MW00358943

SIGNALS AND SYSTEMS FOR BIOENGINEERS

SECOND EDITION

This is a volume in the
ACADEMIC PRESS SERIES IN BIOMEDICAL ENGINEERING

JOSEPH BRONZINO, SERIES EDITOR
TRINITY COLLEGE—HARTFORD, CONNECTICUT

SIGNALS AND SYSTEMS FOR BIOENGINEERS

A MATLAB® -BASED INTRODUCTION

SECOND EDITION

JOHN SEMMLOW

AMSTERDAM • BOSTON • HEIDELBERG • LONDON • NEW YORK • OXFORD
PARIS • SAN DIEGO • SAN FRANCISCO • SINGAPORE • SYDNEY • TOKYO
Academic Press is an imprint of Elsevier

Academic Press is an imprint of Elsevier
225 Wyman Street, Waltham, MA 02451, USA
The Boulevard, Langford Lane, Kidlington, Oxford, OX5 1GB, UK
525 B Street, Suite 1900, San Diego, CA 92101-4495, USA
Radarweg 29, PO Box 211, 1000 AE Amsterdam, The Netherlands

Second Edition 2012

Notices

Knowledge and best practice in this field are constantly changing. As new research and experience broaden our understanding, changes in research methods, professional practices, or medical treatment may become necessary.

Practitioners and researchers must always rely on their own experience and knowledge in evaluating and using any information, methods, compounds, or experiments described herein. In using such information or methods they should be mindful of their own safety and the safety of others, including parties for whom they have a professional responsibility.

To the fullest extent of the law, neither the Publisher nor the authors, contributors, or editors, assume any liability for any injury and/or damage to persons or property as a matter of products liability, negligence or otherwise, or from any use or operation of any methods, products, instructions, or ideas contained in the material herein.

MATLAB® is a trademark of The MathWorks, Inc. and is used with permission. The MathWorks does not warrant the accuracy of the text or exercises in this book. This book's use or discussion of MATLAB® software or related products does not constitute endorsement or sponsorship by The MathWorks of a particular pedagogical approach or particular use of the MATLAB® software.

Library of Congress Cataloging-in-Publication Data

Semmlow, John L.

Signals and systems for bioengineers : a MATLAB-based introduction. / John Semmlow. – 2nd ed.

p. ; cm. – (Biomedical engineering)

Rev. ed. of.: Circuits, signals, and systems for bioengineers / John Semmlow. c2005.

Includes bibliographical references and index.

ISBN 978-0-12-384982-3

1. Biomedical engineering. 2. Signal processing–Digital techniques. I. Semmlow, John L. Circuits, signals and systems for bioengineers. II. Title. III. Series: Academic Press series in biomedical engineering.

[DNLM: 1. Biomedical Engineering. 2. Signal Processing, Computer-Assisted. QT 36]

R856.S453 2012
610'.28–dc23

2011018844

British Library Cataloguing in Publication Data

A catalogue record for this book is available from the British Library

For information on all Academic Press publications visit our website at www.elsevierdirect.com

Typeset by MPS Limited, a Macmillan Company, Chennai, India
www.macmillansolutions.com

Printed and bound in USA

12 13 14 10 9 8 7 6 5 4 3 2 1

Contents

II
SYSTEMS

5. Linear Systems in the Frequency Domain: The Transfer Function

6. Linear Systems Analysis in the Complex Frequency Domain: The Laplace Transform and the Analysis of Transients

7. Linear Systems Analysis in the Time Domain: Convolution and Simulation

8. Linear System Analysis: Applications

III
CIRCUITS

9. Circuit Elements and Circuit Variables

10. Analysis of Analog Circuits and Models

11. Circuit Reduction: Simplifications

Acknowledgements

This text has benefited from the comments of a number of biomedical engineering instructors from around the country. I am indebted to reviewers from a number of prestigious Departments of Biomedical Engineering including:

Brett F. BuSha, Ph.D., The College of New Jersey
Dr. Alan W.L. Chiu, Louisiana Tech University
Delphine Dean, Clemson University
Kenneth Gentry, University of Illinois at Urbana-Champaign
William O. Hancock, Pennsylvania State University
Yingtao Jiang, University of Nevada, Las Vegas
T. Douglas Mast, University of Cincinnati
Michael R. Neuman, Michigan Technological University
Michael Rizk, University of Pennsylvania
Andrew M. Rollins, Case Western Reserve University
Robi Polikar, Rowan University
John H. Schild, Ph.D., IUPUI
Xin Yu, Case Western Reserve University
John X. J. Zhang, The University of Texas, Austin

and two anonymous reviewers. Their contributions were very helpful and any errors or shortcomings that remain are entirely my responsibility. I also wish to thank Susanne Oldham (to whom this book is dedicated) for her patient editing and unwavering support. Thanks also to Lynn Hutchings and Peggy Christ who demonstrated great understanding during the long preparation of this manuscript.

John L. Semmlow, Ph.D.
New Brunswick, N.J., 2011

Preface to the Second Edition

This edition was motivated in large part by the changing educational environment in biomedical engineering. The wide-ranging expertise expected of a competent biomedical engineer continues to grow, putting additional stress on curricula and courses. This text provides a comprehensive introductory coverage of linear systems including circuit analysis. Not all of the material can be covered in a single three-hour course; rather, the book is designed to allow the instructor to choose those topics deemed most essential and be able to find them in a single, coherent text.

NEW TO THIS EDITION

The second edition presents a number of significant modifications over the previous text; in many ways it is a different book. Some of the changes address deficiencies in the first edition, including the lack of biomedical examples and chapter organizational issues. However, the most significant difference is the material that has been added, again with the goal of providing comprehensive coverage of linear systems topics. For example, a chapter has been added on digital filtering, a major section on Simulink®, additional material on the power spectrum including spectral averaging and the short-term Fourier transform, and sections on imaging. Not all instructors will choose to include these topics, but they are relevant and I feel they should be available as options. The chapters on circuits also include sections on lumped-parameter mechanical systems that should be considered optional.

Many biomedical engineering programs cover circuits in a separate course (usually in electrical engineering), so that material has been moved to later chapters. Nonetheless, there may come a time when instructors will want to incorporate this material into their own programs, either to save course time or to provide more effective teaching of this topic. The chapters on circuits could be covered first, although the section on phasor analysis in Chapter 5 will have to be included. Chapter 12 on electronics is not usually included in a linear systems course, but since the students will have the necessary background to analyze useful circuits, I thought it worth including.

What has been retained from the first edition is the strong reliance on MATLAB. In this edition, MATLAB is now considered an essential adjunct and has been embedded in the relevant sections. This text is not meant to be used without this powerful pedagogical tool. Another concept I have retained is the development of some of the deeper concepts, such as the Fourier transform and the transfer function, using an intuitive approach. For example, the Fourier transform is presented in the context of correlation between an unknown function and a number of sinusoidal probing functions. The time-domain concept of convolution is still introduced after covering system characteristics in the frequency domain. I believe this unorthodox arrangement is essential because one needs to understand frequency domain concepts to understand what convolution actually does. In any case, there is no obvious pedagogical reason for covering time-domain concepts first, although in the signals section I do follow the traditional order.

The general philosophy of this text is to introduce and develop concepts though computational methods that allow students to explore operations such as correlations, convolution, the Fourier transform, or the transfer function. With some hesitation, I have added a few examples and problems that follow more traditional methods of mathematical rigor. Again, these examples and related problems should be considered optional. A few more intriguing problems have been sprinkled into the problem sets for added interest and challenge and many more problems are now based on biological examples.

ANCILLARIES

The text comes with a number of educational support materials. For purchasers of the text, a website contains downloadable data and MATLAB functions needed to solve the problems, a few helpful MATLAB routines, and all of the MATLAB examples given in the book. Since many of the problems are extensions or modifications of examples, these files can be helpful in reducing the amount of typing required to carry out an assignment. Please visit *www.elsevierdirect.com*, search on "Semmlow," and click on the "companion site" link on the book's web page to download this material. For instructors, an educational support package is available that contains a set of PowerPoint files that include all of the figures and most of the equations, along with supporting text for each chapter. This package also contains the solutions to the problem sets and some sample exams. The package is available to instructors by registering at: *www.textbooks.elsevier.com*. At Rutgers University, this course is offered in conjunction with a laboratory called "Signals and Measurements," and a manual covering this laboratory is in preparation and should be available shortly.

PART I

SIGNALS

CHAPTER

1

The Big Picture

Bioengineering Signals and Systems

1.1 BIOLOGICAL SYSTEMS

A system is a collection of processes or components that interact with a common purpose. Many systems of the human body are based on function. The function of the cardiovascular system is to deliver oxygen-carrying blood to the peripheral tissues. The pulmonary system is responsible for the exchange of gases (primarily O_2 and CO_2) between the blood and air, while the renal system regulates water and ion balance and adjusts the concentration of ions and molecules. Some systems are organized around mechanism rather than function. The endocrine system mediates a range of communication functions using complex molecules distributed throughout the bloodstream. The nervous system performs an enormous number of tasks using neurons and axons to process and transmit information coded as electrical impulses.

The study of classical physiology and of many medical specialties is structured around human physiological systems. (Here the term *classical physiology* is used to mean the study of whole organs or organ systems as opposed to newer molecular-based approaches.) For example, cardiologists specialize in the cardiovascular system, neurologists in the nervous system, ophthalmologists in the visual system, nephrologists in the kidneys, pulmonologists in the respiratory system, gastroenterologists in the digestive system, and endocrinologists in the endocrine system. There are medical specialties or subspecialties to cover most physiological systems. Another set of medical specialties is based on common tools or approaches, including surgery, radiology, and anesthesiology, while one specialty, pediatrics, is based on the type of patient.

Given this systems-based approach to physiology and medicine, it is not surprising that early bioengineers applied their engineering tools, especially those designed for the analysis of systems, to some of these physiological systems. Early applications in bioengineering research included the analysis of breathing patterns and the oscillatory movements of the iris muscle. Applications of basic science to medical research date from the eighteenth century. In the late nineteenth century, Einthoven recorded the electrical activity of the heart and throughout that century electrical stimulation was used therapeutically (largely to no avail). While early researchers may not have considered themselves engineers, they did draw on the engineering tools of their day.

The nervous system, with its apparent similarity to early computers, was another favorite target of bioengineers, as was the cardiovascular system with its obvious links to hydraulics and fluid dynamics. Some of these early efforts are discussed in this book in the sections on system and analog models. As bioengineering has expanded into areas of molecular biology, systems on the cellular, and even subcellular, levels have become of interest.

Irrespective of the type of biological system, its scale, or its function, we must have some way of interacting with that system. Interaction or communication with a biological system is done through *biosignals*. The communication may only be in one direction, such as when we attempt to infer the state of the system by measuring various biological or physiological variables to make a medical diagnosis. From a systems analytic point of view, changes in physiological variables constitute biosignals. Common signals measured in diagnostic medicine include: electrical activity of the heart, muscles, and brain; blood pressure; heart rate; blood gas concentrations and concentrations of other blood components; and sounds generated by the heart and its valves.

Often, it is desirable to send signals into a biological system for purposes of experimentation or therapy. In a general sense, all drugs introduced into the body can be considered biosignals. We often use the term *stimulus* for signals directed at a specific physiological process, and if an output signal is evoked by these inputs we term it a *response*. (Terms in italics are an important part of a bioengineer's vocabulary.) In this scenario, the biological system is acting like an input-output system, a classic construct or model used in systems analysis, as shown in Figure 1.1.

Classic examples include the knee-jerk reflex, where the input is mechanical force and the output is mechanical motion, and the pupillary light reflex, where the input is light and the output is a mechanical change in the iris muscles. Drug treatments can be included in this input/output description, where the input is the molecular configuration of the drug and the output is the therapeutic benefit (if any). Such representations are further explored in the sections on systems and analog modeling.

Systems that produce an output without the need for an input stimulus, like the electrical activity of the heart, can be considered biosignal *sources*. (Although the electrical activity of the heart can be moderated by several different stimuli, such as exercise, the basic signal does not require a specific stimulus.) Input-only systems are not usually studied, since the purpose of any input signal is to produce some sort of response: even a placebo, which is designed to produce no physiological response, often produces substantive results.

Since all of our interactions with physiological systems are through biosignals, the characteristics of these signals are of great importance. Indeed, much of modern medical

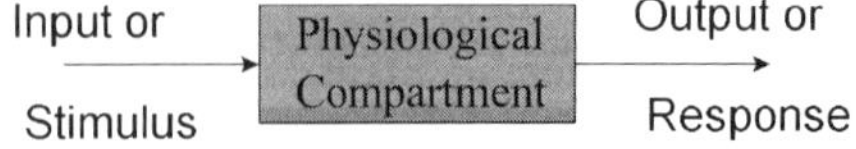

FIGURE 1.1 A classical systems view of a physiological system that receives an external input, or stimulus, which then evokes an output, or response.

technology is devoted to extracting new physiological signals from the body or gaining more information from existing biosignals. The next section discusses some of the basic aspects of these signals.

1.2 BIOSIGNALS

Much of the activity in biomedical engineering, be it clinical or in research, involves the measurement, processing, analysis, display, and/or generation of signals. Signals are variations in energy that carry information. The variable that carries the information (the specific energy fluctuation) depends on the type of energy involved. Table 1.1 summarizes the different energy types that can be used to carry information and the associated variables that encode this information. Table 1.1 also shows the physiological measurements that involve these energy forms, as discussed later.

Biological signals are usually encoded into variations of electrical, chemical, or mechanical energy, although occasionally variations in thermal energy are of interest. For communication within the body, signals are primarily encoded as variations in electrical or chemical energy. When chemical energy is used, the encoding is usually done by varying the concentration of the chemical within a *physiological compartment* for example, the concentration of a hormone in the blood. Bioelectric signals use the flow or concentration of ions, the primary charge carriers within the body, to transmit information. Speech, the primary form of communication between humans, encodes information as variations in pressure.

Outside the body, information is commonly transmitted and processed as variations in electrical energy, although mechanical energy was used in the eighteenth and nineteenth centuries to send messages. The semaphore telegraph used the position of one or more large arms placed on a tower or high point to encode letters of the alphabet. These arm positions could be observed at some distance (on a clear day), and relayed onward if necessary. Information processing can also be accomplished mechanically, as in the early numerical processors constructed by Babbage in the early and mid nineteenth century. In the mid to late twentieth century, mechanically based digital components were attempted using variations in fluid flow.

TABLE 1.1 Energy Forms and Associated Information-Carrying Variables

Energy	Variables (Specific Fluctuation)	Common Measurements
Chemical	Chemical activity and/or concentration	Blood ion, O_2, CO_2, pH, hormonal concentrations, and other chemistry
Mechanical	Position Force, torque, or pressure	Muscle movement Cardiovascular pressures, muscle contractility Valve and other cardiac sounds
Electrical	Voltage (potential energy of charge carriers) Current (charge carrier flow)	EEG, ECG, EMG, EOG, ERG, EGG, GSR
Thermal	Temperature	Body temperature, thermography

Modern electronics provide numerous techniques for modifying electrical signals at very high speeds. The body also uses electrical energy to carry information when speed is important. Since the body does not have many free electrons, it relies on ions, notably Na^+, K^+, and Cl^-, as the primary charge carriers. Outside the body, electrically based signals are so useful that signals carried by other energy forms are usually converted to electrical energy when significant transmission or processing tasks are required. The conversion of physiological energy to an electric signal is an important step, and often the first step, in gathering information for clinical or research use. The energy conversion task is done by a device termed a *transducer*, specifically a *biotransducer*.

A transducer is a device that converts energy from one form to another. By this definition, a lightbulb or a motor is a transducer. In signal processing applications, the purpose of energy conversion is to transfer information, not to transform energy as with a lightbulb or a motor. In physiological measurement systems, all transducers are so-called *input transducers*: they convert nonelectrical energy into an electronic signal. An exception to this is the electrode, a transducer that converts electrical energy from ionic to electronic form. Usually, the output of a biotransducer is a voltage (or current) whose amplitude is proportional to the measured energy. Figure 1.2 shows a transducer that converts acoustic sounds from the heart to electric signals. This *cardiac microphone* uses piezoelectric elements to convert the mechanical sound energy to electrical energy.

The energy that is converted by the input transducer may be generated by the physiological process itself as in the example above, it may be energy that is indirectly related to the physiological process, or it may be energy produced by an external source. In the last case, the externally generated energy interacts with, and is modified by, the physiological process, and it is this alteration that produces the measurement. For example, when externally produced x-rays are transmitted through the body, they are absorbed by the intervening tissue, and a measurement of this absorption is used to construct an image. Most medical imaging systems are based on this external energy approach.

Images can also be constructed from energy sources internal to the body, as in the case of radioactive emissions from radioisotopes injected into the body. These techniques make use of the fact that selected, or tagged, molecules will collect in specific tissue. The areas

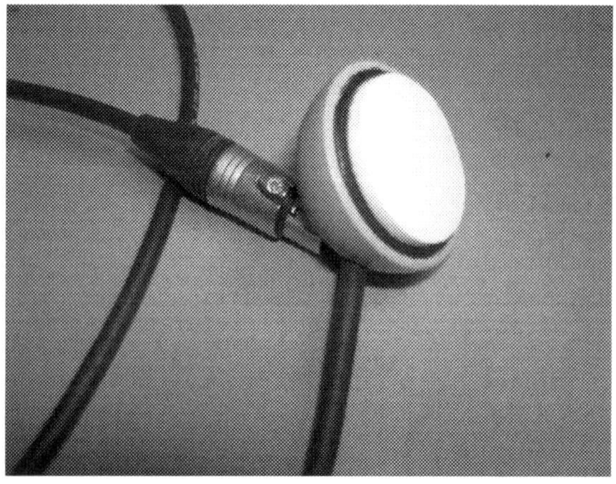

FIGURE 1.2 A cardiac microphone that converts the mechanical sound energy produced by the beating heart into an electrical signal is shown. The device uses a piezoelectric disk that produces a voltage when it is deformed by movement of the patient's chest. The white foam pad covers the piezoelectric disk and is specially designed to improve the coupling of energy between the chest and the piezoelectric disk.

where these radioisotopes collect can be mapped using a gamma camera or, with certain short-lived isotopes, better localized using positron emission tomography (PET).

Many physiological processes produce energy that can be detected directly. For example, cardiac internal pressures are usually measured using a pressure transducer placed on the tip of a catheter and introduced into the appropriate chamber of the heart. The measurement of electrical activity in the heart, muscles, or brain provides other examples of direct measurement of physiological energy. For these measurements, the energy is already electrical and only needs to be converted from ionic to electronic current using an *electrode*. These sources are usually given the term ExG, where the "x" represents the physiological process that produces the electrical energy: ECG—electrocardiogram; EEG—electroencephalogram; EMG—electromyogram; EOG—electrooculogram, ERG—electroretinogram; and EGG—electrogastrogram. An exception to this terminology is the galvanic skin response, GSR, the electrical activity generated by the skin. Typical physiological measurements that involve the conversion of other energy forms to electrical energy are shown in Table 1.1. Figure 1.3 shows the early ECG machine where the interface between the body and the electrical monitoring equipment was buckets filled with saline (E in Figure 1.3).

The *biotransducer* is often the most critical element in the system, since it constitutes the interface between the subject or life process and the rest of the system. The transducer establishes the risk, or invasiveness of the overall system. For example, an imaging system based on differential absorption of x-rays, such as a CT (computed tomography) scanner, is considered more "invasive" than an imaging system based on ultrasonic reflection, since CT uses ionizing radiation that may have an associated risk. (The actual risk of ionizing radiation is still an open question, and imaging systems based on x-ray absorption are considered "minimally invasive.") Both ultrasound and x-ray imaging would be considered less invasive

FIGURE 1.3 An early ECG machine.

than, for example, monitoring internal cardiac pressures through cardiac catheterization in which a small catheter is threaded into the heart chamber. Indeed, many of the outstanding problems in biomedical measurement, such as noninvasive measurement of internal cardiac pressures or intracranial pressure, await a functional transducer mechanism.

1.2.1 Signal Encoding

Given the importance of electrical signals in biomedical engineering, much of the discussion in this text is about electrical or electronic signals. Nonetheless, many of the principles described are general and could be applied to signals carried by any energy form. Irrespective of the energy form or specific variable used to carry information, some type of encoding scheme is necessary. Encoding schemes vary in complexity: human speech is so complex that automated decoding is still a challenge for voice recognition computer programs. Yet the exact same information could be encoded into the relatively simple series of long and short pulses known as Morse code, which is easy for a computer to decode.

Most encoding strategies can be divided into two broad categories or domains: *continuous* and *discrete*. These two domains are also termed *analog* and *digital*. The discrete domain, used in computer-based technology, is easier to manipulate electronically. Within the digital domain many different encoding schemes can be used. For encoding alphanumeric characters, those featured on the keys of a computer keyboard, the ASCII code is used. Here, each letter, the numbers 0 through 9, and many other characters are encoded into an 8-bit binary number. For example, the letters a through z are encoded as 97 through 122, while the capital letters A through Z are encoded by numbers 65 through 90. The complete ASCII code can be found in some computer texts or on the web.

1.2.2 Continuous or Analog Domain Signals

In the continuous analog domain, information is encoded in terms of signal amplitude, usually the intensity of the signal at any given time. Such a time-varying signal is termed an *analog signal* and is mathematically described by the equation:

$$x(t) = f(t) \tag{1.1}$$

where $f(t)$ is some function of time and can be quite complex. For an electronic signal, this could be the value of the voltage or current at a given time. Often it is not possible to find a mathematical $f(t)$ that encompasses the complexity of a signal, and it is presented graphically or by using another method. Note that all signals are by nature *time-varying*, since a time-invariant constant contains no information. Modern information theory makes explicit the difference between information and meaning. The latter depends on the receiver, that is, the device or person for which the information is intended. Many students have attended lectures with a considerable amount of information that, for them, had little meaning. This text strives valiantly for both information and meaning.

Often an analog signal encodes the information it carries as a linear change in signal amplitude. For example, a temperature transducer might encode room temperature into voltage as shown in Table 1.2, below.

TABLE 1.2 Relationship between Temperature and Voltage of a Hypothetical Temperature Transducer

Temperature (°C)	Voltage (volts)
−10	0.0
0.0	5.0
+10	10.0
+20	15.0

Then the encoding equation for temperature would be:
temperature = 2 × *voltage amplitude* −10

Analog encoding is common in consumer electronics such as high-fidelity amplifiers and television receivers, although many applications that traditionally used analog encoding, such as sound and video recording, now primarily use discrete or digital encoding. Nonetheless, analog encoding is likely to remain important to the biomedical engineer because many physiological systems use analog encoding, and most biotransducers generate analog encoded signals.

1.2.3 Discrete-Time or Digital Signals

Signals that originate in the analog domain often need to be represented, processed, and stored in a digital computer. This requires that a continuous analog signal be converted to a series of numbers through a process known as *sampling*. The details of the sampling process and its limitations are discussed later, but essentially the continuous analog signal, $x(t)$, is sliced up into a sequence of digital numbers usually at equal time intervals T_s. T_s is known as the *sampling interval* and is shown in Figure 1.4.

A sampled time signal, also referred to as a *digitized signal* or simply *digital signal*, can be easily stored in a digital computer. A digital signal, $x[k]$, is just a series of numbers: $x[k] = x_1, x_2, x_3, \ldots x_N$. These sequential numbers represent the value of the analog signal at a discrete point in time determined by the sample interval, T_s, or by the sampling frequency, f_s:

$$t = nT_s = n/f_s \tag{1.2}$$

where n is the position of the number in the sequence, t is the signal time represented by the number, and f_s is just the inverse of the sample interval:

$$f_s = 1/T_s \tag{1.3}$$

Usually this series of numbers would be stored in sequential memory locations with $x[1]$ followed by $x[2]$, then $x[3]$, and so on. It is common to use brackets to identify a discrete or digital variable (i.e., $x[k]$) while parentheses are used for continuous variables. Note that the MATLAB® programming language used throughout this text uses parentheses to index a variable (i.e., `x(1)`, `x(2)` `x(3)`...) even though MATLAB variables are clearly digital. MATLAB reserves brackets for concatenation of numbers or variables. In addition, MATLAB uses the integer 1, not zero, to indicate the first number in a variable sequence.

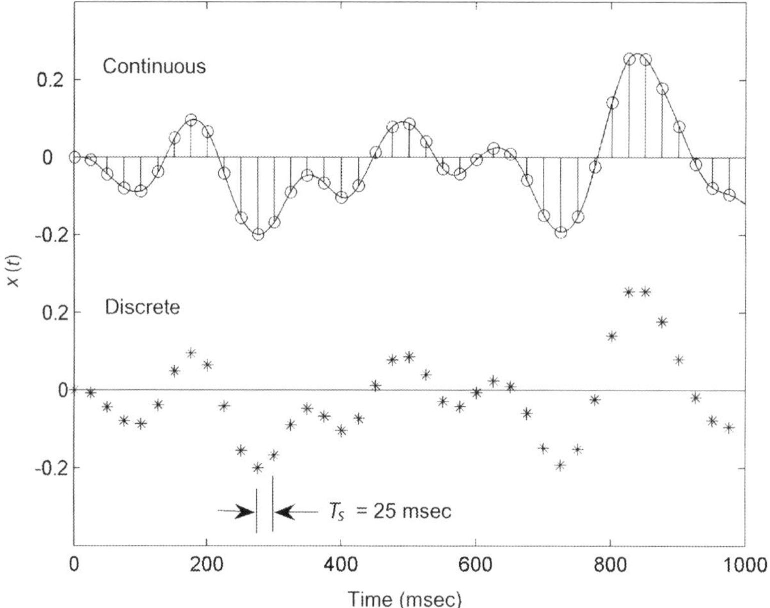

FIGURE 1.4 A continuous time signal, $x(t)$ (upper plot), is converted to a sequence of numbers, $x[k]$ (lower plot), in a process known as sampling. These numbers are just the value of this signal at sequential time intervals spaced $T_s = 0.065$ sec apart (In this figure $x[n] = 0.0288$, 0.8124, 0.9601, 0.4073, −0.3274, −0.6259, −0.3218, 0.1845, 0.3387, −0.0227, −0.5134, −0.5743, −0.0285, 0.7116, 0.9940, 0.5200). After sampling, these numbers can be stored in a digital computer.

This can be a source of confusion when programming some equations that use zero as the index of the first number in a series. Hence x[0] in an equation translates to x(1) in a MATLAB program. (All MATLAB variables and statements used in the book are typeset in a sans serif typeface for improved clarity.)

In some advanced signal processing techniques, it is useful to think of the sequence of n numbers representing a signal as a single vector pointing to the combination of n numbers in an n-dimensional space. Fortunately, this challenging concept (imagine a 256-dimension space) will not be used here except in passing, but it gives rise to use of the term *vector* when describing a series of numbers such as $x[k]$. This terminology is also used by MATLAB, so in this text the term *vector* means "sequence of numbers representing a time signal."

Since digital numbers can only represent discrete or specific amplitudes, the analog signal must also be sliced up in amplitude. Hence, to *"digitize"* or *"sample"* an analog signal requires slicing the signal in two ways: in time and in amplitude.

Time slicing samples the continuous waveform, $x(t)$, at discrete points in time, nT_s, where T_s is the sample interval and $n = 1,2,3,\ldots$. The consequences of time slicing depend on the signal characteristics and the sampling time and are discussed in Section 4.1. Slicing the signal amplitude in discrete levels is termed *quantization*, as shown in Figure 1.5. The equivalent number can only approximate the level of the analog signal, and the degree of approximation will depend on the range of binary numbers and the

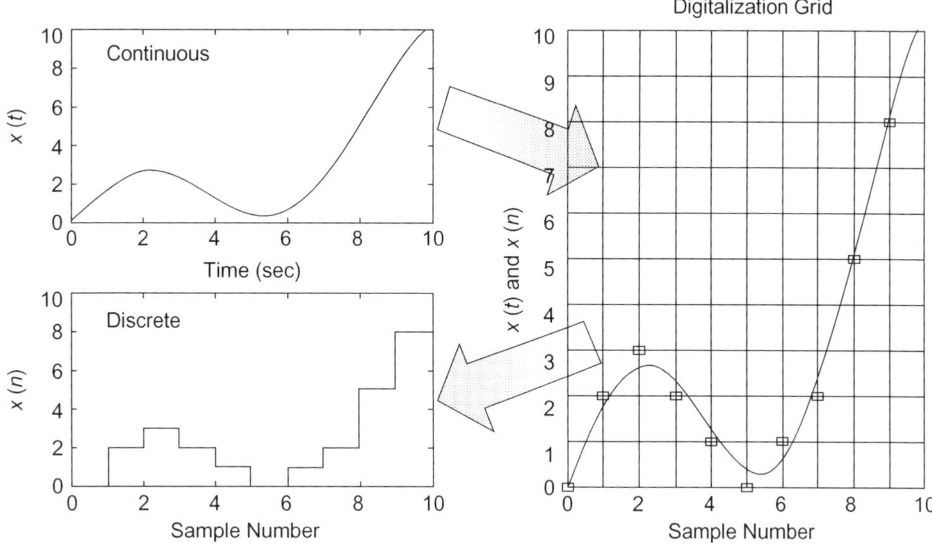

FIGURE 1.5 Digitizing a continuous signal, upper left, requires slicing the signal in time and amplitude, right side. The result is a series of discrete numbers (rectangular points) that approximate the original signal, and the resultant digitized signal, lower left, consists of a series of discrete steps in time and value.

amplitude of the analog signal. For example, if the signal is converted into an 8-bit binary number, this number is capable of 2^8 or 256 discrete values. If the analog signal amplitude ranges between 0.0 and 5.0 volts, then the quantization interval will be 5/256 or 0.0195 volts. If the analog signal is continuous in value, as is usually the case, it can only be approximated by a series of binary numbers representing the approximate analog signal level at discrete points in time, as seen in Figure 1.5. The errors associated with amplitude slicing, or quantization, are described in Chapter 4.

EXAMPLE 1.1

A 12-bit analog-to-digital converter (ADC) advertises an accuracy of ± the least significant bit (LSB). If the input range of the ADC is 0 to 10 volts, what is the accuracy of the ADC in analog volts?

Solution: If the input range is 10 volts then the analog voltage represented by the LSB is:

$$V_{LSB} = \frac{V_{\max}}{2^{\text{Nu bits}} - 1} = \frac{10}{2^{12} - 1} = \frac{10}{4095} = .0024 \text{ volts}$$

Hence the accuracy would be ±.0024 volts.

It is relatively easy, and common, to convert between the analog and digital domains using electronic circuits specially designed for this purpose. Many medical devices acquire the physiological information as an analog signal, then convert it to digital format using an *analog-to-digital converter* (ADC) for subsequent computer processing. For example, the electrical activity produced by the heart can be detected using properly placed electrodes, and the resulting signal, the ECG, is an analog-encoded signal. This signal might undergo some *preprocessing* or *conditioning* using analog electronics, but would eventually be converted to a digital signal using an ADC for more complex, computer-based processing and storage. In fact, conversion to digital format is usually done even when the data are only stored for later use. Conversion from the digital to the analog domain is also possible using a *digital-to-analog converter* (DAC). Most PCs include both ADCs and DACs as part of a sound card. This circuitry is specifically designed for the conversion of audio signals, but can be used for other analog signals. Data transformation cards designed as general-purpose ADCs and DACs are readily available and offer greater flexibility in sampling rates and conversion gains. These cards generally provide multichannel ADCs (using usually 8 to 16 channels) and several DAC channels.

In this text, the basic concepts that involve signals are often introduced or discussed in terms of analog signals, but most of these concepts apply equally well to the digital domain, assuming that the digital representation of the original analog signal is accurate. In this text, the equivalent digital domain equation will be presented alongside the analog equation to emphasize the equivalence. Many of the problems and examples use a computer, so they are necessarily implemented in the digital domain even if they are presented as analog domain problems.

If the digital signal represents a sampled version of an analog signal then we hope there is some meaningful relationship between the two. The conditions under which such a meaningful relationship occurs are described in Chapter 4, but for now we will assume that all computer-based signals used in examples and problems are accurate representations of their associated analog signals. In Chapter 4 we will look at the consequences of the analog-to-digital conversion process in more detail and establish rules for when a digitized signal can be taken as a truthful representation of the original analog signal.

Many of the mathematical operations described here can be implemented either "by hand," that is analytically, or by computer algorithm. Invariably, computer implementation is easier and usually allows the solution of more complicated problems. A solution using a computer algorithm can provide more insight into the behavioral characteristics of a mathematical operation because it is easy to apply that operation to a variety of signals or processes, but computer algorithms do not provide much understanding of the theory behind the operation. Since the mathematical operations in this book are rather fundamental, it is important to understand both theory and application. So we will use analytical examples and problems to gain understanding of the principles and computer implementation to understand their applications. Real-world application of these operations usually relies on computer implementation.

1.3 NOISE

Where there is signal there is noise. Occasionally the noise will be at such a low level that it is of little concern, but more often it is the noise that limits the usefulness of the signal. This is especially true for physiological signals since they are subject to many potential sources of noise. In most usages, *noise* is a general and somewhat relative term: noise is what you do not want and signal is what you do want. This leads to the concept of noise as any form of unwanted variability. Noise is inherent in most measurement systems and is often the limiting factor in the performance of a medical instrument. Indeed, many medical instruments go to great lengths in terms of signal conditioning circuitry and signal processing algorithms to compensate for unwanted noise.

In biomedical measurements, noise has four possible origins: 1) physiological variability; 2) environmental noise or interference; 3) measurement or transducer artifact; and 4) electronic noise. Physiological variability is due to the fact that the information you desire is based on measurements subject to biological influences other than those of interest. For example, assessment of respiratory function based on the measurement of blood pO_2 could be confounded by other physiological mechanisms that alter blood pO_2. Physiological variability can be a very difficult problem to solve, sometimes requiring a total redesign of the approach.

Environmental noise can come from sources external or internal to the body. A classic example is the measurement of the fetal ECG signal where the desired signal is corrupted by the mother's ECG. Since it is not possible to describe the specific characteristics of environmental noise, typical noise reduction approaches such as filtering (described in Chapter 8) are not usually successful. Sometimes environmental noise can be reduced using *adaptive filters* or *noise cancellation* techniques that adjust their filtering properties based on the current environment.

Measurement artifact is produced when the biotransducer responds to energy modalities other than those desired. For example, recordings of electrical signals of the heart, the ECG, are made using electrodes placed on the skin. These electrodes are also sensitive to movement, so-called *motion artifact*, where the electrodes respond to mechanical movement as well as to the electrical activity of the heart. This artifact is not usually a problem when the ECG is recorded in the doctor's office, but it can be if the recording is made during a patient's normal daily living, as in a 24-hour *Holter* recording. Measurement artifacts can sometimes be successfully addressed by modifications in transducer design. For example, aerospace research has led to the development of electrodes that are quite insensitive to motion artifact.

Unlike other sources of variability, electronic noise has well-known sources and characteristics. Electronic noise falls into two broad classes: *thermal* or *Johnson noise*, and *shot noise*. The former is produced primarily in resistors or resistance materials while the latter is related to voltage barriers associated with semiconductors. Both sources produce noise that contains energy over a broad range of frequencies, often extending from DC (i.e., 0 Hz) to 10^{12}–10^{13} Hz. White light contains energy at all frequencies, or at least at all frequencies we can see, so such broad spectrum noise is also referred to as *white noise* since it contains energy at all frequencies, or at least all frequencies of interest to bioengineers. Figure 1.6 shows a plot of the energy in a simulated white noise waveform (actually, an array of random numbers) plotted against frequency. This is similar to a plot of the energy in a beam of light versus wavelength (or frequency), and as with light is also referred to as a *spectrum*, or *frequency plot*, or *spectral plot*, all terms used interchangeably. A method for generating such a frequency plot from the noise waveform, or any other waveform for that matter, is developed in Chapter 3. Note that the energy of the simulated noise is fairly constant across the frequency range.

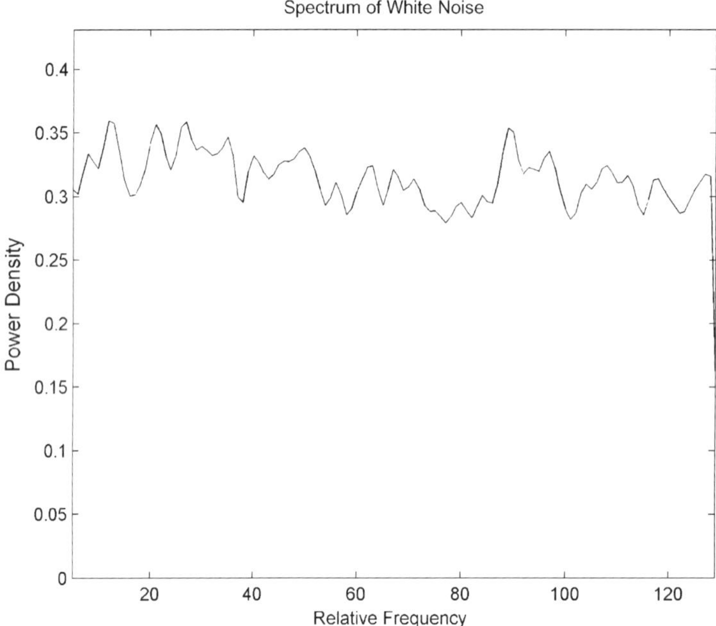

FIGURE 1.6 A plot of the energy in white noise as a function of frequency. The noise has a fairly flat spectral characteristic showing similar energy levels over all frequencies plotted. This equal-energy characteristic gives rise to the use of the term *white noise*. Techniques for producing a signal's spectral plot are covered in Chapter 3.

The various sources of noise or variability along with their causes and possible remedies are presented in Table 1.3. Note that in three out of four instances, appropriate transducer design may aid in the reduction of the variability or noise. This demonstrates the important role of the transducer in the overall performance of the instrumentation system.

1.3.1 Electronic Noise

Johnson or thermal noise is produced by resistance sources, and the amount of noise generated is related to the resistance and to the temperature:

$$VJ = \sqrt{4kT\,R\,BW} \text{ volts} \tag{1.4}$$

where R is the resistance in ohms, T is the temperature in deg Kelvin, and k is Boltzmann's constant ($k = 1.38 \times 10^{-23}$ J/K). A temperature of 310 K is often used as an upper limit for room temperature, in which case $4kT = 1.7 \times 10^{-20}$ J. Here BW is the range of frequencies that is included in the signal. This range of frequencies is termed *bandwidth* and is better defined in Chapter 4. This frequency range is usually determined by the

TABLE 1.3 Sources of Variability

Source	Cause	Potential Remedy
Physiological variability	Measurement indirectly related to variable of interest	Modify overall approach
Environmental (internal or external)	Other sources of similar energy	Apply noise cancellation Alter transducer design
Artifact	Transducer responds to other energy sources	Alter transducer design
Electronic	Thermal or shot noise	Alter transducer or electronic design

characteristics in the measurement system, often the filters used in the system. Since noise is spread over all frequencies, the greater the signal's bandwidth the greater the noise in any given situation.

If noise current is of interest, the equation for Johnson noise current can be obtained from Equation 1.4 in conjunction with Ohm's law:

$$I_J = \sqrt{4kT\,BW/R}\ \text{amps} \tag{1.5}$$

In practice, there will be limits imposed on the frequencies present within any waveform (including noise waveforms), and these limits are used to determine bandwidth. In the problems given here the bandwidth is simply given. Bandwidth, BW, is usually specified as a frequency range and, like all frequency specifications, is given in units of Hertz (Hz), which are actually units of inverse seconds (i.e., 1/sec). In fact, frequencies and bandwidths were given units of "cycles per second" until the 1970s when it was decided that an inverse definition, a "per" definition, was just not appropriate for such an important unit.

Since bandwidth is not always known in advance, it is common to describe a relative noise, specifically the noise that would occur if the bandwidth were 1.0 Hz. Such relative noise specification can be identified by the unusual units required: volts/$\sqrt{\text{Hz}}$ or amps/$\sqrt{\text{Hz}}$.

Shot noise is defined as current noise and is proportional to the baseline current through a semiconductor junction:

$$I_s = \sqrt{2qI_d\,BW}\ \text{amps} \tag{1.6}$$

where q is the charge on an electron (1.662×10^{-19} coulomb which equals 1 amp-sec), and I_d is the baseline semiconductor current. In photodetectors, the baseline current that generates shot noise is termed the *dark current*, hence the letter "d" in the current symbol, I_d, in Equation 1.6. Again, the noise is spread across all frequencies so the bandwidth, BW, must be specified to obtain a specific value or else a relative noise can be specified in amps/$\sqrt{\text{Hz}}$.

When multiple noise sources are present, as is often the case, their voltage or current contributions to the total noise add as the square root of the sum of the squares, assuming that the individual noise sources are independent. For voltages:

$$V_T = \sqrt{V_1^2 + V_1^2 + V_3^2 + \ldots V_N^2} \qquad (1.7)$$

A similar equation applies to current noise.

EXAMPLE 1.2

A 20-ma current flows through both a diode (i.e., a semiconductor) and a 200-Ω resistor, Figure 1.7. What is the net current noise, i_n? Assume a bandwidth of 1 MHz (i.e., 1×10^6 Hz).

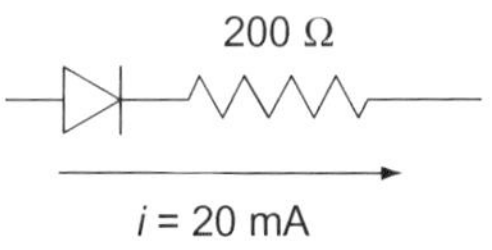

FIGURE 1.7 Simple electric circuit consisting of a resistor and diode used in Example 1.2.

Solution: Find the noise contributed by the diode using Equation 1.6, the noise contributed by the resistor using Equation 1.4, then combine them using Equation 1.7.

$$i_{nd} = \sqrt{2qI_d\,BW} = \sqrt{2(1.66 \times 10^{-19})(20 \times 10^{-3})10^6} = 8.15 \times 10^{-8} \text{ amps}$$

$$i_{nR} = \sqrt{4kT\,BW/R} = \sqrt{1.7 \times 10^{-20}(10^6/200)} = 9.22 \times 10^{-9} \text{ amps}$$

$$i_{nT} = \sqrt{i_{nd}^2 + i_{nR}^2} = \sqrt{6.64 \times 10^{-15} + 8.46 \times 10^{-17}} = 8.20 \times 10^{-8} \text{ amps}$$

Note that most of the current noise is coming from the diode, so the addition of the resistor's current noise does not contribute much to the diode noise current. Also the arithmetic in this example could be simplified by calculating the square of the noise current (i.e., not taking the square roots) and using those values to get the total noise.

1.3.2 Decibels

It is common to compare the intensity of two signals using ratios, V_{Sig1}/V_{Sig2}, particularly if the two are the signal and noise components of a waveform (see next section). Moreover these signal ratios are commonly represented in units of *decibels*. Actually, decibels (dB) are not really units, but are simply a logarithmic scaling of dimensionless ratios. The decibel has several advantageous features: 1) it provides a measurement of the effective power, or power ratio; 2) the log operation compresses the range of values (for example, a range of 1 to 1000 becomes a range of 1 to 3 in log units); 3) when numbers or ratios are to be multiplied, they are simply added if they are in log units; and 4) the logarithmic characteristic is similar to human perception. It was this last feature that motivated Alexander Graham Bell to develop the logarithmic unit the *Bel*. Audio power increments in logarithmic Bels were perceived as equal increments by the human ear.

The Bel turned out to be inconveniently large, so it has been replaced by the *decibel*: dB = 1/10 Bel. While originally defined only in terms of a ratio, *dB* units are also used to express the intensity of a single signal. In this case it actually has a dimension, the dimension of the signal (volts, dynes, etc.), but these units are often unstated.

When applied to a power measurement, the decibel is defined as 10 times the log of the power ratio:

$$P_{dB} = 10\log\left(\frac{P_2}{P_1}\right)\ \text{dB} \tag{1.8}$$

When applied to a voltage ratio, or simply a voltage, the dB is defined as 10 times the log of the RMS value squared, or voltage ratio squared. (RMS stands for "root mean squared" and is found by taking the square root of the mean of the voltage waveform squared. It is formally defined in Equation 1.17.) Since the log is taken, this is just 20 times the unsquared ratio or value ($\log x^2 = 2\log x$). If a ratio of sinusoids is involved, then peak voltages (or whatever units the signal is in) can also be used; they are related to RMS values by a constant (0.707) and the constants will cancel in the ratio (see Example 1.4 below).

$$\begin{aligned} v_{dB} &= 10\log\left(v_2^2/v_1^2\right) = 20\log\left(v_2/v_1\right) \quad \text{or} \\ v_{dB} &= 10\log v_{RMS}^2 = 20\log v_{RMS} \end{aligned} \tag{1.9}$$

The logic behind taking the square of the RMS voltage value before taking the log is that the RMS voltage squared is proportional to signal power. Consider the case where the signal is a time-varying voltage, $v(t)$. To draw energy from this signal, it is necessary to feed it into a resistor, or a resistor-like element, that consumes energy. (Recall from basic physics that resistors convert electrical energy into thermal energy, i.e., heat.) The power (energy per unit time) transferred from the signal to the resistor is given by the equation:

$$P = v_{RMS}^2/R \tag{1.10}$$

where R is the resistance. This equation shows that the power transferred to a resistor by a given voltage depends, in part, on the value of the resistor. Assuming a nominal resistor value of 1 Ω, the power will be equal to the voltage squared; however, for any resistor value, the power transferred will be proportional to the voltage squared. When dB units are used to describe a ratio of voltages, the value of the resistor is irrelevant, since the resistor values will cancel out:

$$v_{dB} = 10\log\left(\frac{v_2^2/R}{v_1^2/R}\right) = 10\log\left(\frac{v_2^2}{v_1^2}\right) = 20\log\left(\frac{v_2}{v_1}\right) \tag{1.11}$$

If dB units are used to express the intensity of a single signal, then the units will be proportional to the log power in the signal.

To convert a voltage from dB to RMS, use the inverse of the defining equation (Equation 1.9):

$$v_{RMS} = 10^{X_{dB}/20} \tag{1.12}$$

Putting units in dB is particularly useful when comparing ratios of signal and noise to determine the *signal-to-noise ratio* (SNR), as described in the next section.

EXAMPLE 1.3

A sinusoidal signal is fed into an *attenuator* that reduces the intensity of the signal. The input signal has a peak amplitude of 2.8 V and the output signal is measured at 2 V peak. Find the ratio of output to input voltage in dB. Compare the power-generating capabilities of the two signals in linear units.

Solution: Convert each peak voltage to RMS, then apply Equation 1.9 to the given ratio. Also calculate the ratio without taking the log.

$$V_{RMS\ dB} = 20\log\left(V_{out\ RMS}/V_{in\ RMS}\right) = 20\log\left(\frac{2.0\times 0.707}{2.8\times 0.707}\right)$$

$$V_{RMS\ dB} = -3\ \mathrm{dB}$$

The power ratio is

$$\text{Power ratio} = \frac{V^2_{out\ RMS}}{V^2_{in\ RMS}} = \frac{(2.0\times .707)^2}{(2.8\times 0.707)^2} = 0.5$$

Analysis: The ratio of the amplitude of a signal coming out of a process to that going into the process is known as the *gain*, and is often expressed in dB. When the gain is <1, it means there is actually a loss, or reduction, in signal amplitude. In this case, the signal loss is 3 dB, so the gain of the attenuator is actually −3 dB. To confuse yourself further, you can reverse the logic and say that the attenuator has an attenuation (i.e., loss) of +3 dB. In this example, the power ratio was 0.5, meaning that the signal coming out of the attenuator has half the power-generating capabilities of the signal that went in. An attenuation of 3 dB is equivalent to a loss of half the signal's energy. Of course, it was not really necessary to convert the peak voltages to RMS since a ratio of these voltages was taken and the conversion factor (0.707) cancels out.

1.3.3 Signal-to-Noise Ratio (SNR)

Most waveforms consist of signal plus noise mixed together. As noted previously, signal and noise are relative terms, relative to the task at hand: the signal is the component interest within the waveform while the noise is everything else. Often the goal of signal processing is to separate a signal from noise, or to identify the presence of a signal buried in noise, or to detect features of a signal buried in noise.

The relative amount of signal and noise present in a waveform is usually quantified by the signal-to-noise ratio, SNR. As the name implies, this is simply the ratio of signal to noise, both measured in RMS (root mean squared) amplitude. This measurement is rigorously defined in the Equation 1.17 in the next section. The SNR is often expressed in dB, as described in the previous section:

$$\text{SNR} = 20\log\left(\frac{\text{signal}}{\text{noise}}\right)\ \text{dB} \tag{1.13}$$

To convert from dB scale to a linear scale:

$$SNR_{Linear} = 10^{dB/20} \tag{1.14}$$

When expressed in dB, a positive number means the signal is greater than the noise and a negative number means the noise is greater than the signal. Some typical dB ratios and their corresponding linear values are given in Table 1.4.

To give a feel for what some SNR dB values actually mean, Figure 1.8 shows a sinusoidal signal with various amounts of white noise. When the SNR is +10 dB the sine wave can be readily identified. This is also true when the SNR is +3 dB. When the SNR

TABLE 1.4 Equivalence between dB and Linear Signal-to-Noise Ratios

SNR in dB	SNR Linear
+20	$v_{signal} = 10\ v_{noise}$
+3	$v_{signal} = 1.414\ v_{noise}$
0	$v_{signal} = v_{noise}$
−3	$v_{signal} = 0.707\ v_{noise}$
−20	$v_{signal} = 0.1\ v_{noise}$

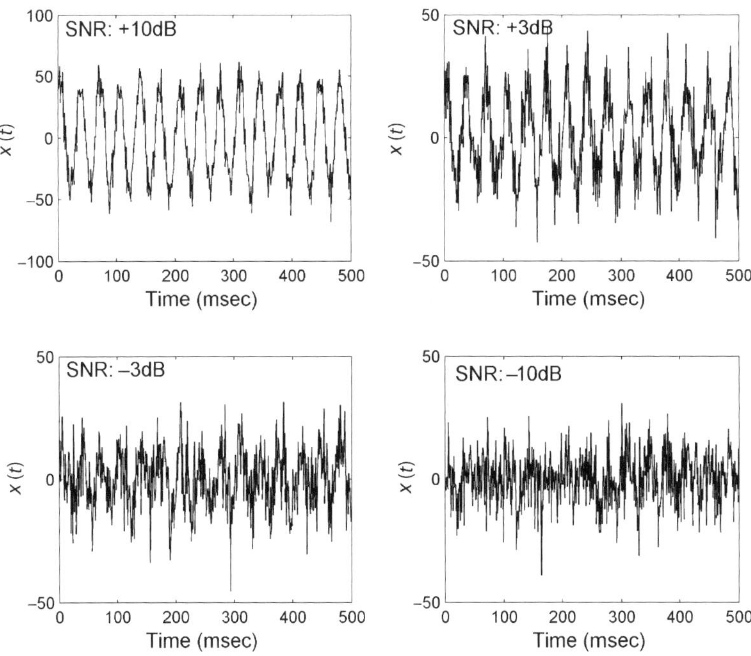

FIGURE 1.8 A 30-Hz sine wave with varying amounts of added noise. The sine wave is barely discernable when the SNR is −3 dB and not visible when the SNR is −10 dB.

is −3 dB you could still spot the signal, at least if you knew what to look for. When the SNR decreases to −10 dB it is impossible to determine if a sine wave is present visually.

1.4 SIGNAL PROPERTIES—BASIC MEASUREMENTS

Biosignals and other information-bearing signals are often quite complicated and defy a straightforward analytical description. An archetype biomedical signal is the electrical activity of the brain as it is detected on the scalp by electrodes termed the *electroencephalogram* or *EEG*, shown in Figure 1.9. While a time display of this signal, as in Figure 1.9, constitutes a unique description, the information carried by this signal is not apparent from the time display, at least not to the untrained eye. Nonetheless physicians and technicians are trained to extract useful diagnostic information by examining the time display of biomedical signals including the EEG. The time display of the ECG signal is so medically useful that it is displayed continuously for patients undergoing surgery or in intensive care units (ICUs). This signal has become an indispensable image in television and movie medical dramas. Medical images, which can be thought of as two-dimensional signals, often need only visual inspection to provide a useful diagnosis.

For some signals, a simple time display provides useful information, but many biomedical signals are not so easy to interpret from their time characteristics alone, and nearly all signals will benefit from some additional processing. For example, the time display of the EEG signal in Figure 1.9 may have meaning for a trained neurologist, but it is likely to be uninterpretable to most readers. A number of basic measurements can be applied to a signal to extract more information, while other types of analyses can be used to probe the signal for specific features. Finally, transformations can be used to provide a different

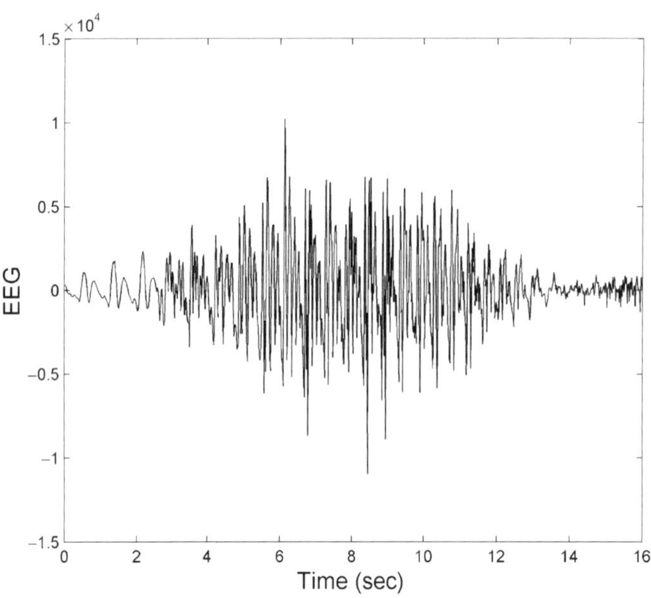

FIGURE 1.9 Segment of an EEG signal. *(From the PhysioNet data bank, Goldberger et al.)*

view of the signal. In this section, basic measurements are described, followed by more complex analyses. In the next chapter some basic transformations are developed.

1.4.1 Mean, Standard Deviation, and Variance

One of the most straightforward of signal measurements is the assessment of its average value. Averaging is most easily described in the digital domain. To determine the average of a series of numbers, simply add the numbers together and divide by the length of the series (i.e., the number of numbers in the series). Mathematically stated:

$$x_{avg} = \bar{x} = \frac{1}{N}\sum_{n=1}^{N} x_n \tag{1.15}$$

where n is an index number indicating a specific number in the series. In this text, the letters k and i are also be used to indicate index number. Also, in some mathematical formulations, it is common to let n (or k or i) range from 0 to $N-1$. In MATLAB the index must be nonzero, so in this text summations will range between 1 and N. The bar over the x in Equation 1.15 stands for "the average of . . ." Equation 1.15 only applies to a digital signal. For a continuous analog signal, $x(t)$, the summation becomes an integration. The average or mean of a continuous signal, the continuous version of Equation 1.15, is obtained by integrating the signal over time and dividing by the length of the signal, which is now a time length.

$$\bar{x}(t) = \frac{1}{T}\int_0^T x(t)\,dt \tag{1.16}$$

Comparing Equations 1.15 and 1.16 shows that the difference between digital and analog domain equations is the conversion of summation to integration, substitution of a time length variable, T, for N, and the use of a continuous variable, t, in place of the discrete integer, n. These conversion relationships are generally applicable, and most digital domain equations can be transferred to continuous or analog equations in this manner. We will usually use these relationships in reverse: the continuous domain equation is developed first, then the corresponding digital domain equation is derived by substituting summation for integration, N for T, and an integer variable for the continuous time variable. Again, the conditions under which a continuous analog signal and a digitized version of that signal can be considered the same are presented in Chapter 4.

Although the average value is a basic property of a signal, it does not provide any information about the variability of the signal. The root mean squared (RMS) value is a measurement that includes both the signal's variability and its average. Obtaining the RMS value of a signal is just a matter of following the measurement's acronym: first squaring the signal, then taking its average (i.e., mean), and finally taking the square root of this average:

$$x(t)_{rms} = \left[\frac{1}{T}\int_0^T x(t)^2\,dt\right]^{1/2} \tag{1.17}$$

The discrete form of the equation is obtained by following the simple rules described above.

$$x_{rms} = \left[\frac{1}{N}\sum_{n=1}^{N} x_n^2\right]^{1/2} \tag{1.18}$$

These equations describe the continuous (Equation 1.17) and discrete (Equation 1.18) versions of the RMS operation and also reflect the two approaches for implementation. Continuous equations require "analytical" solutions implemented by solving equations "by hand." Discrete equations can be solved by hand, but are usually implemented by computer. Example 1.4 uses the analytical approach to find the RMS of a sinusoidal signal while Example 1.5 uses digital implementation to solve the same problem.

EXAMPLE 1.4

Find the RMS value of the sinusoidal signal: $x(t) = A\sin(\omega_p t) = A\sin(2\pi t/T)$ where A is the amplitude and T is the period of the sinusoid.

Solution: Since this signal is periodic, with each period the same as the previous one, it is sufficient to apply the RMS equation over a single period. (This is true for most operations on sinusoids.) Neither the RMS value nor anything else about the signal will change from one period to the next. Applying Equation 1.17:

$$\overline{x}(t)_{rms} = \left[\frac{1}{T}\int_0^T x(t)^2\,dt\right]^{1/2} = \left[\frac{1}{T}\int_0^T \left(A\sin\left(\frac{2\pi t}{T}\right)\right)^2 dt\right]^{1/2}$$

$$= \left[\frac{1}{T}\frac{A^2}{2\pi}\left(-\cos\left(\frac{2\pi t}{T}\right)\sin\left(\frac{2\pi t}{T}\right) + \frac{\pi t}{T}\right)\Bigg|_0^T\right]^{1/2}$$

$$= \left[\frac{A^2}{2\pi}(-\cos(2\pi)\sin(2\pi) + \pi + \cos 0 \sin 0)\right]^{1/2}$$

$$= \left[\frac{A^2\pi}{2\,\pi}\right]^{1/2} = \left[\frac{A^2}{2}\right]^{1/2} = \frac{A}{\sqrt{2}} = 0.707\,A$$

Hence there is a proportional relationship between the amplitude of a sinusoid (A in this example) and its RMS value: specifically, the RMS value is $1/\sqrt{2}$ (rounded here to 0.707) times the amplitude, A, which is half the *peak to peak* amplitude. This is only true for sinusoids. For other waveforms, the application of Equation 1.17 or Equation 1.18 is required.

A statistical measure related to the RMS value is the *variance*, σ^2. The variance is a measure of signal variability irrespective of its average. The calculation of variance for both discrete and continuous signals is given as:

$$\sigma^2 = \frac{1}{T}\int_0^T (x(t) - \overline{x})^2\,dt \tag{1.19}$$

$$\sigma^2 = \frac{1}{N-1}\sum_{n=1}^{N}(x_n - \overline{x})^2 \tag{1.20}$$

where $\overline{x}$ is the mean or signal average (Equation 1.15). In statistics, the variance is defined in terms of an estimator known as the "expectation" operation and is applied to the probability distribution function of the data. Since the distribution of a signal is rarely known in advance, the equations given here are used to calculate variance in practical situations.

The *standard deviation* is another measure of a signal's variability and is simply the square root of the variance:

$$\sigma = \left[\frac{1}{T}\int_0^T (x(t) - \overline{x})^2\, dt\right]^{1/2} \tag{1.21}$$

$$\sigma = \left[\frac{1}{N-1}\sum_{n=1}^{N}(x_n - \overline{x})^2\right]^{1/2} \tag{1.22}$$

In determining the standard deviation and variance from discrete or digital data, it is common to normalize by $1/N-1$ rather than $1/N$. This is because the former gives a better estimate of the actual standard deviation or variance when the data are samples of a larger data set that has a normal distribution (rarely the case for signals). If the data have zero mean, the standard deviation is the same as the RMS value except for the normalization factor in the digital calculation. Nonetheless they come from very different traditions (statistics vs. measurement) and are used to describe conceptually different aspects of a signal: signal magnitude for RMS and signal variability for standard deviation.

Calculating the mean, variance, or standard deviation of a signal using MATLAB is straightforward. Assuming a signal vector[1] x, these measurements are determined using one of the three code lines below.

```
xm = mean(x);     % Evaluate mean of x
xvar = var(x);    % Variance of x normalizing by N-1
xstd = std(x);    % Evaluate the standard deviation of x
```

If x is not a vector, but a matrix, then the output is a row vector resulting from application of the calculation (mean, variance, or standard deviation) to each column of the matrix. The use of these MATLAB routines is demonstrated in the next two examples.

EXAMPLE 1.5

In MATLAB, generate a 4-Hz sine wave having an amplitude of 1.0. Use 500 points to produce the sine wave and assume a sample frequency of 500 Hz. (Hence the sine wave will be

[1]As mentioned above, MATLAB calls numerical strings or arrays "vectors." This term is derived from a mathematical contrivance that views a series of N numbers as a single vector in N-dimensional space. We will not use this feature, but we will still call signal arrays vectors to be compatible with MATLAB terminology.

1 second long and will include 4 cycles.) Plot the sine wave and calculate the RMS value. Compare this answer with the one determined analytically in Example 1.2.

Solution: The biggest challenge in this example is to generate the sine wave. We will take advantage of MATLAB's ability to handle vectors (i.e., strings of numbers) in a single statement. If we construct the proper time vector, `t`, then the 4-Hz sine wave can be constructed using the command: `x = sin(2*pi*f*t);` where `f` = 4-Hz. The vector `t` should consist of 500 points and range from 0 to 1 sec with a sample frequency of 500 Hz. Given the sample frequency and total number of points desired, an easy way to generate the appropriate time vector is: `t = (0:N-1)/fs;` or if `Ts` is given instead: `t = (0:N-1)*Ts;`. Alternatively, the sample interval or frequency could be specified along with the total desired time, `T`, instead of the total number of points, in which case `t` would be constructed as: `t = (0:1/fs:T)` or equivalently, `t = (0:Ts:T);`. These approaches to generating the time vector will be used throughout this text, so it is worthwhile spending an extra moment to study and understand them.

Once we have the time vector, constructing the sine wave is easy: `x = sin(2*pi*f*t);` where `f` is the sine wave frequency and is equal to 4 Hz in this example. Note how the MATLAB code looks just like the mathematical equation. We will see this similarity between the mathematics and MATLAB code on numerous occasions. To find the RMS value of the sine wave, we apply Equation 1.18 directly, taking the square root of the mean of the squared sine wave: `RMS = sqrt(mean(x.^2));`. Note that the .^ operator must be used since we want each point squared separately. The program built around these concepts is given below.

```
% Example 1.5 Program to generate a 500 point vector containing a 4 Hz sine wave
% with a sample frequency of 500 Hz
%
```

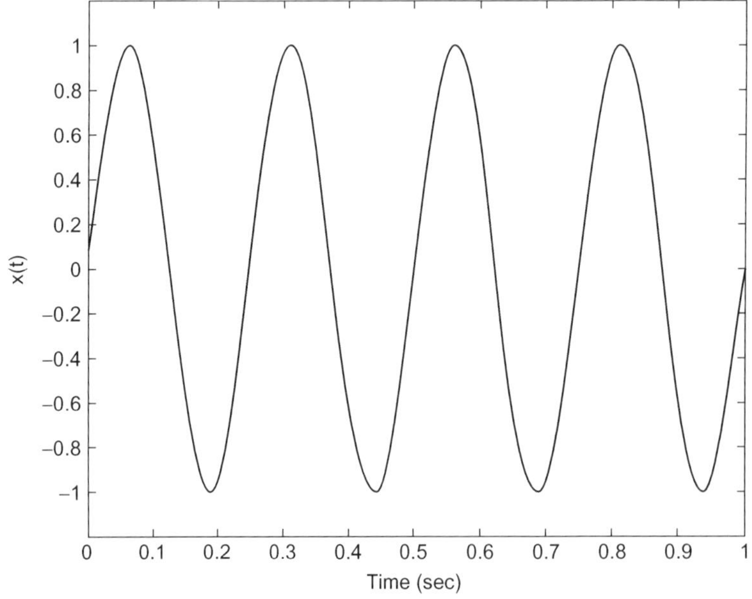

FIGURE 1.10 A 4-Hz sine wave produced by the code in Example 1.5.

```
fs = 500;                                   % Sample interval
N = 500;                                    % Number of points
f = 4;                                      % Frequency of sine wave
t = (0:N-1)/fs;                             % Generate time vector
x = sin(2*pi*f*t);                          % Generate sine wave
RMS = sqrt(mean(x.^2));                     % Calculate RMS
disp (RMS)                                  % and display
plot(t,x,'k','LineWidth',2);                % Plot sine wave
xlabel('Time(sec)'); ylabel('x(t)');        % Label axes
```

Analysis: The calculated RMS value was 0.7071, which is the same as that found in the last example, when the amplitude of the sine wave, A, equals 1.0. The plot generated by this program is shown in Figure 1.10. By plotting `x` against `t` (i.e., `plot(t,x);`), the horizontal axis is correctly scaled in seconds. In addition to being correctly scaled, both axes are labeled. You are always expected to label and scale the axes of your plots correctly as it is not difficult in MATLAB and it significantly improves the communicative power of the graph.

Another MATLAB example that uses some of the statistical measures described above is shown in the next example.

EXAMPLE 1.6

Plot the EEG signal shown in Figure 1.10 and show the range of ± one standard deviation.

Solution: The EEG data are stored as vector `eeg` in the file `eeg_data.mat` which is on this book's accompanying CD. The data were sampled at 50 Hz. The time vector will be constructed using one of the approaches in Example 1.5, but in this case the desired number of points is dictated by the length of the EEG data, which can be determined using the MATLAB length routine. We will calculate the standard deviation using both a direct implementation of Equation 1.22 and MATLAB's `std` routine. To show the limits of one standard deviation, we will plot two horizontal lines: one at the positive value of the calculated standard deviation and the other at the negative value. The mean should be added to the two horizontal lines in case the mean of the data is nonzero. (In this example, the mean of the EEG data is zero so that subtracting the mean is not really necessary, but we often write code to be more general than is needed for the immediate application.)

```
% Example 1.6 Program to plot EEG data and ± one standard deviation
%
load eeg_data;                                        % Load the data (from the CD)
fs = 50;                                              % Sample frequency of 50 Hz
N = length(eeg);                                      % Get data length
t = (0:N-1)/fs;                                       % Generate time vector
plot(t,eeg,'k'); hold on;                             % Plot the data
std_dev = sqrt(sum((eeg - mean(eeg)).^2)/(N-1));      % Eq. 1.22
std_dev1 = std(eeg);                                  % MATLAB standard deviation
avg = mean(eeg);                                      % Determine the mean of EEG
disp([std_dev std_dev1])                              % Display both standard deviations
plot([t(1) t(end)],[std_dev + avg std_dev + avg],':k');
plot([t(1) t(end)],[-std_dev + avg -std_dev + avg],':k');
```

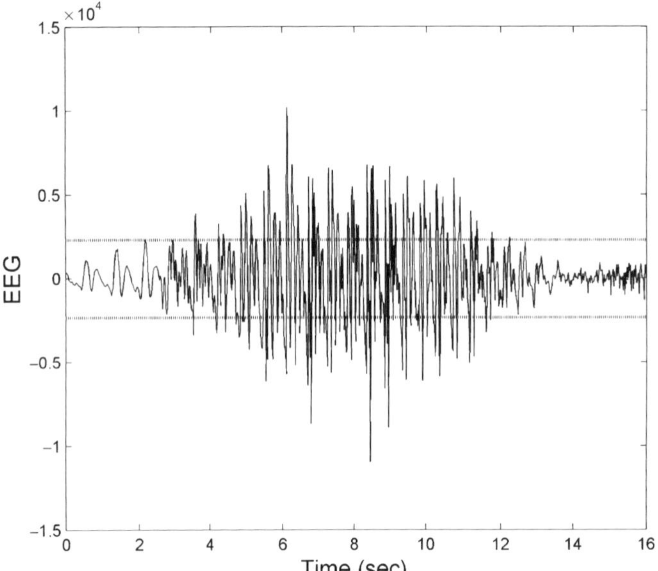

FIGURE 1.11 The segment of an EEG signal with the positive and negative standard deviations indicated by a dotted horizontal line.

The results of this example are shown in Figure 1.11. Both methods for calculating the standard deviation produce identical results: 2.3102.

1.4.2 Averaging and Ensemble Averaging

When multiple measurements are made, multiple values or signals will be generated. If these measurements are combined or added together, the means add so that the combined value or signal has a mean that is the average of the individual means. The same is true for the variance: the variances add and the average variance of the combined measurement is the mean of the individual variances:

$$\overline{\sigma}^2 = \frac{1}{\mathrm{N}}\sum_{n=1}^{\mathrm{N}} \sigma_n^2 \tag{1.23}$$

When a number of signals are added together, it can be shown that the noise components are reduced by a factor equal to $1/N$, where N is the number of measurements that are averaged.

$$\overline{\sigma_{\text{Noise}}} = \sigma/\sqrt{\mathrm{N}} \tag{1.24}$$

In other words, averaging measurements from different sensors, or averaging multiple measurements from the same source, will reduce the standard deviation of the measurement's variability or noise by the square root of the number of averages. For this reason, it is common to make multiple measurements whenever possible and average the results.

Averaging can be applied to entire signals, a technique known as *ensemble averaging*. Ensemble averaging is a simple yet powerful signal processing technique for reducing noise when multiple observations of the signal are possible. Such multiple observations could come from multiple sensors, but in many biomedical applications the multiple observations come from repeated responses to the same stimulus. In *ensemble averaging*, a number of time responses are averaged together on a point-by-point basis; that is, a signal with a lower standard deviation is constructed by taking the average, for each point in time, over all signals in the ensemble. A classic biomedical engineering example of the application of ensemble averaging is the *visual evoked response* (VER) in which a visual stimulus produces a small neural signal embedded in the EEG. Usually this signal cannot be detected in the EEG signal, but by averaging hundreds of observations of the EEG, time-locked to the visual stimulus, the visually evoked signal emerges.

There are two essential requirements for the application of ensemble averaging for noise reduction: the ability to obtain multiple observations, and a reference signal closely time-linked to the response. The reference signal shows how the multiple observations are to be aligned for averaging. Usually a time signal linked to the stimulus is used. An example of ensemble averaging is given in Example 1.7.

EXAMPLE 1.7

Find the average response of a number of individual VER. This is an EEG signal recorded near the visual cortex following a visual stimulus such as a flash of light. Individual recordings have too much noise (probably other brain activity, but noise in this case) to see the neural response produced by the stimulus. But if multiple responses are recorded to the same stimulus, then averaging an ensemble of EEG signals that immediately follow the stimulus can reduce the noise sufficiently so that the signal can be observed. Often 100 or more responses are required. These responses are stored in MATLAB file `ver.mat` which contains variable `ver`, a simulation of the VER. These data were sampled with a sampling interval of 0.005 sec (i.e., 5 msec). Also plot one of the individual responses to show the "raw" EEG signal.

Solution: Use the MATLAB averaging routine `mean`. If this routine is given a matrix variable, it averages each column. Hence if the various signals are arranged as rows in the matrix, the mean routine will produce the ensemble average. To determine the orientation of the data, check the size of the data matrix. Normally the number of signal data points will be greater than the number of signals. Since we want the signals to be in rows, if the number of rows is greater than the number of columns then transpose the data using the MATLAB transposition operator (i.e., the ' symbol).

```
% Example 1.7
load ver;               % Get visual evoked response data;
fs = 1/.005;            % Sample interval = 5 msec
[nu,N] = size(ver);     % Get data matrix size
```

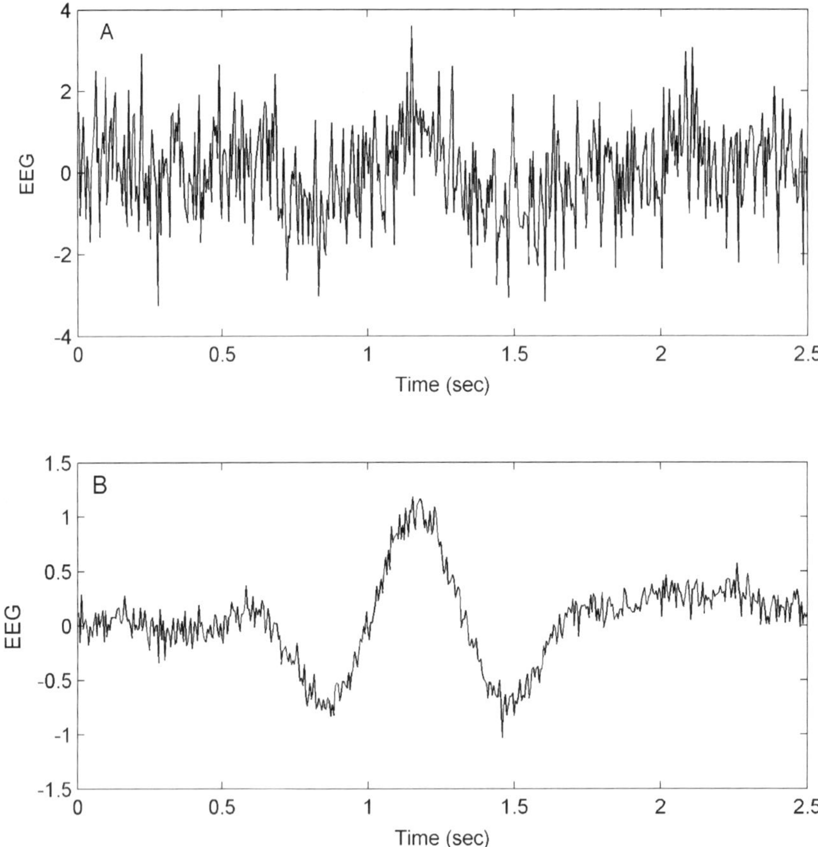

FIGURE 1.12 A) The raw EEG signal showing a single response to an evoked stimulus. B) The ensemble average of 100 individual signals showing the VER.

```
if nu > N
  ver = ver';
  t = (1:nu)/fs;           % Generate time vector
else
  t = (1:N)/fs;            % Time vector if no transpose
end
%
plot(t,ver(1,:));          % Plot individual record
   .......label axes.........
% Construct and plot the ensemble average
avg = mean(ver);           % Calculate ensemble average
figure;
plot(t,avg);               % Plot ensemble average other data
.......label axes.........
```

This program produces the results shown in Figure 1.12. Figure 1.12A shows a single response, and the evoked response is only remotely discernable. In Figure 1.12B the ensemble average of 100 individual responses produced by a light flash is clearly visible.

The next and last example uses the VER data of Example 1.7 to evaluate the noise reduction properties of ensemble averaging and compares the reduction in noise due to averaging with the theoretical prediction of Equation 1.24.

EXAMPLE 1.8

This example evaluates the noise reduction provided by ensemble averaging. Load the VER data used in Example 1.8 along with the actual, noise-free VER (found as variable `actual_ver in file ver.mat`). Construct an ensemble average of 25 and 100 responses. Subtract the noise-free VER from an individual evoked response and the two ensemble averages to get an estimate of just the noise in the three waveforms. Compare the variances of the unaveraged and the two averaged waveforms with that predicted theoretically.

Solution: Load and transpose the visual evoked data used in Example 1.7. Construct two averages: one using the first 25 responses and the other using all 100 responses. Subtract the actual VER from the two averaged waveforms and an unaveraged waveform and take the standard deviation of the result.

```
% Example 1.8 Evaluation of noise reduction from ensemble averaging.
%
  ......Load data and transpose. Same as in Example 1.7........
% Construct the ensemble averages
avg1 = mean(ver);                     % Calculate ensemble avg, all responses
avg2 = mean(ver(1:25,:));             % Calculate ensemble avg, 25 responses
%
% Estimate the noise components
ver_noise = ver(1,:) - actual_ver;    % Single response
avg1_noise = avg1 - actual_ver;       % Overall average response
avg2_noise = avg2 - actual_ver;       % Avg of 25 responses noise component
std_ver = std(ver_noise.^2)           % Output standard deviations
std_avg1 = std(avg1_noise.^2)
std_avg2 = std(avg2_noise.^2)
```

Results: The output from this code was:

```
std_ver = 1.0405; std_avg1 = 0.1005; std_avg2 = 0.1986
```

Based on Equation 1.24, which states that the noise should be reduced by $\sqrt{N}$ where N is the number of responses in the average, we would expect the `std_avg1` value to be $\sqrt{100} = 10$ times less than the noise in an individual record. The standard deviation after averaging was 0.1005, quite close to the expected value of 0.1041. Similarly, the variable `std_avg2` should be $\sqrt{25} = 5$ times less than the unaveraged standard deviation, or 0.2081. The actual value was 0.1986, close, but not exactly as predicted. The values may not be exactly as predicted because noise is a random process and the results of ensemble averaging are subject to some randomness. The noise reduction characteristics of ensemble averaging and its limitations are further explored in Problem 14.

1.5 SUMMARY

From a traditional reductionist viewpoint, living things are described in terms of component systems. Many traditional physiological systems such as the cardiovascular, endocrine, and nervous systems are quite extensive and composed of many smaller subsystems. Biosignals provide communication between systems and subsystems and are our primary source of information on the behavior of these systems. Interpretation and transformation of signals are major focuses of this text.

Biosignals, like all signals, must be "carried" by some form of energy. Common biological energy sources include chemical, mechanical, and electrical energy (in the body, electrical signals are carried by ions, not electrons). In many diagnostic approaches, externally derived energy is used to probe the body as in ultrasound and CT scanning. Measuring a signal, whether of biological or external origin, usually entails conversion to an electric signal, a signal carried by electrons, to be compatible with computer processing. Signal conversion is achieved by a device known as an input transducer, and when the energy is from a living system it is termed a *biotransducer*. The transducer often determines the general quality of a signal and so may be the most important element in a medical device.

Most biological signals exist in the analog domain: they consist of continuous, time-varying fluctuations in energy. These signals are usually converted to the digital domain so they are compatible with digital computers. Conversion of analog to digital (or discrete-time) signals requires slicing the continuous signal into discrete levels of both amplitude and time and is accomplished with an electronic device known, logically, as an analog-to-digital converter. Amplitude slicing adds noise to the signal, and the noise level is dependent on the size of the slice, but typical slices are so small this noise is usually ignored. Time slicing produces a more complicated transformation that is fully explored in Chapter 4.

By definition, signal is what you want and noise is everything else; all signals contain some noise. The quality of a signal is often specified in terms of a ratio of the signal amplitude or power-to-noise amplitude or power, a ratio termed the signal-to-noise (SNR), frequently given in decibels (dB). Noise has many sources and only electronic noise can be easily evaluated based on known characteristics of the instrumentation. Since electronic noise exists at all frequencies, reducing the frequency ranges of a signal is bound to reduce noise. That provides the motivation for filtering a signal, as detailed in Chapter 8.

Some basic measurements that can be made on signals include mean values, standard deviations, and RMS values, all easily implemented in MATLAB. Signal or ensemble averaging is a simple yet powerful tool for reducing noise, but requires multiple observations of a signal, which are not always possible. Isolating evoked neural responses such as the visual evoked response (VER) requires averaging hundreds of responses to a repetitive stimulus. Recovering the very weak evoked response heavily buried in the EEG would not be possible without ensemble averaging.

PROBLEMS

1. Use Equation 1.17 to analytically determine the RMS value of a "square wave" with amplitude of 1.0 volt and a period of 0.2 sec.

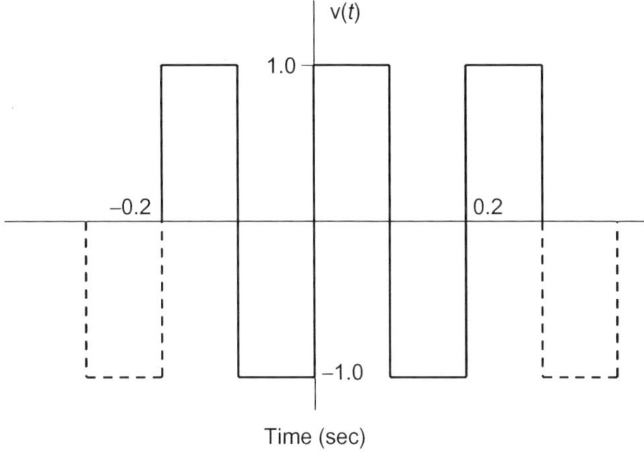

2. Generate one cycle of the square wave similar to the one shown above in a 500-point MATLAB array. Determine the RMS value of this waveform. When you take the square of the data array be sure to use a period before the up arrow so that MATLAB does the squaring point-by-point (i.e., `x.^2`).
3. Use Equation 1.17 to analytically determine the RMS value of the waveform shown below with amplitude of 1.0 volt and a period of 0.5 sec. [*Hint*: You can use the symmetry of the waveform to reduce the calculation required.]

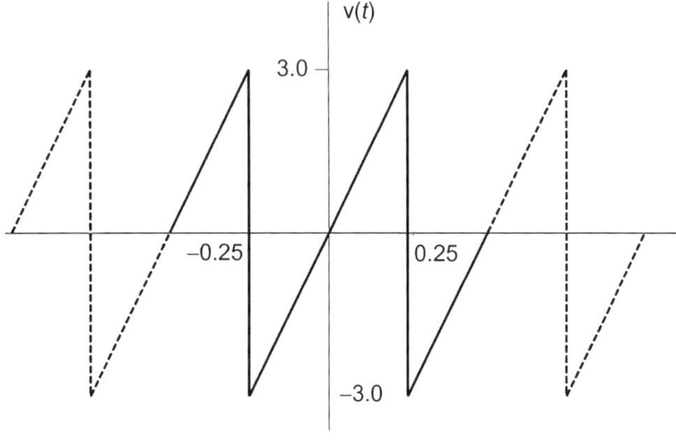

4. Generate the waveform shown for Problem 3 above in MATLAB. Use 1000 points to produce one period. Take care to determine the appropriate time vector. (Constructing the function in MATLAB is more difficult that the square wave of Problem 2, but can still be done in one line of code.) Calculate the RMS value of this waveform as in Problem 2. Plot this function to ensure you have constructed it correctly.
5. Fill a MATLAB vector array with 1000 Gaussianly distributed numbers (i.e., `randn`) and another with 1000 uniformly distributed numbers (i.e., `rand`). Find the mean and standard deviation of both sets of numbers. Modify the array of uniformly distributed numbers to have a mean value of zero. Confirm it has a mean of zero and recalculate the standard deviation.
6. The file on this book's accompanying CD `amplitude_slice.mat` contains a signal, `x`, and an amplitude sliced version of that signal `y`. This signal has been sliced into 16 different levels. In Chapter 4 we will find that amplitude slicing is like adding noise to a signal. Plot the two signals superimposed and find the RMS value of the noise added to the signal by the quantization process. [*Hint*: Subtract the sliced signal from the original and take the RMS value of this difference.]
7. Load the data in `ensemble_data.mat` found in the CD. This file contains a data matrix labeled "data." The data matrix contains 100 responses of a signal in noise. In this matrix each row is a separate response. Plot several randomly selected samples of these responses. Is it possible to identify the signal from any single record? Construct and plot the ensemble average for these data. Also construct and plot the ensemble standard deviation.
8. A resistor produces 10 μV noise (i.e., 10×10^{-6} volts noise) when the room temperature is 310 K and the bandwidth is 1 kHz (i.e., 1000 Hz). What current noise would be produced by this resistor?
9. The noise voltage out of a 1 MΩ (i.e.,10^6 ohm) resistor is measured using a digital voltmeter as 1.5 μV at a room temperature of 310 K. What is the effective bandwidth of the voltmeter?
10. A 3-ma current flows through both a diode (i.e., a semiconductor) and a 20,000-Ω (i.e., 20-kΩ) resistor. What is the net current noise, i_n? Assume a bandwidth of 1 kHz (i.e., 1×10^3 Hz). Which of the two components is responsible for producing the most noise?
11. If a signal is measured as 2.5 volts and the noise is 28 mV (28×10^{-3} volts), what is the SNR in dB?
12. A single sinusoidal signal is found in a large amount of noise. (If the noise is larger than the signal, the signal is sometimes said to be "buried in noise.") If the RMS value of the noise is 0.5 volts and the SNR is 10 dB, what is the RMS amplitude of the sinusoid?
13. The file `signal_noise.mat` contains a variable `x` that consists of a 1.0 volt peak sinusoidal signal buried in noise. What is the SNR for this signal and noise? Assume that the noise RMS is much greater than the signal RMS.
14. This problem evaluates noise reduction in ensemble averaging. Load the `Ver` data, variable `ver`, along with the actual, noise-free VER in variable `actual_ver`. Both these variables can be found in file `Probl_12_data.mat`. The visual response data set consists of 1000 responses, each 100 points long. The sample interval is 5 msec. Construct an

ensemble average of 25, 100, and 1000 responses. The two variables are in the correct orientation and do not have to be transposed. Subtract the noise-free variable `(actual_ver)` from an individual evoked response and the three ensemble averages to get an estimate of the noise in the three waveforms. Compare the standard deviations of the unaveraged with the three averaged waveforms. How does this compare with the reduction predicted theoretically by Equation 1.22? Note that there is a practical limit to the noise reduction that you can obtain by ensemble averaging. Can you explain what factors might limit the reduction in noise as the number of responses in the average gets very large?

CHAPTER

2

Basic Concepts in Signal Processing

2.1 BASIC SIGNALS—THE SINUSOIDAL WAVEFORM

Signals are the foundation of information processing, transmission, and storage. Signals also provide the interface for physiological systems and are the basis for communication between biological processes (see Figure 2.1). Given the ubiquity of signals both within and outside the body, it should be no surprise that understanding basic signal concepts is fundamental to understanding, and interacting with, biological processes.

A few signals are simple and can be defined analytically as mathematical functions. For example, a sine wave signal is defined by the equation:

$$x(t) = A\sin(\omega_p t) = A\sin(2\pi f_p t) = A\sin\left(\frac{2\pi t}{T}\right) \tag{2.1}$$

where A is the amplitude of the signal. Three different methods for describing the sine wave frequency are shown: ω_p is the frequency in rad per sec, f_p is the frequency in Hz, and T is the period in sec. The sine wave is a periodic signal, one that repeats exactly after the time period T. The formal mathematical definition of a periodic signal is:

$$x(t) = x(t+T) \tag{2.2}$$

Recall that frequency can be expressed in either rad or Hz (the units formerly known as "cycles per second") and the two are related by 2π:

$$\omega_p = 2\pi f_p \tag{2.3}$$

Both forms of frequency are used in this text, although frequency in Hz is most common in engineering. The reader should be comfortable with both. The frequency is also the inverse of the period, T:

$$f_p = \frac{1}{T} \tag{2.4}$$

FIGURE 2.1 Signals pass continuously between various parts of the body. These "biosignals" are carried either by electrical energy, as in the nervous system, or by molecular signatures, as in the endocrine system and many other biological processes. Measurement of these biosignals is fundamental to diagnostic medicine and to bioengineering research.

The signal defined by Equation 2.1 is completely quantified by A and f_p (or ω_p or T); once you specify these two terms you have characterized the signal for all time. The sine wave is rather boring: if you've seen one cycle you've seen them all. Worse, the boredom goes on forever since the sine wave of Equation 2.1 is infinite in duration. Since the signal is completely defined by A (amplitude) and f_p (frequency), if neither changes over time, this signal would not convey much information. These limitations notwithstanding, sine waves (and cosine waves) are the foundation of many signal analysis techniques. Some reasons why sine waves are so important in signal processing are given in the next chapter, but part of their importance stems from their simplicity.

Sine-wave-like signals can also be represented by cosines, and the two are related.

$$A\cos(\omega t) = A\sin\left(\omega t + \frac{\pi}{2}\right) = A\sin(\omega t + 90^\circ)$$
$$A\sin(\omega t) = A\cos\left(\omega t - \frac{\pi}{2}\right) = A\cos(\omega t - 90^\circ) \tag{2.5}$$

Note that the right side of these equations (i.e., $A\sin(\omega t + 90^\circ)$ and $A\cos(\omega t - 90^\circ)$) have confused units: the first part of the sine argument, ωt, is in rad while the second part is in deg. Nonetheless, this is fairly common usage, and the form that is used throughout this text. However, when MATLAB computations are involved, the trig function arguments must be in rad. To convert between degrees (deg) and radians (rad), note that 2π rad = 360 deg, so 1 rad = $360/2\pi$ deg and 1 deg = $2\pi/360$ rad.

A general *sinusoid* (as opposed to a pure sine wave or pure cosine wave) has a nonzero phase term, as shown in Equation 2.6,

$$x(t) = A\sin(\omega_p t + \theta) = A\sin(2\pi f_p t + \theta) = A\sin\left(\frac{2\pi t}{T} + \theta\right) \text{ or equivalently}$$
$$x(t) = A\cos(\omega_p t - \theta) = A\cos(2\pi f_p t - \theta) = A\cos\left(\frac{2\pi t}{T} - \theta\right) \tag{2.6}$$

where again the phase θ would be expressed in deg even though the frequency descriptor (ω_p, or $2\pi f_p t$, or $2\pi t/T$) is expressed in rad or Hz. Many of the sinusoidal signals used in this text will follow the format of Equation 2.6.

Example 2.1 uses MATLAB to generate two sinusoids that differ by 60 deg.

EXAMPLE 2.1

Use MATLAB to generate and plot two sinusoids that are 60 deg out of phase.

Solution: We need first to set the sampling frequency and data length. Since no sampling frequency is specified, we arbitrarily assume a sampling frequency of 1 kHz. We then generate a time vector 500 points long; again an arbitrary length, but certainly enough points to generate a smooth plot. Again, since the frequency of the sinusoids was not specified, we construct 2-Hz sinusoids using the MATLAB sine function. The 2-Hz sinusoids give 1 cycle over the 0.5-sec length of the time vector and should make for a nice plot.

```
% Example 2.1 Plot two sinusoids 60 deg out of phase
%
fs = 1000;                      % Assumed sampling frequency
N = 500;                        % Number of points
t = (0:N-1)/fs;                 % Generate time vector
phase = 60*(2*pi/360);          % 60 deg phase; convert to rad
x1 = sin(2*pi*2*t);             % Construct sinusoids
x2 = sin(2*pi*2*t - phase);     % One with a -60 phase shift
hold on;
plot(t,x1);                     % Plot the unshifted sinusoid
plot(t,x2,'- -');               % Plot using a dashed line
plot([0 .5], [0 0]);            % Plot horizontal line
  ......label axes......
```

Analysis: This simple program illustrates how to construct a time vector of a specified length for a given sampling frequency. The graph produced by the example program is shown in Figure 2.2.

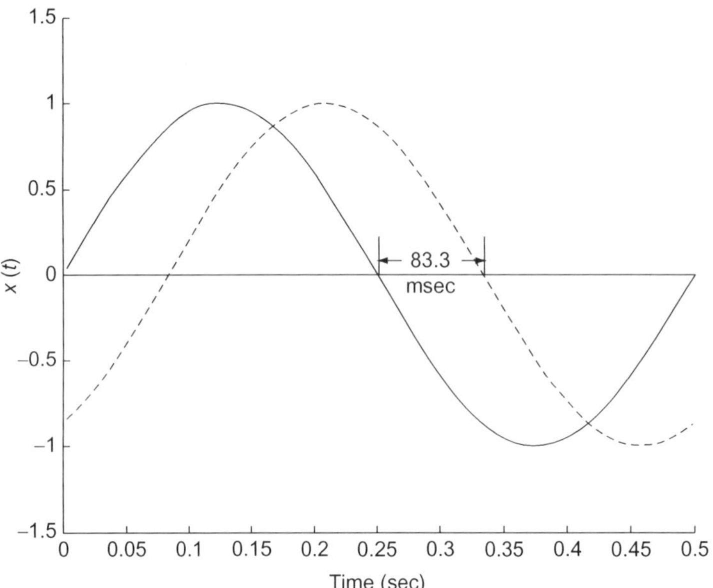

FIGURE 2.2 Two 2-Hz sinusoids that differ in phase by 60 deg. This phase difference translates to a time difference or delay of 83.3 msec, as shown.

To convert the difference in phase angle to a difference in time, note that the phase angle varies through 360 deg during the course of one period, and a period is T sec long. Hence, to calculate the time difference between the two sinusoids given the phase angle θ:

$$t_d = \frac{\theta}{360}T = \frac{\theta}{360f} \tag{2.7}$$

For the 2-Hz sinusoids in Figure 2.2, $T = 0.5$ sec, so substituting into Equation 2.7:

$$t_d = \frac{\theta}{360}T = \frac{60}{360}0.5 = 0.0833 \text{ sec} = 83.3 \text{ msec}$$

EXAMPLE 2.2

Find the time difference between two sinusoids:

$$x_1(t) = \cos(4t + 30), \text{ and } x_2(t) = -2\sin(4t)$$

Solution: This has the added complication that one waveform is defined in terms of the cosine function and the other in terms of the sine function. So we need to convert them both to either a sine or cosine. Here we convert to cosines:

$$x_2(t) = -2\cos(4t - 90)$$

So the total angle between the two sinusoids is $30 - (-90) = -120$ deg, since the period is:

$$T = \frac{1}{f} = \frac{2\pi}{\omega} = \frac{2\pi}{4} = 1.57 \text{ sec}$$

and the time delay is:

$$t_d = \frac{\theta}{360}T = \frac{120}{360}1.57 = 0.523 \text{ sec.}$$

2.1.1 Sinusoidal Arithmetic

Equation 2.6 describes a fairly intuitive way of thinking about a sinusoid as a sine wave with a phase shift. Alternatively, a cosine could just as well be used instead of the sine in Equation 2.6 to represent a general sinusoid, and in this text we use both. Sometimes it is mathematically convenient to represent a sinusoid as a combination of a pure sine and a pure cosine rather than a single sinusoid with a phase angle. This representation can be achieved using the well-known trigonometric identity for the difference of two arguments of a cosine function:

$$\cos(x - y) = \cos(x)\cos(y) + \sin(x)\sin(y) \tag{2.8}$$

Based on this identity, the equation for a sinusoid can be written as:

$$\begin{aligned} C\cos(2\pi ft - \theta) &= C\cos(\theta)\cos(2\pi ft) + C\sin(\theta)\sin(2\pi ft) \\ &= a\cos(2\pi ft) + b\sin(2\pi ft)\text{where} \end{aligned} \tag{2.9}$$

Note that θ is defined as negative in these equations. To convert the other way, from a sine and cosine to a single sinusoid with amplitude C and angle θ, start with Equation 2.9.

If $a = C\cos(\theta)$ and $b = C\sin(\theta)$, then to determine C, square both equations and sum:

$$\begin{aligned} a^2 + b^2 &= C^2(\cos^2\theta + \sin^2\theta) = C^2 \\ C &= \sqrt{a^2 + b^2} \end{aligned} \tag{2.10}$$

The calculation for θ given a and b is:

$$\begin{aligned} \frac{b}{a} &= \frac{C\sin(\theta)}{C\cos(\theta)} = \tan(\theta) \quad \text{and}\ldots \\ \theta &= \tan^{-1}\left(\frac{\mathrm{b}}{\mathrm{a}}\right) \end{aligned} \tag{2.11}$$

Care must be taken in evaluating Equation 2.11 to insure that θ is determined to be in the correct quadrant based on the signs of a and b. Many calculators and MATLAB's `atan` function will evaluate θ as in the first quadrant (0 to 90 deg) if the ratio of b/a is positive, and in the fourth quadrant (270 to 360 deg) if the ratio is negative. If both a and b are positive, then θ is positive, but if b is positive and a is negative, then θ is in the second quadrant between 90 and 180 deg, and if both a and b are negative, θ must be in the third quadrant between 180 and 270 deg. Finally if b is negative and a is positive, then θ must belong in the fourth quadrant between 270 and 360 deg. When using MATLAB it is possible to avoid these concerns by using the `atan2(b,a)` routine which correctly outputs the arctangent of the ratio of b over a (in rad) in the appropriate quadrant.

To add sine waves simply add their amplitudes. The same applies to cosine waves:

$$\begin{aligned} a_1 \cos(\omega t) + a_2 \cos(\omega t) &= (a_1 + a_2) \cos(\omega t) \\ a_1 \sin(\omega t) + a_2 \sin(\omega t) &= (a_1 + a_2) \sin(\omega t) \end{aligned} \tag{2.12}$$

To add two sinusoids (i.e., $C \sin(\omega t + \theta)$ or $C \cos(\omega t - \theta)$), convert them to sines and cosines using Equation 2.9, add sines to sines and cosines to cosines, and convert back to a single sinusoid if desired.

EXAMPLE 2.3

Convert the sum of a sine and cosine wave, $x(t) = -5 \cos(10t) - 3 \sin(10t)$ into a single sinusoid.

Solution: Apply Equations 2.10 and 2.11.

$$a = -5 \quad \text{and} \quad b = -3$$

$$C = \sqrt{a^2 + b^2} = \sqrt{-5^2 - 3^2} = 5.83$$

$$\theta = \tan^{-1}\left(\frac{b}{a}\right) = \tan^{-1}\left(\frac{-3}{-5}\right) = 31 \text{ deg},$$

but θ must be in the 3rd quadrant since both a and b are negative:

$$\theta = 31 + 180 = 211 \text{ deg}.$$

Therefore, the single sinusoid representation would be:

$$x(t) = C \cos(\omega t - \theta) = 5.83 \cos(10t - 211^\circ)$$

Analysis: Using the above equations, any number of sines, cosines, or sinusoids can be combined into a single sinusoid providing they are all at the same frequency. This is demonstrated in Example 2.4.

EXAMPLE 2.4

Combine $x(t) = 4 \cos(2t - 30^\circ) + 3 \sin(2t + 60^\circ)$ into a single sinusoid.

Solution: First expand each sinusoid into a sum of cosine and sine, then algebraically add the cosines and sines, then recombine them into a single sinusoid. Be sure to convert the sine into a cosine (recall Equation 2.5: $\sin(\omega) = \cos(\omega t - 90^\circ)$) before expanding this term.

$4\cos(2t - 30) = a\cos(2t) + b\sin(2t)$

where: $a = C\cos(\theta) = 4\cos(-30) = 3.5$ and $b = C\sin(\theta) = 4\sin(-30) = -2$

$4\cos(2t - 30) = 3.5\cos(2t) - 2\sin(2t)$

Convert the sine to a cosine then decompose the sine into a cosine plus a sine:

$$3\sin(2t + 60) = 3\cos(2t - 30) = 2.6\cos(2t) - 1.5\sin(2t)$$

Combine cosine and sine terms algebraically:

$$4\cos(2t - 30) + 3\sin(2t + 60) = (3.5 + 2.6)\cos(2t) + (-2 - 1.5)\sin(2t)$$
$$= 6.1\cos(2t) - 3.5\sin(2t)$$
$$= C\cos(2t + \theta) \text{ where: } C = \sqrt{6.1^2 + 3.5^2} \text{ and } \theta = \tan^{-1}\left(\frac{-3.5}{6.1}\right)$$

$C = 7$; $\theta = -30$ (Since b is negative, θ is in 4th quadrant.) so $\theta = -30$ deg

$x(t) = 7\cos(2t - 30°)$

This approach can be extended to any number of sinusoids. An example involving three sinusoids is found in Problem 4.

2.1.2 Complex Representation

An even more compact representation of a sinusoid can be expressed if one is willing to use complex notation. A *complex number* combines a *real number* and an *imaginary number*. Real numbers are what we commonly use, while imaginary numbers are the result of square roots of negative numbers and are represented as real numbers multiplied by $\sqrt{-1}$. In mathematics $\sqrt{-1}$ is represented by the letter i, while engineers tend to use the letter j, reserving the letter i for current. So while 5 is a real number, $j5$ (i.e., $5\sqrt{-1}$) is an imaginary number. A complex number is a combination of a real and imaginary number such as $5 + j5$. A *complex variable* simply combines a real and an imaginary variable: $z = x + jy$. The arithmetic of complex numbers and some of their important properties are reviewed in Appendix E. We use complex variables and complex arithmetic extensively in later chapters so it is worthwhile to review these operations.

The beauty of complex numbers and complex variables is that the real and imaginary parts are *orthogonal*. One consequence of orthogonality is that the real and imaginary parts of a complex number (or variable) can be represented by plotting these components perpendicularly, as shown in Figure 2.3.

Orthogonality is discussed in more detail later, but the fundamental importance of orthogonality with regard to complex numbers is that the real number or variable does not interfere or interact with the imaginary number or variable and vice versa. Operations on one component do not affect the other. As a trivial example, if $3 + j4$ is changed to $3 + j5$, the real parts are still the same. This means that a complex number is actually two separate numbers rolled into one and a complex variable is two variables in one. This feature comes in particularly handy when sinusoids are involved, since any sinusoid at a given frequency can be uniquely defined by two variables: its magnitude and phase angle or, equivalently, using Equation 2.9, its cosine and sine magnitudes (a and b). Hence a sinusoid at a given frequency can, in theory, be represented by a single complex number.

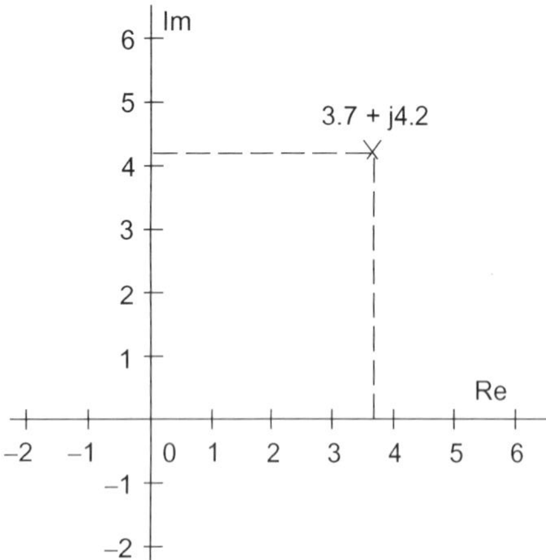

FIGURE 2.3 A complex number represented as an orthogonal combination of a real number on the horizontal axis and an imaginary number on the vertical axis. This graphic representation is useful for understanding complex numbers and aids in the interpretation of some arithmetic operations.

To find the complex representation, we use the equation developed by the Swiss mathematician, Leonhard Euler (pronounced "oiler"). The use of the symbol e for the base of the natural logarithmic system is a tribute to his extraordinary mathematical contributions.

$$e^{\pm jx} = \cos x \pm j \sin x \tag{2.13}$$

The derivation for this equation is given in Appendix A. This equation links sinusoids and exponentials, providing a definition of the sine and cosine in terms of complex exponentials as shown in Appendix C. It also provides a concise representation of a sinusoid, although a few extra mathematical features are required to account for the fact that the second term is an imaginary sine term. This equation will prove useful in two sinusoidally-based analysis techniques: Fourier analysis described in the next chapter and phasor analysis described in Chapter 5.

MATLAB variables can be either real or complex and MATLAB deals with them appropriately. A typical MATLAB number is identified by its real and imaginary part: `x = 2 + 3i` or `x = 2 + 3j`. MATLAB assumes that both `i` and `j` stand for the complex operator unless they are defined to mean something else in your program. Another way to define a complex number is using the complex routine: `z = complex(x,y)` where `x` and `y` are two vectors. The first argument, `x`, is taken as the real part of `z` and the second, `y`, is taken as the imaginary part of z. To find the real and imaginary components of a complex number or vector, use `real(z)` to get the real component and `imag(z)` to get the imaginary component. To find the polar components of a complex variable use `abs(z)` to get the magnitude component and `angle(z)` to get the angle component (in rad).

EXAMPLE 2.5

This example shows a demonstration of the Euler equation in MATLAB. Generate 1 cycle of the complex sinusoid given in Equation 2.13 using a 1.0-sec time vector. Assume $f_s = 500$-Hz.

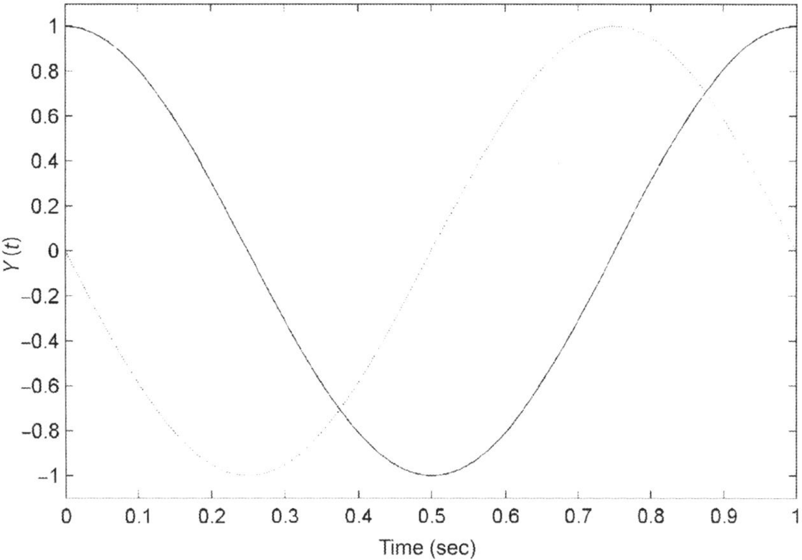

FIGURE 2.4 A one-cycle cosine and sine wave produced by plotting the real and imaginary part of the Euler exponential $e^{(-j2\pi t)}$ where t is a 500-point vector varying between 0 and 1.

Plot the real and imaginary parts of this function superimposed to produce a cosine and sine wave.

Solution: Construct the time vector as described in Example 1.3. Then generate e^{-jx} as in Equation 2.13 where $x = 2\pi t$ so that the complex sinusoid is one cycle (x will range from 0 to 2π as t goes from 0 to 1.0). Plot the real and imaginary part using MATLAB's `real` and `imag` routines.

```
% Example 2.5 Demonstration of the Euler equation in MATLAB
%
Tt = 1;                                  % Total time
fs = 500;                                % Sampling frequency
t = (0:1/fs:Tt);                         % Time vector
z = exp(-j*2*pi*t);                      % Complex sinusoid (Equation 2.13)
plot(t,real(z),'k',t,imag(z),':k');      % Plot result
  .......labels........
```

The sine and cosine wave produced by this program are seen in Figure 2.4.

2.2 MORE BASIC SIGNALS—PERIODIC, APERIODIC, AND TRANSIENT

In the next chapter, we will discover why the sinusoid is the most important type of signal in signal processing. However, there are a number of other signals that are also

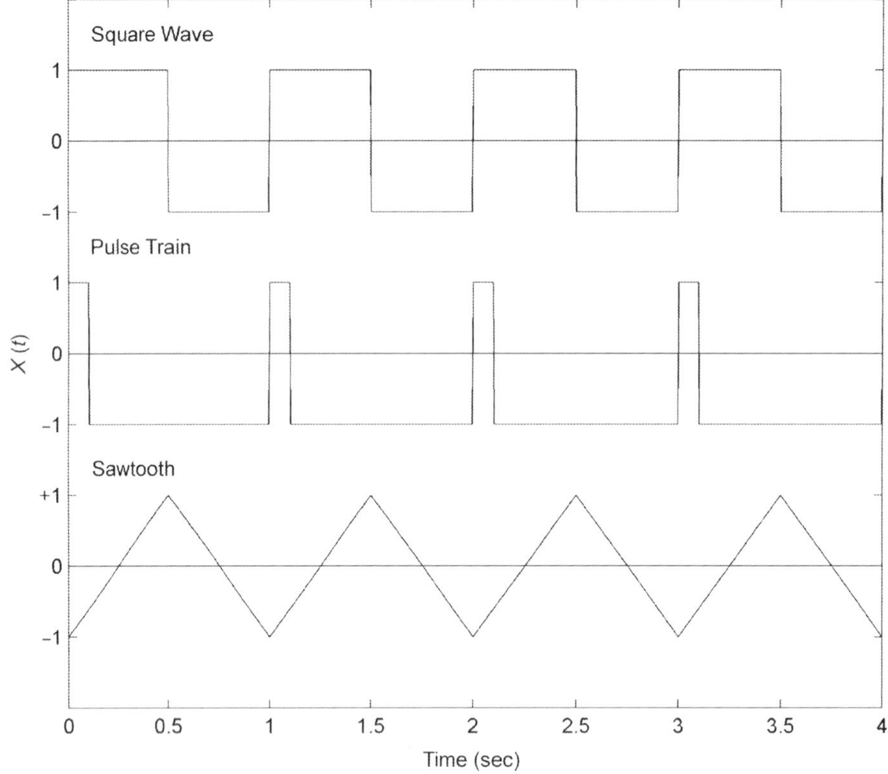

FIGURE 2.5 Three different periodic signals having the same, 1-sec, period: square wave (upper), pulse train, and sawtooth.

important and useful. Signals can be divided into three general classes: periodic, aperiodic, and step-like or transient.[1] The sine wave is an example of a periodic signal, one that repeats in an identical manner following some time period *T*. Other common periodic signals include the square wave, pulse train, and sawtooth as shown in Figure 2.5. Since all periods of a periodic signal are identical, only one period is required to completely define a periodic signal, and any operations on such a signal need only be performed over one cycle. For example, to find the average value of a periodic signal, it is only necessary to determine the average over one period.

Aperiodic signals exist over some finite time frame, but are zero before and after that time frame. Figure 2.6 shows a few examples of aperiodic signals. The assumption is that

[1]All time-varying signals could be called "transient" in the sense that they are always changing. However, when describing signals, the term transient is often reserved for signals that change from one level to another and never return to their initial value.

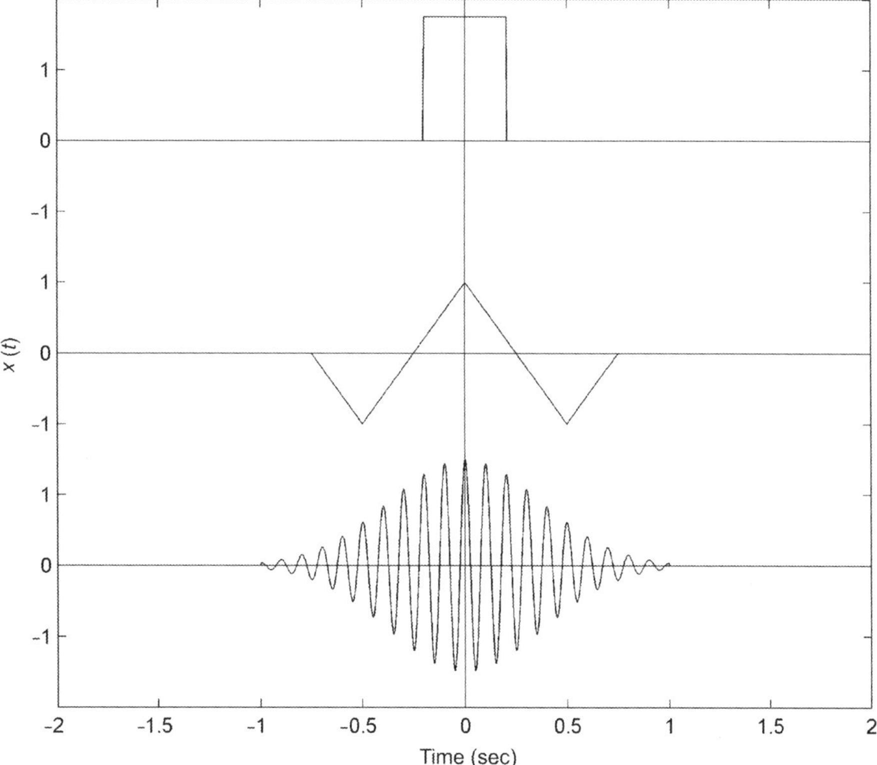

FIGURE 2.6 Three examples of aperiodic signals. It is common to show these signals as centered around $t = 0$, especially when they are symmetrical about a central point.

these signals are zero outside the period shown. A mathematically convenient way to view aperiodic signals is as period signals where the period goes to infinity. Operations on aperiodic signals need only be applied to their finite, nonzero segments.

EXAMPLE 2.6

Find the average value of the aperiodic triangular shaped waveform shown at the center of Figure 2.6.

Solution: Apply Equation 1.16. Since the signal is symmetrical around $t = 0$, we only need to determine the average for positive t. The equation for the signal for $t \geq 0$ is:

$$x(t) = \begin{cases} 1 - 4t & 0 \leq t < 0.5 \\ -1 + 4(t - 0.5) = -3 + 4t & 0.5 \leq t < 0.75 \end{cases}$$

Apply Equation 1.16:

$$Avg = \frac{1}{T}\int_0^T x(t)dt = \frac{1}{0.75}\left(\int_0^{0.5}(1-4t)dt + \int_{0.5}^{0.75}(-3+4t)dt\right)$$

$$Avg = 1.33\left(t\Big|_0^{0.5} - \frac{4t^2}{2}\Big|_0^{0.5} + -3t\Big|_{0.5}^{0.75} + \frac{4t^2}{2}\Big|_{0.5}^{0.75}\right)$$

$$Avg = 1.33((0.5-0)-(0.5-0)-(2.25-1.5)+(1.125-0.5))$$

$$Avg = 1.33(-0.125) = -0.166$$

Transient, or step-like, consist of changes that do not return to an initial level. A few examples are given in Figure 2.7 including the commonly used step signal. These signals do not end at the plot, but continue to infinity. They are the hardest to treat mathematically. They stretch to infinity, but are not periodic, so any mathematical operation must be carried out from zero to infinity. For example, it is impossible to determine an average

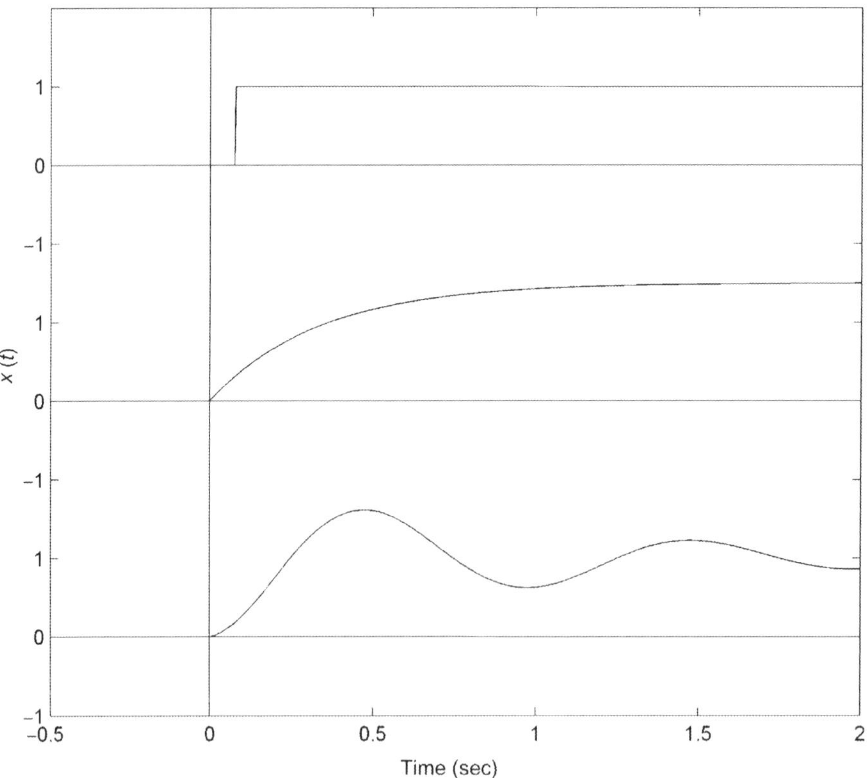

FIGURE 2.7 Three examples of transient or step-like signals. These signals do not end at the end of the plot but continue to infinity. The step signal is a common signal in signal processing.

value for any of the signals in Figure 2.7; since these signals are nonzero to infinity, their average values are also infinite. However, methods for dealing with these signals exist and are discussed in Chapter 6.

2.3 TWO-DIMENSIONAL SIGNALS—IMAGES

The signals we have discussed thus far are one-dimensional functions of time that are represented in MATLAB as arrays or vectors. It is possible to have multiple signals that have some common origin and should be treated together. A common biomedical example is the multilead EEG signal, where the electrical activity of the brain is recorded from 16, 32, or even 64 different locations on the head. Some analyses treat these signals in combination, looking at features common to the signals as well as analyzing them individually. Multiple signals are often represented as a matrix where each row is a different signal.

Images are just two-dimensional signals and can also be represented as matrices. MATLAB has a special toolbox for image processing that provides a number of useful routines for manipulating images, but some taste of image processing can be had using standard MATLAB routines. To generate a simple image, we begin with a sine wave such as that constructed in Example 2.1. This waveform is a single time function, but the sinusoidal variation could also be used to construct an image. The sinusoidal variation in amplitude could be used to fill a matrix where every row was just a copy of the sinusoid. It may not be a very interesting matrix with so much repetition, but it does make an interesting image. When considered and displayed as an image, a matrix of sinusoids would look like a sinusoidally varying pattern of vertical bars. Such an image is called a "sinusoidal grating" and is used in experiments on the human visual system. A sinusoidal grating is constructed and displayed in the next example.

EXAMPLE 2.7

Construct the image of a sine wave grating having 4 cycles across the image. Make the image 100 by 100 pixels. (Pixel is the term used for a two-dimensional image sample.)

Solution: Construct a 4-cycle (Hz) single sine wave having 100 samples. Then construct a matrix having 100 rows where each row contains a copy of the sine wave. Display the matrix as a checkerboard plot using MATLAB's pcolor with the shading option set to interpolate (`shading interp`). This will generate a pseudocolor sine wave grating. Use the bone color map to convert the image to a grayscale image (`colormap(bone)`).

```
% Example 2.7 Construct the image of a sine wave grating having 4 cycles
%
N = 100;                % Number of pixels per line and number of lines
x = (1:N)/N;            % Spatial vector
y = sin(8*pi*x);        % Four cycle sine wave
for k = 1:N
  I(k,:) = y;           % Duplicate 100 times
end
pcolor(I);              % Display image
```

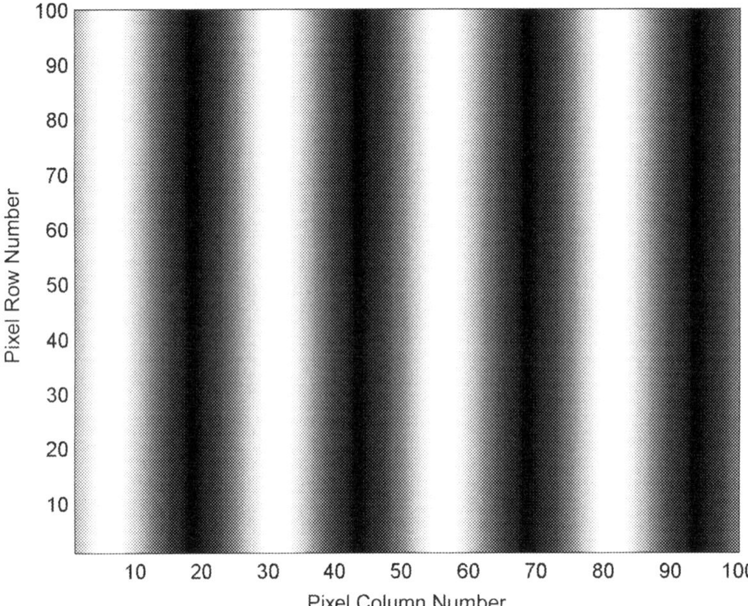

FIGURE 2.8 A sinusoidal grating produced in Example 2.6. Some of the variation in gray level shading is lost in the printing of this image.

```
shading interp;          % Use interpolation
colormap(bone);          % Use a grayscale color map
```

Result: The image produced by this program is shown in Figure 2.8. The sinusoidal gradation in gray levels is compromised somewhat in the printing process.

The grating in Figure 2.8 has a sinusoidal variation in intensity across the image. Since this is an image and not a time plot, the horizontal axis is position, not time, and the variation is a function of spatial position. The frequency of a sine wave that varies with position is called "spatial frequency" and is in units of cycles/distance. If Figure 2.8 is printed to have a width of 4 inches, then the horizontal frequency of the sine wave would be 1.0 cycle/in. There is no variation vertically so the vertical frequency of this image is 0.0 cycles/in. (In Problem 9 at the end of the chapter an image is generated that has a sinusoidal variation in both dimensions.) We will discuss spatial frequency further in Chapter 8 when we apply filters to images.

While MATLAB's Image Processing Toolbox provides a number of different formats for encoding images, the format used here is just the same as that used for standard MATLAB variables: the "double" format. That means that any MATLAB operation can be applied to an image; conceptually the image is just another matrix to MATLAB. The next example shows some basic mathematical and threshold operations applied to an MR image of the brain.

EXAMPLE 2.8

Apply various mathematical and threshold operations to an MR image of the brain. Load the file `brain.mat` which contains the brain image in matrix variable I. Brighten the image and display it, then enhance the contrast of the image and display. Finally, threshold the image and make the darker regions white and all others black (inverted with respect to the normal MATLAB coding of dark and light).

Solution: The file is loaded in the usual way. The image has been scaled to be between 0 (black) and 1.0 (white), a convention used by MATLAB for scaling image data. Since `pcolor` normalizes the data each time it is called, we will fix the display range to be between 0 and 1.0 using the `caxis` routine (`caxis([0 1])`). After displaying the original image as in the preceding example, we will lighten the image by increasing all the pixel values in the image by a value of 0.3 (remember 1.0 is white). After displaying this image, we multiply the original image by 1.75 to increase its contrast and display. To apply the threshold operation, we examine every pixel and generate a new image that contains a 1.0 (i.e., white) in the corresponding pixel if the original pixel is below 0.25 and a 0 (i.e., black) otherwise. This image is then displayed.

```
% Example 2.8 Example to apply some mathematical and threshold operations to an MR
image of the brain.
%
load brain;              % Load image
subplot(2,2,1);          % Display the images 2 by 2
pcolor(I);               % Display original image
shading interp;          % Use interpolation
colormap(bone);          % Grayscale color map
caxis([0 1]);            % Fix pcolor scale
title('Original Image','FontSize',12);
%
subplot(2,2,2);
I_1 = I + .3;            % Brighten the image by adding 0.3
 .......displayed and titled as above.......
%
subplot(2,2,3);
I_1 = 1.75*I;            % Increase image contrast by multiplying by 1.75
 .......displayed and titled as above.......
%
subplot(2,2,4);
thresh = 0.25;           % Set threshold
for k = 1:r
  for j = 1:c
    if I(k,j) < thresh   % Test each pixel separately
      I_1(k,j) = 1;      % If low make corresponding pixel white (1)
    else
      I_1(k,j) = 0;      % Otherwise make it black (0)
    end
  end
end
 .......displayed and titled as above.......
```

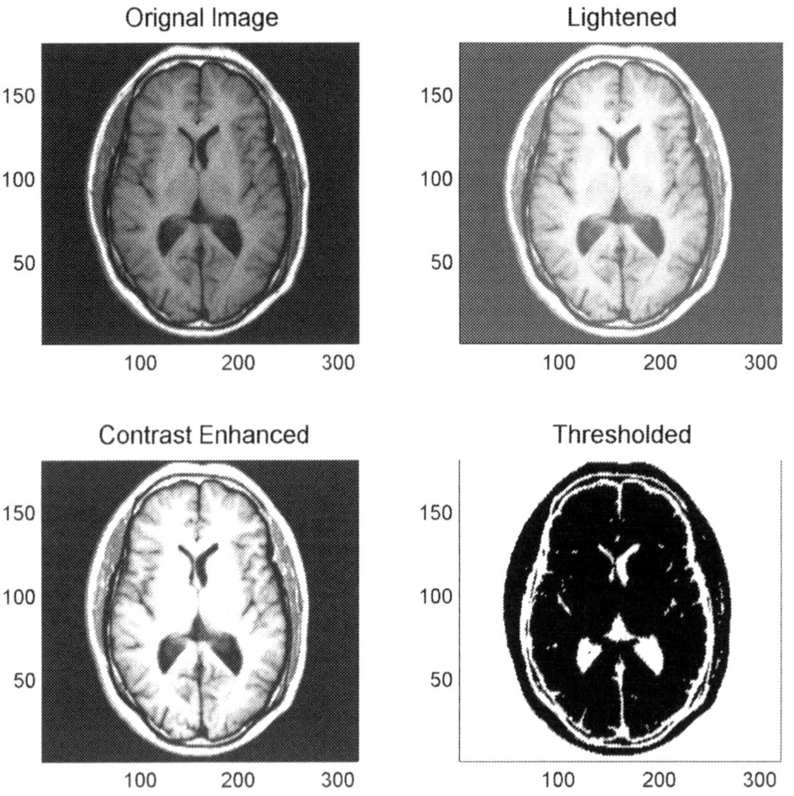

FIGURE 2.9 An MR image of the brain that has been brightened, contrast-enhanced, and thresholded. In thresholding, the darker regions are shown as white and all other regions as black.

Results: The four images produced by this are shown in Figure 2.9. The changes produced by each of these operations are clearly seen. Many other operations could be performed on this, or any other image, using standard MATLAB routines. A few are explored in the Problems at the end of this chapter, and others in later chapters.

2.4 SIGNAL COMPARISONS AND TRANSFORMATIONS

The basic measurements of mean, variance, standard deviation, and RMS described in Chapter 1 are often not all that useful in capturing the important features of a signal. For example, given the EEG signal shown in Figure 1.5, we can easily compute its mean, variance, standard deviation, and RMS value, but these values do not tell us much about either the EEG signal or the processes that created it. More insight might be gained by a more complicated analysis where the EEG signal is compared to some reference signal(s) or converted into something that is more meaningful. Signal conversions are better known as transformations and are at the heart of many signal processing operations. In signal processing, transformations are used to remove noise or make the signal more meaningful.

Transformations are also used later in linear systems analysis to make the descriptive equations easier to solve.

Comparisons between the signal of interest and one or more reference signals are often useful in themselves, but are also at the heart of many transformations. In fact comparisons form the basis of all the transformations described in this text: a signal is compared with a reference or, more often, a family of related reference functions. (Again the terms "signal," "waveform," and "function" can be used interchangeably.) When a family of reference functions is used, you get a number of comparisons. If the reference function is shorter than the signal, then the comparison is sometimes made for a number of different alignments between the two functions: the reference function can be thought of as sliding along the signal function making comparisons as it goes.

Usually the reference function or family is much less complicated than the signal, for example, a sine wave or series of sine waves. (As shown later, a sine wave is about as simple as it gets.) A quantitative comparison can tell you how much your complicated signal is like a simpler, easier to understand, reference function or family. If enough comparisons are made with a well-chosen family, these comparisons taken together can provide an alternative representation of the signal. These comparisons produce only an approximate representation of the signal but one that, as we engineers like to say, "for all practical purposes" is good enough. When this happens, the alternate representation by the family of comparisons can be viewed as a *transformation* of the original signal. Sometimes this new representation of the signal is more informative or enlightening than the original.

2.4.1 Correlation

One of the most common ways of quantitatively comparing two functions is through correlation. Correlation seeks to quantify how much one function is like another. Mathematical correlation does a pretty good job of describing similarity, but once in a while it breaks down so that some signals that are very much alike have a mathematical correlation of zero, as shown in Figure 2.10.

Correlation between different signals is illustrated in Figure 2.10, which shows various pairs of waveforms and the correlation between them. Note that a sine and a cosine have zero correlation even though the two are alike in the sense that they are both sinusoids, shown in Figure 2.10 (upper plot). This lack of correlation is also demonstrated in Example 2.8. Intuitively we see that this is because any positive correlation between them over one portion of a cycle is canceled by a negative correlation over the rest of the cycle. This shows that correlation does not always measure general similarity: a sine and a cosine of the same frequency are by this mathematical definition as unalike as possible, even though they have similar behavioral patterns.

The linear correlation between two functions or signals can be obtained using the Pearson correlation coefficient defined as:

$$r_{xy} = \frac{1}{(N-1)\sigma_x\sigma_y}\sum_{n=1}^{N}(x_n - \overline{x})(y_n - \overline{y}) \tag{2.14}$$

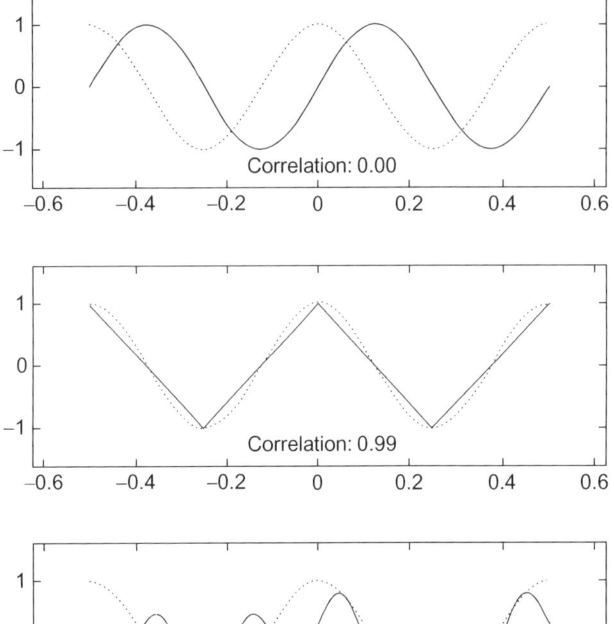

FIGURE 2.10 Three pairs of signals and the correlation between them, as given by Equation 2.14. The high correlation between the sine and triangle waves in the center plot correctly expresses the general similarity between them. However, the correlation between the sine and cosine in the upper plot is zero, even though they are both sinusoids at the same frequency.

where r_{xy} is a common symbol in signal analysis to represent the correlation between x and y; the variables x and y represent the signal and reference functions; $\bar{x}$ and $\bar{y}$ are the means of x and y, and σ_x and σ_y are the standard deviations of x and y as defined in Equation 1.22. This equation scales the correlation coefficient to be between ± 1, but if we are not concerned with the scale then it is not necessary to normalize by the standard deviations. Further, if the means of the signal and reference functions are zero, correlations can be determined using a simpler equation:

$$r_{xy} = \frac{1}{N}\sum_{n=1}^{N} x[n]y[n] = \frac{1}{N}(x[1]y[1] + x[2]y[2] + x[3]y[3] + \ldots x[N]y[N]) \tag{2.15}$$

where r_{xy} is the unscaled correlation between x and y. If continuous functions are involved, the summation becomes an integral and the discrete functions, $x[n]$ and $y[n]$, become continuous functions, $x(t)$ and $y(t)$:

$$r_{xy} = \frac{1}{T}\int_0^T x(t)y(t)dt \tag{2.16}$$

The integration (or summation) and scaling (dividing by T or N) simply takes the average of the product over its range. The correlation values produced by Equations 2.15 and 2.16 will not range between ± 1, but they do give relative values that are proportional to the *linear correlation* of x and y. The correlation of r_{xy} will have the largest possible positive value when the two functions are identical and the largest negative value when the two functions are exact opposites of one another (i.e., one function is the negative of the other). The average product produced in these equations will be zero when the two functions are mathematically completely dissimilar.

Note that the relationship between the Pearson correlation coefficient given in Equation 2.14 and the correlation given in Equation 2.15 is just the linear correlation normalized by the standard deviations. This will make the correlation value equal to $+1$ when the two signals are identical and -1 if they are exact opposites:

$$Corr_{normalized} = \frac{r_{xy}}{\sqrt{\sigma_1^2 \sigma_2^2}} \tag{2.17}$$

where the variances, σ^2, are defined in Equations 1.20. The term *correlation coefficient* implies this normalization while the term *correlation* is used more loosely and could mean normalized or unnormalized correlation.

Recall that in Chapter 1 it is mentioned that a signal sequence made up of n numbers could be thought of as a vector in n-dimensional space. In this concept, correlation becomes the *projection* of one signal vector upon the other: it is the mathematical equation for projecting an n-dimensional x vector upon an n-dimensional y vector, Equation 2.15 or 2.16. When two functions are uncorrelated, they are also said to be orthogonal, a term that derives from the n-dimensional vector concept since orthogonal vectors have a projection of zero on one another. In fact, a good way to test if two functions are orthogonal is to assess their correlation. In advanced signal processing, correlation is viewed as projecting one signal upon another and evaluating how much they overlap. When this concept of vector projection is used, it is common to refer to a family of reference functions as a *basis* or *basis functions*. Hence "projecting the signal on a basis" is the same as correlating the signal with a reference family.

Covariance computes the variance that is shared between two (or more) signals. Covariance is usually defined in discrete format as:

$$\sigma_{xy} = \frac{1}{N-1}\sum_{n=1}^{N}(x_n - \overline{x})(y_n - \overline{y}) \tag{2.18}$$

The equation for covariance is similar to the discrete form of correlation except that the mean values of the signals have been subtracted. Of course, if the signals have average values of zero, then the two operations, unnormalized correlation and covariance, are the same.

As seen from the equations above, correlation operations have both continuous and discrete versions so that both analytical and computer solutions are possible. The next two examples solve for the correlation between two waveforms using both the continuous, analytical approach and a computer algorithm.

EXAMPLE 2.9

Use Equation 2.16 (continuous form) to find the correlation (unnormalized) between the sine wave and the square wave shown in Figure 2.11. Both have amplitudes of 1.0 volt (peak-to-peak) and periods of 1.0 sec.

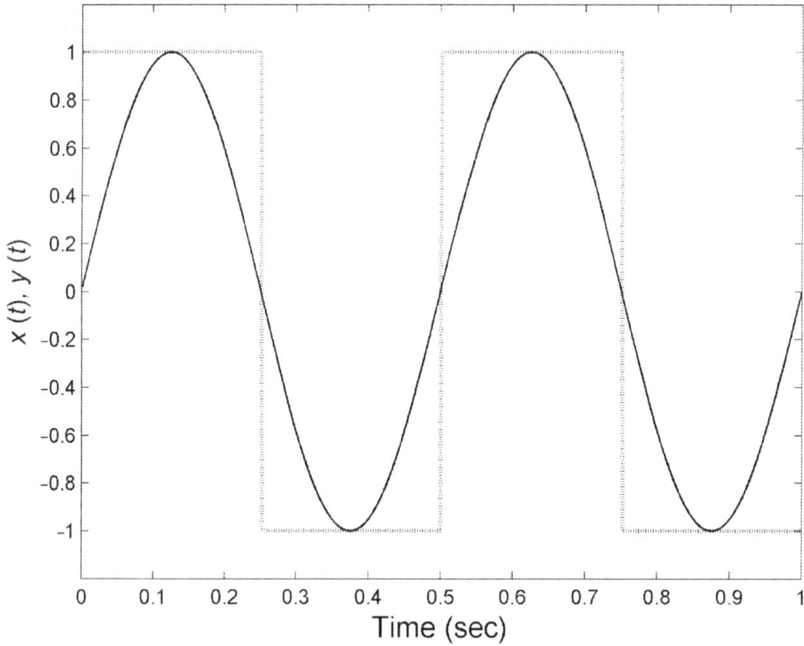

FIGURE 2.11 Waveforms for correlation analysis in Examples 2.9 and 2.11.

Solution: Apply Equation 2.16. By symmetry, the correlation in the second half of the 1-second period equals the correlation in the first half, so it is only necessary to calculate the correlation period in the first half period.

$$Corr = \frac{1}{T}\int_0^T x(t)y(t)dt = \frac{2}{T}\int_0^{T/2}(1)\sin\left(\frac{2\pi t}{T}\right)dt = \frac{2}{T}\frac{T}{2\pi}\left(-\cos\left(\frac{2\pi t}{T}\right)\right)\Bigg|_0^{T/2}$$

$$Corr = \frac{1}{\pi}(-\cos(\pi) - -\cos(0)) = \frac{2}{\pi}$$

EXAMPLE 2.10

Find the correlation between the cosine wave and square wave shown in Figure 2.11.
Solution: Apply Equation 2.16 as in the last example.

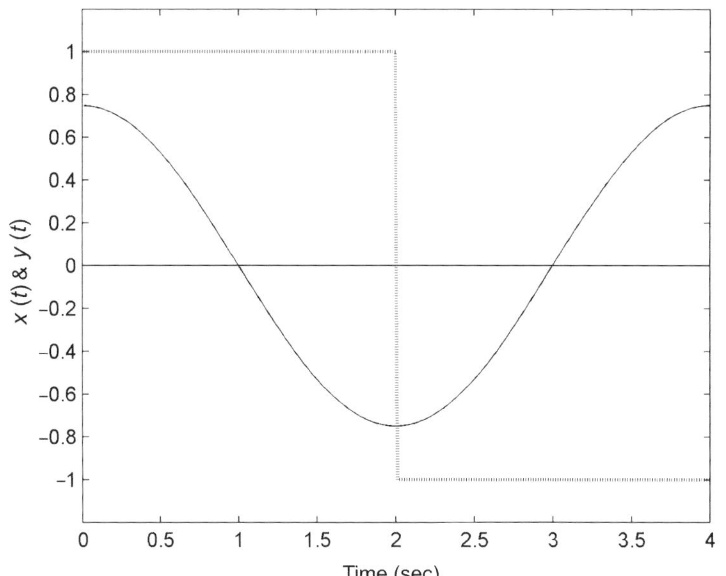

FIGURE 2.12 Cosine wave and square wave used in Example 2.10.

$$Corr = \frac{1}{T}\int_0^T x(t)y(t)dt = \frac{2}{4}\int_0^2 (1)\cos\left(\frac{2\pi t}{4}\right)dt + \frac{2}{4}\int_2^4 (-1)\cos\left(\frac{2\pi t}{4}\right)dt$$

$$Corr = \frac{2}{4}\frac{4}{\pi}\left(\sin\left(\frac{2\pi t}{4}\right)\right)\Bigg|_0^2 - \frac{2}{4}\frac{4}{\pi}\left(\sin\left(\frac{2\pi t}{4}\right)\right)\Bigg|_2^4$$

$$Corr = \frac{1}{\pi}(\sin(\pi) - \sin(0) - \sin(\pi) + \sin(2\pi)) = \frac{1}{\pi}(\sin(\pi) - \sin(\pi)) = 0$$

So there is no correlation between the two waveforms in Figure 2.12. Note that the product of the two waveforms over the first half period is cancelled by the product over the second half waveform.

EXAMPLE 2.11

Use Equation 2.15 and MATLAB to confirm the result given in Example 2.9; also find the correlation between a sine wave and a cosine wave.

Solution: Construct a time vector 500 points long ranging from 0.0 to 1.0 sec. Use the time vector to construct a 1-Hz sine wave and a cosine wave. Then evaluate the correlation between the two by direct application of Equation 2.15.

```
N = 500;                    % Number of points
Tt = 1.0                    % Desired total time
fs = N/Tt;                  % Calculate sampling frequency
t = (0:N-1)/fs;             % Time vector from 0 (approx.) to 1 sec
x = sin(2*pi*t);            % 1 Hz sine wave
```

```
y = cos(2*pi*t);                       % 1 Hz cosine wave
z = [ones(1,N/2) -ones(1,N/2)];        % 1 Hz square wave
rxz = mean(x.*z);                      % Correlation (Eq. 2.15) x and z
rxy = mean(x.*y);                      % Correlation x and y
disp([rxz rxy])                        % Output correlations
```

Analysis: In Example 1.5 we learned how to construct a time vector given the number of points and sampling frequency (N and f_s) or the total time and sampling frequency (T_t and f_s). Here we are presented with a third possibility, the number of points and total time (N and T_t). To calculate this time vector, the appropriate sampling frequency was determined from the total points and total time: $f_s = N/T_t$. Note that Equation 2.15 takes a single line of MATLAB code.

Result: The correlation of the sine wave and square wave, rxz, is 0.6366 which is equivalent to $2/\pi$, the result found in Example 2.9. The correlation of the sine wave and cosine wave was 0.0 as expected. The fact that sines and cosines are uncorrelated turns out to be useful in some analysis procedures as described in the next chapter.

2.4.2 Orthogonal Signals and Orthogonality

Orthogonal signals and functions are very useful in a variety of signal processing tools. In common usage, "orthogonal" means perpendicular: if two lines are orthogonal they are perpendicular. In the graphical representation of complex numbers shown in Figure 2.3, the real and imaginary components are perpendicular to one another; hence they are also orthogonal. But what of signals: what makes two signals orthogonal? The formal definition for orthogonal signals is that their inner product (also called the dot product) is zero:

$$\int_{-\infty}^{\infty} x(t)y(t)dt = 0 \tag{2.19}$$

For discrete signals, this becomes:[2]

$$\sum_{n=1}^{N} x[n]y[n] = 0 \tag{2.20}$$

These equations are the same as the correlation equations (Equation 2.15 and 2.16) except for normalizing by $1/N$. So signals that are orthogonal are uncorrelated and vice versa, and either the standard correlation equations or their unnormalized versions above could be used to test for orthogonality.

An important characteristic of signals that are orthogonal/uncorrelated is that when they are combined or added together they do not interact with one another. This becomes of particular value in the next chapter when a complicated signal is represented by summations of simpler signals. The amount of each orthogonal signal in the summation can be determined without regard to the other orthogonal signals in the summation. Orthogonality simplifies

[2]A footnote in Chapter 1 notes that a digital signal composed of N samples can be considered a single vector in N-dimensional space. In this representation, two orthogonal signals have vectors that are actually perpendicular. So the concept of orthogonal meaning perpendicular holds even for signals.

many calculations in a similar manner and some analyses could not be done, at least not practically, without orthogonal signals. Orthogonality is not limited to two signals. Whole families of signals can be orthogonal to all other members in the family. Such families of orthogonal signals are called *orthogonal sets.*

2.4.3 Matrix of Correlations

MATLAB has functions for determining the correlation and/or covariance that are particularly useful when a number of signals are compared. A matrix of correlations or covariances between different combinations of signals can be calculated using the `corrcoef` or `cov` functions respectively. For zero mean signals, the matrix of covariances is just the signal correlations normalized only by N as given by Equation 2.15, while the matrix of correlations is the covariances normalized as in Equation 2.17. This normalization means that the matrix of correlations runs between ± 1 with 1.0 indicating identical signals and -1.0 indicating a pair of the same signals but with one inverted with respect to the other. The calls are similar for both functions:

```
Rxx = corrcoef(X);  % Signal correlations
S = cov(X);         % Signal covariances
```

where `X` is a matrix that contains the various signals to be compared in columns. Some options are available as explained in the associated MATLAB help file. The output, `Rxx`, of the `corrcoef` routine will be an n-by-n matrix where n is the number of signals (i.e., columns). The diagonals of this matrix represent the correlation of the signals with themselves and therefore will be 1. The off-diagonals represent the correlation coefficients of the various combinations. For example, r_{12} is the correlation between signals 1 and 2. Since the correlation of signal 1 with signal 2 is the same as signal 2 with signal 1, $r_{12} = r_{21}$, and in general $r_{m,n} = r_{n,m}$, the matrix will be symmetrical about the diagonals:

$$r_{xx} = \begin{bmatrix} r_{1,1} & r_{1,2} & \cdots & r_{1,N} \\ r_{2,1} & r_{2,2} & \cdots & r_{2,N} \\ \vdots & \vdots & \ddots & \vdots \\ r_{N,1} & r_{N,2} & \cdots & r_{N,N} \end{bmatrix} \tag{2.21}$$

The `cov` routine produces a similar output, except the diagonals are the variances of the various signals and the off-diagonals are the covariances, the same correlation values given by Equation 2.15. However, it is easier to use `cov` if multiple correlations are needed.

$$S = \begin{bmatrix} \sigma_{1,1}^2 & \sigma_{1,2}^2 & \cdots & \sigma_{1,N}^2 \\ \sigma_{2,1}^2 & \sigma_{2,2}^2 & \cdots & \sigma_{2,N}^2 \\ \vdots & \vdots & \ddots & \vdots \\ \sigma_{N,1}^2 & \sigma_{N,2}^2 & \cdots & \sigma_{N,N}^2 \end{bmatrix} \tag{2.22}$$

Example 2.12 uses covariance and correlation analysis to determine if sines and cosines of different frequencies are orthogonal. Recall that two signals that are orthogonal will have zero correlation. Either covariance or correlation could be used to determine if signals are orthogonal. Example 2.12 uses both.

EXAMPLE 2.12

Determine if a sine wave and cosine wave at the same frequency are orthogonal and if sine waves at harmonically related frequencies are orthogonal. The term "harmonically related" means that sinusoids are related by frequencies that are multiples. Thus the signals sin(2*t*), sin (4*t*), and sin(6*t*) are harmonically related. Also determine if sawtooth and sine waves at the same frequency are orthogonal.

Solution: Generate a 500-point, 1.0-sec time vector as in the last example. Use this time vector to generate a data matrix where the columns represent 1.0-, 2-, and 2.5-Hz cosine and sine waves. Apply the covariance and correlation MATLAB routines (i.e., `cov` and `corrcoef`) and display results.

```
% Example 2.12
% Application of the covariance matrices to sinusoids that
% are orthogonal and a sawtooth
%
N = 1000;                  % Number of points
Tt = 2;                    % desired total time
fs = N/Tt;                 % Calculate sampling frequency
t = (0:N-1)/fs;            % Time vector from 0 (approx.) to 2 sec
x(:,1) = cos(2*pi*t)';     % Generate a 1 Hz cosine
x(:,2) = sin(2*pi*t)';     % Generate a 1 Hz sine
x(:,3) = cos(4*pi*t)';     % Generate a 2 Hz cosine
x(:,4) = saw(1,2,fs)';     % Generate a 1 Hz sawtooth
%
S = cov(x)                 % Print covariance matrix
Rxx = corrcoef(x)          % and correlation matrix
```

Analysis: The program defines a time vector in the standard manner. The program then generates the three sinusoids using this time vector in conjunction with `sin` and `cos` functions, arranging the signals as columns of x. The program then uses the `saw` routine (included on this book's accompanying CD) to generate a 1-Hz sawtooth having the same number of samples as the sinusoids. The four signals are placed in a single matrix variable `x`. The program then determines the covariance and correlation matrices of `x`.

Results: The output from this program is a covariance and a correlation matrix. The two matrices are shown in Table 2.1.

The diagonals of the covariance matrix give the variance of the 4 signals. These are consistent for the sinusoids and slightly reduced for the sawtooth. The correlation matrix shows similar results except that the diagonals are now 1.0 since these reflect the correlation of the signals with themselves.

The covariance and correlation between the various signals are given by the off-diagonals. The off-diagonals show that the sinusoids have no correlation with one another, a reflection of

TABLE 2.1 Covariance and Correlations of Signals in Example 2.11

Covariance Matrix (S)				Correlation Matrix (Rxx)			
0.5005	−0.0000	−0.0000	−0.4057	1.0000	−0.0000	−0.0000	−0.9927
−0.0000	0.5005	−0.0000	0.0025	−0.0000	1.0000	0.0000	0.0062
−0.0000	−0.0000	0.5005	0.0000	−0.0000	−0.0000	1.0000	0.0000
−0.4057	0.0025	0.0000	0.3337	−0.9927	0.0062	0.0000	1.0000

their orthogonal relationship. The 1-Hz cosine wave is highly correlated with the sawtooth and there is some correlation between the 1-Hz sine wave and the sawtooth; however, no correlation is found between the 2-Hz cosine wave and the sawtooth.

2.4.4 Multiple Correlations

Returning to the EEG signal presented in Chapter 1 (Figure 1.9), we can use our new-found correlation tool to ask questions such as "how much is this EEG signal like a sine wave, or diamond-shaped wave, or other less complicated function?" In Example 2.13, we pursue this idea by comparing the EEG signal with two sinusoids at two different frequencies: 6.5 and 14-Hz. The problem with such a comparison is that it is poorly defined: we are not asked to find the similarity between the EEG signal and a sine wave (or cosine wave), but rather between the signal and a couple of apparently arbitrary "sinusoidal waveforms." What does this mean exactly? (Exam questions sometimes appear like this: not always definitive.) Figure 2.13 illustrates the problem finding the correlation between a sine and cosine; it depends on their relative positions. In Figure 2.13A they are not shifted and have zero correlation, but when the sine is shifted by 45 deg (Figure 2.13B) the correlation coefficient is 0.71, and at a 90-deg shift, the correlation coefficient is 1.0 (Figure 2.13C). Figure 2.13D shows that the correlation as a function of shift is itself a sine wave ranging between ± 1. Let us assume that what is wanted is the best match with any sinusoidal waveform at the two frequencies. That means we will have to shift the sine wave to find the best match. For a continuous function, there is an infinite number of different sinusoidal waveforms at any one frequency, each with a different phase shift. Since it is impossible to do the comparison with an infinite number of sinusoids, we limit it to 360 sinusoids, each having a 1.0-deg difference in phase shift. (Later we show that this is far more comparisons than necessary, but it does work.)

EXAMPLE 2.13

Find the correlation between the EEG signal given in Figure 1.5 and a 6.5-Hz sinusoid. Is it more similar to a 14-Hz sinusoid or a 6.5-Hz sinusoid? (Why these two particular frequencies? Because, as we will see later, the EEG signal "contains" different amounts of these two sinusoids and thus exhibits differing correlations.)

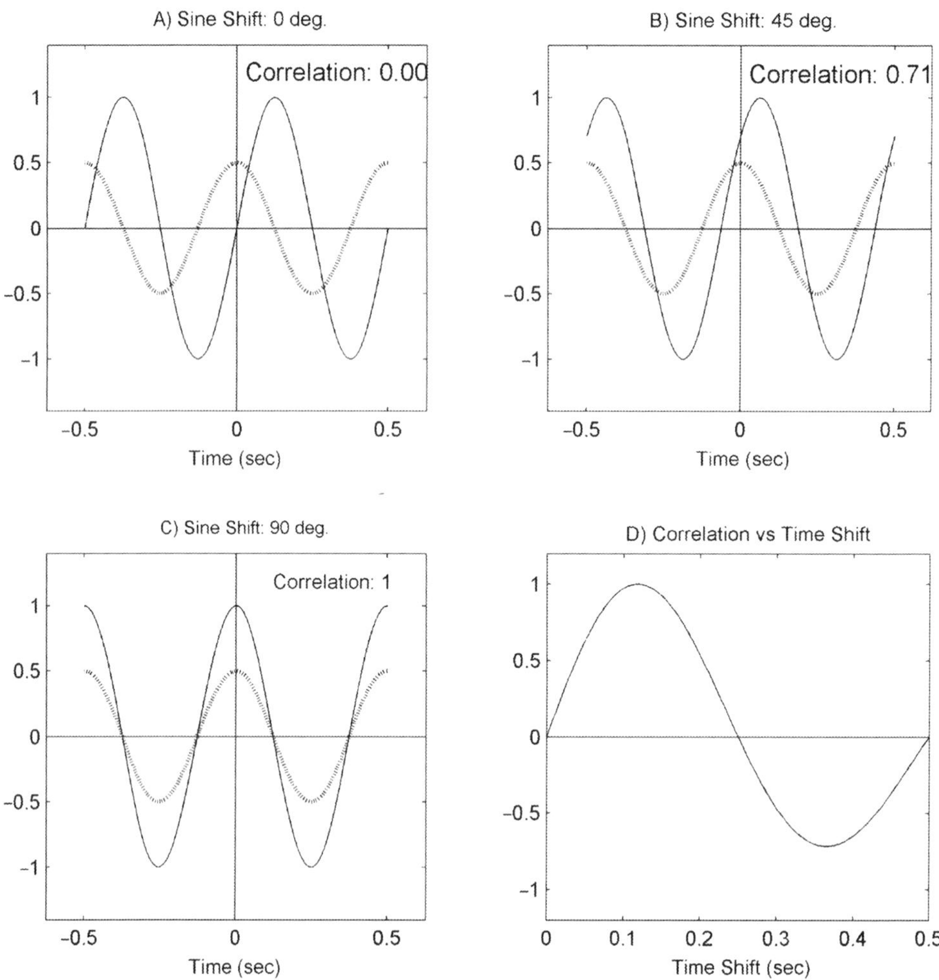

FIGURE 2.13 A) The correlation between a 2-Hz cosine reference (dashed) and an unshifted 2-Hz sine is 0.0. B) When the sine wave is time shifted by the equivalent of 45 deg, the correlation is 0.71. C) When the sine is shifted by 90 deg, the two functions are identical in shape and the correlation is 1.0. D) The correlation coefficient as a function of the sine wave phase (or time) shift is itself sinusoidal. A peak value of 1.0 occurs at a shift of 0.125 sec corresponding to a shift of 90 deg, zero at 0.25 sec corresponding to a shift of 180 deg, and a correlation of −1.0 at a time shift of 0.375 sec corresponding to a shift of 270 deg. The changes in correlation can be viewed as a function of phase shift or time shift, as shown here.

Solution: For each frequency, compare the EEG with 360 sine waves, each with a different phase shift incremented in 1 deg intervals. Since MATLAB's trigonometric functions assume that input arguments are in rad, we have to convert deg to rad. We then select the correlation having the highest value. This is 360 individual correlations, essentially implementing Equation 2.15 360 times, but

MATLAB hardly breaks a sweat. We also normalize the correlation to be between ±1 using Equation 2.17 to provide a better feel for the similarities.

We could make the program much more efficient. If we make use of the fact that shifting a sine wave by 180 deg is the same as changing its sign, we could cut the number of correlations in half: shift between 0 and 180 deg and take the maximum positive or negative value. We could reduce the number of correlations even further if we insisted that the phase shift be at least one sample interval. A 12-Hz signal has a period of 83.3 msec and the 1 deg shift is 83.3/360 = 0.23 msec while the sample interval is only $T_s = 1/f_s = 1/50 = 20$ msec! So we could increase the incremental shift from 1 deg to 20 deg, reducing the number of correlations accordingly, and still get the same accuracy. Increasing the incremental shift is a modification left for Problem 14, but in this example the inefficient code remains as is.

```
% Example 2.13 Find the correlation between the EEG signal and both a
% 8-Hz and 12-Hz sinusoid.
%
load eeg_data                          % Loads data array 'eeg'
fs = 50;                               % Sampling freq of EEG data
t = (1:length(eeg))/fs;                % Time vector from 0 to 1 sec
f = [6.5 14];                          % Frequencies used
for k = 1:2                            % Loop for different frequencies
  for m = 1:360                        % Loop for phases
    phase = (m-1)*pi/180;              % Change phase, convert to rad
    x = sin(2*pi*f(k)*t + phase);      % Construct sinusoid
    rxy(m) = mean(x.*eeg);             % Eq. 2.15
  end
  subplot(1,2,k);                      % Plot correlations versus phase
  rxy = rxy/sqrt(var(x)*var(eeg));     % Normalize (Eq. 2.17)
  plot((1:360)-1,rxy,'k');             % Corr as a function of phase
  .......axis labels.......
  title(['Max correl: ',num2str(max(abs(rxy)),2)]);
end
```

Analysis: The time vector is constructed to be the same length as the data. This insures that the sinusoidal waveform used in the comparison is the same length as the EEG signal. Also by dividing the time vector by the sampling frequency (50-Hz), the time intervals have the correct spacing. The outer loop repeats the correlation analysis for the two different frequencies while the inner loop does the phase shifting. The phase ranges between 0 and 359 deg (`m-1`) and is converted to rad. The phase adds to the time vector argument of the `sin` function.

Results: The results produced by this program are shown in Figure 2.14. The two curves show the correlations as a function of phase shift and the maximum correlations (absolute values) are given in the graph titles. The 6.5-Hz sinusoid shows a much higher correlation reaching a value of 0.16. Figure 2.15 shows a 1-second time plot of the EEG signal (solid line) and the 6.5-Hz sinusoid at the optimum phase shift (dashed line). A modest similarity between the two signals is apparent. Of course, there are probably other sinusoidal frequencies that produce an even higher correlation and it is tempting to do this comparison for a wider range of frequencies. It would be easy to change the code in this example to include any number of frequencies; however, there are better strategies for implementing this comparison as described next.

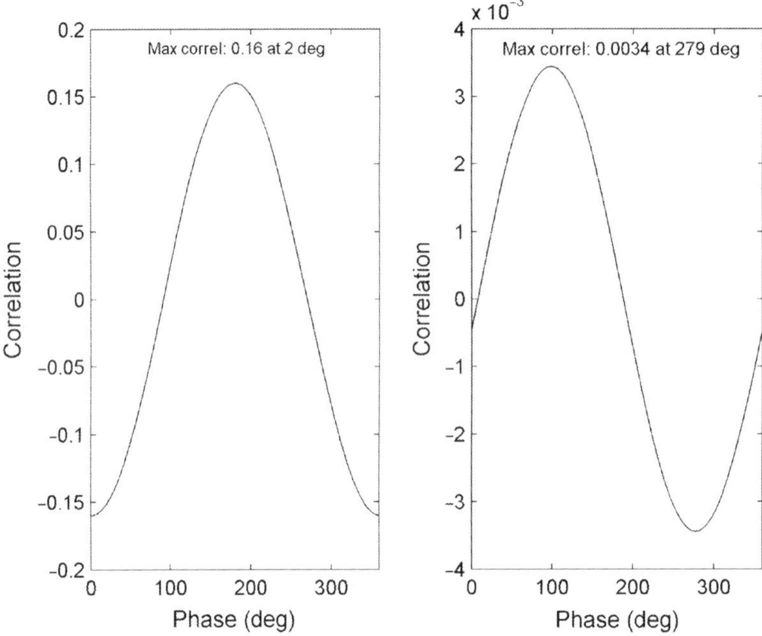

FIGURE 2.14 Correlation between a 6.5-Hz (left) and 14-Hz (right) sine wave and an EEG signal (Figure 1.9) as a function of sine wave phase shift. The maximum correlation is also given.

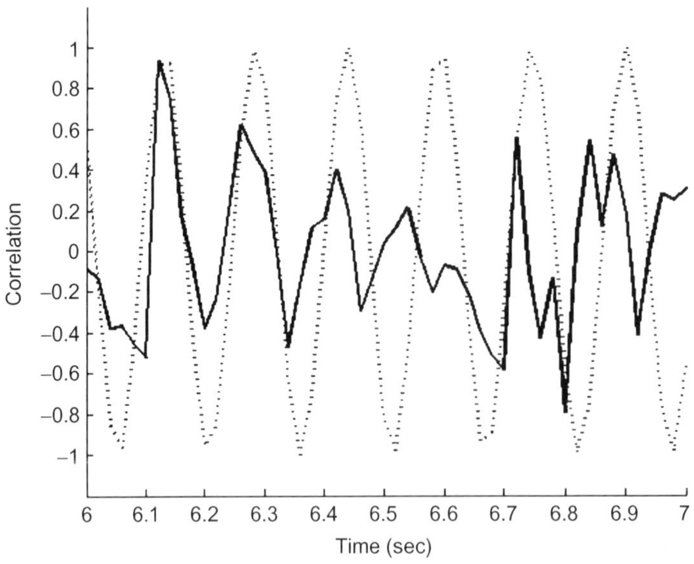

FIGURE 2.15 A time plot of a 1-sec segment of the EEG signal and the 6.5-Hz sine wave shifted to give the highest correlation between the two signals. Some similarity between the two signals can be observed. The correlation coefficient for these two signals is 0.16.

2.4.5 Crosscorrelation

The mathematical dissimilarity between a sine and a cosine is disconcerting and a real problem if you are trying to determine if a signal has general sinusoidal-like features. Usually you just want to know if a signal has sinusoidal components at a given frequency, irrespective of the phase of those sinusoidal components. Correlating a sine wave with an unknown function might lead you to believe it is not sinusoidal even if it is a perfect cosine wave, as in Figure 2.13A. In Example 2.13, we got around this problem by performing the correlation for many different phase shifts. However, it would have been much more efficient to shift the sine function one data point rather than 1.0 deg since the latter produces a number of redundant correlations. (A 1.0-deg shift of an 8-Hz sinusoid sampled at 50-Hz results in over 55 redundant correlations.) Also, shifting one sample point is more general and could be applied to nonsinusoidal functions. A running comparison between two waveforms when one of the waveforms is shifted one data sample per correlation is known as *crosscorrelation*.

The equation for crosscorrelation can be derived from the basic correlation equation, Equation 2.15, by introducing a variable shift into one of the two functions. It does not matter which function is shifted with respect to the other, the results would be the same, although usually the reference function is shown as shifted. The correlation operation of Equation 2.15 then becomes a series of correlations over different time shifts, ℓ, and the result is not a single number, but a series of numbers for different values of ℓ:

$$r_{xy}[\ell] = \frac{1}{N}\sum_{n=1}^{N} y[n]x[n+\ell] \qquad \ell = 0, 1, 2, \ldots L \tag{2.23}$$

where ℓ is the shift and ranges between 0 and L, where L depends on how the endpoints are treated. Note that this means correlating the signal and reference L times, where L could be quite large, but MATLAB really does not mind the work. This shift ℓ is often termed *lags* and specifies the number of samples shifted for a given correlation. If the crosscorrelated signals were originally time functions, the lags may be converted to time shifts in secs. The value of L could be just the length of the data sets, but often the data are extended with zeros to enable correlations at all possible shift positions: positive and negative. In such cases, L would be the combined length of the two data sets minus one. Extending a data set with zeros to permit calculation beyond its nominal end is called *zero padding* and is discussed in the next chapter. The multiple correlations produce a *crosscorrelation function*, a function of the shift ℓ. This function is often abbreviated as r_{xy}, where x and y are the two signal functions being correlated.

For continuous signals, the time shifting is continuous and the correlation becomes a continuous function of a continuous time shift. This leads to an equation for crosscorrelation that is an extension of Equation 2.16 adding a time shift variable, τ:

$$Crosscorrelation \equiv r_{xy}(\tau) = \frac{1}{T}\int_0^T y(t)x(t+\tau)dt \tag{2.24}$$

where variable τ is a continuous variable of time that specifies the time shift of $x(t)$ with respect to $y(t)$. The variable τ is analogous to the lag variable ℓ in Equation 2.23. It is the time variable for the crosscorrelation function r_{xy}, but the signals themselves need a different time variable, which is t. The τ time variable is sometimes referred to as a *dummy time variable*. The continuous crosscorrelation function also requires multiple operations, in this case one integration for every value of τ, in theory an infinite number of integrations. An example of evaluating Equation 2.24 analytically is provided for a very similar operation in Example 7.1 of Chapter 7.

Fortunately, most analyses that involve crosscorrelation are done in the digital domain where software like MATLAB does all the work. However it is instructive to do at least one digital evaluation manually because it provides insight into the algorithm used to implement the digital crosscorrelation equation, Equation 2.23.

EXAMPLE 2.14

Evaluate the crosscorrelation of the two short digital signals, $x[n]$ and $y[n]$, shown in Figure 2.16.

Solution: If the crosscorrelation is to include all possible shifts, we need to determine what to do when a shifted signal falls outside the range of the other signal. The common approach is to add zeros to the signals so that the signals always overlap. Since $y[n]$ contains four samples and $x[n]$ contains three samples, it is only necessary to add two zeros to each side of $y[n]$, assuming that we shift the shorter signal, $x[n]$. The discrete crosscorrelation equation (Equation 2.23) begins with one sample of $x[n]$ overlapping one sample of $y[n]$, Figure 2.17 (upper left). The calculation continues as $x[n]$ is shifted leftward one position, as shown left to right and top to bottom in Figure 2.17. In this

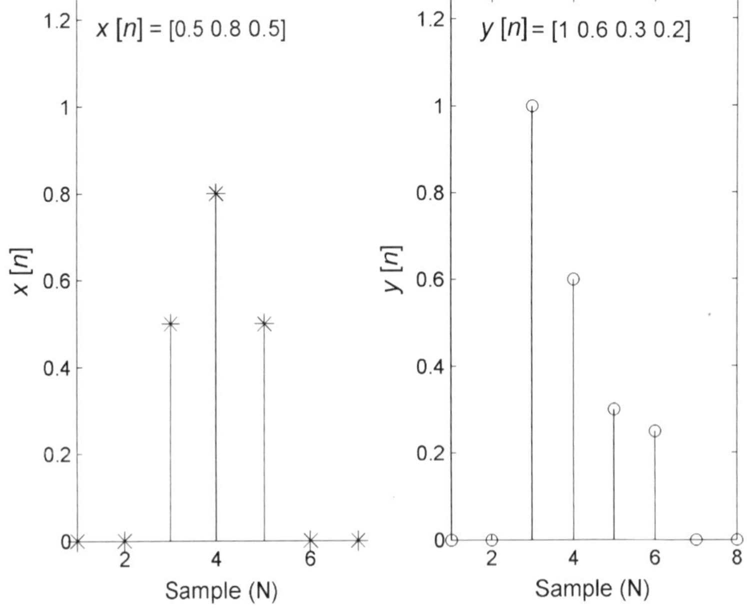

FIGURE 2.16 Two very short digital signals used in Example 2.14 to show manual calculation of the crosscorrelation function given in Equation 2.23. The signal $x[n]$ contains three samples and the $y[n]$ has four samples.

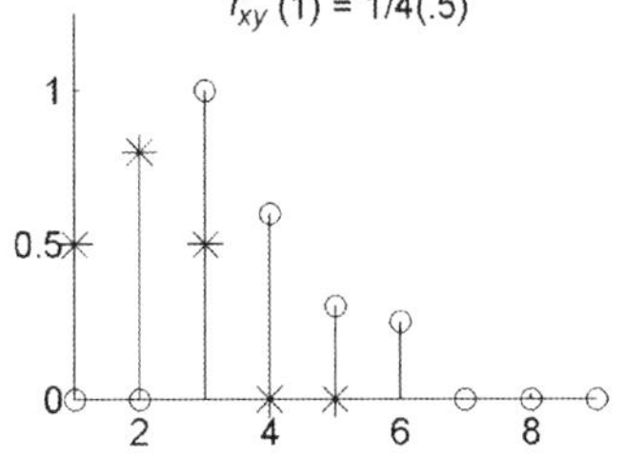

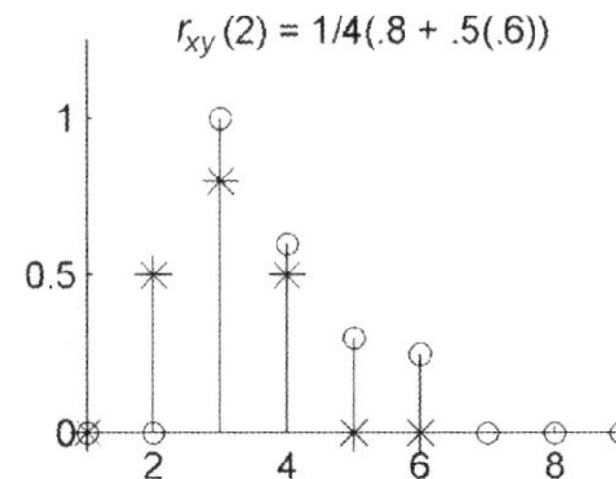

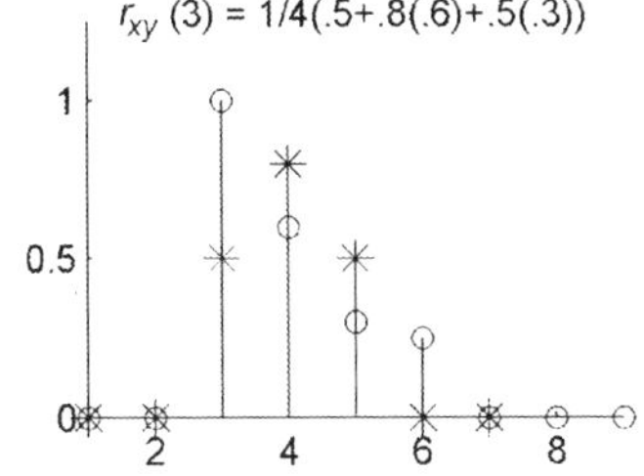

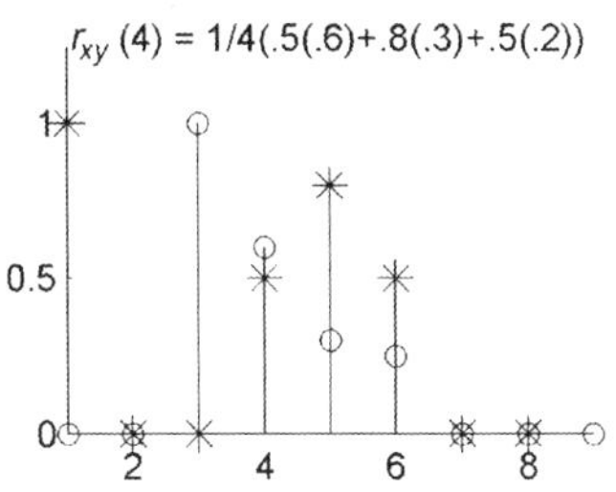

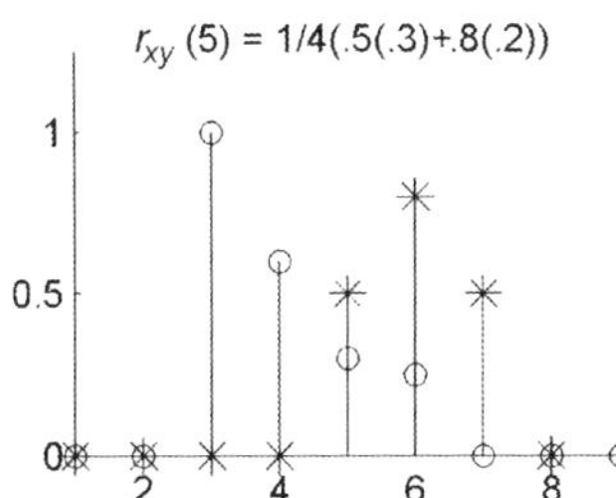

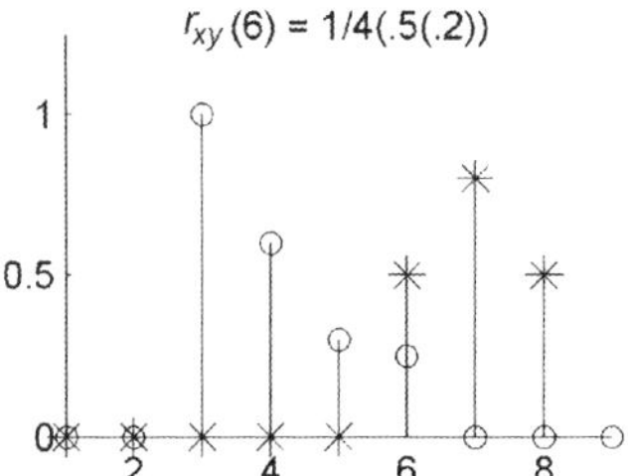

FIGURE 2.17 Manual calculation of the crosscorrelation of the two short signals shown in Figure 2.16. In this case $x[n]$ (*points) is shifted from left to right beginning with a one-sample overlap in the upper left plot.

case N, the normalization value, is taken as 4, the length of the longer signal. The resulting equations are given in Figure 2.17 and below.

$$r[1] = \frac{1}{4}(0.5(1)) = 0.125$$

$$r[2] = \frac{1}{4}(0.8(1) + 0.5(0.6)) = 0.275$$

$$r[3] = \frac{1}{4}(0.5(1) + 0.8(0.6) + 0.5(0.3)) = 0.2825$$

$$r[4] = \frac{1}{4}(0.5((0.6) + 0.8(0.3) + 0.5(0.2))0.160$$

$$r[5] = \frac{1}{4}(0.5(0.3) + 0.8(0.2)) = 0.0775$$

$$r[6] = \frac{1}{4}(0.5(0.2)) = 0.025$$

The maximum correlation occurs at a shift of 3 and is 0.2825 for this normalization. Figure 2.17 indicates that this shift does correspond to the greatest similarly between the two signals.

In MATLAB, crosscorrelation could be implemented by modifying and extending Example 2.13, and this has been done in the routine axcor which can be found on this book's accompanying CD:

```
[rxy,lags] = axcor(x,y);
```

where x and y are the data to be crosscorrelated and rxy is the crosscorrelation vector and is normalized to ±1 using Equation 2.17. Since the correlation is performed at all possible shift combinations, the length of rxy is the length of x plus the length of y minus one. The optional output vector lags is the same length as rxy and shows the actual shift position, positive and negative, corresponding to a given rxy value. This function shifts x with respect to y so a positive lag value means x is shifted to the left with respect to y, and vice versa. The lag vector can be helpful in scaling the horizontal axis when plotting the crosscorrelation against lag. If the MATLAB Signal Processing Toolbox is available, a MATLAB routine called xcorr is available that does the same thing, but features a wider range of options.

Crosscorrelation can be used for the problem in Example 2.13 to determine the similarity between a signal and a reference waveform when the relative position or phase shift for the best match is unknown. Simply take the maximum value of the crosscorrelation function. Another application for crosscorrelation is shown in the next example: finding the time delay between two signals.

EXAMPLE 2.15

File neural_data.mat contains two waveforms, x and y, that were recorded from two different neurons in the brain with a sampling interval of 0.2 msec. They are believed to be involved in the same function, but separated by one or more neuronal junctions that would impart a delay to the signal. Plot the original data, determine if they are related and, if so, the time delay between them.

Solution: Take the crosscorrelation between the two signals. Find the maximum correlation and the time shift at which that maximum occurs. The former will tell us if they are related and the latter the time delay.

```
load neural_data.mat;                      % Load data
fs = 1/0.0002                              % Sampling freq.
t = (1:length(x))/fs;                      % Calc. time vector
plot(t,y,'k',t,x,':');                     % Plot data
  ....... labels......
[rxy,lags] = axcor(x,y);                   % Compute crosscorrelation
figure;
plot(lags/fs,rxy);                         % and plot crosscorrelation
  .......labels.......
[max_corr, max_shift] = max(rxy);
disp([max_corr, lags(max_shift)/fs])       % Output delay in sec
```

Analysis: Finding the maximum correlation is straightforward: apply MATLAB's max operator to the crosscorrelation function. Finding the time at which the maximum value occurs is a bit more complicated. The max operator provides the index of the maximum value, labeled here as max_shift. To find the actual shift corresponding to this index, we need to find the lag value at

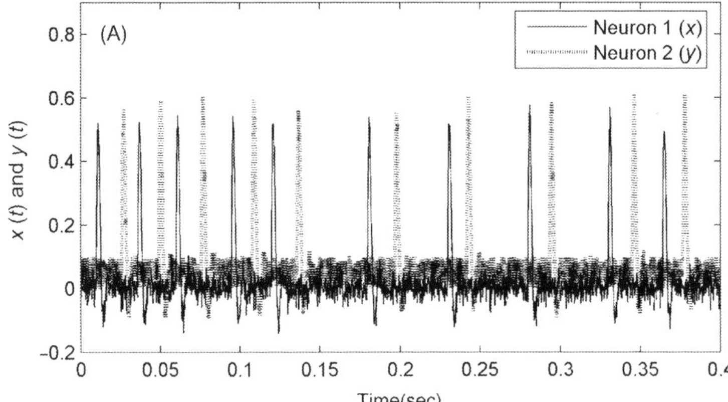

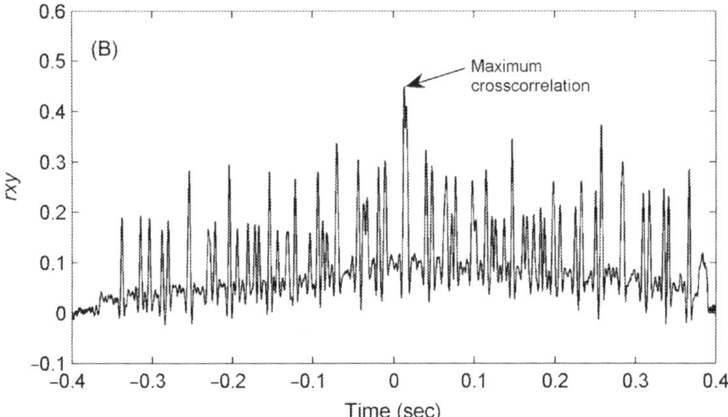

FIGURE 2.18 A) Recordings made from two different neurons believed to be involved in the same neural process, but delayed by intervening synapses (simulated data). Some background noise is seen in both recordings. B) Crosscorrelation between the two neural recordings showing a peak at 0.013 sec.

this shift, that is, `lags(max_shift)`. This lag value then needs to be converted to a corresponding time shift by dividing by the sampling frequency, fs. In the program, these operations are done in the display command: `lags(max_shift)/fs`.

Result: The two signals are shown in Figure 2.18A. The dashed signal, x, follows the solid line signal, y. The crosscorrelation function is seen in Figure 2.18B. While the time at which the peak occurs is difficult to determine visually due to the large scale of the horizontal axis, the peak is found to be 0.013 sec using the method described above. The maximum correlation at this lag is 0.45.

Unlike correlation, crosscorrelation can be done using signals that have different lengths and is often used for probing waveforms with short reference signals. For example, given the waveform seen in Figure 2.19A, we might ask: "does any portion of this signal contain something similar to the short reference waveform shown on the left side?" The search

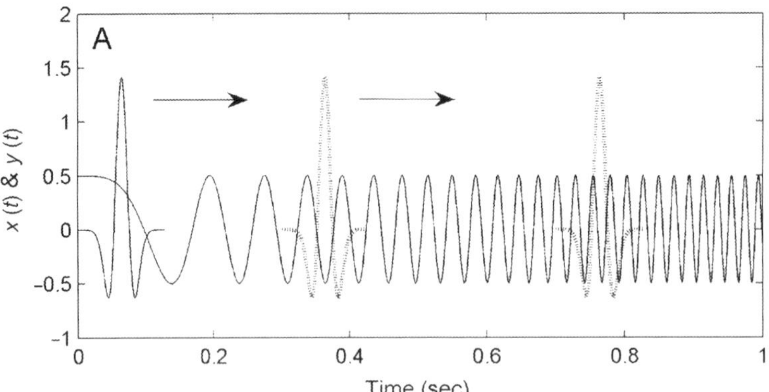

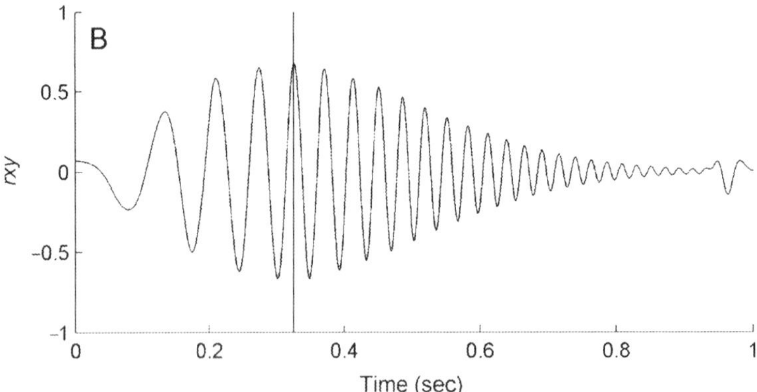

FIGURE 2.19 A) The short reference waveform on the left is to be compared to the longer signal. The longer signal continuously increases its frequency with time and is termed a *chirp* signal. Crosscorrelation can be used to perform a running correlation. B) Crosscorrelation between the two signals. A maximum correlation of 0.677 was found at 0.325 sec (vertical line).

for segments of the original signal that may correlate with the shorter reference waveform requires performing the correlation analysis at all possible relative positions and shifting the reference signal one sample interval at a time; crosscorrelation can be used for this sliding correlation. Peak correlation values indicate where the original signal is most similar to the reference function. An example of this application of crosscorrelation is shown in Example 2.16.

EXAMPLE 2.16

Evaluate the signal shown in Figure 2.19A to determine if it contains segments similar to the short reference signal also shown in Figure 2.19A. Determine at what time the most significant match occurs. Both waveforms are found as x (signal) and y (reference) in file `chirp_signal.mat` and have a sampling interval of 1.0 msec.

Solution: Load the waveforms and apply crosscorrelation between the signal and reference. Find the maximum correlation and time of correlation.

```
fs=1/.001  % Sampling frequency (1/Ts)
load chirp_signal;                    % Load data
[rxy,lags]=axcor(y,x);               % Crosscorrelate: shift reference (y)
plot(lags/fs,rxy,'k');               % Plot data
  .......label and axis.......
[max_corr max_shift]=max(rxy);       % Find max values
disp([max_corr lags(max_shift)/fs])  % Display max values
```

Result: The crosscorrelation of the original chirp signal and the shorter reference shows a maximum correlation of 0.677 occurring at 0.325 sec. As shown in Figure 2.19B, the chirp signal appears most similar to the reference at this time point (vertical line).

For a final crosscorrelation example, let us return to the EEG signal and the sinusoidal comparison used in Example 2.13. Rather than compare the EEG signal with just two sinusoids, we use crosscorrelation to compare it with a number of sinusoidal frequencies ranging between 1.0 and 25-Hz using 1.0-Hz intervals between them. Using crosscorrelation will ensure that we correlate with a sinusoid having all possible phase shifts.

EXAMPLE 2.17

Find the similarity between the EEG signal and sinusoid over a range of frequencies from 1.0 to 25-Hz. The sinusoidal frequencies should be in 1.0-Hz increments.

Solution: Use a loop to generate a series of sine waves from 1 to 25-Hz in 1.0-Hz increments. (Cosine waves work just as well since the crosscorrelation covers all possible phase shifts.) Crosscorrelate these sine waves with the EEG signal and find the maximum crosscorrelation. Plot this maximum correlation as a function of the sine wave frequency.

```
% Example 2.17 Comparison of sinusoids at different frequencies with the EEG signal
% using crosscorrelation.
%
load eeg_data;              % Get EEG data
fs=50;                      % Sampling frequency
t=(1:length(eeg))/fs;       % Time vector
%
for i=1:25
  f(i)=i;                   % Frequency range: 1-25 Hz
  x=sin(2*pi*f(i)*t);       % Generate sine
  rxy=axcor(eeg,x);         % Perform crosscorrelation
  rmax(i)=max(rxy);         % Store max value
end
plot(f,rmax,'k');           % Plot max values as function of freq.
  .......labels.......
```

Result: The result of the multiple crosscorrelations is seen in Figure 2.20, and an interesting structure emerges. Some frequencies show much higher correlation between the sinusoid and

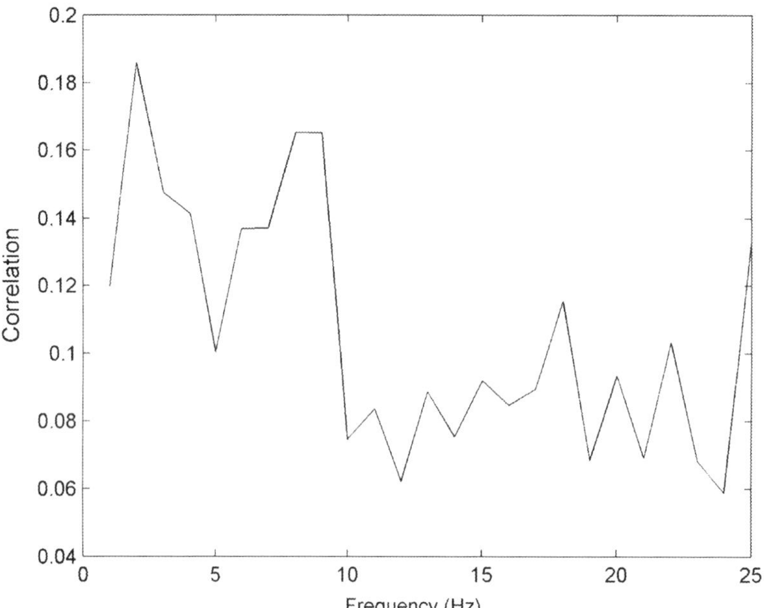

FIGURE 2.20 The maximum crosscorrelation between sine waves and the EEG signal plotted as a function of the sine wave frequency. Several peaks are seen, including a peak between 7 and 9-Hz which indicates the presence of an oscillatory pattern known as the "alpha wave."

the EEG. A particularly strong peak is seen in the region of 7 to 9-Hz, indicating the presence of an oscillatory pattern known as the "alpha wave." A more efficient method for obtaining the same information is given in section 3.2.

2.4.6 Autocorrelation

It is also possible to correlate a signal with other segments of itself. This can be done by performing crosscorrelation on two identical signals, a process called *autocorrelation*. Autocorrelation is easy to implement—simply apply crosscorrelation with the same signal as both original signal and reference (e.g., axcor(x,x))—but it is harder to understand what the result signifies. Basically, the autocorrelation function describes how well a signal correlates with shifted versions of itself. This could be useful in finding segments of a signal that repeat. Another way of looking at autocorrelation is that it shows how the signal correlates with neighboring portions of itself. As the shift increases, the signal is compared with more distant neighbors. Determining how neighboring segments of a signal relate to one another provides some insight into how the signal was generated or altered by intervening processes. For example, a signal that remains highly correlated with itself over a period of time must have been produced, or modified, by some process that took into account previous values of the signal. Such a process can be described as having *memory*, since it must remember past values of the signal and use this information to shape the signal's current values. The longer the memory, the more the signal remains partially correlated with shifted versions of itself. Just as memory tends to fade over time, the autocorrelation function usually goes to zero for large enough time shifts.

To derive the autocorrelation equation, simply substitute the same variable for x and y in Equation 2.23 or Equation 2.24:

$$r_{xx}[\ell] = \frac{1}{N}\sum_{n=1}^{N} x[n]x[n+\ell] \quad \ell = 0, 1, 2, \ldots L \tag{2.25}$$

$$r_{xx}(\tau) = \frac{1}{T}\int_0^T x(t)x(t+\tau)dt \tag{2.26}$$

where r_{xx} is the autocorrelation function.

Figure 2.21 shows the autocorrelation of several different waveforms. In all cases, the correlation has a maximum value of one at zero lag (i.e., no time shift) since when the lag (τ or ℓ) is zero, this signal is being correlated with itself. It is common to normalize the autocorrelation function to 1.0 at lag 0. The autocorrelation of a sine wave is another sinusoid, as shown in Figure 2.21A, since the correlation varies sinusoidally with the lags, or phase shift. Theoretically the autocorrelation should be a pure cosine, but because the correlation routine adds zeros to the end of the signal in order to compute the autocorrelation at all possible time shifts, the cosine function decays as the shift increases.

A rapidly varying signal, as in Figure 2.21C, *decorrelates* quickly: the correlation of neighbors falls off rapidly for even small shifts of the signal with respect to itself. One could say that this signal has a poor memory of its past values and was probably the product of a process with a short memory. For slowly varying signals, the correlation falls slowly, as in Figure 2.21B. Nonetheless, for all of these signals there is some time shift for which the signal becomes completely decorrelated with itself. For a random signal, the correlation falls to zero instantly for all positive and negative lags, as in Figure 2.21D. This indicates that each instant of the random signal (each instantaneous time point) is completely uncorrelated with the next instant. A random signal has no memory of its past and cannot be the product of a process with memory.

Since shifting the waveform with respect to itself produces the same results no matter which way the function is shifted, the autocorrelation function will be symmetrical about lag zero. Mathematically, the autocorrelation function is an even function:

$$r_{xx}(-\tau) = r_{xx}(\tau) \tag{2.27}$$

The maximum value of r_{xx} clearly occurs at zero lag, where the waveform is correlated with itself. If the autocorrelation is normalized by the variance, which is common, the value will be 1 at zero lag. (Since in autocorrelation the same function is involved twice, the normalization equation given in Equation 2.17 reduces to $1/\sigma^2$.)

When autocorrelation is implemented on a computer, it is usually considered just a special case of crosscorrelation. That is the case here where `axcor` is used with a single input argument.

```
[rxx, lags]=axcor(x)  % Autocorrelation
```

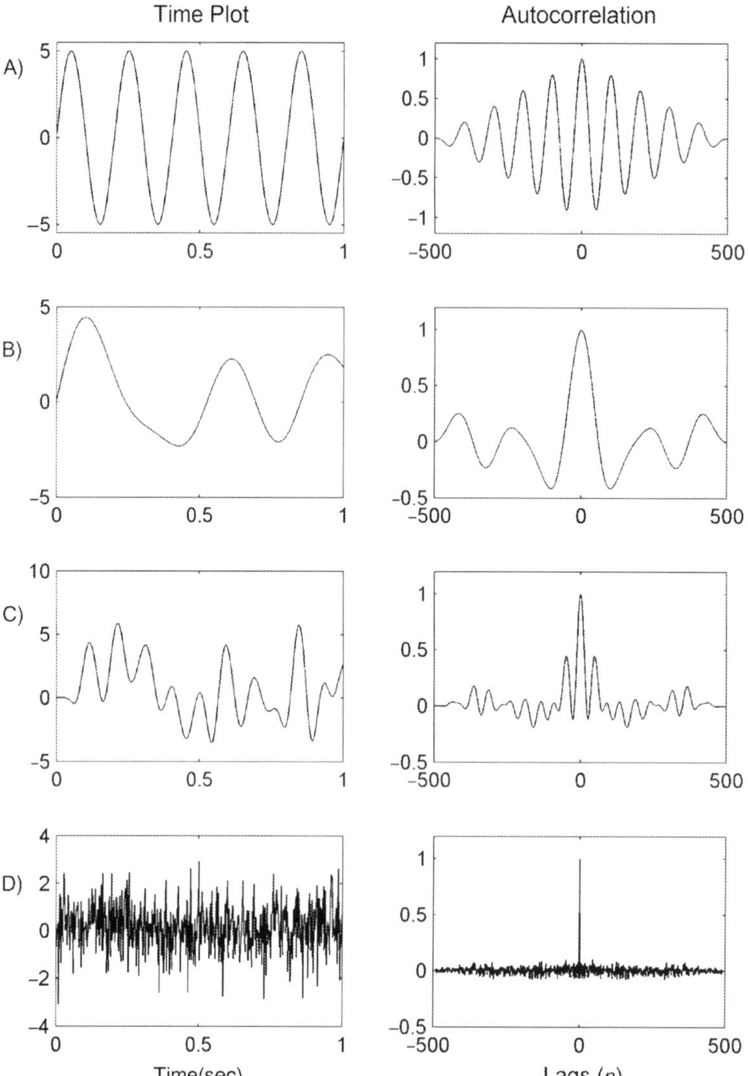

FIGURE 2.21 Four different signals (left side) and their autocorrelation functions (right side): A) A truncated sinusoid. The reduction in amplitude of the autocorrelation is due to the finite length of the signal. A true (i.e., infinite) sinusoid would have a nondiminishing cosine wave as its autocorrelation function. B) A slowly varying signal. C) A rapidly varying signal. D) A random signal.

When only a single input argument is present, axcor performs autocorrelation and normalizes the resultant function to 1.0 at zero shift (i.e., `lags = 0`) when the signal is identically correlated with itself. A simple application of autocorrelation is shown in the next example.

EXAMPLE 2.18

Evaluate and plot the autocorrelation function of the EEG signal. To better view the decrease in correlation at small shifts, plot only shifts between ± 1 sec.

Solution: Load the EEG signal and apply axcor with the EEG signal as the only input argument. Plot the result using the lags output scaled by the sampling frequency to provide the delay (i.e., lags) in sec. Recall the sampling frequency of the EEG signal is 50-Hz. Scale the x axis to show only the first ± 1 sec lags.

```
load eeg_data;
fs = 50;                              % Sample frequency 50 Hz
[rxx,lags] = axcor(eeg);              % Autocorrelation
plot(lags/fs, rxx); hold on;          % Plot autocorrelation function
plot([lags(1) lags(end)], [0 0]);     % Plot zero line
axis([-1 1 -.5 1.2]);                 % Scale x-axis to be between ± 1 sec
  ......title and labels......
```

Result: The autocorrelation function of the EEG signal is shown in Figure 2.22. The signal decorrelates quickly, reaching a value of zero correlation after a time shift of approximately 0.08 sec. The EEG signal is likely to be contaminated with noise and since noise decorrelates instantly (Figure 2.21D), some of the rapid decorrelation seen in Figure 2.22 is likely due to noise.

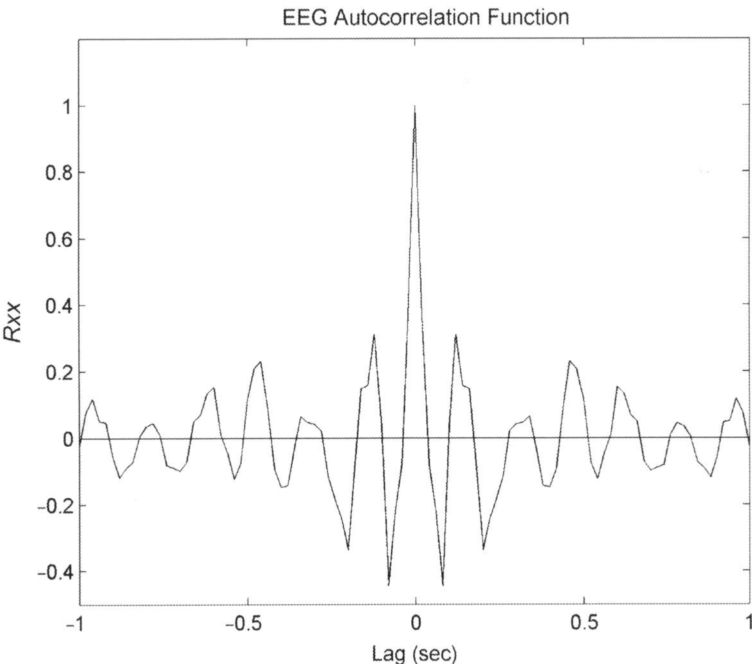

FIGURE 2.22 Autocorrelation function of the EEG signal. The autocorrelation function decorrelates rapidly probably due to noise added to the signal. Nonetheless, some correlation is seen up to 1.0 sec.

A common approach to estimating the autocorrelation of the signal without noise is to draw a smooth curve across the peaks and use that curve as the estimated autocorrelation function of signal without noise. From Figure 2.22 we see that such an estimated function decorrelates to a low value after around 0.6 to 1.0 sec.

2.4.7 Autocovariance and Crosscovariance

Two operations closely related to autocorrelation and crosscorrelation are autocovariance and crosscovariance. The relationship between correlation and covariance functions is similar to the relationship between standard correlation and covariance given in the last section. Covariance and correlation functions are the same except that in covariance the means have been subtracted from the input signals, $x(t)$ and $y(t)$ (or just $x(t)$ in the case of autocovariance):

$$\begin{aligned} Autocovariance \equiv C_{xx}(\tau) &= \frac{1}{T}\int_0^T [x(t) - \overline{x(t)}][x(t+\tau) - \overline{x(t)}]dt \\ C_{xx}[\ell] &= \frac{1}{N}\sum_{n=1}^{N}(x[n] - \overline{x})(x[n+\ell] - \overline{x}) \quad \ell = 0,1,2,\ldots L \end{aligned} \tag{2.28}$$

$$\begin{aligned} Crosscovariance \equiv C_{xy}[\tau] &= \frac{1}{T}\int_0^T [y(t) - \overline{y(t)}][x(t+\tau) - \overline{x(t)}]dt \\ Crosscovariance \equiv C_{xy}[\ell] &= \frac{1}{N}\sum_{n=1}^{N}(y[n] - \overline{y[n]})(x[n+\ell] - \overline{x[n]}) \end{aligned} \tag{2.29}$$

The autocovariance function can be thought of as measuring the memory or self-similarity of the deviation of a signal about its mean level. Similarly, the crosscovariance function is a measure of the similarity of the deviation of two signals about their respective means. An example of the application of the autocovariance to the analysis of heart rate variability is given in Example 2.19.

Recall that auto- and crosscovariance are the same as auto- and crosscorrelation if the data have zero means. Hence autocovariance or crosscovariance can be determined using `axcor` with the data means subtracted out.

```
[c,lags]=axcor(x-mean(x), y-mean(x)); % Crosscovariance
```

The autocorrelation and autocovariance functions describe how one segment of data is correlated, on average, with adjacent segments. As mentioned previously, such correlations could be due to memory-like properties in the process that generated the data. Many physiological processes are repetitive, such as respiration and heart rate, yet vary somewhat on a cycle-to-cycle basis. Autocorrelation and crosscorrelation can be used to explore this variation. For example, considerable interest revolves around the heart rate and its

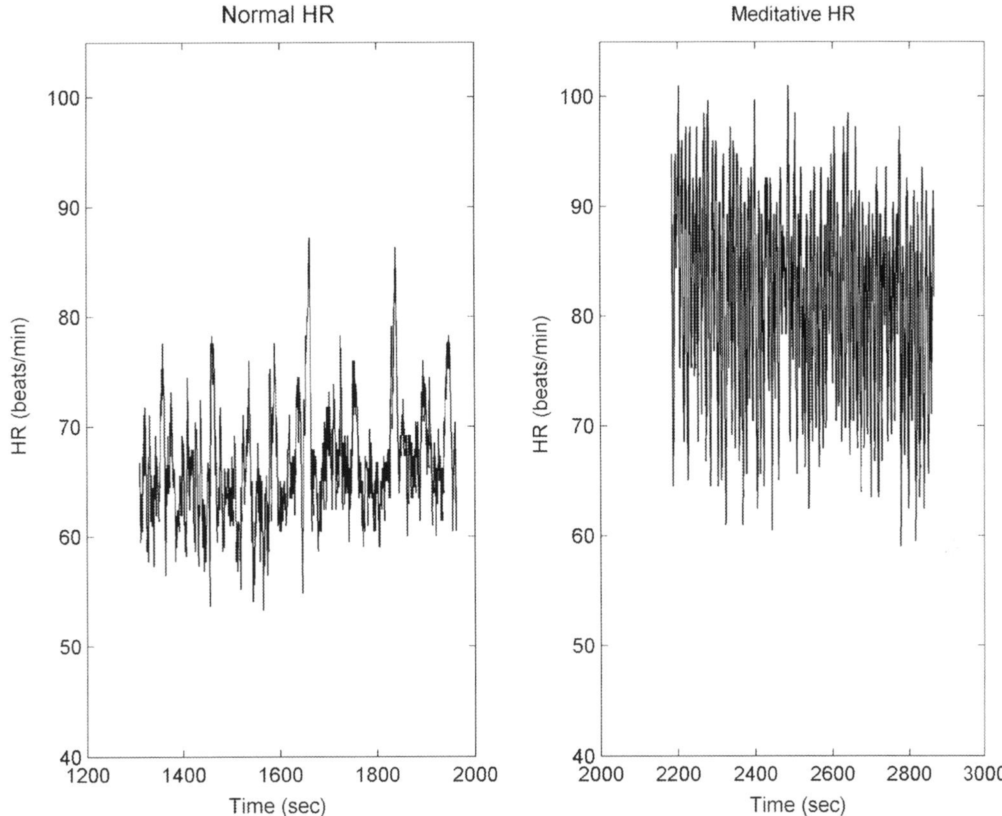

FIGURE 2.23 Approximately 10 minutes of instantaneous heart rate data taken from a normal subject and one who is meditating. The meditating subject shows a higher overall heart rate and much greater beat-to-beat fluctuation.

beat-to-beat variations. Autocovariance can be used to tell us if these variations are completely random or if there is on average some correlation between beats or over several beats. In this instance, we use autocovariance, not autocorrelation, since we are interested in correlation of heart rate variability, not the correlation of heart rate per se. (Remember that autocovariance will subtract the mean value of the heart rate from the data and analyze only the variation.)

Figure 2.23 shows the time plots of instantaneous heart rate in beats per minute taken under normal and meditative conditions. These data are found on this book's accompanying CD as `HR_pre.mat` (preliminary) and `HR_med.mat` (meditative) and were originally from the PhysioNet database (Goldberger et al., 2000). Clearly the meditators have a higher average heart rate, a different structure, and their variation is greater. These differences will be explored further in the next chapter. Example 2.19 analyzes the normal heart rate data of Figure 2.23 to determine the correlation over successive beats.

EXAMPLE 2.19

Determine if there is any correlation in the variation between the timing of successive heartbeats under normal resting conditions.

Solution: Load the heart rate data taken during normal conditions. The file `Hr_pre.mat` contains the variable `hr_pre`, the instantaneous heart rate. However, the heart rate is determined each time a heartbeat occurs so it is not evenly time-sampled and a second variable `t_pre` contains the time at which each beat is sampled. For this problem, we will determine the autocovariance as a function of heart beat and we will not need the time variable. We can determine the autocovariance function using axcor by first subtracting the mean heart rate. The subtraction can be done within the `axcor` input argument. We then plot the resulting autocovariance function and limit the x axis to $\pm$ 30 successive beats to better evaluate the decrease in covariance with successive beats.

```
% Example 2.19 Use of autocovariance to determine the correlation
% of heart rate variation between heart beats
%
load Hr_pre;                                          % Load normal HR data
[cov_pre,lags_pre] = axcor(hr_pre - mean(hr_pre));    % Auto-covariance
plot(lags_pre,cov_pre,'k'); hold on;                  % Plot normal auto-cov
plot([lags_pre(1) lags_pre(end)], [0 0],'k');         % Plot a zero line
axis([-30 30 -0.2 1.2]);                              % Limit x-axis to ±30 beats
```

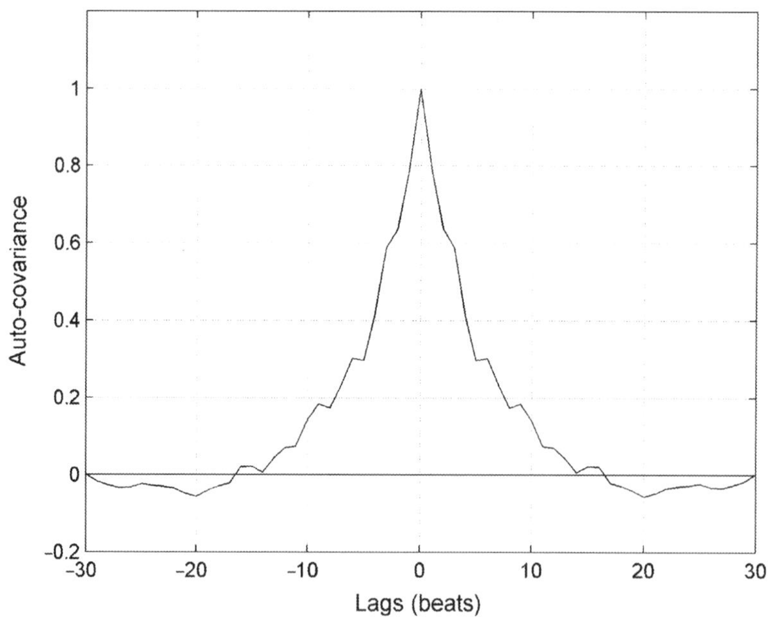

FIGURE 2.24 Autocovariance function of the heart rate under normal resting conditions (solid line). Some correlation is observed in this individual over approximately 10 successive heartbeats.

Results: The results in Figure 2.24 show that there is some correlation between adjacent heartbeats all the way out to 10 beats.

2.5 SUMMARY

The sinusoidal waveform is arguably the single most important waveform in signal processing. Some of the reasons for this importance are provided in the next chapter. Because of their importance, it is essential to know the mathematics associated with sines, cosines, and general sinusoids, including the complex representations.

Several basic measurements discussed in the last chapter apply to any signal including mean value, RMS value, and variance or standard deviation. While these measurements provide some essential basic information, they do not provide much information on signal content or meaning. A common approach to obtain more information is to probe a signal by correlating it with one or more reference waveforms. One of the most popular probing signals is the sinusoid, and sinusoidal correlation is covered in detail in the next chapter. Sometimes a signal will be correlated with another signal in its entirety, a process known as correlation or the closely related covariance. Zero correlation between signal and reference does not necessarily mean they have nothing in common, but it does mean the two signals are mathematically orthogonal. Signals and families of signals that are orthogonal are particularly useful in signal processing because when they are used in combination, each orthogonal signal can be treated separately: it does not interact with the other signals.

If the probing signal is short or the correlation at a number of different relative positions is desired, a running correlation known as crosscorrelation is used. Crosscorrelation not only quantifies the similarity between the reference and the original, but also shows where that match is greatest. A signal can be correlated with shifted versions of itself, a process known as autocorrelation. The autocorrelation function describes the time period for which a signal remains partially correlated with itself, and this relates to the structure of the signal. A signal consisting of random noise decorrelates immediately, whereas a slowly varying signal will remain correlated over a longer period. Correlation, crosscorrelation, autocorrelation, and the related covariances are all easy to implement in MATLAB.

PROBLEMS

1. Two 10-Hz sine waves have a relative phase shift of 30 deg. What is the time difference between them? If the frequency of these sine waves doubles, but the time difference stays the same, what is the phase difference between them?
2. Convert $x(t) = 6\sin(5t) - 5\cos(5t)$ into a single sinusoid; i.e., $A\sin(5t + \theta)$.
3. Convert $x(t) = 30\sin(2t + 50)$ into sine and cosine components. (Angles are in deg unless otherwise specified.)
4. Convert $x(t) = 5\cos(10t + 30) + 2\sin(10t - 20) + 6\cos(10t + 80)$ into a single sinusoid, as in Problem 2.
5. Find the delay between $x_1(t) = \cos(10t + 20)$ and $x_2(t) = \sin(10t - 10)$.
6. Equations 2.9 and 2.10 were developed to convert a sinusoid such as $\cos(\omega t - \theta)$ into a sine and cosine wave. Derive the equations to convert a sinusoid based on the sine function, $\sin(\omega t + \theta)$, into a sine and cosine wave. [*Hint*: Use the appropriate identity from Appendix C.]

7. A) Using the values for a and b in Example 2.3 (-5 and -3), find the angle using the MATLAB `atan` function. Remember MATLAB trig. functions use rad for both input and output. Note the quadrant-related error in the result. B) Now use the `atan2` functions to find the angle. The arguments for this function are `atan2(b,a)`. C) Finally, put a and b into complex form: $C = a + jb$. Find the angle of the complex number C using the MATLAB angle function (i.e., `angle(C)`). Note that the angle is the same as that found by the `atan2` function. Also evaluate the magnitude of C using the MATLAB `abs` function (i.e., `abs(C)`) and note that it is the same as the value found in Example 2.3 although the angle is given as negative instead of positive.
8. Modify the complex exponential in Example 2.5 to generate and plot a cosine wave of amplitude 5.0 that is shifted by 45 deg. [*Hint*: Since the cosine is desired, you only need to plot the real part and modify the magnitude and phase of the exponential. Remember MATLAB uses rad.]
9. Image generation. Using the approach shown in Example 2.7, construct a sinusoidal grating having a spatial frequency of 5 cycles per horizontal distance. Transpose this image and display. You should have a sinusoidal grating with horizontal strips. Then multiply each row in the transposed image by the original sine wave and display. This should generate a checkerboard pattern.
10. Load the image of the brain found in `brain.mat`, display the original, and apply several mathematical operations. A) Invert the image: make black white and white black. B) Apply a nonlinear transformation: make a new image that is the square root of the pixel value of the original. (Note that this requires only 1 line of MATLAB code.) C) Create a doubly thresholded image. Set all values below 0.25 in the original image to zero (black), all values above 0.5 to 1.0 (white), and anything in between to 0.5 (gray). (In this exercise, it is easier not to use `caxis` to set the grayscale range. Just let the `pcolor` routine automatically adjust to the range of your transformed images.) These images should be plotted using the `bone` colormap to reproduce the grayscale image with accuracy, but after plotting out the figure you could apply other colormaps such as `jet`, `hot`, or `hsv` for some interesting pseudocolor effects. Just type `colormap(hot)`, and so on.
11. Use Equation 2.15 to show that the correlation between $\sin(2\pi t)$ and $\cos(2\pi t)$ is zero.
12. Use Equation 2.15 to find the normalized correlation between a cosine and a square wave, as shown in the figure below. This is the same as Example 2.10 except that the sine has been replaced by a cosine.

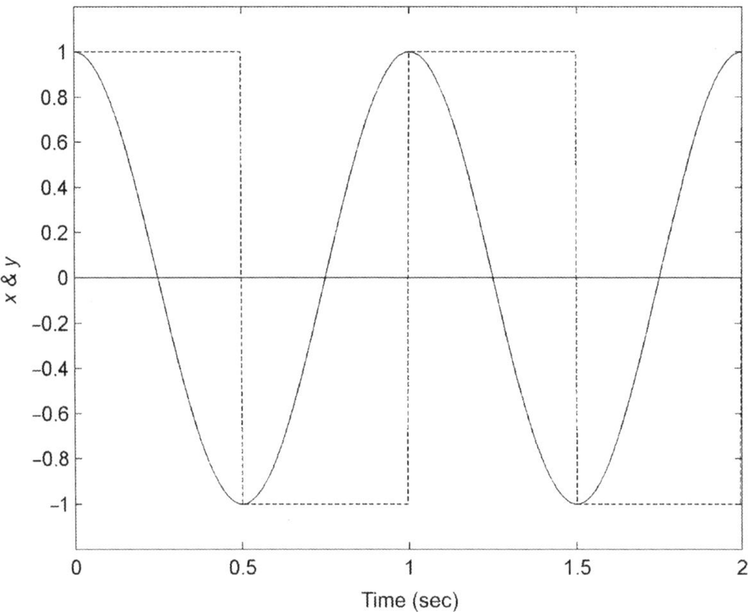

13. Use Equation 2.15 to find the correlation between the two waveforms shown in the figure below.

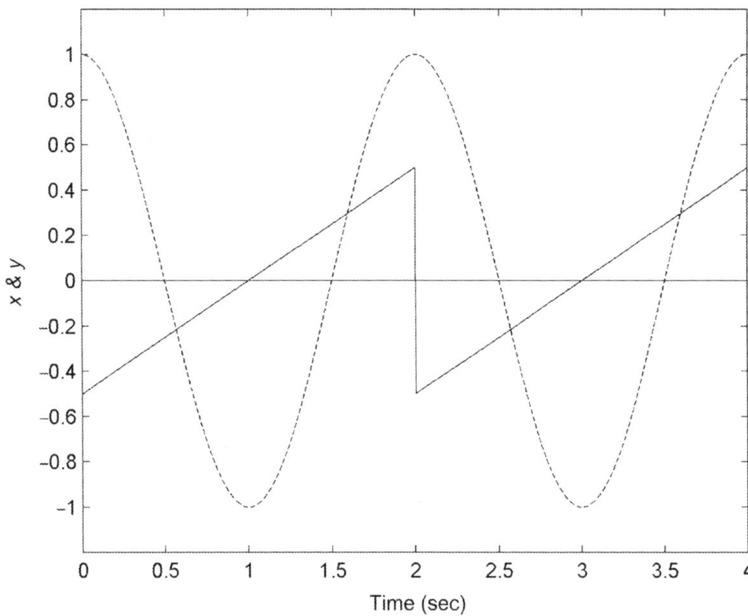

14. Modify Example 2.17 to use incremental shifts of 1, 2, 10, 20, and 50 degrees. Use a loop to implement each of the phase shifts. Do not plot the correlations, just print out the maximum correlation for each frequency and for each shift. What is the difference in the resulting maximum correlations for the various shifts? What is the difference in the phase values at which the maximum shift is measured? [*Hint*: To find the phase, save the various phase shifts generated in the loop, then use the index obtained from the max function to find the phase; i.e., `max_phase = phase(index)` converted to deg.

15. Use MATLAB and crosscorrelation to find the delay between the 10-Hz sine waves described in Problem 1. Use a sample frequency of 2 kHz and total time of 0.5 sec. Remember MATLAB trig. functions use rad. (Note that the second part of Problem 1, finding the phase when the frequency doubled would be difficult to do in MATLAB. There are occasions where analytical solutions are preferred.)
16. Use MATLAB and crosscorrelation to find the delay in the two sinusoids of Problem 5. (You choose the sampling frequency and number of data points.)
17. Use MATLAB and crosscorrelation to find the phase shift between $x(t)$ in Problem 4 and a sine wave of the same frequency. Plot $x(t)$ and the crosscorrelation function and find the lag at which the maximum (or minimum) correlation occurs. You choose the sample frequency and number of points. [*Hint*: Generate the two signals and find the delay between $x(t)$ and $\sin(10t)$ using crosscorrelation. You will have to follow the approach used in Example 2.15 to get the correct time shift. To convert that time shift to a phase shift, note that the period of $x(t)$ is $1/(2\pi/10)$ sec and the phase is the ratio of the time delay to this period, times 360.]
18. The file `two_var.mat` contains two variables `x` and `y`. Is either of these variables random? Are they orthogonal to each other? (Use any valid method to determine orthogonality.)
19. Modify Example 2.17 so that the maximum correlation is found between the EEG signal and a number of sinusoids ranging in frequency between 1 and 25-Hz. Increase the sinusoid frequencies in 0.5-Hz increments using an additional loop—you will need to repeat the inner loop 50 times for 50 different frequencies. Also limit the phase shift to between 0 and 179 deg, and find the maximum positive or negative correlation. Plot only one graph: the maximum correlation as a function of the sinusoidal frequency. Compare the output with the results from Example 2.17, presented in Figure 2.20.
20. The file `prob2_20_data.mat` contains a signal `x` ($f_s = 500$-Hz). Determine if this signal contains a 50-Hz sine wave and if so at what time(s).
21. The file `prob2_21_data.mat` contains a variable `x` that is primarily noise but may contain a periodic function. Plot `x` with the correct time axis ($f_s = 1$ kHz). Can you detect any structure in this signal? Apply autocorrelation and see if you can detect a periodic process. If so, what is the frequency of this periodic process? It may help to expand the x axis to see detail. Chapter 3 presents a more definitive approach for detecting periodic processes and their frequencies.
22. Develop a program along the lines of Example 2.19 to determine the correlation in heart rate variability during meditation. Load file `Hr_med.mat`, which contains the heart rate in variable `hr_med` and the time vector in variable `t_med`. Calculate and plot the autocovariance. The result will show that the heart rate under meditative conditions contains some periodic elements. Can you determine the frequency of these periodic elements? A more definitive approach is presented in Chapter 3, which shows how to identify the frequency characteristics of any waveform.

CHAPTER

3

Fourier Transform: Introduction

3.1 TIME- AND FREQUENCY-DOMAIN SIGNAL REPRESENTATIONS

Both analog and digital signals can be represented in either the time or the frequency domain. Time-domain representations are usually intuitive and are easy to interpret and understand. The discussion of signals thus far has been in the time domain. The frequency-domain representation is an equally valid representation of a signal and consists of two components: magnitude and phase. These two components are discussed in detail below, but essentially the components show how the signal energy is distributed over a range of frequencies. The frequency components of a signal are called the *signal spectrum*. Hence, the words *spectrum* or *spectral* are inherently frequency-domain terms. The popular word *bandwidth* is also a frequency-domain term. The two representations are completely interchangeable. You can go from the time to the frequency domain and back again with no constraints; you just have to follow the mathematical rules.

Sometimes the spectral representation of a signal is more understandable than the time-domain representation. The EEG time-domain signal, such as shown in Figure 1.9, is quite complicated. The time-domain signal may not be random, but whatever structure it contains is not apparent in the time response. When the EEG signal is transformed in the frequency domain, organization emerges. In fact, clinicians who use the EEG, so-called electroneurophysiologists, define EEG waves by their frequencies: the alpha wave ranges between 8 and 12-Hz, the beta wave between 12 and 30-Hz, and the theta wave between 4 and 7-Hz.

In Chapter 5, we find that systems as well as signals can be represented and thought of in either the time or frequency domain. Since systems operate on signals, both time- and frequency-domain representations show how the system affects a signal. With systems, the frequency domain is the more intuitive: it shows how the system alters each frequency component in the signal. The *spectrum of a system* shows if the signal energy at a given frequency is increased, decreased, or left the same and how it might be shifted in time. For linear systems, those are the only choices: altering the energy that is already in the signal. However, nonlinear systems can produce a signal with energy at frequencies that were not in the original signal. The same mathematical tools used to convert signals between the time and frequency domains are used to flip between the two system representations and, again, there are no constraints.

3.1.1 Frequency Transformations

One approach to finding the frequency-domain equivalent of a time-domain signal is to use crosscorrelation with sine waves. As shown below, sine waves have some special properties that allow them to be used as a gateway into the frequency domain. In Example 2.17, the EEG signal was correlated against a reference family of sinusoids and we found that sinusoids at some frequencies showed much higher correlation with the EEG signal than at other frequencies (Figure 2.20). This crosscorrelation approach can be used to examine the structure of any waveform using sinusoids as a reference waveform. When a high correlation is found between the signal and the reference waveform, we could say the signal "contains" some of that reference waveform. The higher the correlation, the more the reference signal is part of the signal being analyzed.

There are two problems in using this approach to analyze the structure of a signal. First, it is computationally intensive, although this is not much of a problem with modern high-speed computers. Second, you have to know what you are looking for, or at least have an idea of the general shape. If you are probing with sinusoids, you have to have some idea about what sinusoidal frequencies might be present. For example, suppose you are examining a signal that is made up of three sinusoids of different frequencies. If your crosscorrelation search includes the three frequencies, you will get an accurate picture of the three sinusoids contained in the signal. But what if your search does not include the exact frequencies? Perhaps they are close, but not identical, to the ones contained in the signal. As will be demonstrated in Example 3.1, you would likely get an inaccurate picture of the sinusoidal composition of the signal.

EXAMPLE 3.1

Use crosscorrelation to probe two signals, each containing a mixture of three sinusoids. One signal is made up of sines at 100, 200, and 300-Hz while the other contains sines at 100, 204, and 306-Hz. Begin your search at 10.0-Hz and continue up to 500-Hz crosscorrelating every 10-Hz. Assume a sampling frequency of 1.0 kHz to generate the two waveforms. In constructing the test waveform, use different sine-wave amplitudes to test the ability of the crosscorrelating analysis to find the amount of sinusoidal signal in the waveform.

Solution: Modify the MATLAB code in Example 2.17 to include the generation of the two sine-wave mixtures. Plot out the maximum correlation for the two mixtures side-by-side.

```
% Example 3.1
% Correlation analysis of two signals each containing three sinusoids
%
fs = 1000;                          % Sample frequency
N = 2000;                           % Number of points in the test signals
t = (1:N)/fs;                       % Time vector
f = [100 200 300];                  % Test signal frequencies
```

```
%
% Generate the test signal as a mixture of 3 sinusoids at different freq.
x = 1.5*sin(2*pi*f(1)*t) + 1.75*sin(2*pi*f(2)*t) + 2.0*sin(2*pi*f(3)*t);
%
for i = 1:50                          % Analysis loop
  f(i) = i*10;                        % Frequency range: 10-500 Hz
  y = cos(2*pi*f(i)*t);               % Generate sinusoid
  [r,lags] = axcor(x,y);              % Crosscorrelate
  [rmax(i),ix(i)] = max(r);           % Find maximum value
end
subplot(1,2,1); plot(f,rmax,'k');     % Plot and label crosscorrelation results
......label and title......
%
% Redo for a test signal having slightly different frequencies.
f = [100 204 306];                    % New test signal frequencies
x = 1.5*sin(2*pi*f(1)*t) + 1.75*sin(2*pi*f(2)*t) + 2.0*sin(2*pi*f(3)*t);
......same code as above......
```

Results: As shown in Figure 3.1A, the crosscorrelation analysis correctly identifies the three sinusoids found in the first mixture. The peaks of the three components accurately reflect the change in amplitude of the three sine waves (from 1.5 at 100-Hz, 1.75 at 200-Hz and 2.0 at 300-Hz). However, when the frequencies of two of the component sine waves are shifted just a few Hz, the analysis fails to find them. This is because the crosscorrelation is searching in increments of 10-Hz and does not compare the test signal with sinusoids at 204 and 306-Hz frequencies. (Moreover, this program takes over 10 sec to run on a typical personal computer.)

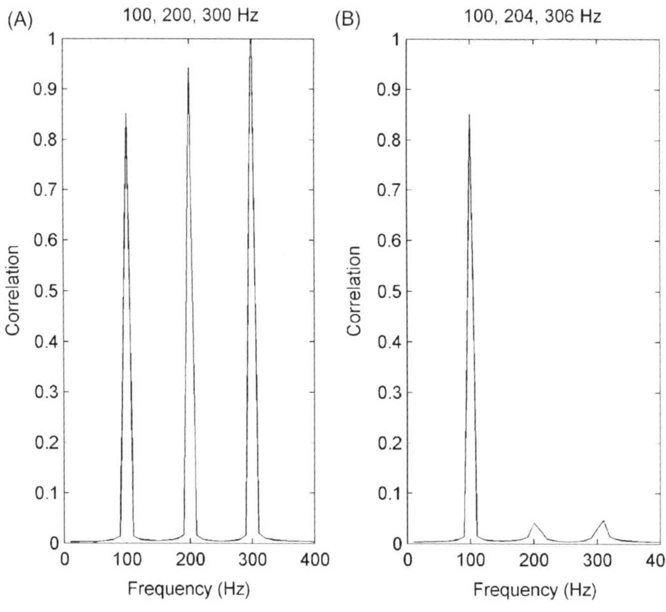

FIGURE 3.1 Crosscorrelation analysis of two very similar signals, both containing mixtures of three sinusoids. The amplitudes of the three sine waves are 1.5, 1.75, and 2.0 in both mixtures. A) The analysis of a mixture containing 100, 200, and 300-Hz sine waves clearly identifies the three sinusoids at the correct frequencies. B) When two of the sinusoids are shifted slightly in frequency, from 200 to 204-Hz and from 300 to 306-Hz, the analysis fails to find them.

From Example 3.1 we can see that using crosscorrelation to probe the contents of a signal works well if we know the specific pattern of interest, but we often do not know what pattern(s) a signal may contain. Of course we can decrease the frequency increment and use more sinusoids to probe the signals, but this increases the analysis time and we may still miss some important frequencies. The question is: when probing a signal whose characteristics are unknown, what probing frequencies should we use? If we are probing with sinusoids (a common probing signal for reasons described next) and the signal we are probing is periodic, or taken to be periodic, the answer to this question of what frequencies we should use is found in an important theorem known as the *Fourier series theorem*. Before getting into Fourier's theorem, let us look at some of the remarkable properties embodied in the sinusoid.

3.1.2 Useful Properties of the Sinusoidal Signal

Despite appearances, sinusoids are the most basic of waveforms and that makes them of particular value in signal analysis. Sinusoids have six properties that we will find extraordinarily useful for signal and systems analysis. We will make use of these properties in the next few chapters.

1. Sinusoids have energy at only one frequency: the frequency of the sinusoid. This property is unique to sinusoids. Because of this property, sinusoids are sometimes referred to as "pure": the sound a sinusoidal wave produces is perceived as pure or basic. (Of the common orchestral instruments, the flute produces the most sinusoidal-like tone.)
2. Sinusoids are completely described by three parameters: amplitude, phase, and frequency. If frequency is fixed, then only two variables are required, amplitude and phase. Moreover, if you use complex variables, these two parameters can be rolled into a single complex variable.
3. Thanks to Fourier's theorem, discussed below, many signals can be broken down into an *equivalent* representation as a series of sinusoids (see Figure 3.2). The only constraint on the signal is that it must be periodic, or assumed to be periodic. Recall that a periodic signal repeats itself exactly after some period of time, T: $x(t) = x(t + T)$. Moreover, the sinusoidal components will be at the same frequency or integer multiples of the frequency of the periodic signal.
4. Looking at it the other way around, you can always find some combination of harmonically related sinusoids that, when added together, reconstruct any periodic signal. You may need a large number of such sinusoids to represent a signal, but it can always be done.[1]
5. In other words, the sinusoidal representation of a signal is complete and works in both directions: signals can be decomposed into some number of sinusoids and can also be accurately reconstructed from that series (see Figure 3.2). Because it works both ways, this decomposition is known mathematically as an *invertible transform*.

[1] In a few cases, such as the *square wave*, you need an infinite number of sine waves. As shown in Figure 3.6, any finite series will show an inappropriate oscillatory behavior.

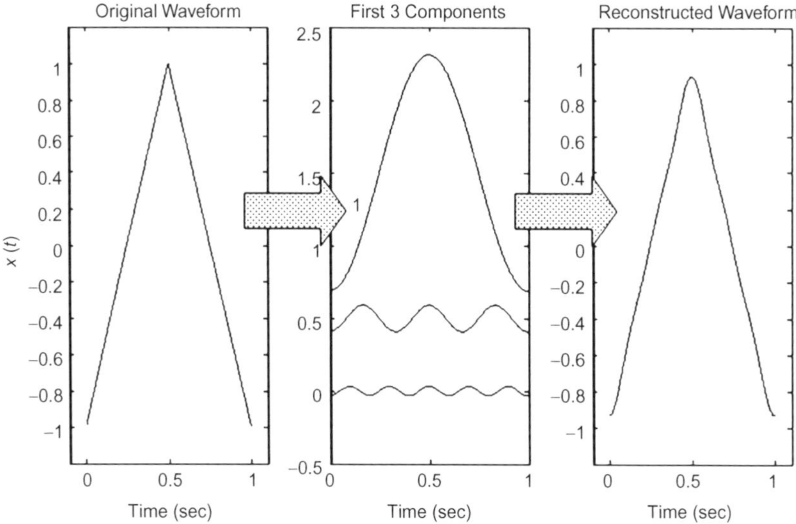

FIGURE 3.2 A periodic signal decomposed into a series of sinusoids which when summed will reconstruct the original signal. In this figure, a triangle wave is deconstructed into three sinusoidal components which are then used to reconstruct the original waveform. Even with only three components, the reconstruction is quite good. More components would lead to a more accurate representation and a better reconstruction (see Figure 3.6).

6. Since sinusoids have energy at only one frequency, they can easily be represented in what is termed the *frequency domain*. If for any frequency a sinusoid is completely described by two variables, amplitude and phase, then it can be represented accurately using two plots: one plot specifying amplitude and the other phase at the frequency of the sinusoid, as shown in Figure 3.3. (This property is really just a combination of Properties 1 and 3 above, but it is so useful that it is listed separately.)
7. Combining the decomposition properties described in Property 3 with the frequency representation of Property 4 gives us a technique to represent any periodic signal in the frequency domain. If a waveform can be decomposed into equivalent sinusoids, and each sinusoid plots directly to the magnitude and phase plots of frequency domain, then sinusoids can be used as a gateway between any periodic function and its frequency representation. The frequency representation of a signal is also referred to as its *spectrum*. Using this approach, any periodic signal can be represented by an equivalent magnitude and phase spectrum (Figure 3.4), and that frequency representation is just as valid as the time representation.
8. The calculus operations of differentiation and integration change only the magnitude and phase of a sinusoid. The result of either operation is still a sinusoid at the same frequency. For example, the derivative of a sine is a cosine, which is just a sine with a 90-degree phase shift; the integral of a sine is a negative cosine or a sine with a −90-degree phase shift. The same rules apply to any general sinusoid (i.e., $\cos(2\pi ft + \theta)$).

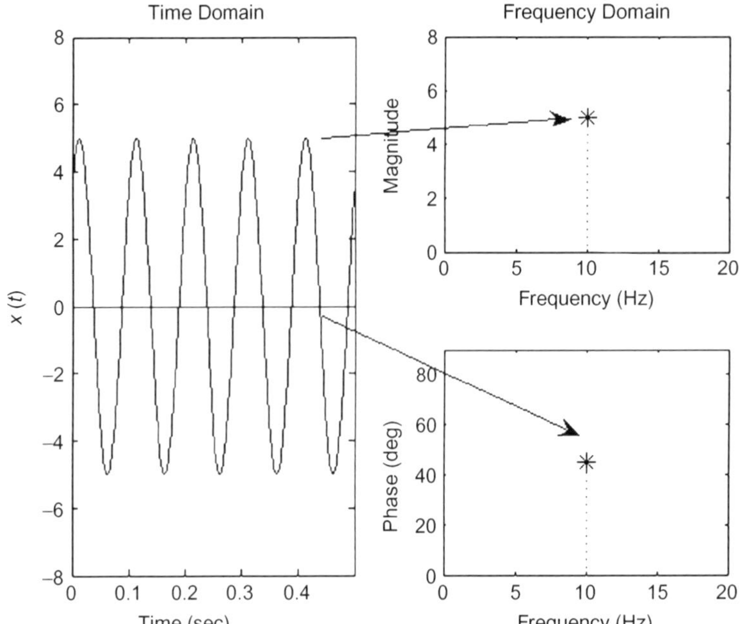

FIGURE 3.3 A sinusoid is completely represented by its magnitude and phase at a given frequency. These defining characteristics can be represented as plots of magnitude and phase against frequency. If a series of sinusoids is involved, each sinusoid contributes one point to the magnitude and phase plot.

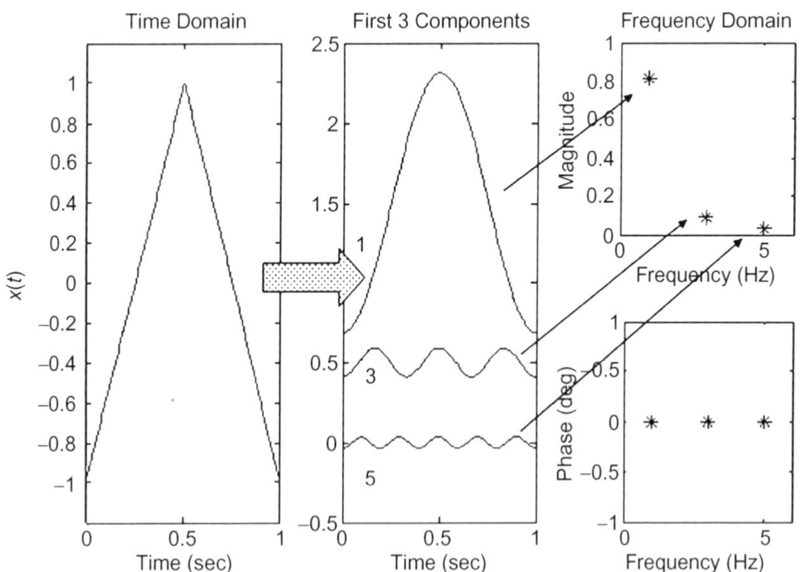

FIGURE 3.4 By combining the simplicity of a sinusoid's frequency representation (a magnitude and a phase point on the frequency plot) with its decomposition ability, sinusoids can be used to convert any periodic signal from a time-domain plot to a frequency-domain plot. If enough sinusoidal components are used, the frequency-domain representation is equivalent to the time-domain representation.

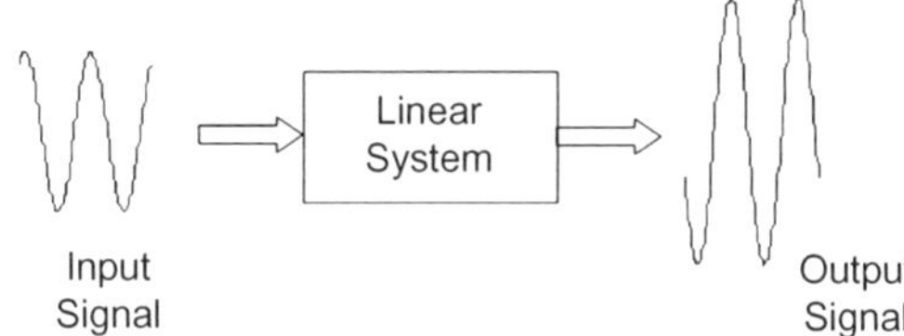

FIGURE 3.5 The output of any linear system driven by a sinusoid is a sinusoid at the same frequency. Only the magnitude and phase of the output sinusoid differs from the input.

9. If the input to any linear system is a sinusoid, the output is just another sinusoid at the same frequency irrespective of the complexity of the system. (Linear systems are defined and studied in Chapter 5.) The only difference between the input and output is the magnitude and phase of the sinusoid (Figure 3.5). This is because linear systems are composed of elements that can be completely defined by scaling, derivative, or integral operations and, as noted in Property 8, above, these operations modify only the magnitude and phase of a sinusoid.
10. Harmonically related sinusoids are orthogonal, as illustrated in Example 2.11. If a periodic waveform is decomposed into harmonically related sinusoids, then each component in the decomposition will be independent from all other components. Practically this means that if we decide to decompose a waveform into 10 harmonically related sinusoids, but later decide to decompose it into 12 sinusoids to attain more accuracy, the addition of more components will not change the value of the 10 components we already have.

3.2 FOURIER SERIES ANALYSIS

The Fourier series theorem is one of the most important and far-reaching concepts presented in this book. Its several versions will be explored using both analytical continuous domain implementations and computer algorithms. Conceptually, the Fourier series theorem is just a mathematical statement of Property 3 above: any periodic signal, no matter how complicated, can be represented by a sum of sinusoids; specifically, a series of sinusoids that are the same as, or multiples of, the signal's frequency. In other words, any periodic signal can be *equivalently* represented by sinusoids that are harmonically related to the base frequency of the signal. For example, if a periodic signal repeats every 10 seconds, that signal can be completely represented by sinusoids having frequencies of 0.1-Hz (i.e., the base frequency = 01/10 = 0.1-Hz), 0.2-Hz, 0.3-Hz, 0.4-Hz, and so on. In theory, we may have a large, possibly infinite, number of sinusoids, but for most real signals, the magnitude of the sinusoidal components eventually becomes negligibly small as frequency increases. Biological signals are not usually periodic but, as described in more detail later, can usually be taken as periodic for purposes of analysis.

To put the Fourier series theorem in mathematical terms, note that if the time period of a periodic function is T, then the base or *fundamental* frequency is:

$$f_1 = \frac{1}{T} \tag{3.1}$$

The base cosine wave, the cosine at the fundamental frequency, becomes:

$$x(t) = \cos(2\pi f_1 t) \tag{3.2}$$

and the series of harmonically related cosine waves is written as:

$$\text{Harmonic series} = \cos(2\pi m f_1\, t) = \cos\left(\frac{2\pi mt}{T}\right) \qquad m = 1, 2, 3, \ldots \tag{3.3}$$

where m is a constant that multiplies the base frequency, f_1, to obtain the higher harmonic frequencies. This constant, m, is called the *harmonic number*. The Fourier series theorem states that a series of sinusoids is necessary to represent most periodic signals, not just a cosine (or a sine) series. Therefore, we need to add a phase term, θ, to the argument of the cosines. The theorem also allows for each sinusoid in the series to be different, not just in frequency where they are harmonically related, but also in amplitude and phase. Using the letter A for the amplitude term and θ for the phase term, the series would be stated mathematically as:

$$\text{Harmonic series} = A_m \cos\left(\frac{2\pi mt}{T} - \theta_m\right) \qquad m = 1, 2, 3, \ldots \tag{3.4}$$

The Fourier series theorem simply states that any periodic function can be *completely and equivalently* represented by a summation of a sinusoidal series along with a possible constant term. The Fourier series is defined mathematically as:

$$\begin{aligned} x(t) &= \frac{A_0}{2} + \sum_{m=1}^{\infty} A_m \cos\left(\frac{2\pi mt}{T} - \theta_m\right) \qquad m = 1, 2, 3, \ldots \text{ or in terms of } f_1 \\ &= \frac{A_0}{2} + \sum_{m=1}^{\infty} A_m \cos(2\pi m f_1\, t - \theta_m) \qquad m = 1, 2, 3, \ldots \end{aligned} \tag{3.5}$$

where $x(t)$ is a periodic function with a period of T, and the first term, $A_0/2$, accounts for any nonzero mean value of the signal. This term is also known as the "DC" term. If the signal has zero mean, as is often the case, then this term will be zero. For reasons explained below, A_0 is calculated as *twice* the mean value of $x(t)$ which is why it is divided by 2 in Equation 3.5.

$$A_0 = \frac{2}{T}\int_0^T x(t)\, dt \tag{3.6}$$

Sometimes $2\pi m f_1$ is stated in terms of radians, where $2\pi m f_1 = m\omega_1$. Using frequency in radians makes the equations look cleaner, but in practice frequency is usually measured in Hz. Both are used here. Another way of writing $2\pi m f_1$ is to combine the mf_1 into a single term, f_m, so the sinusoid is written as $A_m \cos(2\pi f_m t - \theta_m)$ or in terms of radians as $A_m \cos(\omega_m t - \theta_m)$. Using this format for a sinusoid, the Fourier series equation becomes:

$$x(t) = \frac{A_0}{2} + \sum_{m=1}^{\infty} A_m \cos(\omega_m\, t - \theta_m) \tag{3.7}$$

The more sinusoids that are included in the summations of Equations 3.5 or 3.7, the better the representation of the signal $x(t)$. For an exact representation, the summation is theoretically infinite, but in practice the number of sine and cosine components that have

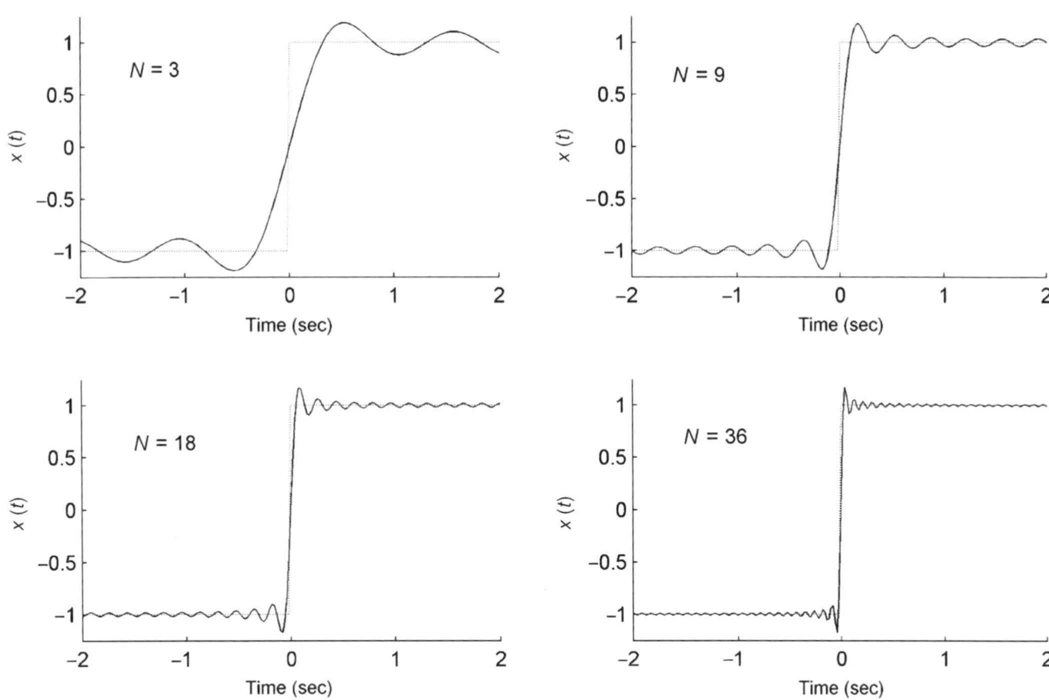

FIGURE 3.6 Reconstruction of a square wave using 3, 9, 18, and 36 sinusoids. The square wave is one of the most difficult signals to represent with a sinusoidal series. The oscillations seen in the sinusoidal approximations are known as Gibbs artifacts, and they increase in frequency, but do not diminish in amplitude, as more sinusoids are added to the summation.

meaningful amplitudes is limited. Often only a few sinusoids are required for a decent representation of the signal. Figure 3.6 shows the reconstruction of a square wave by 4 different series each containing a different number of sinusoids, from 3 to 36 sinusoids. The square wave is one of the most difficult waveforms to represent using a sinusoidal series because of the sharp transitions. Figure 3.6 shows that the reconstruction is fairly accurate even when the summation contains only 9 sinusoids.

The square wave reconstructions shown in Figure 3.6 become sharper as more sinusoids are added, but still contain oscillations. These oscillations, termed *Gibbs artifacts*, occur when a continuous, finite sinusoidal series is used to represent a discontinuity. They do not decrease when more sinusoidal terms are added, but increase in frequency and, because of this increase in frequency, the largest overshot moves closer to the time of the discontinuity. Gibbs artifacts are due to truncation of what should be an infinite series and are sometimes called *truncation artifacts*. Gibbs artifacts occur in a number of circumstances that involve truncation. Gibbs artifacts can be found in MR images when there is a sharp transition in the image and the resonance signal is truncated during data acquisition. They are seen as subtle dark ridges adjacent to high contrast boundaries as shown in the MR image of Figure 3.7. We encounter Gibbs artifacts again due to truncation of digital filter coefficients in Chapter 8.

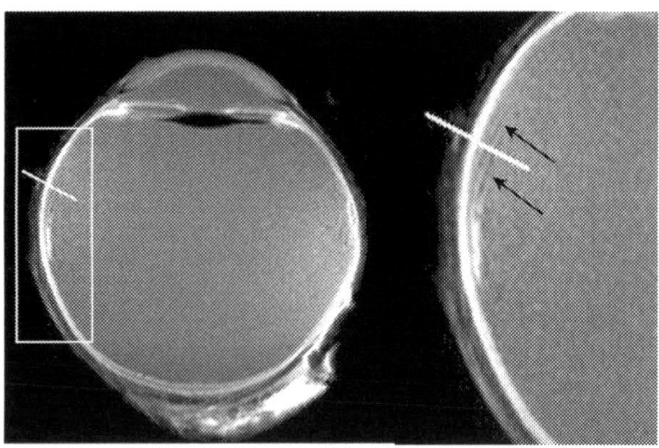

FIGURE 3.7 MR image of the eye showing Gibbs artifacts near the left boundary of the eye. This section is enlarged on the right side to better show the artifacts (arrows). *Image courtesy of Susan and Lawrence Strenk of MRI Research, Inc.*

To simplify manual calculations, the sinusoidal series based on cosines in Equations 3.5 and 3.7 can also be represented in terms of a sine and cosine series using Equation 2.8. The Fourier series equation of Equation 3.5 can also be written as:

$$x(t) = \frac{a_0}{2} + \sum_{m=1}^{\infty} a_m \cos(2\pi m f_1\, t) + \sum_{m=1}^{\infty} b_m \sin(2\pi m f_1\, t) \tag{3.8}$$

where $a_0 \equiv A_0$ in Equation 3.5; $a_m = A_m \cos(\theta)$; and $b_m = B_m \sin(\theta)$, as given in Equation 2.9. In Equation 3.8, the a and b coefficients are just the amplitudes of the cosine and sine components of $x(t)$. The Fourier series equations using a and b coefficients are referred to as the *rectangular representation* of the Fourier series, while the equations using A and θ are termed the *polar representation*. Rectangular to polar conversion is done using Equations 2.10 and 2.11; these equations are repeated here:

$$A_m = \sqrt{a_m^2 + b_m^2} \tag{3.9}$$

$$\theta_m = \tan^{-1}\left(\frac{b_m}{a_m}\right) \tag{3.10}$$

Since the Fourier series theorem says that periodic functions can be completely represented by sinusoids at the same and multiple frequencies, if we are looking for sinusoids in a signal we need only correlate the periodic signal with sinusoids at those frequencies. In other words, we only need to do the correlation at the fundamental and harmonic frequencies of the signal. To solve for the a and b coefficients in Equation 3.8, we simply correlate our signal, $x(t)$, with the cosine and sine terms in Equation 3.8:

$$a_m = \frac{2}{T}\int_0^T x(t)\cos(2\pi m f_1\, t)\, dt \qquad m = 1, 2, 3\ldots \tag{3.11}$$

$$b_m = \frac{2}{T}\int_0^T x(t)\sin(2\pi m f_1\, t)\, dt \qquad m = 1, 2, 3\ldots \tag{3.12}$$

These correlations must be carried out for each value of m to obtain a series of as and bs representing the cosine and sine amplitudes at frequencies $2\pi m f$. The series of a_ms and b_ms are just a string of constants known as the *Fourier series coefficients*. A formal derivation of Equations 3.11 and 3.12 based on Equation 3.8 is given in Appendix A.2. The factor of 2 is required because the Fourier series coefficients, a_m and b_m, are defined in Equation 3.8 as amplitudes, not correlation values. Unfortunately there is no agreement about the best way to scale the Fourier equations, so you might find these equations with other scalings. The MATLAB routine that implements Equations 3.11 and 3.12 bypasses the issue of scaling by computing the Fourier series coefficients using no scaling. So for the MATLAB Fourier series routine, Equations 3.11 and 3.12 would be written without the $2/T$ in front of the integral sign, and it is up to you to scale the output as you wish.

Equations 3.11 and 3.12 are also sometimes written in terms of a period normalized to 2π. This can be useful when working with a generalized $x(t)$ without the requirement for defining a specific period, T.

$$a_m = \frac{1}{\pi}\int_{-\pi}^{\pi} x(t)\ \cos(mt)\,dt \qquad m = 1, 2, 3 \ldots \tag{3.13}$$

$$b_m = \frac{1}{\pi}\int_{-\pi}^{\pi} x(t)\sin(mt)\,dt \qquad m = 1, 2, 3 \ldots \tag{3.14}$$

Here we always work with time functions that have a specific period to make our analyses correspond more closely to real-world applications.

Equations such as 3.11 and 3.12 (or Equations 3.13 and 3.14) calculate the Fourier series coefficients from $x(t)$ and are termed *analysis equations*. Equations 3.5, 3.7 and 3.8 work in the other direction, generating $x(t)$ from the *a*s and *b*s, and are termed *synthesis equations*.

The a_0 term in Equation 3.8 is the same as A_0 in Equations 3.5 and 3.7 and is determined the same way as twice the mean value of $x_T(t)$.

$$a_0 = \frac{2}{T}\int_0^T x(t)\,dt \tag{3.15}$$

The reason a_0 and A_0 are calculated as twice the average value is to be compatible with the Fourier analysis equations of Equations 3.9 and 3.10, which also involve a factor of 2. To offset this doubling, a_0 and A_0 are divided by 2 in the synthesis Equations 3.7 and 3.8.

In order to carry out the correlation and integration in Equations 3.11 through 3.14, there are a few constraints on $x(t)$. First, $x(t)$ must be capable of being integrated over its period; that is:

$$\int_0^T |x(t)|dt < \infty \tag{3.16}$$

Second, while $x(t)$ can have discontinuities, they must be finite in number and have finite amplitudes. Finally, the number of maxima and minima must also be finite. These three criteria are sometimes referred to as the *Dirichlet conditions* and are met by real-world signals.

The fact that periodic signals can be represented completely by a sinusoidal series has implications beyond simply limiting the appropriate sinusoids for probing a periodic signal. It implies that any periodic signal, $x_T(t)$, is as well-represented by the sinusoidal series as by the time signal itself (assuming there are sufficient terms in the series). In other words, the string of constants, A_ms and θ_ms in Equation 3.5 or 3.7 (or the equivalent a_ms and b_ms in Equation 3.8), is as good a representation of $x(t)$ as $x(t)$ itself. This is implied by the fact that given only the A_ms and θ_ms (or equivalently a_ms and b_ms) and the fundamental frequency, you could reconstruct $x(t)$ using Equation 3.5 or 3.8.

Converting a signal into its sinusoidal equivalent is known as a *transformation* since it transforms $x(t)$ into an alternative representation. This transformation based on sinusoids is often referred to as the *Fourier transform*, but this term really should be reserved for *aperiodic* or *transient* signals, as described below. It is important to use technical terms carefully (at least if you are an engineer), so here we use the term *Fourier series analysis* to describe the transformation between the time and frequency representations of a periodic signal. In general, a transformation can be thought of as a remapping of the original signal into an alternative form in the hope that the new form will be more useful or have more meaning. Since it is possible to reverse or invert the transformation and go from the sinusoidal series back to the original function, the Fourier series analysis is referred to as an *invertible transformation*.

3.2.1 Symmetry

Some waveforms are symmetrical or antisymmetrical about $t = 0$, so that one or the other of the components, a_m or b_m in Equations 3.7 and 3.8, is zero. Specifically, if the

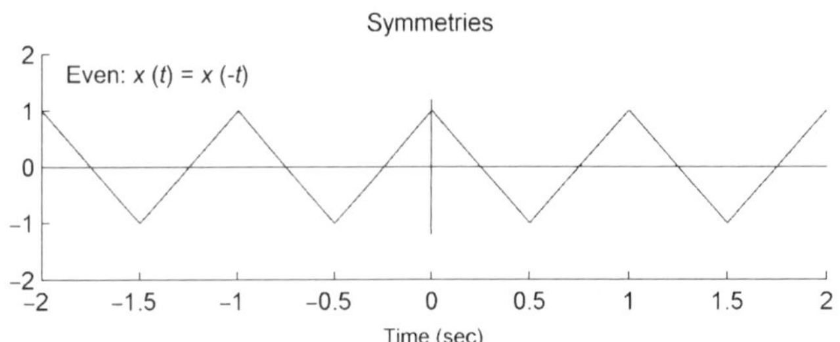

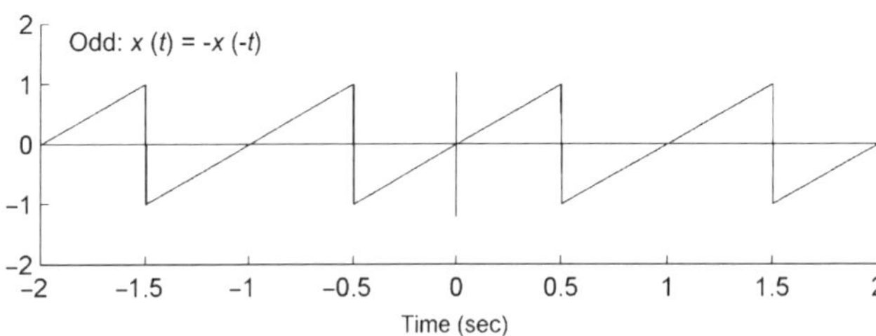

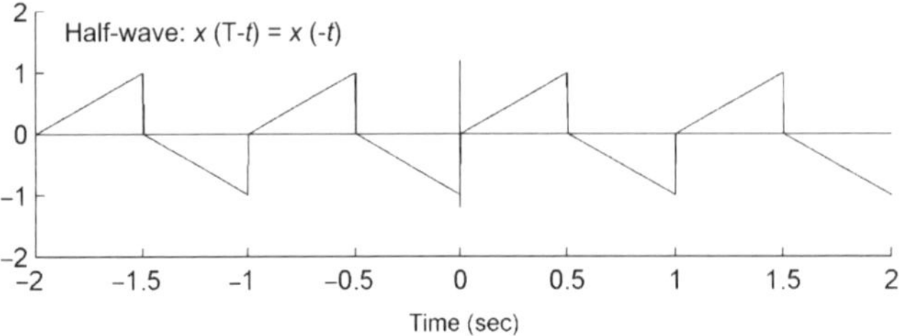

FIGURE 3.8 Waveform symmetries. Each waveform has a period of one second. *Upper plot*: even symmetry; *middle plot*: odd symmetry; *lower plot*: half-wave symmetry.

TABLE 3.1 Function Symmetries

Function Name	Symmetry	Coefficient Values
Even	$x(t) = x(-t)$	$b_m = 0$
Odd	$x(t) = -x(-t)$	$A_m = 0$
Half-wave	$x(t) = x(T-t)$	$A_m = b_m = 0$; for m even

waveform has mirror symmetry about $t = 0$, that is, $x(t) = x(-t)$ (Figure 3.8, upper plot), then multiplications with all sine functions will be zero, so the b_m terms are all zero. Such mirror symmetry functions are termed *even* functions. If the function has antisymmetry, $x(t) = -x(-t)$ (Figure 3.8, middle plot), it is termed an *odd* function and all multiplications with cosines will be zero, so the a_m coefficients are all zero. Finally, functions that have *half-wave* symmetry will have no even coefficients, so both a_m and b_m are zero for $m =$ even. These are functions where the second half of the period looks like the first half, but inverted, that is, $x(T - t) = -x(t)$ (Figure 3.7, lower plot). These symmetries are useful for simplifying the task of solving for the coefficients manually, but are also useful for checking solutions done on a computer. Table 3.1 and Figure 3.8 summarize these properties.

3.3 FREQUENCY REPRESENTATION

Functions other than sinusoids can be, and are, used to perform transformations, but sinusoids are especially useful because of their unique frequency characteristics: a sinusoid contains energy at only one frequency (Figure 3.3). The sinusoidal components of a signal map directly to its frequency characteristics or *spectrum*. A complete description of a waveform's frequency characteristics consists of two sets of data points (or two plots): a set of sinusoidal magnitudes versus frequency, A_m, and a set of sinusoidal phases versus frequency, θ_m. We could also use plots of a_m and b_m, but these are harder to interpret; it is easier to think in terms of the magnitude (A_m) and phase (θ_m) of a sinusoid than to think about the cosine (a_m) and sine (b_m) coefficients. Although both magnitude and phase plots are necessary to represent the signal completely and to convert the frequency representation back into a time representation, often only the magnitude spectrum is of interest in signal analysis.

The specific frequency related to a given component, the relationship between the frequency of a component and its harmonic number (m), is:

$$f = \frac{m}{T} = mf_1 \tag{3.17}$$

For example, each A_m in Equation 3.5 appears as a single point on the *magnitude* (upper) plot while each θ_m shows as a single point on the *phase* (lower) curve, as shown in Figure 3.3. These frequency plots are referred to as the *frequency-domain representation*, just as the original waveform plot of $x(t)$ is called the *time-domain representation*.

The Fourier series analysis, as described by Equations 3.11 and 3.12, can be used to convert a periodic time-domain signal into its equivalent frequency-domain representation.

Indeed, this time-to-frequency conversion is the primary *raison d'etre* for Fourier series analysis. The Fourier series analysis and related Fourier transform are not the only paths to a signal's frequency characteristics or spectrum, but they constitute the most general approach; they make the fewest assumptions about the signal. The digital version[s] of Equation 3.11 and 3.12 can be calculated with great speed using an algorithm known as the *fast Fourier transform* (*FFT*). To go in the reverse direction, from the frequency to the time domain, the Fourier series equation 3.5 or 3.8 is used. In the digital domain this transformation is known as the *inverse fast Fourier transform* (*IFFT*).

The resolution of a spectrum can be loosely defined as the difference in frequencies that a spectrum can resolve, that is, how close two frequencies can get and still be identified as two frequencies. This resolution clearly depends on the frequency spacing between harmonic numbers, which in turn is equal to $1/T$ (Equation 3.17); the longer the signal period, the better the spectral resolution. Later we find this holds for digital data as well.

EXAMPLE 3.2

Find the Fourier series of the triangle waveform shown in Figure 3.9. The equation for this waveform is:

$$x(t) = \begin{cases} t & 0 < t \le 0.5 \\ 0 & 0.5 < t \le 1.0 \end{cases}$$

Find the first four components (i.e., $m = 1, 2, 3, 4$). Construct frequency plots of the magnitude and phase components.

Solution: Use Equations 3.11 and 3.12 to find the cosine (a_m) and sine (b_m) coefficients. Then convert to magnitude (A_m) and phase (θ_m) using Equations 3.9 and 3.10. Plot A_m and θ_m against the associated frequency, $f = mf_1$.

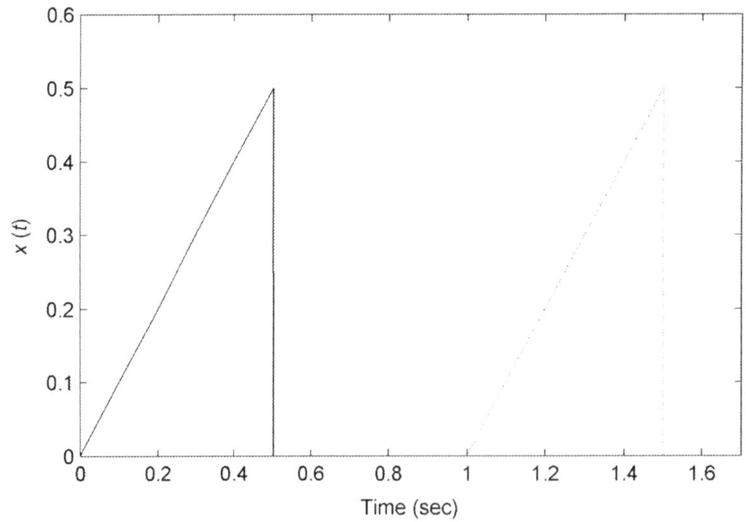

FIGURE 3.9 A half-triangle waveform used in Example 3.2.

We could start by evaluating either the sine or cosine coefficients. We begin with the sine coefficients using Equation 3.12:

$$b_m = \frac{2}{T}\int_0^T x(t)\sin(2\pi m f_1\, t)dt = \frac{2}{T}\int_0^{0.5} t\sin(2\pi m t)dt$$

$$\frac{2}{4\pi^2 m^2}[\sin(2\pi m t) - 2\pi m t\cos(2\pi m t)]_0^{0.5}$$

$$\frac{1}{2\pi^2 m^2}[sin(\pi m) - \pi m\cos(\pi m)] = \frac{-1}{2\pi m}[\cos(\pi m)]$$

$$= \frac{1}{2\pi};\quad \frac{-1}{4\pi};\quad \frac{1}{6\pi};\quad \frac{-1}{8\pi}\ldots = 0.159;\quad -0.080;\quad 0.053;\quad -0.040$$

To find the cosine coefficients, use Equation 3.11:

$$a_m = \frac{2}{T}\int_0^T x(t)\cos(2\pi m f_1\, t)dt = \frac{2}{T}\int_0^{0.5} t\cos(2\pi m t)dt$$

$$\frac{2}{4\pi^2 m^2}[\cos(2\pi m t) - 2\pi m t\sin(2\pi m t)]_0^{0.5}$$

$$\frac{1}{2\pi^2 m^2}[\cos(\pi m) - \pi m\sin(\pi m) - 1] = \frac{-1}{2\pi^2 m^2}[\cos(\pi m) - 1]$$

$$= \frac{-1}{\pi^2};\quad 0;\quad \frac{-1}{9\pi^2};\quad 0\ldots = -0.101;\quad 0;\quad -0.0118;\quad 0$$

To express the Fourier coefficients in terms of magnitude (C_m) and phase (θ_m) use Equations 3.9 and 3.10. Care must be taken in computing the phase angle to insure that it represents the proper quadrant:[2]

$$A_m = \sqrt{a_m^2 + b_m^2} = .187,\quad .080,\quad .054,\quad .040$$

$$\theta_m = tan^{-1}\left(\frac{b_m}{a_m}\right) = 58\ deg(2^{nd}),\ 90\ deg(3^{rd}),\ 77\ deg(2^{nd}),\ 90(3^{rd})deg$$

$$\theta_m = (180 - 58) = 122\ deg,\ (-90) = -90deg,\ (180 - 77) = 103\ deg,\ (-90) = -90\ deg$$

For a complete description of $x(t)$ we need more components and also the a_0 term. The $a_0/2$ term is the average value of $x(t)$ which can be obtained using Equation 3.15:

$$\frac{a_0}{2} = \overline{x}(t) = \frac{1}{T}\int_0^T x(t)\,dt = \frac{1}{1}\int_0^{.5} tdt = \left.\frac{t^2}{2}\right|_0^{0.5} = .125$$

[2]The arctangent function on some calculators does not take into account the signs of b_m and a_m so it is up to you to figure out the proper quadrant. This is also true of MATLAB's `atan` function. However, MATLAB does have a function, `atan2(b,a)`, that takes the signs of `b` and `a` into account and produces an angle (in radians) in the proper quadrant.

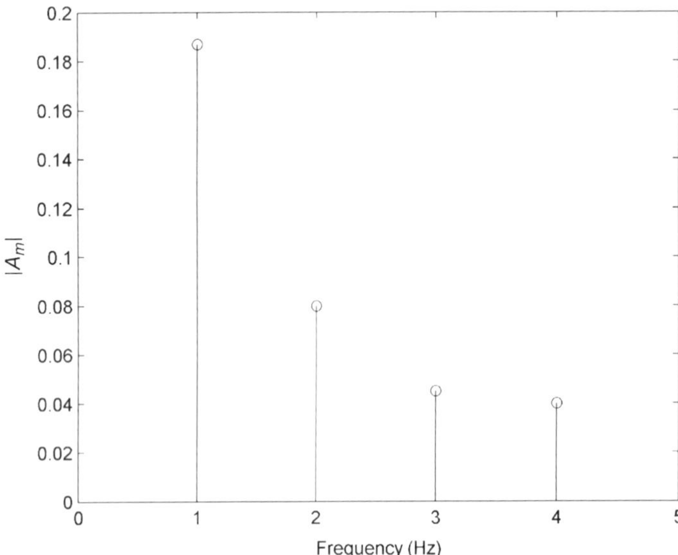

FIGURE 3.10 Magnitude plot (circles) of the frequency characteristics of the waveform given in Example 3.2. Only the first four components are shown because these are the only ones calculated in the example. These components were determined by taking the square root of the sum of squares of the first four cosine and sine terms (*a*s and *b*s).

To plot the spectrum of this signal, or rather a partial spectrum, we just plot the sinusoid coefficients, A_m and θ_m, against frequency. Since the period of the signal is 1.0 second, the fundamental frequency, f_1, is 1-Hz. Therefore the first four values of m represent 1-Hz, 2-Hz, 3-Hz, and 4-Hz. The magnitude plot is shown in Figure 3.10.

The four components computed in Example 3.2 are far short of that necessary for a reasonable frequency spectrum, but just these four components do provide a rough approximation of the signal, as shown in the next example.

EXAMPLE 3.3

Use MATLAB to reconstruct the waveform of Example 3.2 using only the first four components found analytically in that example. Use the magnitude and phase format of the Fourier series equation, Equation 3.5. Plot the waveform.

Solution: Apply Equation 3.5 directly using the four magnitude and phase components found in the last example. Remember to add the DC components, a_0. Plot the result of the summation.

```
fs = 500;                              % Assumed sample frequency
N = 500;                               % Number of points for 1 sec
t = (1:N)/N;
A = [0.187 0.08 0.054 0.04];           % Component magnitudes
theta = [122 - 90 103 -80]*2*pi/360;   % Component phase (in radians)
%
x = zeros(1,N);
```

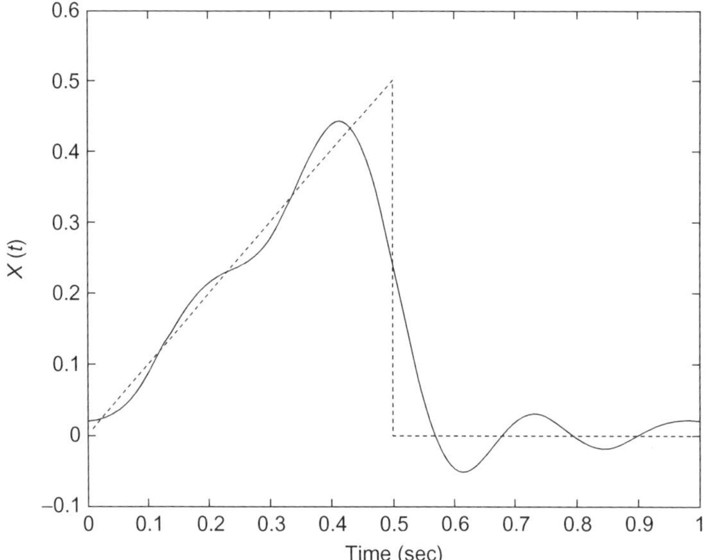

FIGURE 3.11 The waveform produced by applying the magnitude and phase version of the Fourier series equation (Equation 3.5) to the four magnitude and phase components found analytically in Example 3.3. The Fourier summation (solid line) produces a fair approximation of the original waveform (dotted line) even though only four components are used.

```
for f = 1:4                                  % Add the 4 terms of Equation 3.5
  x = x + A(f)*cos(2*pi*f*t - theta(f));     % using appropriate A and theta
end
x = x + 0.125;                               % Add the DC term
plot(t,x,'k');                               % Plot the result
  ......labels......
```

The result produced by this program is shown in Figure 3.11. It is a fair approximation of the original signal despite being constructed from only 4 components.

In Example 3.1, we used crosscorrelation to search for the sinusoidal components of a waveform and found that it failed to find two of three sinusoids at certain "unusual" frequencies. The next example examines how the Fourier series analysis deals with this data set.

EXAMPLE 3.4

Write a program in MATLAB to determine the spectrum of the three sine-wave signals used in Example 3.1. Analyze only the second waveform with sinusoidal frequencies of 100-Hz, 204-Hz, and 306-Hz. Use the basic equations, Equations 3.11 and 3.12, to find the a and b coefficients, and Equation 3.13 to combine them into the A coefficients. Plot only the magnitude spectrum, A_m, as in Figure 3.1.

Solution: The most difficult part of the solution is determining the fundamental frequency. After generating the second waveform of Example 3.1, we determine the period, T, from the number of data points, N, by going back to the relationship between the signal time and a number's sequence position given in Chapter 1. Equation 1.2 states that $t = n/fs$, so the equivalent time length of a signal with N digitized numbers is $T = N/f_s$. If we assume that the data in the computer represent one period of a periodic waveform then, using Equation 3.1, the fundamental frequency is $f_1 = 1/T = 1/N/f_s = f_s/N$.

From here the solution is straightforward: apply Equations 3.11 and 3.12 to find the cosine (a) and sine (b) coefficients, then use Equation 3.9 to convert the as and bs to magnitude components (i.e., As). Plot each A_m against its associated frequency, mf_1. We do not need to find the phase, as only the magnitude plot was requested. Since the signal is composed of three sinusoids all with zero mean, we do not need to calculate the a_0 term (i.e., the DC term) since it will be zero.

The one question remaining is how many components should we evaluate: what should be the range of m? Later we find that there is a definitive answer to the number of coefficients required when analyzing digitized data. For now let us just use a large number, say, 1000.

```
% Example 3.4 Evaluate the 3 sine wave signals of Example 3.1 using
% the Fourier series analysis equations.
%
fs = 1000;                              % Sampling freq.
N = 2000;                               % Number of points in the test signal
T = N/fs;                               % Period from Equation 1.2
f1 = 1/T;                               % Fundamental frequency from Equation 3.1
t = (1:N)/fs;                           % Time vector (for generating the signal)
% Construct waveform with 3 sinusoids
fsig = [100 204 306];                   % Test signal frequencies (from Example 3.1)
x = 1.5*cos(2*pi*fsig(1)*t) + 1.75*sin(2*pi*fsig(2)*t) + 2.0*sin(2*pi*fsig(3)*t);
%
% Calculate the spectrum from the basic equation
for m = 1:1000
  f(m) = m*f1;                          % Construct the frequency vector (Eq. 3.17)
  a(m) = (2/N)*sum(x.*(cos(2*pi*m*f1*t))); % Equation 3.11
  b(m) = (2/N)*sum(x.*(sin(2*pi*m*f1*t))); % Equation 3.12
  A(m) = sqrt(a(m).^2 + b(m).^2);       % Equation 3.11
end
plot(f,A);                              % Plot Am against f
  ......label axes......
```

Results: This program produces the plot shown in Figure 3.12. All three sinusoids are now found at the correct amplitudes and correct frequencies. The frequency resolution also looks better as the peaks are much sharper than those of Figure 3.1. From Equation 3.17, we know that frequency resolution is equal to mf_1 and in this case f_1 as found from the period T is much smaller than that used in Example 3.1: 0.5-Hz versus 10-Hz in Example 3.1. If we had used a frequency increment of 0.5-Hz in Example 3.2, we would have obtained the same result as shown in Figure 3.12, so what is the big deal about the Fourier approach? Well, for one it tells you what frequencies to search for and for another the Fourier series equations can be computed much faster than the crosscorrelation equations. For digitized data, a special algorithm exists for solving the Fourier series equations that is so fast it is called, as mentioned previously, the fast Fourier transform or, simply, the FFT. This approach uses complex numbers to represent the as and bs (or As and θs). Since a complex number is actually two numbers rolled into one, using complex numbers means that only a single complex coefficient is required to represent each sinusoidal component in the series. The complex representation of the Fourier series is described next.

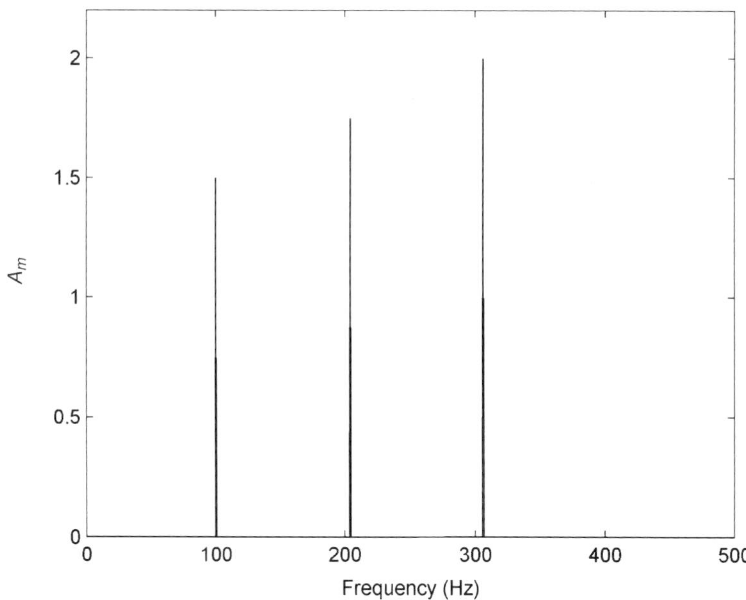

FIGURE 3.12 The spectrum obtained using Fourier series analysis applied to a waveform consisting of three sinusoids at 100, 204, and 306-Hz. This is the same waveform that is analyzed using correlation in Example 3.1 and displayed in Figure 3.1. The three sinusoids are correctly determined by the Fourier analysis.

3.4 COMPLEX REPRESENTATION

The complex representation of the Fourier series analysis can be derived from Equations 3.7 and 3.8 using only algebra and Euler's equation (Equation 2.12). We start with the exponential definitions of the sine and cosine functions:

$$\cos(2\pi m f_1\, t) = \frac{1}{2}(e^{+j2\pi m f_1\, t} + e^{-j2\pi m f_1\, t}) \text{ and}$$
$$\sin(2\pi m f_1\, t) = \frac{1}{j2}(e^{+j2\pi m f_1\, t} - e^{-j2\pi m f_1\, t}) \tag{3.18}$$

Recall the definition of the Fourier series given in Equation 3.8 and repeated here:

$$x(t) = \frac{a_0}{2} + \sum_{m=1}^{\infty} a_m \cos(2\pi m f_1\, t) + \sum_{m=1}^{\infty} b_m \sin(2\pi m f_1\, t)$$

Substituting the complex definitions for the sine and cosines into the Fourier series equation and expanding:

$$x(t) = \frac{a_0}{2} + \sum_{n=1}^{\infty}\left(\frac{a_m}{2}e^{j2\pi m f_1\, t} + \frac{a_m}{2}e^{-j2\pi m f_1\, t} + \frac{b_m}{2j}e^{j2\pi m f_1\, t} - \frac{b_m}{2j}e^{-j2\pi m f_1\, t}\right)$$

Moving the j term to the numerator using the fact that $1/j = -j$:

$$x(t) = \frac{a_0}{2} + \sum_{n=1}^{\infty}\left(\frac{a_m}{2}e^{j2\pi m f_1 t} + \frac{a_m}{2}e^{-j2\pi m f_1 t} - \frac{jb_m}{2}e^{j2\pi m f_1 t} + \frac{jb_m}{2}e^{-j2\pi m f_1 t}\right)$$

Rearranging:

$$x(t) = \frac{a_0}{2} + \sum_{n=1}^{\infty}\left(\left(\frac{a_m}{2} - \frac{jb_m}{2}\right)e^{j2\pi mf_1 t} + \left(\frac{a_m}{2} + \frac{jb_m}{2}\right)e^{-j2\pi mf_1 t}\right)$$

Collecting terms into positive and negative summations of m:

$$x(t) = \frac{a_0}{2} + \sum_{m=1}^{\infty}\left(\frac{a_m - jb_m}{2}\right)e^{j2\pi mf_1\, t} + \sum_{m=-\infty}^{-1}\left(\frac{a_{-m} + jb_{-m}}{2}\right)e^{j2\pi mf_1\, t} \tag{3.19}$$

This can be combined into a single equation going from negative to positive infinity:

$$x(t) = \sum_{m=-\infty}^{m=\infty} C_m e^{j2\pi mf_1\, t} \tag{3.20}$$

where the new coefficient, C_m is defined as:

$$C_{+m} = \frac{a_m - jb_m}{2};\, C_{-m} = \frac{a_m + jb_m}{2};\ \text{and } C_0 = a_0 \tag{3.21}$$

Equation 3.21 is known as the complex Fourier series synthesis equation. Note the DC term, a_0, is incorporated into the complex number series C_m.

To find the Fourier series analysis equation, substitute the original correlation equations for a_m and b_m (Equations 3.11 and 3.12) into Equation 3.20 noting the relationships between C_m and a_m and b_m given by Equation 3.21:

$$C_m = \frac{2}{2T}\int_0^T x(t)\cos(2\pi mf_1\, t)\,dt - \frac{2j}{2T}\int_0^T x(t)\sin(2\pi mf_1\, t)\,dt$$

Combining: (3.22)

$$C_m = \frac{1}{T}\int_0^T x(t)\left[\cos(2\pi mf_1\, t) - j\sin(2\pi mf_1\, t)\right]dt$$

The term in the brackets in Equation 3.22 has the form $\cos(x) - j\sin(x)$. This is just equal to e^{-x} as given by Euler's equation (Equation 2.13). Hence the term in the brackets can be replaced by a single exponential which leads to the complex form of the Fourier series:

$$C_m = \frac{1}{T}\int_0^T x(t)e^{-j2\pi mf_1\, t}dt \qquad m = 0, \pm1, \pm2, \pm3\ldots \tag{3.23}$$

where $f_1 = 1/T$. In the complex representation, the sinusoidal family members must range from minus to plus infinity as in Equation 3.23, essentially to account for the negative signs introduced by the complex definition of sines and cosines (Equation 3.18). However, the $a_0/2$ term (the average value or DC term in Equation 3.8) does not require a separate equation, since when $m = 0$ the exponential becomes $e^{-j0} = 1$, and the integral computes the average value.

The negative frequencies implied by the negative m terms are often dismissed as mathematical contrivances since negative frequency has no meaning in the real world. However, for a digitized signal, these negative frequencies do have consequences as shown later. Equation 3.23 shows C_m is symmetrical about zero frequency so the negative frequency components are just mirror images of the positive components. Since they are the same components in reverse order, they do not provide additional information. However, adding the mirror image negative components to the positive frequency components does double their magnitude, so the normalization of the complex analysis equation is now $1/T$, not the $1/2T$ used for the noncomplex analysis equations (Equations 3.11 and 3.12). The complex Fourier transform equations are more commonly used and they are commonly normalized by $1/T$ so the noncomplex forms must be normalized by $1/2T$ if both normalizations are to produce the same results. For the same reason, the magnitude of C_m does not need to be scaled by 2 to get the amplitude. (Recall that different normalization schemes are sometimes used and, although MATLAB uses the complex form to compute the Fourier transform, it provides no normalization so the magnitude of C_m is relative to the number of points analyzed. You are expected to apply whatever normalization you wish. Example 3.10 shows how to scale the values of C_m produced by MATLAB's FFT routine to match that produced by the noncomplex equations.)

From the definition of C_m (Equation 3.21) it can be seen that this single complex variable combines the cosine and sine coefficients. To get the a_ms and b_ms from C_m, we use Equation 3.21 which states that the a_ms are twice the real part of C_m while the b_ms are twice the imaginary part of C_m.

$$a_m = 2Re\,(C_m);\, b_m = 2Im\,(C_m) \tag{3.24}$$

The magnitude A_m and phase θ_m representation of the Fourier series can be found from C_m using complex algebra:

$$|C_m| = \sqrt{\mathrm{Re}\,(C_m)^2 + \mathrm{Im}\,(C_m)^2} = \sqrt{\frac{a_m^2 + b_m^2}{2^2}} = \frac{1}{2}\sqrt{a_m^2 + b_m^2} = A_m/2 \tag{3.25}$$

$$Angle\,(C_m) = \tan^{-1}\left(\frac{\mathrm{Im}\,(C_m)}{\mathrm{Re}\,(C_m)}\right) = \tan^{-1}\left(\frac{b_m/2}{a_m/2}\right) = \tan^{-1}\left(\frac{b_m}{a_m}\right) = \theta_m \tag{3.26}$$

Hence the magnitude of C_m is equal to $1/2\ A_m$, the magnitude of the sinusoidal components, and the angle of C_m is equal to the phase, θ_m, of the sinusoidal components. Not only does the complex variable C_m contain both sine and cosine coefficients, but these components can be readily obtained in either the polar form (As and θs, most useful for plotting) or the rectangular form (as and bs). Note that the factors of 2 in Equation 3.24 and 3.25 are based on the normalization strategy used in this book. When MATLAB is used and normalized by N, the number of points in the signal, the half magnitude component, $A_m/2$, is equal to $|C_m|$. Hence, if the values of $|C_m|$ produced by MATLAB are divided by $N/2$, then these values are equal to A_m, the peak-to-peak amplitude of the associated sinusoidal component. Similarly, to find the DC component (i.e., $a_0/2$), C_0 should be divided by N. C_0 will be real (the imaginary part is zero). MATLAB normalization techniques are demonstrated in Example 3.10.

EXAMPLE 3.5

Find the Fourier series of the "pulse" waveform shown in Figure 3.12 with a period T and a pulse width of W. Use Equation 3.23, the complex equation.

Solution: Apply Equation 3.23 directly, except since the signal is an even function ($x(t) = x(-t)$), it is easier to integrate from $t = -T/2$ *to* $+T/2$.

$$C_m = \frac{1}{T}\int_{-T/2}^{T/2} x(t)e^{-j2\pi mf_1 t}dt = \frac{1}{T}\int_{-W/2}^{W/2} V_p e^{-j2\pi mf_1 t}dt = \frac{V_p}{T(-j2\pi mf_1)}\left[e^{-j2\pi mf_1 W/2} - e^{j2\pi mf_1 W/2}\right]$$
$$= \left[\frac{V_p}{\pi mf_1 T}\right]\frac{[e^{j2\pi mf_1 W/2} - e^{-j2\pi mf_1 W/2}]}{2j}$$

where the second term in brackets is $\sin(2\pi mf_1 W/2)$. So the expression for C_m becomes:

$$C_m = \frac{V_p}{\pi mf_1 T}\sin(\pi mf_1 W/2) \quad m = 0, \pm1, \pm2, \pm3, \ldots$$

It is common to rearrange this equation so that the second term has the format:

$$\frac{\sin(x)}{x} = \mathrm{sinc}(x).$$

$$C_m = \left(\frac{V_p W}{T}\right)\frac{\sin(\pi mf_1 W/2)}{2\pi mf_1 W/2} \quad m = 0, \pm1, \pm2, \pm3, \ldots$$

Using specific values for T and W it is possible to solve for the values of C_m. In the next example, MATLAB is used to plot C_m for two different values of W. A plot of the magnitude of C_m is given in Figure 3.13. In general, it is easier to solve for the Fourier series analytically (i.e., by hand) using the sine and cosine equations (Equations 3.11 and 3.12) as was done in Example 3.2, but the complex representation is favored when talking about Fourier series analysis because of its succinct presentation (i.e., only one equation). It is also the approach used in MATLAB and other computer algorithms and is the format cited in journal articles that involve Fourier series analysis.

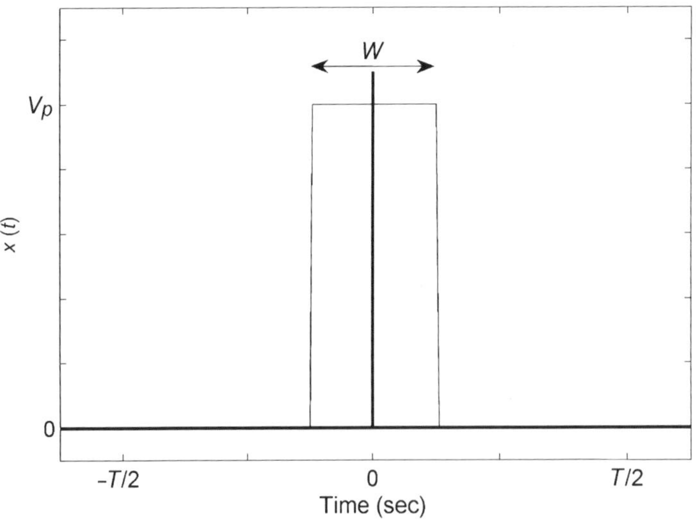

FIGURE 3.13 Pulse wave used in Example 3.5.

EXAMPLE 3.6

Use MATLAB and the complex form to find the spectrum of the pulse-like waveform, such as that used in the last example. Assume pulse amplitude V_p of 1.0, a period, T, of 1 sec and use a sampling frequency, f_s, of 500-Hz. Determine and plot the magnitude spectrum for two pulse widths: 0.005 and 0.05 sec.

Solution: Calculate the number of points needed to produce the specified T and f_s using the method first described in Example 1.3. Generate the pulse waveform of the proper width and correlate it with the complex exponential in Equation 3.23. Find the magnitude spectrum using Equation 3.25 and plot.

```
% Example 3.6 Find the magnitude spectrum of two pulse waveforms
% using the complex form of the Fourier series analysis (Equation 3.23)
%
T = 1.0;                                    % Period (sec)
fs = 500;                                   % Sampling frequency (Hz)
f1 = 1/T;                                   % Fundamental frequency
N = round(T*fs);                            % Calculate the number of points
                                            needed
t = (1:N)/fs;                               % Time vector for complex sinusoids
x = zeros(1,N);                             % Generate pulse waveform
PW = round(.005*fs);                        % PW = 0.005 sec
x(1:PW) = 1;                                % Amplitude = 1.0
%
%Apply complex Fourier series analysis
for m = 1:round(N/2);                       % Make m half the number of points
                                            % (explained later)
  f(m) = m*f1;                              % Frequency vector for plotting
  C(m) = mean(x.*exp(-j*2*pi*m*f1*t));      % Find complex components (Equation
                                            % 3.23)
  Cmag(m) = sqrt(real(C(m))^2+imag(C(m))^2); % Find magnitude (Equation 3.25)
end
subplot(2,2,1);
  plot(t,x,'k','LineWidth',2); % Plot pulse
  ......labels, axes and titles......
subplot(2,2,2);
  plot(f,Cmag,'.');                         % Plot |Cm| as points
  ......labels and titles; repeat for longer pulse width......
```

Analysis: After the pulse waveform is generated, the complex components C_m are determined by correlation of the waveform with the complex exponential $e^{-j2\pi f1t}$ as called for by Equation 3.23. The magnitude component was found by direct application of Equation 3.25; however, an easier approach would be to apply MATLAB's `abs` function which does the same thing: `Cmag(m)=2*abs(C(m));`.

Result: The spectra produced by the program in Example 3.6 are shown in Figure 3.14. Note that the spectra consist of individual points (filled circles) spaced $1/T$ or 1.0-Hz apart; however, in the upper plot the spectrum looks like a continuous curve because the points are so closely spaced. Both curves have the form of the magnitude of function $\sin(x)/x$. This is the function found analytically in Example 3.5 where $x = 2\pi f_1 W/2$. The function $\sin(x)/x$ is also known as $\text{sinc}(x)$. Note the *inverse relationship* between pulse width and spectrum width: a wider time domain pulse produces a narrower frequency domain spectrum. This inverse relationship is

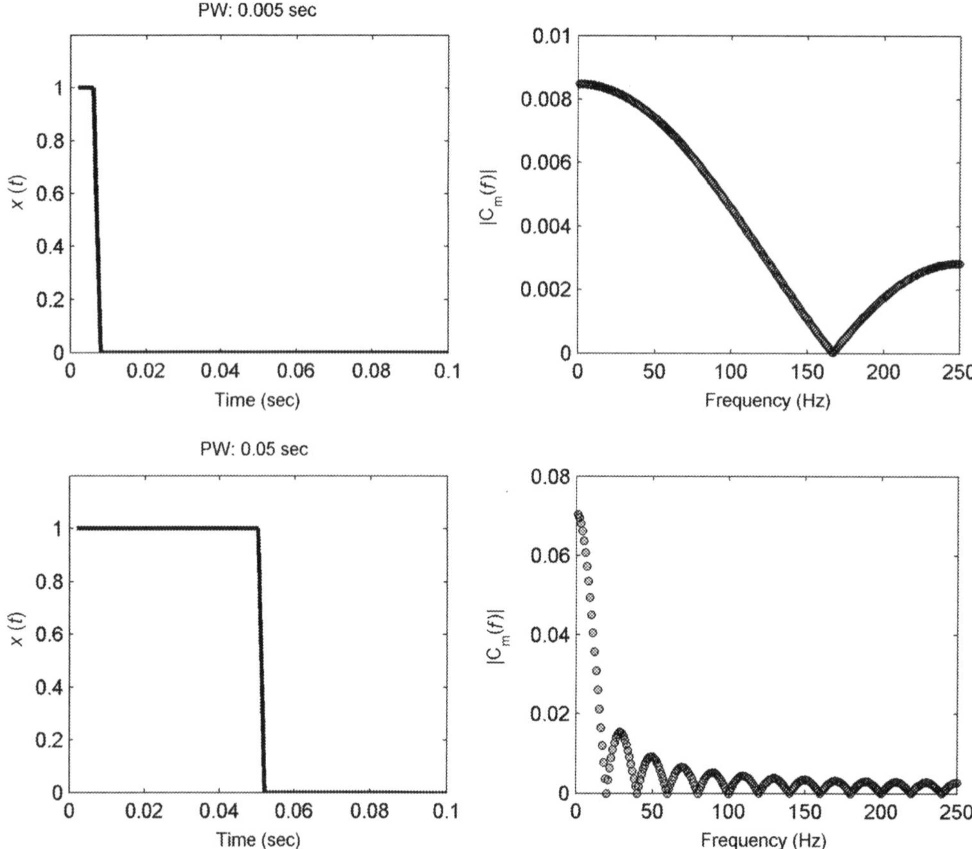

FIGURE 3.14 The frequency spectra of two pulse functions having different pulse widths found using the complex Fourier series analysis. The spectra consist of individual points spaced $1/T = 2.0$-Hz apart. The longer pulse produces a spectrum with a shorter peak. Both spectra have the shape of $|\sin(x)/x|$. The $\sin(x)/x$ function is also known as the *sinc* function. Note the inverse relationship between pulse width and the width of the spectrum: the narrower the pulse the broader the spectrum and vice versa.

typical of time-frequency conversions. Later we see that in the limit, as $PW \to 0$, the spectrum becomes infinitely broad, that is, a flat horizontal line.

EXAMPLE 3.7

Find the Fourier series of the waveform shown in Figure 3.15 using the complex form of the equation.

Solution: Apply Equation 3.23 directly, noting that $x(t) = e^{2.4t} - 1$ for $0 < t \leq 1$ and $f_1 = 1/T = 1$.

$$C_m = \frac{1}{T}\int_0^T x(t)e^{-j2\pi m f_1 t}dt = \int_0^1 (e^{2.4t} - 1)e^{-j2\pi m t}dt$$

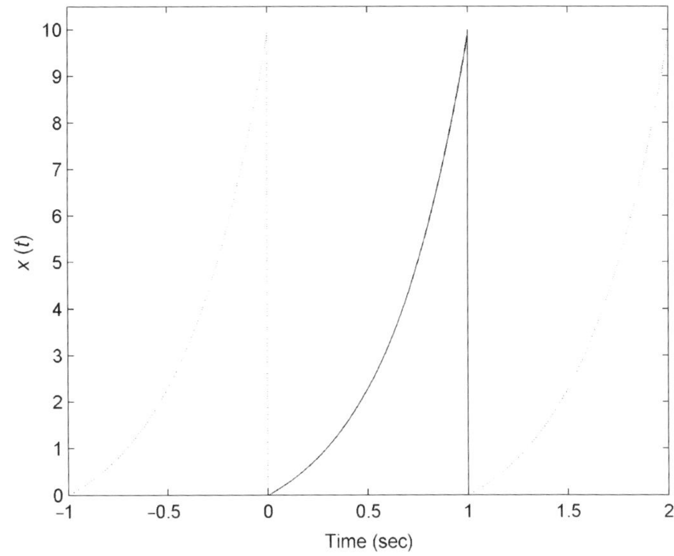

FIGURE 3.15 Periodic waveform used in Example 3.7. The equation for this waveform is $x(t) = 5e^{2.45t} - 1$ for $0 < t < 1$. The period, T, of this waveform is 1 sec.

$$C_m = \int_0^1 (e^{2.4t}e^{-j2\pi mt} - e^{-j2\pi mt})dt = \int_0^1 (e^{2.4t-j2\pi mt} - e^{-j2\pi mt})dt$$

$$C_m = \left.\frac{e^{2.4t-j2\pi mt}}{2.4 - j2\pi m}\right|_0^1 - \left.\frac{e^{-j2\pi mt}}{2\pi m}\right|_0^1$$

$$C_m = \frac{e^{(2.4-j2\pi m)} - 1}{2.4 - j2\pi m} - \frac{e^{-j2\pi m} - 1}{2\pi m}$$

Note that $e^{-j2\pi m} = 1$ for $m = 0, \pm 1, \pm 2, \pm 3, \ldots$ (This can be seen by substituting in cosine and sine from Euler's identity: $e^{-j2\pi m} = \cos(2\pi m) + j\sin(2\pi m)$; hence, the sine term is always zero and the cosine term always 1 if m is an integer or zero.)

$$C_m = \frac{e^{2.4}e^{-j2\pi m}}{2.4 - j2\pi m} - \frac{e^{-j2\pi m} - 1}{2\pi m} = \frac{e^{2.4} - 1}{2.4 - j2\pi m} - \frac{1 - 1}{2\pi m} = \frac{e^{2.4} - 1}{2.4 - j2\pi m} \quad m = 0, \pm 1, \pm 2, \pm 3, \ldots$$

The problems at the end of the chapter contain additional applications of the complex Fourier series equation.

3.5 THE CONTINUOUS FOURIER TRANSFORM

The Fourier series analysis is a good approach to determining the frequency or spectral characteristics of a periodic waveform, but what if the signal is not periodic? Most real signals are not periodic, and for many physiological signals, such as the EEG signal introduced in the first chapter, only a part of the signal is available. The segment of EEG signal shown previously is just a segment of the actual recording, but no matter how long the record, the EEG signal existed before the recording and will continue after the recording session ends (unless the EEG recording session was so traumatic as to cause untimely

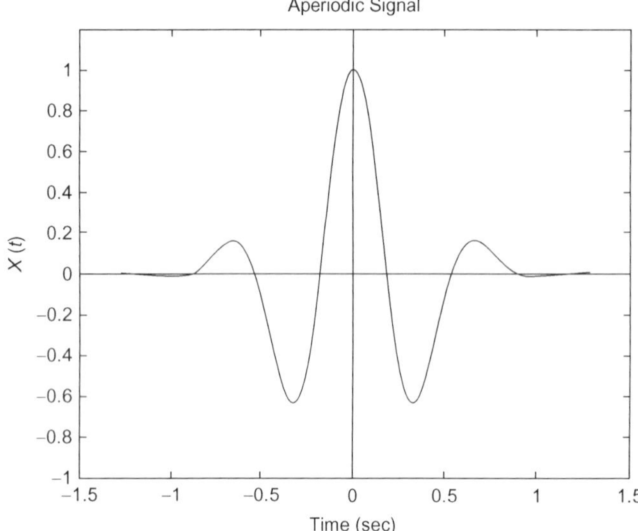

FIGURE 3.16 An aperiodic function exists for a finite length of time and is zero everywhere else. Unlike with a sine wave, you are seeing the complete signal here; this is all there is!

death!). Dealing with very long signals generally entails estimations or approximations, but if the signal is *aperiodic*, an extension of the Fourier series analysis can be used. An aperiodic signal is one that exists for a finite period of time and is zero at all other times (see Figure 3.16). Aperiodic signals are sometimes referred to as *transient* signals.

To extend the Fourier series analysis to aperiodic signals, these signals are treated as periodic, but with a period that goes to infinity, (i.e., $T \rightarrow \infty$). If the period becomes infinite, then $f_1 = 1/T \rightarrow 0$; however, mf_1 does not go to zero since m goes to infinity. This is a case of limits: as T gets longer and longer, f_1 becomes smaller and smaller as does the increment between harmonics (Figure 3.17). In the limit, the frequency increment mf_1 becomes a continuous variable f. All the various Fourier series equations described above remain pretty much the same for aperiodic functions, only the $2\pi mf_1$ goes to $2\pi f$. If radians are used instead of Hz, then $m\omega_1$ goes to ω in these equations. The equation for the continuous Fourier transform in complex form becomes:

$$\lim_{\substack{m \to \infty \\ f_1 \to 0}} C_m = \int_0^T x(t)e^{-j2\pi mf_1 t}dt = \int_{-\infty}^{\infty} x(t)e^{-j2\pi ft}dt$$

$$\left.\begin{matrix} X(f) \equiv C(f) \\ X(\omega) \equiv C(\omega) \end{matrix}\right\} = \int_{-\infty}^{\infty} x(t)e^{-j2\pi ft}dt = \int_{-\infty}^{\infty} x(t)e^{-j\omega t}dt \tag{3.27}$$

Or for the noncomplex equations in terms of the sine and cosine:

$$\left.\begin{matrix} A(f) \\ A(\omega) \end{matrix}\right\} = \int_{-\infty}^{\infty} x(t)\cos(2\pi ft)dt = \int_{-\infty}^{\infty} x(t)\cos(\omega t)dt$$

$$\left.\begin{matrix} B(f) \\ B(\omega) \end{matrix}\right\} = \int_{-\infty}^{\infty} x(t)\sin(2\pi ft)dt = \int_{-\infty}^{\infty} x(t)\sin(\omega t)dt \tag{3.28}$$

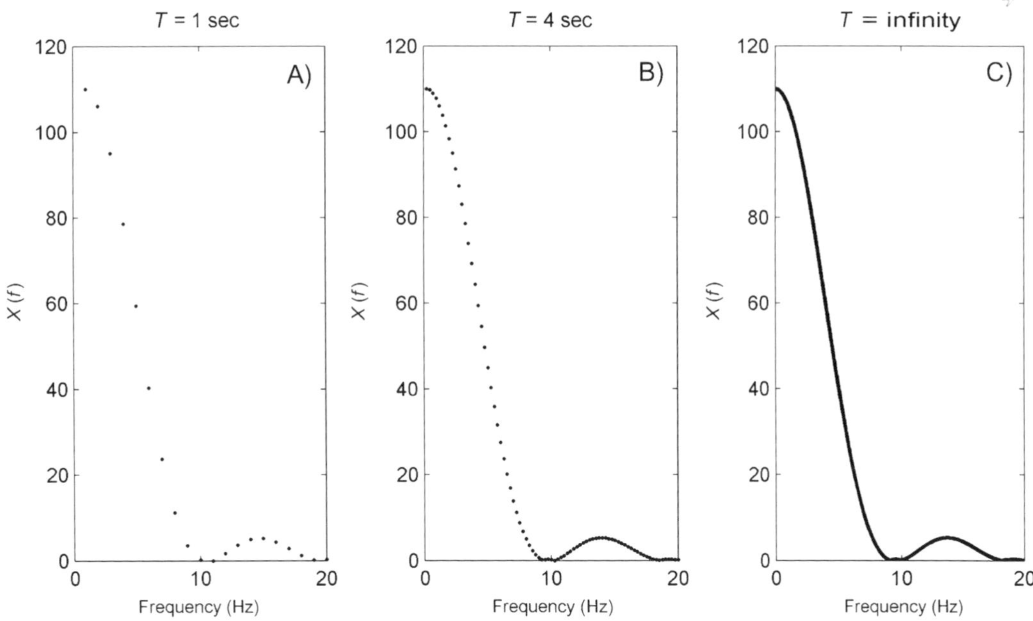

FIGURE 3.17 The effect of increasing the period, T, on the frequency spectrum of a signal. A) When the period is relatively short (1 sec), the spectral points are widely spaced at 1-Hz. B) When the period becomes longer (4 sec), the points are spaced closer together at intervals of 0.25-Hz. C) When the period becomes infinite, the points become infinitely close together and the spectrum becomes a continuous curve.

These transforms produce a continuous function as an output and it is common to denote these terms with capital letters. Also the transform equation is no longer normalized by the period since $1/T \rightarrow 0$. Although the transform equation is integrated over time between plus and minus ∞, the actual limits will only be over the nonzero values of $x(t)$.

Although it is rarely used in analytical computations, the inverse continuous Fourier transform is given below. This version is given in radians, ω, instead of $2\pi f$ just for variety:

$$x(t) = \frac{1}{2\pi}\int_{-\infty}^{\infty} X(\omega)e^{j\omega t}d\omega \tag{3.29}$$

where ω is frequency in radians.

EXAMPLE 3.8

Find the Fourier transform of the pulse in Example 3.4 assuming the period, T, goes to infinity.

Solution: We could apply Equation 3.27, but since $x(t)$ is an even function we could also use just the cosine equation of the Fourier transform in Equation 3.28 knowing that all of the sine terms will be zero for the reasons given in the section on symmetry.

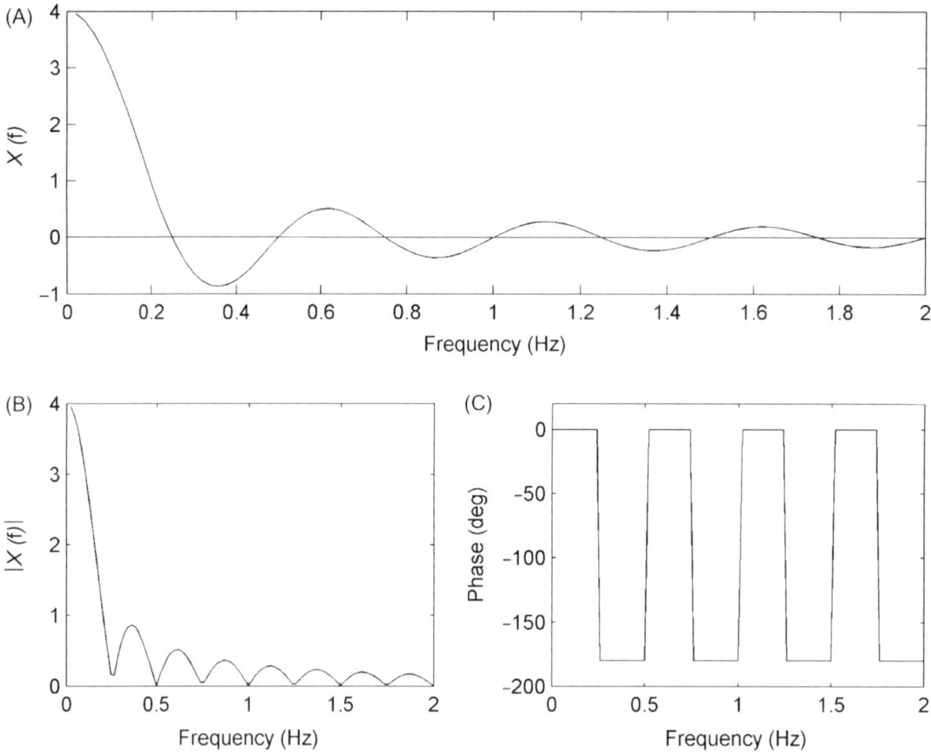

FIGURE 3.18 Upper plot) The complex spectrum of an aperiodic pulse as determined in Example 3.8. In this case the complex function *X(f)* is shown in the upper plots. Lower plots) The magnitude and phase are shown in the lower plots.

$$X(f) = \int_{-\infty}^{\infty} x(t)\cos(2\pi ft)dt = \int_{-W/2}^{W/2} V_p \cos(2\pi ft)dt = \frac{V_p}{2\pi f}\sin(2\pi f)\Big|_{t=-W/2}^{t=W/2}$$

$$X(f) = \frac{V_p}{2\pi f}(\sin(2\pi fW/2) - \sin(-2\pi fW/2)) = \frac{V_p}{2\pi f}(\sin(\pi fW) + \sin(\pi fW))$$

$$X(f) = \frac{V_p}{2\pi f}(2\sin(\pi fW)) = \frac{V_p \sin(\pi fW)}{\pi f} \tag{3.30}$$

Result: A plot of *X(f)* is shown in Figure 3.18 upper plot. Although this particular function is real and does not have an imaginary component, it still represents both magnitude and phase components. The magnitude is the absolute value of the function (Figure 3.16 lower plots), while the phase alternates between 0 degrees when the function is positive and 180 degrees when the function is negative.

A special type of pulse plays an important role in systems analysis: a pulse which is very short but maintains a pulse area of 1.0. For this special pulse, termed an "impulse function," the pulse width, *W*, theoretically goes to 0 but the amplitude of the pulse becomes infinite in such a way that the area of the pulse stays at 1.0. The spectrum of an impulse can be determined from Equation 3.30. For small angles, $\sin x \approx x$, so in the solution above, as W gets small:

$$\sin(\pi f W) \rightarrow \pi f W$$

Thus the frequency spectrum of this special pulse becomes:

$$X(f) = \frac{V_p}{\pi f} \pi f W = VW$$

However, VW is the area under the pulse and from the definition of an impulse function the area of an impulse equals 1.0.

$$X(f) = VW = 1 \tag{3.31}$$

Thus the frequency spectrum of this special pulse, the impulse function, is a constant value of 1.0 for all frequencies. You might imagine that a function with this spectrum would be particularly useful. You might also speculate that a pulse with these special properties – infinitely short, infinitely high, yet still with an area of 1.0 – could not be realized, and you would be right. However, it is possible to generate a real-world pulse that is short enough and high enough to function as an impulse at least for all practical purposes. The approach for generating a real-world impulse function in specific situations is provided in Chapter 7.

3.6 DISCRETE DATA: THE DISCRETE FOURIER SERIES AND DISCRETE FOURIER TRANSFORM

Most Fourier analysis is done using a digital computer and is applied to discrete data. Discrete data differ from continuous periodic data in two fundamental ways: they are time and amplitude sampled, as discussed in Chapter 1, and they are always finite. The discrete version of the Fourier analysis equation is termed the *discrete time Fourier series* and is given as:

$$X[m] = \sum_{n=1}^{N} x[n] e^{-\frac{j2\pi mn}{N}} \quad m = 1, 2, 3, \ldots, M \tag{3.32}$$

where n is the index of the data array of length N, and m is the harmonic number. Equation 3.32 is the discrete version of the Fourier series and is the equation implemented on a computer. In the popular FFT algorithm, the number of components evaluated, M, is equal to the number of points in the data, N.

As with the continuous version of the Fourier series, Equation 3.32 produces a series of complex numbers that describe the amplitude and phase of a harmonic series of sinusoids. The fundamental period of this series is still $1/T$, which can also be given in terms of the harmonic number, M, and the sampling frequency, f_s:

$$f_1 = \frac{1}{T}; \quad T = \frac{M}{f_s}; \quad f_1 = \frac{f_s}{M} \tag{3.33}$$

Since the discrete time Fourier series is applied to finite data, the underlying assumption about the data set must be that it represents one period of a periodic function, although this is almost never the case in real situations. Then the Fourier series output is a set of numbers representing frequency components spaced $1/T$ apart. After the Fourier series is calculated, we sometimes pretend that the data set in the computer is actually an aperiodic function: that the original time series is zero for all time except for that time segment captured in the computer. Under this aperiodic assumption, we can legitimately connect the points of the Fourier series together when plotting, since the difference between points goes to zero. This produces a smooth, apparently continuous curve, and gives rise to an alternate term for the operation in Equation 3.32, the *discrete Fourier transform* or *DFT*. Nonetheless Equation 3.32 assumes periodicity and produces a Fourier *series* irrespective of what it is called; however, the aperiodic assumption is often employed and the term *discrete Fourier transform* commonly used. No matter how we interpret the operation of Equation 3.32, it is implemented using a high-speed algorithm known as the *fast Fourier transform* (*FFT*). (Note the use of the term "Fourier transform" even though the FFT actually calculates the Fourier series.)

In some theoretical analyses, infinite data are assumed, in which case the n/N in Equation 3.32 does become a continuous variable f as $N \to \infty$ (at least theoretically). In these theoretical analyses the data string, $x[n]$, really is aperiodic and the modified equation that results is called the *discrete time Fourier transform* (the *DTFT* as opposed to just the *DFT*, i.e., discrete Fourier transform, for periodic data). Since no computer can hold infinite data, the discrete time Fourier transform is of theoretical value only and is not used in calculating the real spectra.

The number of different Fourier transforms has led to some confusion regarding terminology and a tendency to call all these operations "Fourier transforms" or, even more vaguely, "FFTs." Using appropriate terminology will reduce confusion about what you are actually doing, and may even impress less-informed bioengineers. To aid in being linguistically correct, Table 3.2 summarizes the terms and abbreviations used to describe the various aspects of Fourier analyses equations, and Table 3.3 does the same for the Fourier synthesis equation.

Tables 3.2 and 3.3 show a potentially confusing number of options for converting between the time and frequency domains. Of these options, there are only two real-world

TABLE 3.2 Analysis Equations (Forward Transform)[1]

$x(t)$ periodic and continuous: Fourier series		$x(n)$ periodic and discrete: discrete Fourier transform (DFT) or discrete time Fourier series[2,3]	
$X(m) = \frac{1}{T}\int_0^T x(t)e^{-j2\pi m f_1 t}dt$ $m = 0, \pm1, \pm2, \pm3 \ldots f_1 = 1/T$	(3.34)	$X[m] = \sum_{n=0}^{N-1} x[n]e^{-j2\pi mn/N}$	(3.35)
$x(t)$ aperiodic and continuous: Fourier transform (FT) or continuous time Fourier transform		$x[n]$ aperiodic and discrete: discrete time Fourier transform (DTFT)[4]	
$X(f) = \int_{-\infty}^{\infty} x(t)e^{-j2\pi ft}\,dt$	(3.36)	$X[f] = \sum_{n=-\infty}^{\infty} x[n]e^{-j2\pi nf}$	(3.37)

[1] *Often ω is substituted for 2πf to make the equation shorter. In such cases the Fourier series integration is carried out between 0 and 2π or from −π to +π.*

[2] *Alternatively, 2πmnT/N is sometimes used instead of 2πmn/N, where T is the sampling interval.*

[3] *Sometimes this equation is normalized by 1/N and then no normalization term is used in the inverse DFT.*

[4] *The DTFT cannot be calculated with a computer since the summations are infinite. It is used in theoretical problems as an alternative to the DFT.*

TABLE 3.3 Synthesis Equations (Reverse Transform)

Inverse Fourier series

$$x(t) = \sum_{n=-\infty}^{\infty} X(m)e^{j2\pi f_1 t} \quad m = 0, \pm 1, \pm 2, \pm 3, \ldots \tag{3.38}$$

Inverse discrete Fourier transform or inverse discrete time Fourier series[1,2]

$$x[n] = \frac{1}{N}\sum_{n=0}^{N-1} X[m]e^{j2\pi mn/N} \tag{3.39}$$

Inverse Fourier transform or inverse continuous time Fourier transform

$$x(t) = \int_{-\infty}^{\infty} X(f)e^{j2\pi f t}\, df \tag{3.40}$$

Inverse discrete time Fourier transform[3,4]

$$x[n] = \frac{1}{2\pi}\int_{0}^{2\pi} X[f]e^{j2\pi nf}\, df \tag{3.41}$$

[1]*Alternatively, 2πmnT/N is sometimes used instead of 2πmn/N, where T is the sampling interval.*
[2]*Sometimes this equation is normalized by 1/N and then no normalization term is used in the inverse DFT.*
[3]*The DTFT cannot be calculated with a computer since the summations are infinite. It is used in theoretical problems as an alternative to the DFT.*
[4]*Evaluation of the integral requires an integral table because the summations that produce X[f] are infinite in number.*

TABLE 3.4 Real-World Application of Fourier Transform Equations

Analysis Type (Equation)	Application
Continuous Fourier series (3.34)	Analytical (i.e., textbook) problems involving periodic continuous signals
Discrete Fourier transform (DFT) (3.35)	*All real-world problems* (implemented on a computer using the FFT algorithm)
Continuous Fourier transform (3.36)	Analytical (i.e., textbook) problems involving aperiodic continuous signals
Discrete time Fourier transform (DTFT) (3.37)	None; theoretical only
Inverse Fourier series (3.38)	Analytical (i.e., textbook) problems to reconstruct a signal from the Fourier series; not common even in textbooks (one partial example, Example 3.3, in this book)
Inverse discrete Fourier transform (IDFT) (3.39)	*All real world problems* (implemented on a computer using the inverse FFT algorithm)
Inverse Fourier transform (3.40)	Analytical (i.e., textbook) problems to reconstruct a signal from the continuous Fourier transform; not common even in textbooks (no applications in this book)
Inverse discrete time Fourier transform (4.41)	None; theoretical only

choices, both Fourier transformations, one forward and one reverse. In the interest of simplicity, the eight options are itemized in Table 3.4 above (*sans* equations) with an emphasis on when they are applied.

Note that the two equations that are used in real-world problems are the discrete Fourier series, usually just called the discrete Fourier transform or DFT, and its inverse. That is because these are the only equations that can be implemented on time-frequency transformations. However, you need to know about all the versions, because they are sometimes referenced in engineering articles. If you do not understand these variations, you do not really understand the Fourier transform.

The discrete Fourier transform (DFT) is implemented on a computer and very occasionally in textbooks as an exercise. For some students, it is helpful to see a step-by-step implementation of the DFT, so the next example is dedicated to them.

EXAMPLE 3.9

Find the discrete Fourier transform of the discrete periodic signal shown in Figure 3.19 manually.

Solution: The signal has a period of 7 points beginning at $n=-3$ to $n=+3$. To find the Fourier series, apply Equation 3.32, noting that $N=7$ and $x[n]$ is defined by Figure 3.19. Specifically:

$$x[-3]=x[3]=0;\, x[-2]=\ 1;\, x[-1]=0.5;\, x[0]=0;\, x[1]=-0.5;\, x[2]=-1;$$

Substituting into Ex. 3.32 and expanding:

$$X[m]=\sum_{n=1}^{N}x[n]e^{-\frac{j2\pi mn}{N}}=x[-3]e^{j2\pi m3/7}+x[-2]e^{j2\pi m2/7}+x[-1]e^{j2\pi m1/7}$$
$$+x[0]e^{j2\pi m0}+x[1]e^{-j2\pi m1/7}+x[2]e^{-j2\pi m2/7}+x[3]e^{-j2\pi m3/7}$$
$$=0e^{j\pi m6/7}+1e^{j\pi m4/7}+0.5e^{j\pi m2/7}+0e^{j\pi m0}-0.5e^{-j\pi m2/7}-1e^{-j\pi m4/7}+0e^{-j\pi m6/7}$$

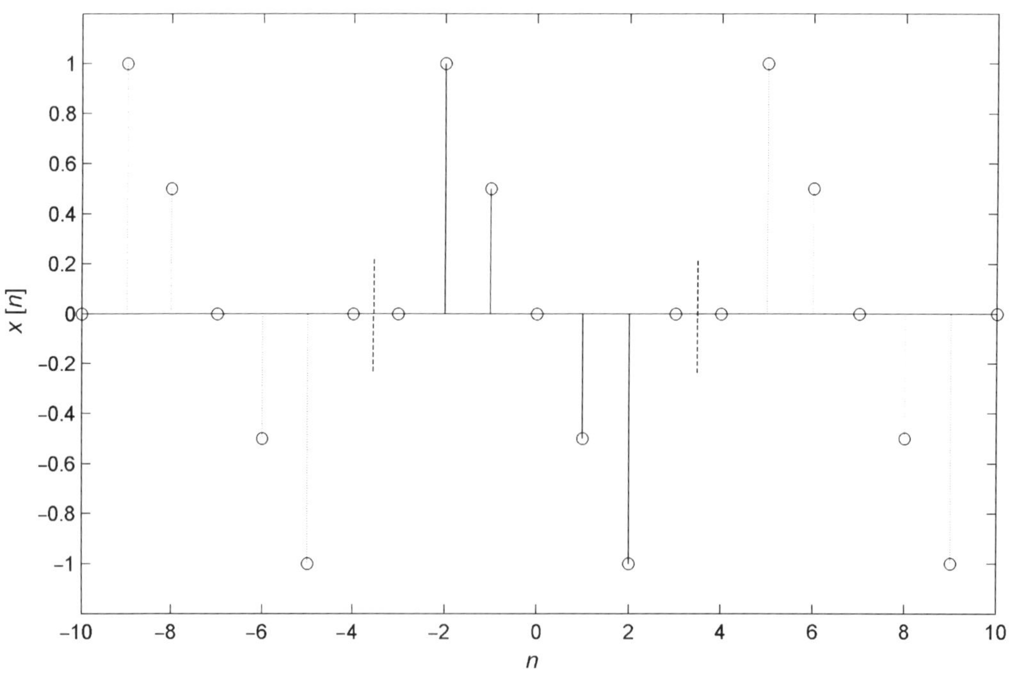

FIGURE 3.19 Discrete periodic waveform used in Example 3.9. The signal consists of discrete points indicated by the circles. The lines are simply to improve visibility. Discrete data are frequently drawn in this manner.

Collecting terms and rearranging:

$$X[m] = e^{j\pi m4/7} - e^{-j\pi m4/7} + 0.5(e^{j\pi m2/7} - e^{-j\pi m2/7})$$

Noting that:

$$\sin(x) = \frac{e^{jx} - e^{-jx}}{2j}$$

$$X[m] = 2j\sin(4\pi m/7) + j\sin(2\pi m/7)$$

The problems at the end of this chapter offer a couple of additional opportunities to solve for the DFT manually.

3.7 MATLAB IMPLEMENTATION OF THE DISCRETE FOURIER TRANSFORM (DFT)

MATLAB provides a variety of methods for calculating spectra, particularly if the Signal Processing Toolbox is available. The basic Fourier transform routine is implemented as:

```
Xf = fft(x,n)   % Calculate the Fourier transform
```

where x is the input waveform and Xf is a complex vector providing the sinusoidal coefficients. (It is common to use capital letters for the Fourier transform variable.) The first term of Xf is real and is the *unnormalized* DC component; you need to divide by the length of x to get the actual DC component. Hence the second term in Xf is the complex representation of the fundamental sinusoidal component; the third term represents the second harmonic; and so on. The argument n is optional and is used to modify the length of data analyzed: if n is less than the length of x, then the analysis is performed over the first n points. If n is greater than the length of x, then the signal is padded with trailing zeros to equal n. The fft routine implements Equation 3.32 above and employs a high-speed algorithm. The FFT algorithm requires the data length to be a power of 2 and, while the MATLAB routine will interpolate, calculation time becomes highly dependent on data length. The algorithm is fastest if the data length is a power of 2, or if the length has many prime factors. For example, on a slow machine a 4096-point FFT takes 2.1 seconds, but requires 7 seconds if the sequence is 4095 points long, and 58 seconds if the sequence is 4097 points long. If at all possible, it is best to stick with data lengths that are powers of 2.

The magnitude of the frequency spectra is obtained by applying the absolute value function, abs, to the complex output Xf:

```
Magnitude = abs(Xf);   % Take the magnitude of Xf
```

This MATLAB function simply takes the square root of the sum of the real part of Xf squared and the imaginary part of Xf squared. The phase angle of the spectra can be obtained by application of the MATLAB angle function:

```
Phase = angle(Xf)   % Find the angle of Xf
```

The angle function takes the arctangent of the imaginary part divided by the real part of Xf. Unlike most handheld calculators and the MATLAB atan function, the angle routine, like the `atan2` function, does take note of the signs of the real and imaginary parts and will generate an output in the proper quadrant. The magnitude and phase of the spectrum can then be plotted using standard MATLAB plotting routines. The angle routine gives phase in radians so to convert to the more commonly used degrees, scale by $360/(2\pi)$.

EXAMPLE 3.10

Construct the waveform used in Example 3.2 (repeated below) and determine the Fourier transform using both the MATLAB `fft` routine and a direct implementation of the defining equations (Equations 3.11 and 3.12).

Solution: We need a total time of 1 sec; let us assume a sample frequency of 256-Hz. This leads to a value of N that equals 256, which is a power of 2 as preferred by the `fft` routine. The MATLAB `fft` routine does no scaling, so its output should be multiplied by $2/N$, where N is the number of points, to get the correct coefficient value in RMS. To get the peak-to-peak values, the output would have to be scaled further by dividing by 0.707.

```
% Example 3.10 Find the Fourier transform of half triangle waveform
% used in Example 3.2. Use both the MATLAB fft and a direct
% implementation of Equations 3.11 and 3.12
%
T = 1;                                    % Total time
fs = 256;                                 % Assumed sample frequency
N = fs*T;                                 % Calculate number of points
t = (1:N)*T/N;                            % Generate time vector
f = (1:N)*fs/N;                           % and frequency vector for plotting
x = [t(1:N/2) zeros(1,N/2)];              % Generate signal, ramp-zeros
%
Xf = fft(x);                              % Take Fourier transform, scale
Mag = abs(Xf(2:end))/(N/2);               % and remove first point (DC value)
Phase = -angle(Xf(2:end))*(360/(2*pi));
%
plot(f(1:20),Mag(1:20),'xb'); hold on;    % Plot magnitude only lower frequencies
xlabel('Frequency (Hz)','FontSize',12); ylabel('|X(f)|','FontSize',12);
%
% Calculate discrete Fourier transform using basic equations
for m = 1:20
  a(m) = (2/N)*sum(x.*(cos(2*pi*m*t)));         % Equation 3.11
  b(m) = (2/N)*sum(x.*(sin(2*pi*m*t)));         % Equation 3.12
  C(m) = sqrt(a(m).^2 + b(m).^2);               % Equation 3.9
  theta(m) = (360/(2*pi)) * atan2(b(m),a(m));   % Equation 3.10
end
disp([a(1:4)' b(1:4)' C(1:4)' Mag(1:4)' theta(1:4)' Phase(1:4)'])
plot(f(1:20),C(1:20),'sr');               % Plot squares superimposed
```

Results: The spectrum produced by the two methods is identical, as seen by the perfect overlap of the x points and square points in Figure 3.20.

The numerical values produced by this program are given in Table 3.5.

Analysis: Both methods produce identical magnitude spectra and similar phase spectra (compare C_m with Mag(fft) and Theta with Phase(fft)). The angles calculated from the a and b coefficients

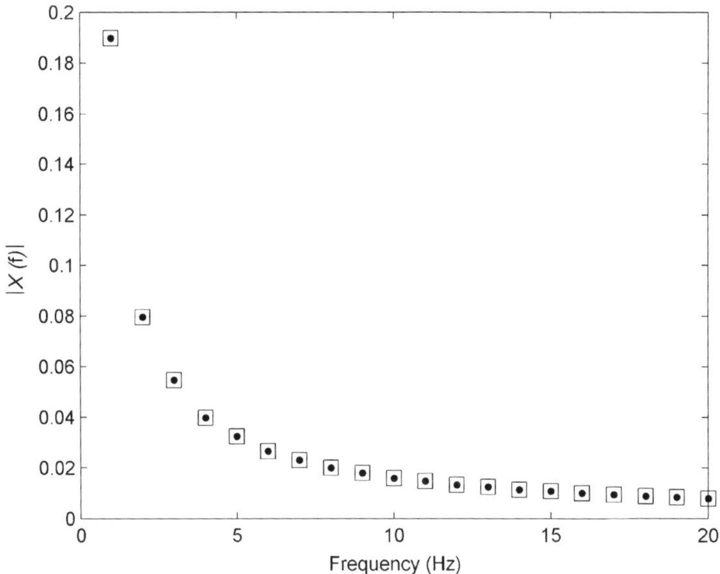

FIGURE 3.20 Magnitude frequency spectra produced by the MATLAB `fft` routine (dots) and a direct implementation of the Fourier transform equations, Equations 3.11 and 3.12 (squares).

TABLE 3.5 Summary of Results from Example 3.2 and Example 3.10

m	a_m	b_m	C_m	Mag(fft)	Theta	Phase (fft)
1	− 0.1033	0.1591	0.1897	0.1897	122.9818	121.5756
2	0.0020	− 0.0796	0.0796	0.0796	− 88.5938	− 91.4063
3	− 0.0132	0.0530	0.0546	0.0546	103.9947	99.7760
4	0.0020	− 0.0398	0.0398	0.0398	− 87.1875	− 92.8125

are in the correct quadrant because the MATLAB `atan2` function was used. This would not be the case if the `atan` function was used. Note that both the `atan2` and `angle` routines give the angle in radians; so to convert to degrees the program scales this output by $360/2\pi$.

The magnitudes and phases found by both methods closely match the values determined analytically in Example 3.2. However, the values for a_2 and a_4 are not exactly zero as found analytically. Because of these small computational errors, the phase angle for the second and fourth components is not exactly -90 degrees as determined analytically. This shows that analytical solutions done by hand can occasionally be more accurate than computer results!

An example applying the MATLAB `fft` to a waveform containing sinusoids and white noise is provided below, along with the resultant spectra in Figure 3.21. Other applications are explored in the problems at the end of this chapter. Example 3.11 uses a special routine, `sig_noise`, found on this book's accompanying CD. This routine generates data consisting of sinusoids and noise and can be useful in evaluating spectral analysis algorithms. The calling structure for `sig_noise` is:

```
[x,t] = sig_noise([f],[SNR],N);   % Generate a signal in noise
```

where f specifies the frequency of the sinusoid(s) in Hz, SNR specifies the desired noise associated with the sinusoid(s) in dB, and N is the number of points. If f is a vector, then a number of sinusoids is generated, each with a signal-to-noise ratio in dB specified by SNR, assuming it is a vector. If SNR is scalar, its value is used for the SNR of all the frequencies generated. The output waveform is in x and t is a time vector useful in plotting. The routine assumes a sample frequency of 1 kHz.

EXAMPLE 3.11

Use fft to find the magnitude spectrum of a waveform consisting of a single 250-Hz sine wave and white noise with an SNR of −14 dB. Calculate the Fourier transform of this waveform and plot the magnitude spectrum.

Solution: Use sig_noise to generate the waveform, take the Fourier transform using fft, obtain the magnitude using abs, and plot. Since we are only interested in the general shape of the curve, we will not bother to normalize the magnitude values.

```
% Example 3.11 Determine the magnitude spectrum of a noisy waveform.
%
N = 1024;  % Number of data points
fs = 1000;  % 1 kHz fs assumed by sig_noise.
f = (0:N-1)*fs/(N-1);  % Frequency vector for plotting
% Generate data using "sig_noise"
% 250 Hz sin plus white noise; N data points; SNR = -14 dB
[x,t] = sig_noise (250,-14,N);  % Generate signal and noise
%
Xf = fft(x);  % Calculate FFT
Mf = abs(Xf);  % Calculate the magnitude
plot(f,Mf);  % Plot the magnitude spectrum
  ......label and title......
```

Analysis: The program is straightforward. After constructing the waveform using the routine sig_noise, the program takes the Fourier transform with fft and then plots the magnitude (constructed using abs) versus frequency using a frequency vector to correctly scale the frequency axis. The frequency vector ranges from 0 to N − 1 to include the DC term (which is zero in this example.)

Results: The spectrum is shown in Figure 3.21 and the peak related to the 250-Hz sine wave is clearly seen. For reasons described in the next section, the spectrum above $f_s/2$ (i.e., 500-Hz) is a mirror image of the lower half of the spectrum. Ideally the background spectrum should be a constant value since the background noise is white noise, but the background produced by noise is highly variable with occasional peaks that could be mistaken for signals. A better way to determine the spectrum of white noise is to use an averaging strategy, described in the next chapter.

The fft routine produces a complex spectrum that includes both the magnitude and phase information, but often only the magnitude spectrum is of interest. Calculating and plotting the phase from the complex spectrum using Equation 3.26 or the equivalent MATLAB angle routine should be straightforward, but a subtle problem can arise. The

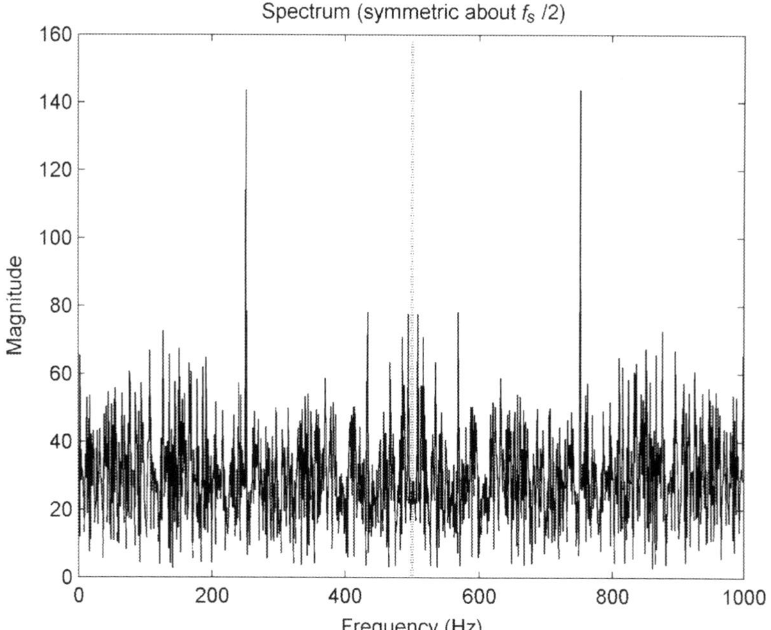

FIGURE 3.21 Plot produced by the MATLAB program above. The peak at 250-Hz is apparent. The sampling frequency of these data is 1 kHz and the spectrum is symmetric about $f_s/2$ (dashed line at 500-Hz) for reasons described later. Normally only the first half of this spectrum would be plotted. (f_{sine} = 250-Hz; SNR = −14 dB; N = 1024.)

output of the `angle` routine is limited to $\pm 2\pi$; larger values "wrap around" to fall within that limit. Thus a phase shift of $3\pi/2$ would be rendered as $\pi/2$ by the `angle` routine, as would a phase shift of $5\pi/2$. When a signal's spectrum results in large phase shifts, this wrap-around characteristic produces jumps on the phase curve. An example of this is seen in Figure 3.22, which shows the magnitude and phase spectra of a delayed pulse waveform shown in Figure 3.23. This signal consists of a periodic 20-msec pulse with a period of 4 sec (Figure 3.23). However, the pulse is delayed by 100 msec and this delay leads to large phase shifts in the phase spectrum. (The effect of delays on magnitude and phase spectra is examined further in the problems at the end of this chapter and in the next chapter.) The phase plot (Figure 3.22B), shows a sharp transition of approximately 2π whenever the phase reaches $-\pi$ (lower line, Figure 3.22B).

The correct phase spectrum could be recovered by subtracting 2π from the curve every time there is a sharp transition. This is sometimes referred to as "unwrapping" the phase spectrum. It would not be difficult to write a routine to check for large jumps in the phase spectrum and subtract 2π from the rest of the curve whenever they are found. As is so often the case, it is not necessary to write such a routine since MATLAB has already done it. The MATLAB routine `unwrap` checks for jumps greater than π and does the subtraction whenever they occur. (The threshold for detecting a discontinuity can be modified as described in the `unwrap` help file.) The next example illustrates the use of `unwrap` for unwrapping the phase spectrum.

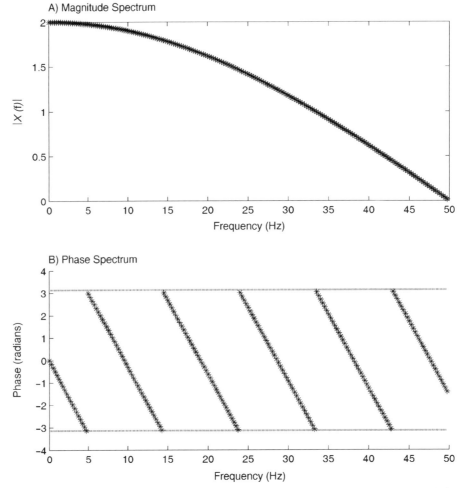

FIGURE 3.22 A) Magnitude spectrum of the delayed pulse waveform shown in Figure 3.23. B) Phase spectrum in radians of the delayed pulse waveform showing upward jumps of approximately 2π whenever the phase approaches $-\pi$. The dashed lines represent $\pm\pi$.

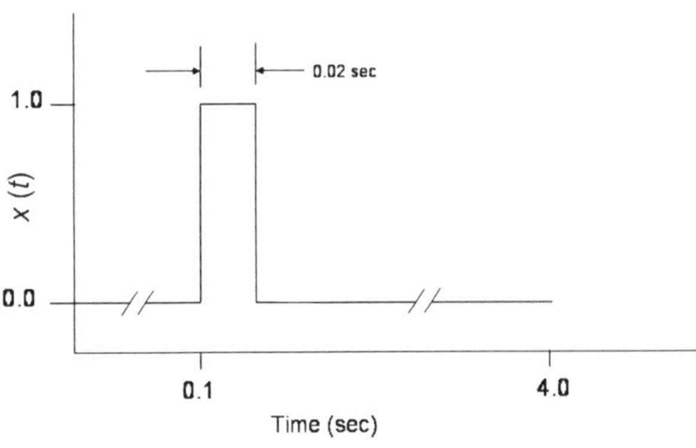

FIGURE 3.23 The time function of the delayed pulse waveform whose magnitude and phase spectra are shown in Figure 3.22. The 20-msec pulse is periodic with a period of 4 sec, but the pulse is delayed by 100 msec from the start of the period.

EXAMPLE 3.12

Calculate the magnitude and phase spectrum of the triangle waveform used in Examples 3.2 and 3.10, but extend the period with zeros to 2 sec. (In the next chapter we find that adding zeros increases the apparent resolution of the spectrum.) As the magnitude spectrum decreases rapidly with increasing frequency, plot only the first 20 points in the spectrum plus the DC term. Plot the phase with and without unwrapping.

Solution: The approach is the same as in the last example: a time vector is constructed and used to generate the signal; a frequency vector ranging from 0 to −1 is constructed for plotting; the Fourier transform is calculated using `fft`, and the magnitude and phase are determined and plotted. In addition, the `unwrap` routine is applied to the phase component and the result compared with the original phase.

```
% Example 3.12 Evaluate the spectrum of a triangle waveform
%
T = 2;                                        % Total time (2 sec in this example)
fs = 256;                                     % Assumed sample frequency
N = fs*T;                                     % Calculate number of points
t = (1:N)*T/N;                                % Generate time vector
f = (0:N-1)*fs/(N-1);                         % and frequency vector for plotting
x = [t(1:N/4) zeros(1,3*N/4)];                % Generate signal, ramp-zero
%
X = fft(x);                                   % Calculate complex spectrum
Mag = abs(X);                                 % Get magnitude
Phase = angle(X)*360/(2*pi);                  % Get phase in degrees
Phase_unwrap = unwrap(angle(X))*360/(2*pi);;  % Unwrapped phase in degrees
plot(f(1:21),Mag(1:21),'*k');                 % Plot magnitude spectrum
  .......labels and title.......
figure;
subplot(2,1,1);
plot(f(1:21),Phase(1:21),'*k');
  ........label and title.......
subplot(2,1,2);
plot(f(1:21),Phase_unwrap(1:21),'*k');
  .......label and title........
```

Results: The magnitude spectrum is shown in Figure 3.24 and the two phase spectra are shown in Figure 3.25. In Figure 3.25A, the phase spectra are seen to transition upward sharply whenever they approach −180 degrees as expected. The unwrapped spectrum is a straight line that decreases smoothly to −1500 degrees at 10-Hz.

This last example applies the DFT to a signal related to respiration and uses the spectrum to find the breathing rate.

EXAMPLE 3.13

Find the magnitude and phase spectrum of the respiratory signal found in variable `resp` of file `Resp.mat` (data from PhysioNet, http://www.physionet.org, Golberger et al., 2000). The data

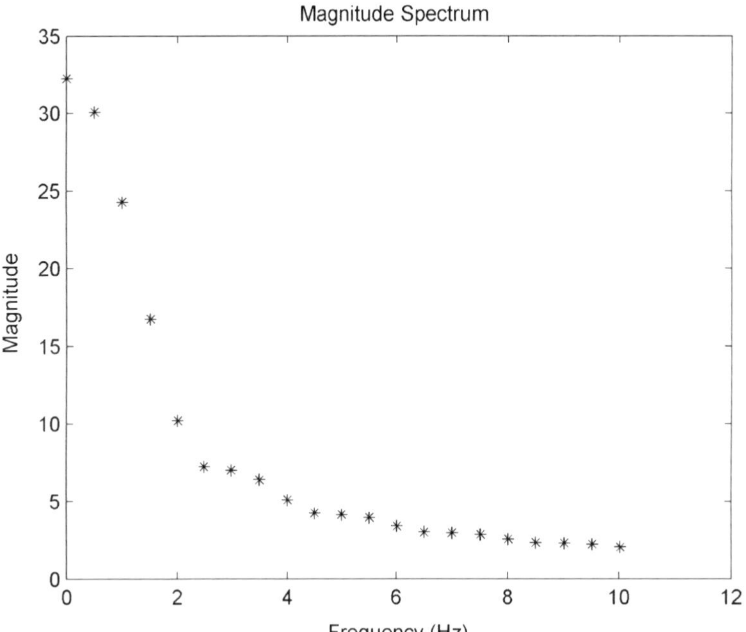

FIGURE 3.24 The magnitude spectrum of a triangle as shown in Figure 3.9 but with a period of 2 sec obtained by extending the zero portion of the signal.

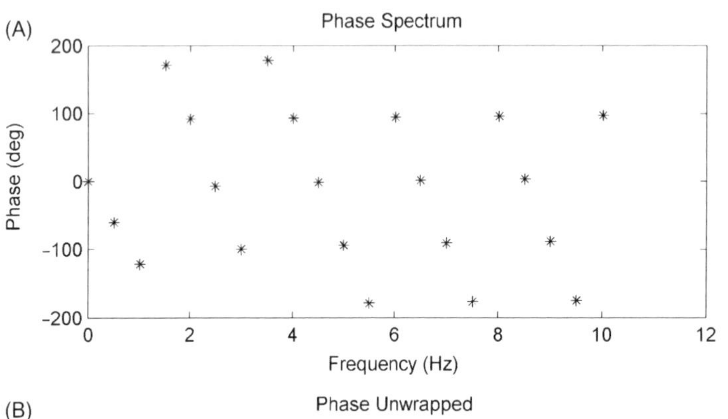

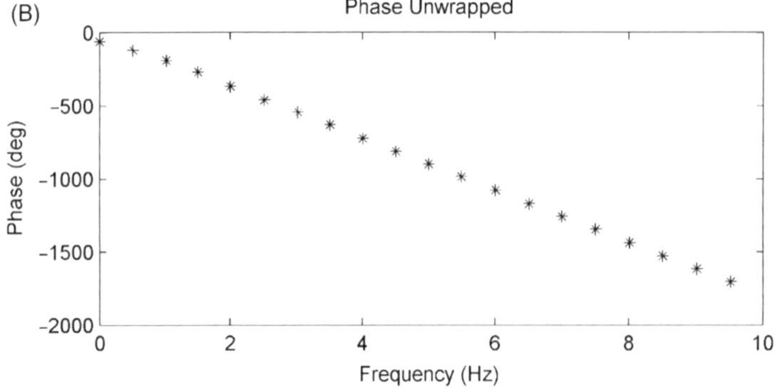

FIGURE 3.25 The phase spectra of the triangle signal. A) Without phase unwrapping. B) With phase unwrapping. This is the correct spectrum: a smooth straight line that decreases continually with increasing frequency.

have been sampled at 125-Hz. Plot only frequencies between the fundamental and 2-Hz. From the spectrum's magnitude peak, estimate the breaths per minute of this subject.

Solution: Loading the data and taking the DFT are straightforward, as is finding the spectral peak. We then need to calculate the spectral point corresponding to 2-Hz. From Equation 3.17, $f = m/T$ and $T = N/f$ where N is the total number of points. So 2-Hz corresponds to point $m = fT = 2T = 2N/f_s$. To find the peak frequency, first find the value of m that corresponds to the peak frequency using MATLAB's `max` routine (the second output argument of this routine). The value of the frequency vector at index m will be the corresponding frequency, f_{peak} and $1/f_{peak}$ will be the corresponding time, t_{peak}. This represents the average time between each breath so the breaths per min will be this time divided into 60: $60/t_{peak}$.

```
% Example 3.13 Example applying the FT to a respiration signal.
%
load Resp                                   % Get respiratory signal
fs = 125;                                   % Sampling frequency in Hz
max_freq = 2;                               % Max desired plotting frequency in Hz
N = length(resp);                           % Length of respiratory signal
f = (1:N)*fs/N;                             % Construct frequency vector
t = (1:N)/fs                                % Construct the time vector
plot(t,resp)                                % Plot the time signal
  ......labels......
X = fft(resp);                              % Calculate the spectrum
m_plot = round(2/(fs/N));                   % Find m for 2 Hz
subplot(2,1,1);
plot(f(1:m_plot-1),abs(X(2:m_plot)));       % Plot magnitude spectrum
  ......labels......
phase = unwrap(angle(X));                   % Calculate phase spectrum
phase = phase * 360/(2*pi);                 % Unwrap phase spectrum
%
subplot(2,1,2);
plot(f(1:m_plot-1),phase(2:m_plot));        % Plot phase spectrum
  ......labels......
[peak m_peak] = max(abs(X(2:m_plot)));      % Find m at max magnitude peak
max_freq = f(m_peak)                        % Calculate and display frequency
max_time = 1/max_freq                       % Calculate and display max time
breath_min = 60/max_time                    % Calculate and display breath/min
```

Results: The time data are shown in Figure 3.26 to be a series of peaks corresponding to maximum inspiration. The spectrum plot produced by the program is shown in Figure 3.27. A peak is shown around 0.3-Hz with a second peak around twice that frequency, a harmonic of the first peak. The frequency of the peak was found to be 0.3125-Hz corresponding to an inter-breath interval of 3.2 sec giving rise to a respiratory rate of 18.75 breaths/min. Of course the respiratory rate could have been found from the time domain data by looking at intervals between the peaks in Figure 3.26, then taking averages, but this would have been more difficult and prone to error.

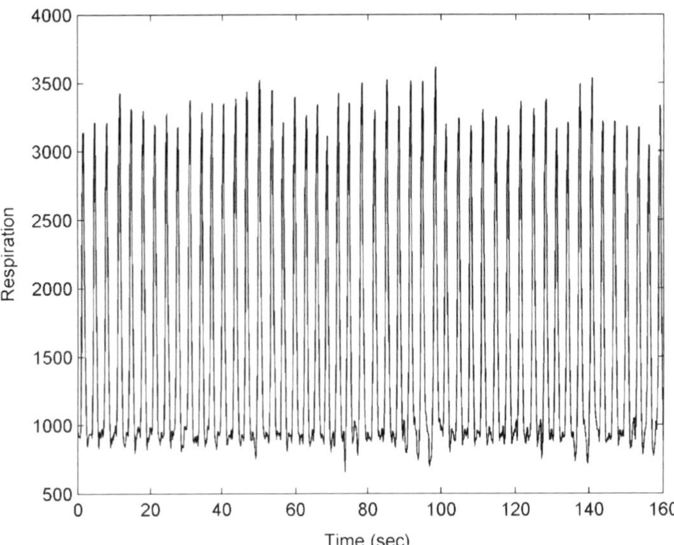

FIGURE 3.26 Respiratory signal used in Example 3.13. The peaks correspond to maximum inspiration.

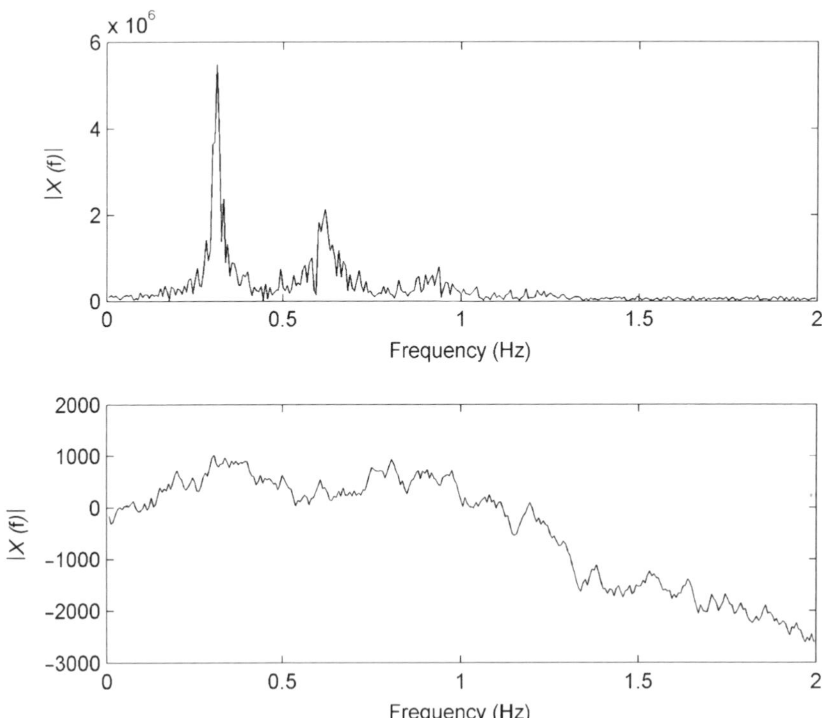

FIGURE 3.27 The magnitude and phase spectrum of the respiratory signal seen in Figure 3.26. The peak around 0.3-Hz corresponds to the interval between breaths. This is equivalent to 18.75 breaths/min.

3.8 SUMMARY

The sinusoid (i.e., $A\cos(\omega t + \theta)$) is a unique signal with a number of special properties. A sinusoid can be completely defined by three values: its amplitude A, its phase θ, and its frequency ω (or $2\pi f$). Any periodic signal can be broken down into a series of harmonically related sinusoids, although that series might have to be infinite in length. Reversing that logic, any periodic signal can be reconstructed from a series of sinusoids. Thus any periodic signal can be equivalently represented by a sinusoidal series. A sinusoid is also a pure signal in that it has energy at only one frequency, the only waveform to have this property. This means that sinusoids can serve as intermediaries between the time domain representation of a signal and its frequency domain representation. When the input to a linear system is a sinusoid, the output is also a sinusoid at the same frequency. Only the magnitude and phase of a sinusoid can be altered by a linear system. Finally, harmonically related sinusoids are orthogonal so they are independent and do not influence one another during mathematical operations.

The technique for determining the sinusoidal series representation of a periodic signal is known as Fourier series analysis. To determine the Fourier series, the signal of interest is correlated with sinusoids at harmonically related frequencies. This correlation provides the amplitude and phase of a sinusoidal series that represents the periodic signal, and these can be plotted against frequency to describe the frequency-domain composition of the signal. Fourier series analysis is often described and implemented using the complex representation of a sinusoid.

If the signal is not periodic, but exists for a finite time period, Fourier decomposition is still possible by assuming that this aperiodic signal is in fact periodic, but that the period is infinite. This approach leads to the Fourier transform where the correlation is now between the signal and infinite number of sinusoids having continuously varying frequencies. The frequency plots then become continuous curves. The inverse Fourier transform can also be constructed from the continuous frequency representation. The Fourier transform and Fourier series analysis are a linear operations. This property and other important properties of the Fourier transform are described in Appendix B.

Fourier decomposition applied to digitized data is known as the discrete Fourier series or the discrete Fourier transform (DFT). In fact, real-world Fourier analysis is always done on a computer using a high-speed algorithm known as the fast Fourier transform (FFT). The discrete equations follow the same pattern as those developed for the continuous time signal, except integration becomes summation and both the sinusoidal and signal variables are discrete. There is a discrete version of the continuous Fourier transform known as the discrete time Fourier transform (DTFT), but this equation and its inverse are of theoretical interest only.

PROBLEMS

1. Find the Fourier series of the square wave below using noncomplex analytical methods (i.e., Equations 3.11 and 3.12). [*Hint*: Take advantage of symmetries to simplify the calculation.]

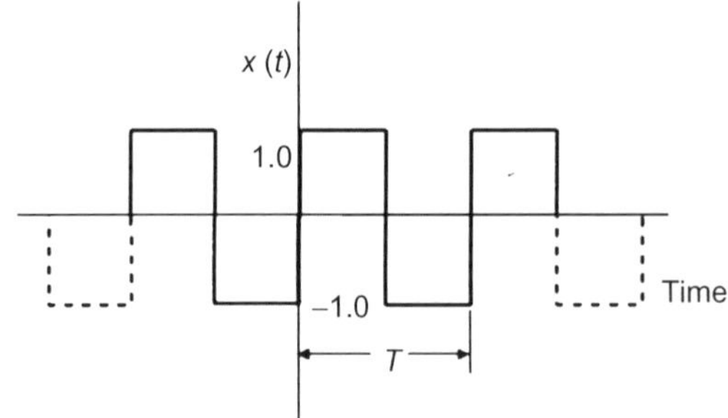

2. Find the Fourier series of the waveform below using noncomplex analytical methods. The period, T, is 1 sec.

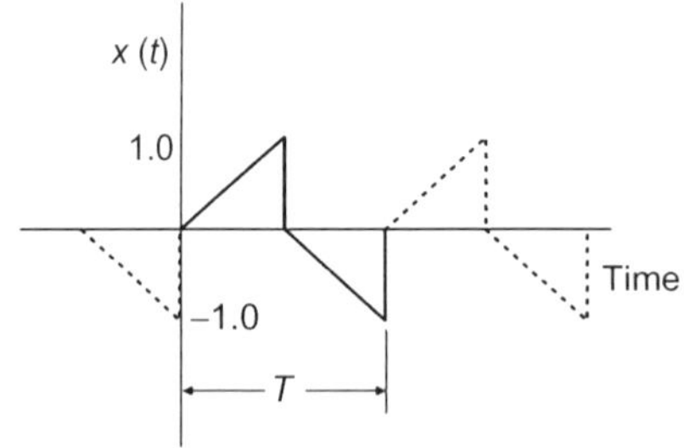

3. Find the Fourier series of the "half-wave" rectified sinusoidal waveform below using noncomplex analytical methods. Take advantage of symmetry properties.

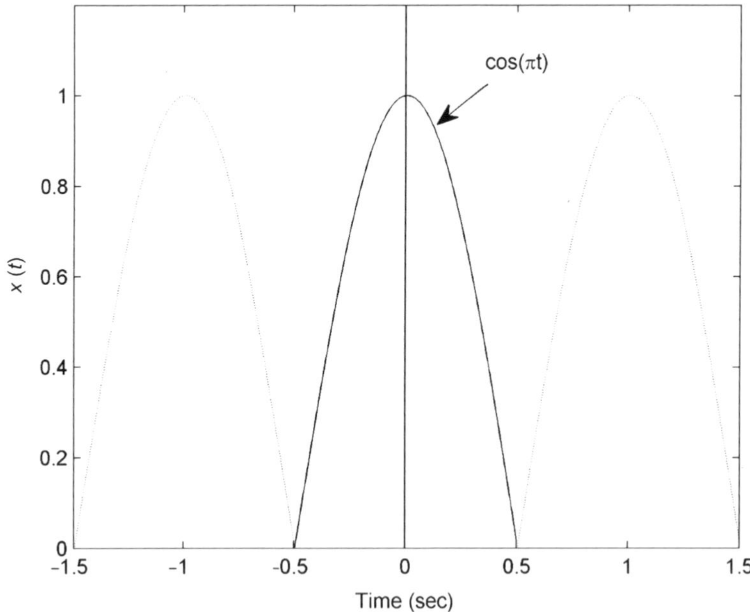

4. Find the Fourier series of the "sawtooth" waveform below using noncomplex analytical methods.

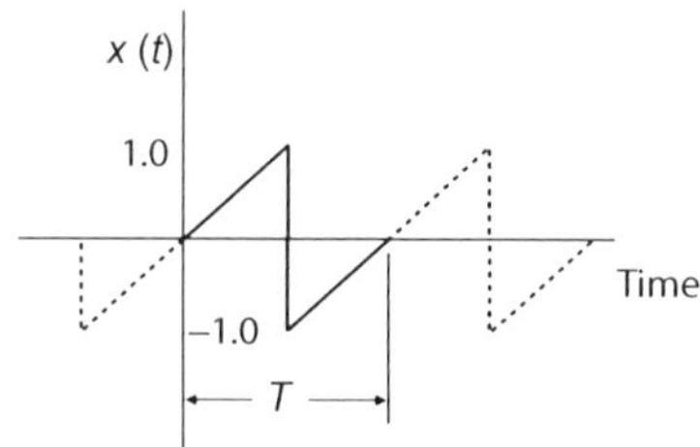

5. Find the Fourier series of the periodic exponential waveform shown below where $x(t) = e^{-2t}$ for $0 < t \leq 2$ (i.e., $T = 2$ sec). Use the complex form of the Fourier series in Equation 3.23.

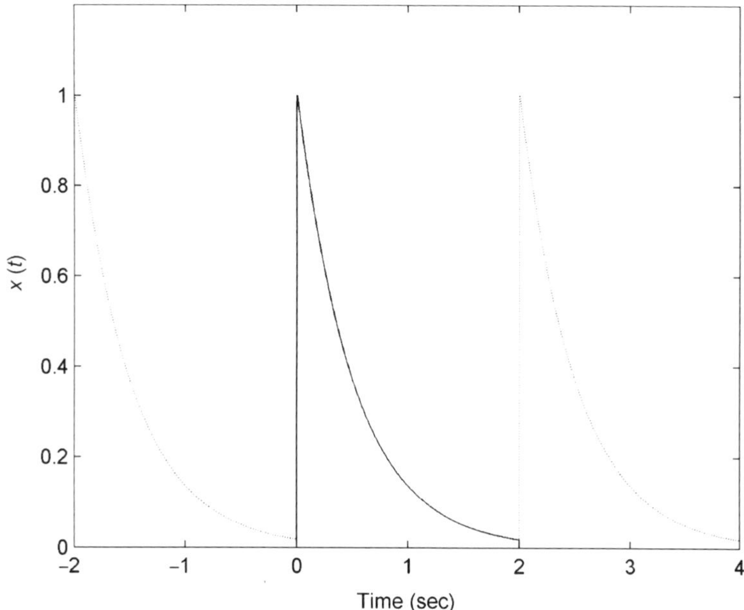

6. Find the continuous Fourier transform of an aperiodic pulse signal given in Example 3.5 using the noncomplex equation, Equation 3.28 and symmetry considerations.
7. Find the continuous Fourier transform of the aperiodic signal shown below, using the complex equation, Equation 3.27.

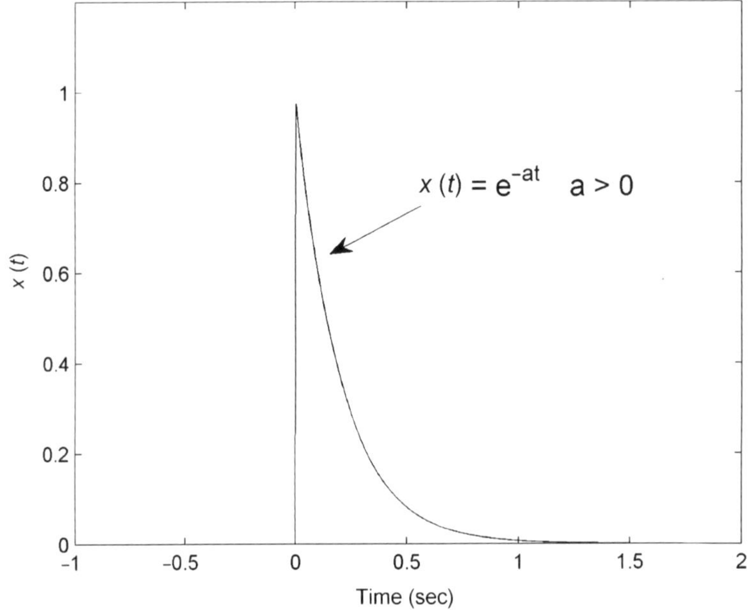

8. Find the continuous Fourier transform of the aperiodic signal shown below. This is easier to solve using the noncomplex form, Equation 3.13 and 3.14, in conjunction with symmetry considerations.

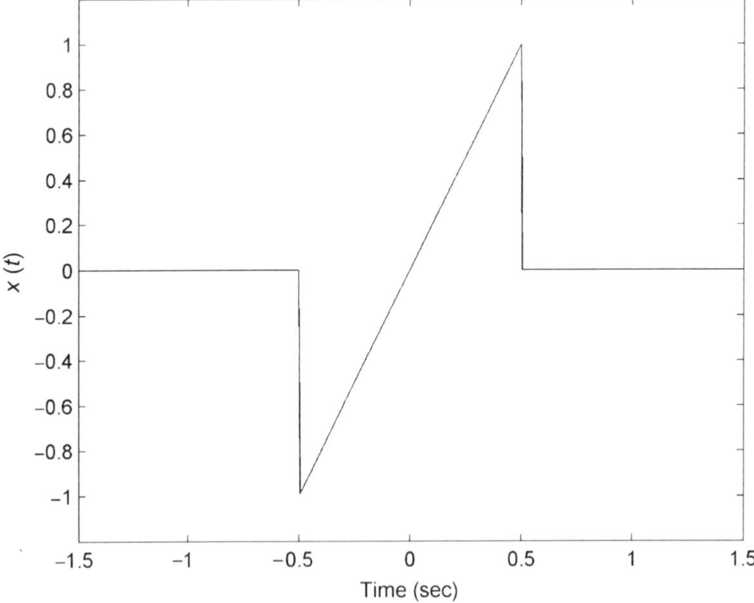

9. Use the MATLAB Fourier transform routine to find the spectrum of a waveform consisting of two sinusoids at 200 and 400-Hz. First generate the waveform in a 512-point array assuming a sampling frequency of 1 kHz. Take the Fourier transform of the waveform and plot the magnitude of the full spectrum (i.e., 512 points). Generate a frequency vector as in Example 3.10 so the spectrum plot has a properly scaled horizontal axis. Since the DC term will be zero, there is no need to plot it. [*Hint*: To generate the waveform, first construct a time vector `t`, then generate the signal using the code: `x=sin(2*pi*f1*t)+sin(2*pi*f2*t)` where `f1` = 200, and `f2` = 400.]
10. Use the routine `sig_noise` to generate a waveform containing 200- and 400-Hz sine waves as in Problem 9, but add noise so that the signal-to-noise ratio (SNR) is −8 dB; i.e., `x=sig_noise([200 400], -8,N)` where `N` = 512. Plot the magnitude spectrum, but only plot the nonredundant points (2 to N/2) and do not plot the DC term, which again is zero. Repeat for an SNR of −16 dB. Note that the two sinusoids are hard to distinguish at the higher (−16 dB) noise level.
11. Use the routine `sig_noise` to generate a waveform containing 200- and 400-Hz sine waves as in Problems 9 and 10 with an SNR of −12 dB with 1000 points. Plot only the nonredundant points (no DC term) in the magnitude spectrum. Repeat for the same SNR, but for a signal with only 100 points. Note that the two sinusoids are hard to distinguish with the smaller data sample. Taken together, Problems 10 and 11 indicate that both data length and noise level are important when detecting sinusoids (also known as *narrowband* signals) in noise.
12. Generate the signal shown in Problem 2 and use MATLAB to find both the magnitude and phase spectrum of this signal. Assume that the period, T, is 1 sec and assume a sampling frequency of 500-Hz; hence you will need to use 500 points to generate the signal. Plot the time domain signal, $x(t)$, then calculate and plot the spectrum. The spectrum of this curve falls off rapidly, so plot only the first 20 points plus the DC term and plot them as discrete points, not as a line plot. Note that every other frequency point is zero, as predicted by the symmetry of this signal.
13. Repeat Problem 12 using the signal in Problem 4, but generate the signal between $-T/2$ and $+T/2$ (i.e., `x=t;` for `t=(-N/2:N/2);`). Also plot the phase in degrees, that is, scale by $360/(2\pi)$. Again plot only the first 20 values (plus the DC term) as discrete points.
14. Repeat Problem 13 for the signal in Problem 4, but generate the signal between 0 and T. Unwrap the phase specturm, plot in degrees and plot only the first 20 points (plus DC term) as in Problem 13. Note the difference in phase, but not magnitude, from the plots of Problem 13.
15. Find the discrete Fourier transform for the signal shown below using the manual approach of Example 3.9. The dashed lines indicate one period.

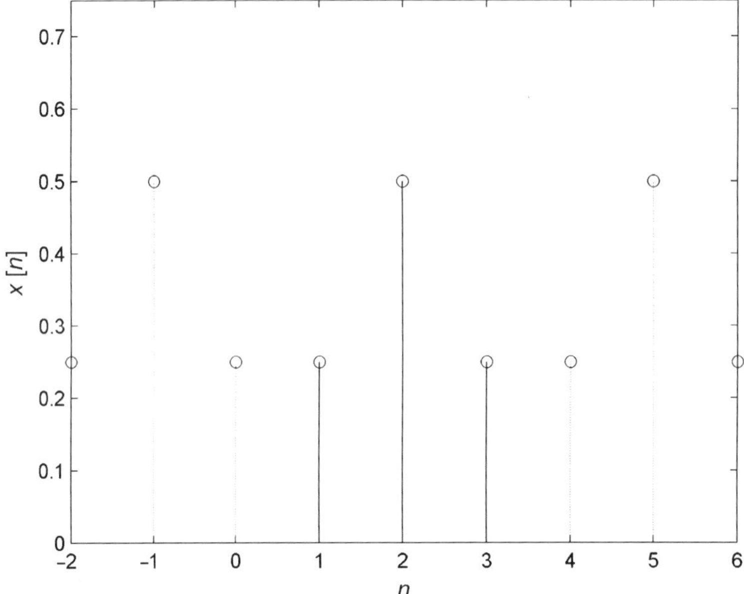

16. Find the discrete Fourier transform for the signal shown below using the manual approach of Example 3.9. The dashed lines indicate one period. [*Hint*: You can just ignore the zeros on either side, except for the calculation of N.]

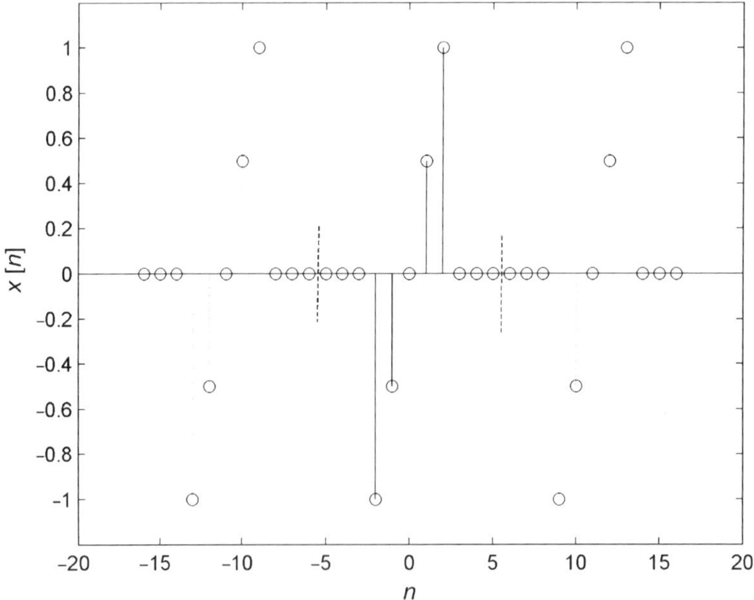

17. Plot the magnitude and phase components of the ECG signal found as variable `ecg` in file `ECG.mat`. Plot only the nonredundant points and do not plot the DC component (i.e., the first point in the Fourier series). Also plot the time function and correctly label and scale the time and frequency axes. The sampling frequency was 125-Hz. Based on the magnitude plot, what is the bandwidth (i.e., range of frequencies) of the ECG signal?
18. The data file `pulses.mat` contains 3 signals: `x1`, `x2`, and `x3`. These signals are all 1 sec in length and were sampled at 500-Hz. Plot the 3 signals and show that each contains a single 40-msec pulse, but at 3 different delays: 0, 100, and 200 msec. Calculate and plot the spectra for the 3 signals superimposed on a single magnitude and single phase plot. Plot only the first 20 points as discrete points plus the DC term using a different color for each signal's spectra. Apply the `unwrap` routine to the phase data and plot in degrees. Note that the 3 magnitude plots are identical, while the phase plots are all straight lines but with radically different slopes.
19. Load the file `chirp.mat` which contains a sinusoidal signal, `x`, that increases its frequency linearly over time. The sampling frequency of this signal is 5000-Hz. This type of signal is called a *chirp* signal because of the sound it makes when played through an audio system. If you have an audio system, you can listen to this signal after loading the file using the MATLAB command: `sound(x,5000);`. Take the Fourier transform of this signal and plot magnitude and phase (no DC term). Note that the magnitude spectrum shows the range of frequencies that are present but there is no information on the timing of those frequencies. Actually, information on signal timing is contained in the phase plot but, as you can see, this plot is not easy to interpret. Advanced signal processing methods known as *time-frequency methods* are necessary to recover the timing information.
20. Load the file `ECG_1min.mat` that contains 1 minute of ECG data in variable `ecg`. Take the Fourier transform. Plot both the magnitude and phase (unwrapped) spectrum up to 20-Hz and do not include the DC term. Find the average heart rate using the strategy found in Example 3.13. The sample frequency is 250-Hz. [*Note*: The spectrum will have a number of peaks; however, the largest low-frequency peak (which is also the largest overall) will correspond to the cardiac cycle.]

CHAPTER

4

The Fourier Transform and Power Spectrum: Implications and Applications

4.1 DATA ACQUISITION AND STORAGE

4.1.1 Data Sampling—The Sampling Theorem

Discrete time data differ from continuous time data in two important ways: the data are time and amplitude sampled, and the data are shortened to fit within the finite constraints of computer memory. Since most real-world time and frequency operations are carried out on discrete time data, it is important to understand how these operations influence the data.

Slicing the signal into discrete time intervals (usually evenly spaced) is a process known as sampling and is described in Chapter 1 (see Figures 1.4 and 1.5). Sampling is a nonlinear process and has some peculiar effects on the signal's spectrum. Figure 4.1A shows an example of a magnitude frequency spectrum of a hypothetical 1.0-second periodic, continuous signal, as may be found using continuous Fourier series analysis. The fundamental frequency is $1/T = 1$-Hz and the first 10 harmonics are plotted out to 10-Hz. For this particular signal, there is little energy above 7.0-Hz. One of the peculiarities of the sampling process is that it produces many additional frequencies that are not in the original signal. Figure 4.1B shows just a portion of the theoretical magnitude spectrum of the signal whose spectrum is shown in Figure 4.1A, but after it is sampled at 20-Hz. Only a portion can be shown because the spectrum of the sampled signal is infinite. The new spectrum is itself periodic, with a period equal to the sample frequency, f_s (in this case 20-Hz). The spectrum also contains negative frequencies (it is, after all, a theoretical spectrum). It has even symmetry about $f = 0.0$-Hz as well as about both positive and negative multiples of f_s. Finally,

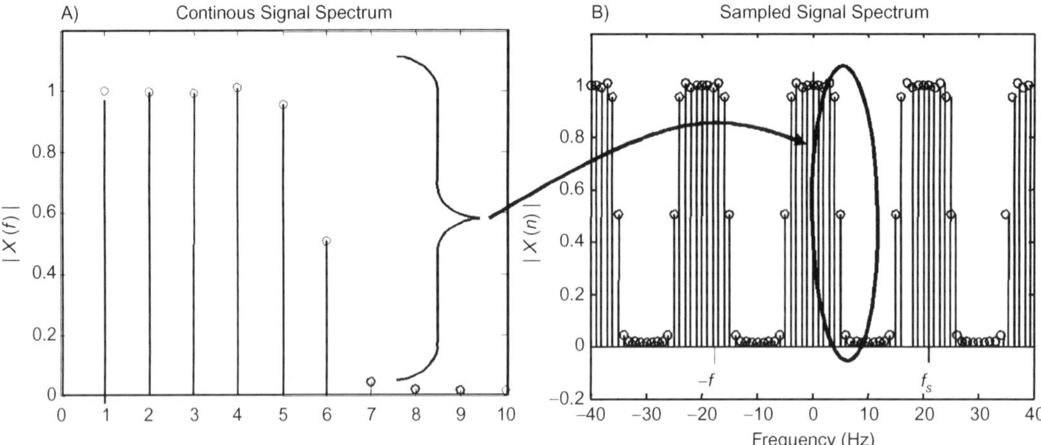

FIGURE 4.1 A) The spectrum of a continuous signal. B) The spectrum of this signal after being sampled at $f_s = 20$-Hz. Sampling produces a large number of frequency components not in the original spectrum, even components having negative frequency. The sampled signal has a spectrum that is periodic at the sampling frequency (20-Hz) and has an even symmetry about 0.0-Hz, and positive and negative multiples of the sampling frequency, f_s. Since the sampled spectrum is periodic, it goes on forever and only a portion of it can be shown. The spectrum is just a series of points; the vertical lines are drawn to improve visualization.

the portion between 0 and 20-Hz also has even symmetry about the center frequency, $f_s/2$ (in this case 10-Hz).

The spectrum of the sampled signal is certainly bizarre, but comes directly out of the mathematics of sampling as touched on in Chapter 7. When we sample a continuous signal, we effectively multiply the original signal by a periodic impulse function (with period f_s) and that multiplication process produces all those additional frequencies. Even though the negative frequencies are mathematical constructs, their effects are felt because they are responsible for the symmetrical frequencies above $f_s/2$.

If the sampled signal's spectrum is different from the original signal's spectrum, it stands to reason that the sampled signal is different from the original. If the sampled signal is different from the original and if we cannot recover the original signal from the sampled signal, then digital signal processing is a lost cause. We would be processing something unrelated to the original signal. The critical question is: given that the sampled signal is different from the original, can we find some way to reconstruct the original signal from the sampled signal? The frequency domain version of that question is: can we reconstruct the unsampled spectrum from the sampled spectrum? The definitive answer is: maybe, but fortunately that maybe depends on things we can measure and understand.

Figure 4.2 shows just one period of the spectrum shown in Figure 4.1B, the period between 0 and f_s Hz. In fact, this is the only portion of the spectrum that can be calculated by the discrete Fourier transform; all the other frequencies shown in Figure 4.1B are theoretical (but not, unfortunately, inconsequential). Comparing this spectrum to the spectrum of the original signal (Figure 4.1A), we see that the two are the same for the first half of the spectrum up to $f_s/2$, and the second half is just the mirror image (a manifestation of

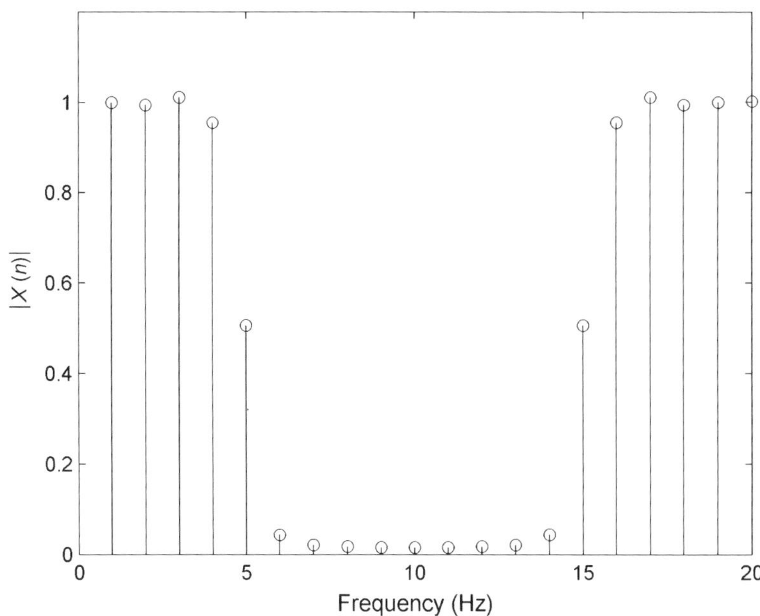

FIGURE 4.2 A portion of the spectrum of the sampled signal whose unsampled spectrum is shown in Figure 4.1A. There are more frequencies in this sampled spectrum than in the original, but they are distinct from, and do not overlap, the original frequencies.

those theoretical negative frequencies). It would appear that we could obtain a frequency spectrum that was identical to the original if we somehow got rid of all frequencies above $f_s/2$. In fact, we can get rid of the frequencies above $f_s/2$ by an approach known as filtering. Just knowing we can get back to the original spectrum is sufficient to justify our sampled computer data and we can ignore the frequencies above $f_s/2$. The frequency $f_s/2$ is so important it has its own name: the *Nyquist*[1] *frequency*.

This strategy of just ignoring all frequencies above the Nyquist frequency ($f_s/2$) works well and is the approach that is commonly adopted. But it can only be used if the original signal does not have spectral components at or above $f_s/2$. Consider a situation where 4 sinusoids with frequencies of 100, 200, 300, and 400-Hz are sampled at a sampling frequency of 1000-Hz. The spectrum produced after sampling actually contains 8 frequencies (Figure 4.3A): the 4 original frequencies plus the 4 mirror image frequencies reflected about $f_s/2 = 500$-Hz. As long as we know, in advance, that the sampled signal does not contain any frequencies above the Nyquist frequency (500-Hz), we will not have a problem: We know that the first 4 frequencies are those of the signal and the second 4, above the Nyquist frequency, are the reflections which can be ignored. However, a problem occurs if the original signal contains frequencies higher than the Nyquist frequency. The frequencies above $f_s/2$ will be reflected back into the lower half of the spectrum. This is shown in Figure 4.3B where the signal now contains 2 additional frequencies at 650 and 850-Hz. These frequency components have their reflections in the lower half of the

[1]Nyquist was one of many prominent engineers to hone his skills at the former Bell Laboratories during the first half of the twentieth century. He was born in Sweden, but received his education in the United States.

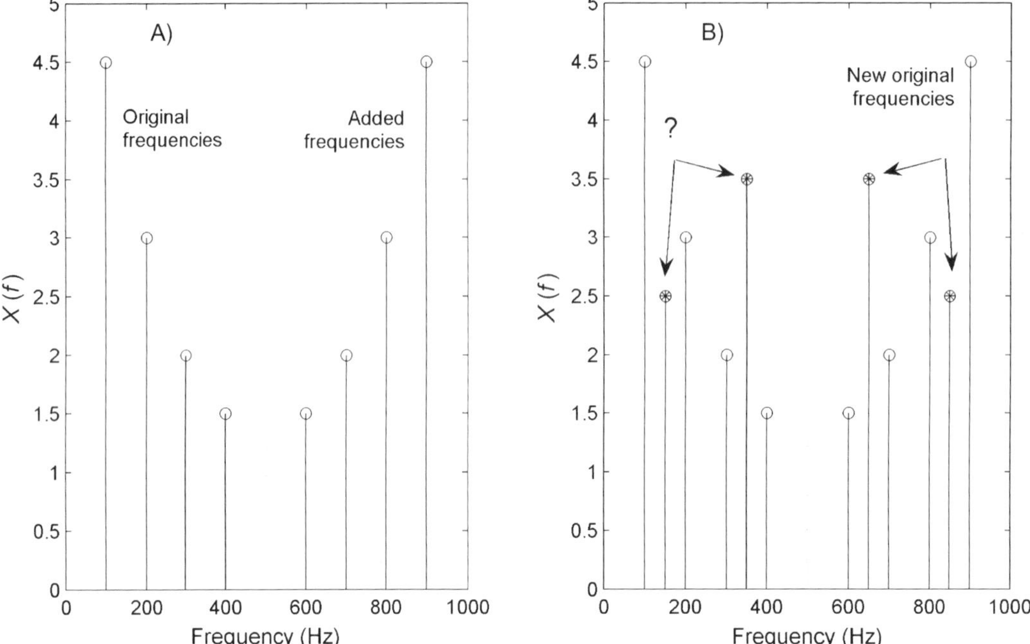

FIGURE 4.3 A) Four sine waves between 100 and 400-Hz are sampled at 1-kHz. Only one period of the sampled spectrum is shown, the period between 0 and f_s Hz. Sampling essentially produces new frequencies not in the original signal. Because of the periodicity and even symmetry of the sampled spectrum, the additional frequencies are a mirror image reflection around $f_s/2$, the Nyquist frequency. If the frequency components of the sampled signal are all below the Nyquist frequency as shown here, the upper frequencies do not interfere with the lower spectrum and can simply be ignored. B) If the sampled signal contains frequencies above the Nyquist frequency, they are reflected into the lower half of the spectrum (circles with "*"). It is no longer possible to determine which frequencies belong where, an instance of *aliasing*.

spectrum: at 350 and 150-Hz, respectively. It is now no longer possible to determine if the 350-Hz and 150-Hz signals are part of the true spectrum of the signal (i.e., the spectrum of the signal before it was sampled) or whether these are reflections of signals with frequency components greater than $f_s/2$ (which in fact they are). Both halves of the spectrum now contain mixtures of frequencies above and below the Nyquist frequency, and it is impossible to know where they really belong. This confusing condition is known as *aliasing*. The only way to resolve this ambiguity is to insure that all frequencies in the original signal are less than the Nyquist frequency.

If the original signal contains frequencies above the Nyquist frequency, then you cannot determine the original spectrum from what you have in the computer and you cannot reconstruct the original analog signal from the one currently stored in the computer. The frequencies above the Nyquist frequency have hopelessly corrupted the signal stored in the computer. Fortunately, the converse is also true. If there are no corrupting frequency components in the original signal (i.e., the signal contains no frequencies above half the sampling frequency), the spectrum in the computer can be adjusted to match the original

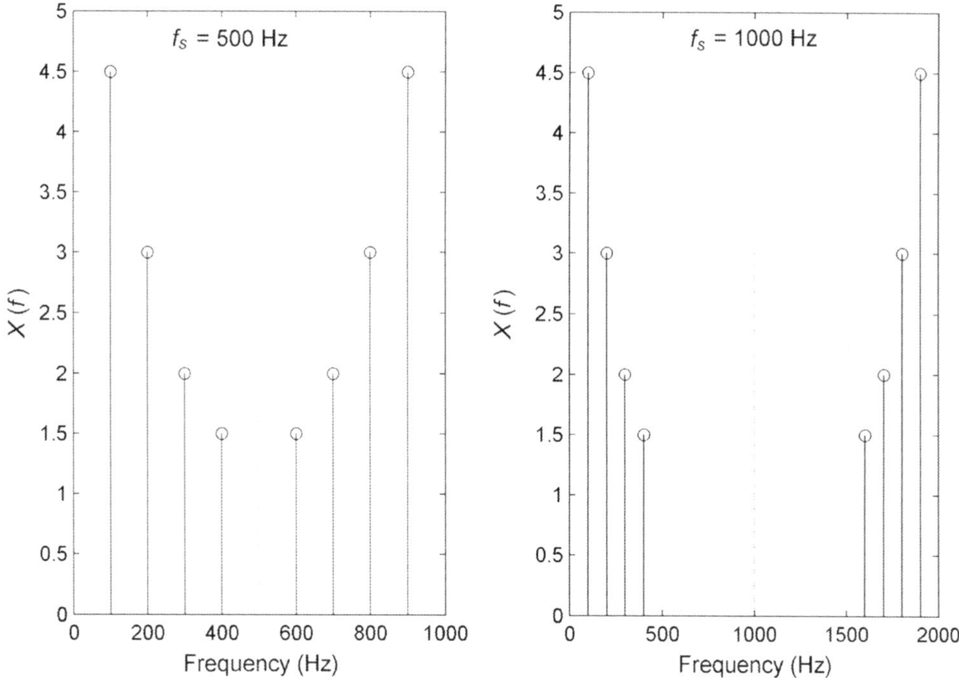

FIGURE 4.4 The same signal is sampled at two different sampling frequencies. The higher sampling frequency (right hand plot) provides much greater separation between the original spectral components and those produced by the sampling process.

signal's spectrum if we eliminate or disregard the frequencies above the Nyquist frequency. (Elimination of frequencies above the Nyquist frequency can be achieved by lowpass filtering, and the original signal can be reconstructed.) This leads to Shannon's famous sampling theorem: The original signal can be recovered from a sampled signal provided the sampling frequency is more than twice the maximum frequency contained in the original (Equation 4.1).

$$f_s > 2f_{max} \tag{4.1}$$

Usually the sampling frequency is under software control, and it is up to the biomedical engineer doing the sampling to ensure that f_s is high enough. To make elimination of the unwanted higher frequencies easier, it is common to sample at 3 to 5 times f_{max}. This increases the spacing between the frequencies in the original signal and those generated by the sampling process (Figure 4.4). The temptation to set f_s higher than is really necessary is strong, and it is a strategy often pursued. However, excessive sampling frequencies leads to large data storage and processing requirements that could overtax the computer system.

The concepts behind sampling and the sampling theorem can also be described in the time domain. Consider a single sinusoid as an example waveform. Since all waveforms can be broken into sinusoids, the sampling constraints on a single sinusoid can be extended to cover any general waveform. In the time domain, Shannon's sampling theorem states that a sinusoid can be accurately reconstructed as long as two or more (evenly spaced) samples are taken over its period. This is equivalent to saying that f_s must be greater than 2 $f_{sinusoid}$. Figure 4.5 shows a sine wave (solid line) defined by two samples per cycle, (black circles). The Shannon sampling theorem states that no sinusoids of a lower frequency can pass through these two points: these two frequencies uniquely define one sine wave that is less than $f_s/2$. But there are many higher frequency sine waves that can pass cleanly through these two points: two examples of such higher frequency sine waves are shown in Figure 4.5. In fact there is an infinite number of higher frequency sine waves that can pass though the two points: all the higher harmonics, for example. These and others give rise to the added points in the sample spectrum, as shown in Figure 4.1

Looking at this from the perspective of the sampled signal, frequencies generated by aliasing are illustrated in Figure 4.6. A 5-Hz sine wave is sampled by 7 samples during the 1 sec shown, giving an $f_s/2 = 3.5$-Hz. A 2-Hz sine wave (dotted line) also passes through the points as predicted by aliasing, since the 5-Hz sine is 1.5-Hz above the 3.5-Hz Nyquist frequency and the 2-Hz sine is 1.5-Hz below the Nyquist frequency.

Aliasing can also be observed in images. Figure 4.7 shows two image pairs: the left images are correctly sampled and the right images are undersampled. The upper images are of a sine wave that increases in spatial frequency going from left to right. This is the image version of a "chirp" signal that increases in frequency over time giving the appearance of a vertical strip that becomes narrower going from left to right. The left image

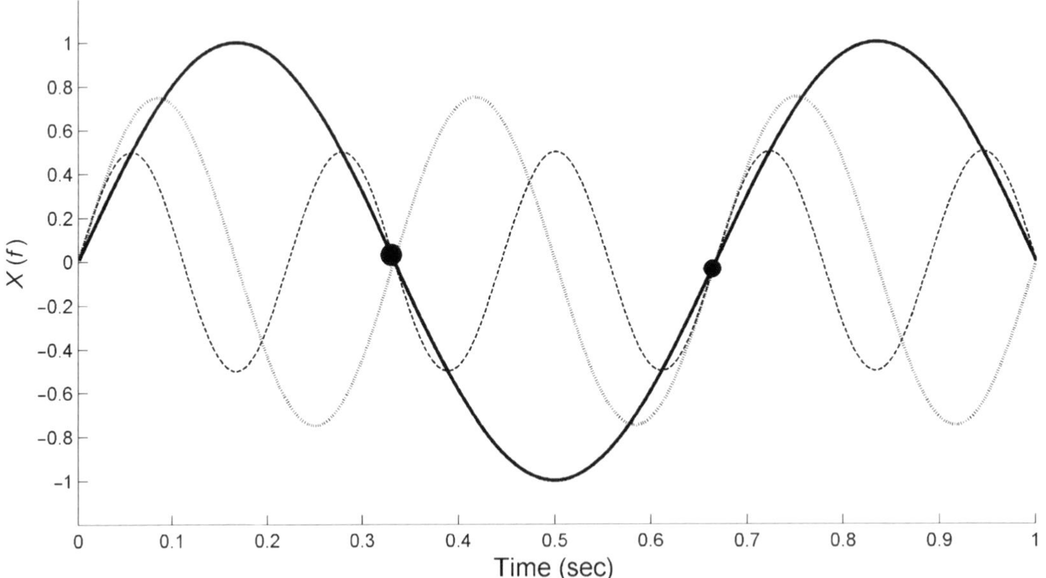

FIGURE 4.5 A sine wave (solid line) is sampled at two locations (black circles) within one period. The time domain interpretation of Shannon's sampling theorem states that no sine wave of a lower frequency can pass though these two points and this lowest frequency sine wave is uniquely defined. However, an infinite number of higher frequency sine waves can pass though those two points (all the harmonics for example) and two of these are shown. These higher frequency sine wave contributes the additional points shown on the spectrum of Figure 4.1 but can be eliminated by filtering.

shows a smooth progression of increasingly narrow strips, but the undersampled image shows a disruption in the progression as wider strips (i.e., lower frequencies) are created due to aliasing. The lower images are MR images of the brain and the undersampled image shows jagged diagonals and a type of moiré pattern that is characteristic of undersampled images.

4.1.2 Amplitude Slicing—Quantization

By selecting an appropriate sampling frequency, it is possible to circumvent problems associated with time slicing but, as mentioned in Chapter 1, the digitization process also requires that the data be sliced in amplitude as well as in time. Amplitude resolution is given in terms of the number of bits in the binary output with the assumption that the least significant bit (LSB) in the output is accurate (which may not always be true). Typical analog-to-digital converters (ADCs) feature 8-, 12-, and 16-bit outputs with 12 bits presenting a good compromise between conversion resolution and cost. In fact, most biological signals do not have sufficient signal-to-noise ratio to justify a higher resolution; you are simply obtaining a more accurate conversion of the noise.

The number of bits used for conversion sets an upper limit on the resolution, and determines the quantization error, as shown in Figure 4.8. The more bits used to represent the signal, the finer the resolution of the digitized signal and the smaller the quantization error. The quantization error is the difference between the original continuous signal value and the digital representation of that level after sampling. This error can be thought of as some sort of noise superimposed on the original signal. If a sufficient number of quantization levels exist (say $N > 64$, equivalent to 7 bits), the distortion produced by quantization

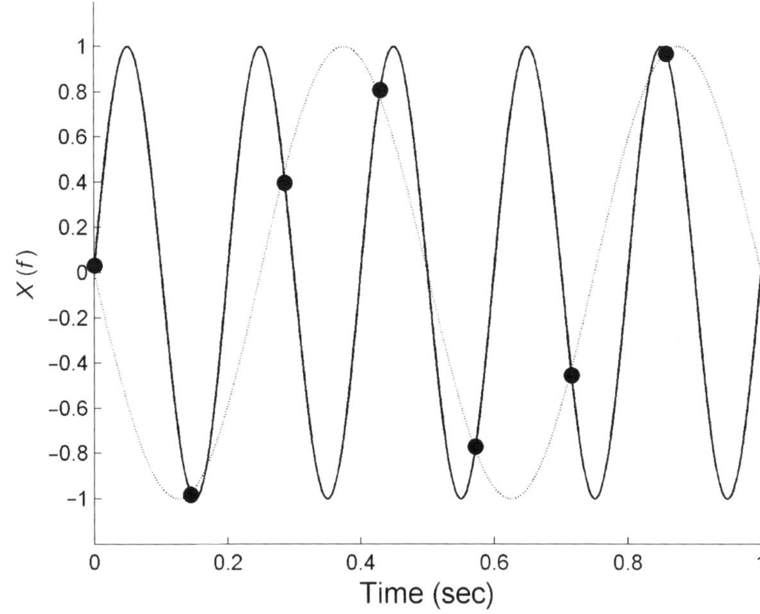

FIGURE 4.6 A 5-Hz sine wave (solid line) is sampled at 7-Hz (7 samples over a 1-sec period). The 7 samples also fit a 2-Hz sine wave, as is predicted by aliasing.

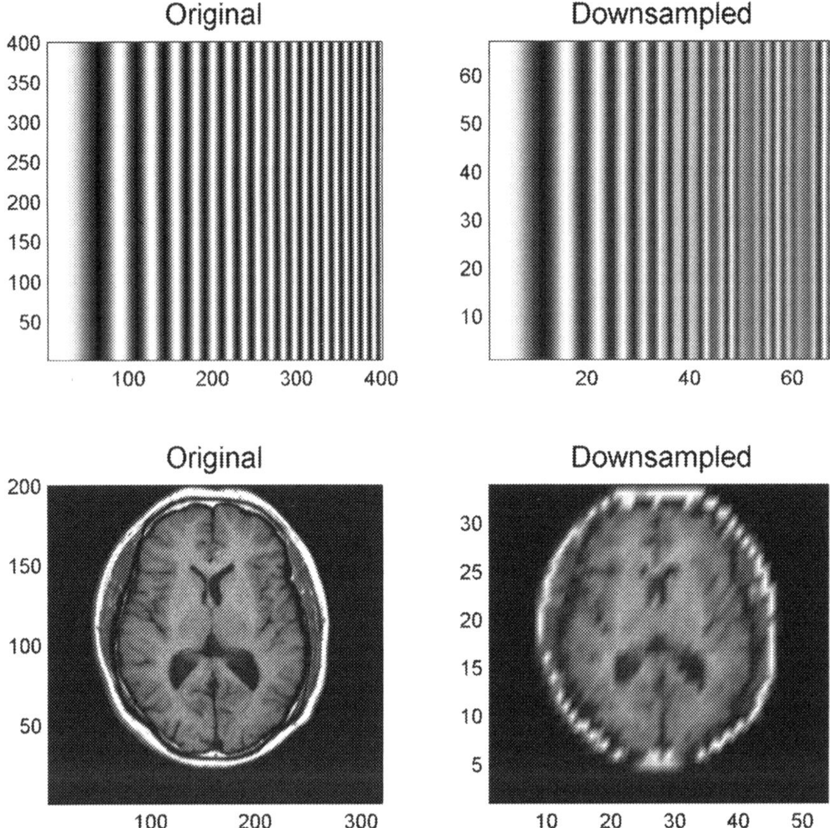

FIGURE 4.7 Two images that are correctly sampled (left side) and undersampled (right side). The upper pattern is sinusoidal with increasing spatial frequencies going from left to right. The undersampled image shows additional sinusoidal frequencies folded into the original pattern due to aliasing. The lower image is an MR image of the brain and the undersampled image has jagged diagonals and a moiré pattern, both characteristics of undersampled images.

error may be modeled as additive independent white noise with zero mean and a variance determined by the quantization step size, q. As described in Example 1.1, the quantization step size, q (V_{LSB} in Example 1.1), is the maximum voltage the ADC can convert divided by the number of quantization levels, which is $2^N - 1$; hence, $q = V_{MAX}/2^N - 1$. The variance or mean square error can be determined using the expectation function from basic statistics:

$$\sigma^2 = \overline{\sigma^2} = \int_{-\infty}^{\infty} e^2 PDF(e)de \tag{4.2}$$

where $PDF(e)$ is the uniform probability density function and e is the error voltage (the bottom trace of Figure 4.5). The $PDF(e)$ for a uniform probability distribution that ranges

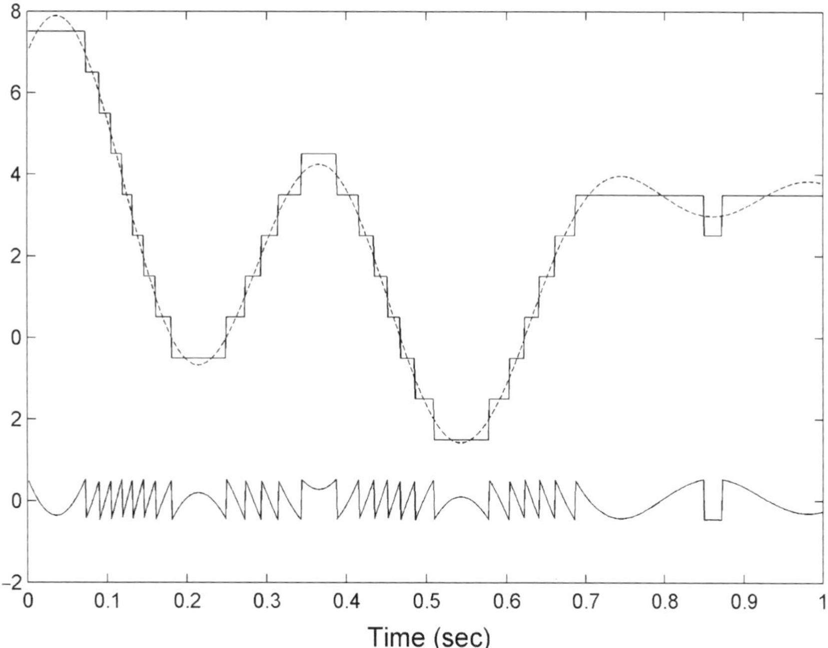

FIGURE 4.8 Quantization (amplitude slicing) of a continuous waveform. The lower trace shows the error between the quantized signal and the input.

between $-q/2$ to $+q/2$ is simply $1/q$. Substituting $1/q$ for $PDF(e)$ in Equation 4.2:

$$\sigma^2 = \int_{-q/2}^{q/2} e^2\left(\frac{1}{q}\right)de = \frac{(1/q)e^3}{3}\Bigg|_{-q/2}^{q/2} = \frac{q^2}{12} = \frac{V_{MAX}^2}{12(2^N-1)^2} \tag{4.3}$$

where V_{MAX} is the maximum voltage the ADC can convert and N is the number of bits out of the ADC. Assuming a uniform distribution for the quantization noise, the RMS value of the noise would be approximately equal to the standard deviation, σ, as shown in Section 1.4.1 of Chapter 1.

EXAMPLE 4.1

What is the equivalent quantization noise (RMS) of a 12-bit analog-to-digital converter that has an input voltage range of ± 5.0 volts? Assume the converter is perfectly accurate.

Solution: Apply Equation 4.3. Note that V_{max} would be 10 volts since the range of the ADC is from -5 to $+5$ volts

$$\sigma^2 = \frac{V_{MAX}^2}{12(2^N-1)^2} = \frac{10^2}{12(2^{12}-1)^2} = \frac{100}{12(4095)^2} = 4.97 \times 10^{-7}$$

$$V_{RMS} \cong \sigma = \sqrt{4.974 \times 10^{-7}} = 0.705 \text{ mV}$$

4.1.3 Data Length—Truncation

Given the finite memory capabilities of computers, it is usually necessary to digitize a shortened version of the signal. For example, only a small segment of a subject's ongoing EEG signal can be captured for analysis on the computer. The length of this shortened data segment is determined during the data acquisition process and is usually a compromise between the desire to acquire a large sample of the signal and the need to limit computer memory usage. Taking just a segment of a much longer signal involves "truncation" of the signal, and this has an influence on the spectral resolution of the stored signal. As described next, both the number of data points and the manner in which the signal is truncated will alter the spectral resolution.

4.1.3.1 Data Length and Spectral Resolution

The length of the data segment determines the apparent period of the data and hence the frequency range of the discrete Fourier transform (DFT). Recall that the frequency components obtained by the Fourier series analysis depend on the period, specifically:

$$f = \frac{m}{T} = mf_1 \tag{4.4}$$

where m is the harmonic number and f_1 is the fundamental frequency, which also equals $1/T$. For a discrete signal where N is the total number of points in the digitized signal, the equivalent signal period is the total number of points, N, multiplied by the sample interval, T_s, or divided by the sample frequency, f_s:

$$T_{effective} = \frac{N}{f_s} \tag{4.5}$$

and the equivalent frequency, f, of a given harmonic number, m, can be written as:

$$f = \frac{m}{T_{effective}} = \frac{m}{N/f_s} = \frac{mf_s}{N} \tag{4.6}$$

As noted previously, this equation is often used in MATLAB programs to generate the horizontal axis of a frequency plot (i.e., `f = (0:N-1)*fs/(N-1)`), where `N` is the length of the signal. (See Example 3.11.) The last data point from the DFT has a frequency value of f_s since:

$$f_{\text{Last}} = f|_{m=N} = \frac{Nf_s}{N} = f_s \tag{4.7}$$

The frequency resolution of a spectral plot is the difference in frequency between harmonics, which according to Equation 4.6 is just f_s/N:

$$f_{\text{Resolution}} = \frac{f_s}{N} \tag{4.8}$$

Just as with the continuous Fourier transform, frequency resolution depends on the period, that is, the time length of the data. For the DFT, the resolution is equal to f_s/N, Equation 4.8. For a given sampling frequency, the larger the number of samples in the signal, N, the smaller the frequency increment between successive DFT data points: the more points sampled, the higher the spectral resolution.

Once the data have been acquired, it would seem that the number of points representing the data, N, is fixed, but there is a trick that can be used to increase the data length *post hoc*. We can increase N simply by tacking on constant values, usually zeros. This may sound like cheating, but it is justified by the underlying assumption that the signal is zero outside of the data segment in the computer. In signal processing, adding zeros to extend the period of a signal is known as *zero-padding*. (Other, more complicated, "padding" techniques can be used, such as extending the last values of the signal on either end, but zero-padding is by far the most common strategy for extending data.)

Zero-padding gives the appearance of a spectrum with higher resolution. Example 4.2 and Figure 4.9 show what appear to be advantages of extending the period by adding zeros to the data segment. The signal is a triangle wave with an actual period of 1 sec but

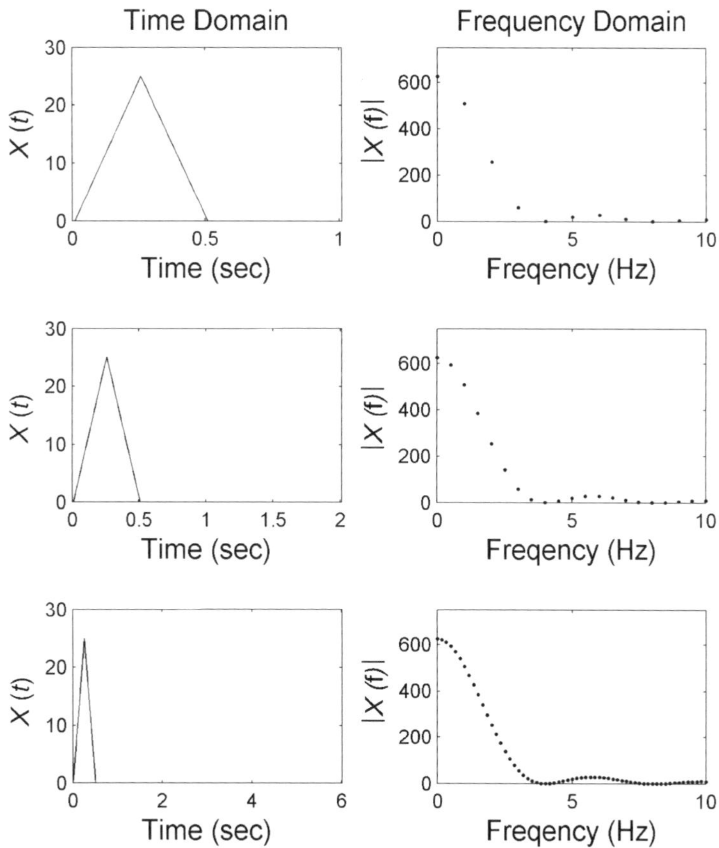

FIGURE 4.9 A waveform having a period of 0.5 sec (upper left) and its associated frequency spectrum (upper right). Extending the period to 2 and 6 seconds (middle and lower plots) by adding zeros decreases the spacing between frequency points, producing a smoother-looking frequency curve.

which has been extended to 2 and 6 seconds with zero-padding. Zero-padding the data seems to improve resolution, as the more closely-spaced frequency points show more of the spectrum's details. In fact, extending the period with zeros does *not* increase the information in the signal, and the resolution of the signal is really not any better. Zero-padding provides an interpolation of the points in the unpadded signal: it fills in the gaps of the original spectrum using an estimation process. Overstating the value of zero-padding is a common mistake of practicing engineers. Zero-padding does not increase the resolution of the spectrum, only the apparent resolution. However, the interpolated spectrum will certainly look better when plotted.

EXAMPLE 4.2

Generate a 1 sec wave with a half second triangle wave assuming a sample frequency of 100-Hz, so $N = 100$ points. Calculate and plot the magnitude spectrum. Zero-pad the signal so the period is extended to 2 and 6 seconds and recalculate and plot the magnitude spectrum. Limit the spectral plot to a range of 0 to 20-Hz.

Solution: First generate the 0.5-sec (50-point) triangle waveform with an additional 50 zeros to make a 1 sec period. Since $f_s = 100$-Hz, the signal should be padded by 100 and 500 additional points to extend the 1.0-sec signal to 2 and 6 sec. Calculate the spectrum using `fft`. Use a loop to calculate and plot the spectra of the 3 signals.

```
% Example 4.2
%
clear all, close all;
fs = 100;                                % Sampling frequencies
N1 = [0 100 500];                        % N values for padding
x = [ (0:25) (24:-1:0) zeros(1,50) ];   % Generate triangle signal, 100 pts long
for k = 1:3
   y = [x zeros(1,N1(k))];               % Zero-pad signal with 0, 100, and 500 zeros
   N = length(y);                        % Get data length
   t = (1:N)/fs;                         % Construct time vector for plotting
   f = (0:N-1)*fs/(N-1);                 % Construct frequency vector for plotting
   subplot(3,2,k*2-1);
   plot(t,y,'k');                        % Plot the signal
    ...... Labels and titles.........
   subplot(3,2,k*2);
   Y = abs(fft(y));                      % Calculate the magnitude spectrum
   plot(f, Y,'.k');                      % Plot spectrum using points
     ......Labels and titles.........
end
```

Results: The time and magnitude spectrum plots are shown in Figure 4.6. The spectral plots all have the same shape, but the points are more closely spaced with the zero-padded data. As shown in Figure 4.6, simply adding zeros to the original signal produces a better-looking curve which explains the popularity of zero-padding (even if no additional information is produced). MATLAB makes zero-padding during spectral analysis easy, as the second argument of the `fft` routine (i.e., `fft(x,N)`) will add zeros to the original signal to produce a waveform of length `N`. The second argument can also be used to shorten the data set, but this usage is less common.

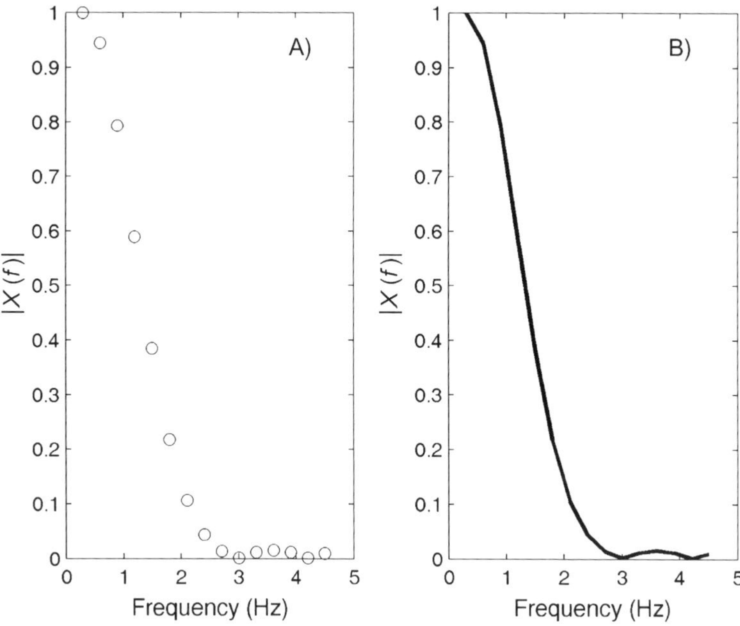

FIGURE 4.10 A) The DFT assumes the signal is periodic and computes a spectrum consisting of discrete points. B) Making the assumption that the signal really is aperiodic, having an infinite period, allows you to connect the discrete points to form a smooth curve. This assumption is commonly made when plotting frequency spectra.

The fast Fourier transform method operates under the assumption that the signal is periodic with a period of $T = N/f_s$ (Equation 4.5) and the frequency spectrum produced is a set of numbers spaced f_s/N apart (Equation 4.8). However, if we make the assumption that our signal is actually aperiodic (i.e., that $T \rightarrow \infty$), then the spacing between points goes to zero and the points become a continuous line. This is the assumption that is commonly made when we plot the spectrum not as a series of points, but as a smooth curve (Figure 4.10). Although this is commonly done when plotting spectra, you should be aware of the underlying truth: an implicit assumption is being made that the signal is actually aperiodic, and that the calculated spectrum is really a series of discrete points that have been joined together because of this assumption.

4.1.4 Data Truncation—Window Functions

The digitized waveform must necessarily be truncated to the length of the memory storage array, a process described as *windowing*. The windowing process can be thought of as multiplying the data by some window shape. If the waveform is simply truncated and no further shaping is performed on the resultant digitized waveform, then the window shape is rectangular by default (Figure 4.11). Note that a rectangle will usually produce abrupt changes or discontinuities at the two endpoints.

Other window shapes can be imposed on the data by multiplying the truncated waveform by the desired shape. There are many different window shapes and all have the effect of tapering the two ends of the data set toward zero. An example of a tapering window is the Hamming window shown in Figure 4.12. This window is often used by MATLAB as a default window in routines that generate short data sets. The Hamming

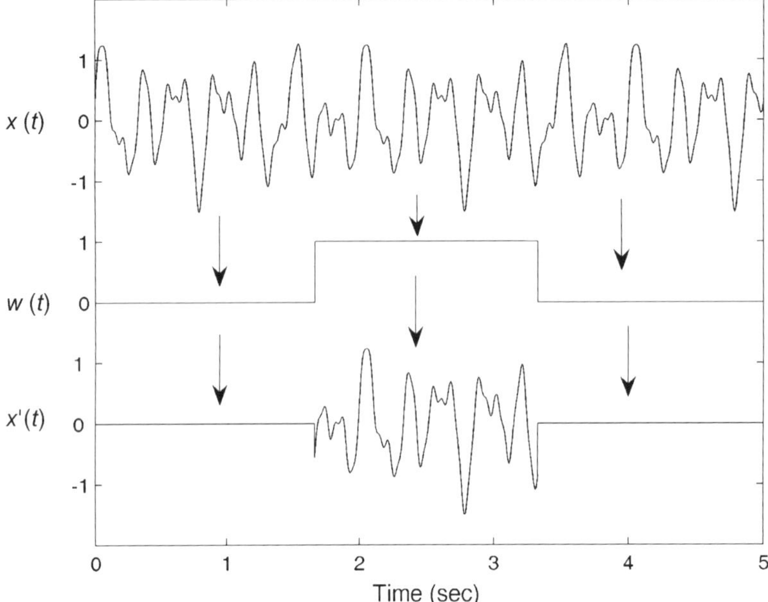

FIGURE 4.11 Data truncation or shortening can be thought of mathematically as multiplying the original data, $x(t)$ (upper curve), by a window function, $w(t)$ (middle curve), to produce the truncated data that are in the computer, $x'(t)$ (lower curve). The window function has a value of 1.0 over the length of the truncated data and 0.0 everywhere else.

window has the shape of a raised half sine wave, and when the truncated signal is multiplied by this function the endpoint discontinuities are reduced. The equation for the Hamming function is:

$$w[n] = 0.5 - 0.46\cos\left(\frac{2\pi n}{N}\right) \tag{4.9}$$

where $w[n]$ is the Hamming window function and n ranges between 1 and $N+1$ the window length.

Multiplying your signal by a function like that shown in Figure 4.12 (middle curve) may seem a bit extreme, and there is some controversy about the use of nonrectangular windows; however, the influence of windows such as the Hamming window is generally quite subtle. Except for very short data segments, a rectangular window (i.e., simple truncation) is usually the best option. Applying the Hamming window to the data will result in a spectrum with slightly better noise immunity, but overall frequency resolution is reduced. Figure 4.13 shows the magnitude spectra of two closely-spaced sinusoids with noise that were obtained using the FFT after application of two different windows: a rectangular window (i.e., simple truncation) and a Hamming window. The spectra are really quite similar, although the spectrum produced after applying a Hamming window to the data shows a slight loss in resolution as the two frequency peaks are not as sharp. The peaks due to background noise are also somewhat reduced by the Hamming window.

If the data set is fairly long (perhaps 256 points or more), the benefits of a nonrectangular window are slight. Figure 4.14 shows the spectra obtained with and without the Hamming window to be nearly the same except for a scale difference produced by

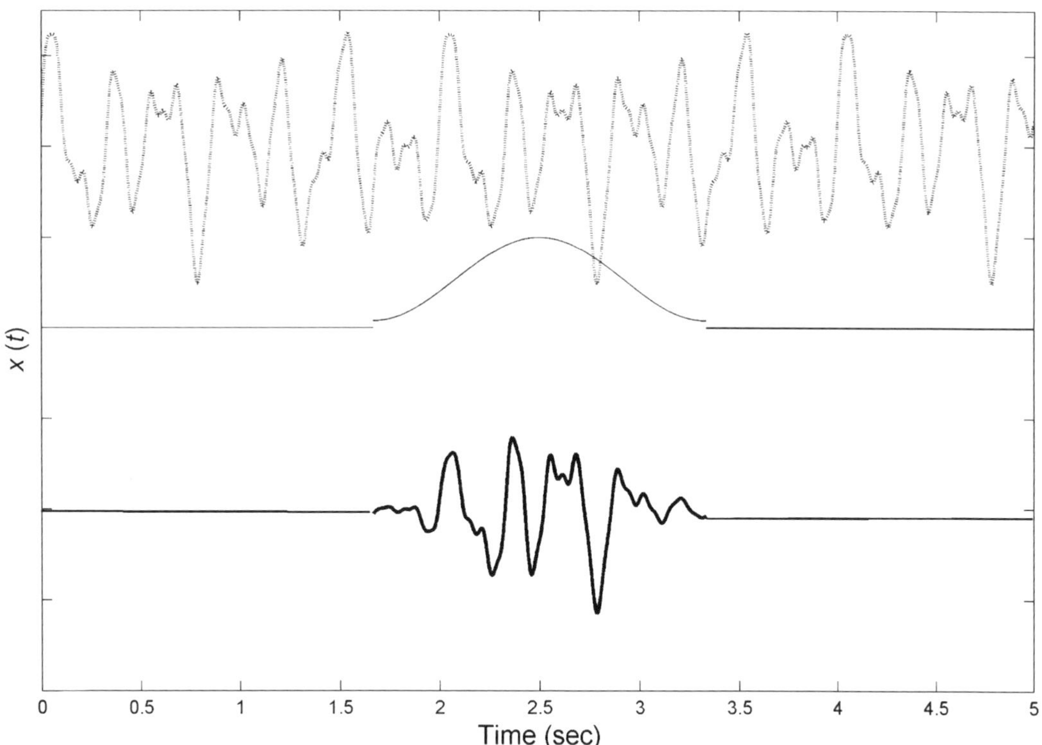

FIGURE 4.12 The application of a Hamming window to truncate the data shown in Figure 4.11. This window is similar to a half sine function and, when multiplied with the data set, tapers the two ends of the signal toward zero. The application of such a tapering window decreases the resultant spectral resolution but also reduces the influence of noise on the spectrum.

the Hamming window. Semmlow (2009) provides a thorough discussion of the influence of various nonrectangular window functions.

4.2 POWER SPECTRUM

The power spectrum is commonly defined as the Fourier transform of the autocorrelation function. In continuous and discrete notations the power spectrum equation becomes:

$$PS(f) = \frac{1}{T}\int_0^T r_{xx}(t)e^{-j2\pi mf_1 t}\,dt \qquad m = 0, 1, 2, 3\ldots \tag{4.10}$$

$$PS[m] = \sum_{n=1}^{N} r_{xx}[n]e^{-\frac{j2\pi mn}{N}} \qquad m = 0, 1, 2, 3\ldots N \tag{4.11}$$

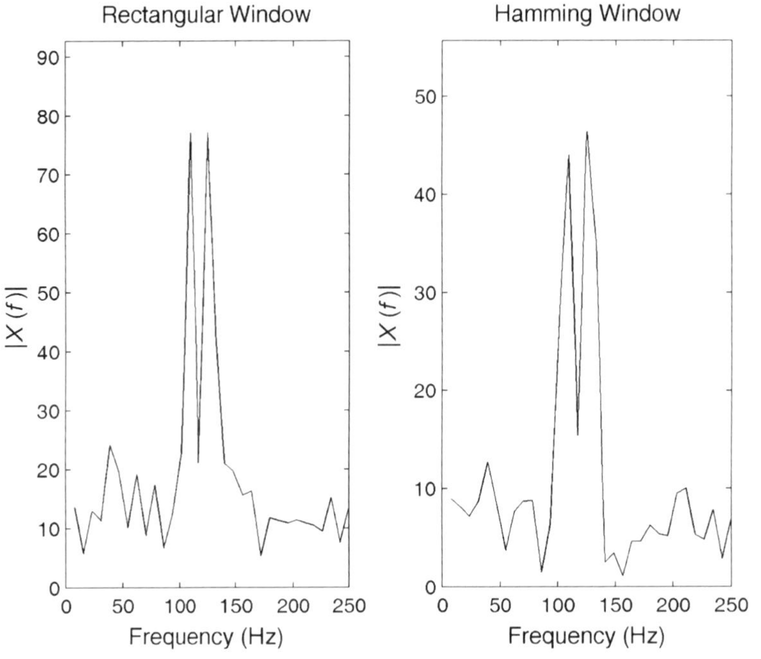

FIGURE 4.13 The spectrum from a short signal ($N = 128$) containing two closely spaced sinusoids (100 and 120-Hz). A rectangular and a Hamming window have been applied to the signal to produce the two spectra. The spectrum produced after application of the Hamming window shows a slight loss of spectral resolution as the two peaks are not as sharp; however, background spectral peaks due to noise have been somewhat reduced.

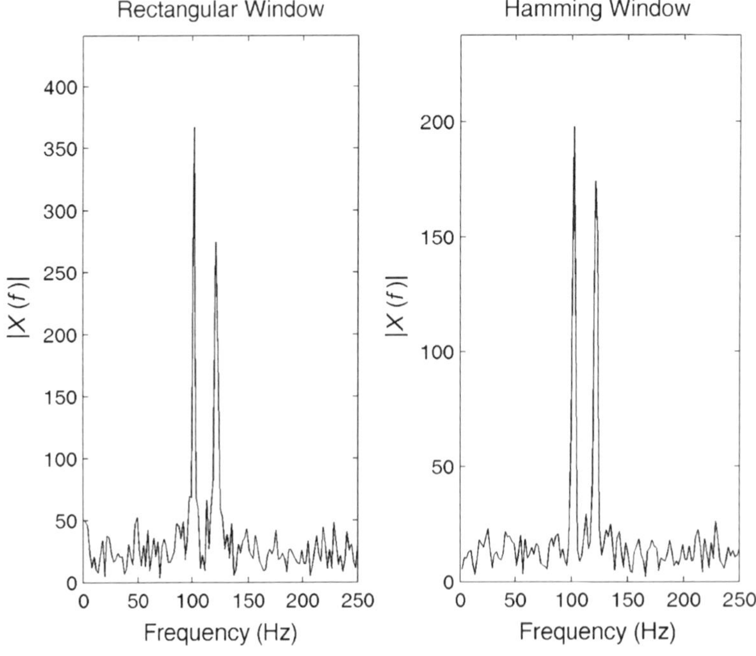

FIGURE 4.14 A spectrum from a relatively long signal ($N = 512$) containing two closely spaced sinusoids (100 and 120-Hz). Rectangular and Hamming windows have been applied to the signal to produce the two spectra. When the data set is long, the window function has little effect on the resultant spectrum and you may as well use a rectangular window; i.e., simple truncation.

where $r_{xx}(t)$ and $r_{xx}[n]$ are autocorrelation functions, as described in Chapter 2. Since the autocorrelation function has even symmetry, the sine terms of the Fourier series will all be zero (see Table 3.1), and the two equations can be simplified to include only real cosine terms:

$$PS(f) = \frac{1}{T}\int_0^T r_{xx}(t)\cos(2\pi mft)\,dt \qquad m = 0, 1, 2, 3, \ldots \tag{4.12}$$

$$PS[m] = \sum_{n=0}^{N-1} r_{xx}[n]\cos\left(\frac{2\pi nm}{N}\right) \qquad m = 0, 1, 2, 3, \ldots N \tag{4.13}$$

Equations 4.12 and 4.13 are sometimes referred to as *cosine transforms*.

A more popular method for evaluating the power spectrum is the *direct approach*. The direct approach is motivated by the fact that the energy contained in an analog signal, $x(t)$, is related to the magnitude of the signal squared integrated over time:

$$E = \int_{-\infty}^{\infty} |x(t)|^2\,dt \tag{4.14}$$

By an extension of a theorem attributed to Parseval it can be shown that:

$$\int_{-\infty}^{\infty} |x(t)|^2\,dt = \int_{-\infty}^{\infty} |X(f)|^2\,df \tag{4.15}$$

Hence $|X(f)|^2$ equals the energy density function over frequency, also referred to as the *energy spectral density*, the *power spectral density* (PSD), or simply the *power spectrum* (PS). In the "direct approach," the power spectrum is calculated as the magnitude squared of the Fourier transform (or Fourier series) of the waveform of interest:

$$PS(f) = |X(f)|^2 \tag{4.16}$$

This direct approach of Equation 4.16 has displaced the cosine transform for determining the power spectrum because of the efficiency of the fast Fourier transform. (A variation of this approach is still used in some advanced signal processing techniques involving time and frequency transformation.) Problem 7 compares the power spectrum obtained using the direct approach of Equation 4.16 with the traditional cosine transform method represented by Equation 4.11 and, if done correctly, shows them to be identical.

Unlike the Fourier transform, the power spectrum does not contain phase information, so the power spectrum is not an invertible transformation: It is not possible to reconstruct the signal from the power spectrum. However, the power spectrum has a wider range of applicability and can be defined for some signals that do not have a meaningful Fourier transform (such as those resulting from random processes). Since the power spectrum does not contain phase information, it is applied in situations where phase is not considered useful or to data that contain a lot of noise, since phase information is easily corrupted by noise.

An example of the descriptive properties of the power spectrum is given using the heart rate data shown in Figure 2.23. The heart rates show major differences in the mean and standard deviation of the rate between meditative and normal states. Applying the autocovariance to the meditative heart rate data (Example 2.19 and Problem 22 in Chapter 2) indicates a possible repetitive structure for the variation in heart rate during the meditative average rate and variance. The next example uses the power spectrum to search for structure in the frequency characteristics of both normal and meditative heart rate data.

EXAMPLE 4.3

Determine and plot the power spectra of heart rate variability data recorded during both normal and meditative states.

Solution: The power spectrum can be evaluated through the Fourier transform using the direct method given in Equation 4.16. However, the heart rate data should first be converted to evenly sampled time data, and this is a bit tricky. The data set obtained by a download from the PhysioNet data base provides the heart rate at unevenly spaced times, where the sample times are provided as a second vector. These interval data need to be rearranged into evenly spaced time positions. This process, known as *resampling*, will be done through interpolation using MATLAB's `interp1` routine. This routine takes in the unevenly spaced $x-y$ pairs as two vectors along with a vector containing the desired evenly spaced x_i values. The routine then uses linear interpolation (other options are possible) to approximate the y_i values that match the evenly spaced x values. Details can be found in the MATLAB help file for `interp1`.

In the program below, the uneven $x-y$ pairs for the normal conditions are in vectors `t_pre` and `hr_pre`, respectively, both in the MATLAB file `Hr_pre`. (For meditative conditions, the vectors are named `t_med` and `hr_med` and are in file `Hr_med`.) The evenly spaced time vector is `xi`, and the resampled heart rate values are in vector `yi`.

```
% Example 4.3
% Frequency analysis of heart rate data in the normal and meditative state
%
fs = 100;                                        % Sample frequency (100 Hz)
ts = 1/fs;                                       % Sample interval
load Hr_pre;                                     % Load normal and meditative data
%
% Convert to evenly spaced time data using interpolation; i.e., resampling
% First generate and evenly space time vectors having one second
% intervals and extending over the time range of the data
%
xi = (ceil(t_pre(1)):ts:floor(t_pre(end)));      % Evenly spaced time vector
yi = interp1(t_pre,hr_pre,xi');                  % Interpolate
yi = yi - mean(yi);                              % Remove average
N2 = round(length(yi)/2);
f = (1:N2)*fs/N2);                               % Vector for plotting
%
% Now determine the Power Spectrum
YI = abs((fft(yi)).^2);                          % Direct approach (Eq. 4.16)
subplot(1,2,1);
plot(f,YI(2:N2 + 1,'k');                         % Plot spectrum, but not DC value
axis([0 .15 0 max(YI)*1.25]);                    % Limit frequency axis to 0.15 Hz
  ......label and axis......
%
% Repeat for meditative data
```

Analysis: To convert the heart rate data to a sequence of evenly spaced points in time, a time vector, `xi`, is first created that increases in increments of 0.01 second ($1/f_s = 1/100$) between the lowest and highest values of time (rounded appropriately) in the original data. A 100-Hz resampling frequency was chosen because this is common in heart rate variability studies that use certain nonlinear methods, but in this example a wide range of resampling frequencies give the same result. Evenly spaced time data, `yi`, were generated using the MATLAB interpolation routine `interp1`. The example requested the power spectrum of heart rate *variability*, not heart rate per se; in other words, the changes in beat-to-beat rate, not the beat-to-beat rate itself. To get this change, we simply subtract out the average heart rate before evaluating the power spectrum.

After interpolation and removal of the mean heart rate, the power spectrum is determined using `fft` then taking the square of the magnitude component. The frequency plots in Figure 4.15 are limited to a range between 0.0 to 0.15-Hz since this is where most of the spectral energy is to be found.

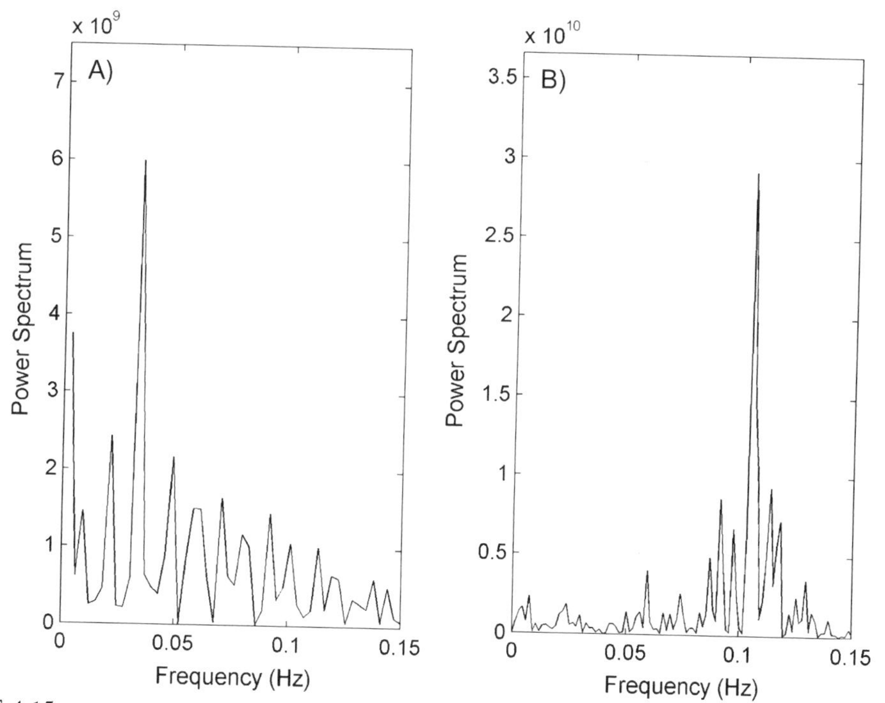

FIGURE 4.15 A) Power spectrum of heart rate variability under normal conditions. The power decreases with frequency. B) Power spectrum of heart rate variability during meditation. Strong peaks in power around 0.12-Hz are seen indicating that much of the variation in heart rate is organized around these frequencies. Note the larger scale of the meditative power spectrum.

Results: The power spectrum of normal heart rate variability is low and decreases with frequency, showing little energy above 0.1-Hz (Figure 4.15A). The meditative state (Figure 4.15B), shows large peaks at around 0.1 to 0.12-Hz, indicating that some resonant process is active at these frequencies, corresponding to a timeframe of around 10 seconds. Speculation as to the mechanism behind this heart rate rhythm is left to the reader.

4.3 SPECTRAL AVERAGING

While the power spectrum is usually calculated using the entire waveform, it can also be applied to isolated segments of the data. The power spectra determined from each of these segments can then be averaged to produce a spectrum that better represents the broadband, or "global," features of the spectrum. This approach is popular when the available waveform is only a sample of a longer signal. In such situations, spectral analysis is necessarily an estimation process, and averaging improves the statistical properties of the result. When the power spectrum is based on a direct application of the Fourier transform followed by averaging, it is referred to as an *average periodogram*.

Averaging is usually achieved by dividing the waveform into a number of segments, possibly overlapping, and evaluating the power spectrum on each of these segments (Figure 4.16). The final spectrum is constructed from the *ensemble average* of the power spectra obtained from each segment. Note that this averaging approach can only be applied to the power spectrum because the power spectrum, like the magnitude spectrum of the Fourier transform, is not sensitive to time translation (recall Problem 19 in Chapter 3). Applying this averaging technique to the standard Fourier transform would not make

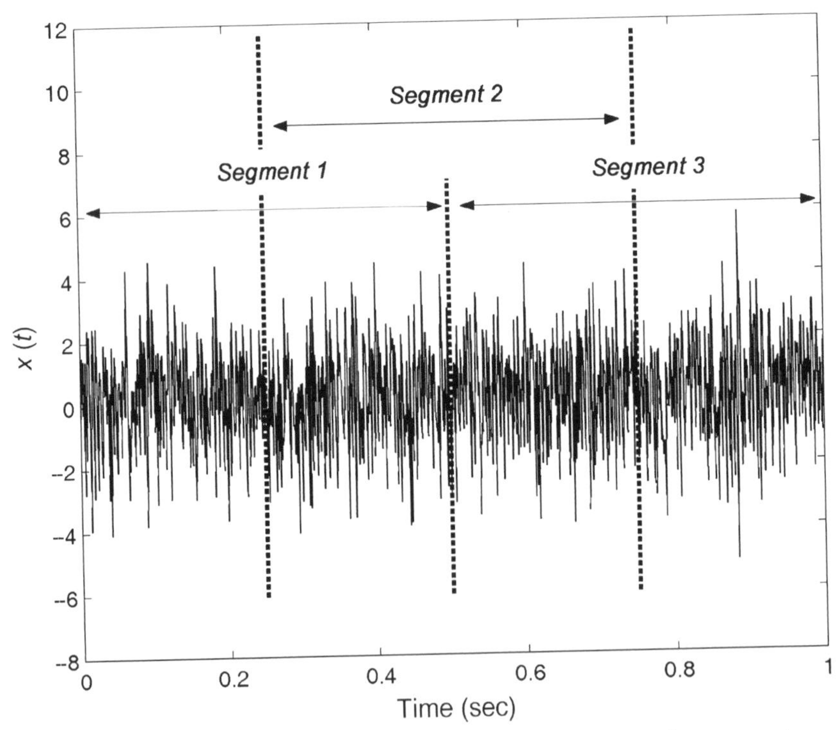

FIGURE 4.16 A waveform is divided into three segments with a 50% overlap between each segment. In the *Welch* method of spectral analysis, the power spectrum of each segment is computed separately and an average of the three transforms gives the final spectrum.

sense because the phase spectrum is sensitive to segment position. Averaging phases obtained for different time positions would be meaningless.

One of the most popular procedures to evaluate the average periodogram is attributed to Welch and is a modification of the segmentation scheme originally developed by Bartlett. In this approach, overlapping segments are used and a shaping window (i.e., a nonrectangular window) is sometimes applied to each segment. More heavily overlapped segments can be averaged for a given segment and data length. Averaged periodograms obtained from noisy data traditionally average spectra from half-overlapping segments: segments that overlap by 50%. Higher amounts of overlap have been recommended in applications when computing time is not a factor. Maximum overlap occurs when the windowed segment is shifted by only one sample.

Segmenting the data reduces the number of data samples being analyzed in each segment: each segment contains the original N divided by the number of segments. Equation 4.8 states that frequency resolution is proportional to f_s/N where N is now the number of samples in a segment. So, averaging produces a trade-off between spectral resolution, which is reduced by averaging, and statistical reliability. Choosing a short segment length (a small N) will provide more segments for averaging and improve the reliability of the spectral estimate, but will also decrease frequency resolution. This trade-off is explored in the next example.

EXAMPLE 4.4

Evaluate the influence of averaging power spectra on a signal made up of a combination of broadband and narrowband signals along with added noise. The data can be found in file `broadband1.mat` which contains a white noise signal filtered to be between 0 and 300-Hz and two closely spaced sinusoids at 390 and 410-Hz.

Solution: Load the test data file `broadband1` containing the narrowband and broadband processes. First calculate and display the unaveraged power spectrum. Then apply power spectrum averaging using an averaging routine. Use a segment length of 128 points and the maximum overlap of 127 points. To implement averaging, write a routine called `welch` that takes in the data, segment size, and the number of overlapping points and produces the averaged power spectrum.

```
% Example 4.4 Investigation of the use of averaging to improve
% broadband spectral characteristics in the power spectrum.
%
load broadband1;                        % Load data (variable x)
fs = 1000;                              % Sampling frequency
nfft = 128;                             % Segment length
%
% First calculate the unaveraged spectrum using the direct method of Eq. 4.17
%
PS = abs((fft(x)).^2)/length(x);        % Calculate un-averaged PS
half_length = fix(length(PS)/2);        % Find data length/2 and eliminate redundant
                                        % pts
f = (1:half_length)* fs/(2*half_length); % Frequency vector for plotting
subplot(1,2,1)
plot(f,PS(1:half_length),'k');          % Plot unaveraged Power Spectrum
```

```
  ......labels and title......
%
[PS_avg,f]=welch(x,nfft, nfft-1,fs);      % Calculate periodogram, max. overlap
%
subplot(1,2,2)
plot(f,PS_avg,'k');                       % Plot periodogram
  ......labels and title......
```

The welch routine takes in the sampling frequency, an optional parameter which is used to generate a frequency vector useful for plotting. MATLAB power spectrum routines generally include this feature and we need to make our programs as good as MATLAB's. This program outputs only the nonredundant points of the power spectrum (i.e., up to $f_s/2$) and the frequency vector. Finally, it will check the number of arguments and use defaults if necessary (another common MATLAB feature). The program begins with a list and description of input and output parameters, then sets up defaults as needed. This is the text that will be printed if someone types in "help welch."

```
function [PS,f]=welch(x,nfft,noverlap,fs);
% Function to calculate averaged spectrum
% [ps,f]=welch(x,nfft,noverlap,fs);
% Output arguments
%     sp  spectrogram
%     f frequency vector for plotting
% Input arguments
%     x data
%     nfft window size
%     noverlap number of overlapping points in adjacent segments
%     fs sample frequency
%   Uses Hamming window
%
N=length(x);                              % Get data length
half_segment=round(nfft/2);               % Half segment length
if nargin < 4                             % Test inputs and set defaults
  fs=2*pi;                                % if needed. Default fs is pi
end
if nargin < 3 || isempty(noverlap) == 1
  noverlap=half_segment;                  % Set default overlap at 50%
end
%
% Defaults complete. The routine now calculates the appropriate number of points
%   to shift the window and the number of averages that can be done given
%   the data length (N), window size (nff) and overlap (noverlap).
f=(1:half_segment)* fs/(nfft);            % Calculate frequency vector
increment=nfft - noverlap;                % Calculate window shift
nu_avgs=round(N/increment);               % Determine the number of segments
%
% Now shift the segment window and calculate the PS using Eq. 4.16
for k=1:nu_avgs                           % Calculate spectra for each data point
  first_point=1+(k - 1) * increment;
  if (first_point+nfft -1) > N            % Check for possible overflow
```

```
      first_point = N - nfft + 1;            % Shift last segment to prevent overflow
    end
    data = x(first_point:first_point + nfft -1);
  % Get data segment
  % MATLAB routines would add a nonrectangular window here, the default being
  % a Hamming window. This is left as an exercise in Problem 13.
    if k == 1
      PS = abs((fft(data)).^2);              % Calculate power spectrum first time
    else
      PS = PS + abs((fft(data)).^2);         % Calculate power spectrum and add to
                                             average
    end
  end
  % Scale average and remove redundant points. Also do not include DC term
  PS = PS(2:half_segment + 1)/(nu_avgs*nfft/2);
```

Analysis: Spectral averaging is done in the `welch` routine, which first checks if the sampling frequency and desired overlap are specified. If not, the routine sets the sampling frequency to π and the overlap to a default value of 50% (i.e., half the segment length: `nfft/2`). Then a frequency vector, `f`, is generated from 1 to `fs/2` or π. (Setting the frequency vector from 1 to π is another common MATLAB default.) Next, the number of segments to be averaged is determined based on the segment size (i.e., `nfft`) and the overlap (`noverlap`). A loop is used to calculate the power spectrum using the direct method of Equation 4.16, and individual spectra are summed. Finally, the power spectrum is shortened to eliminate redundant points and the DC term, and then is normalized by both the number of points and the number of spectra in the sum to generate the average spectrum.

This example uses the `welch` routine to determine the averaged power spectrum, but also calculates the unaveraged spectrum. For the averaged spectrum or *periodogram*, a segment length of 128 (a power of 2) was chosen along with maximal overlap. In practice, the selection of segment length and averaging strategy is usually based on experimentation with the data.

Results: In the unaveraged power spectrum (Figure 4.17A), the two sinusoids at 390 and 410-Hz are clearly seen; however, the broadband signal is noisy and poorly defined. The periodogram produced from the segmented and averaged data in Figure 4.17B is much smoother, better reflecting the constant energy of white noise, but the loss in frequency resolution is apparent as the two high-frequency sinusoids are barely visible. This demonstrates one of those all-so-common engineering compromises. Spectral techniques that produce a good representation of "global" features such as broadband features are not good at resolving narrowband or "local" features such as sinusoids and vice versa.

EXAMPLE 4.5

Determine and plot the frequency characteristics of heart rate variability during both normal and meditative states using spectral averaging. Divide the time data into 8 segments and use a 50% overlap (the default). Plot using both a linear horizontal axis and an axis scaled in dB.

Solution: Construct the time data by resampling at 100-Hz, as was done in Example 4.3. Find the segment length by dividing the total length by 8 and round this length downward. Use the `welch` routine with the default 50% overlap to calculate the average power spectrum and plot.

```
% Example 4.5 Influence of averaging on heart rate data.
  ......Data loading and reorganization as in Example 4.3......
%
nfft = floor(length(yi)/8);              % Calculate segment length
[PS_avg,f] = welch(yi,nfft,[ ],fs);      % Calculate periodogram
subplot(1,2,1)
plot(f,PS_avg,'k');                      % Plot periodogram
  ......label and axis......
  ......Repeat for meditative data......
figure ;                                 % Replot in dB
subplot(1,2,1)
plot(f,20*log(PS_avg));                  % Take 20*log(power spectrum)
```

Results: The `welch` routine developed in the last example is used to calculate the average power spectrum. The sample segment length with a 50% overlap was chosen empirically as it produced smooth spectra without losing the peak characteristic of the meditative state.

The results in Figure 4.18 show much smoother spectra than those of Figure 4.15, but they also lose some of the detail. The power spectrum of heart rate variability taken under normal conditions now appears to decrease smoothly with frequency. It is common to plot

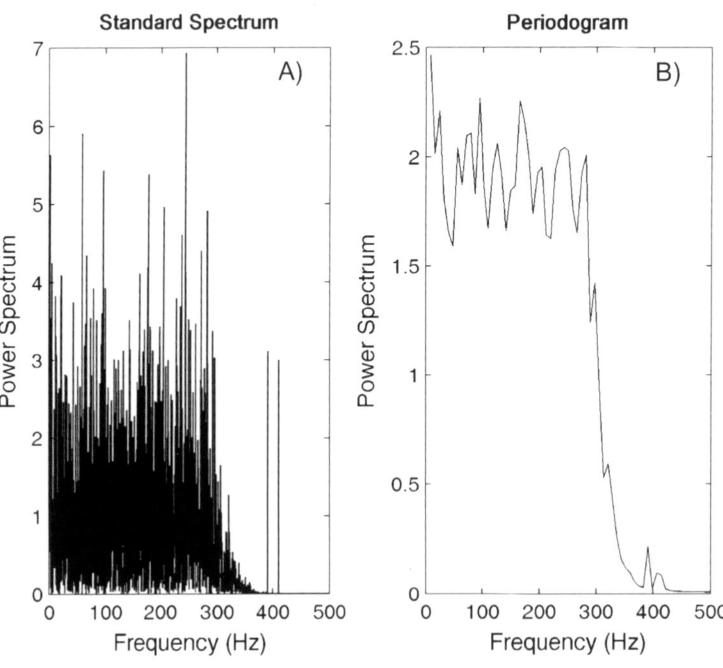

FIGURE 4.17 Power spectra obtained from a waveform consisting of a broadband signal having energy between 0 and 300-Hz and two sine waves at 390 and 410-Hz. The sine waves are also buried in white noise (SNR = −16 dB). A) The unaveraged spectrum clearly shows the 390- and 410-Hz components, but the features of the broadband signal are unclear. B) The averaged spectrum barely shows the two sinusoidal components, but produces a smoother estimate of the broadband spectrum which should be flat up to around 300-Hz. In the next figure, averaging is used to estimate the power spectrum of the normal and meditative heart rate variability data.

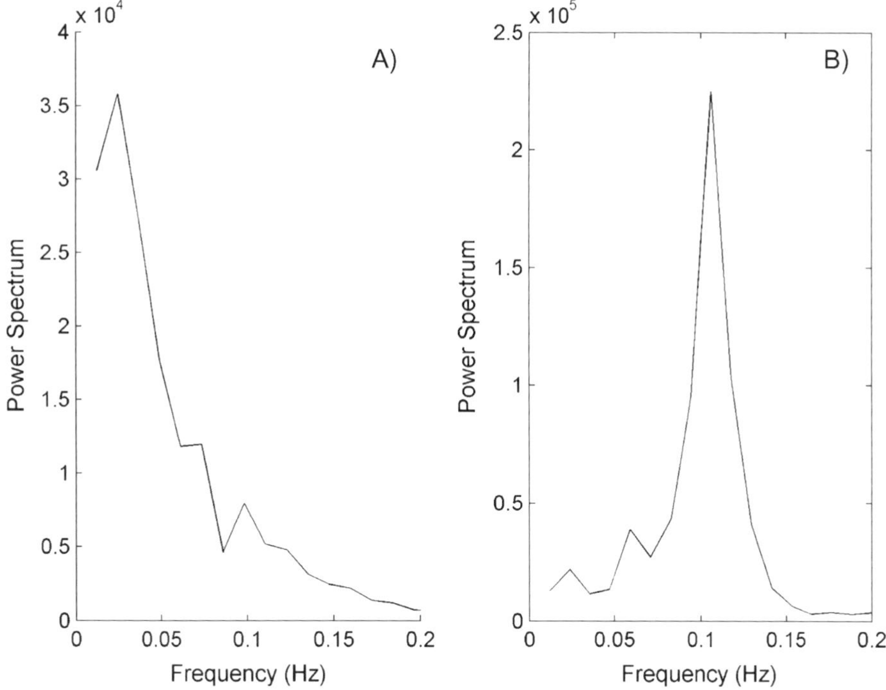

FIGURE 4.18 Power spectra taken from the same heart rate variability data used to determine the spectra in Figure 4.15, but constructed using an averaging process. The spectra produced by averaging are considerably smoother. A) Normal conditions. B) Meditative conditions.

power spectra in dB. Since we are dealing with power, we apply Equation 1.8 directly; i.e., 10 log(*PS*). We use 10 log instead of 20 log because we have a spectrum in power; the square of the magnitude spectrum has already been taken. Replotting in dB shows that the normal spectrum actually decreases linearly with frequency (Figure 4.19A). The principal feature of the dB plot of meditative data is the large peak at 0.12-Hz. Note that this peak is also present, although much reduced, in the normal data (Figure 4.19A). This suggests that meditation enhances a process that is present in the normal state as well, some neurological feedback process with an 8-sec delay ($T = 1/0.12$).

4.4 STATIONARITY AND TIME-FREQUENCY ANALYSIS

All Fourier analyses are done on a block of data so the result represents signal spectra over some finite period of time. An underlying and essential assumption is that the data have consistent statistical properties over the length of the data set being analyzed. Data that meet this criterion are known as *stationary* data, and the processes that produce such data are termed *stationary processes*. There is a formal mathematical definition of stationary based on consistency of the data's probability distribution function, but for our purposes,

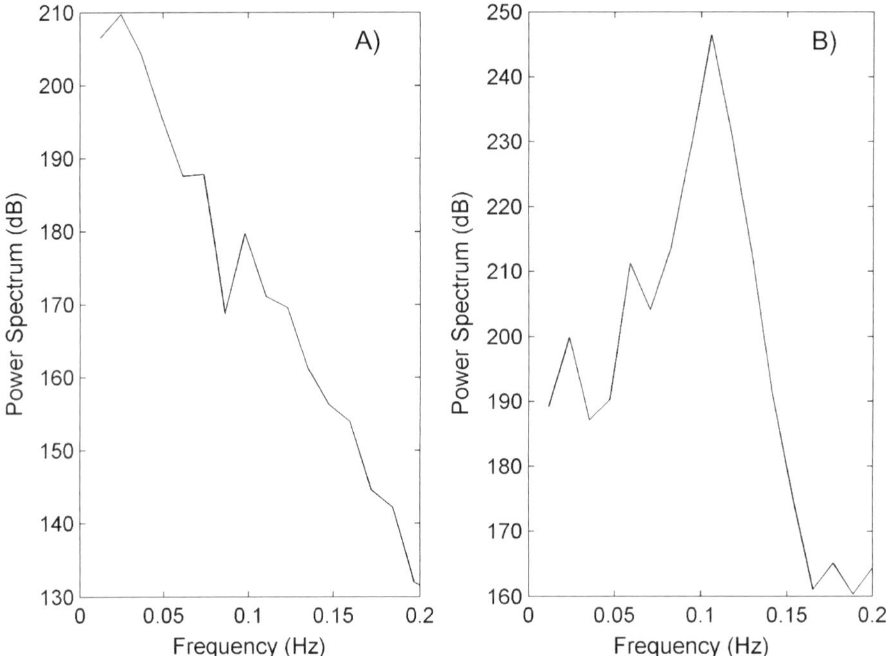

FIGURE 4.19 Normal and meditative spectra of heart rate variability data replotted in dB by taking 10 log of the power spectral curves in Figure 4.18. A) The spectrum obtained under normal conditions is seen to decrease linearly with frequency. B) The spectrum obtained under meditative conditions shows a large peak at 0.12-Hz. A reduced version of this peak is also seen in the normal data.

a stationary signal is one that does not change its mean, standard deviation, or autocorrelation function.

Unfortunately, most biological signals are not stationary, so in order to use Fourier analyses and many other signal processing techniques some additional data transformation is required. If the nonstationarity exists as a variation in mean, *detrending* techniques that subtract out the time-varying mean can be used. Sometimes just taking the difference between successive data points (i.e., the derivative) will eliminate the nonstationarity. However, most of the time we just try to limit the data to time periods short enough to be stationary (or assumed to be stationary). If the data set is long, we might break it up into shorter segments and analyze each set separately.

Breaking up a long data set into (we hope) stationary segments and performing a Fourier analysis on each segment is known as *time-frequency analysis*. Each spectrum is a function of frequency, but it is also a function of the time segment from which it was obtained. For example, if a 1-hour data set was divided into 60 1-min segments and a Fourier analysis performed on each segment, the resulting data would be 60 functions of frequency, each also a function of a different time period.

There are many different approaches to performing time-frequency analyses, but the one based on the Fourier series is the best understood and perhaps most common. Since it is based on dividing a data set into shorter segments, sometimes quite short, it is called the *short-term Fourier transform* approach. Since the data segments are usually short, a window function such as the Hamming window is usually applied to each segment before taking the Fourier transform. The next section provides an application of the short-term Fourier transform.

EXAMPLE 4.6

Load the respiratory signal found as variable `resp` in file `Resp_long.mat`. This signal is similar to one used in Example 3.11 except that it is 10 min long. The sampling frequency is 125-Hz. Find the power spectra for 1-minute segments that overlap by 30 sec (i.e., 50% resulting in 20 segments in total). Plot these spectra in 3 dimensions using MATLAB's `mesh` routine. Limit the frequency range to 2-Hz and do not plot the DC component.

Solution: Simply modify the `welch` routine so that it saves the individual spectra in a matrix rather than computing an average spectrum. Also add code to compute the relative time corresponding to each segment. The main program is given below followed by the modified `welch` routine now called `stft` for the short-term Fourier transform.

```
% Example 4.6 Example of the short-term Fourier transform time-frequency analysis
%
load Resp;                                    % Get respiratory data
N = length(resp);                             % Determine data length
fs = 125;                                     % Sampling frequency
nfft = fs*60;                                 % 1 min of samples
noverlap = round(nfft/2);                     % Use 50% overlap
m_plot = round(2/(fs/nfft));                  % Find m for 2 Hz
[PS,f,t] = stft(resp',nfft,noverlap,fs);      % Calculate the STFT
PS1 = PS(:,2:m_plot);                         % Resize power spectra to range to
                                              2 Hz
f1 = f(1:m_plot-1);                           % Resize frequency vector for
                                              plotting
mesh(f1,t,PS1);                               % Plot in 3D
view([17 30]);                                % Adjust view for best perspective
  ......labels......


The modified welch routine is:
function [PS,f,t] = stft(x,nfft,noverlap,fs);
% Function to calculate short-term Fourier transform
%
  ......same as welch.m up to the calculation of the power spectra......
% Data have been windowed
PS_temp = abs((fft(data)).^2)/N1;                % Calculate power spectrum
                                                 (normalized)
PS(k,:) = PS_temp(2:half_segment + 1)/(nfft/2);  % Remove redundant points
t(k) = (k - 1)*increment/fs;                     % Construct time vector
end                                              % Program end here
```

Results: The results from the short-term Fourier transform are shown in Figure 4.20. The view was initially adjusted interactively to find a perspective that shows the progression of power spectra. In Figure 4.20 each timeframe shows a spectrum that is similar to the spectrum in

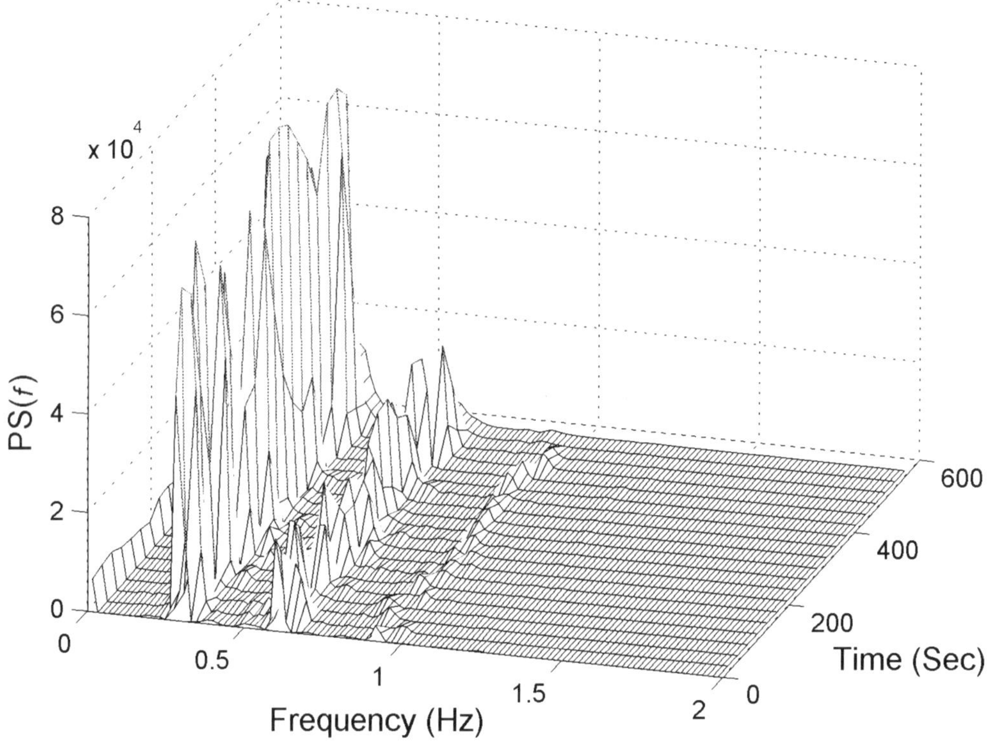

FIGURE 4.20 Time-frequency plot of a 10 minute respiratory signal. This plot was constructed by segmenting the 10 minute signal into 1-minute increments and taking the power spectrum of each increment separately rather than averaging as in the `welch` method. Segments overlap by 50%.

Figure 3.27 with a large peak around 0.35-Hz and a smaller peak around 0.7-Hz. There is some variation in the amplitude of the peaks with time.

4.5 SIGNAL BANDWIDTH

The concept of representing a signal in the frequency domain brings with it additional concepts related to a signal's spectral characteristics. One of the most important of these new concepts is signal bandwidth. In the next chapter the concept of bandwidth is extended to systems as well as signals, and we find that the definitions related to bandwidth are essentially the same. The modern everyday use of the word "bandwidth" relates to a signal's general ability to carry information. Here we define the term more specifically with respect to the range of frequencies found in the signal. Figure 4.21A shows the spectrum of a hypothetical signal which contains energy at frequencies in equal measure up to 200-Hz, labeled f_c, above which no energy is found. We would say that this signal has a *flat* spectrum up to f_c, the *cutoff frequency*, above which it has no energy. Since the signal

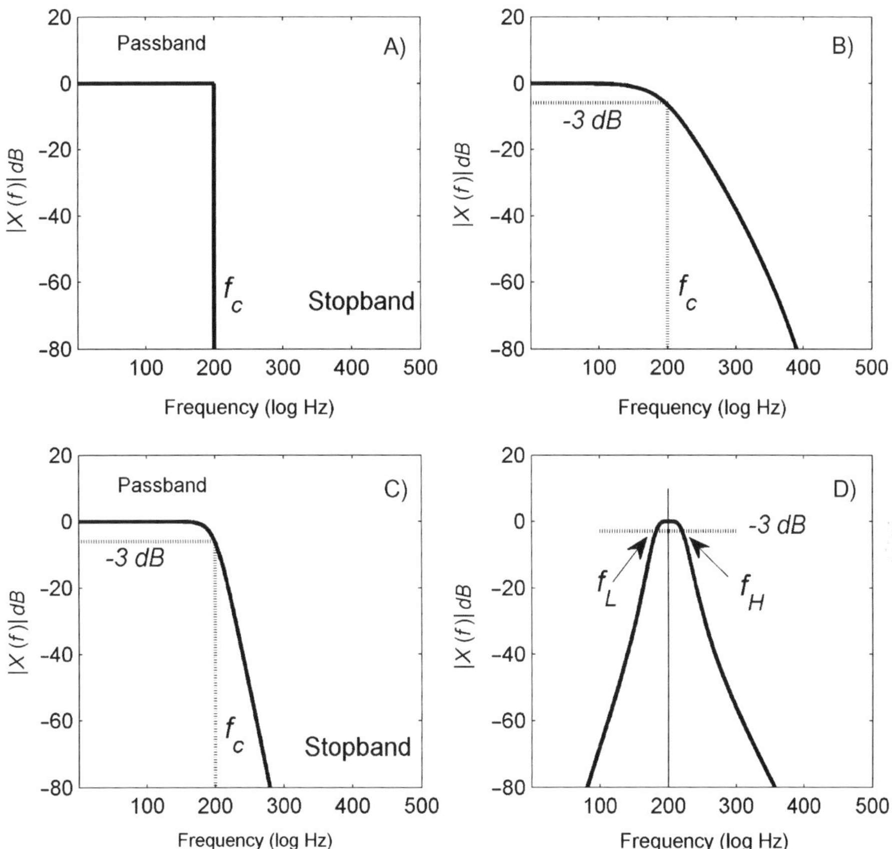

FIGURE 4.21 Frequency characteristics of ideal and realistic signals. The frequency plots shown here are plotted in dB. The horizontal axis is in log frequency. A) The spectrum of a signal with an idealized and perfectly defined frequency range: 0 to 200-Hz. B) A more realistic signal spectrum where signal energy decreases gradually at higher frequencies. C) A realistic signal spectrum where signal energy decreases more sharply at higher frequencies. D) A realistic signal spectrum where the signal energy decreases both above and below a *center frequency* of 200-Hz.

contains energy only between 0 and f_c Hz, the bandwidth of this signal would be defined here as from 0 to 200-Hz, or simply 200-Hz. The frequency range below 200-Hz is called the *passband*, while the frequency range about 200-Hz is called the *stopband*.

Although some real signals can be quite flat over selected frequency ranges, they are unlikely to show such an abrupt cessation of energy above a given frequency, as seen in Figure 4.21A. Figure 4.21B shows the spectrum of a more realistic signal, where the energy begins to decrease at a specific frequency but decreases gradually. When the decrease in signal energy takes place gradually, as in Figure 4.21B, defining the signal bandwidth is problematic. If we want to attribute a single bandwidth value for this signal, we need to define some sort of *cutoff frequency*: a frequency defining a boundary between the region of substantial energy and the region of minimal energy. Such a boundary frequency has been arbitrarily defined as the frequency when the signal's value has declined by 3 dB with respect to its average unattenuated RMS value. (In Figure 4.21, the nominal unattenuated value for all signals is normalized to 0 dB.) The 3-dB boundary is not entirely arbitrary. If we convert −3 dB to linear units using Equation 1.14 we get $10^{(-3\text{dB}/20)} = 0.707$. Thus, when the amplitude of a signal is reduced 3 dB, it has a value of 0.707 of its unattenuated value. Its power at that value is proportional to the amplitude squared, and $0.707^2 = 0.5$. So when the amplitude of the signal is attenuated by 3 dB, or decreased in amplitude by 0.707, it has lost half its power. (Recall that the dB scale is logarithmic, so −3 dB means a reduction of 3 dB; e.g., see Table 1.4.)

When the signal magnitude spectrum (the magnitude spectrum, not the power spectrum) is reduced by 3 dB, the power in the signal is half the nominal value, so this boundary frequency is also known as the *half-power point*. In terms of defining the cutoff frequency, the *3-dB point* and *half-power point* are synonymous. In Figure 4.17B, the signal would also have a bandwidth of 0.0 to 200-Hz, or simply 200 (or f_c) Hz based on the −3-dB cutoff frequency. The signal in Figure 4.21C has a sharper decline in energy, referred to as the *frequency rolloff*, but it still has a bandwidth of 200-Hz based on the −3-dB point.

It is possible that a signal *rolls off* or *attenuates* at both the low-frequency and high-frequency ends as shown in Figure 4.21D. In this case, the signal's frequency characteristic has two cutoff frequencies, one labeled f_L and the other f_H. For such signals, the bandwidth is defined as the range between the 2 cutoff frequencies (or 3-dB points), that is, $BW = f_H - f_L$ Hz.

EXAMPLE 4.7

Find the effective bandwidth of the noisy signal, x, in file `Ex4_7_data.mat`. The sampling frequency for this signal is 500-Hz.

Solution: To find the bandwidth we first get the spectrum using the Fourier transform. Since the signal is noisy, we will likely want to use spectral averaging to get a relatively clean spectrum. We can then use MATLAB's `find` routine to search the magnitude spectrum for the first and last points that are greater than 0.707, or 0.5 if we use the power spectrum.

```
% Example 4.7 Find the effective bandwidth of the signal x in file ex4_6_data.mat
%
load Ex4_7_data.mat;                  % Load the data file. Data in x
fs = 500;                             % Sampling frequency (given)
[PS1,f1] = welch(x,length(x),0,fs);   % Determine and plot the unaveraged
                                      spectrum
  ......plot, axes labels and new figure......
%
nfft = 256;                           % Power spectrum window size
[PS,f] = welch(x,nfft,nfft-1,fs); % Compute averaged spectrum, maximum overlap
PS = PS/max(PS);                      % Normalize peak spectrum to 1.0
plot(f,PS,'k'); hold on;              % Plot the normalized, average
                                      spectrum
```

```
    ......axes labels......
%
i_fl = find(PS > .5, 1, 'first');           % Find index of low freq. cutoff
i_fh = find(PS > .5, 1, 'last');            % Find index of high freq. cutoff
f_low = f(in_fl);                           % Convert low index to cutoff freq.
f_high = f(in_fh);                          % Convert high index to cutoff freq.
    ......plot markers......
BW = f_high - f_low;                        % Calculate bandwidth
title(['Bandwidth: ',num2str(BW,3),' Hz']); % Display bandwidth in title
```

Results: The unaveraged power spectrum was obtained using the `welch` routine developed in Example 4.4 by making the window size the same as the data length. Alternatively the square of the magnitude Fourier transform could have been used, but the `welch` routine is convenient as it also provides the frequency vector. The unaveraged power spectrum is quite noisy (see Figure 4.22), so we do want to use spectral averaging. Using the `welsh` routine with a window size of 256 samples produces a spectrum that is fairly smooth, but still has a reasonable spectral resolution (Figure 4.23). (The effect of changing window size is explored in Problem 16.)

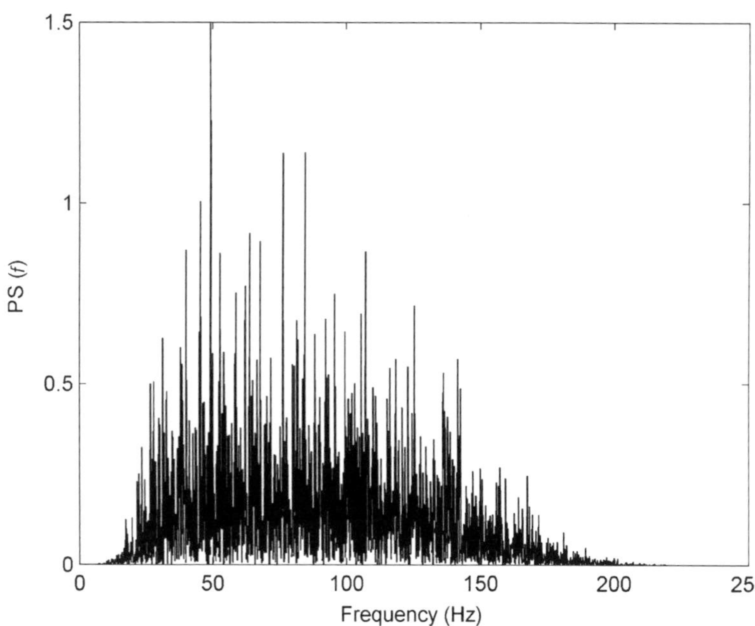

FIGURE 4.22 The spectrum of the noisy signal used in Example 4.7. Spectral averaging will be used to produce a smoother spectrum that makes it easier to identify the half-power (i.e., −3-dB) points and calculate the bandwidth.

The `find` routine with the proper options gives us the indices of the first and last points above 0.5, and these are plotted superimposed on the spectral plot (large dots in Figure 4.23). These high and low sample points can be converted to frequencies using the frequency vector produced by the `welch` routine. The difference between the two cutoff frequencies is the bandwidth

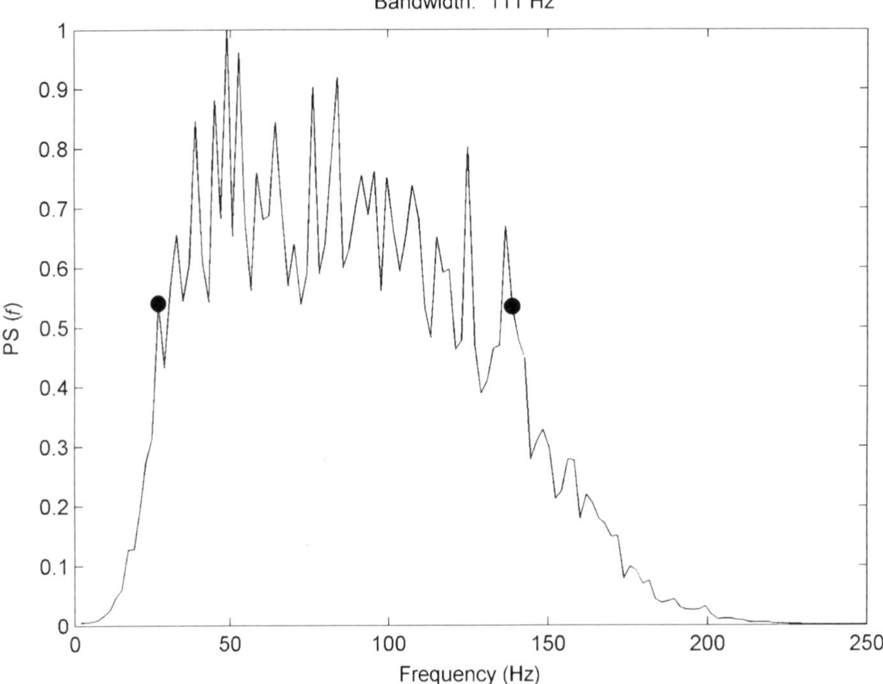

FIGURE 4.23 The spectrum of the signal in Example 4.6 obtained using spectral averaging. The window size was selected empirically to be 256 samples. This size appears to give a smooth spectrum while maintaining good resolution. The half-power points which are the same as the −3-dB points in the magnitude spectrum are determined by searching for the first and last points in the spectrum that are greater than 0.5. MATLAB's `find` routine was used to locate these points, which are identified as large dots on the spectral curve. The frequency vector is used to convert the index of these points to equivalent frequency and the bandwidth calculated from these frequencies is shown in the figure title.

and is displayed in the title of the spectral plot (Figure 4.23). The estimation of the high and low cutoff frequencies could have been improved by interpolating between the spectral frequencies on either side of the 0.5 values. This is done in Problem 17.

4.6 SUMMARY

The discrete Fourier transform can be used to understand the relationship between a continuous time signal and the sampled version of that signal. This frequency-based analysis shows that the original, unsampled signal can be recovered if the sampling frequency is more than twice the highest frequency component in the unsampled signal.

The Fourier series or Fourier transform can be used to construct the power spectrum of a signal. The power spectral curve describes how the signal's power varies with frequency. The power spectrum is particularly useful for random data where phase characteristics have little meaning. By dividing the signal into a number of possibly overlapping segments and averaging the spectrum obtained from each segment, a smoothed power spectrum can be obtained. The resulting frequency curve will emphasize the broadband or general characteristics of a signal's spectrum, but will lose some of the fine detail.

Bandwidth is a term used commonly to describe the capability of a communication channel to carry information. Bandwidth is specifically defined here as the range of frequencies included in a signal. To be included, the energy associated with a given frequency must be greater than half that of the signal's nominal values. This means that the signal amplitudes of included frequencies must be greater than 0.707 of the nominal signal amplitude or, equivalently, attenuated no more than 3 dB from the nominal RMS value.

PROBLEMS

1. This problem demonstrates aliasing. Generate a 512-point waveform consisting of 2 sinusoids at 200 and 400-Hz. Assume a sampling frequency of 1 kHz. Generate another waveform containing frequencies at 200 and 900-Hz. Take the Fourier transform of both waveforms and plot the magnitude of the spectrum up to $f_s/2$. Plot the 2 spectra superimposed, but in different colors to highlight the additional peak due to aliasing at 100-Hz. [*Hint*: To generate the sine waves, first construct a time vector, `t`, then generate the signal using: `x = sin(2*pi*f1*t) + sin(2*pi*f2*t)` where `f1` = 200 for both signals, while `f2` = 400 for one waveform and `f2` = 900 for the other.]
2. Load the chirp signal, `x`, in file `chirp.mat` and plot the magnitude spectrum (f_s = 5000 Hz). Now decrease the sampling frequency by a factor of 2 by removing every other point, and recalculate and replot the magnitude spectrum. Note the distortion of the spectrum produced by aliasing. Do not forget to recalculate the new frequency vector based on the new data length and sampling frequency. (The new sampling frequency is effectively half that of the original signal.) Decreasing the sampling frequency is termed *downsampling* and can be done easily in MATLAB: `x1 = x(1:2:end)`.
3. Load the file `sample_rate.mat`, which contains signals `x` and `y`. Is either of these signals oversampled (i.e., $f_s/2 \leq f_{max}$)? Alternatively, could the sampling rate of either signal be safely reduced? Justify your answer.
4. The file labeled `quantization.mat` contains 3 vector variables: `x`, `y`, and `z`. This file also contains a time vector, `t`, useful for plotting. Variable `x` represents an original signal. Variable `y` is the original signal after it has been sampled by a 6-bit ADC, and variable `z` is the original signal after sampling by a 4-bit ADC. Both converted signals have been scaled to have the same range as `x`, 0 to 5 volts maximum. On one plot show the 3 variables superimposed and on another plot show the error signals `x – y` and `x – z`. Then calculate the RMS error between `x` and `y` and `x` and `z`. Also calculate the theoretical error for the 2 ADCs based on Equation 4.3. (Recall, the RMS value is approximately equal to the square root of the variance as given by Equation 4.3)
5. The file `short.mat` contains a very short signal of 32 samples. Plot the magnitude spectrum as discrete points obtained with and without zero-padding. Zero-pad out to a total of 256 points.

6. Construct 2 arrays of white noise using `randn`: one 128 points in length and the other 1024 points in length. Take the FT of both. Does increasing the length improve the spectral estimate of white noise, which should be flat? (Eliminate the first point—the average or DC term—when you plot the spectra and plot only nonredundant points.)
7. Use `sig_noise` to generate a 256-point waveform consisting of a 300-Hz sine wave with an SNR of −12 dB (`x = sig_noise(300, −12,256);`). Calculate and plot the power spectrum using 2 different approaches. First use the direct approach: take the Fourier transform and square the magnitude function. In the second approach use the traditional method defined by Equation 4.11: take the Fourier transform of the autocorrelation function. Calculate the autocorrelation function using `axcor`, then take the absolute value of the `fft` of the autocorrelation function. You should only use the second half of the autocorrelation function (those values corresponding to positive lags). Plot the power spectrum derived from both techniques. The scales will be different because the MATLAB `fft` routine does not normalize the output.
8. The variable `x` in file `prob4_8_data.mat` contains a combination of 200- and 300-Hz sine waves with SNRs of −6 dB. The sampling frequency is 1000-Hz and the data segment is fairly short, 64 samples. Plot the magnitude spectrum obtained with and without a Hamming window. Use Equation 4.9 to generate a Hamming window function 64 points long and multiply point-by-point with the signal variable `x`. Note that the difference is slight, but could be significant in certain situations.
9. Use MATLAB routine `sig_noise` to generate two arrays, one 128 points long and the other 512 points long. Include 2 closely spaced sinusoids having frequencies of 320 and 340-Hz with an SNR of −12 dB. The MATLAB call should be:
   `x = sig_noise([320 340],-12,N);` where `N` = either 128 or 512.
   Calculate and plot the (unaveraged) power spectrum. How does data length affect the spectra?
10. Use `sig_noise` to generate a 128-point array containing 320- and 340-Hz sinusoids as in Problem 9. Calculate and plot the unaveraged power spectrum of this signal for an SNR of −12 dB, −14 dB, and −16 dB. How does the presence of noise affect the ability to detect and distinguish between the two sinusoids?
11. Load the file `broadband2` which contains variable `x`, a broadband signal with added noise. Assume a sample frequency of 1 kHz. Calculate the averaged Power Spectrum using the `welch` routine. Evaluate the influence of segment length using segment lengths of `N/4` and `N/16`, where `N` is the length of the date of variable, `x`. Use the default overlap.
12. Load the file `eeg` that contains EEG data similar to that shown in Figure 1.9. Analyze these data using the unaveraged power spectral technique and an averaging technique using the `welch` routine. Find a segment length that smoothes the background spectrum, but still retains any important spectral peaks. Use a 99% overlap.
13. Modify the `welsh` routine to create a routine `welch_win` that applies a Blackman window to the data before taking the Fourier transform. The Blackman window adds another cosine term to the equation of the Hamming window giving:

$$w[n] = 0.41 - 0.5\cos\left(\frac{2\pi n}{N}\right) + 0.08\cos\left(\frac{4\pi n}{N}\right)$$

Load the file broadband3.mat which contains a broadband signal and a sinusoid at 400-Hz in variable x. Analyze the broadband/narrowband signal using both welsh and welch_win with segment lengths of $N/4$ and $N/16$, where N is the length of the data of variable, x. Use the default overlap. Note that the Blackman window provides slightly more smoothing than the rectangular window used by welch.

14. Load the file ECG_1hr which contains 1 hour of ECG signal in variable ecg. Plot the time-frequency spectra of this signal using the short-term Fourier transform approach shown in Example 4.6 (stft.m is on this book's accompanying CD). Limit the frequency range to 3-Hz and do not include the DC term. The sample frequency is 250-Hz. [*Hint*: You can improve the appearance of the higher frequency components by limiting the z axis using ylim and the color axis using caxis.]

15. Repeat Problem 14 but change the segment length to 30 sec and the overlap to 20 sec (2/3). Also change the upper limit on the frequency range to 8-Hz but again do not include the DC term. Use pcolor with shading set to interp to plot the time-frequency data and correctly label and scale the 2 axes. Limit the color range to 200. You will see a light-colored wavy line around 1-Hz that reflects the variation in the heart rate with time. Now lower the upper limit of the color range to around 20 and note the chaotic nature of the higher frequency components. [*Hint*: If you make the lower limit of the color range slightly negative, it lightens the background and improves the image.]

16. Find the effective bandwidth of the signal x in file broadband2.mat. Load the file broadband2.mat and find the bandwidth using the methods of Example 4.7. Find the power spectrum using welch and a window size ranging between 100 and 1000-Hz. Find the window size that appears to give the best estimate of bandwidth.

17. Load the file broadband2.mat and find the bandwidth using the methods of Example 4.7, but this time using interpolation. Determine the power spectrum using welch with a window size of 50 and maximum overlap. This will give a highly smoothed estimate of the power spectrum but a poor estimate of bandwidth due to the small number of points. Interpolation can be used to improve the bandwidth estimate from the smoothed power spectrum. The easiest way to implement interpolation in this problem is to increase the number of points in the power spectrum using MATLAB's interp routine. (For example: PS1 = interp(PS,5) increases the number of points in PS by a factor of 5.) Estimate the bandwidth using the approach of Example 4.7 before and after expanding the power spectrum by a factor of 5. Be sure to expand the frequency vector (f in Example 4.7) by the same amount.

PART II

SYSTEMS

CHAPTER

5

Linear Systems in the Frequency Domain: The Transfer Function

5.1 LINEAR SIGNAL ANALYSIS—AN OVERVIEW

In the first four chapters we studied signals. Signals come from, and go through, systems. A system either acts on a signal to produce a modified signal or it is the origin of the signal. From a mechanistic point of view, all living things are a collection of systems. These systems[1] act or interact through signals. Such biological signals are generated by the manipulation of molecular mechanisms, chemical concentrations, ionic electrical current, and/or mechanical forces and displacements. A physiological system performs some operation or manipulation in response to one or more inputs and gives rise to one or more outputs. Terminology aside, our main concern with any system is to be able to determine the behavior of the system in response to a range of stimuli. Ideally we would like to be able to calculate a system's response to any input no matter how complicated the stimulus, or how complicated the system.

To be able quantitatively to predict the behavior of complex processes in response to complex stimuli, we usually impose rather severe simplifications and/or assumptions. The most common assumptions are: 1) that the process behaves in a linear manner, and 2) that its basic characteristics do not change over time. Combined, these two assumptions are referred to as a *linear time-invariant* (LTI) system. A formal, mathematical definition of an LTI system is given in the next section, but the basic idea is intuitive: Such systems exhibit proportionality of response (double the input/double the output) and they are stable over time. These assumptions allow us to apply a powerful array of mathematical tools and are known collectively as *linear systems analysis*. Of course, living systems change over time, they are adaptive, and they are often nonlinear. Nonetheless, the power of

[1]Sometimes the term *system* is reserved for larger structures and the term *process* for smaller systems with fewer elements, but the two terms are often used interchangeably, as they are here. Hence, the term *physiological process* is identical to the term *physiological system*.

linear systems analysis is sufficiently seductive that the necessary assumptions or approximations are often made so that these tools can be applied. Linearity can be approximated by using small-signal conditions since, when a range of action is restricted, many systems behave more-or-less linearly. Alternatively, piecewise linear approaches can be used where the analysis is confined to operating ranges over which the system behaves linearly. In Chapter 7 we learn simulation techniques that allow us to analyze systems with certain kinds of nonlinearity. To deal with processes that change over time, we can limit our analysis to a timeframe that is short enough that the process can be considered time invariant.

A major property of linear systems is that they can be represented solely by linear differential equations. This means that any linear system, no matter how complex, can be represented by combinations of at most four element types: arithmetic (addition and subtraction), scaling (i.e., multiplication by a constant), differentiation, and integration. (Using some simple mathematical manipulations, it is possible to limit the differential equations to only integrators or only differentiators, but we will allow for both elements here.) A typical system might consist of only one of these four basic elements, but is usually made up of a combination of basic operators. A summary of these basic element types is provided a little later in this section.

5.1.1 Analog and System Representations of Linear Processes

There are several approaches to the study of linear systems. In this text, two different approaches are developed and explored: *analog analysis* using *analog models*, and *systems analysis* using *systems models*. There is potential confusion in this terminology. While analog analysis and systems analysis are two different approaches, both are included as tools of *linear systems analysis*. Hence, linear systems analysis includes both analog and systems analysis.

The primary difference between analog and systems models is the way the underlying processes are represented. In analog analysis, individual components are represented by analogous elements. Only three basic elements are used in analog models: scaling, differentiation, and integration. The arithmetic operations are determined by the way the elements are arranged in the model. An electric circuit is an example of an analog model. Figure 5.1 shows what appears to be a simple electric circuit, but it is actually an analog model of the cardiovascular system known as the *windkessel model*. In this circuit, voltage represents blood pressure, current represents blood flow, R_P and C_P are the resistance and compliance of the systemic arterial tree, and Z_o is the characteristic impedance of the proximal aorta. In Chapters 10 and 11 we study analog models and we find that using an electrical network to represent what is basically a mechanical system is mathematically

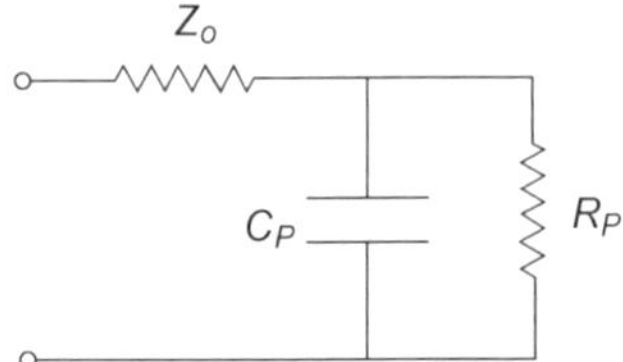

FIGURE 5.1 An early analog model of the cardiovascular system that uses electrical elements to represent mechanical processes. In this model, voltage is equivalent to blood pressure and current to blood flow.

appropriate. In this analog model, the elements are not really very elemental as they represent processes distributed throughout various segments of the cardiovascular system; however, the model can be, and has been, expanded to represent the system in greater detail.

With *systems models*, the process is represented as an input/output relationship—transforming an input signal or stimulus into a response. Figure 5.2 shows a model of the neural pathways that mediate the vergence eye movement response, the processes used to turn the eyes inwa rd to track visual targets at different depths. The model shows three different neural paths converging on the elements representing the oculomotor plant (the two right-most system elements). Neural processes in the upper two pathways provide a velocity-dependent signal to move the eyes quickly to approximately the right position. The lower pathway represents the processes that use visual feedback to fine-tune the position of the eyes more slowly and attain an accurate final position.

Systems models and concepts are very flexible: inputs can take many different energy forms (chemical, electrical, mechanical, or thermal), and outputs can be of the same or different energy forms. While the elements in analog models more closely relate to the underlying physiological processes, system representations provide a better overall view of the system under study. System representations can also be more succinct. The next few chapters address system representation of physiological or other processes while Chapters 10 and 11 are devoted to analog models and circuits.

5.1.2 Linear Elements—Linearity, Time Invariance, Causality

Irrespective of the type of model representation used, all linear systems are composed of linear elements. The concept of linearity has a rigorous definition, but the basic concept is one of proportionality of response. If you double the stimulus into a linear system, you will get twice the response. One way of stating this proportionality property mathematically is: If

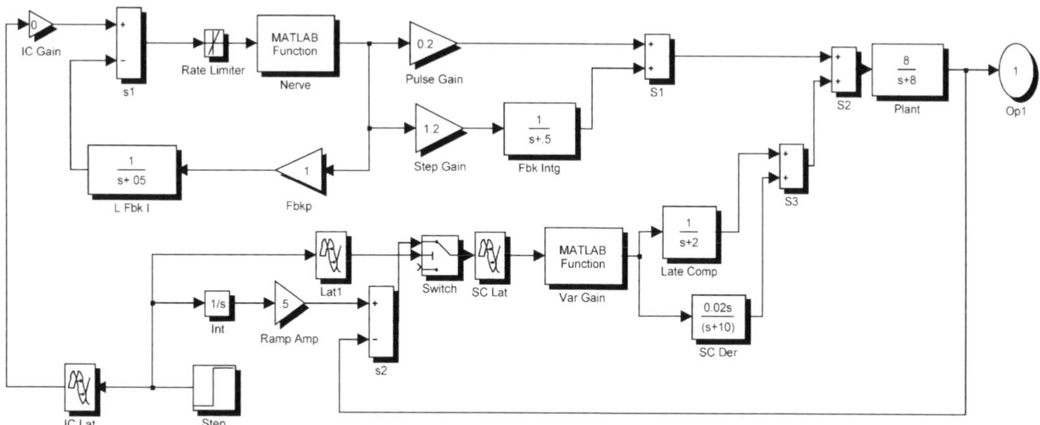

FIGURE 5.2 A system model of the vergence eye movement neural control system showing some of the control details. This description emphasizes the system's role in producing a response to a given stimulus: the input/output relationship. This model can be simulated using MATLAB's Simulink, as described in Chapter 7.

the independent variables of a linear function are multiplied by a constant, k, the output of the function is simply multiplied by k. Assuming:

$$\begin{aligned} &y = f(x) \quad \text{where } f \text{ is a linear function, then:} \\ &ky = f(kx) \end{aligned} \tag{5.1}$$

Also if f is a linear function:

$$f(x_1(t)) + f(x_2(t)) = f(x_1(t) + x_2(t)) \tag{5.2}$$

In addition, if: $y = f(x)$ and

$$z = \frac{df(x)}{dx}$$

then

$$\frac{df(kx)}{dx} = k\left(\frac{df(x)}{dx}\right) = kz$$

Similarly, if: $y = f(x)$ and

$$z = \int f(x)dx$$

then

$$\int f(kx)dx = k\int f(x)dx = kz$$

Derivation and integration are linear operations. When they are applied to linear functions, these operations preserve linearity.

If the system has the same behavior at any given time, that is, its basic response characteristics do not change over time, it is said to be *time-invariant*. Time invariance could be thought of as a stricter version of stationarity, since a time-invariant system would also be stationary. The mathematical definition of a time-invariant function, f, is given as:

$$\begin{aligned} &y = f(x) \quad \text{where } f \text{ is a linear function, then:} \\ &y(t-T) = f(x(t-T)) \end{aligned} \tag{5.3}$$

As mentioned above, systems that are both linear and time-invariant are, not surprisingly, referred to as *linear time-invariant* systems, usually abbreviated as *LTI* systems. If a system responds only to current and past inputs, the system is termed *causal*. Real-world systems are causal and you might think that all systems must be causal; how can a system possibly respond to a future input or stimulus? If the system is represented in a computer, such a *simulated* system could be made to produce an output that, for any given simulated time period, acts on both past and futures values of a signal. Examples of such noncausal systems are shown in Chapter 8.

If a system only responds to its current input and not to inputs past, it is said to be *memoryless*. (Such a memoryless system is inevitably doomed to repeat its mistakes.) Conversely, a system whose current output depends to some extend on past inputs is said to possess *memory*. Examples of both types are given below.

5.1.3 Superposition

Response proportionality, or linearity, is required for the application of an important concept known as *superposition*. Superposition states that when two or more influences are acting on some process, the resultant behavior is the same as the summation of the process's response to each influence as if it were acting alone. In systems analysis, the principle of superposition means that when two or more inputs are active in a system, a valid solution can be obtained by solving for each source as if it were the only one in the system, then algebraically summing these partial solutions. In fact, the sources could be anywhere in the system, and often are at different locations in analog models. In system representations, the multiple sources are usually at one location, defined as the input.

In system models, these *multiple sources* could be the sinusoidal components of a single, more complicated input signal. If superposition holds, then the response of any LTI system can be found to any periodic or aperiodic input by: 1) decomposing the signal into its sinusoidal components using the Fourier series or Fourier transform; 2) finding the system's response to each sinusoid in the decomposition; and 3) summing the individual response sinusoids using the Fourier series equation. This decomposition strategy may seem like a lot of work were it not for the fact that step 2, finding the response of any linear system to a sinusoidal input, can be accomplished using only algebra. Even for systems that contain calculus operations like integration or differentiation, techniques exist that convert these operations to algebraic operation. This strategy, combining decomposition with the ease of algebraic solutions, motivates many of the concepts developed in the next few chapters.

You can also determine the response of a linear system in the time domain by decomposing the signal using standard calculus tricks: 1) divide the signal into infinitely small segments; 2) find the response of the system to each segment; and 3) sum the segments. This is just a time-domain analogy to the frequency-domain strategy described in the last paragraph. This time-domain strategy is developed in Chapter 7.

The superposition principle is a consequence of linearity and only applies to LTI systems. It makes the tools that we have studied, and study later in this book, much more powerful since it enables decomposition strategies, either in the frequency or time domain. Examples using the principle of superposition are found throughout this text.

5.1.4 Systems Analysis and Systems Models

Systems models usually represent processes using so-called *black box* components. Each element of a systems model consists only of a mathematically defined input-output relationship and is represented by a geometrical shape, usually a rectangle. No effort is made to determine what is actually inside the box, hence the term black box: Only the element's relationship between input and output is of importance.

A typical element in a systems model is shown graphically as a box, or sometimes as a circle when an arithmetic process is involved. Two such elements are shown in Figure 5.3. The inputs and outputs of all elements are signals with a well-defined direction of flow or influence. These signals and their direction of influence are shown by lines and arrows connecting the system elements.

The letter G in the right-hand element of Figure 5.3 represents the mathematical description of the element: the mathematical operation that converts the input signal into an output signal. Stated mathematically:

$$Output = G(Input) \tag{5.4}$$

Note that Equation 5.4 represents a very general concept: G could be any linear relationship and the terms *Input* and *Output* could represent a variety of signal modalities of any complexity. Also, we can use any letter, but usually capital letters are used. Rearranging this basic equation, G can be expressed as the ratio of output to input:

$$G = \frac{Output}{Input} \tag{5.5}$$

The term G in Equation 5.5 could be quite complex if it is describing a complex system. Because it relates the *Output* to the Input of a system (Equation 5.5), it is called the *transfer function*. The transfer function is covered in some length below and in the next chapter. It is a critical concept in linear systems analysis because it defines the input/output relationship of a system, the only concern in a classical system model. Although the transfer function concept is sometimes used very generally, in the context of linear systems analysis it is an algebraic multiplying function. Linear system elements can be represented algebraically and this representation leads to a transfer function that is also algebraic. When a number of systems elements are connected together, the overall input/output relationship can be determined from the input/output relationship of each element using algebra. The next section of this chapter shows how these relationships can be transformed into algebraic functions even if they contain calculus operators, at least for sinusoidal signals.

If the transfer function is an algebraic multiplier, then when two systems are connected in series, as in Figure 5.4, the transfer function of the overall paired systems is just the product of the two individual transfer functions:

$$\begin{aligned} &Out_2 = G_2 \times In_2 \text{ and } Out_1 = In_2 = G_1 \times In_1, \textit{then:} \\ &Out_2 = G_2 \times (G_1 \times In_1) = G_2 G_1 In_1. \end{aligned} \tag{5.6}$$

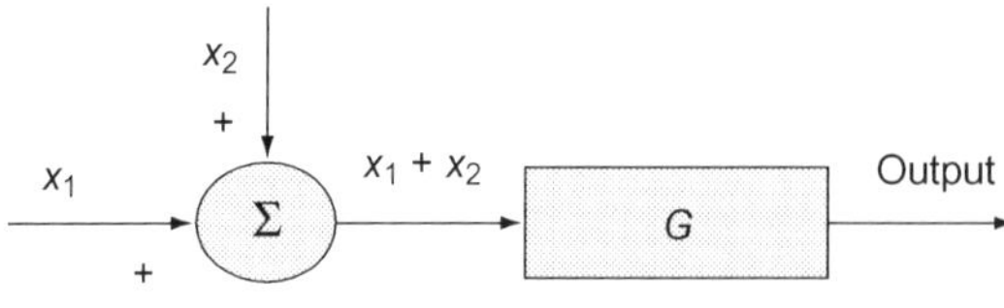

FIGURE 5.3 Typical elements of a system model. The left-hand element is an *adder* whose output signal is the sum of the two inputs x_1 and x_2. The right-hand element is a general element that takes the input ($x_1 + x_2$ from the adder in this case) and operates on it with the mathematical operation G to produce the output signal.

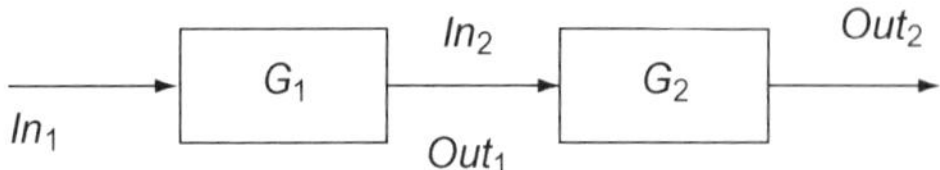

FIGURE 5.4 A system with two system elements in series. The transfer function of the combined elements Out_2/Out_1 is the product of the two individual transfer functions, G_1 G_2 (Equation 5.7).

and the overall transfer function for the two series elements is:

$$\frac{Out_2}{In_1} = G_1 G_2 \tag{5.7}$$

Note that the overall transfer function of the combined systems in Figure 5.4 would be the same even if the order of the two systems were reversed. This is a property termed *associativity*. We can extend this concept to any number of system elements in series:

$$Output = Input \prod_i G_i \tag{5.8}$$

The overall transfer function is the product of the individual series element transfer functions:

$$\frac{Output}{Input} = \prod_i G_i \tag{5.9}$$

In practice, determining the transfer function of most biological systems is challenging and usually involves extensive empirical observation of the input/output relationship. However, once the relationships for the individual elements have been determined, finding the input/output relationship of the overall system is straightforward, as suggested by the simple system analysis problem in Example 5.1.

EXAMPLE 5.1

Find the transfer function for the systems model in Figure 5.5. The mathematical description of each element is either an algebraic term or an arithmetic operation. G and H are assumed to be linear algebraic functions so they produce an output that is the product of function times the input (i.e., Equation 5.2). The left hand element is an arithmetic element that performs subtraction. The system shown is a classic feedback system because the output is coupled back to the input via the lower pathway. In this system, the upper pathway is called the *feedforward pathway* because it moves the signal toward the output. The lower pathway is called the *feedback pathway* because it moves its input signal (which is actually the output signal) away from the output and toward the system input. Note that the signal in this feedback pathway becomes mixed with the input signal through the subtraction operator. To finish off the terminology lecture, G in this configuration is called the *feedforward gain* and H is called the *feedback gain*.

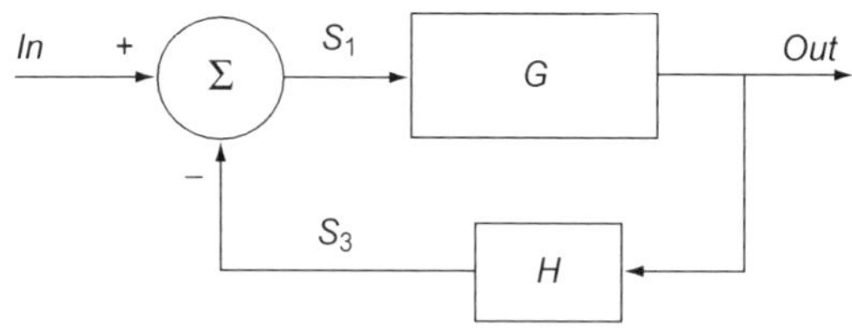

FIGURE 5.5 A system model consisting of simple *gain* type elements and an arithmetic operation (subtraction). The gain elements are completely described in system model terms by a constant (G or H). In this system, G is called the feedforward gain and H is called the feedback gain. In complex systems, these two elements could represent quite complicated operations, but the system would still be termed a *feedback system*.

Solution: Generate an algebraic equation based on the configuration of the system and the fact that the output of each process is the input multiplied by the associated gain term: $Output = G \times Input$.

$$\text{For the upper box: } G = \frac{Out}{S_1};$$

$$\text{and for the lower box: } \quad H = \frac{S_3}{Out}$$

$$\text{and for the arithmetic element: } \quad S_1 = In - S_3$$

$$\text{Rearranging: } \quad S_1 = \frac{Out}{G}; \quad S_3 = Out \times H$$

$$\text{Since } S_1 = In - S_3 \quad \text{Substituting in the above: } \quad \frac{Out}{G} = In - Out \times H$$

$$\text{Rearranging: } \quad Out = In \times G - Out \times GH; \quad Out(1 + GH) = In \times G$$

$$\frac{Out}{In} = \frac{G}{1 + GH} \tag{5.10}$$

Analysis: The solution given in Equation 5.10 is known as the *feedback equation* and is used in later analyses of more-complex systems. In this example, the two elements, G and H, were constants, but could have been anything as long as they can be treated algebraically. When the individual elements contain differential or integral operations in their input/output relationships, the techniques described below are required to encode these calculus operations into algebraic manipulations, but this equation, and the algebraic manipulations used in this example, still apply.

Systems models are evaluated by analyzing their response to different stimulus conditions. Like the Fourier transform this analysis can be done analytically, but like the Fourier transform it is easiest to do on a computer. MATLAB's Simulink model simulator described in Chapter 7 is specifically designed for analyzing a system model's response to a range of stimuli, a process called *simulation*. Simulation not only provides a reality check—do the simulations produce responses similar to that of the real system?—but also permits evaluation of internal components and signals not available to the experimentalist. For example, what would happen in the vergence model of Figure 5.2 if the neural components responsible for the pulse signal (upper pathway) were not functioning, or functioning erratically perhaps due to a brain tumor? Simulink simulations of the vergence model have produced responses that are very close to actual vergence eye movements. Although Simulink is only developed for systems models, it is easy to convert analog models into system formats so this software can be used to simulate analog models as well. Alternatively, there are programs such as PSpice specifically designed to simulate electronic circuits that can also be applied to analog models.

Figures 5.4 and 5.5 illustrate another important property of systems models. The influence of one process on another is explicitly stated and indicated by the line connecting two processes. This line has a direction usually indicated by an arrow, which implies that the influence or information flow travels only in that direction. If there is also a reverse flow of information, such as in feedback systems, this must be explicitly stated in the form of an additional connecting line showing information flow in the reverse direction.

The next example has some of the flavor of the simulation approach, but does not require the use of Simulink.

EXAMPLE 5.2

There is a MATLAB function on this book's accompanying CD that represents a system. The function is called `process_x` and it takes an input signal, `x`, and generates an output signal `y` (as in: `y = process(x)`). We are to determine if `process_x` is a linear process over the range of 0 to ±100 signal units. We can input to the process any signal we desire and examine the output.

Solution: The basic strategy is to input a number of signals with different amplitudes and determine if the outputs are proportional. But what is the best signal to use? The easiest approach might be to input two or three signals that have a constant value; for example, $x = 1$, then $x = 10$, then $x = 100$, along with the negative values. The output should be proportional. But what if the process contains a derivative operation? While the derivative is a linear operation, the derivative of a constant is zero, and so we would get zero output for all three signals. Similarly, if the process contains integrations, the output to a constant input signal is a ramp and is more difficult to interpret. A sinusoid would be ideal because the derivative or integral of a sinusoid is still a sinusoid, only the amplitude and phase would be different. This approach is illustrated in Figure 5.6.

So our strategy will be to input a number of sine waves having different amplitudes. If the output signals are all proportional to the input amplitudes, we predict that `process_x` is a linear process, at least over the range of values tested. Since the work is done on a computer, we can use any number of sine-wave inputs, so let us use 100 different input signals ranging in amplitude from ±1 to ±100. If we plot the peak amplitude of the sinusoidal output, it will plot as a straight line if the system is linear or some other curve if it is not. The MATLAB code for this evaluation is given below.

```
% Example 5.2 Example to evaluate whether an unknown process is linear.
%
t = 0:2*pi/500:2*pi;       % Sine wave time function, 500 points
for k = 1:100              % Amplitudes (k) will vary from 1 to 100
  x = k*sin(t);            % Generate a 1 cycle sine wave
  y = process_x(x);        % Input sine to process
  output(k) = max(y);      % Save max value of output
end
plot(output);              % Output, horizontal scale will be in amplitude
  ......label axes......
```

Analysis: The first instruction generates the time vector. Here we use a series of 500 numbers ranging from 0 to 2π. A "for-loop" is used to generate sine waves with amplitudes ranging from ±1 to ±100. Each sine wave, `x`, is used as the input signal to `process_x`. The function `process_x` produces an output signal, `y`, and the peak value of the output is determined using MATLAB's max routine and saved in variable array, `output`. These peak values are then plotted.

The plot of maximum output values (i.e., variable output) is a straight line, as shown in Figure 5.7, and indicates a linear relationship between the input and output amplitudes. Whatever the process really is, it does appear to be linear.

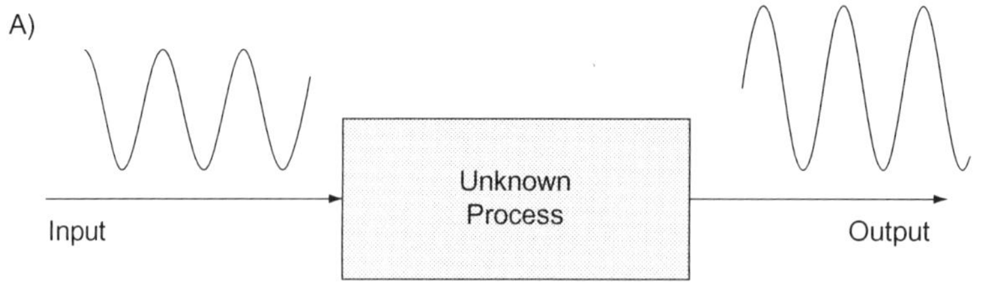

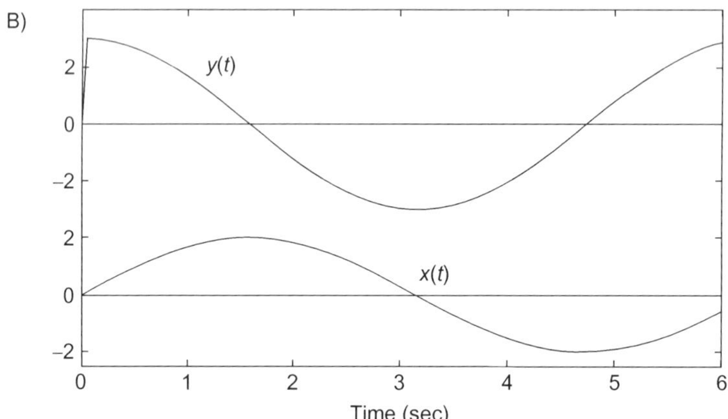

FIGURE 5.6 A) Many different sine waves are used as input signals to the "Unknown Process" and the output sine waves are analyzed for linearity. B) An example of one of the input signals, $x(t)$, and the corresponding output signal, $y(t)$, produced by `process_x`. The input signal is a sine wave as expected, since that is what was produced by the MATLAB code. The output looks like a cosine wave that is slightly larger than the input signal. This suggests that the process does contain a derivative operation, but in this example we are only interested in whether or not the process is linear.

Deriving the exact input/output equation for an unknown process solely by testing it with external signals can be a challenge. The field of *system identification* deals with approaches to obtain a mathematical representation for an unknown system by probing it with external signals. Comparing the input and output sinusoids, it looks like `process_x` contains a derivative and a multiplying factor that increases signal amplitude, but more input-output combinations using different frequency sine waves would have to be evaluated to confirm this guess.

5.1.5 Systems and Analog Analysis—Summary

Analog models represent the physiological process using elements that are, to some degree, analogous to those in the actual process. Good analog models can represent the system at a lower level, and in greater detail, than systems models, but not all analog models offer such detail. Analog models provide better representation of secondary

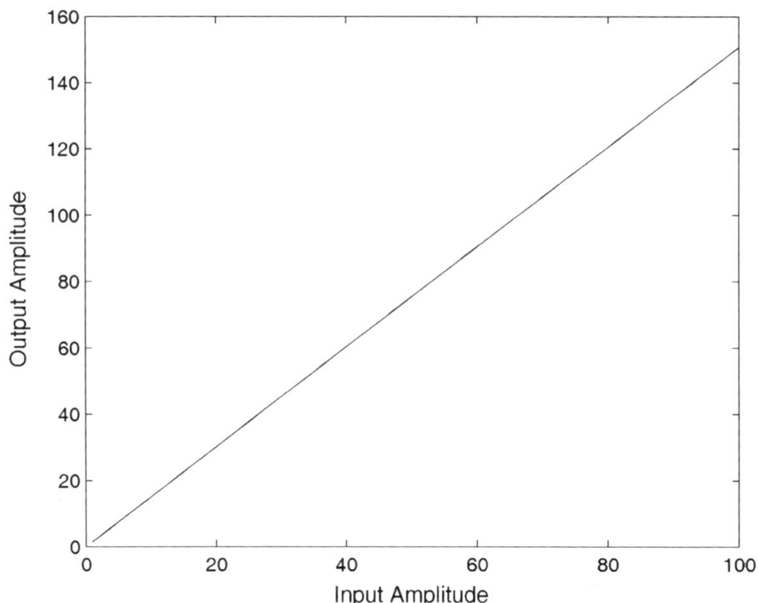

FIGURE 5.7 The relationship between the input and output amplitudes. The relationship is a straight line indicating that the system is linear.

features such as energy use, which is usually similar between analog elements and the actual components they represent. However, in analog models, the interaction between components may not be obvious from inspection of the model.

System models emphasize component interaction, particularly with regard to information flow, and clearly show the overall organization. For example, in the vergence system model of Figure 5.2, the fact that the ocular response also influences the neural controller is explicitly shown by a feedback pathway. This can be of great benefit in clarifying the control structure of a complex system. Perhaps the most significant advantage of the systems approach is what it does not represent: it allows the behaviors of biological processors to be quantitatively described without requiring the modeler to know the details of the underlying physiological mechanism.

5.2 THE RESPONSE OF SYSTEM ELEMENTS TO SINUSOIDAL INPUTS—PHASOR ANALYSIS

If the signals or variables in a system are sinusoidal or can be converted to sinusoids using the Fourier series or Fourier transform, then a technique known as phasor analysis can be used to convert calculus operations into algebraic operations. As used here, the term *phasor analysis* is considerably more mundane than the name implies: the analysis of phasors such as used on Star Trek is, unfortunately, beyond the scope of this text. The phasor analysis we will learn combines complex representation of sinusoids with the fact that calculus operations (integration and differentiation) change only the magnitude and phase of a sinusoid. This analysis assumes the signals are in sinusoidal steady state which means they always have been and always will be sinusoidal.

One of the great virtues of the sinusoid is that its frequency is not modified by calculus operations (see Property 8 in Chapter 3's "Useful Properties of the Sinusoidal Signal"), nor is it modified by scaling or arithmetic operations. That means that in a linear system, if the input is a sinusoidal steady-state signal, then *all* system variables including the output will be sinusoidal (see Property 9 in Chapter 3's "Useful Properties of the Sinusoidal Signal") at the same frequency. All signals in such an LTI system can be described by the same equation:

$$x(t) = A\cos(\omega t + \theta) = A\cos(2\pi f t + \theta) \tag{5.11}$$

where the values of A and θ can be modified by the system elements, but the value of ω (or f) will be the same throughout the system.

Sinusoids require just three variables for complete description: amplitude, phase, and frequency (see Property 2 in Chapter 3's "Useful Properties of the Sinusoidal Signal"). However, if the frequency is always the same, as it would be for any variable of an LTI system, then we really only need to keep track of two variables: amplitude[2] and phase. This suggests that complex variables and complex arithmetic may be of use in simplifying the mathematics of sinusoids, since a single complex variable is actually two variables rolled into one (i.e., $a + jb$). A single complex number or variable should be able to describe the amplitude and phase of a sinusoid.

To find a way to represent a sinusoid by a single complex variable, we return to the complex representation of sinusoids given by Euler's equation:

$$e^{jx} = \cos(x) + j\sin(x) \tag{5.12}$$

or in terms of the general equation for a sinusoid, Equation 5.11:

$$Ae^{j(\omega t + \theta)} = A\cos(\omega t + \theta) + jA\sin(\omega t + \theta) \tag{5.13}$$

Comparing the basic sinusoid equation, Equation 5.11, with Equation 5.13 shows that only the real part of e^{jx} is needed to represent a sinusoid:

$$A\cos(\omega t + \theta) = \mathrm{Re}\, Ae^{j\theta}e^{j\omega t} \tag{5.14}$$

If all variables in an equation contain the real part of the complex sinusoid, the real terms can be dropped, since if:

$$\begin{aligned} &\mathrm{Re}\, Ae^{j\theta_1}e^{j\omega t} = \mathrm{Re}\, Be^{j\theta_2}e^{j\omega t} \quad \text{for all } t \\ &\text{then} \\ &Ae^{j\theta_1}e^{j\omega t} = Be^{j\theta_2}e^{j\omega t} \end{aligned} \tag{5.15}$$

[2]There is a tendency to use the word "amplitude" when referring to the peak value of the sinusoid and the word "magnitude" when referring to the RMS value. Here the words *amplitude* and *magnitude* are used interchangeably.

In general, if Re A = Re B, A does not necessarily equal B. However the only way the Re $Ae^{j\omega t}$ can equal the Re $Be^{j\omega t}$ at all values of t is if $A = B$. Appendix E presents a review of complex arithmetic.

Since all variables in a sinusoidally-driven LTI system are the same except for amplitude and phase they will all contain the "Re" operator, and these terms can be removed from the equations as was done in Equation 5.15. They do not actually cancel; they are simply unnecessary since the equality stands just as well without them. Similarly, since all variables will be at the same frequency, the identical $e^{j\omega t}$ term will appear in each variable and will cancel after the "Re"s are dropped. Therefore, the general sinusoid of Equation 5.11 can be represented by a single complex number:

$$A\cos(\omega t + \theta) \Leftrightarrow Ae^{j\theta} \tag{5.16}$$

where $Ae^{j\theta}$ is the *phasor* representation of a sinusoid.

Equation 5.16 does not indicate a mathematical equivalence, but a transformation from the standard sinusoidal representation to a complex exponential representation without loss of information. In the phasor representation, the frequency, ω, is not explicitly stated, but is understood to be associated with every variable in the system (sort of a "virtual variable"). Note that the phasor variable, $Ae^{j\theta}$, is in polar form as opposed to the rectangular form of complex variable, $a + jb$, and may need to be converted to a rectangular representation in some calculations. Since the phasor, $Ae^{j\theta}$, is defined in terms of the cosine (Equation 5.16), sinusoids defined in terms of sine waves must be converted to cosine waves when using this analysis.

If phasors (i.e., $Ae^{j\theta}$) only offered a more succinct representation of a sinusoid, their usefulness would be limited. It is their "calculus-friendly" behavior that endears them to engineers. To determine the derivative of the phasor representation of a sinusoid, we return to the original complex definition of a sinusoid (i.e., Re $Ae^{j\theta}e^{j\omega t}$):

$$\frac{d(\text{Re}\, Ae^{j\theta}e^{j\omega t})}{dt} = \text{Re}\, j\omega Ae^{j\theta}e^{j\omega t} \tag{5.17}$$

The derivative of a phasor is the original phasor, but multiplied by $j\omega$. Hence, in phasor representation, taking the derivative is accomplished by multiplying the original term by $j\omega$, and a calculus operation has been reduced to a simple arithmetic operation:

$$\frac{d}{dt} \Leftrightarrow j\omega \tag{5.18}$$

Similarly, integration can be performed in the phasor domain simply by dividing by $j\omega$:

$$\int \text{Re}\, Ae^{j\theta}e^{j\omega t}dt = \text{Re}\frac{Ae^{j\theta}e^{j\omega t}}{j\omega} \tag{5.19}$$

and the operation of integration becomes an arithmetic operation:

$$\int dt \Leftrightarrow \frac{1}{j\omega} \tag{5.20}$$

The basic rules of complex arithmetic are covered in Appendix E; however, a few properties of the complex operator j are noted here. Note that $1/j$ is the same as $-j$, since:

$$\frac{1}{j} = \frac{1}{\sqrt{-1}} = \frac{-\sqrt{-1}}{(-\sqrt{-1})(-\sqrt{-1})} = \frac{-\sqrt{-1}}{-(-1)} = -\sqrt{-1} = -j \tag{5.21}$$

So Equation 5.20 could also be written as:

$$\int dt \Leftrightarrow -j\omega \tag{5.22}$$

Multiplying by j in complex arithmetic is the same as shifting the phase by 90 deg, which follows directly from Euler's equation:

$$je^{jx} = j(\cos x + j\sin x) = j\cos(x) - \sin(x) = -\sin(x) + j\cos(x)$$

Substituting in $\cos(x + 90)$ for $-\sin(x)$, and $\sin(x + 90)$ for $\cos(x)$, je^{jx} becomes:

$$je^{jx} = \cos(x + 90) + j\sin(x + 90)$$

This is the same as e^{jx+90} which equals $e^{jx}\,e^{90}$

$$je^{jx} = e^{jx}\,e^{90} \tag{5.23}$$

Similarly, dividing by j is the equivalent of shifting the phase by -90 deg:

$$\frac{e^{jx}}{j} = \frac{\cos(x) + j\sin(x)}{j} = \frac{\cos x}{j} + \sin(x) = -j\cos(x) + \sin(x)$$

Substituting in $\cos(x - 90)$ for $\sin(x)$, and $\sin(x - 90)$ for $-\cos(x)$; $\frac{e^{x/j}}{j}$ becomes:

$$\frac{e^{jx}}{j} = \cos(x - 90) + j\sin(x - 90) = e^{jx}e^{-90} \tag{5.24}$$

Equations 5.18, and 5.20 demonstrate the benefit of representing sinusoids by phasors: the calculus operations of differentiation and integration become the algebraic operations of multiplication and division. Moreover, the bilateral transformation that converts between the time and phasor representation domain (Equation 5.16), is easy to implement going in either direction.

EXAMPLE 5.3

Find the derivative of $x(t) = 10\cos(2t + 20)$ using phasor analysis.

Solution: Convert $x(t)$ to phasor representation (represented as $x(j\omega)$), multiply by $j\omega$, then take the inverse phasor transform:

$$10\cos(2t + 20) \Leftrightarrow 10e^{j20}$$

$$\frac{dx(j\omega)}{dt} \Leftrightarrow j\omega(10e^{j20}) = j2(10e^{j20}) = j20e^{j20}$$

$$j20e^{j20} = 20e^{j20}e^{j90} \Leftrightarrow 20\cos(2t + 20 + 90) = 20\cos(2t + 110)$$

Since $\cos(x) = \sin(x + 90) = -\sin(x - 90)$, this can also be written as $-20\sin(2t + 20)$, which is obtained from straight differentiation.

A shorthand notation is common for the phasor description of a sinusoid. Rather than write $Ve^{j\theta}$, we simply write $V\angle\theta$ (stated as V and at an angle of θ). When a time variable such as $v(t)$ is converted to a phasor variable, it is common to write it as a function of ω using capital letters: $V(\omega)$. This acknowledges the fact that phasors represent sinusoids at a specific frequency even though the sinusoidal term, $e^{j\omega t}$, is not explicitly shown in the phasor itself. When discussing phasors and implementing phasor analysis, it is common to represent frequency in radians/sec, ω, rather than in Hz, f, even though Hz is more likely to be used in practical settings. Hence, the time-phasor transformation for variable $v(t)$ can be stated as:

$$\mathrm{v}(t) \Leftrightarrow V(\omega) = V\angle\theta \tag{5.25}$$

In this notation, the phasor representation of $20\cos(2t + 110)$ is written as $20\ \angle 110$ rather than $20e^{j110}$. Sometimes, the phasor representation of a sinusoid expresses the amplitude of the sinusoid in RMS values rather than peak-to-peak values, in which case the phasor representation of $20\cos(2t + 110)$ is written as $(0.707)\ 20\angle 110 = 14.14\ \angle 110$. In this text, peak-to-peak values are used. It really does not matter, as long as we are consistent.

The phasor approach is an excellent method for simplifying the mathematics of LTI systems. It can be applied to all systems that are driven by sinusoids or, with the use of the Fourier series analysis or the Fourier transform, to systems driven by any periodic signal.

5.3 THE TRANSFER FUNCTION

The transfer function introduced above gets its name because it describes how an input, *Input*(ω), is "transferred" to the output, *Output*(ω). The concept of the transfer function is so compelling that it has been generalized to include many different types of processes or systems with different types of inputs and outputs. In the phasor domain, the transfer function is a modification of Equation 5.5:

$$Transfer\ Function(\omega) = \frac{Output(\omega)}{Input(\omega)} \tag{5.26}$$

Frequently both *Input*(ω) and *Output*(ω) are signals and have units in volts, but for now we will continue to use general terms. By strict definition, the transfer function should always be a function of ω, f, or, as shown in the next chapter, the Laplace variable, s, but the idea of expressing the behavior of a process by its transfer function is so powerful that it is sometimes used in a nonmathematical, conceptual sense.

EXAMPLE 5.4

The system shown in Figure 5.8 is a simplified linearized model of the Guyton-Coleman body fluid balance system presented by Rideout (1991). This model describes how the arterial blood pressure P_A in mmHg responds to a small change in fluid intake, F_{IN} in ml/min. Find the transfer function for this system, $P_A(\omega)/F_{IN}(\omega)$, using phasor analysis.

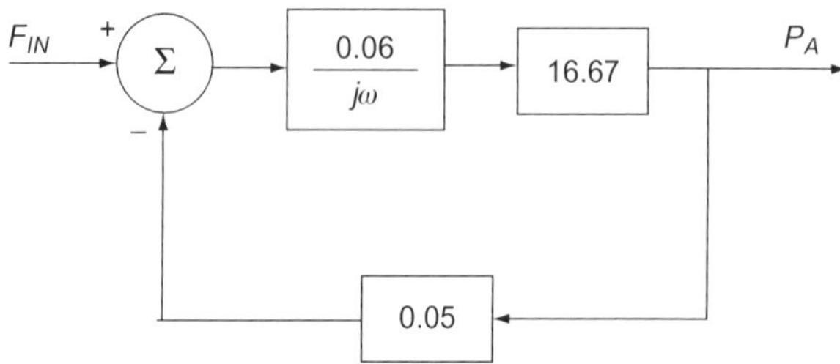

FIGURE 5.8 A simplified, linearized model of the Guyton-Coleman body fluid balance system that describes how the arterial blood pressure P_A responds to a small change in fluid intake, F_{IN}. If the two elements in the upper path are combined, this system has the same configuration as that shown in Figure 5.5.

Solution: The two elements in the upper path can be combined into a single element using Equation 5.7. The transfer function of the combined element is just the product of the two individual transfer functions:

$$G(\omega) = G_1(\omega)G_2(\omega) = \frac{0.06}{j\omega}(16.67) = \frac{1}{j\omega}$$

The resulting system has the same configuration as the feedback system in Figure 5.5, which has already been solved in Example 5.1. We could go though the same algebraic process, but it is easier to use the fact that the system in Figure 5.8 has the same configuration as the feedback system in Figure 5.5, where the transfer function of G is $1/j\omega$ and the feedback gain, H, is 0.05.

Substituting the expressions for G and H into the feedback equation (Equation 5.10), gives the transfer function of this system:

$$\frac{Out(\omega)}{In(\omega)} = \frac{G}{1+GH} = \frac{1/j\omega}{1+0.05(1/j\omega)} = \frac{1}{0.05+j\omega} = \frac{20}{1+j20\omega}\frac{\text{mmHg}}{\text{ml/min}}$$

Analysis: This transfer function applies to any input signal at any frequency, ω, as long as it is sinusoidal steady state. Note that the denominator consists of a real and an imaginary part, and that the constant term is normalized to 1.0. This is the common format for transfer function equations: the lowest power of ω, usually a constant term, is normalized to 1.0.

Systems having transfer functions with a $1 + jk\omega$ term in the denominator, where k is a constant, are called *first-order* systems and are discussed later in this chapter. The output of this system to a specific sinusoidal input is determined in the next example.

Many of the transfer functions encountered in this text have the same units for both numerator and denominator terms (e.g., volts/volts) so these transfer functions are dimensionless. However, when this approach is used to represent physiological systems, the numerator and denominator often have different units, as is the case here (i.e., the transfer function has the dimensions of mmHg/ml/min).

EXAMPLE 5.5

Find the output of the system in Figure 5.8 if the input signal is $F_{IN}(t) = 0.5\sin(0.3t + 20)$ ml/min. The model parameters in Figure 5.8, have been determined assuming time, t, is in hours even though the input is in ml/min.

Solution: Since phasors are based on cosine, we first need to convert $F_{IN}(t)$ to a cosine wave. From Appendix C, Equation 4: $\sin(\omega t) = \cos(\omega t - 90)$, so $F_{IN}(t) = 0.5\cos(0.3t - 70)$ ml/min.

In phasor notation the input signal is:

$$F_{IN}(t) = 0.5\cos(0.1t - 70) \Leftrightarrow F_{IN}(\omega) = 0.5\ \angle -70\ ml/min$$

From the solution to Example 5.4, the transfer function is:

$$\frac{Out(\omega)}{In(\omega)} = \frac{20}{1 + j20\omega}$$

Solving for $Out(\omega)$ and then substituting in $0.5\angle -70$ for $In(\omega)$ and letting $\omega = 0.3$

$$Out(\omega) = \frac{20}{1 + j20}\quad In(\omega) = \frac{20}{1 + j20(0.3)}(0.5\angle -70) = \frac{10\angle -70}{1 + j6}$$

The rest of the problem is just working out the complex arithmetic. To perform division (or multiplication), it is easiest to have a complex number in polar form. The number representing the input, which is also the numerator, is already in polar form ($5\angle -70$), so the denominator needs to be converted to polar form:

$$1 + j6 = \sqrt{1^2 + 6^2}\angle \tan^{-1}\left(\frac{6}{1}\right) = 6.08\angle 80$$

Substituting and solving:

$$P_A(\omega) = \frac{10\angle -70}{6.08\angle 80} = 1.64\angle -150$$

Converting to the time domain:

$$P_A(t) = 1.64\cos(0.3t - 150)\ \text{mmHg}$$

So the arterial pressure response to this input is a sinusoidal variation in blood pressure of 1.64 mmHg. That variation has a phase of 150 deg, which is shifted 80 deg from the input phase. The corresponding time delay between the stimulus and response sinusoids can be found from Equation 2.7 noting that time is in hours:

$$t_d = \frac{\theta}{360f} = \frac{\theta}{360(\omega/2\pi)} = \frac{80}{360(0.3/2\pi)} = 4.65\ \text{hrs}$$

So the response is delayed 4.65 hrs from the stimulus due to the long delays in transferring fluid into the blood. As with many physiological systems, generating a sinusoidal stimulus, in this case a sinusoidal variation in fluid intake, is difficult. In the next chapter we see how to apply these same techniques to stimuli that are more commonly used with living systems. The next example shows the application of phase analysis to a more complicated system.

EXAMPLE 5.6

Find the transfer function of the system shown in Figure 5.9. Find the time-domain output if the input is 10 cos(20*t* + 30).

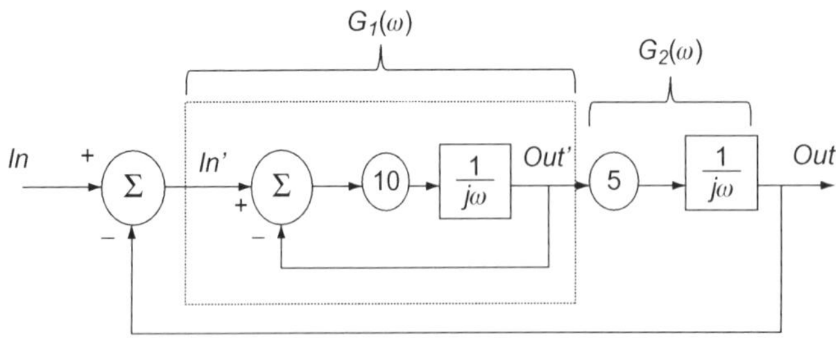

FIGURE 5.9 Configuration of basic elements in the system used in Example 5.6. This system features three of the four basic elements that can be found in a linear system: arithmetic, scalar, and integrator. In this "system diagram," the transfer function of each element is indicated inside the black box representing that element.

Solution: Since we do not already have a transfer function for this configuration of elements, we need to write the equations for the elements or combination of elements and solve for the overall output in terms of the input. In all of these problems, the best strategy is to find the equivalent feedforward gain, *G*, and then apply the feedback equation to find the overall transfer function. To facilitate this strategy, some internal signals are labeled for reference. Note that the elements inside the rectangle constitute an internal feedback system with input *In′* and output *Out′*. So we can use the basic feedback equation (Equation 5.10) to get the transfer function of this internal subsystem. For this subsystem, the feedforward gain is a combination of the scalar and integrator:

$$G' = 10\left(\frac{1}{j\omega}\right)$$

and since the all of the output, *Out*, is feedback to the subtractor, $H = 1$. So the transfer function for the subsystem becomes:

$$G_1 = \frac{Out'}{In'} = \frac{10/j\omega}{1 + 10/j\omega} = \frac{10}{j\omega + 10} = \frac{1}{1 + j0.1\omega}$$

As noted above, it is common to normalize the denominator of these transfer functions so that the constant term (or lowest power of ω) is equal to 1 as was done here.

The rest of the feedforward path consists of another scalar and integrator which can be represented by transfer function G_2.

$$G_2 = 5\left(\frac{1}{j\omega}\right) = \frac{5}{j\omega}$$

The transfer function of the feedforward gain is the product of G_1 and G_2:

$$G = G_1 G_2 = \left(\frac{5}{j\omega}\right)\left(\frac{1}{1 + j0.1\omega}\right) = \frac{5}{-0.1\omega^2 + j\omega}$$

We apply the feedback equation yet again, this time to find the transfer function of the overall system. Substituting G above into the feedback equation and letting $H = 1$:

$$\frac{Out}{In} = \frac{G}{1 + GH} = \frac{\frac{5}{-0.1\omega^2 + j\omega}}{1 + \frac{5}{-0.1\omega^2 + j\omega}} = \frac{5}{-0.1\omega^2 + j\omega + 5} = \frac{1}{1 - (0.1/5)\omega^2 + j\omega/5}$$

Rearranging the denominator to collect the real and imaginary parts, the system transfer function becomes:

$$\frac{Out}{In} = \frac{1}{1 - 0.02\omega^2 + j0.2\omega}$$

This type of transfer function is known as a second-order function, as the polynomial in the denominator is a second-order frequency term, ω^2.

Finding the output given the input is only a matter of complex algebra. In phasor notation the input signal is $10\angle 30$ with $\omega = 20$. The output in phasor notation becomes:

$$Out = TF\ In = \frac{1(10\angle 30)}{1 - 8 + j4} = \frac{10\angle 30}{-7 + j4} = \frac{10\angle 30}{8\angle 150} = 1.25\angle -120$$

Converting to the time domain: $out(t) = 1.25\cos(20t - 120)$.

5.3.1 The Spectrum of a Transfer Function

One of the best ways to examine transfer functions and the behavior of the systems they represent is the same approach as that used for signals: a plot of their spectral characteristics. The transfer function specifies how a system influences input signals at different frequencies, so the spectrum of the transfer function shows how frequencies passing through a system are altered. In the phasor domain, the transfer function is a function of ω, so at any specific frequency it will resolve to a single number, probably a complex number. As with the Fourier series of a signal, this number can be determined for a range of frequencies and plotted as magnitude and phase at each frequency.

The signal spectrum describes the energy levels in a signal as a function of frequency. The system spectrum describes how the system changes the energy levels from input to output, also as a function of frequency. For example, Figure 5.10 shows the spectrum of a system where the lower frequencies have a magnitude ratio of 1.0. This means that frequencies in this range pass through the system without change: They emerge at the same amplitude as they went in. Unlike a signal spectrum, this does not mean that the output of the system will contain energy at those lower frequencies; the output will contain energy at these frequencies only if the input contains energy at those frequencies. At the higher frequencies the system spectrum is reduced, so any energy in the input signal at these frequencies will be reduced in the output signal. For example, at around 100-Hz the transfer function has a magnitude value of around 0.707. That means that if the input signal has energy at 100-Hz, the output signal will have half the energy at that frequency. (Recall that when the magnitude is reduced by 0.707, the energy is reduced by 0.707^2 or 0.5.) Following the same logic, the phase plot does not show the phase angle of the output as it would in the signal spectrum, but rather the change in phase angle induced by the system.

If the entire spectral range of a system is explored, then the transfer function spectrum provides a complete description of the way in which the system alters all possible input signals. A complete system spectrum is a complete description of the system, at least from

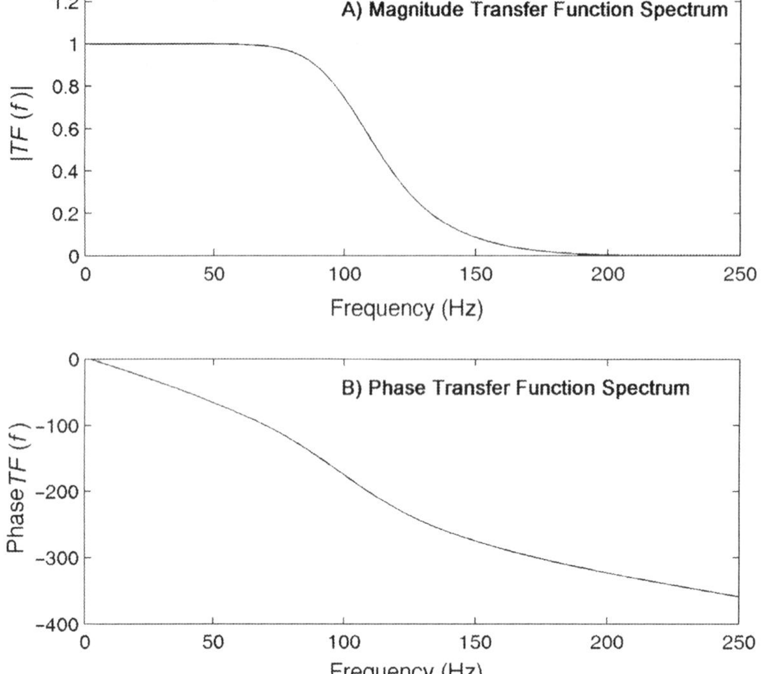

FIGURE 5.10 The spectrum of a system showing how the system alters the frequency components of an input signal in both magnitude (upper plot) and phase (lower plot). For this system, the lower frequencies pass though the system unaltered in magnitude but slightly shifted in phase. Higher frequencies are significantly reduced in amplitude and greatly shifted in phase.

the systems analysis point of view where only the input/output characteristics are of concern. Chapter 3 shows us that a signal could be represented equally well in either the time or the frequency domain. The same is true of a system: the frequency characteristics of the transfer function provide a complete description of the system's input-output characteristics. In fact, the input-output or transfer properties of systems are most often represented in the frequency domain. It is easy to generate the frequency characteristics of any transfer function using MATLAB, as shown in Example 5.7.

EXAMPLE 5.7

The transfer function in this example represents the relationship between applied external pressure $P(\omega)$ and airway flow $Q(\omega)$ for a person on a respirator in an intensive care unit (ICU). This transfer function was derived by Khoo (2000) for typical lung parameters, but could be modified for specific patients. (The original model was given in the Laplace domain as covered in the next chapter, but has been modified to the frequency domain for this example.) We want to find the magnitude and phase spectrum of this airway system. We use MATLAB to plot the magnitude in dB and phase spectrum in degrees, both against log frequency in Hz. The transfer function is:

$$\frac{Q(\omega)}{P(\omega)} = \frac{9.52j\omega(1 + j0.0024\omega)}{1 - 0.00025\omega^2 + j0.155\omega} \quad \frac{l/\text{min}}{\text{mmHg}}$$

where $Q(\omega)$ is airway flow in l/min and $P(\omega)$ is the pressure applied by the respirator in mmHg. Again transfer functions derived from biological systems usually have inputs and outputs in different units, as is the case here, so the transfer function has a dimension.

Solution: To plot this in MATLAB, we need to construct a frequency vector and substitute it for ω in the transfer function equation. The only question is what frequency ranges should we include in the vector? Since MATLAB will do all the work, we might as well use a large-frequency vector and reduce if it seems too extensive. The original range that was tried was from 0.1 to 1000 rad/sec in steps of 0.1 radians, but was extended to 0.001 to 10,000 in steps of 0.01 to better cover the lower frequencies. In this model, time is in minutes not seconds, so the plotting frequency needs to be modified accordingly (to cycles/min). Since there are 60 seconds in a minute, the plotting frequency should be divided by 60 (as well as by 2π to convert to cycles/sec; i.e., Hz).

```
% Example 5.7 Use MATLAB to plot the given transfer function
%
w = .001:0.001:1000;                        % Construct frequency vector
TF = (9.52*j*w.*(1+j*w*.0024))./(1 - (w.^2)*.00025 + .155*j*w);
% Trans func
Mag = 20*log10(abs(TF));                    % Magnitude of TF in dB
Phase = unwrap(angle(TF))*360/(2*pi);       % Phase of TF in deg
subplot(2,1,1);
semilogx(w/(2*pi),Mag);                     % Plot |TF| in dB against log freq in Hz.
  ......labels, gird......
subplot(2,1,2);
semilogx(w/2*pi),Phase,'k','LineWidth',2);  % Plot phase in deg, log freq in Hz
  ......labels, gird......
```

The spectral plots for this respiratory-airway system are shown in Figure 5.11. The flow is constant for a given external pressure for frequencies around 0.1-Hz (corresponding to 6- breaths/

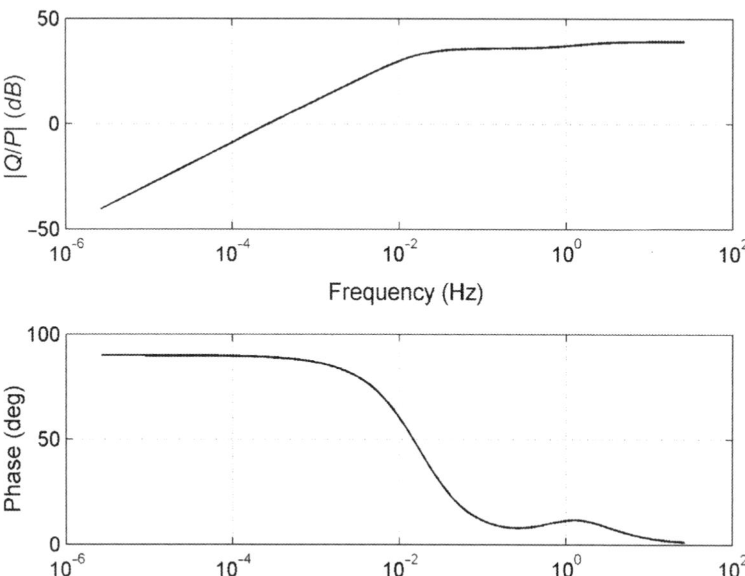

FIGURE 5.11 The magnitude and phase plots of the transfer function found for the respiratory-airway system in Example 5.7. For a given external pressure, flow is reduced for frequencies below approximately 0.1-Hz (6- breaths/min) but is constant for higher frequencies.

min) but decreases for slower breathing. However, around the normal respiratory range (11 breaths/min = 0.18-Hz), the magnitude transfer function is close to maximum.

Relegating the problem to the computer is easy but sometimes more insight is found by working it out manually. To really understand the frequency characteristics of transfer functions, it is necessary to examine the typical components of a general transfer function. In so doing, we learn how to plot transfer functions without the aid of a computer. More importantly, by examining the component structure of a typical transfer function, we also gain insight into what the transfer function actually represents. This knowledge is often sufficient to allow us to examine a process strictly within the frequency domain and learn enough about the system that we need not even bother with time-domain responses.

5.4 TRANSFER FUNCTION SPECTRAL PLOTS—THE BODE PLOT

To demonstrate how to determine the spectral plot from the transfer function (without using MATLAB), we start with the four basic elements. We then look at a couple of popular configurations of these elements. The arithmetic element (summation and subtraction) does not really have a spectrum, as it performs its operations the same way at all frequencies. Learning the spectra of the various elements is a bit tedious, but results in increased understanding of systems and their design.

5.4.1 Constant Gain Element

The constant gain element shown in Figure 5.12 has already been introduced in previous examples. This element simply scales the input to produce the output and the transfer function for this element is a constant called the *gain*. When the gain is greater than 1, the output is larger than the input and vice versa. In electronic systems, a gain element is called an *amplifier*.

The transfer function for this element is:

$$TF(\omega) = G \tag{5.27}$$

The output of a gain element is $y(t) = Gx(t)$ and hence only depends on the instantaneous and current value of t. Since the output does not depend on past values of time, it is a memoryless element.

In the spectrum plot, the transfer function magnitude is usually plotted in dB so the transfer function magnitude equation is:

$$|TF(\omega)|_{dB} = |20 \log TF(\omega)| = 20 \log G \tag{5.28}$$

The magnitude spectrum of this element is a horizontal line at 20 log G. If there are other elements contributing to the spectral plot, it is easiest simply to rescale the vertical

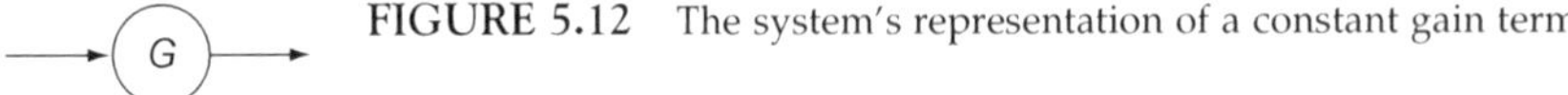

FIGURE 5.12 The system's representation of a constant gain term.

axis so that the former zero line equals 20 log G. This rescaling is shown in the examples below.

The phase of this element is zero since the phase angle of a real constant is zero. The phase spectrum of this element plots as a horizontal line at a level of zero and essentially makes no contribution to the phase plot.

$$\angle TF(\omega) = \angle G = 0 \tag{5.29}$$

5.4.2 Derivative Element

The derivative element (Figure 5.13A), has somewhat more interesting frequency characteristics. The output of a derivative element depends on both the current and past input values, so it is an example of an element with memory. This basic element is sometimes referred to as an *isolated zero* for reasons that become apparent later. The transfer function of this element is:

$$TF(\omega) = j\omega \tag{5.30}$$

The magnitude of this transfer function in dB is 20 log($|j\omega|$) = 20 log(ω), which is a logarithmic function of ω, and would plot as a curve on a linear frequency axis. However, it is common to plot transfer function spectra using the log of frequency on the horizontal axis. A log frequency axis is useful because frequencies of interest often span several orders of magnitude. The function log(ω) plots as a straight line when plotted against log frequency since it is log(ω) against log(ω). To find the intercept, note that when $\omega = 1$, 20 log(ω) = 0 since the log(1) = 0. So the magnitude plot of this transfer function is a straight line that intersects the 0 dB line at $\omega = 1$ rad/sec.

To find the slope of this line, note that when $\omega = 10$ the transfer function in dB equals 20 log(10) = 20 dB, and when $\omega = 100$ it equals 20 log(100) = 40 dB. So for every order of magnitude increase in frequency there is a 20-dB increase in the value of the Equation 5.30. This leads to unusual dimensions for the slope: specifically 20 dB/decade, but this is the result of the logarithmic scaling of the horizontal and vertical axes. The magnitude spectrum plot is shown in Figure 5.14 along with that of the integrator element.

The angle of $j\omega$ is +90 deg irrespective of the value of ω:

$$\angle j\omega = 90 \text{ deg.} \tag{5.31}$$

So the phase spectrum of this transfer function is a straight line at +90 deg. In constructing spectral plots manually, this term is usually not plotted, but is used to rescale the phase plot after the phase characteristics of the other elements are plotted.

FIGURE 5.13 A) The system representation of the derivative element. An alternate more commonly used representation is given in Chapter 6 using Laplace notation. B) The system representation of an integrator element.

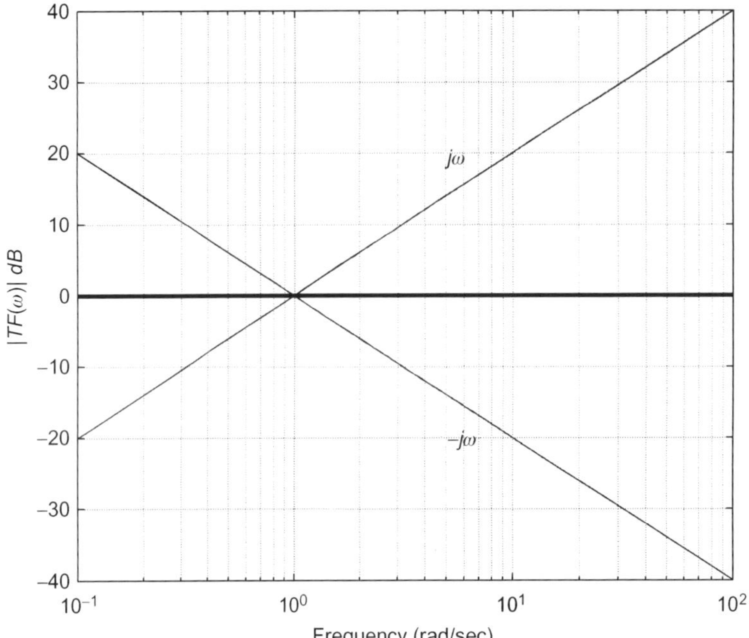

FIGURE 5.14 The magnitude spectra of the differentiator and integrator elements. The two elements have similar plots with opposite slopes.

5.4.3 Integrator Element

The last of the basic elements is the integrator element (Figure 5.13B), with a transfer function that is the inverse of the derivative element:

$$TF(\omega) = \frac{1}{j\omega} \tag{5.32}$$

This element also depends on current and past values of the input so is an element with memory. The derivative element is sometimes referred to as an *isolated pole* for reasons explained later. The magnitude spectrum in dB is $20\log\left|\frac{1}{j\omega}\right| = -20\ \log|j\omega| = -20\ \log(\omega)$, which plots as a straight line when the frequency axis is in log ω. The line intercepts 0 dB at $\omega = 1$ since $-20\ \log(1) = 0$ dB. This is similar to the derivative element described above, but unlike the derivative element, this element has the opposite slope: −20 dB/decade. The magnitude plot of this transfer function is shown in Figure 5.14.

The phase spectrum of this transfer function is:

$$\angle\frac{1}{j\omega} = -90\ \text{deg} \tag{5.33}$$

Again this is usually not plotted (since it is just a straight line), but rescales the phase axis after the plot is complete.

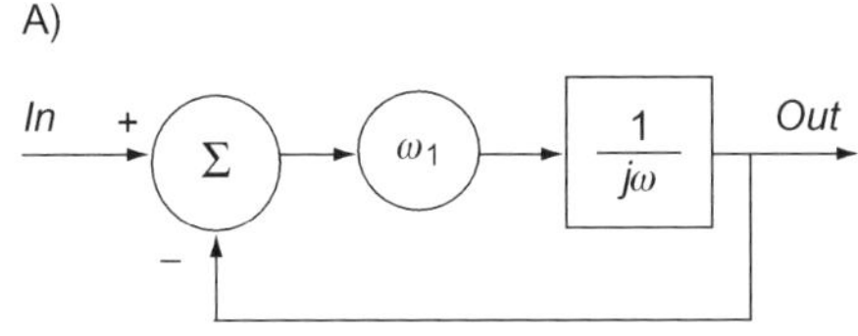

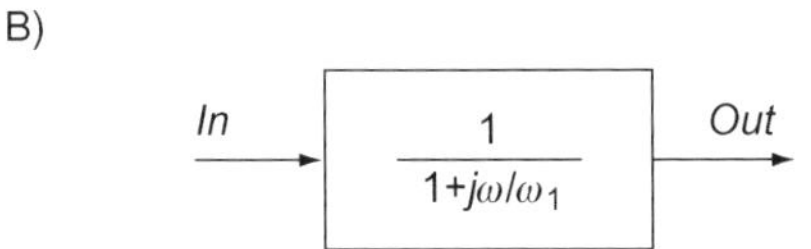

FIGURE 5.15 A) A system consisting of an integrator and constant elements enclosed with a feedback path. This gives rise to a first-order element. B) The single-element version of a first-order element.

5.4.4 First-Order Element

We can now plot both the magnitude and phase spectra of four basic elements, and none of them is particularly interesting. However, if we start putting some of these elements together, we can construct more interesting, and useful, spectra. The first-order element can be constructed by placing a negative feedback path around a gain term and an integrator (Figure 5.15A). The gain term is given the symbol ω_1 for reasons that will be apparent, but it is still only a constant.

The transfer function for a first-order element can be found easily from the feedback equation, Equation 5.10.

$$TF(\omega) = \frac{G}{1+GH} = \frac{\omega_1(1/j\omega)}{1+\omega_1(1/j\omega)} = \frac{\omega_1}{\omega_1 + j\omega} = \frac{1}{1+j\omega/\omega_1} \tag{5.34}$$

This combination is often presented as a single element, as shown in Figure 5.15B. It is even given its own name, a *first-order element*, so-called because the denominator of its transfer function is a first-order polynomial of the frequency variable, ω. This is also true of the integrator element (Equation 5.32), but it is an isolated term and not truly a polynomial. Since the first-order element contains an integrator and an integrator element has memory, the first-order element also has memory; its output depends on current and past values of the input.

The magnitude spectrum of the first-order element is established using an asymptote technique: the high-frequency asymptote and low-frequency asymptote are first plotted. Here high frequency means frequencies where $\omega >> \omega_1$ and low frequency means frequencies where $\omega << \omega_1$. To find the low-frequency asymptote in dB, take the limit as $\omega << \omega_1$ of the absolute value of 20 log of Equation 5.34:

$$|TF(\omega_{low})| = \lim_{\omega << \omega_1} 20\log\left|\frac{1}{1+j\,\omega/\omega_1}\right| = 20\log\left(\frac{1}{1+j0}\right) = 0\text{ dB} \tag{5.35}$$

The low-frequency asymptote given by Equation 5.35 plots as a horizontal line at 0 dB. The high-frequency asymptote is obtained in the same manner except that the limit taken is $\omega >> \omega_1$:

$$|TF(\omega_{high})| = \lim_{\omega>>\omega_1}\left[20\log\left|\frac{1}{1+j\,\omega/\omega_1}\right|\right] = 20\log\left|\frac{1}{j\,\omega/\omega_1}\right| = -20\log\left(\frac{\omega}{\omega_1}\right)\text{dB} \tag{5.36}$$

The high-frequency asymptote, $-20\log(\omega/\omega_1)$, is a logarithmic function of ω and plots as a straight line when frequency is plotted against log frequency since, again, it is log (ω/ω_1) against $\log(\omega)$. It has the same slope as the integrator element: -20 dB/decade. This line intersects the 0 dB line at $\omega = \omega_1$ since $-20\log(\omega_1/\omega_1) = -20\log(1) = 0$ dB.

Errors between the actual curve and the asymptotes occur when the asymptotic assumptions are no longer true: when the frequency, ω, is neither much greater than nor much less than ω_1. The biggest error will occur when ω exactly equals ω_1. At that frequency the magnitude value is:

$$|TF(\omega = \omega_1)| = 20\log\left|\frac{1}{1+j\,\omega_1/\omega_1}\right| = 20\log\left|\frac{1}{1+j}\right| = -20\log(\sqrt{2}) = -3\text{ dB} \tag{5.37}$$

To approximate the spectrum of the first-order element, we first plot the high- and low-frequency asymptotes, as shown in Figure 5.16 (dashed lines). Then we find the -3-dB point at $\omega = \omega_1$, Figure 5.16 'x' point. Finally, we draw a freehand curve from asymptote to asymptote through the -3 dB (Figure 5.16, solid line). In Figure 5.16, MATLAB was used to plot the actual curve, and the deviation from the two asymptotes is small. Often the asymptotes alone provide sufficient accuracy for this spectral curve.

The phase plot of the first-order element can be estimated using the same approach: taking the asymptotes and the worst case point, $\omega = \omega_1$. The low-frequency asymptote is:

$$\angle TF(\omega_{low}) = \lim_{\omega<<\omega_1}\left[\angle\left(\frac{1}{1+j\,\omega/\omega_1}\right)\right] = \angle 1 = 0\text{ deg.} \tag{5.38}$$

This is a straight line at 0 deg. In this case, the assumption "much much less than" is taken to mean one order of magnitude less, so that the low-frequency asymptote is assumed to be valid from 0.1 ω_1 on down, Figure 5.17 dashed line. The high-frequency asymptote is determined to be -90 deg in Equation 5.39 below, and by the same reasoning is assumed to be valid from 10 ω_1 and up, Figure 5.17 dashed line.

$$\angle TF(\omega_{high}) = \lim_{\omega>>\omega_1}\left[\angle\left(\frac{1}{1+j\,\omega/\omega_1}\right)\right] = \angle\left(\frac{1}{j\,\omega/\omega_1}\right) = -90\text{ deg.} \tag{5.39}$$

Again the greatest difference between the asymptotes and the actual curve is when ω equals ω_1:

$$\angle(TF(\omega = \omega_1)) = \angle\left(\frac{1}{1+j\,\omega_1/\omega_1}\right) = \angle\left(\frac{1}{1+j}\right) = -45\text{ deg.} \tag{5.40}$$

This value, 45 deg, is exactly halfway between the high- and low-frequency asymptotes, Figure 5.17 'x' point. What is usually done is to draw a straight line between the high end of the low-frequency asymptote at 0.1 ω_1 and the low end of the high-frequency asymptote

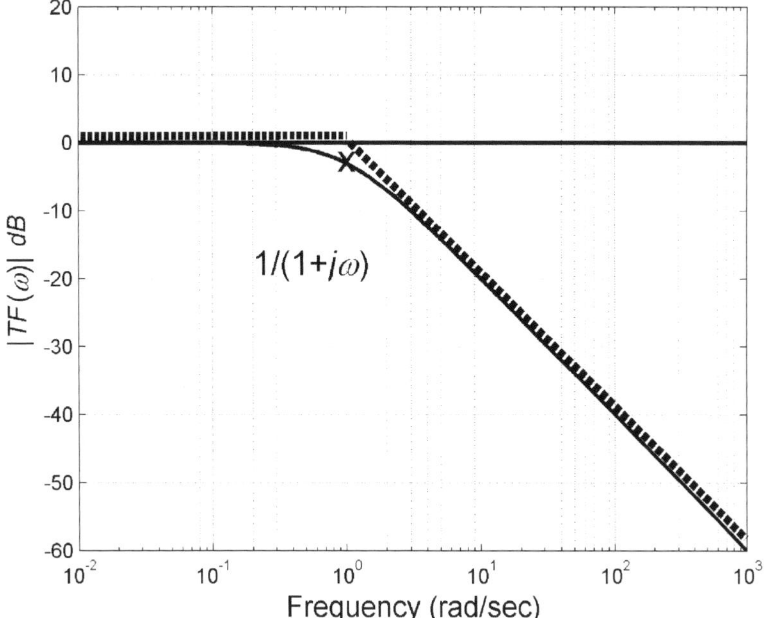

FIGURE 5.16 Magnitude spectrum of a first-order element. In this plot ω_1 has been normalized to 1.0 rad/sec. The low-frequency asymptote is valid for frequencies $\omega << \omega_1$ and is a straight line at 0 dB. The high-frequency asymptote is valid for frequencies $\omega >> \omega_1$ and plots as a line with a slope of -20 dB/decade. The high-frequency asymptote intersects the 0 dB line at $\omega = \omega_1$. Thus the two curves intersect at $\omega = \omega_1$ (in this case $\omega_1 = 1.0$ rad/sec). The greatest deviation between the curves and their asymptotes occurs at ω_1 and this error is small (-3 dB), 'x' point.

at 10 ω_1 passing through 45 deg, Figure 5.17 dashed line. Although the phase curve is nonlinear in this range, Figure 5.17 solid line, the error induced by a straight line approximation is surprisingly small, as shown in Figure 5.17. To help you to plot these spectral curves by hand, the CD that accompanies this book contains files semilog.pdf which can be printed out to produce semilog graph paper.

There is a variation of the transfer function described in Equation 5.34 that, while not as common, is sometimes found as part of the transfer function of other elements. One such variation has the first-order polynomial in the numerator and no denominator:

$$TF(\omega) = 1 + j\frac{\omega}{\omega_1} \tag{5.41}$$

The asymptotes for this transfer function are very similar to that of Equation 5.34; in fact, the low-frequency asymptote is the same.

$$|TF(\omega_{low})| = \lim_{\omega<<\omega_1} 20\log\left|1 + j\frac{\omega}{\omega_1}\right| = 20\log(1) = 0 \text{ dB} \tag{5.42}$$

The high-frequency asymptote has the same intercept, but the slope is positive, not negative:

$$|TF(\omega_{high})| = \lim_{\omega>>\omega_1}\left[20\log\left|1 + j\frac{\omega}{\omega_1}\right|\right] = 20\log\left|j\frac{\omega}{\omega_1}\right| = 20\log\left(\frac{\omega}{\omega_1}\right)\text{dB} \tag{5.43}$$

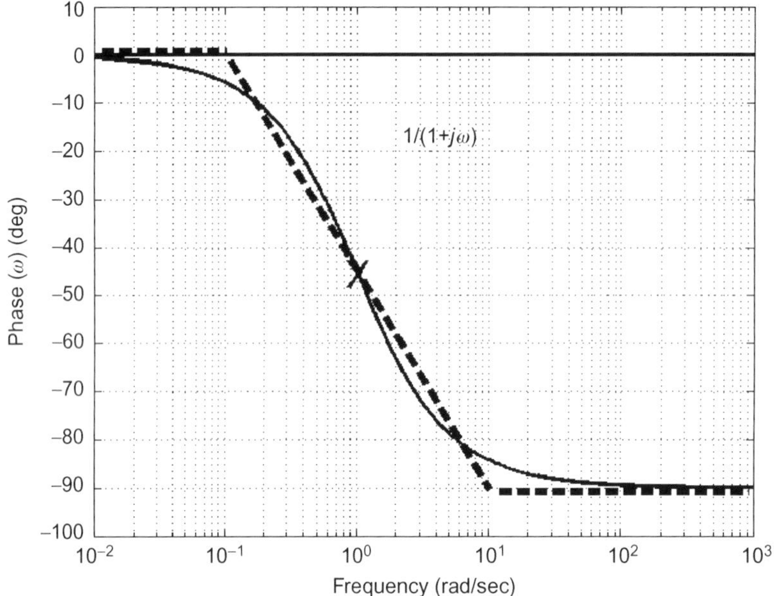

FIGURE 5.17 Phase spectrum of a first-order element. For this plot, ω_1 has been normalized to 1.0 rad/sec. The low-frequency asymptote is 0.0 deg up to 0.1 ω_1 and the high-frequency asymptote is −90 deg from 10 ω_1 on up, dashed lines. The asymptote between 0.1 ω_1 and 10 ω_1 is drawn as a straight line that passes through −45 deg at $\omega = \omega_1$, dashed line. This intermediate asymptote approximates the actual curve quite well, solid line.

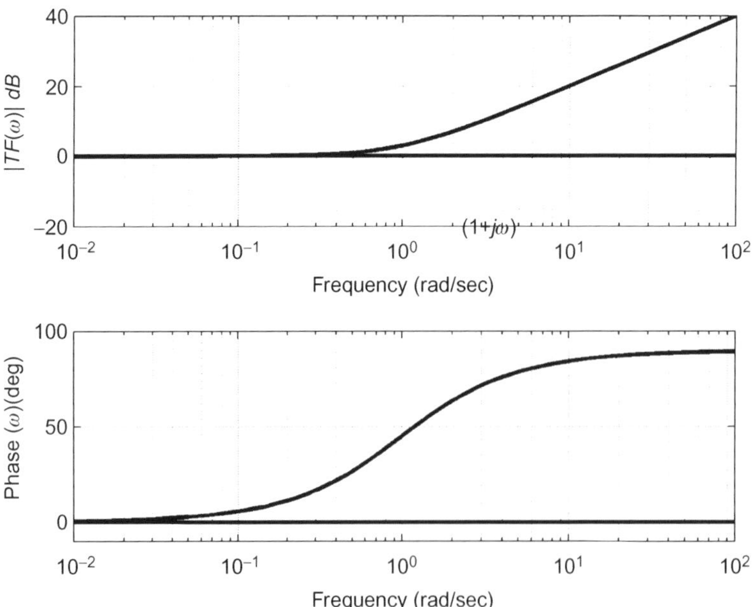

FIGURE 5.18 The magnitude and phase spectrum for the first-order transfer function when the frequency polynomial is in the numerator (Equation 5.41).

Similarly the value of the transfer function when $\omega = \omega_1$ is +3 dB instead of −3 dB:

$$|TF(\omega = \omega_1)| = 20 \log\left|1 + j\frac{\omega_1}{\omega_1}\right| = 20 \log(\sqrt{2}) = 3 \text{ dB} \tag{5.44}$$

The phase plots are also the same except that the change in angle with frequency is positive rather than negative. Demonstration of this is left as an exercise in Problem 12 at the end of this chapter. The magnitude and phase spectra of the transfer function given in Equation 5.41 is shown in Figure 5.18.

EXAMPLE 5.8

Plot the magnitude and phase spectrum of the transfer function below:

$$TF(\omega) = \frac{100}{(1 + j0.1\omega)}$$

Solution: This transfer function is actually a combination of two elements: a gain term of 100 and a first-order term where ω_1 is equal to $1/0.1 = 10$. The easiest way to plot the spectrum of this transfer function is to plot the first-order term first, then rescale the vertical axis of the magnitude spectrum so that 0 dB is equal to 20 log(100) = 40 dB. The rescaling will account for the gain term. The gain term has no influence on the phase plot so this may be plotted directly.

Results: Figure 5.19 shows the resulting magnitude and phase spectral plots. For the magnitude plot, the spectrum was drawn freehand through the −3-dB point. In the phase plot, usually only the asymptotes are used, but in this case a freehand curve was drawn.

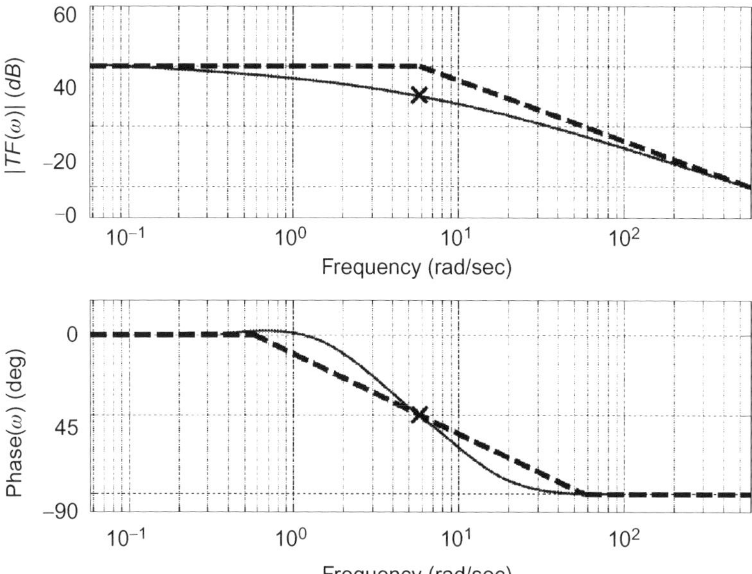

FIGURE 5.19 The magnitude and phase spectrum of the transfer function given in Example 5.8. The magnitude spectrum features a hand-drawn approximation that traverses the −3-dB point and the horizontal axis has been rescaled so that 0 dB is now 40 dB. In the phase plot, a freehand curve is also show but usually only the asymptotes are plotted. The magnitude spectrum is identical to the one in Figure 5.16 except for the location of the −3-dB point and the vertical scale of the magnitude spectrum.

5.4.5 Second-Order Element

A second-order element has, as you might expect, a transfer function that includes a second-order polynomial of ω. A second-order element contains two integrators so it also has memory: its output at any given time depends on both current and past values of the input. One way to construct a second-order system is to put two first-order systems in series, but other configurations also produce a second-order equation as long as they contain two integrators. For example, the system shown in Figure 5.9 was found to have a second-order ω term in the denominator. The general equation for a second-order system is given as:

$$TF(\omega) = \frac{1}{1 - (\omega/\omega_n)^2 + j2\delta\,\omega/\omega_n} \tag{5.45}$$

where ω_n and δ are two constants associated with the second-order polynomial. These constants have a more complicated relationship to the frequency (i.e., ω) coefficients, but they have direct relationships to both the time behavior and the spectrum. The parameter δ is called the *damping factor* and ω_n the *undamped natural frequency*. These names relate to time-domain behaviors, so they will not be meaningful until we look at the time-domain behavior in Chapter 7. We find that ω_n and δ are useful, so we want to determine their value from a second-order polynomial, as shown in the example below.

EXAMPLE 5.9

Find the constants ω_n and δ in the second-order transfer function below:

$$TF(\omega) = \frac{1}{1 - .03\omega^2 + j.01\omega} \tag{5.46}$$

Solution: We can obtain the equivalent constants ω_n and δ by equating coefficients between the denominator terms of Equation 5.45 and Equation 5.46:

$$\left(\frac{1}{\omega_n}\right)^2 = 0.03; \quad \omega_n = \frac{1}{\sqrt{0.03}} = 5.77 \text{ rad/sec}$$

$$\frac{2\delta}{\omega_n} = 0.1; \quad \delta = \frac{0.01\omega_n}{2} = \frac{(5.77)\,0.01}{2} = 0.029$$

If the roots of the denominator polynomial are real, it can be factored into two first-order terms and then separated into two first-order transfer functions (Equation 5.34) using partial fraction expansion. This is done in Section 6.6 of Chapter 6, but for spectral plotting, the second-order term will be treated as is.

Not surprisingly, the second-order element is more challenging to plot that the previous transfer functions. Basically the same strategy is used as in the first-order function except special care must be taken when $\omega = \omega_n$. We begin by finding the high- and low-frequency asymptotes. These now occur when ω is either much greater or much less than ω_n. When $\omega << \omega_n$, both the ω and the ω^2 terms go to 0 and the denominator goes to 1.0:

$$|TF(\omega_{low})| = \lim_{\omega<<\omega_n}\left[20\log\left|\frac{1}{1-(\omega/\omega_1)^2+j2d\ \omega/\omega_1}\right|\right] = 20\log(1) = 0\ \text{dB} \tag{5.47}$$

The low-frequency asymptote is the same as for the first-order element, namely a horizontal line at 0 dB.

The high-frequency asymptote occurs when $\omega >> \omega_n$, where the ω^2 term in the denominator dominates:

$$|TF(\omega_{high})| = \lim_{\omega>>\omega_n}\left[20\log\left|\frac{1}{1-(\omega/\omega_n)^2+j2\delta\ \omega/\omega_n}\right|\right] = 20\log\left(\frac{1}{\omega/\omega_n}\right)^2 = -40\log(\omega/\omega_n)\ \text{dB} \tag{5.48}$$

The high-frequency asymptote is also similar to that of the first-order element, but with double the slope: −40 dB/decade instead of −20 dB/decade.

A major difference between the first- and second-order terms occurs when $\omega = \omega_n$:

$$|TF(\omega=\omega_n)| = 20\log\left|\frac{1}{1-(\omega/\omega_n)^2+j2\delta\ \omega/\omega_n}\right| = 20\log\left|\frac{1}{j2\delta}\right| = -20\log(2\delta)\ \text{dB} \tag{5.49}$$

The magnitude spectrum at $\omega = \omega_n$ is not a constant, but depends on δ: specifically, it is 20 log(2δ). If δ is less than 1.0 then log(2δ) will be negative and the transfer function will be positive. In fact, the value of δ can radically alter the shape of the magnitude curve and must be taken into account when plotting. Usually the magnitude plot is determined by calculating the value of $TF(\omega)$ at $\omega = \omega_n$ using Equation 5.49, plotting that point, then drawing a curve freehand through that point, converging smoothly with the high- and low-frequency asymptotes above and below ω_n. The second-order magnitude plot is shown in Figure 5.20 for several values of δ. This plot assumes the second-order function is in the denominator.

The phase plot of a second-order system is also approached using the asymptote method. For phase angle the high- and low-frequency asymptotes are given as:

$$\angle TF(\omega_{low}) = \lim_{\omega<<\omega_n}\angle\left[\frac{1}{1-(\omega/\omega_n)^2+j2\delta\omega/\omega_n}\right] = \angle 1 = 0\ \text{deg.} \tag{5.50}$$

$$\angle TF(\omega_{high}) = \lim_{\omega>>\omega_n}\angle\left[\frac{1}{1-(\omega/\omega_n)^2+j2\delta\omega/\omega_n}\right] = \angle\left(\frac{1}{-\omega^2}\right) = -180\ \text{deg.} \tag{5.51}$$

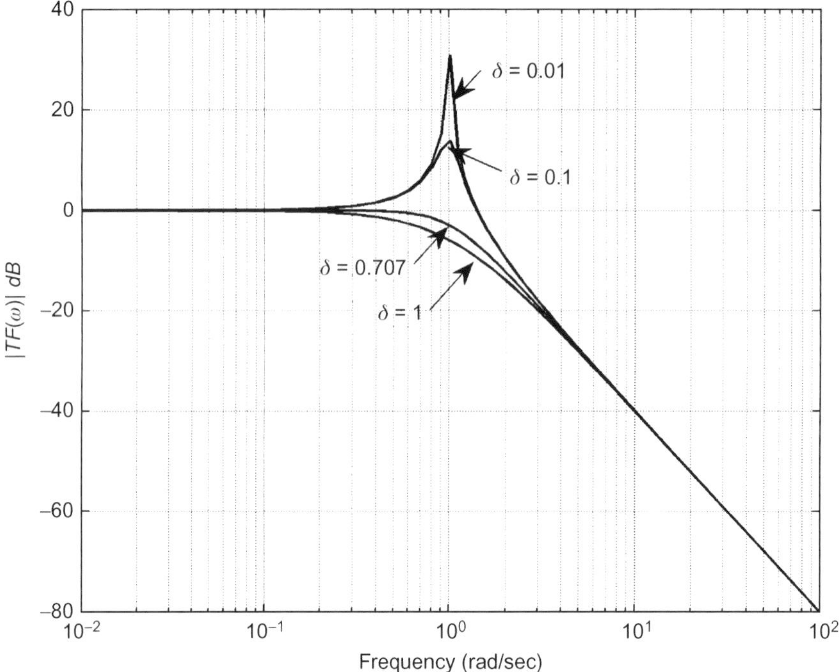

FIGURE 5.20 Magnitude frequency plot for a second-order element for several values of δ. The value of ω_n has been normalized to 1.0 rad./sec. The values of δ are 0.01, 0.1, 0.707, and 1.0.

This is similar to the asymptotes of the first-order process except the high-frequency asymptote is at −180 deg instead of −90 deg. The phase angle when $\omega = \omega_n$ can easily be determined:

$$\angle TF(\omega_n) = \angle\left(\frac{1}{1 - (\omega_n/\omega_n)^2 + j2\delta\omega_n/\omega_n}\right) = \angle\left(\frac{1}{j2\delta}\right) = -90 \text{ deg.} \tag{5.52}$$

So the phase at $\omega = \omega_n$ is −90 deg, halfway between the two asymptotes. Unfortunately, the shape of the phase curve between 0.1 ω_n and 10 ω_n is a function of δ and can no longer be approximated as a straight line except at larger values of δ (>2.0). Phase curves are shown in Figure 5.21 for the same range of values of δ used in Figure 5.20. Again, the plot is only for a second-order denominator term. The curves for low values of δ have steep transitions between 0 and −180 deg, while the curves for high values of δ have gradual slopes approximating the phase characteristics of a first-order element, except for the larger phase change. Hence if δ is 2.0 or more, a straight line between the low-frequency asymptote at 0.1 ω_n and the high-frequency asymptote at 10 ω_n works well. If δ is much less than 2.0, the best that can be done using this manual method is to approximate, freehand, the appropriate curve in Figure 5.21.

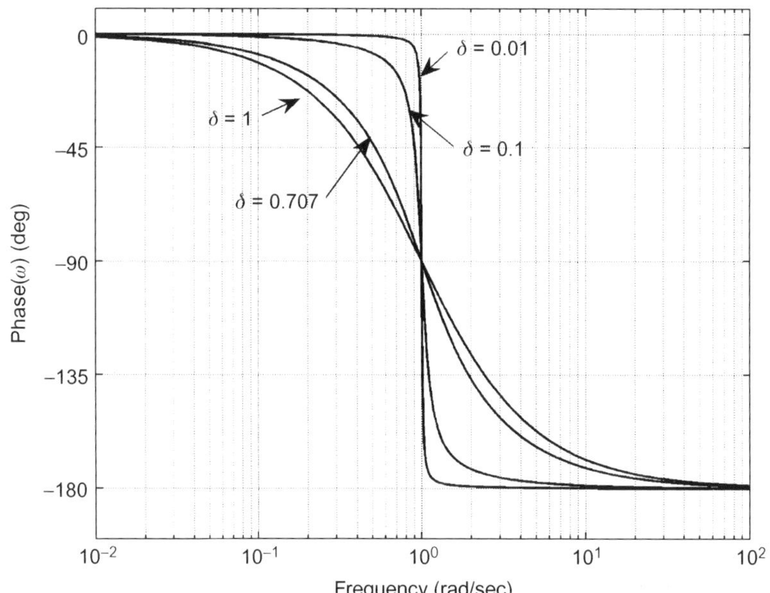

FIGURE 5.21 Phase curves for a second-order term in the denominator. The value of ω_n has been normalized to 1.0 rad/sec. The values of δ are the same as in Figure 5.20.

EXAMPLE 5.10

Plot the magnitude and phase spectra of the second-order transfer function used in Example 5.9 (i.e., Equation 5.46). Confirm the results obtained manually with those obtained using MATLAB.[3]

Solution: The equivalent values of ω_n and δ are found in Example 5.9 to be 5.77 rad/sec and 0.029, respectively. The magnitude and phase spectra can be plotted directly using the asymptotes, noting that the magnitude spectrum equals $-20\log(2\delta) = -4.7$ dB when $\omega = \omega_n$.

To plot the transfer function using MATLAB, first construct the frequency vector. Given that $\omega_n = 5.77$ rad/sec, use the frequency vector range between .01 $\omega_n = .0577$ rad/sec and 100 $\omega_n = 577$ rad/sec in increments of, say, 0.1 rad/sec. Then evaluate the function in MATLAB and take 20 times the log of the magnitude and the angle in degrees. Plot these functions against frequency in radians using a semilog plot.

[3]MATLAB has a function, `bode.m`, that will plot the Bode plot of any linear system. The system is defined from the transfer using another MATLAB routine, `tf.m`. Both these routines are in the Control System Toolbox. However, as shown in the following examples, it is not that difficult to generate Bode plots using standard MATLAB.

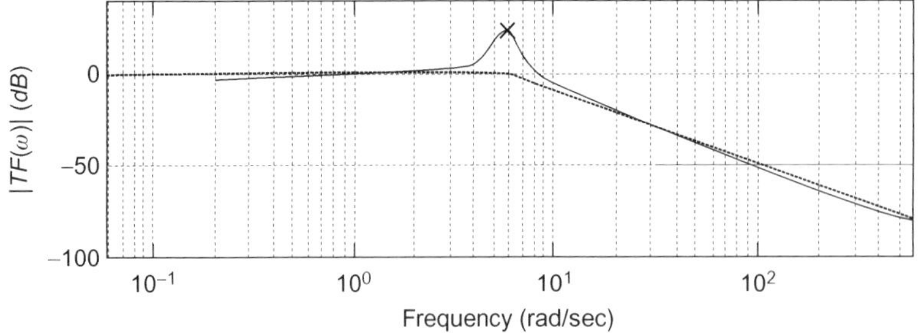

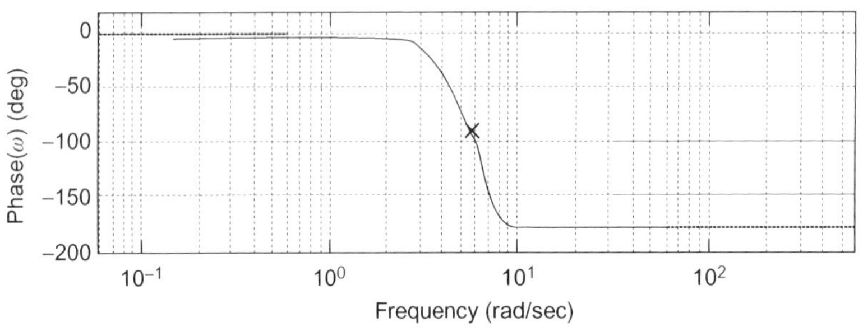

FIGURE 5.22 The magnitude and phase spectra of the second-order transfer function given in Equation 5.46 showing the asymptotes, $\omega = \omega_n$ point, and freehand spectral plots.

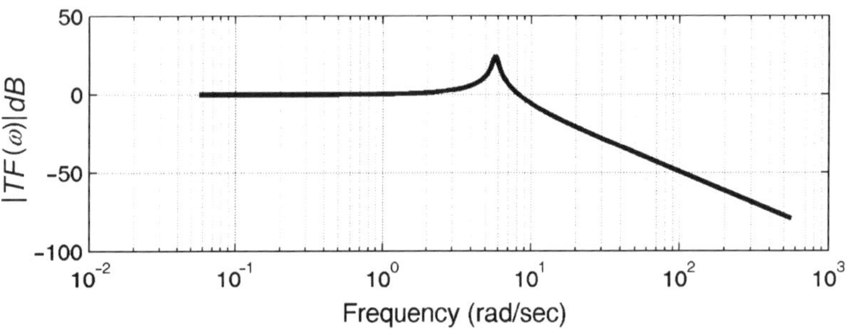

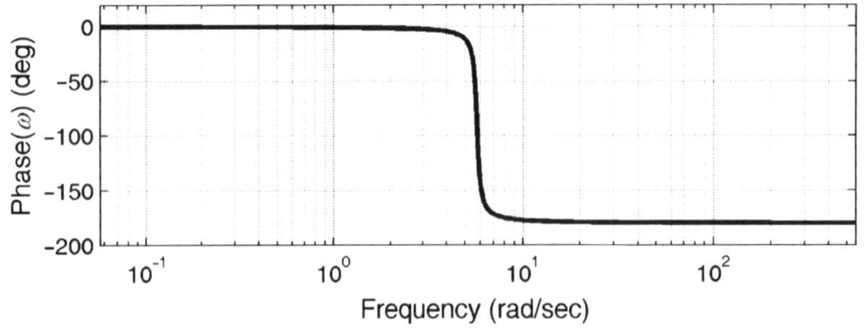

FIGURE 5.23 The magnitude and phase spectra of the second-order transfer function given in Equation 5.46 obtained using a MATLAB program. While the hand-drawn spectrum of Figure 5.22 is not as polished (nor as accurate), it does capture the essential spectral features.

```
% Example 5.10 Use MATLAB to plot the transfer function given in Equation 5.46
%
w = .057:.1:557;                     % Define frequency vector
TF = 1./(1 - .03*w.^2 + .01*j*w);   % Transfer function
Mag = 20*log10(abs(TF));             % Magnitude in dB
Phase = angle(TF)*360/(2*pi);        % Phase in deg
subplot(2,1,1);
semilogx(w,Mag,'k');                 % Plot as log frequency
  ......labels......
subplot(2,1,2);
semilogx(w,Phase,'k');
  ......labels......
```

Results: The asymptotes, the point where $\omega = \omega_n$, and freehand spectral plots are shown in Figure 5.22, and the results of the MATLAB code are shown in Figure 5.23. The ease of plotting the transfer function using MATLAB is seductive, but the manual method provides more insight into the system. It also allows us to go in the reverse direction: from a spectrum to an equivalent transfer function, as shown later. This is sometimes an important step in system design.

Occasionally the second-order term is found in the numerator as part of a more complex transfer function. In this case, both the magnitude and phase spectra are inverted so the phase spectrum goes from 0 to +180 deg passing through +90 deg when $\omega = \omega_n$. Modifying the equations to demonstrate this is given in Problem 13 at the end of this chapter.

Table 5.1 summarizes the various elements (also referred to as Bode plot primitives), providing alternate names where applicable, and Table 5.2 brings together their magnitude and phase spectral characteristics.

TABLE 5.1 Transfer Function Elements

	Name(s)	
Basic Equation	**Numerator**	**Denominator**
G	Constant or gain	———
$j\omega$	Differentiator* Isolated zero	Integrator* Isolated pole
$1 + j\omega/\omega_1$	Real zero or just zero Lead element	First-order element* Real pole or just pole Lag element
$1 - (\omega/\omega_n)^{+} + j2\delta\omega/\omega_n$	Complex zeros[+]	Complex poles[+] Second-order element*

Name most commonly used in this text.

[+] *Depends on the value of δ.*

TABLE 5.2 Bode Plot Primitives

Denominator Term	Magnitude Plot	Phase Plot
Constant Gain	20 log G	---
$j\omega$	0.0; 1.0; −20 dB/decade	−90 deg
$1 + j\dfrac{\omega}{\omega_1}$	0.0; −20 dB/decade; ω_1	0.0; .1 ω_1; −45; 90; ω_1; 10 ω_1
$1 - \left(\dfrac{\omega}{\omega_n}\right)^2 + j\dfrac{2\delta\omega}{\omega_n}$	20 log (2δ); δ = 0.01; 0.0; δ = 1.0; -40 dB/decade; ω_n	0.0; δ = 0.01; Sharpness depends on δ; δ = 1.0; -180; .1 ω_n; ω_n; 10 ω_n

5.5 BODE PLOTS COMBINING MULTIPLE ELEMENTS

Complex systems may contain a number of the elements describe above. By putting the magnitude of the transfer function in dB, determining the spectra of such functions from the individual transfer function components is not difficult. Assuming that a general transfer function consists of combinations of the elements described in the last section, the general transfer function becomes:

$$TF(\omega) = \frac{Gj\omega(1 + j\,\omega/\omega_1)(1 - (\omega/\omega_{n1})^2 + j\,2\delta_1\omega/\omega_{n1})\ldots}{j\omega(1 + j\,\omega/\omega_2)(1 - (\omega/\omega_{n2})^2 + j\,2\delta_2\omega/\omega_{n2})\ldots} \tag{5.53}$$

This assumes that terms higher than second order can be factored into first- or second-order terms. Polynomial factoring can be tedious, but is easily accomplished using MATLAB's `roots` routine. An example involving a fourth-order denominator polynomial is found in Example 6.11 at the end of the next chapter. The constants for first- and second-order elements are described above. This is a general equation illustrating the various possibilities; if the same elements actually appear in both numerator and denominator, as does the $j\omega$ element in Equation 5.53, of course they would cancel.

To convert Equation 5.53 to a magnitude function in dB, we take 20 log of the absolute value

$$TF(\omega) = 20\log\left|\frac{Gj\omega(1 + j\,\omega/\omega_1)(1 - (\omega/\omega_{n1})^2 + j\,2\delta_1\omega/\omega_{n1})\ldots}{j\omega(1 + j\,\omega/\omega_2)(1 - (\omega/\omega_{n2})^2 + j\,2\delta_2\omega/\omega_{n2})\ldots}\right| \tag{5.54}$$

Note that multiplication and division in Equation 5.53 becomes addition and subtraction after taking the log, so Equation 5.54 can be expanded to:

$$\begin{aligned}|TF(\omega)|_{\mathrm{dB}} = {} & 20\log(G) + 20\log|j\omega| + 20\log|1 + j\omega/\omega_1| + 20\log|1 - (\omega/\omega_{n1})^2 + j2\delta_1\omega/\omega_{n1}| \\ & + \ldots -20\log|j\omega| - 20\log|1 + j\omega/\omega_2| - 20\log|1 - (\omega/\omega_{n2})^2 + j2\delta_2\,\omega/\omega_{n2}| - \ldots\end{aligned} \tag{5.55}$$

where the first line in Equation 5.55 is the expanded version of the numerator in Equation 5.54 and the second line is the expanded version of the denominator. Aside from the constant term, the first and second lines in Equation 5.55 have the same form except for the sign: numerator terms are positive and denominator terms are negative. Moreover, each term in the summation is one of the element types described in the last section. This suggests that the magnitude spectrum of any transfer function can be plotted by plotting each individual element then adding the curves graphically. This is not as difficult as it first appears. Usually only the asymptotes and a few other important points (such as the value of a second-order term at $\omega = \omega_n$) are plotted, then the overall curve is completed freehand by connecting the asymptotes and critical points. Aiding this procedure is the fact that most real transfer functions do not contain a large number of elements. This approach is termed the *Bode plot technique*, and the individual format types are called *Bode plot primitives* (see Table 5.2). Although the resulting Bode plot is only approximate and often somewhat crude, it is usually sufficient to represent the general spectral characteristics of the transfer function.

The phase portion of the transfer function can also be dissected into individual components:

$$\angle TF(\omega) = \frac{\angle[Gj\omega(1 + j\,\omega/\omega_1)(1 + (\omega/\omega_{n1})^2 + j\,2\delta_1\omega/\omega_{n1})]}{\angle[j\omega(1 + j\,\omega/\omega_2)(1 + (\omega/\omega_{n2})^2 + j\,2\delta_2\omega/\omega_{n2})]} \tag{5.56}$$

By the rules of complex arithmetic, the angles of the individual elements simply add if they are in the numerator or subtract if they are in the denominator:

$$\begin{aligned}\angle TF(\omega) = {} & \angle G + \angle j\omega + \angle(1 + j\omega/\omega_1) + \angle(1 + (\omega/\omega_{n1})^2 + j\,2\delta_1\omega/\omega_{n1}) \\ & - \angle j\omega - \angle(1 + j\,\omega/\omega_2) - \angle(1 + (\omega/\omega_{n2})^2 + j\,2\delta_2\omega/\omega_{n2})\end{aligned} \tag{5.57}$$

The phase transfer function also consists of individual components that add or subtract, and these elements again correspond to the elements described in the last section and in Table 5.2. Like the magnitude spectrum, the phase spectrum of any general transfer function can be constructed manually by plotting the spectrum of each element and adding the curves graphically. Again, we usually start with the asymptotes and fill in freehand, and like the magnitude spectrum the result is frequently a crude, but usually adequate, approximation of the phase spectrum. As with signal spectra, often only the magnitude plot is of interest and it is not necessary to construct the phase plot.

The next two examples demonstrate the Bode plot approach for transfer functions of increasing complexity.

EXAMPLE 5.11

Plot the magnitude and phase spectra using Bode plot methods for the transfer function:

$$TF(\omega) = \frac{100j\omega}{(1 + j1\omega)(1 + j.1\omega)}$$

Solution: The transfer function contains four elements: a constant, an isolated zero (i.e., $j\omega$ in the numerator), and two first-order terms in the denominator. For the magnitude curve, plot the asymptotes for all but the constant term. Add these asymptotes together graphically to get an overall asymptote. Finally, use the constant term to scale the value of the vertical axis. For the phase plot, construct the asymptotes for the two first-order denominator elements, then rescale the axis by +90 deg to account for the $j\omega$ in the numerator. Recall that the constant term does not contribute to the phase plot.

The general form for the first two elements is

$$\frac{1}{1 + j\,\omega/\omega_1}$$

where ω_1 is the point where the high- and low-frequency asymptotes intersect. In this transfer function, these two frequencies, ω_1 and ω_2, are $1/1 = 1.0$ rad/sec and $1/0.1 = 10$ rad/sec. Figure 5.24 shows the asymptotes obtained for the two first-order primitives plus the $j\omega$ primitive.

Graphically adding the three asymptotes shown in Figure 5.24 gives the curve consisting of three straight lines shown in Figure 5.25. Note that the numerator and denominator asymptotes cancel out at $\omega = 1.0$ rad/sec, so the overall asymptote is flat until the additional downward asymptote comes in at $\omega = 10$ rad/sec. The actual magnitude transfer function is also shown in Figure 5.25 and closely follows the overall asymptote. A small error (3 dB) is seen at the two *breakpoints*: $\omega = 1.0$ and $\omega = 10$. A final step in constructing the magnitude curve is to rescale the vertical axis so that 0 dB corresponds to 20 log(100) = 40 dB. (The original values are given on the right vertical axis.)

The asymptotes of the phase curve are shown in Figure 5.26 along with the overall asymptote that is obtained by graphical addition. Also shown is the actual phase curve which, as with the magnitude curve, closely follows the overall asymptote. As a final step the vertical axis of this plot has been rescaled by +90 deg on the right side to account for the $j\omega$ term in the numerator.

In both the magnitude and phase plots, the actual curves follow the overall asymptote fairly closely for this and other transfer functions that have terms no higher than first-order. Tracing freehand through the −3-dB points would further improve the match, but often the asymptotes are sufficient. As we will see in the next example, this is not true for transfer functions that contain second-order terms, at least when the damping factor, δ, is small.

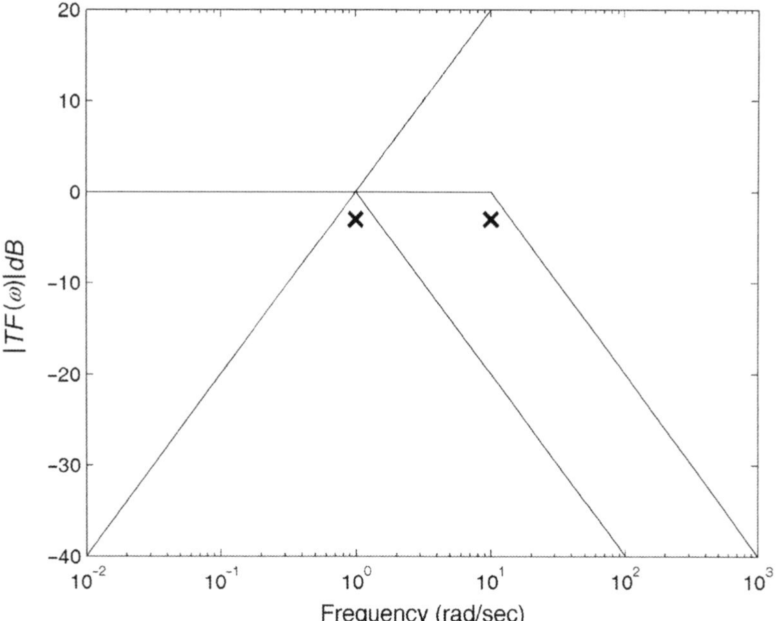

FIGURE 5.24 Asymptotes for the magnitude transfer function given in Example 5.11 along with the −3 dB points. These asymptotes are graphically added to get an overall asymptote in Figure 5.25.

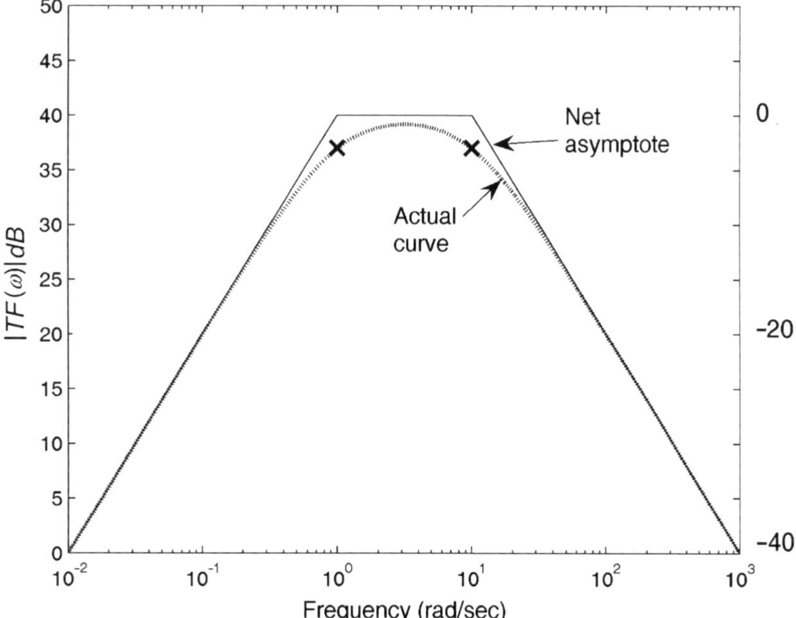

FIGURE 5.25 Magnitude plot for the transfer function given in Example 5.11 constructed from the asymptotes. The solid line is the addition of the individual asymptotes shown in Figure 5.24. The dashed line shows the actual magnitude transfer function. The vertical axis has been rescaled by 40 dB on the left side to account for the 100 in the numerator: 20 log(100) = 40 dB.

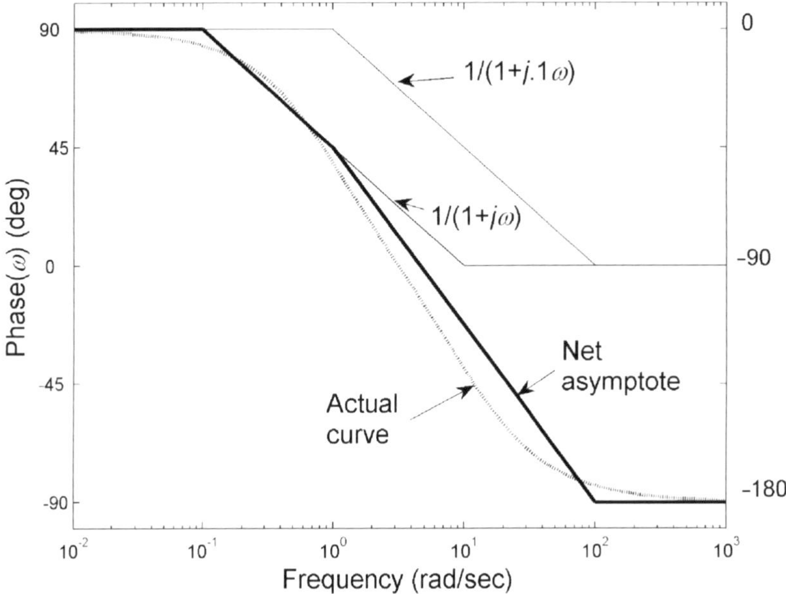

FIGURE 5.26 Phase frequency characteristics for the transfer function given in Example 5.11. The two first-order terms in the denominator produce downward curves that cross 45 deg at $\omega = 1.0$ and $\omega = 10$. The actual curve closely follows the overall asymptote that is created by graphically adding the two first-order curves. To complete the plot, the vertical axis has been rescaled by +90 deg to account for the $j\omega$ term in the numerator, as shown on the left axis. (The original axis values are given on the right vertical axis.)

EXAMPLE 5.12

Find the magnitude and phase spectra using Bode plot methods and MATLAB code for the transfer function:

$$TF(\omega) = \frac{10(1 + j2\omega)}{j\omega(1 - 0.04\omega^2 + j0.04\omega)}$$

Solution: This transfer function contains four elements: a constant, a numerator first-order term, an isolated pole (i.e., $j\omega$) in the denominator, and a second-order term in the denominator. For the magnitude curve, plot the asymptotes for all primitives except the constant, then add these up graphically. Lastly, use the constant term to scale the value of the vertical axis. For the phase curve, plot the asymptotes of the first-order and second-order terms, then rescale the vertical axis by −90 deg to account for the $j\omega$ term in the denominator. To plot the magnitude asymptotes, it is first necessary to determine ω_1, ω_n, and δ from the associated coefficients:

$$\omega_1: \quad 1/\omega_1 = 2; \quad \omega_1 = 0.5 \text{ rad/sec}$$

$$\omega_n: \quad 1/\omega_n^2 = 0.04; \quad \omega_n = \frac{1}{\sqrt{0.04}} = 5 \text{ rad/sec}$$

$$\delta: \quad 2\delta/\omega_n = 0.04; \quad \delta = 0.04\omega_n/2 = 0.04(5)/2 = 0.1$$

$$-20\log(2\delta) = 14 \text{ dB}$$

Results: Note that the second-order term is positive 14 dB when $\omega = \omega_n$. This is because 2δ is less than 1 and so the log is negative and the two negatives make a positive. Using these values and including the asymptotes leads to the magnitude plot shown in Figure 5.27.

The log frequency scale can be a little confusing, so the position of 0.5 rad/sec and 5.0 rad/sec are indicated by arrows in Figure 5.27. The overall asymptote and the actual curve for the magnitude transfer function are shown in Figure 5.28. The vertical axis has been rescaled by 20 dB to account for the constant term in the numerator. The second-order denominator system will reach a value that is 14 dB above the net asymptote. The net asymptote at $\omega = \omega_n$ is approximately 5 dB, so the actual peak, including the rescaling, will be around $14 + 5 + 20 = 39$ dB.

The individual phase asymptotes are shown in Figure 5.29 while the overall asymptote and the actual phase curve are shown in Figure 5.30. Note that the actual phase curve is quite different from the overall asymptote because the second-order term has a fairly small value for δ. This low value of δ means the phase curve has a very sharp transition between 0 and -180 deg and will deviate substantially from the asymptote. The vertical axis is rescaled by -90 deg to account for the $j\omega$ term in the denominator. The original axis is shown on the right side.

The MATLAB code required for this example is a minor variation of that required in previous examples. Only the range of the frequency vector and the MATLAB statement defining the transfer function have been changed. Note that the "period" point-by-point operators, .* and ./, are required whenever vector terms are being multiplied. The plots produced by this code are shown as lighter dashed lines in Figures 5.28 and 5.30.

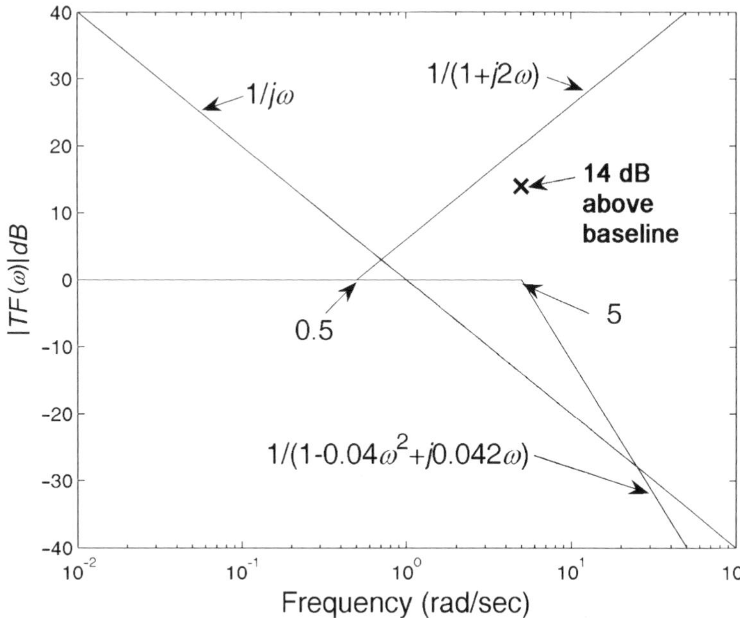

FIGURE 5.27 The individual asymptotes for the magnitude plot of the transfer function given in Example 5.12. The value of the second-order term at $\omega = \omega_n$ is also shown. The vertical axis will be rescaled in the final plot. The position of 0.5 rad/sec and 5.0 rad/sec are indicated by the arrows.

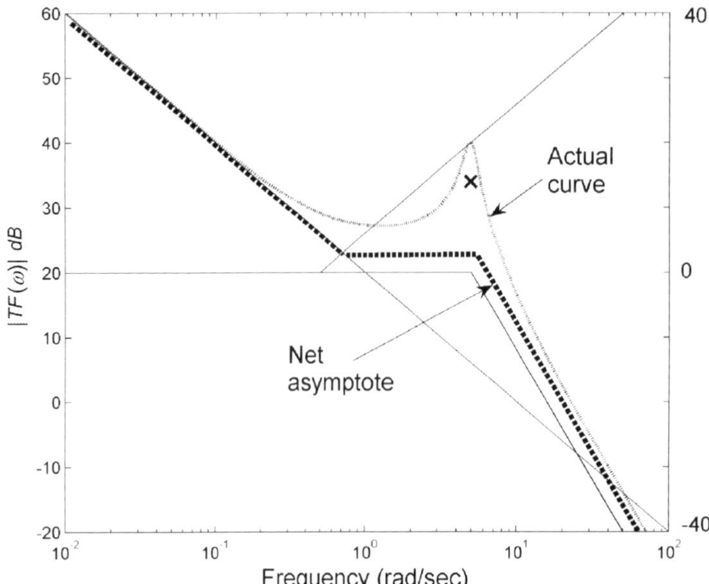

FIGURE 5.28 Overall asymptote (heavy dashed line) of the magnitude frequency characteristic of the transfer function given in Example 5.12. The curve follows the overall asymptote fairly closely except for the region on either side of ω_n. To find the actual peak at ω_n where the second-order term has a 34-dB peak on the rescaled axis (x point), the value of the overall asymptote at that point (about 5 dB) must be added in. This gives a total peak value of 39 dB on the rescaled axis, which matches the actual curve. The vertical axis is scaled with the original scale shown on the right side.

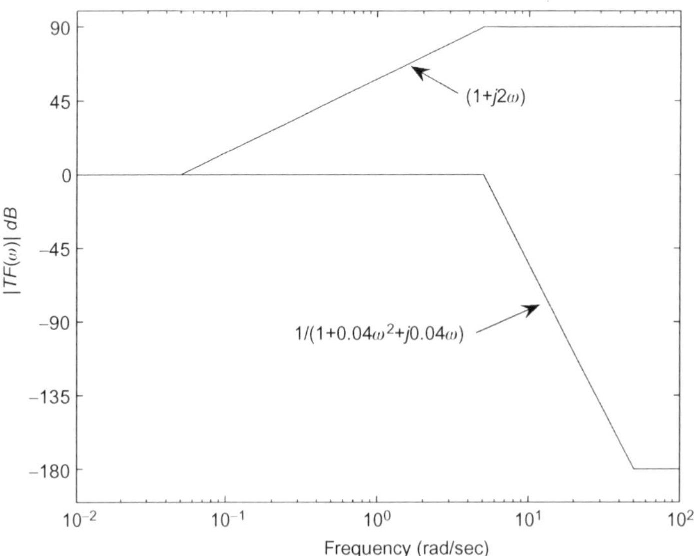

FIGURE 5.29 Asymptotes for the phase frequency characteristics of the transfer function in Example 5.12. Only two terms, the first-order term in the numerator and the second-order term in the denominator, contribute to the phase curve. The overall asymptote and actual phase curve are shown in Figure 5.30.

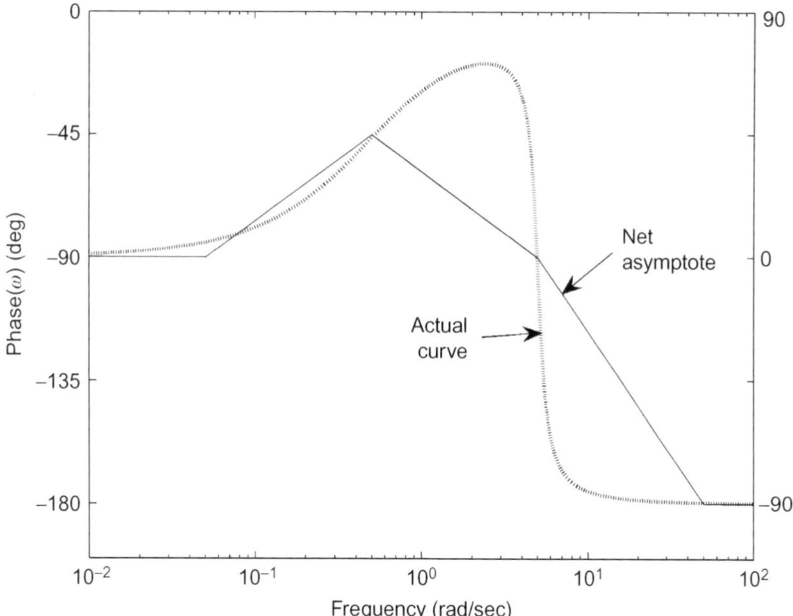

FIGURE 5.30 Overall asymptote and actual phase curve for the transfer function given in Example 5.12. The actual curve deviates from the asymptote due to the phase characteristics of the second-order term. The $j\omega$ term in the denominator scales the axis. The original axis is shown on the right side.

```
% Example 5.12 Use MATLAB to plot the transfer function given in this example
%
w = .005:.1:500;                                        % Define frequency vector
TF = 10*(1+j*2*w)./(j*w.*(1 - .04*w.^2 + j*.04*w));    % Define transfer function
  ......the rest of the code is the same as Example 5.7......
```

Clearly the code in this example could be easily modified to plot any transfer function of any complexity, and some of the problems at the end of this chapter demonstrate this. The Bode plot approach may seem like a lot of effort to achieve a graph that could be done better with a couple of lines of MATLAB code. However, these techniques provide us with a mental map between the transfer function and the system's spectrum. Also Bode plot techniques can help go the other way, from frequency curve to transfer function. Given a desired frequency response curve, we can use Bode plot methods to construct the transfer function that will produce that frequency curve. It is then only one more step to design a process (e.g. a electric circuit) that has this transfer function and this step is covered in the later chapters. If we can get the frequency characteristics of a physiological process, we ought to be able to quantify its behavior with a transfer function and ultimately design a quantitative model for this system. Examples of these concepts are touched on in the next section.

5.6 THE TRANSFER FUNCTION AND THE FOURIER TRANSFORM

The transfer function can be used to find the output to any input as long as that input can be decomposed into a series of sinusoids using the Fourier transform. This example shows how to combine the transfer function with Fourier decomposition to find the output of a complex signal to a particular transfer function. More examples are explored in later chapters.

EXAMPLE 5.13

Find the output of the system having the transfer function shown below when the input is the EEG signal first shown in Figure 1.9 in Chapter 1. Plot both input and output signals in both the time and frequency domain.

$$TF(\omega) = \frac{V_{out}(\omega)}{V_{in}(\omega)} = \frac{1}{1 - .05\omega^2 + j.1\omega}$$

Solution: In a MATLAB program, this transfer function is applied to the EEG data by first decomposing it into individual sinusoids using the MATLAB `fft` command, giving us $V_{in}(\omega)$. The output from all these sinusoids is then determined from the transfer function: $V_{out}(\omega) = TF(\omega)\ V_{in}(\omega)$. Then this frequency-domain output is converted back into the time domain using the inverse Fourier transform command, `ifft`. Finally, the program plots the time-domain reconstruction, $v_{out}(t)$. This approach is summarized in Figure 5.31 followed by the MATLAB code. The three steps are shown bordered by horizontal lines. This decomposition approach invokes the principle of superposition.

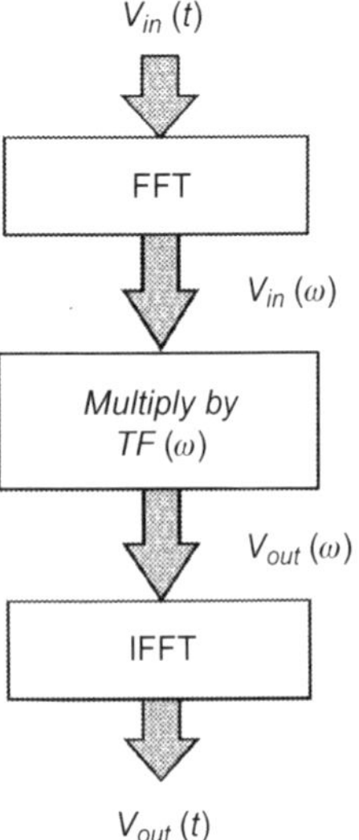

FIGURE 5.31 The steps involved in determining the output of a system to any input given the transfer function of that system. Although they may seem complicated, these steps are easy to implement in MATLAB.

```
% Example 5.13 Apply a specific transfer function to EEG data
%
load eeg_data;                                   % Load EEG data (in variable eeg)
N = length(eeg);
fs = 50;                                         % Sample frequency is 50 Hz
t = (1:N)/fs;                                    % Construct time vector for plotting
f = (0:N-1)*fs/N;
  ......plot and label EEG time data......
% -------------------------------
Vin = (fft(eeg));                                % Decompose data
Vout = Vin./(1-.002*(2*pi*f).^2 + j*.003*2*pi*f); % Get output using Fourier components
vout = ifft(Vout);                               % Convert output to time domain
% -------------------------------
  ......plot and label system output in the time domain......
  ......plot signal spectrum of input (Vin)......
  ......plot signal spectrum of output (Vout)......
```

The program loads the data, constructs time and frequency vectors based on the sampling frequency, and plots the EEG time data. Next the `fft` routine converts the EEG time data to the frequency domain, then the frequency components are used as input to the transfer function. The output of the transfer function is converted back into the time domain using the inverse Fourier transform program, `ifft`. The last section plots the magnitude spectra and reconstructed time data. The Fourier series conversion, multiplication by the transfer function, and the inverse Fourier series conversion each require only one line of MATLAB code.

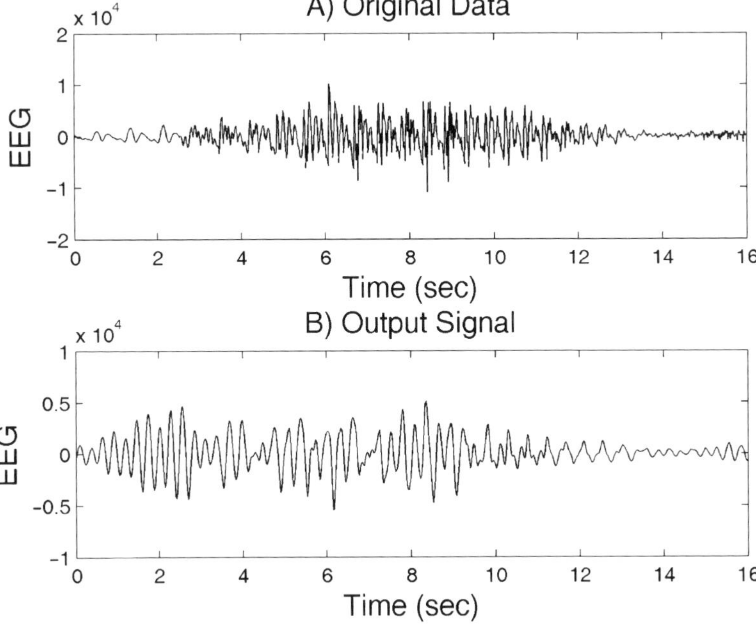

FIGURE 5.32 EEG data before (upper plot) and after (lower plot) processing by the transfer function given in Example 5.13. Note the strong oscillatory behavior seen in the filtered data.

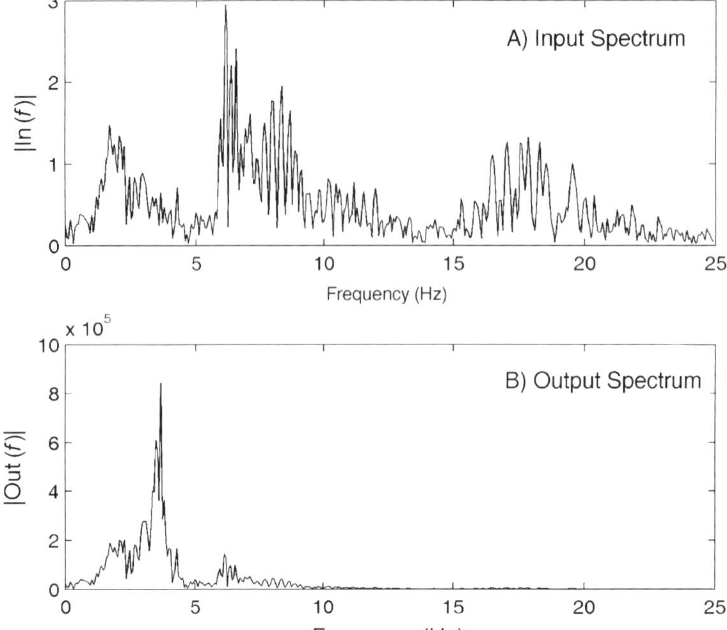

FIGURE 5.33 The frequency characteristics of the original EEG signal (upper plot) and of the signal coming out of the transfer function (lower plot). A comparison of the two spectra shows the transfer function acts as a filter removing frequencies outside a small passband range between 3- and 10-Hz.

The plots generated by this program are shown in Figures 5.32 and 5.33. It is clear from the frequency plots that this circuit acts as a narrowband bandpass filter with a peak frequency around 4-Hz. This filter emphasizes the EEG activity in this frequency region as shown in the time-domain plots. Note how this filtering significantly alters the appearance of the EEG waveform in Figure 5.32.

5.7 SUMMARY

Linear systems can be represented by differential equations and transfer functions that are combinations of four basic elements. If signals are restricted to steady-state sinusoids, then phasor techniques can be used to represent these elements by algebraic equations that are functions only of frequency. Phasor techniques represent steady-state sinusoids as a single complex number. With this representation, the calculus operation of differentiation can be implemented in algebra simply as multiplication by $j\omega$, where $j = \sqrt{-1}$and ω is frequency in radians. Similarly, integration becomes division by $j\omega$.

If all the elements in a system can be represented by algebraic operations, they can be combined into a single equation termed the *transfer function*. The transfer function relates the output of a system to the input through a single equation. As presented here, the transfer function can only deal with signals that are sinusoidal or can be decomposed into sinusoids, but the transfer function concept is extended to a wider class of signals in the next chapter.

The transfer function not only offers a succinct representation of the system, it also gives a direct link to the frequency spectrum of the system. Frequency plots can be constructed directly from the transfer function using MATLAB, or they can be drawn by hand using a graphical technique based on Bode plot methods. With the aid of these Bode plot methods, we can also go the other way: from a given frequency plot to the corresponding transfer function.

Using the Bode plot approach, we should be able to estimate a transfer function to represent the system as is shown in later chapters. The transfer function can then be used to predict the system's behavior to other input signals, including signals that might be difficult to generate experimentally. Thus the transfer function concept, coupled with Bode plot methods, can be a powerful tool for determining the transfer function representation of a wide range of physiological systems and for predicting their behavior to an equally broad range of stimuli.

The transfer function can provide the output of the system to any input signal as long as that signal can be decomposed into sinusoids. We first transform the signal into its frequency-domain representation (i.e., its sinusoidal components) using Fourier series decomposition. Then we multiply these sinusoidal components by the transfer function to obtain the frequency-domain representation of the output signal. Finally, to reconstruct the time-domain signal, we apply the inverse Fourier transform to the output signal spectrum.

PROBLEMS

1. Assume that the feedback control system presented in Example 5.1 is in a steady state or static condition. If $G = 100$ and $H = 1$ (i.e., a unity gain feedback control system), find the output if the input equals 1. Find the output if the input is increased to 10. Find the output if the input is 10 and G is increased to 1000. Note how the output is proportional to the input, which accounts for why this system is sometimes termed a *proportional control system*.
2. In the system given in Problem 1, the input is changed to a signal that smoothly goes from 0.0 to 5.0 in 10 seconds (i.e., $In(t) = 0.5\,t$ sec). What will the output look like? (Note G and H are simple constants so Equation 5.10 still holds.)
3. Convert the following to phasor representation:
   a. $10\cos(10t)$
   b. $5\sin(5t)$
   c. $6\sin(2t + 60)$
   d. $2\cos(5t) + 4\sin(5t)$ [*Hint*: see Example 2.4.]
   e. $\int 5\cos(20t)dt$
   f. $\frac{d}{dt}(2\cos(20t + 30))$

4. Add the following real, imaginary, and complex numbers to form a single complex number:
   a. $6, j10, 5+j12, 8+j3, 10\angle 0, 5\angle -60, 1/(j.1)$
5. Evaluate the following expressions to produce a single complex number:
   a. $(10+j6) + 10\angle -30 - 10\angle 30$
   b. $6\angle -120 + \frac{5-j10}{j4}$
   c. $\frac{10+j5}{15-j6} - \frac{8-j8}{12+j4}$.
   d. $\frac{(6+j5)(3-j4)j}{(8+j3)(10\angle 260)}$
6. Find the transfer function, $Out(\omega)/In(\omega)$, using phasor notation for the system shown below.

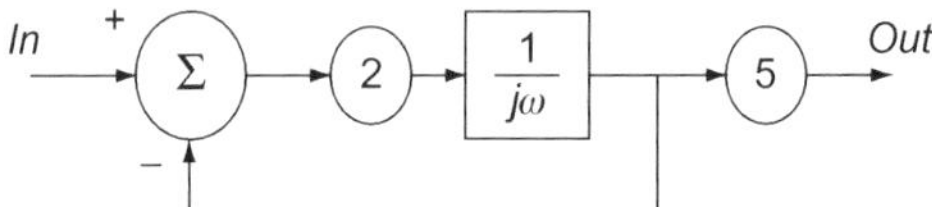

7. Find the transfer function, $Out(\omega)/In(\omega)$, for the system below. Find the time function, $out(t)$ if $in(t) = 5\cos(5t - 30)$. [*Hint*: You could still use the feedback equation to find this transfer function, but the G term contains an arithmetic operator; specifically, it contains a summation operation.]

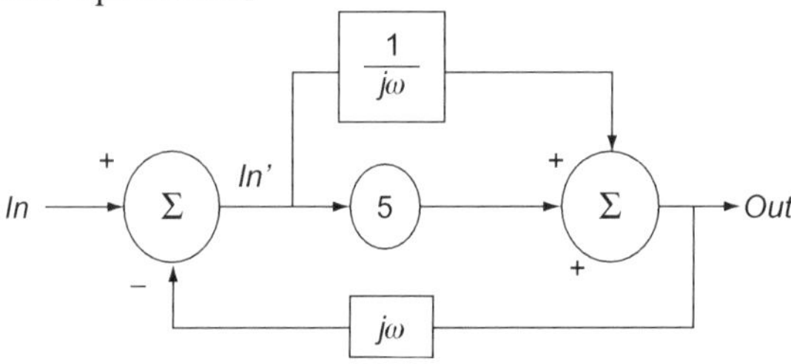

8. Using the transfer function of the respirator-airway pathway given in Example 5.7, find the airway pressure, $Q(\omega)$ if the respirator produces an external pressure of 2.0 mmHg at 11 breaths per minute. [*Hint*: Generate the input function $P(\omega)$, then solve using phasor analysis. Recall the transfer function is in minutes, which needs to be taken into account for the calculation of $P(\omega)$.]
9. Find the transfer function of a modified version of the Guyton-Colemen fluid balance system shown below. Find the time function, $P_A(t)$ if $F_{IN}(t) = 5\sin(5t - 30)$.

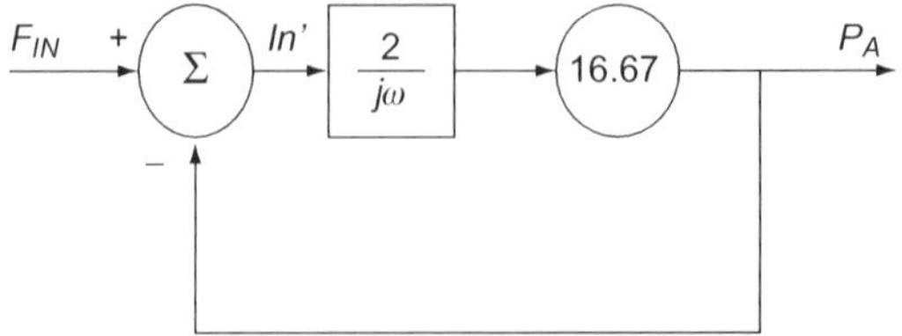

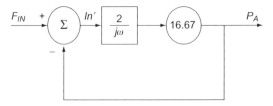

10. Find the transfer function of the circuit below. Find the time function, $out(t)$ if $in(t) = 5\sin(5t - 30)$. [*Hint*: This system is a feedback system containing another feedback system in the feedforward path so the feedback equation can be applied twice.]

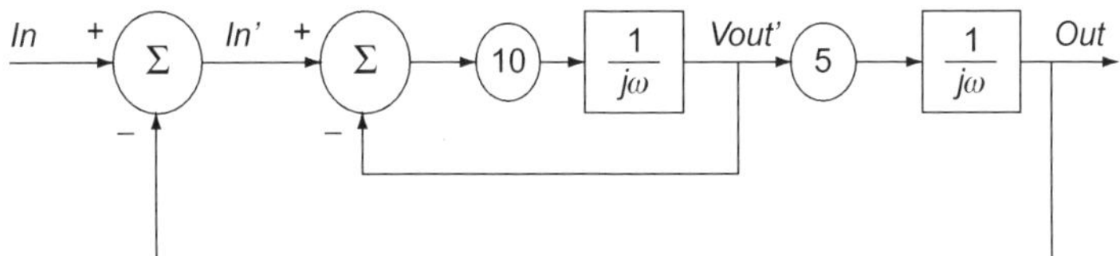

11. Plot the magnitude spectra of the transfer functions in Problems 6 and 7. What is the difference between the two?
12. Modify Equations 5.38 through 5.40 to find the equations for the asymptotes and worst-case angle for the phase spectrum of the transfer function: $TF(\omega) = 1 + j\,\omega/\omega_1$.
13. Modify Equations 5.47 through 5.49 to show that when the second-order term appears in the numerator, the magnitude and phase spectra are inverted. Use MATLAB to plot the magnitude and phase spectra of the second-order transfer function used in Example 5.9 but with the second-order term in the numerator (i.e., $TF(\omega) = 1 - 0.03\omega^2 + j0.1\omega$).
14. Plot the Bode plot (magnitude and phase) for the transfer function below using graphical techniques.

$$TF(\omega) = \frac{100(1 + j.05\omega)}{(1 + j.01\omega)}$$

15. Plot the Bode Plot (magnitude and phase) for the transfer function below using graphical techniques:

$$TF(\omega) = \frac{100(1 + j.001\omega)}{(1 + j.005\omega)(1 + j0.0002\omega)}$$

16. Plot the Bode plot (magnitude and phase) of the transfer function below using graphical techniques.

$$TF(\omega) = \frac{10\,j\omega}{(1 - 0.0001\omega^2 + j0.002\omega)}$$

17. Plot the Bode plot for the transfer function of the respirator-airway system given in Example 5.7 using graphical methods. Compare with the spectrum generated by MATLAB given in Figure 5.11.
18. Use MATLAB to plot the transfer functions (magnitude and phase) given in Problems 14 and 15.
19. Plot the Bode plot (magnitude and phase) of the transfer function given below using graphical techniques.

$$TF(\omega) = \frac{100(1 + j.1\omega)}{(1 - 0.0004\omega^2 + j0.028\omega)}$$

20. Use MATLAB to plot the transfer functions (magnitude and phase) given in Problems 16 and 19.

21. In the system used in Problem 10, find the effective values of ω_n and δ.

22. For the respirator-airway system of Example 5.7, find the effective values of ω_n and δ. Remember that time for this model is in minutes so this has an effect on the value of ω_n. Convert ω_n to Hz and determine the breaths per minute that correspond to that frequency.

23. In the system below, find the value of k that makes $\omega_n = 10$ rad/sec.

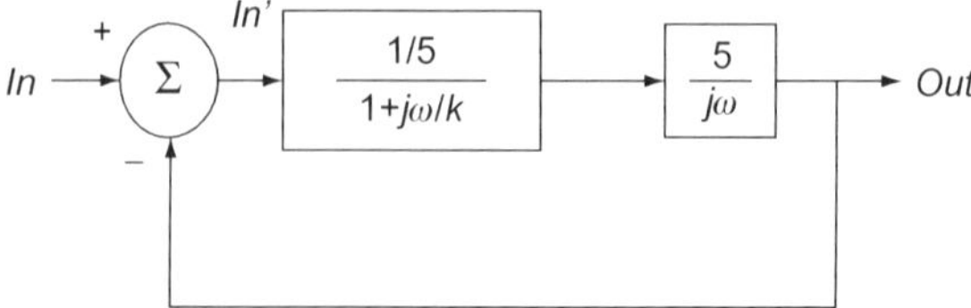

24. In the system below, find the maximum value of k so that the roots of the transfer function are real (i.e., the range of values of k for which $\delta \geq 1$).

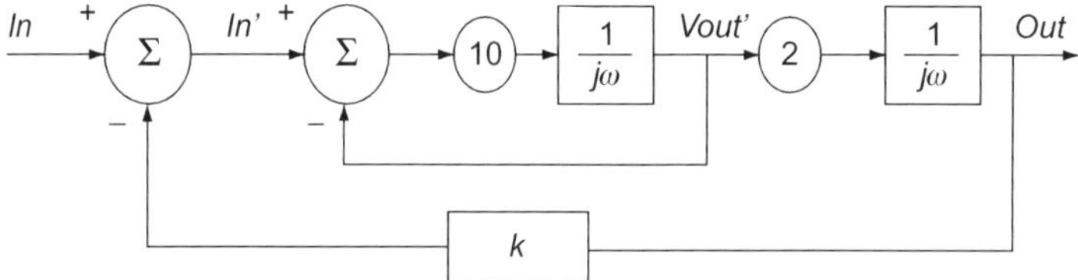

25. Find the transfer function that produces the magnitude spectrum curve shown below. Plot the phase curve of your estimated transfer function

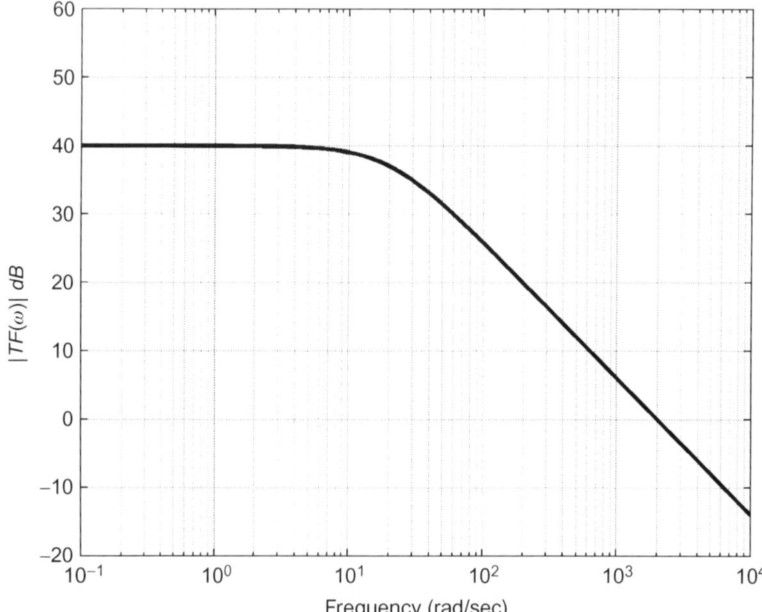

26. Find the transfer function that produces the magnitude frequency curve shown below. Plot the spectrum of your estimated transfer function using MATLAB and compare.

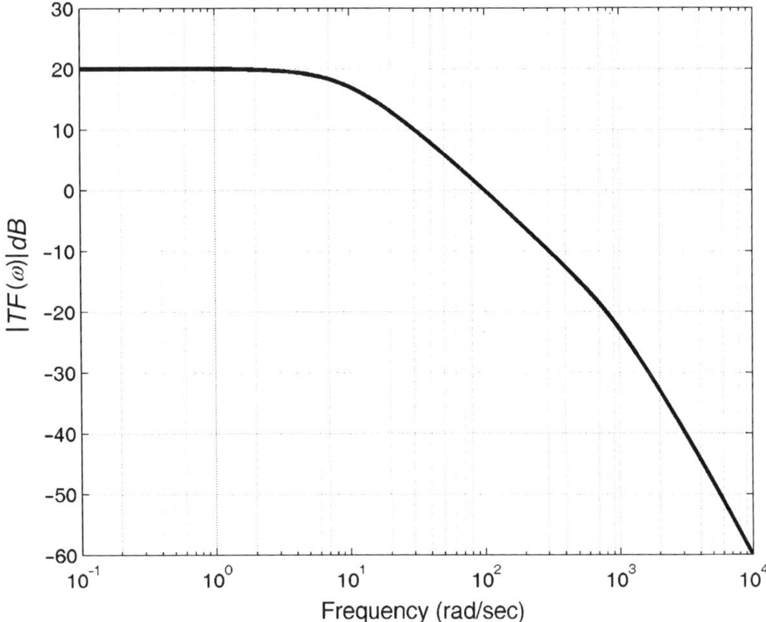

27. Find the transfer function that produces the magnitude frequency curve shown below.

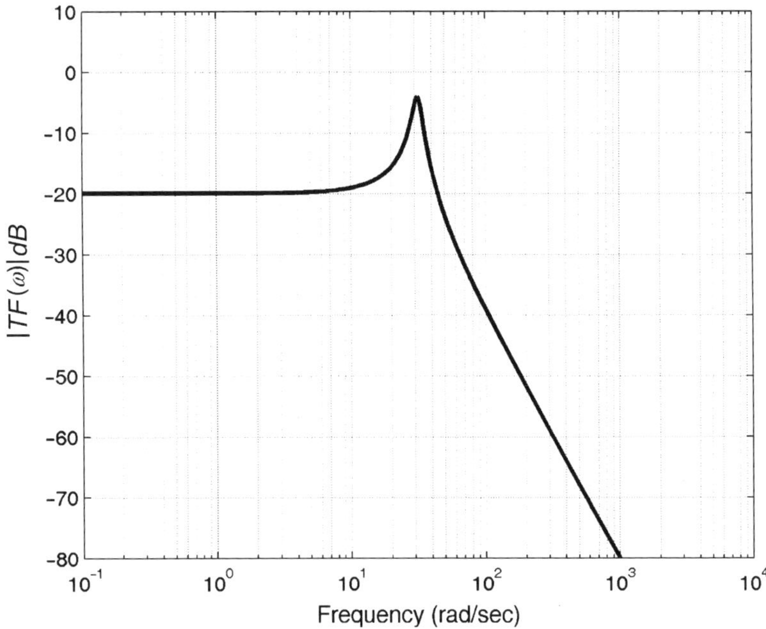

28. Find the transfer function that produces the magnitude frequency curve shown below.

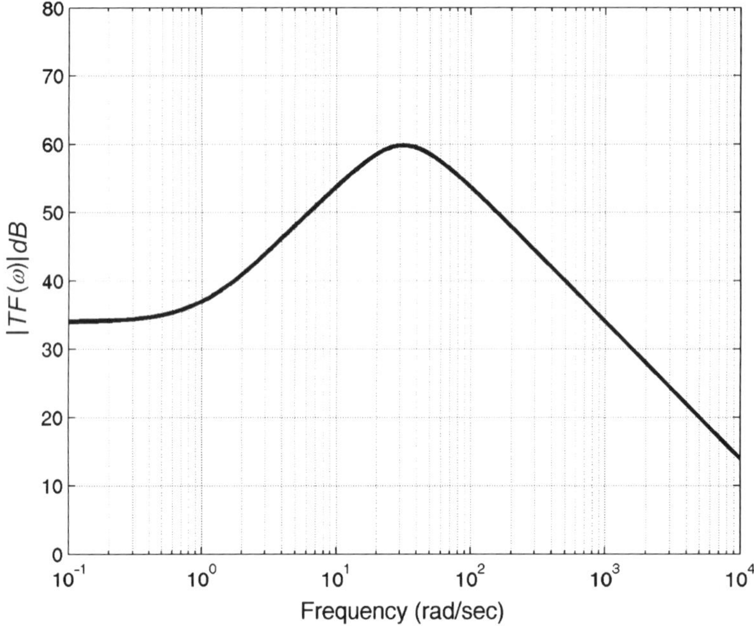

29. The MATLAB file `unknown.m` found on this book's accompanying CD represents a biological system that is known to be linear. The input is the input argument, `x`, and the output is the output argument, `y`; i.e., `y = unknown(x);`. Determine the magnitude spectrum of the transfer function for this unknown process by using a sine wave as `x` and measuring the RMS of the output. Vary the range of frequencies of the sine wave between 0 and 1-kHz. If the amplitude of the sine wave is 1.414, then it will have an RMS value of 1.0, so you can plot the output RMS directly on the spectral curve. Plot in dB versus log frequency.

CHAPTER

6

Linear Systems Analysis in the Complex Frequency Domain: The Laplace Transform and the Analysis of Transients

6.1 THE LAPLACE TRANSFORM

The frequency domain transfer function discussed in the last chapter allows us to determine the output of any linear system to any input that can be decomposed using the Fourier series, that is, any input that is periodic or can be taken to be periodic. A signal is broken down into sinusoidal components using Fourier series analysis, and the principle of superposition allows us to analyze each component separately, then add all the individual contributions to get the total response. This is the approach taken in the last example of Chapter 5 (Example 5.13). This approach can be extended to aperiodic functions through the application of the continuous Fourier transform. What the frequency domain transfer function cannot handle are waveforms that suddenly change and never return to a baseline level. A classic example is the *step function*, which changes from one value to another at one point in time (often taken as $t = 0$) and remains at the new value for all eternity (Figure 6.1). Other examples of these signals were presented in Figure 2.7 in Chapter 2.

One-time changes, or changes that do not return to some baseline level, are common in nature and even in everyday life. For example, this text should leave the reader with a lasting change in his or her knowledge of biosystems and biosignals, one that does not return to the baseline, precourse level. (However, this change is not expected to occur instantaneously as with the step function.) An input signal could change a number of times, either upward or downward, and might even show a partial return to its original starting point (probably a better model for this course!). Nonetheless, if a signal is not

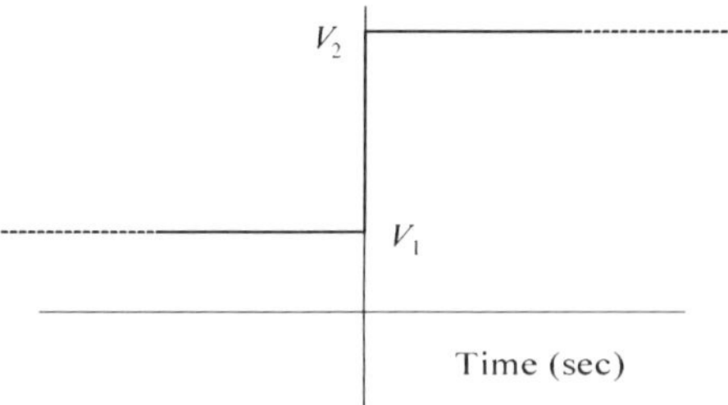

FIGURE 6.1 Time plot of a step function which changes from value V_1 to value V_2 at $t = 0.0$ sec. It remains at this value for all time.

periodic, or does not fully return to a baseline level so it can be taken as aperiodic, we simply do not have the tools to determine the output of systems to such signals. As noted in Chapter 2, signals that change and never return to baseline are sometimes referred to as "transient" signals. Of course all signals vary in time and therefore could be called transient signals, but this term is often reserved for signals that have one-time or step-like changes. This is another linguistic problem in engineering where context can vary meaning.

The *Laplace transform* can be used to develop a transfer function similar to the frequency domain transfer function of the last chapter. However, a transfer function in the Laplace domain can represent processes having inputs that change and do not return to a baseline level. With this tool, we can extend all of the techniques developed in the last chapter to systems exposed to this wider class of signals. It can also be used to analyze systems that begin with nonzero initial conditions, a situation we have managed to avoid thus far, but one that does occur in the real world.

6.1.1 Definition of the Laplace Transform

Transfer functions written in Laplace notation differ only slightly from the frequency domain transfer function used in the last chapter. In fact, it is trivial to convert from a Laplace transfer function to a frequency domain transfer function, although not the other way around. In this sense, the Laplace transfer function is more general than the frequency domain transfer function. The reason that the frequency domain transfer function cannot be used when signals change in a step-like manner is simply that these functions cannot be decomposed into sinusoidal components using either Fourier series analysis or the Fourier transform. Consider a function similar to that shown in Figure 6.1, a function that is 0.0 for a $t \leq 0$ and 1.0 for all $t > 0$:

$$x(t) = \begin{cases} 0 & t \leq 0 \\ 1 & t > 0 \end{cases} \tag{6.1}$$

The function shown in Figure 6.1 is known as the *unit step function* since it begins at zero and jumps to 1.0 at $t = 0$. In general, a step function can begin at any level and jump to any other level. The Fourier transform of this function would be determined using:

$$FT(\omega) = \int_0^\infty x(t)e^{-j\omega t}dt = \int_0^\infty 1e^{-j\omega t}dt \Rightarrow \infty \tag{6.2}$$

The problem is that because $x(t)$ does not return to its baseline level (zero in this case), the limits of the integration are infinite. Since the sinusoidal function $e^{-j\omega t}$ has nonzero values out to infinity, the integral goes to infinity. In the past the input signals had a finite life span and the Fourier transform integral need only be taken over the time period when the function was nonzero, but for transient signals this integral cannot be computed.

The trick used to solve this infinite integral problem is to modify the exponential function so that it converges to zero at large values of t even if the signal, $x(t)$, does not. This can be accomplished by multiplying the sinusoidal term, $e^{-j\omega t}$, by a decaying exponential such as $e^{-\sigma t}$ where σ is some positive real variable. In this case the sinusoid term in the Fourier transform becomes a complex sinusoid, or rather a sinusoid with a complex frequency:

$$e^{-j\omega t}e^{-\sigma t} = e^{-(\sigma + j\omega)t} = e^{-st} \tag{6.3}$$

where $s = \sigma + j\omega$ and is termed the *complex frequency* because it is a complex variable, but plays the same role as frequency, ω, in the Fourier transform exponential. The complex variable, s, is also known as the *Laplace variable* since it plays a critical role in the Laplace transform. A modified version of the Fourier transform can now be constructed using complex frequency, in place of regular frequency, that is, $s = \sigma + j\omega$ instead of just $j\omega$. This modified transform is termed the Laplace transform:

$$X(s) = \mathscr{L}x(t) = \int_0^\infty x(t)e^{-st}\,dt \tag{6.4}$$

where the script $\mathscr{L}$ indicates the Laplace transformation.

The trick is to use complex frequency, with its decaying exponential component, to cause convergence for functions that would not converge otherwise. Not all functions when multiplied by $e^{-(\sigma + j\omega)t}$ converge to zero as $t \to \infty$, in which case the Laplace transform does not exist.[1] Some advanced signal processing gurus have spent a lot of time worrying about such functions: which functions converge, their ranges of convergence, or how to get them to converge. Fortunately, such matters need not concern us since most common real-world functions, including the step function, do converge, at least for some values of σ, and so they do have a Laplace transform. The range of values of σ for which a given product of $x(t)$ and $e^{-(\sigma + j\omega)t}$ converges is another occupation of signal processing theoreticians, but again is not something bioengineering signal processors worry about. If a signal has a Laplace transform (and all the signals we will ever use do), then the product $x(t)e^{-st}$ must converge as $t \to \infty$.

The Laplace transform has two downsides: It cannot be applied to functions for negative values of t, and it is difficult to evaluate using Equation 6.4 for any but the simplest of functions. The restriction against representing functions for negative time values comes

[1]For example, the function $x(t) = e^{t^2}$ will not converge as $t \to \infty$ even when multiplied by e^{-st}, so it does not have a Laplace transform.

from the fact that e^{-st} becomes a positive exponential and will go to infinity as t goes to large negative values. (For negative t, the real part of the exponential becomes $e^{+\sigma t}$ and does exactly the opposite of what we want it to do: it forces divergence rather than convergence.) The only way around this is to limit our analyses to $t > 0$, but this is not a severe restriction. The other downside, the difficulty in evaluating Equation 6.4, stems from the fact that s is complex, so although the integral in Equation 6.4 does not look so complicated, the complex integration becomes very involved for all but a few simple functions of $x(t)$. To get around this problem, we use tables that give the Laplace transform of frequently used functions. Such a table is given in Appendix B, and a more extensive list can be easily found on the internet. The Laplace transform table is used to calculate both the Laplace transform of a signal and, by using the table in reverse, the inverse Laplace transform. The only difficulty in finding the inverse Laplace transform is rearranging the output Laplace function into one of the formats found in the table.

EXAMPLE 6.1

Find the Laplace transform of the step function in Equation 6.1.

Solution: The step function is one of the few functions that can be evaluated easily using the basic defining equation of the Laplace transform (Equation 6.4).

$$X(s) = \int_0^\infty x(t)e^{-st}dt = \int_0^\infty 1e^{-st}dt = -\frac{e^{-st}}{s}\bigg|_0^\infty = 0 - \left(-\frac{1}{s}\right)$$

$$X(s) = \frac{1}{s}$$

6.1.2 Laplace Transform Representation of Elements—Calculus Operations in the Laplace Domain

The calculus operations of differentiation and integration can be reduced to algebraic operations in the Laplace domain. The Laplace transform of the derivative operation can be determined from the defining equation (Equation 6.4).

$$\mathcal{L}\frac{dx(t)}{dt} = \int_0^\infty \frac{dx(t)}{dt}e^{-st}dt$$

Integrating by parts:

$$\mathcal{L}\frac{dx(t)}{dt} = x(t)e^{-st}\bigg|_0^\infty + s\int_0^\infty x(t)e^{-st}dt$$

From the definition of the Laplace transform $\left(\int_0^\infty x(t)e^{-st}dt = X(s)\right)$, so the right term in the summation is just: $sX(s)$, and the equation becomes:

$$\begin{aligned}\mathscr{L}\frac{dx(t)}{dt} &= x(\infty)e^{-\infty} - x(0)e^{-0} + sX(s)\\ \mathscr{L}\frac{dx(t)}{dt} &= sX(s) - x(0-)\end{aligned} \tag{6.5}$$

Equation 6.5 shows that in the Laplace domain, differentiation becomes multiplication by the Laplace variable s with the additional subtraction of the value of the function at $t = 0$. The value of the function at $t = 0$ is known as the *initial condition*. This value can be used, in effect, to account for all negative time history of $x(t)$. In other words, all of the behavior of $x(t)$ when t was negative can be lumped together into a single initial value at $t = 0$. This trick allows us to include some aspects of the system's behavior over negative values of t even if the Laplace transform does not itself apply to such time values. If the initial condition is zero, as is frequently the case, then differentiation in the Laplace domain is simply multiplication by s.

Multiple derivatives can also be taken in the Laplace domain, although this is not such a common operation. Multiple derivatives involve multiplication by s n-times, where n is the number of derivative operations, and taking the derivatives of the initial conditions:

$$\mathscr{L}\frac{d^n x(t)}{dt^n} = s^n X(s) - s^{n-1}x(0-) - s^{n-2}\frac{dx(0-)}{dt}\dots\frac{d^{n-1}x(0-)}{dt} \tag{6.6}$$

Again, if there are no initial conditions, taking n derivatives becomes just a matter of multiplying $x(t)$ by s^n.

If differentiation is multiplication by s in the Laplace domain, it is not surprising that integration is just a matter of dividing by s. Again, the initial conditions need be taken into account:

$$\mathscr{L}\left[\int_0^T x(t)dt\right] = \frac{1}{s}X(s) + \frac{1}{s}\int_{-\infty}^{0} x(t)dt \tag{6.7}$$

If there are no initial conditions, then integration is accomplished by dividing by s. The second term of Equation 6.7 is a direct integral that accounts for the initial conditions and for the negative time history of the system.

6.1.3 Sources—Common Signals in the Laplace Domain

In the Laplace domain both signals and systems are represented by functions of s. As mentioned above, the Laplace domain representation of signals are determined from a table. While it might seem that there could be a large variety of such signals, which would require a very large table, in practice only a few signal types commonly occur in Laplace analysis. The signal most frequently encountered in Laplace analysis is the step function

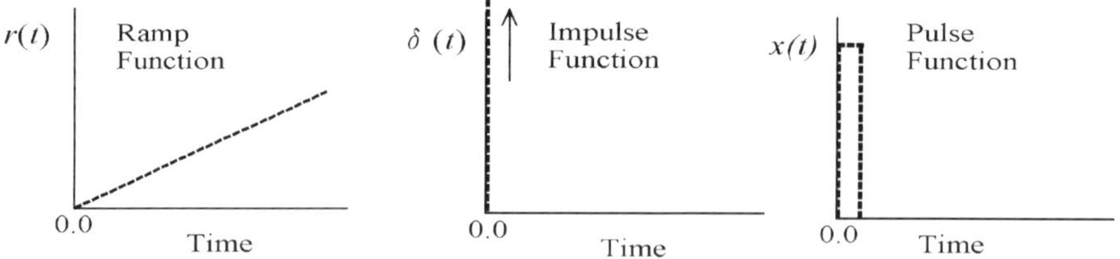

FIGURE 6.2 The ramp and impulse are two functions related to the step function and commonly encountered in Laplace analysis. The ideal impulse occurs at $t = 0$ and is infinitely narrow and infinitely tall. In the real world, impulse functions are approximated by short pulses (right-hand plot).

shown in Figure 6.1, or its more constrained version, the unit step function given in Equation 6.1 and repeated here:

$$x(t) = u(t) = \begin{cases} 0 & t \leq 0 \\ 1 & t > 0 \end{cases} \tag{6.8}$$

The symbol u is used frequently to represent the unit step function. The Laplace transform of the step function is found in Example 6.1 and repeated here:

$$X(x) = U(s) = \mathscr{L}\mu(t) = \frac{1}{s} \tag{6.9}$$

As with the Fourier transform, it is common to use capital letters to represent the Laplace transform of a time function. Two functions closely related to the unit step function are the ramp and impulse functions (Figure 6.2).

These functions are related to the step function by differentiation and integration. The unit ramp function is a straight line with slope of 1.0.

$$r(t) = \begin{cases} t & t > 0 \\ 0 & t \leq 0 \end{cases} \tag{6.10}$$

Since the unit ramp function is the integral of the unit step function, its Laplace transform will be that of the step function divided by s:

$$R(s) = \mathscr{L}r(t) = \frac{1}{s}\left(\frac{1}{s}\right) = \frac{1}{s^2} \tag{6.11}$$

The impulse function is the derivative of a unit step, which leads to one of those mathematical fantasies: a function that becomes infinitely short, but as it does, its amplitude becomes infinite so the area under the function remains 1.0, as quantified in Equation 6.12 below.

$$x(t) = \delta(t) = \lim_{a \to 0} \frac{1}{a} \quad \frac{-a}{2} \leq t \leq \frac{a}{2} \tag{6.12}$$

In practice a short pulse is used as an impulse function, a pulse with decidedly finite amplitude and a short, but finite width (see Figure 6.2, right-hand plot). In real situations, it is only necessary that the pulse be much shorter (say, an order of magnitude) than the fastest process in the system that is driven by this signal. The method to determine when a pulse can be taken as a "practical impulse function" is demonstrated in Example 7.4 in Chapter 7.

Since the impulse response is the derivative of the unit step function, its Laplace transfer function is that of a unit step multiplied by s:

$$\Delta(s) = \mathscr{L}\delta(t) = s\left(\frac{1}{s}\right) = 1 \tag{6.13}$$

Hence the Laplace transform of an impulse function is a constant, and if it is a unit impulse (the derivative of a unit step) then that constant is 1. As you might guess, this fact will be especially useful in the analysis of Laplace transfer functions. The Laplace transforms of other common signal functions are given in Appendix B.

6.1.4 Converting the Laplace Transform to the Frequency Domain

Remember that s is a complex frequency and equals a real term plus the standard imaginary frequency term: $s = \alpha + j\omega$. To convert a Laplace function to the frequency domain, we simply substitute $j\omega$ for s. Essentially we are agreeing to restrict ourselves to sinusoidal steady-state signals, so the real component of s is no longer needed. Its only purpose is to insure convergence in the integral for transient signals that would not normally converge. Converting the Laplace transform function to the frequency domain allows us to determine the frequency characteristics of that function using Bode plot techniques or MATLAB. This approach is used in the next section to evaluate the frequency characteristics of a time-delay process.

6.1.5 The Inverse Laplace Transform

Working in the Laplace domain is pretty much the same as working in the frequency domain: the math is still algebra. The transfer function of system elements will be in Laplace notation as described below. The input signal is converted to Laplace notation using the table in Appendix B. Combining the input signal with the transfer function provides the output response, but as a function of s not $j\omega$. Sometimes the Laplace representation of the solution or even just the Laplace transfer function is sufficient, but if a time-domain solution is desired, then the inverse Laplace transform must be determined. The equation for the inverse Laplace transform is given as:

$$x(t) = \mathscr{L}^{-1}X(x) = \frac{1}{2\pi}\int_{\sigma-j\infty}^{\sigma+j\infty} X(s)e^{st}ds \tag{6.14}$$

Unlike the inverse Fourier transform, this equation is quite difficult to solve even for simple functions. So to evaluate the inverse Laplace transform, we use the Laplace transform table in Appendix B in the reverse direction: Find a transformed function (on the right side of the table) that matches your Laplace output function and convert it to the equivalent time domain function. The trick is usually in rearranging the Laplace output function to conform to one of the formats given in the table. Methods for doing this are described next and specific examples given.

6.2 LAPLACE ANALYSIS—THE LAPLACE TRANSFER FUNCTION

The analysis of systems using the Laplace transform is no more difficult than in the frequency domain, except that there may be the added task of accounting for initial conditions. In addition, to go from the Laplace domain back to the time domain may require some algebraic manipulation of the Laplace output function.

The transfer function introduced in Chapter 5 is ideally suited to Laplace domain analysis, particularly when there are no initial conditions. In the frequency domain, the transfer function is used primarily to determine the spectrum of a system or system element. It can also be used to determine the system's output or response to any input, provided that input can be expressed as a sinusoidal series or is aperiodic. In the Laplace domain, the transfer function can be used to determine a system's output to a broader class of inputs. The Laplace domain transfer function is widely applied in biomedical engineering and in other areas of science and engineering. The Laplace domain transfer function is similar to its cousin in the frequency domain, except the frequency variable, ω, is replaced by the complex frequency variable, s:

$$TF(s) = \frac{Output(s)}{Input(s)} \tag{6.15}$$

Like its frequency domain cousin, the general Laplace domain transfer function consists of a series of polynomials such as Equation 5.53, repeated here:

$$TF(\omega) = \frac{Gj\omega(1 + j(\omega/\omega_1))(1 - (\omega/\omega_{n1})^2 + j(2\delta_1\omega/\omega_{n1}))\ldots.}{j\omega(1 + j(w/w_2))(1 - (\omega/\omega_{n2})^2 + j(2\delta_2\omega/\omega_{n2}))\ldots} \tag{6.16}$$

The general form of the Laplace transfer function differs in three ways: 1) the frequency variable $j\omega$ is replaced by the complex frequency variable s; 2) the polynomials are normalized so that the highest order of the frequency variable is normalized to 1.0 (as opposed to the lower order of $j\omega$); and 3) the order of the frequency terms is reversed with the highest order term (now normalized to 1) coming first:

$$TF(s) = \frac{Gs(s + \omega_1)(s^2 + 2\delta_1\omega_{n1}s + \omega_{n1}^2)\ldots\ldots}{s(s + \omega_2)(s^2 + 2\delta_2\omega_{n2}s + \omega_{n2}^2)\ldots\ldots} \tag{6.17}$$

The constant terms, δ, ω_1, ω_n, and so on, have exactly the same definitions given in Chapter 5. Even the terminology used to describe the elements is the same. For example: $\frac{1}{s+\omega_1}$ which is equivalent to $\frac{1}{1+(j\omega/\omega_1)}$ is called a first-order element, and $\frac{1}{s^2+2\delta\omega_n s+\omega_n^2}$ is known as a second-order element. Note that in Equation 6.17 the constants δ and ω_n have a more orderly arrangement in the transfer function equation. As described in section 6.1.4 above, it is possible to go from the Laplace domain transfer function to the frequency representation simply by substituting $j\omega$ for s, but you should also rearrange the coefficients and frequency variable to fit the frequency domain format. As in the frequency domain, the assumption is that any higher-order polynomial (third-order or above) can be factored into the first- and second-order terms shown.

As with our discussion of the frequency domain transfer function, each of the various element types are discussed independently. First the gain and first-order elements are covered, then the intriguing behavior of second-order elements is explored in greater detail. We also add a new element, the time-delay element common in physiological systems.

6.2.1 Time-Delay Element—The Time-Delay Theorem

Many physiological processes have a delay before they begin to respond to a stimulus. Such physiological delays are termed *reaction time*, or *response latency*, or simply *response delay*. In these processes, there is a period of time between the onset of the stimulus and the beginning of the response. In systems involving neurological control, this delay is due to processing delays in the brain. The Laplace transform can be used to represent such delays.

The *Time-Delay* theorem can be derived from the defining equation of the Laplace transform (Equation 6.4). Assume a function $x(t)$ that is zero for negative time, but normally changes at $t = 0$ sec. If this function is delayed from $t = 0$ by a delay time of T sec, then the delayed function would be $x(t - T)$. From the defining equation (Equation 6.4), the Laplace transform of such a delayed function would be:

$$\mathscr{L}[x(t-T)] = \int_0^\infty x(t-T)e^{-st}dt$$

Defining a new variable: $\gamma = t - T$

$$\mathscr{L}[x(t-T)] = \int_0^\infty x(\gamma)e^{-s(T+\gamma)}d\gamma = \int_0^\infty x(\gamma)e^{-sT}e^{-s\gamma}d\gamma = e^{-sT}\int_0^\infty x(\gamma)e^{-s\gamma}d\gamma$$

But the integral in the right-hand term is the Laplace transform of the function, $x(t)$, before it was shifted; that is, the right-hand integral is just $\mathscr{L}[x(t)]$. Hence the Laplace transform of the shifted function becomes:

$$\mathscr{L}[x(t-T)] = e^{-sT}\mathscr{L}[x(t)] \tag{6.18}$$

Equation 6.18 is the Time-Delay theorem, which can also be used to construct an element that represents a pure time delay, specifically an element that has a transfer function:

$$TF(s) = e^{-sT} \tag{6.19}$$

where T equals the delay, usually in seconds. Thus a system element having a Laplace transfer function of $TF(s) = e^{-sT}$ would be a pure time delay of T seconds. Such an element is commonly found in systems models of neurological control processes, where it represents neural processing delays.

EXAMPLE 6.2

Find the spectrum of a system time-delay element for two values of delay: 0.5 sec and 2.0 sec. Use MATLAB to plot the spectral characteristics of this element.

Solution: The Laplace transfer function of a pure time delay of 2.0 sec would be:

$$TF(s) = e^{-2s}$$

To determine the spectrum of this element, we need to convert the Laplace transfer function to the frequency domain. This is done simply by substituting $j\omega$ for s in the transfer function equation:

$$TF(\omega) = e^{-j2\omega}$$

The magnitude of this frequency domain transfer function is:

$$|TF(\omega)| = |e^{-j2\omega}| = 1 \tag{6.20}$$

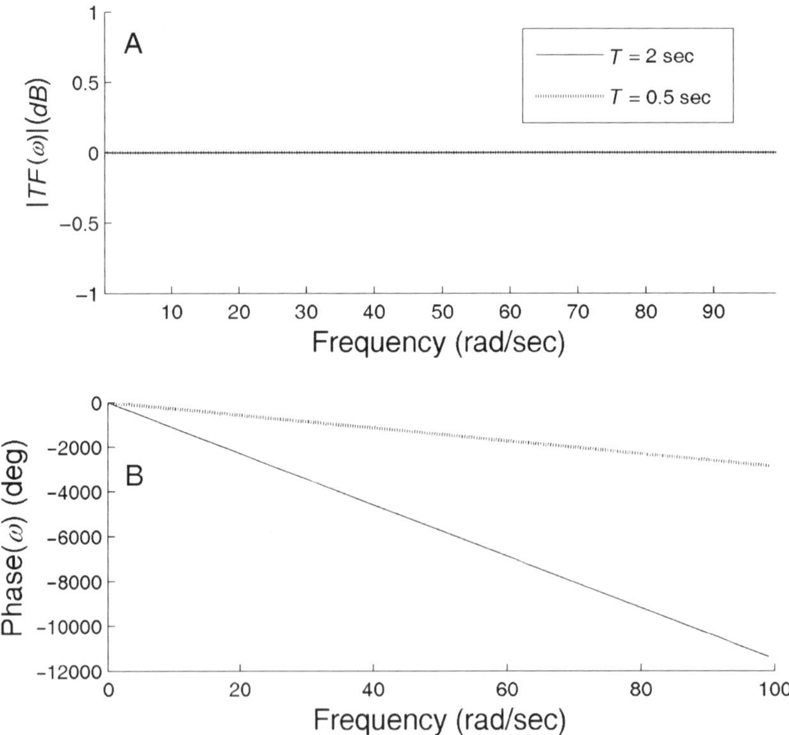

FIGURE 6.3 The magnitude and phase spectrum of a time-delay element with two time delays: 2 sec and 0.5 sec. In both cases the magnitude spectrum is 0 dB as expected and the phase spectrum decreases linearly with frequency. Note that when $\omega = 100$ rad/sec, the phase angle is close to $-12{,}000$ degrees for the 2-sec time delay, as predicted by Equation 6.21 ($-2*100*360/2\pi = -11{,}459$).

for all values of ω since changing ω (or the constant 2 for that matter) only changes the angle of the imaginary exponential, not its magnitude. The phase of this transfer function is:

$$\angle TF(\omega) = \angle e^{-j2\omega} = -2\omega \text{ rad} \tag{6.21}$$

Since the magnitude of the transfer function is 1.0, the output signal of a time-delay process has the same magnitude frequency characteristics as the input signal. The time-delay process does have a frequency-dependent phase term, one that increases linearly with frequency (-2ω).

For a 0.5-sec delay, $\angle TF(\omega) = \angle e^{-j2\omega} = -0.5$ rad so the phase curve would be -0.5ω. A pure time delay decreases the phase component of a signal's spectrum in a linear manner (Figure 6.3). A time-delay process is also explored in Problem 14 at the end of the chapter.

The MATLAB code to generate the magnitude and phase spectrum is a minor variation of Example 5.7. The plotting is done using linear frequency rather than log, since the phase spectrum is a linear function of frequency.

```
% Example 6.2 Use MATLAB to plot the transfer function of a time delay
%
T = 2;                                   % Time delay in sec.
w = .1:1:100;                            % Frequency vector
TF = exp(-j*T*w);                        % Transfer function
Mag = 20*log10(abs(TF));                 % Calculate magnitude spectrum
Phase = unwrap(angle(TF))*360/(2*pi);    % Calculate phase spectrum
  ......Repeat for T = 0.5 and plot and label......
```

The results produced by this program are shown in Figure 6.3.

6.2.2 Constant Gain Element

The gain element is not a function of frequency, so its transfer function is the same irrespective of whether standard or complex frequency is involved:

$$TF(s) = G \tag{6.22}$$

The system representation for a gain element is the same as in the frequency domain representation (Figure 5.12).

6.2.3 Derivative Element

Equation 6.5 shows that the derivative operation is the Laplace domain, implemented simply by multiplication with the Laplace variable s. This is analogous to this operation in the frequency domain, where differentiation is accomplished by multiplying by $j\omega$. So in the absence of initial conditions, the Laplace transfer function for a derivative element is:

$$TF(s) = s \tag{6.23}$$

The Laplace system representation of this element is shown in Figure 6.4A.

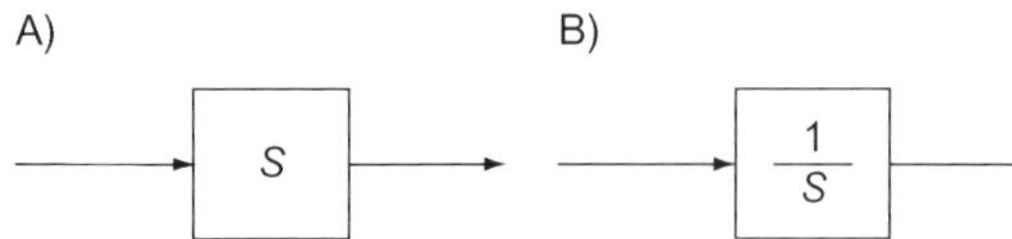

FIGURE 6.4 A) Laplace system representation of a derivative element. B) Laplace system representation of an integrator element.

6.2.4 Integrator Element

As shown in Equation 6.6, integration in the Laplace domain is accomplished by dividing by s. So in the absence of initial conditions, the system representation of integration in the Laplace domain has a transfer function of

$$TF(s) = 1/s \tag{6.24}$$

This parallels the representation in the frequency domain where $1/s$ becomes $1/j\omega$. The representation of this element is shown in Figure 6.4B.

EXAMPLE 6.3

Find the Laplace and frequency domain transfer function of the system in Figure 6.5. The gain term k is a constant.

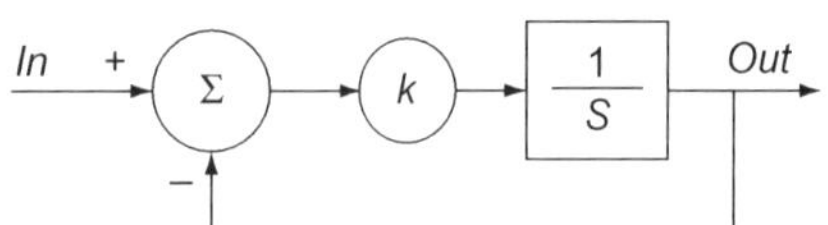

FIGURE 6.5 Laplace representation of a system used in Examples 6.3 and 6.4.

Solution: The approach to finding the transfer function of a system in the Laplace domain is exactly the same as in the frequency domain used in the last chapter. Here we are asked to determine both the Laplace and frequency domain transfer functions. First we find the Laplace transfer function, then substitute $j\omega$ for s and rearrange the format.

We could solve this several different ways, but since this is clearly a feedback system, we first find the equivalent feedforward gain function, $G(s)$, then the equivalent feedback function, $H(s)$, and apply the feedback equation (Equation 5.10). The feedback function is $H(s) = 1$ since all of the output feeds back to the input in this system. Again, such systems are known as *unity gain feedback systems* since the feedback gain, $H(s)$, is 1.0. The feedforward gain is the product of the two elements:

$$G(s) = k\frac{1}{s}$$

Substituting $G(s)$ and $H(s)$ into the feedback equation:

$$TF(s) = \frac{G(s)}{1 + G(s)H(s)} = \frac{\frac{k}{s}}{1 + \frac{k}{s}} = \frac{k}{s + k}$$

Again, the Laplace domain transfer function has the highest coefficient of complex frequency, in this case, s, normalized to 1. Substituting $s = j\omega$ and rearranging the normalization:

$$TF(\omega) = \frac{k}{j\omega + k} = \frac{1}{1 + j(\omega/k)}$$

This has the same form as a first-order transfer function given in Equation 5.34, where the frequency term $\omega_1 = k$.

6.2.5 First-Order Element

First-order processes contains an $(s + \omega_1)$ term in the denominator. Later we find that these systems contain a single energy storage device. The Laplace transfer function for a first-order process is:

$$TF(s) = \frac{1}{s + \omega_1} = \frac{1/\tau}{s + (1/\tau)} \tag{6.25}$$

In this transfer function equation, a new variable is introduced, τ, which is termed the *time constant* for reasons that will become apparent. Equation 6.25 shows that this time-constant variable, τ, is just the inverse of the frequency, ω_1:

$$\tau = \frac{1}{\omega_1} \tag{6.26}$$

Chapter 5 shows that the frequency ω_1 is where the magnitude spectrum is -3 dB (Equation 5.37), and the phase spectrum is -45 degrees (Equation 5.40). Next we find that the time constant τ has a direct relationship to the time behavior of a first-order element.

Now that the first-order transfer function is in the Laplace domain, we can explore the response of this element using input signals other than sinusoids. Two popular signals that are used are the step function (Figure 6.1), and the impulse function (Figure 6.2). Typical impulse and step responses of a first-order system are presented in the next example.

EXAMPLE 6.4

Find the arterial pressure response of the linearized model of the Guyton-Coleman body fluid balance system (Rideout, 1991) to a step increase of fluid intake of 0.5 ml/min. The frequency domain version of this model is shown in Figure 5.8 in Chapter 5. Also find the pressure response to fluid intake of 250 ml administered as an impulse.[2] This is equivalent to drinking approximately 1 cup of fluid quickly. Use MATLAB to plot the outputs in the time domain for 2 time-constant values: $\tau = 0.5$ sec and $\tau = 2$ sec.

Solution, step response: The first step is to convert to Laplace notation from the frequency notation used in Figure 5.8. The feedforward gain becomes $G(s) = 16.67(0.06/s) = 1/s$ and the

[2]Logically, the response to a step change in input is called the *step response* and the response due to an impulse is termed the *impulse response*.

feedback gain becomes $H(s) = 0.05$. Applying the feedback equation, the transfer function is:

$$TF(s) = \frac{P_A(x)}{F_{IN}(s)} = \frac{G(s)}{1 + G(s)H(s)} = \frac{1/s}{1 + 0.05(1/s)} = \frac{1}{s + 0.05}$$

where P_A is arterial blood pressure in mmHg and F_{IN} is the change in fluid intake in ml/min.

To find the step response, note that from the basic definition of the transfer function:

$$P_A(s) = F_{IN}(s)TF(s) = \frac{1}{s}TF(s) \tag{6.27}$$

Substituting in for $TF(s)$, Equation 6.27 becomes:

$$P_A(s) = \frac{1}{(s + 0.05)}\left(\frac{1}{s}\right) = \frac{1}{s(s + 0.05)} \tag{6.28}$$

Now we need to rearrange the resulting Laplace output function so that it has the same form as one of the functions in the Laplace transform table of Appendix B. Often this can be the most difficult part of the problem. Fortunately, in this problem, we see that the right-hand term of Equation 6.28 matches the Laplace function in entry number 4) of the table. Thus, the inverse Laplace transform for $V_{out}(s)$ in Equation 6.28 can be obtained as:

$$\frac{1}{s(s + 0.05)} \text{ has the form: } (1/\alpha)\left[\frac{\alpha}{s(s + \alpha)}\right] \Leftrightarrow (1 - e^{-\alpha t})/\alpha$$

where $\alpha = 0.05$. From the definition of the Laplace transform (Equation 6.4), any constant term can be removed from the integral, so the Laplace transform of a constant times a function is just the constant times the Laplace transform of the function. Similarly, the inverse Laplace transform of a constant times a Laplace function is just the constant times the inverse Laplace transform. Stating these two characteristics formally:

$$\mathscr{L}[kx(t)] = k\mathscr{L}x(t) \tag{6.29}$$

$$\mathscr{L}^{-1}[kX(s)] = k\mathscr{L}^{-1}[X(s)] \tag{6.30}$$

Hence the step response in the time domain for this system is the exponential function given in entry number 4) multiplied by $1/0.05 = 20$:

$$p_A(t) = \frac{1}{0.05}(1 - e^{-0.05}) = 20(1 - e^{-0.05}) \text{ mmHg}$$

Solution, impulse response: Solving for the impulse response is even easier since for the impulse function, $F_{IN}(s) = 1$. So the impulse response of a system in the Laplace domain is the transfer function itself. Therefore, the impulse response in the time domain is the inverse Laplace transform of the transfer function:

In this example the impulse is scaled by 250 ml, so $F_{IN}(s) = 250$ and $P_A(s) = 250\ TF(s)$:

$$P_A(s) = \frac{250}{s + 0.05} \text{ mmHg}$$

The Laplace output function is matched by entry 3) in the Laplace transform table except for the constant term.

So the time-domain solution for $P_A(s)$ can be obtained as:

$$V_{out}(s) = 250\left(\frac{1}{s + 0.05}\right) \text{ which has the same form as: } k\frac{1}{s + \alpha} \Leftrightarrow ke^{-\alpha t}$$

$$v_{out}(t) = 250e^{-0.05} \text{ mmHg}$$

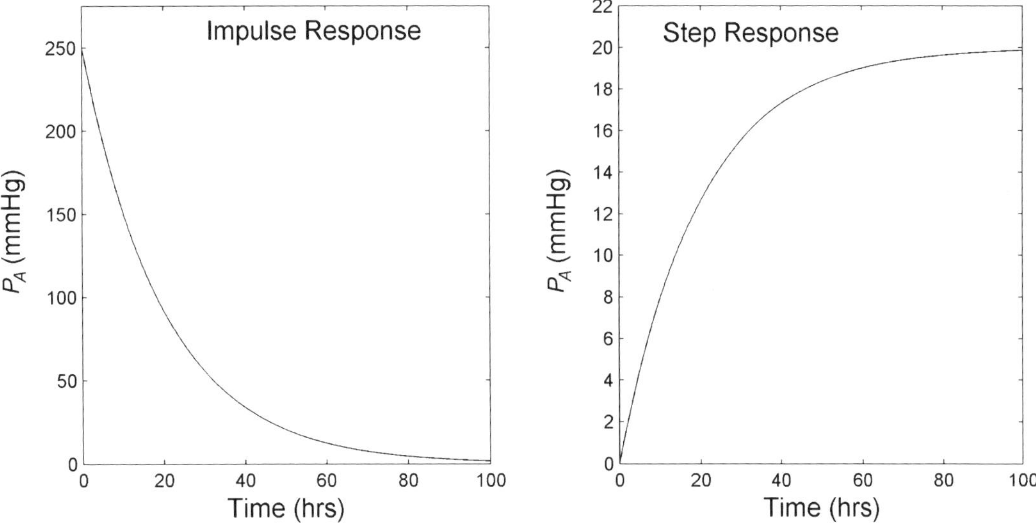

FIGURE 6.6 Response of a linearized model of the Guyton-Coleman body fluid balance system. The graph shows the change in arterial pressure to an increase of fluid intake, either taken in a step-like manner (right) or as an impulse (left). Note the long time periods involved.

The MATLAB program to plot these results is given below.

```
% Example 6.4 First-order system impulse and step response for two time constants
%
tau = .5;                    % Time constant value
t = 0:.01:10;                % Define time vector: 0 to 10 sec
x = 250*exp(−0.05);          % Impulse response
plot(t,x,'k');               % Plot impulse response
  .........labels, title, and other text......
x = 20*(1 − exp(−0.05));     % Step response.
plot(t,x,'k');               % Plot step response
  .........plot, labels, title, and other text......
```

Results: The system responses are shown in Figure 6.6. Both the impulse and step responses are exponential with time constants of $1/0.05 = 20$ hrs. When time equals the time constant (i.e., $t = \tau$), the value of the exponential becomes: $e^{-t/\tau} = e^{-1} = 0.37$. Hence at time $t = \tau$, the exponential is within 37% of its final value. An alternative, equivalent statement is that the exponential has attained 63% of its final value in one time constant.

The time constant makes a good measure of the relative "speed" of an exponential response. As a rule of thumb, an exponential is considered to have reached its final value when $t > 5\tau$, although in theory the final value is reached only when $t = \infty$.

There are many other possible input waveforms in addition to the step and impulse function. As long as the waveform has a Laplace representation, the response to any signal can be found using the equivalent of Equation 6.27.[3] However, the step and/or impulse responses usually provide the most insight into the general behavior of the system.

First-order transfer functions can be slightly different than that of Equation 6.25. There can be a single s or even an $(s + 1/\tau)$ in the numerator. What bonds all first-order systems is the denominator term, $(s + 1/\tau)$, or $(s + \omega_1)$. All first-order systems will contain this term in the denominator, although they might have different numerator terms. In the next section, we find that all second-order systems contain a second-order s in the denominator of the transfer function, usually as a quadratic polynomial of s. The close link between the denominator term and the system order explains why the denominator of the transfer function is sometimes called the *characteristic equation*: The characteristic equation defines the general type of system described by the transfer function. The characteristic equation also tells us something about the behavior of the system even without evaluating the time-domain solution. For example, a characteristic equation such as $(s + 3)$, tells us: 1) the system is a first-order system, and 2) the system's response will include an exponential having a time constant of 1/3 or 0.33 sec. Second-order characteristic equations are even more informative, as the next section shows.

EXAMPLE 6.5

Find the transfer function of the system shown in Figure 6.7.

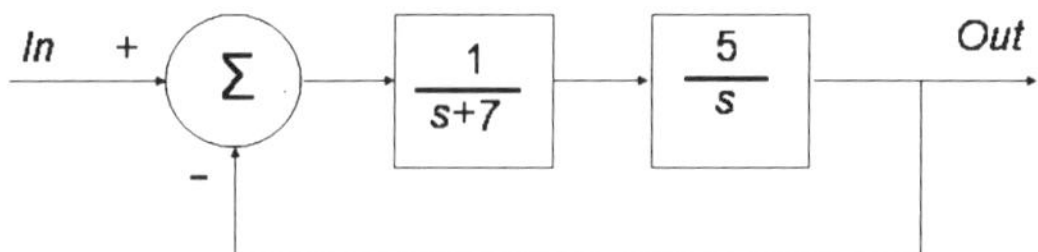

FIGURE 6.7 System used in Example 6.5. The transfer function of this system is to be determined.

Solution: As in Example 6.3 and previous examples, find $G(s)$ and $H(s)$ and apply the feedback equation (Equation 5.10). Again $H(s) = 1$ since all of the output feeds back to the input. The feedforward gain function is just the product of the two feedforward elements:

$$G(s) = \frac{1}{s+7}\left(\frac{5}{s}\right) = \frac{5}{s^2 + 7s}$$

Substituting into the feedback equation:

$$TF(s) = \frac{G(s)}{1 + G(s)H(s)} = \frac{\frac{5}{s^2+7s}}{1 + \frac{5}{s^2+7s}} = \frac{5}{s^2 + 7s + 5}$$

The fact that the denominator has a quadratic polynomial of s indicates that this transfer function represents a second-order system. This is indeed the case, and these systems are covered in the next section.

[3]Even a sine wave has a Laplace transform, but it is complicated; it is much easier to use the frequency domain techniques developed in Chapter 5 for sine waves that are in steady state. If the sine wave starts at a particular time, say $t = 0$, then it is not in steady state and Laplace techniques including the Laplace transform of a sine wave must be used.

6.2.6 Second-Order Element

Second-order processes will have quadratic polynomials of s in the denominator of the transfer function:

$$TF(s) = \frac{1}{s^2 + 2\delta\omega_n s + \omega_n^2} = \frac{1}{s^2 + bs + c} \tag{6.31}$$

The denominator of the right-hand term is the familiar notation for a standard quadratic equation and can be factored using Equation 6.32 below.

One method for dealing with second-order terms would be to factor them into two first-order terms.[4] We could then isolate the two factors in the denominator, which would have the form $(s + \alpha)$, as first-order terms using partial fraction expansion. Indeed this method is perfectly satisfactory if the factors, the roots of the quadratic equation, are real. Examination of the classic quadratic equation demonstrates when this approach will work. Since the coefficient of the s^2 term is always normalized to 1.0, the a coefficient in the quadratic is always 1.0 and the roots of the quadratic equation become:

$$r_1, r_2 = \frac{-b}{2} \pm \frac{1}{2}\sqrt{b^2 - 4c} \tag{6.32}$$

If $b^2 \geq 4c$, then the roots are real and the quadratic can be factored into two first-order terms. However, if $b^2 < 4c$, both roots are complex and have real and imaginary parts:

$$r_1 = \frac{-b}{2} + j\frac{1}{2}\sqrt{4c - b^2} \quad \text{and} \quad r_2 = \frac{-b}{2} - j\frac{1}{2}\sqrt{4c - b^2} \tag{6.33}$$

If the roots are complex, they both have the same real part $(-b/2)$, while the imaginary parts also have the same values but with opposite signs. Complex number pairs that feature this relationship, the same real part but oppositely signed imaginary parts, are called *complex conjugates*. Whether or not the roots of a second-order characteristic equation are real or complex has important consequences in the behavior of the system. Sometimes all we want to know about a second-order system is the type of roots in the characteristic equation. This saves the effort of finding the inverse Laplace transform.

The second-order transfer function variables δ and ω_n are first introduced in Chapter 5. Recall that the parameter δ is called the *damping factor* while ω_n is called the *undamped natural frequency*. As with the term *time constant*, introduced for first-order systems, these names relate to the step and impulse response behavior of the second-order systems. As is shown later, second-order systems with low damping factors respond with an exponentially decaying oscillation, and the smaller the damping factor the longer the oscillation persists. The frequency of oscillation is related to ω_n. Specifically, the frequency of oscillation, ω_d, of systems with small damping factors is:

$$\omega_d = \omega_n\sqrt{1 - \delta^2} \tag{6.34}$$

So as δ becomes smaller and smaller, the oscillation frequency ω_d approaches ω_n. When δ equals 0.0, the system is "undamped" and the oscillation continues forever at frequency ω_n.

[4]This also applies to frequency domain equations, but is not as useful as it will be here.

Hence the term *undamped natural frequency* for ω_n: It is the frequency at which the system oscillates if it has no damping, that is, $\delta \to 0$.

Here we equate the variables ω_n and δ to coefficients a and b in Equation 6.31 and insert them into the solution to the quadratic equation:

$$r_1, r_2 = \frac{-2\delta\omega_n}{2} \pm \frac{\sqrt{4\delta^2\omega_n^2 - 4\omega_n^2}}{2} = -\delta\omega_n \pm \omega_n\sqrt{\delta^2 - 1} = -\delta\omega_n \pm \omega_d \tag{6.35}$$

From this equation we see that the damping factor δ alone determines if the roots will be real or complex. Specifically, if $\delta > 1$, then the constant under the square root is positive and the roots will be real. Conversely, if $\delta < 1$, the square root will be a negative number and the roots will be complex. If $\delta = 1$, the two roots are also real, but are the same: both roots are equal to $-\omega_n$.

Again, the behavior of the system is quite different if the roots are real or imaginary, and the form of the inverse Laplace transform is also different. Accordingly, it is best to examine the behavior of a second-order system with real roots separately from that of a system with complex roots, acting as if the second-order system were two different animals.

6.2.6.1 Second-Order Elements with Real Roots

The first step in evaluating a second-order transfer function is to determine if the roots are real or imaginary. If $\delta > 1$, the system is said to be *overdamped* because it does not exhibit oscillatory behavior in response to a step or impulse input. Such systems have responses that consist of double exponentials.

The best way to analyze overdamped second-order systems is to factor the quadratic equation into two first-order terms each with its own time constant, τ_1 and τ_2:

$$TF(s) = \frac{1}{s^2 + 2\delta\omega_n s + \omega_n^2} = \frac{1}{(s + (1/\tau_1))(s + (1/\tau_2))} \tag{6.36}$$

These two time constants are just the negative of the inverse of the root of the quadratic equation, as given by Equation 6.33: $\tau_1 = -1/r_1$ and $\tau_2 = -1/r_2$. Like the first-order transfer function, the second-order transfer function can also have numerator terms other than 1, but the analysis strategy does not change. Typical overdamped impulse and step responses will be shown in the example below.

After the quadratic equation is factored, the next step is either to separate this function into two individual first-order terms, $\frac{k_1}{s + (1/\tau_1)} + \frac{k_2}{s + (1/\tau_2)}$, using partial fraction expansion (see below), or find a Laplace transform that matches the form of the unfactored equation in Equation 6.36. If the numerator is a constant, then entry 9 in the Laplace transform table (Appendix B) matches the unfactored form. If the numerator is more complicated, a match may not be found and partial fraction expansion will be necessary. Figure 6.8 shows the response of a second-order system to a unit step input. The step responses shown are for an element with four different combinations of the two time constants: τ_1, $\tau_2 = 1$, 0.2 sec; 1, 2 sec; 4, 0.2 sec; and 4, 2 sec. The time it takes for the output of this element to reach its final value depends on both time constants, but is primarily a factor of the

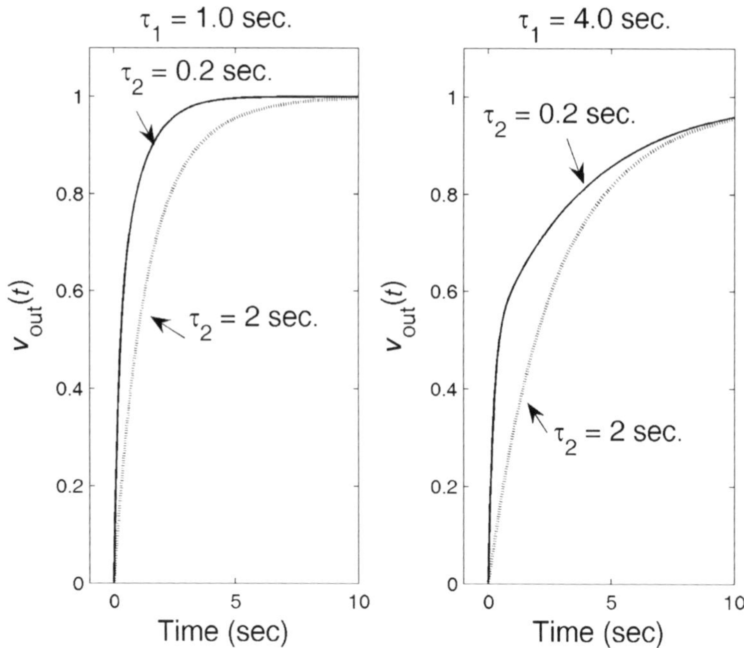

FIGURE 6.8 Typical step responses of a second-order system with real roots. These responses are termed overdamped because of the exponential-like behavior of the response. Four different combinations of τ_1 and τ_2 are shown. The speed of the response depends on both time constants.

longest time constant. The next example illustrates the use of Laplace transform analysis to determine the impulse and step responses of a typical second-order overdamped system.

EXAMPLE 6.6

Find the impulse and step responses of the second-order transfer function given below. Use Laplace analysis to find the time functions and MATLAB to plot the two time responses.

$$TF(s) = \frac{25}{s^2 + 12.5s + 25} \tag{6.37}$$

Solution, impulse response: We can find the values of δ and ω_n by equating coefficients with the basic equation (Equation 6.31):

$$\omega_n^2 = 25; \quad \omega_n = 5$$

$$2\delta\omega_n = 12.5; \quad \delta = \frac{12.5}{2\omega_n} = \frac{12.5}{10} = 1.25$$

Since $\delta > 1$, the roots are real. In this case, the next step is to factor the denominator using the quadratic equation (Equation 6.33):

$$r_1, r_2 = \frac{-b}{2} \pm \frac{1}{2}\sqrt{b^2 - 4c} = \frac{-12.5}{2} \pm \frac{1}{2}\sqrt{12.5^2 - 4(25)} = -6.25 \pm 3.75$$

$$r_1, r_2 = -10.0 \quad \text{and} \quad -2.5$$

For an impulse function input, the output $V_{out}(s)$ is the same as the transfer function since the Laplace transform of the input is $V_{in}(s) = 1$:

$$V_{out}(s) = \frac{25}{(s+10)(s+2.5)} = \left(\frac{25}{7.5}\right)\frac{7.5}{(s+10)(s+2.5)}$$

The output function can be rearranged to match entry 9) in the Laplace transform table (Appendix B):

$$\frac{\gamma - \alpha}{(s+\alpha)(s+\gamma)} \Leftrightarrow e^{-\alpha t} - e^{-\gamma t}$$

where $\alpha = 2.5$ and $\gamma = 10$.

So $v_{out}(t)$ becomes:

$$v_{out}(t) = \left(\frac{25}{7.5}\right)(e^{-2.5t} - e^{-10t}) = 3.33(e^{-2.5t} - e^{-10t}) \tag{6.38}$$

Solution, step response: The first step, factoring the denominator into two first-order terms, has already been done. To find $V_{out}(s)$, follow Equation 6.27 and multiply the transfer function by the Laplace transform of the step function, $1/s$:

$$V_{out}(s) = \left(\frac{1}{s}\right)\frac{25}{(s+2.5)(s+10)} = \frac{25}{s(s+2.5)(s+10)} \tag{6.39}$$

With the extra s added to the denominator, this function no longer matches any in the Laplace transform table. However, we can expand this function using partial fraction expansion into the form:

$$V_{out}(s) = \frac{k_1}{s} + \frac{k_2}{s+2.5} + \frac{k_3}{s+10}$$

6.2.6.2 Partial Fraction Expansion

Partial fraction expansion is the opposite of finding the common denominator: Instead of trying to combine fractions into a single fraction, we are trying to break them apart. We want to determine what series of simple fractions will add up to the fraction we are trying to decompose. The technique described here is a simplified version of partial fraction expansion that deals with distinct linear factors, that is, denominator components of the form $(s - p)$. Moreover, this analysis will be concerned only with single components, not multiple components such as $(s - p)^2$.

Under these restrictions, the partial fraction expansion can be defined as:

$$TF(s) = \frac{N(s)}{(s-p_1)(s-p_2)(s-p_3)\ldots.} = \frac{k_1}{s-p_1} + \frac{k_2}{s-p_2} + \frac{k_3}{s-p_3} + \cdots. \tag{6.40}$$

where:

$$k_n = (s - p_n)TF(s)\big|_{s=p_n} \tag{6.41}$$

Since the constants in the Laplace equation denominator will always be positive, the values of p will always be negative. Applying partial fraction expansion to the Laplace function of Equation 6.39, the values for p_1, p_2, and p_3 are, respectively, −0, −2.5, and −10, which produces the numerator terms k_1, k_2, and k_3:

$$k_1 = (s+0)\frac{25}{s(s+2.5)(s+10)}\bigg|_{s=-0} = \frac{25}{2.5(10)} = 1.0$$

$$k_2 = (s+2.5)\frac{25}{s(s+2.5)(s+10)}\bigg|_{s=-2.5} = \frac{25}{-2.5(-2.5+10)} = -1.33$$

$$k_3 = (s+10)\frac{25}{s(s+2.5)(s+10)}\bigg|_{sS=-10} = \frac{25}{-10(2.5-10)} = 0.33$$

This gives rise to the expanded version of Equation 6.39:

$$V_{out}(s) = \frac{25}{s(s+2.5)(s+10)} = \frac{1}{s} - \frac{1.33}{s+2.5} + \frac{0.33}{s+10} \tag{6.42}$$

Each of the terms in Equation 6.42 has an entry in the Laplace table (Appendix B). Taking the inverse Laplace transform of each of these terms separately gives:

$$v_{out}(t) = 1.0 - 1.33e^{-2.5t} + 0.33e^{-10t} \text{ volts} \tag{6.43}$$

MATLAB is used as in Example 6.4 to plot the two resulting time-domain equations (Equations 6.38 and 6.43), in Figure 6.9.

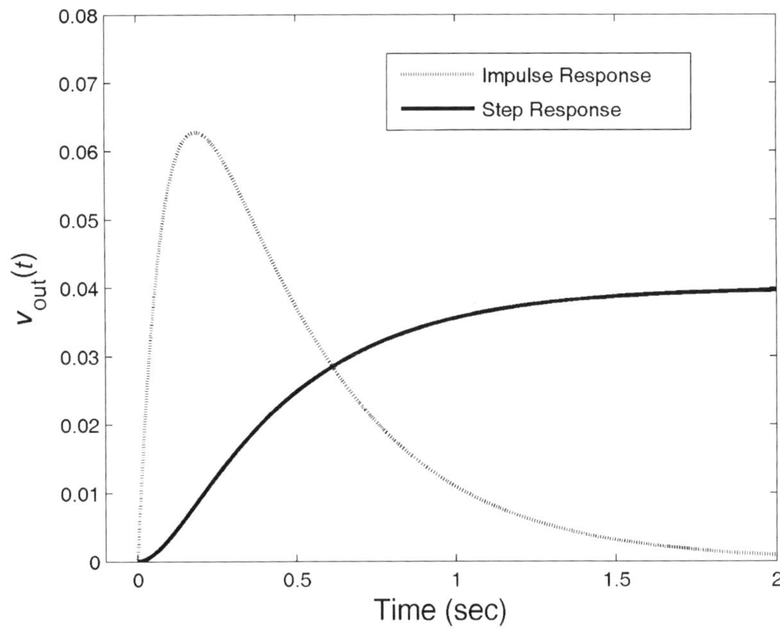

FIGURE 6.9 The impulse (dashed curve) and step (solid curve) responses of the transfer function used in Example 6.6.

EXAMPLE 6.7

Find the impulse response of the system having the transfer function:

$$TF(s) = \frac{s/4}{s^2 + 15s + 50} \tag{6.44}$$

Since $In(s) = 1/s$, the output in Laplace notation becomes:

$$Out(s) = \left(\frac{1}{s}\right)\frac{s/4}{s^2 + 15s + 50} = \frac{.25}{s^2 + 15s + 50}$$

Next, we evaluate the value of δ by equating coefficients:

$$2\delta\omega_n = 15; \quad \delta = \frac{15}{2\omega_n} = \frac{15}{2\sqrt{50}} = 1.06$$

Since $\delta = 1.06$, the roots will be real and the system will be overdamped. The next step is to factor the roots using the quadratic equation (Equation 6.30):

$$r_1, r_2 = \frac{-15}{2} \pm \frac{1}{2}\sqrt{15^2 - 4(50)} = -7.5 \pm 2.5 = -10.0, \ -5.0$$

and the output becomes: $Out(s) = \dfrac{.25}{(s + 10)(s + 5)}$

The inverse Laplace transform for this equation can be found in Appendix B (entry 9) where: $\gamma = 10$ and $\alpha = 5$. To get the numerator constants to match, multiply top and bottom by $10 - 5 = 5$:

$$Out(s) = \left(\frac{0.25}{5}\right)\frac{5}{(s + 10)(s + 5)} = (0.05)\frac{5}{(s + 10)(s + 5)}$$

This leads to the time function:

$$out(t) = 0.05(e^{-5t} - e^{-10t}) \tag{6.45}$$

Alternatively, we could decompose this second-order polynomial in the denominator into two first-order terms using partial fraction expansion.

$$k_1 = (s + 10)\frac{.25}{(s + 10)(s + 5)}\bigg|_{s=-10} = \frac{.25}{(-10 + 5)} = -.05$$

$$k_2 = (s + 5)\frac{.25}{(s + 10)(s + 5)}\bigg|_{s=-5} = \frac{.25}{(-5 + 10)} = .05$$

$$Out(s) = \frac{0.05}{s + 5} - \frac{0.05}{s + 10}$$

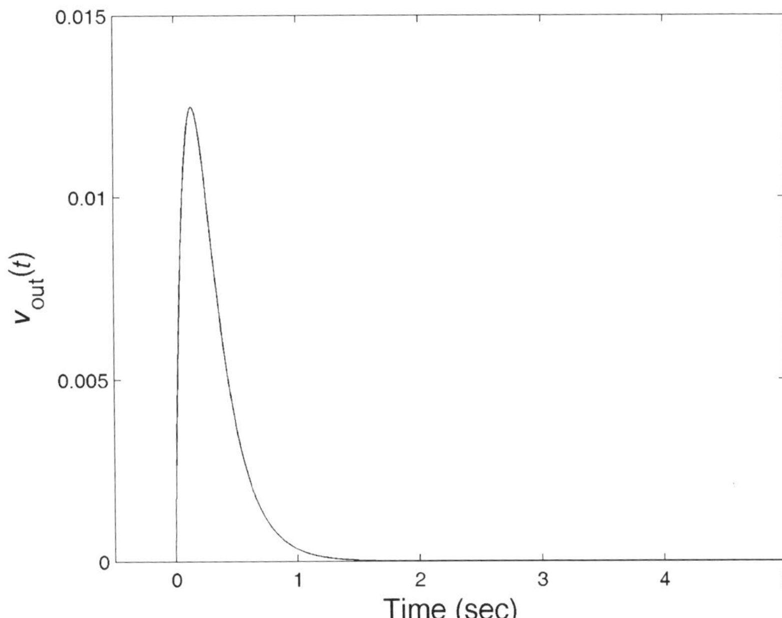

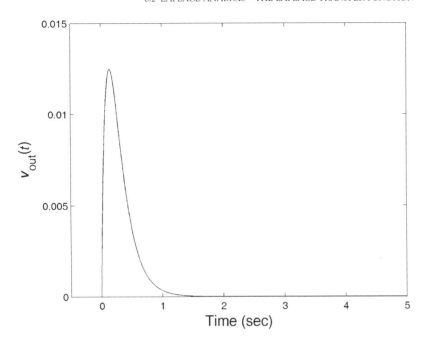

FIGURE 6.10 The impulse response of the second-order system in Example 6.7.

Taking the inverse Laplace transform of the two first-order terms leads to the same time function found using the unfactored transfer function. Figure 6.10 shows *out*(t) plotted using MATLAB code.

6.2.6.3 Second-Order Processes with Complex Roots

If the damping factor, δ, of a second-order transfer function is <1, then the roots of the characteristic (i.e., denominator) equation will be complex and the step and pulse responses will have the behavior of a damped sinusoid: a sinusoid that decreases in amplitude exponentially with time (i.e., of the general form $e^{-\alpha t} \sin \omega_n t$). Second-order systems that have complex roots are said to be *underdamped*, a term relating to the oscillatory behavior of such systems. This terminology can be a bit confusing. *Underdamped* systems respond with *damped* sinusoidal behavior while *overdamped* responses have double exponential responses and no sinusoidal-like behavior.

If the roots are complex, the quadratic equation is not factored, but an inverse Laplace transform can usually be determined directly. Inverse Laplace transforms for second-order underdamped responses are provided in the table in terms of ω_n and δ and in terms of general coefficients: numbers 13–16 in Appendix B. Usually the only difficulty in finding the inverse Laplace transform of these systems is in matching coefficients and scaling the transfer function to match the constants in the table. The next example demonstrates the solution of a second-order underdamped system.

EXAMPLE 6.8

In an experiment performed by Lawrance Stark in the early 1960s, a subject gripped a handle while an impulse of torque was applied. (To create the torque impulse, an unseen pendulum struck a lever arm attached to the handle.) The resulting rotation of the wrist under relaxed conditions was measured and it was determined that this response could be represented by the second-order equation given below (Equation 6.46). As shown in Chapter 10, such an equation is typical of mechanical systems that contain a mass, elasticity, and friction or viscosity. In this example, we solve for the angular rotation, $\theta(t)$.

$$TF(s) = \frac{\theta(s)}{T(s)} = \frac{150}{s^2 + 6s + 310} \tag{6.46}$$

where $T(s)$ is the torque input and $\theta(s)$ is the rotation of the wrist in radians.

Solution: The values for δ and ω_n are found from the quadratic equation's coefficients.

$$\omega_n = \sqrt{310} = 17.6 \quad \text{and} \quad \delta \text{ is}: \; 2\delta\omega_n = 6; \quad \delta = \frac{6}{2\omega_n} = \frac{6}{2(17.6)} = 0.17$$

Since $\delta < 1$, the roots are complex.

The input $T(s)$ was an impulse of torque with a value of 2×10^{-2} newton-meters, so in the Laplace domain: $T(s) = 2 \times 10^{-2}$. $\theta(s)$ becomes:

$$\theta(s) = (2 \times 10^{-2})\frac{150}{s^2 + 6s + 310} = \frac{3}{s^2 + 6s + 310} \tag{6.47}$$

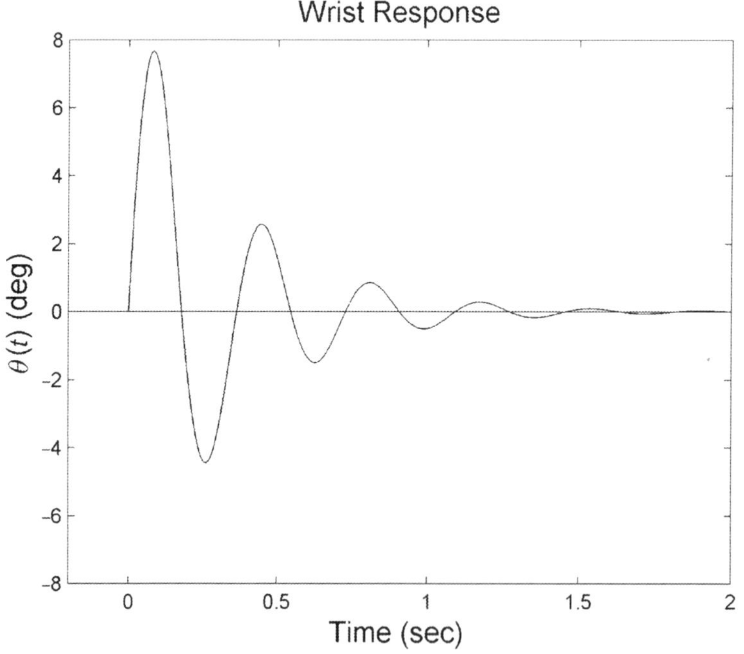

FIGURE 6.11 The step response of the relaxed human wrist to an impulse of torque showing an underdamped response.

Two of the transforms given in the Laplace tables will work (number 13 and 15): entry 13) is used here:

$$e^{-\alpha t}\left(\frac{c-b\alpha}{\beta}\sin(\beta t)+b\cos(\beta t)\right) \Leftrightarrow \frac{bs+c}{s^2+2\alpha s+\alpha^2+\beta^2}$$

Equating coefficients with Equation 6.47 we get:

$$c=3;\quad b=0;\quad \alpha=\frac{6}{2}=3;\quad \alpha^2+\beta^2=310;\quad \beta=\sqrt{310-\alpha^2}=\sqrt{310-9}=17.3$$

Substituting these values into the time-domain equivalent on the left side gives $\theta(t)$:

$$\theta(t)=e^{-3t}\left(\frac{3}{17.3}\sin(17.3t)\right)=0.173e^{-3t}(\sin(17.3t))$$

This response, scaled to degrees, is plotted in Figure 6.11 using MATLAB code. Other examples of second-order underdamped responses are presented in the next chapter, which analyzes system response behavior in the time domain.

6.3 NONZERO INITIAL CONDITIONS—INITIAL AND FINAL VALUE THEOREMS

6.3.1 Nonzero Initial Conditions

The Laplace analysis method cannot deal with negative values of time but, as mentioned above, it can handle elements that have a nonzero condition at $t=0$. So one way of dealing with systems that have a history for $t<0$ is to summarize that history as an initial condition at $t=0$. To evaluate systems with an initial condition, the full Laplace domain equations for differentiation and integration must be used. These equations (Equations 6.5 and 6.7), are repeated here.

$$\mathscr{L}\frac{dx(t)}{dt}=sX(s)-x(0-) \tag{6.48}$$

$$\mathscr{L}\left[\int_0^T x(t)dt\right]=\frac{1}{s}X(s)+\frac{1}{s}\int_{-\infty}^{0} x(t)dt \tag{6.49}$$

With the derivative, the initial value of the variable, $x(0-)$, is subtracted from the standard (i.e., zero initial condition) Laplace domain derivative term. With integration, a constant divided by s is added to the zero initial condition Laplace term. The constant is the integral of the past history of the variable from minus infinity to zero. This is because the current state of an integrative process is the integration of all past values.

To analyze systems with nonzero initial conditions, it is only necessary to apply both terms in Equations 6.48 and 6.49. An example is given in the solution of a one-compartment diffusion model. This model is a simplification of diffusion in biological compartments as the cardiovascular system. The one-compartment model would apply to large molecules in the blood that cannot diffuse into tissue and are only slowly eliminated. Moreover, the compartment is assumed to provide perfect mixing: mixing of the blood and substance is instantaneous and complete.

EXAMPLE 6.9

Find the concentration, $c(t)$, of a large molecule solute delivered as a step input to the blood compartment. Assume an initial concentration of $c(0)$ and a diffusion coefficient, K.

The kinetics of a one-compartment system with no outflow is given by the differential equation

$$V\frac{dc(t)}{dt} = F_{in}(t) - Kc(t)$$

where V is the volume of the compartment, K is a diffusion coefficient, and $F_{in}(t)$ is the input, which is assumed to be an impulse of a given amount A: $F_{in}(t) = A\mu(t)$. Converting this to the

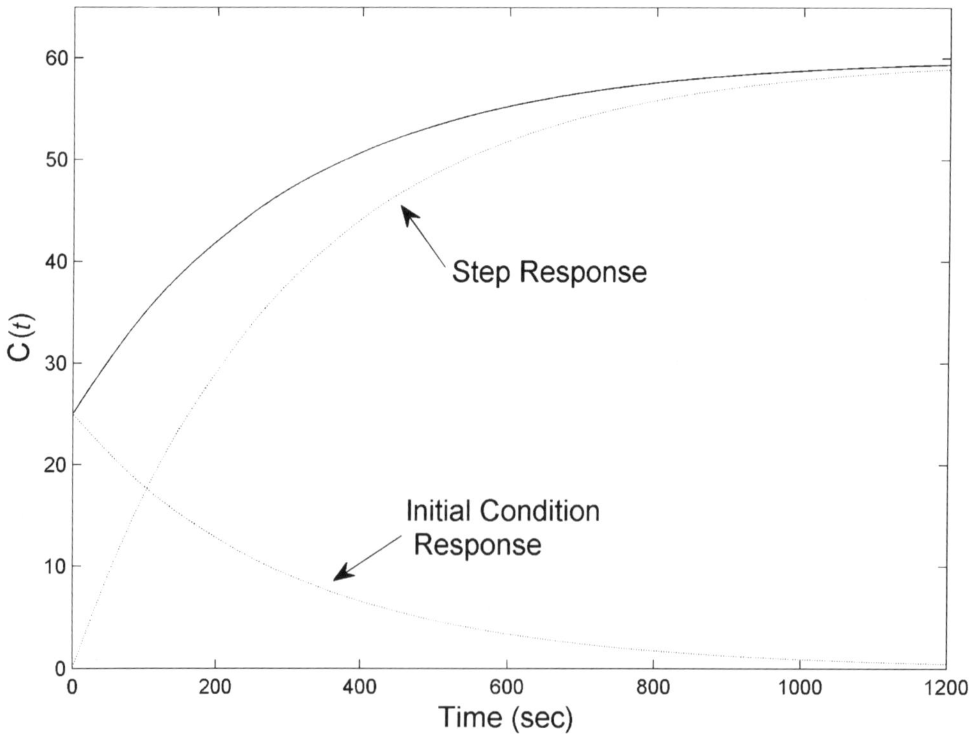

FIGURE 6.12 The diffusion of a large molecule solute delivered as a step input into the blood compartment. The molecule had an initial concentration of 25 gm/ml. Perfect mixing in the blood compartment is assumed.

Laplace domain, $F_{in}(s)$ becomes A/s. Using both terms of Equation 6.48 for the derivative operation:

$$V(sC(s) - c(0)) = (A/s) - KC(s)$$

Solving for the concentration, $C(s)$:

$$C(s)(Vs + K) = Vc(0) + (A/s)$$

$$C(s) = \frac{A}{s(Vs+K)} + \frac{Vc(0)}{Vs+K} = \frac{A/V}{s(s+(K/V))} + \frac{c(0)}{s+(K/V)}$$

Taking the inverse Laplace transform:

$$c(t) = \frac{A/V}{K/V}(1 - e^{-tK/V}) + c(0)e^{-tK/V} = (A/K)(1 - e^{-tK/V}) + c(0)e^{-tK/V}$$

Hence the concentration increases exponentially from an initial value of $c(0)$ to A/K while the initial value decays exponentially to zero. Using typical values from Rideout (1991) of $A = 0.6$ gm, $V = 3.0$ ml, $K = 0.01$, and $c(0) = 25.0$ gm/ml results in the plot shown in Figure 6.12 (solid line).

Nonzero initial conditions also occur in circuits where preexisting voltages exist or in mechanical systems which have a nonzero initial velocity; more examples of nonzero initial condition problems are found in Chapter 9.

6.3.2 Initial and Final Value Theorems

The time representation of a Laplace function is obtained by taking the inverse Laplace transform using tables such as those found in Appendix B. Sometimes we are only looking for the value of the function at the very beginning of the stimulus, $out(t = 0)$, or at its very end, $out(t \to \infty)$. Two useful theorems that can easily supply us with this information are the *Initial Value Theorem* and *Final Value Theorem*. As their names imply, these theorems give us the initial and the final output values without the need for taking the inverse Laplace transform. The Initial Value Theorem provides us with the value of the function at $t = 0$, while the Final Value Theorem, as you might expect, gives us the value of the function as $t \to \infty$.

Since the Laplace variable, s, is a form of complex frequency, and frequency is inversely related to time, it is not to unreasonable to find that the value of a Laplace function at $t = 0$ can be obtained by letting $s \to \infty$. In particular, the Initial Value Theorem states:

$$x(0+) = \lim_{s\to\infty} sX(s) \tag{6.50}$$

The Final Value Theorem follows the same logic, except since $t \to \infty$, s goes to 0. Specifically, the Final Value Theorem states:

$$\lim_{t\to\infty} x(0+) = \lim_{s\to 0} sX(s) \tag{6.51}$$

The application of either theorem is straightforward, as is shown in the following example.

EXAMPLE 6.10

Use the Final Value Theorem to find the final value of $x(t)$, the step response of the system whose transfer function is given below. Also find the final value the hard way: by determining $x(t)$ from the inverse Laplace transform, then letting $t \to \infty$.

$$TF(s) = \frac{.28s + .23}{s^2 + 0.3s + 2} \tag{6.52}$$

Solution: Using the inverse Laplace transform method find the output Laplace function $X(s)$ by multiplying $TF(s)$ by the impulse function in the Laplace domain:

$$X(s) = \frac{1}{s} TF(s) = \frac{.28s + 0.92}{s(s^2 + 0.3s + 2)} \tag{6.53}$$

Next we find the full expression for $x(t)$ by finding the inverse Laplace transform. First we need to calculate the value of δ to determine if the roots are real or complex:

$$2\delta\omega_n = 0.3; \quad \delta = \frac{0.3}{2\omega_n} = \frac{0.3}{2\sqrt{2}} = 0.106 < 1$$

So the system is clearly underdamped and the appropriate transfer function should be used to match $X(s)$ (numbers 13 through 16 in the Laplace transform table). The function $X(s)$ matches entry number 14 in the Laplace transform table, but requires some rescaling to match the numerator.

$$1 - e^{-\alpha t}\left(\frac{\alpha - b}{\beta}\right)\sin\beta t + b\cos\beta t \Leftrightarrow \frac{bs + \alpha^2 + \beta^2}{s(s^2 + 2\alpha s + \alpha^2 + \beta^2)}$$

Considering only the dominator, the sum $\alpha^2 + \beta^2$ should equal 2, but then the constant in the numerator which is also $\alpha^2 + \beta^2$ must also equal 2. To make the numerator match, we need to multiply it by: $2/0.92 = 2.17$. Multiplying top and bottom by 2.17, the rescaled Laplace function becomes:

$$X(s) = \left(\frac{1}{2.17}\right)\frac{0.61s + 2}{s(s^2 + 0.3s + 2)} = \frac{0.46(0.61s + 2)}{s(s^2 + 0.3s + 2)}$$

Now equating coefficients with entry 14:

$$b = .61; \quad \alpha = \frac{0.3}{2} = 0.15; \quad \alpha^2 + \beta^2 = 2; \quad \beta^2 = 2 - 0.15^2; \quad \beta = \sqrt{1.98} = 1.41$$

the inverse Laplace transform becomes:

$$x(t) = 0.46\left[1 - e^{-.15t}\left(\frac{.15 - .61}{1.41}\sin(1.41t) + \cos(1.41t)\right)\right]$$

$$x(t) = 0.46(1 - e^{-.15t}(-0.33\sin(1.41t) + \cos(1.41t)))$$

Now letting $t \to \infty$, the exponential term goes to zero and the final value becomes:

$$x(t) = 0.46$$

Solution, Final Value Theorem: This approach is much easier. Substitute the output Laplace function given in Equation 6.52 into Equation 6.51:

$$\lim_{s\to 0} sX(s) = \lim_{s\to 0} s\left[\frac{0.28s + 0.92}{s(s^2 + 0.3s + 2)}\right] = \lim_{s\to 0}\left[\frac{0.28s + 0.92}{s^2 + 0.3s + 2}\right] = \frac{0.92}{2} = 0.46$$

This is the same value that is obtained when letting $t \to \infty$ in the time solution $x(t)$.

An example of the application of the Initial Value Theorem is found in Problem 16 at the end of the chapter.

6.4 THE LAPLACE DOMAIN AND THE FREQUENCY DOMAIN

Since s is a complex frequency variable, there is a relationship between the Laplace domain and the frequency domain. Given a Laplace transfer function, it is easy to find the frequency domain equivalent by substituting $s = j\omega$. Then, after renormalizing the coefficients so the constant term equals 1, the frequency plot can be constructed using Bode plot techniques (or MATLAB). Some of the relationships between the Laplace transfer function and the frequency domain characteristics have already been mentioned, and these depend largely on the characteristic equation. A first-order characteristic equation gives rise to first-order frequency characteristics such as those shown in Figures 5.16 and 5.17.

Second-order frequency characteristics, like second-order time responses, are highly dependent on the value of the damping coefficient, δ. As shown in Figure 5.20, the frequency curve develops a peak for values of $\delta < 1$, and the height of that peak increases as δ decreases. In the time domain, $\delta < 1$ corresponds to an overshoot response. In the frequency domain, the peak in the frequency curve occurs at frequency ω_d, which for small values of δ is close to ω_n, the undamped natural frequency (Equation 6.34). In the time domain, the response will oscillate at frequency ω_d. As indicated by Equation 6.34, the oscillating frequency is not quite the same as the undamped frequency, the resonant frequency at which the system would like to oscillate if no damping was present. This is because the decay in the oscillation due to the exponential damping lowers the actual resonant frequency to ω_d. As the damping factor, δ, decreases, the resonant frequency, ω_d, approaches the undamped natural frequency, ω_n.

The next example uses MATLAB to compare the frequency and time characteristics of a second-order system for various values of damping.

EXAMPLE 6.11

Define a second-order transfer function with an undamped natural frequency of 100 rad/sec and 3 values of damping factor: 0.05, 0.1, and 0.5. Plot the frequency spectrum (i.e., Bode plot) of the time domain transfer function and the step response of the system for the 3 damping factors.

Use the Laplace transform to solve for the time response and MATLAB for calculation and plotting.

Solution: The second-order transfer function with the desired ω_n is:

$$TF(s) = \frac{1}{s^2 + 2(10^2)\delta s + 10^4}$$

The step response can be obtained by multiplying the transfer function by $1/s$, then determining the inverse Laplace transform leaving δ as a variable:

$$V_{out}(s) = \frac{1}{s(s^2 + 2(10^2)\delta s + 10^4)}$$

This matches entry number 16 (Appendix B) for $\delta < 1$, although a minor rescaling is required:

$$V_{out}(s) = \left(\frac{1}{10^4}\right)\frac{10^4}{s(s^2 + 200\delta s + 10^4)}$$

$$v_{out}(t) = \left(\frac{1}{10^4}\right)\left(1 - \frac{e^{-\delta\omega_n t}}{\sqrt{1-\delta^2}}\sin(\omega_n\sqrt{1-\delta^2}\,t + \theta)\right); \quad \theta = \tan^{-1}\left(\frac{\sqrt{1-\delta^2}}{\delta}\right)$$

where:

$$\omega_n^2 = 10^4; \quad \omega_n = 100;$$

The equivalent time response will be programmed directly into the MATLAB code.

To find the frequency response, convert the Laplace transfer function to a frequency domain transfer function by substituting $j\omega$ of s and rearranging it into the frequency domain format where the constant is normalized to 1.0:

$$TF(\omega) = \frac{1}{(j\omega)^2 + 200\delta j\omega + 10^4} = \frac{1}{1 - 10^{-4}\omega^2 + 0.02\delta j\omega}$$

This equation can also be programmed directly into MATLAB. The resulting program is shown below, and the time and frequency plots are shown in Figures 6.9 and 6.10.

```
% Ex 6.11 Comparison of time and frequency characteristics of a second-order system
with % different damping factors.
%
w = 1:1:10000;                                   % Frequency vector: 1 to 10,000 rad/sec
t = 0:.0001:.4;                                  % Time vector: 0 to 0.4 sec.
delta = [0.5 0.1 0.05];                          % Damping factors
% Calculate spectra
for k = 1:length(delta)                          % Repeat for each damping factor
  TF = 1./(1 -10.^ -4*w.^2 + j*delta(k)*.02*w);  % Transfer function
  Mag = 20*log10(abs(TF));                       % Take magnitude in dB
  semilogx(w,Mag);                               % Plot semilog.
end
```

```
.........labels and title......
% Calculate and plot time characteristics
figure ;
wn = 100;                                     % Undamped natural frequency
for k = 1:length(delta)
  c = sqrt(1 - delta(k)^2);
  theta = atan(c/delta(k));
  xt = (1 - (1/c)*(exp(-delta(k)*wn*t)).*...
  (sin(wn*c*t + theta)))/10000;               % Time function
  plot(t,xt)
end
......label, title, and text......
```

The results are shown in Figures 6.13 and 6.14. The correspondence between the frequency characteristics and the time responses is evident. A system that has a spectrum with a peak corresponds to a time domain response with an overshoot. Note that the frequency peak associated with a δ of 0.5 is modest, but the time domain response still has some overshoot. As shown here, the larger the spectral peak the greater the overshoot.

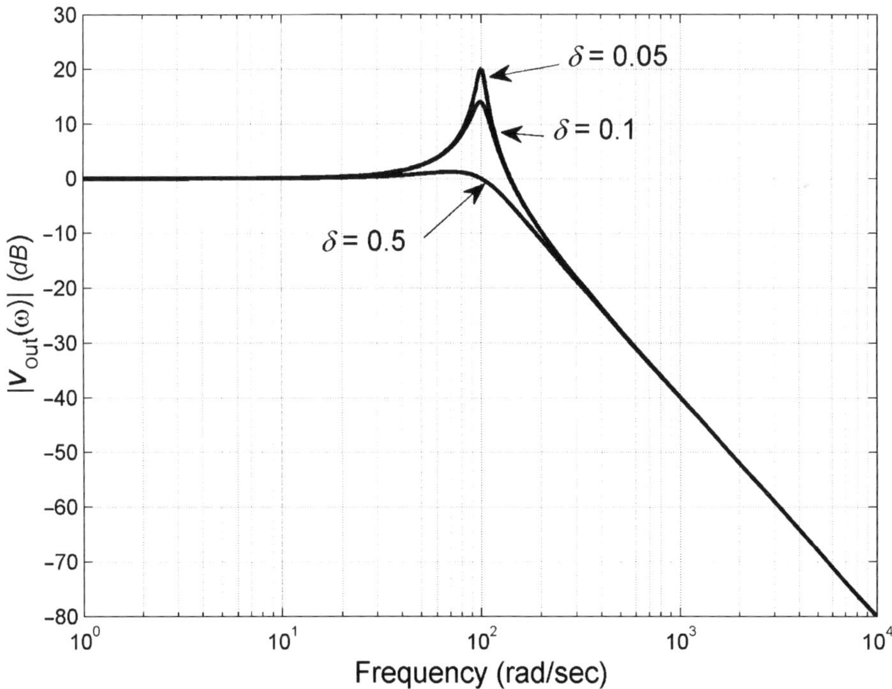

FIGURE 6.13 Comparison of magnitude spectra of the second-order system presented in Example 6.10 having three different damping coefficients: $\delta = 0.05$, 0.1, and 0.05.

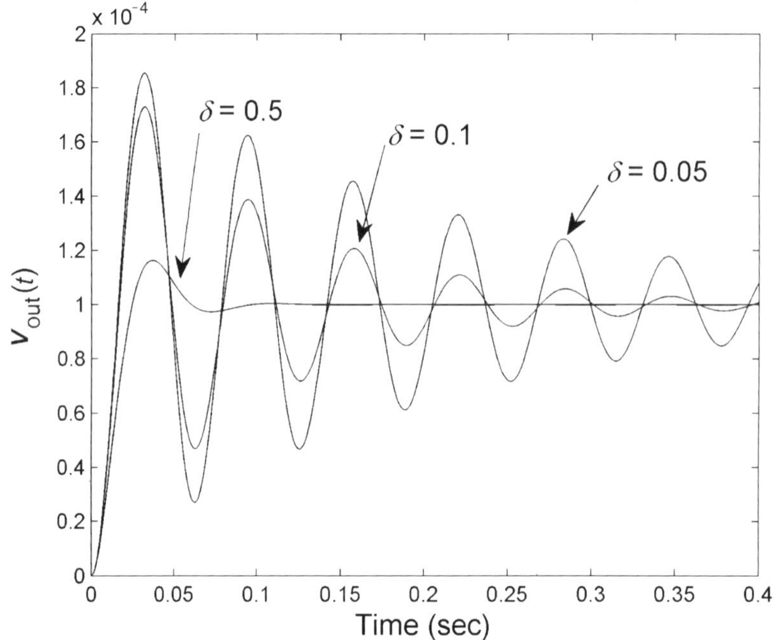

FIGURE 6.14 Comparison of step responses for a second-order system presented in Example 6.10 having the same three different damping factors used in Figure 6.13.

Often it is not practical to simulate biological systems sinusoidally, but it is possible to generate an impulse input. For example, it is impractical to simulate a pharmacological system sinusoidally, but a sudden infusion of a drug, a drug *bolus*, is a common input. The knee-jerk response is another example of the use of an impulse input.

The final example of this chapter finds the Bode plot of a transfer function with a fourth-order polynomial in the denominator. This example uses MATLAB to factor the denominator polynomial and to generate the plot, but the plotting could also have been done by hand using the Bode plot methods developed in Chapter 5.

EXAMPLE 6.12

Factor the transfer function shown below into first- and second-order terms. Write out the factored transfer function and plot the Bode plot (magnitude and phase). Use MATLAB to factor the numerator and denominator, rearrange into first- and second-order terms, and generate the Bode plot.

$$TF(s) = \frac{s^3 + 5s^2 + 3s + 10}{s^4 + 12s^3 + 20s^2 + 14s + 10}$$

Solution: The numerator contains a third-order term and the denominator contains a fourth-order term. We use the MATLAB `roots` routine to factor these two polynomials and rearrange into first- and second-order terms, then substitute $s = j\omega$ and plot. Of course, we could plot the transfer function directly without factoring, but the factors provide some insight into the system. The first part of the code factors the numerator and denominator.

```
% Example 6.12 Find the Bode plot of a higher order transfer function.
%
% Define numerator and denominator
num = [1 5 3 10];          % Define numerator polynomial
den = [1 12 20 14 10];     % Define denominator polynomial
n_root = roots(num)        % Factor numerator and display
d_root = roots(den)        % Factor denominator and display
  The results of this program are:
n_root = -4.8086
   -0.0957 + 1.4389i
   -0.0957 - 1.4389i
d_root = -10.1571
   -1.4283
   -0.2073 + 0.8039i
   -0.2073 - 0.8039i
```

Thus the numerator factors into a single root and a complex pair indicating a second-order term with complex roots, and the denominator factors into two single roots and one complex pair. To find the second-order polynomials, we could multiply the complex roots:

$$(s + 0.0957 + j1.4389)(s + 0.0957 - j1.4389)$$

or save ourselves some complex algebra by using the MATLAB `poly` routine with the complex congregate as arguments:

```
nun_sec_order = poly( n_root(2:3,:) )  % Construct second-order numerator
den_sec_order = poly( d_root(3:4,:) )  % Construct second-order denominator
This gives the output:
nun_sec_ord = 1.0000 0.1914 2.0796
den_sec_order = 1.0000 0.4146 0.6893
```

Results: The factored transfer function is written as:

$$TF(s) = \frac{(s + 4.81)(s^2 + 0.19s + 2.08)}{(s + 10.2)(s + 1.43)(s^2 + 0.41s + 0.69)}$$

The Bode plot is obtained by substituting in $s = j\omega$ in this equation and evaluating over ω where now each of the terms is identifiable as one of the Bode plot elements presented in Chapter 5. In this example, we will use MATLAB to plot this function so there is no need to rearrange this equation into the standard Bode plot format (i.e., with the constant terms equal to 1). The resulting plot is shown in Figure 6.15.

```
%
% Plot the Bode Plot
w = 0.1:.05:100;                                   % Define the freq. vector
TF = (j*w + 4.81).*(- w.^2 + j*0.19*w + 2.08)...
```

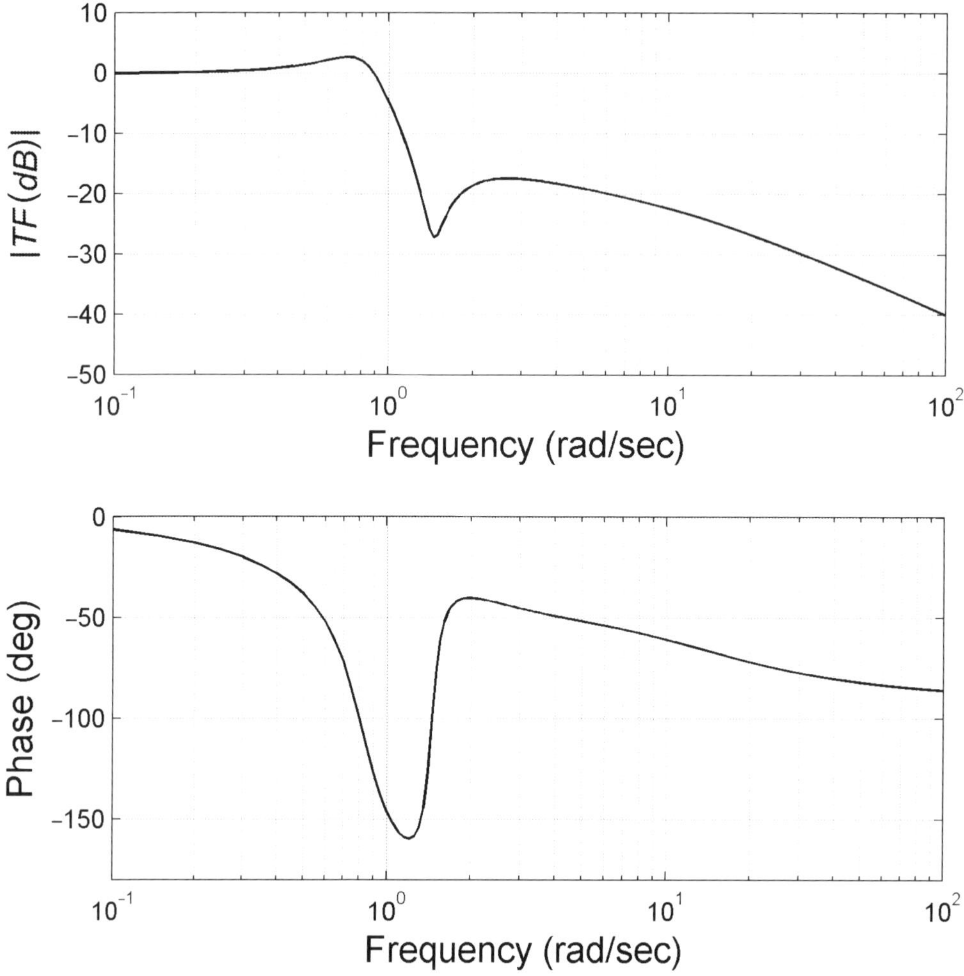

FIGURE 6.15 The magnitude and phase frequency characteristic (i.e., Bode plot) of the higher-order transfer function used in Example 6.11.

```
  ./((j*w+10.2).*(j*w+1.43).*(-w.^2+j*0.41*w+0.69));   % Define TF
subplot(2,1,1);
  semilogx(w,20*log10(abs(TF)));                       % Plot magnitude spectrum
                                                       % in dB
  ......labels......
subplot(2,1,2);
  semilogx(w,angle(TF)*360/(2*pi));                    % Plot the phase spectrum
  .......labels......
```

6.5 SUMMARY

With the Laplace transform, all of the analysis tools developed in Chapter 5 and other tools developed later can be applied to systems exposed to a broader class of signals. Transfer functions written in terms of the Laplace variable s (i.e., complex frequency) serve the same function as frequency domain transfer functions, but to a broader class of signals. Here only the response to the step and impulse signals is explored because these are the two stimuli that are most commonly used in practice. Their popularity stems from the fact that they provide a great deal of insight into system behavior, and they are usually easy to generate in practical situations. However, responses to other signals such as ramps or exponentials, or any signal that has a Laplace transform, can be analyzed using these techniques. Laplace transform methods can also be extended to systems with nonzero initial conditions, a useful feature explored in Chapter 9.

The Laplace transform can be viewed as an extension of the Fourier transform where complex frequency, s, is used instead of imaginary frequency, $j\omega$. With this in mind, it is easy to convert from the Laplace domain to the frequency domain by substituting $j\omega$ for s in the Laplace transfer functions. Bode plot techniques can be applied to these converted transforms to construct the magnitude and phase spectra. Thus the Laplace transform serves as a gateway into both the frequency domain and the time domain through the inverse Laplace transform.

PROBLEMS

1. Find the Laplace transform of the following time functions:

a.

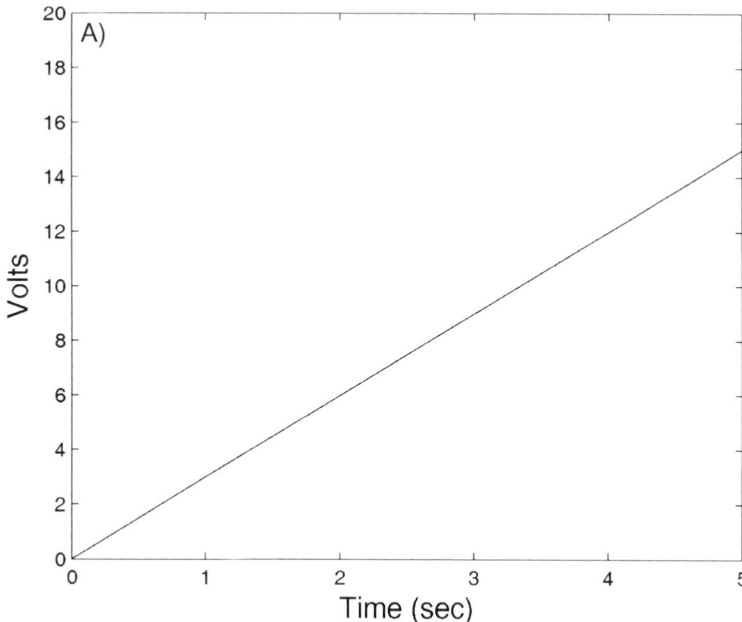

b.

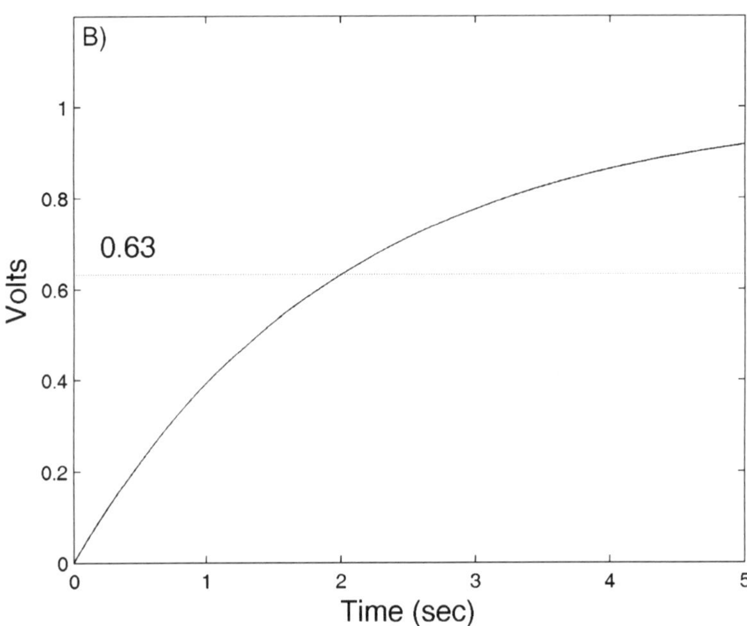

c. $(e^{-2t} - e^{-5t})$
d. $2e^{-3t} - 4e^{-6t}$
e. $5 + 3e^{-10t}$

2. Find the inverse Laplace transform of the following Laplace functions:
a. $\frac{10}{s+5}$
b. $\frac{10}{s(s+5)}$
c. $\frac{5s+4}{s^2+5s+20}$ [*Hint*: Check roots]
d. $\frac{5s+4}{s(s^2+5s+20)}$

3. Use partial fraction expansion to find the inverse Laplace transform of the following functions:
a. $\frac{s+4}{s^2+10s+10}$
b. $\frac{10}{s(s^2+4s+3)}$

4. Find the time response of the system below if the input is a step from 0 to 5.

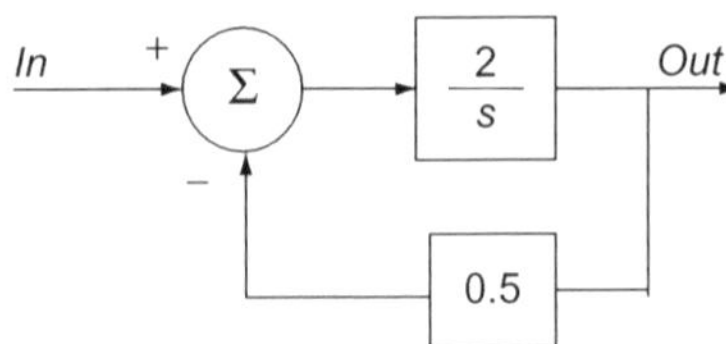

5. Find the time response of the system below to a unit step function. Use Laplace methods to solve for the time response as a function of k. Then use MATLAB to plot the time function for $k = 0.1$, 1, and 10.

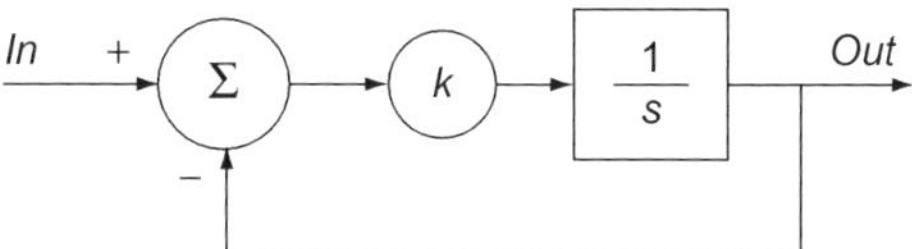

6. Find the time response of the two systems below if the input is a step from 0 to 8. Use MATLAB to plot the time responses.

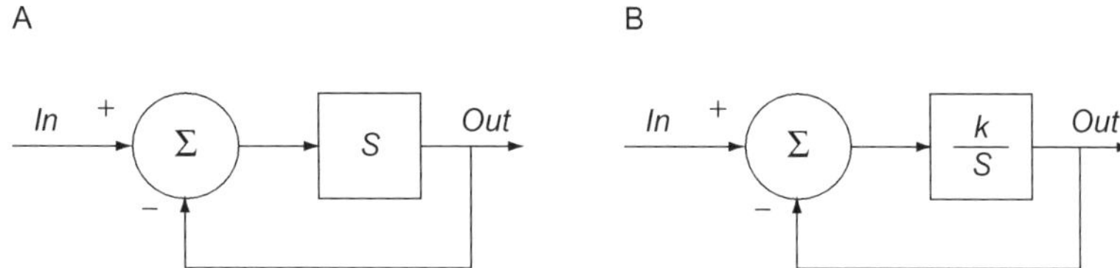

7. Solve for the Laplace transfer function of the system below. Find the time response to a step from 0 to 4 and a ramp having a slope of 4/sec. [*Hint*: You can apply the feedback equation twice.]

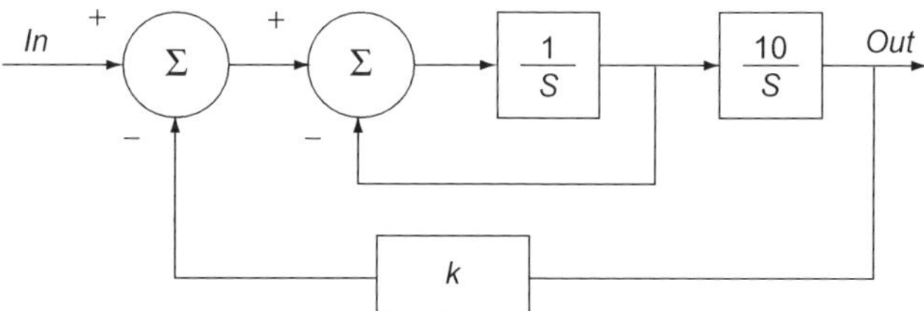

8. Find the time response of the system below if the input is a unit step. Also find the response to a unit impulse function. Assume $k = 5$.

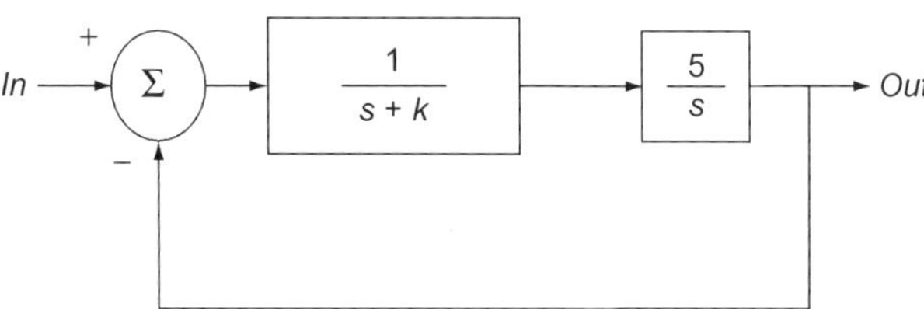

9. Repeat Problem 8 assuming $k = 0.2$. How does decreasing k change the output behavior?
10. Find the impulse response to the system shown below for $k = 1$ and $k = 0.1$. Use MATLAB to plot the two impulse responses. How does decreasing k change the output behavior?

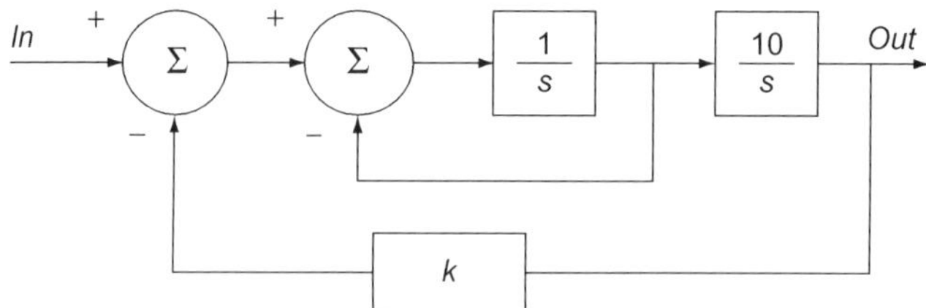

11. Find the impulse response to the system shown below for $k = 1$ and $k = 0.1$. Use MATLAB to plot the two impulse responses. How does decreasing k change the output behavior? Compare with the results in Problem 10 for the same values of k.

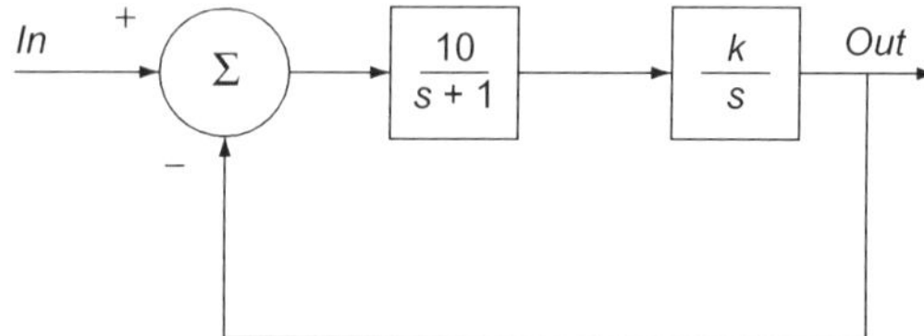

12. Use MATLAB to plot the magnitude spectra for the two systems shown in Problem 6.

13. Use MATLAB to plot the magnitude spectra for the system in Problem 10 for the two values of k. Repeat for the system in Problem 11.

14. Use Laplace analysis to find the transfer function of the system below that contains a time delay of 0.01 sec ($e^{-0.01s}$). Find the Laplace domain response of a step from 0 to 10. Assume $k = 1.0$. Then convert to the frequency domain and use MATLAB to find the magnitude and phase spectrum of the response. Note that time delays are common in biological systems. Plot the system spectrum over a frequency range of 1 to 200 rad/sec. Also note that the phase curve with the delay will exceed −180 degrees and will wrap around, so use the MATLAB `unwrap` routine.

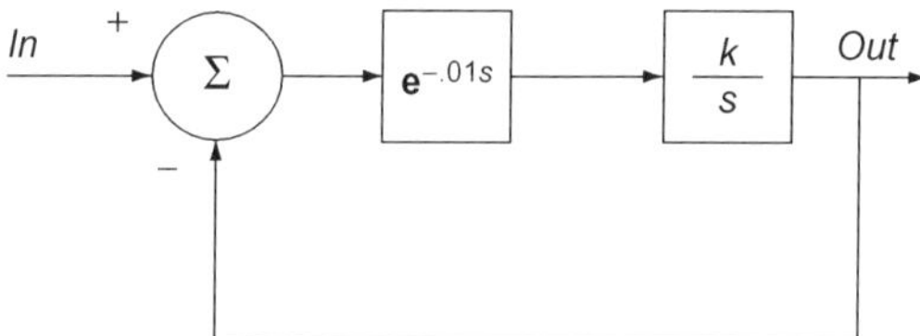

15. Demonstrate the effect of a 0.2-sec delay on the frequency characteristics of a second-order system. The system should have a ω_n of 10 rad/sec and a δ of 0.7. Plot the magnitude and phase with and without the delay. [*Hint*: Use MATLAB to plot the spectrum of the second-order system by substituting $j\omega$ for s. Then replot adding an $e^{-.2s} = e^{-j.2\omega}$ to the transfer function.] Plot for a frequency range of 1 to 100 rad/sec. Again the `unwrap` routine should be used as the phase plot will exceed −180 degrees.

16. The impulse response of a first-order system is:

$$V_{out}(s) = \left(\frac{1}{\tau}\right)\left(\frac{1}{s + (1/\tau)}\right)$$

Use the Initial Value Theorem to find the system's output at $t = 0$ (i.e., $v_{out}(t = 0)$) for the first-order system.

17. The transfer function of an electronic system has been determined as:

$$TF(s) = \frac{5s + 4}{s^2 + 5s + 20}$$

Use the Final Value Theorem to find the value of this system's output for $t \to \infty$ if the input is a step function that jumps from 0 to 5 at $t = 0$.

CHAPTER

7

Linear Systems Analysis in the Time Domain: Convolution and Simulation

7.1 LINEAR SYSTEMS

The systems model is a process-oriented representation that emphasizes the influences, or flow, of information between modules. A systems model describes how processes interact and what operations these processes perform, but it does not go into details as to how these processes are implemented. The basic element of the systems model in this text is a block enclosing a transfer function. Some very general systems-type models use only descriptive boxes as elements i.e., boxs without even transfer functions. The purpose of such models is to show the general relationship and flow of influence between processes; no effort is made to develop them for mathematical analysis.

A system's model elements are sometimes referred to as *black boxes* because they are defined only by the transfer function, which is not concerned with, and does not reveal, the inner working of the box. Both the greatest strength and the greatest weakness of systems model are found in their capacity to ignore the details of mechanism and emphasize the interactions between the elements.

One of the primary concerns of system analysis is to find the output of a system to any possible input. In the frequency domain, we use the Fourier transform to decompose a signal into sinusoids (Chapter 5) or the Laplace transform to decompose it into complex sinusoids (Chapter 6). We then apply the transfer function equation $Out = In\ TF$ (where the three terms are functions of either ω or s) to find the outputs to all of the individual input components. Inherent in this decompose–recompose approach is the principle of superposition, so these approaches only work on linear, time-invariant (LTI) systems. Between the two approaches (Fourier and Laplace), the output of any LTI system can be determined for any input.

There are also two time-domain approaches can be used to find the output of an LTI system to almost any input. Both also decompose the signal, but into slices of time instead of frequency. Slicing the input signal into small (or even infinitesimal) segments is reminiscent of the fundamental calculus trick of dividing a function into small segments, finding the function's behavior to each segment, then adding them up to get the output function.

The first approach is termed *convolution*, which applies this time-slice trick to the entire system in one operation. In convolution, these signal segments can be infinitesimal and integration can be used to sum up the contributions of each segment. However, if digital signals are involved then the minimum segment size is one sample, a time slice determined by the sample interval. So for digital signals, the time slices are usually small, but not infinitesimal.

The second time-domain approach is a computer-based method called *simulation*. It again divides the input into small but finite segments and finds the output response to each individual time slice one at a time. Beginning with the first time slice, the output of the input element(s) is found which then becomes the input to the next element(s) in the system. Responses are found for all elements in turn, working through from the first input element to the last output element. The process repeats for the second time slice, then the third, until the entire input signal has been processed. This approach is more computationally intensive but works for any input signal; and it can even be used on systems that contain nonlinearities. Convolution is described first, followed by simulation. The latter will rely on MATLAB's powerful simulation software, Simulink.

7.2 THE CONVOLUTION INTEGRAL

The output to any input in the frequency domain is found in Chapter 6 by multiplying the transfer function with the frequency domain input signal (i.e., the input spectrum). This requires transferring the signal into the frequency domain using the Fourier (or Laplace) transform and converting the solution back into the time domain with the appropriate inverse transform. If we want to stay in the time domain, we might expect the procedure to be a bit more complicated than multiplication, and so it is. Dividing the input signal into infinitesimal segments is appealing because we already know how to find a system's response to an infinitesimal signal segment. An infinitesimal signal segment is equivalent to an impulse function, so the response to an infinitesimal time slice is just the impulse response. The impulse representing the segment has the same shape as the unit impulse, but is scaled by the value of the signal at a given time segment. The contribution of such a scaled impulse response to the output is shifted to correspond to its time slot in the input signal (Figure 7.1). Since LTI systems are, by definition, time invariant, time shifting the signal does not alter the impulse response of the shifted signal. When superposition is applied, the output signal is the sum of all those contributing scaled and shifted impulse responses.

For a given time slot, the impulse response is scaled by the value of the input signal at that time slot, as shown in Figure 7.1. When implementing convolution, we reverse the input signal, because the way signals are graphed, the left-hand side of a time plot is actually the *low-time* side. It is the low-time side of the signal that enters the system first. So, from the point of view of the system, the left side of the input signal is encountered first

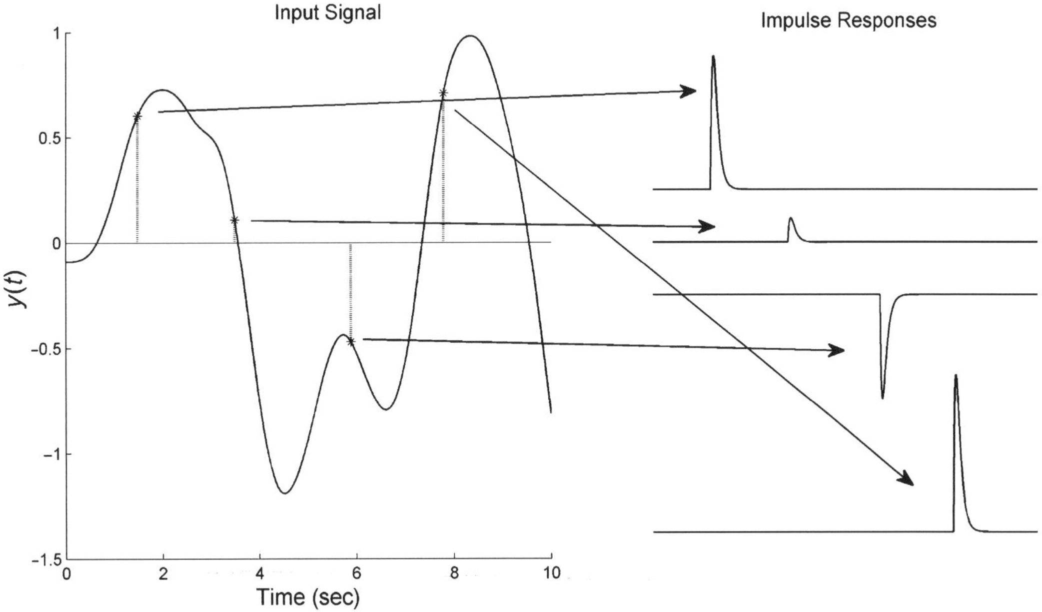

FIGURE 7.1 Each segment of a signal can be represented by an impulse and produces an impulse response in the system's output. The amplitude of the impulse response depends on the value of the signal segment that is represented, but its shape is always the same.

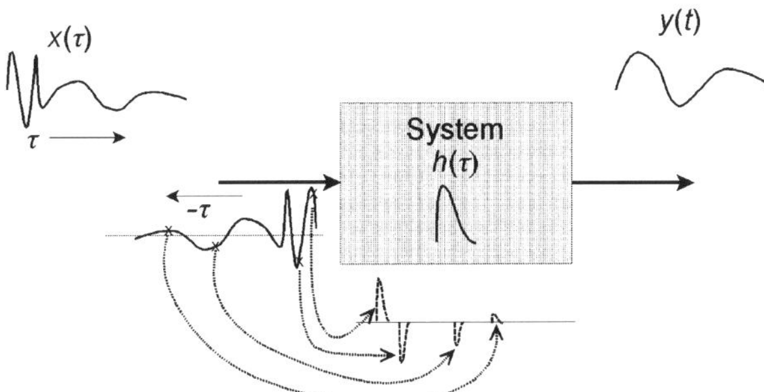

FIGURE 7.2 An input signal first enters its target system at its smallest (i.e., most negative) time value. So the first impulse generated by the input signal comes from the left side of the signal as it is normally plotted. As it proceeds in time through the system backwards, it generates a series of impulse responses that are scaled by the value of the input signal at any given time.

and the left-most segment produces the first impulse response (Figure 7.2). The input signal then proceeds through the system backward from the way it is normally plotted.

Since the segments are infinitesimal, there are an infinite number of impulse responses at various time positions, so the summation has to be done by integration. Standard calculus is well-equipped to describe this operation, although implementation is another matter. Scaling and summing the time-shifted impulse responses generated by a reversed input signal leads to the *convolution integral* equation.

$$y(t) = \int_{0}^{\infty} h(\tau)x(t-\tau)d\tau \tag{7.1}$$

where x is the input signal, h is the system's impulse response, and y is the output. It is the same as the crosscorrelation equation (Equation 2.24), except for two alterations. First, the role of t and τ are switched so that t is now the time-shift variable and τ is now the time variable for the input and impulse response functions. This modification is a matter of convention and is trivial. Second, there is a change in the sign between the time variables of x: $x(t+\tau)$ changes to $x(t-\tau)$. This is because the input signal, $x(\tau)$, must be reversed, as described above, and the $-\tau$ accounts for this reversal. A third difference not reflected in the equations is that the intent of convolution is very different from that of crosscorrelation. Crosscorrelation is used to compare two signals, while convolution is used to find the output of an LTI system to any input.

As with crosscorrelation, it is just as valid to shift the impulse response instead of the input signal. This leads to an equivalent representation of the convolution integral equation:

$$y(t) = \int_{0}^{\infty} x(\tau)h(t-\tau)d\tau \tag{7.2}$$

The solution of Equation 7.1 or 7.2 requires multiple integrations, one integration for every value of t. Since t is continuous, that works out to be an infinite number of integrations. Fortunately, except for one example and couple of easy problems, we only solve the convolution equation in the discrete domain. In the discrete domain, t takes on discrete values at sample intervals T_s and integration becomes summation, so solving the convolution equation is just a matter of performing a large number of summations. This would still be pretty tedious except that, as usual, MATLAB does all the work.

An intuitive feel for convolution is provided in Figures 7.3 through 7.5 as more and more segments are added to the process. Figure 7.3 shows the impulse response of an example system in the left plot and the input to that system in the right plot. Note the different time scales: the impulse response is much shorter than the input signal, as is generally the case.

In Figure 7.4A, the impulse responses to four signal segments at 2, 4, 6, and 8 sec are shown. Each segment produces an impulse response that is shifted to the same time slot as the impulse, and the impulse response is scaled by the amplitude of the input signal at that time. Note that the input signal has been reversed as explained above. Some responses are larger, one is scaled negatively, but they all have the basic shape as the impulse response shown in Figure 7.3A. A larger number (50) of these scaled impulse responses is shown in Figure 7.4B.

The 50 impulse responses in Figure 7.4 begin to suggest the shape of the output signal. A better picture is given in Figure 7.5, which shows 150 impulse responses, the summation of those responses (dashed line), and for comparison the actual output response (solid line) obtained by convolution. The summation of 150 impulse responses looks very similar to the actual output. (The actual output signal is scaled down to aid comparison.) As mentioned above, convolution of a continuous signal requires an infinite number of segments, but the digital version is limited to one for each data sample, as discussed below. (The input signal used in the figures above has 1000 samples.)

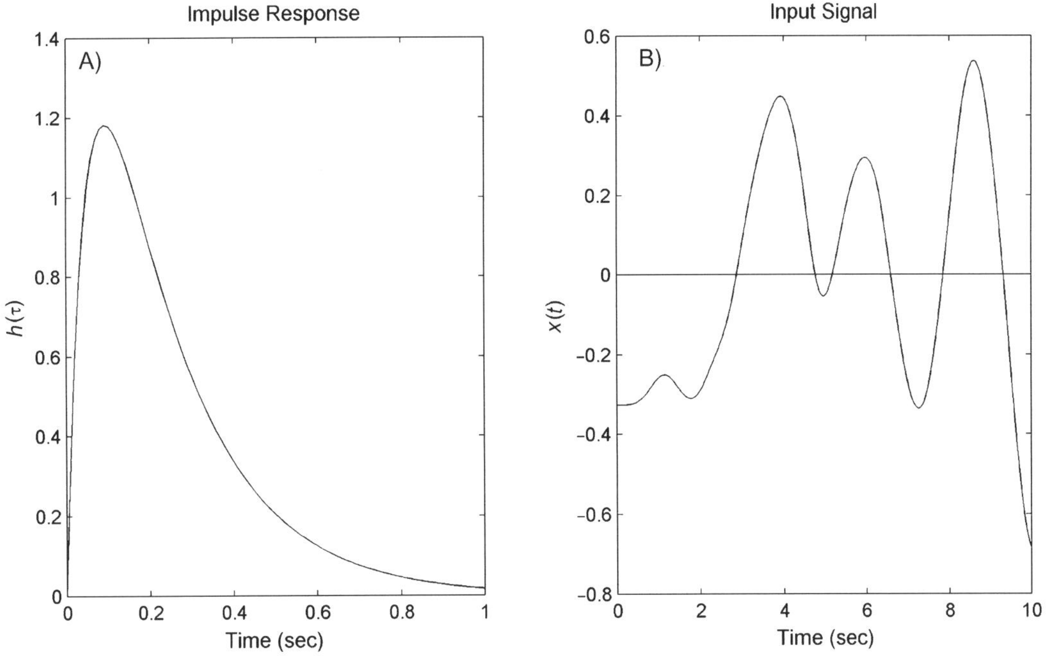

FIGURE 7.3 A) The impulse response of a hypothetical system used to illustrate convolution. B) The input signal to a system having the impulse response shown in A. The impulse response has a much shorter duration than the input signal: 1 sec versus 10 sec. The impulse response is usually much shorter than the input signal.

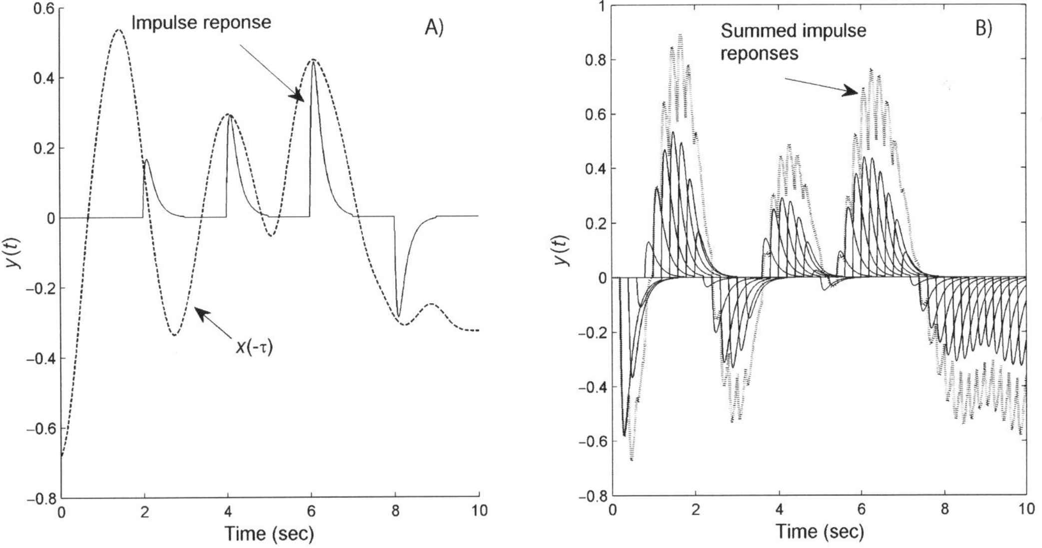

FIGURE 7.4 A) The output response produced by four time slices of the input signal shown in Figure 7.3B. Impulse responses from input segments at 2, 4, 6, and 8 sec are shown (solid curves) along with the reversed input signal $x(-\tau)$. B) 50 impulse responses for the same signal (solid curve). The sum of these signals is also shown (dotted curve).

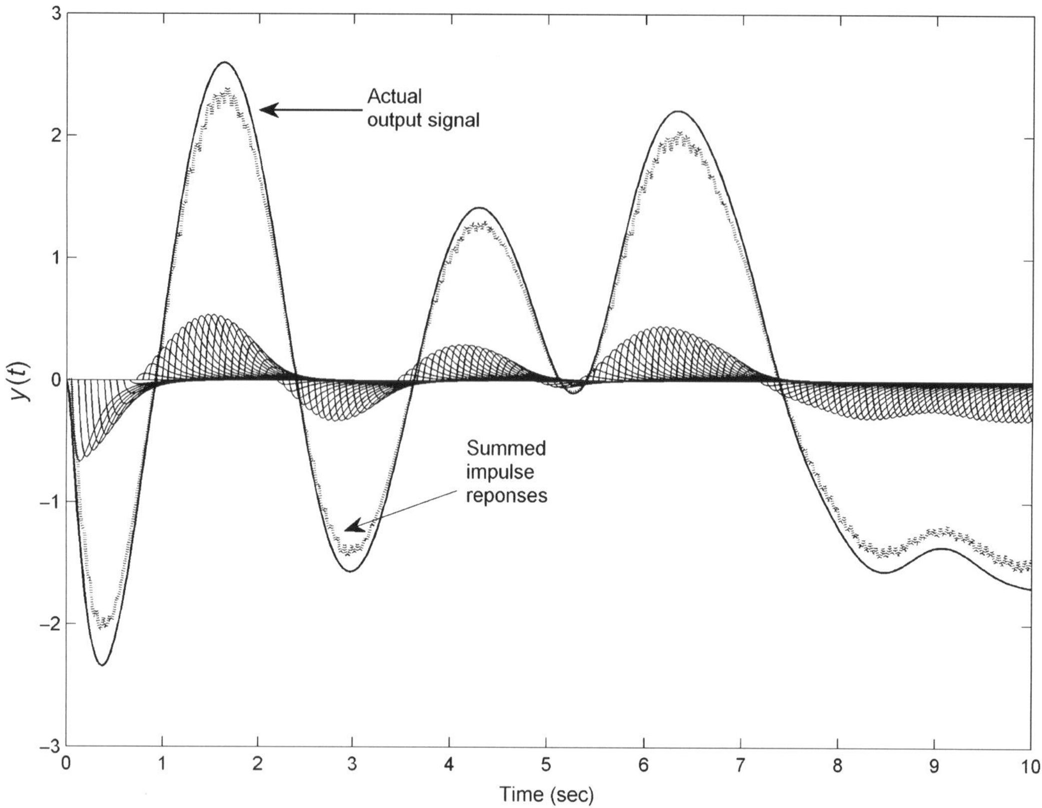

FIGURE 7.5 The summation of 150 impulse responses (dashed curve) from input segments spaced evenly over the input signal now closely resembles the actual output curve (solid line). The actual output is obtained from implementation of the digital convolution equation (Equation 7.3).

If the impulse extends over a long period of time at high values, then a large contribution comes from former portions of the input signal. Such a system is said to have more memory. If the impulse response is short, then very little comes from past inputs and more of the output is shaped by the instantaneous input at a given time. In the limit, if the impulse response is itself an impulse, then nothing comes from past inputs—all of the output comes from the current instantaneous input, and the output looks just like the input. The only difference might be the scale of the output, which would be altered by the height of the impulse. For example, if the impulse response of a system is an impulse with an area of −1.0 (i.e., a negative impulse function), then the output is the same as the input inverted. Such a system is memoryless.

For discrete signals, the integration becomes a summation and the discrete version of the convolution equation, sometimes called the convolution *sum*, becomes:

$$y[n] = \sum_{k=0}^{K-1} x[k]h[n-k] = \sum_{k=0}^{K-1} h[k]x[n-k] \qquad (7.3)$$

where K is the length of the shorter function, usually $h[k]$, and n represents the time-shift variable. Again, it does not matter whether $h[k]$ or $x[k]$ is shifted; the net effect is the same. Since both $h[n]$ and $x[n]$ must be finite (they are stored in finite memory), the summation is also finite. The convolution operation may also be abbreviated by using an "*" between the two signals:

$$y(t) = \int_0^{\infty} h(\tau)x(t-\tau)d\tau \equiv h(t)*x(t) \tag{7.4}$$

Or:

$$y[n] = \sum_{n=0}^{N-1} x[k]h[n-k] \equiv x[n]*h[n] \tag{7.5}$$

Unfortunately the * symbol is broadly used to represent multiplication in many computer languages, including MATLAB, so its use to represent the convolution operation can be a source of confusion.

Superposition is a fundamental assumption of convolution. The impulse response describes how a system responds to a small (or infinitely small) signal segment. Each of these small (or infinitesimal) segments of an input signal generates its own little impulse response scaled by the amplitude of the segment and shifted in time to correspond with the segment's time slot. The output is obtained by summing (or integrating) the multitude of impulse responses, but this is valid only if superposition holds. Since convolution invokes superposition, it can only be applied to LTI systems where superposition is valid.

Convolution is frequently used to implement linear systems. Given the impulse response of the linear system and some input signal, convolution is used to determine an output signal. We see numerous examples of this in the next chapter, where convolution is used to implement digital filters. In the real world, convolution is always done on a computer; digital signals are involved, and Equation 7.3 or 7.5 is applied. In the textbook world, convolution may be done manually in both the continuous and digital domains. Sometimes these exercises provide insight and, in that hope, the following two examples of manual convolution are presented along with a few problems.

EXAMPLE 7.1

Find the result of convolving the two continuous signals shown in Figure 7.6A and B. Use direct application of Equation 7.2.

Solution: First, one of the signals has to be flipped, then a sliding integration can be performed. In this example, we flip the impulse response as shown in Figure 7.6C. So the basic signals in Equation 7.2 are:

$$x(\tau) = \begin{cases} 1 & 0 < \tau < 2 \\ 0 & \text{otherwise} \end{cases}; \quad h(\tau) = e^{-\tau} \quad t > 0; \quad h(t-\tau) = \begin{cases} e^{-(t-\tau)} & 0 < \tau < 1.5 \\ 0 & \text{otherwise} \end{cases}$$

To solve the integral equation, we use the classic calculus trick of dividing the problem into segments where the relationship between the two signals is consistent. For these two signals,

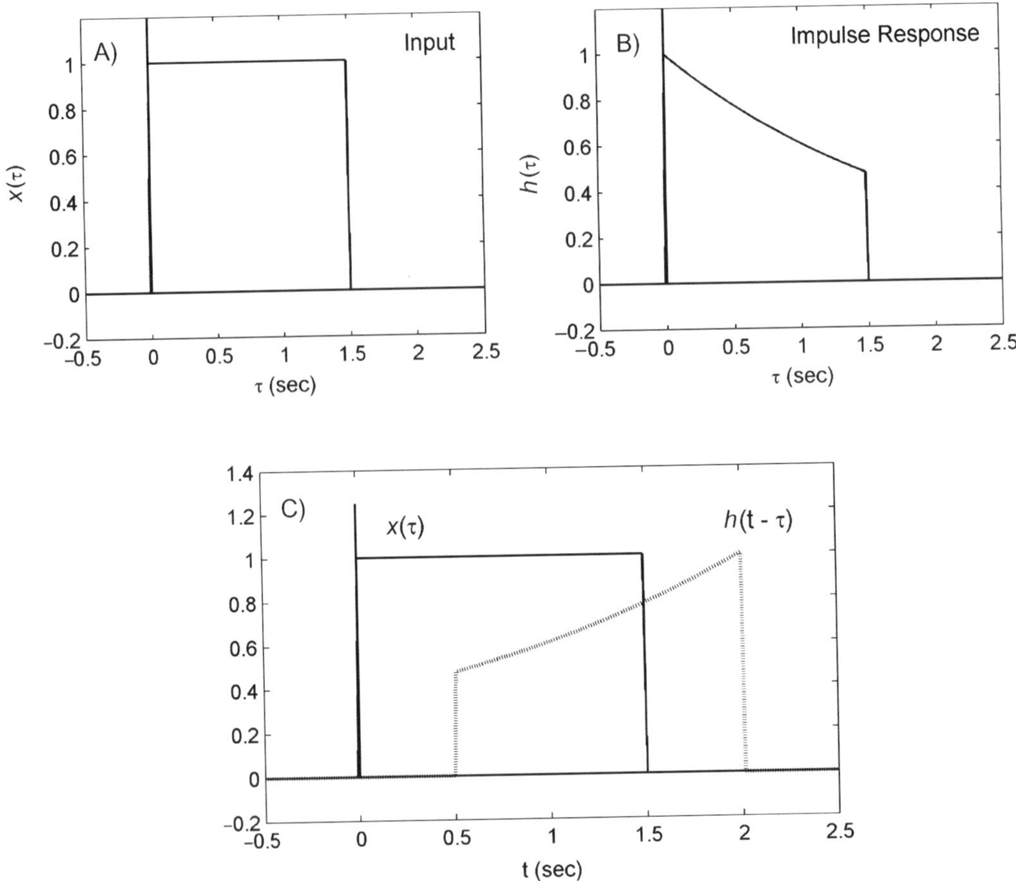

FIGURE 7.6 A) Input signal and B) impulse response used in Example 7.1. C) The impulse response is flipped before it is multiplied by the input and integrated.

there are three time periods during which the mathematical statement for $x(t)h(t-\tau)$ remains consistent (see Figure 7.7).

In the first region, for shifts where $t<0$, there is no overlap between the two signals (Figure 7.7A). So their product is zero and the integral is also 0:

$$y(t)=0 \text{ for } t<0$$

When $t>0$ sec, the functions overlap to an extent determined by the impulse response (Figure 7.7B).

$$y(t)=\int_0^{\infty} x(\tau)h(t-\tau)d\tau=\int_0^{t} 1e^{-(t-\tau)}d\tau=e^{-t}(e^{\tau}|_0^t)=1-e^{-t} \qquad 0\le t<2$$

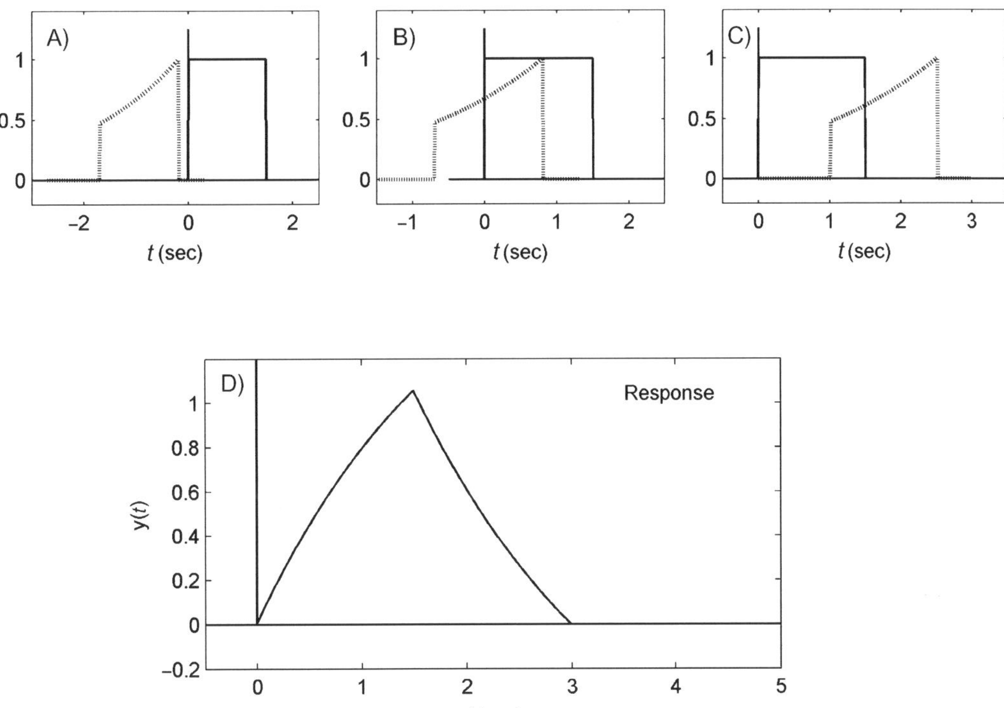

FIGURE 7.7 A) One of the three regions where the integral convolution of the two signals in Figure 7.6 holds consistent. In this case $t < 0$ and there is no overlap of the function. B) In the region where $0 \leq t < 2$, the convolution integral is consistent and the upper limit is set by t, which is at the leading edge of the flipped impulse response. C) In this region, $t > 2$, the representation of the functions is the same as for B, but the upper limit on the integral is set by the input signal at 2 sec. D) The output signal found by convolution.

This relationship continues until $t > 2$ sec when the overlap between the two signals is determined by the pulse function. For this region the integral is the same, except that the upper limit of integration is 2 sec:

$$y(t) = \int_0^2 1e^{-(t-\tau)}d\tau = e^{-t}(e^{\tau}|_0^2) = (e^2 - 1)e^{-t} \qquad t < 2$$

Combining these three solutions gives:

$$y(t) = \begin{cases} 0 & t < 0 \\ 1 - e^{-t} & 0 \leq t < 2 \\ (e^2 - 1)e^{-t} & t \geq 2 \end{cases}$$

A time plot of the signal resulting from this convolution is shown in Figure 7.7D. An example of manual convolution with a step function that is 0 for $t < 0$ and a constant for all $t > 0$ is given in Problem 1 at the end of this chapter. The next example applies convolution to a digital signal.

EXAMPLE 7.2

Find the result of convolving the two discrete signals shown in Figure 7.8. Use the direct application of the convolution sum in Equation 7.3.

Solution: This example is similar to the digital crosscorrelation shown in Example 2.14 except that one of the signals is flipped. The signal values for $x[k]$ and h[k] are given in Figure 7.8 as: $x[0]=0.2$; $x[1]=0.4$; $x[2]=0.6$; $x[3]=0.2$ and $h[0]=0.2$; $h[1]=0.5$; $h[2]=0.75$. After reversing $h[k]$, we substitute these values into Equation 7.3 for various values of n as was done in Example 2.14. The result is illustrated in Figure 7.9 as the flipped impulse response is shifted across the signal $x[k]$:

$$y[n]=\sum_{k=0}^{K-1}x[k]h[n-k]$$
$$y[1]=0.2(0.2)=0.04$$
$$y[2]=0.5(0.2)+0.2(0.4)=0.18$$
$$y[3]=0.75(0.2)+0.5(0.4)+0.2(0.6)=0.47$$
$$y[4]=0.75(0.4)+0.5(0.6)+0.2(0.2)=0.64$$
$$y[5]=0.75(0.6)+0.5(0.2)=0.55$$
$$y[6]=0.75(0.2)=0.15$$

The output signal $y[n]$ is shown in Figure 7.10

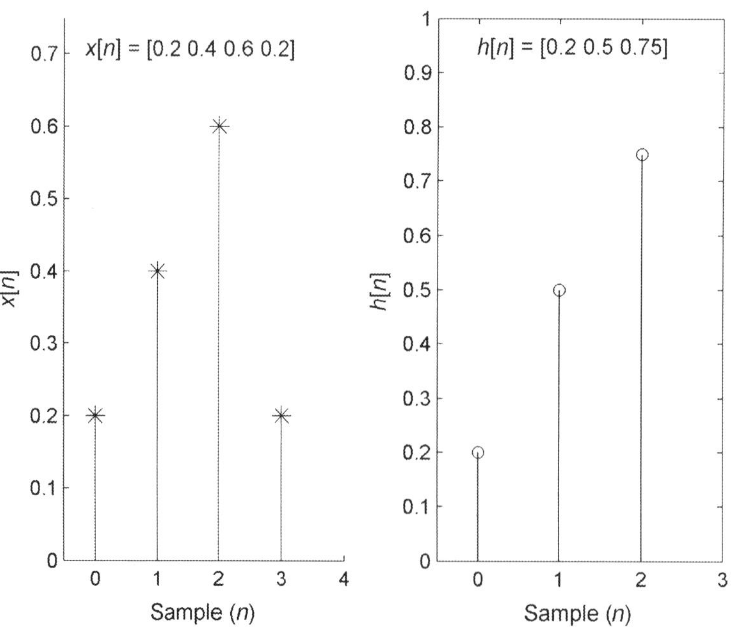

FIGURE 7.8 The discrete input and impulse response signals used in Example 7.2.

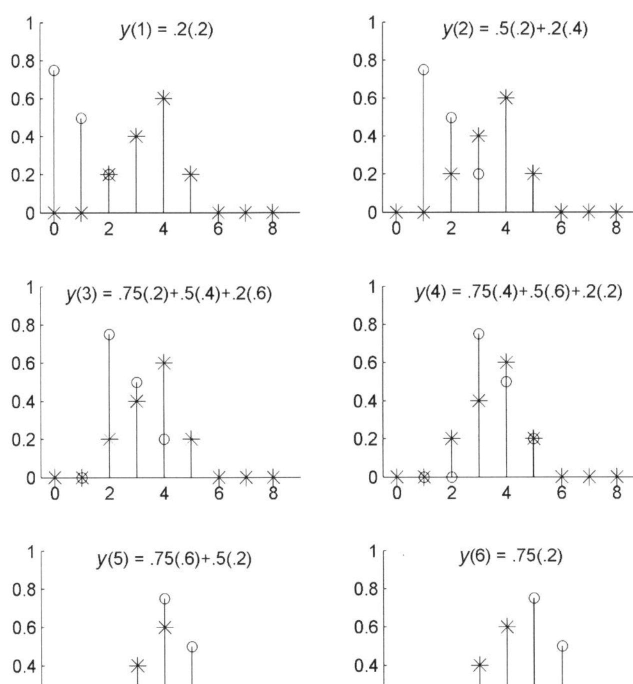

FIGURE 7.9 Progression of steps, left to right, top to bottom, in evaluating the convolution sum in Example 7.2.

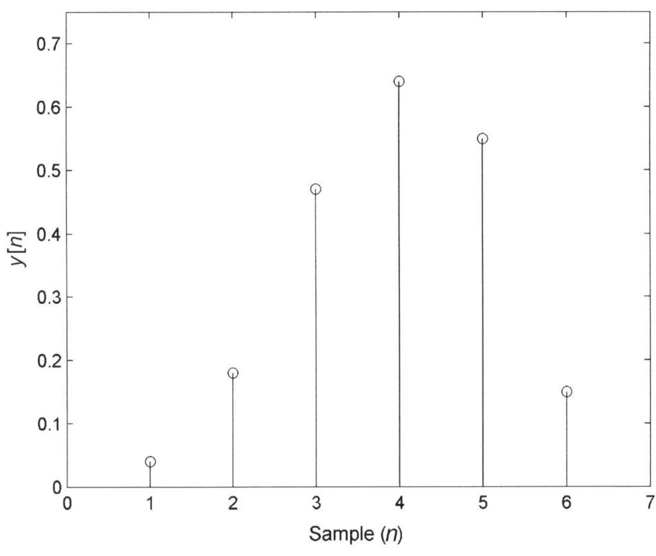

FIGURE 7.10 The response signal produced by discrete convolution of the signal and impulse response shown in Figure 7.8.

7.2.1 Determining the Impulse Response

A quantitative description of the impulse response of a system is necessary in order to use convolution to find the system's output to any input. There are several approaches to determining the impulse response. One route is through the transfer function. As shown in Chapter 6, the Laplace transform of an impulse function is 1.0 (Equation 6.13: $\mathscr{L}\delta(t) = \Delta(s) = 1$). Since the Laplace transfer function is $TF(s) = Out(s)/In(s)$, then if the input, $In(s)$, is an impulse:

$$TF(s) = \frac{Out(s)}{\Delta(s)} = Out(s) \text{ when } In(s) = \Delta(s) \tag{7.6}$$

But the output of a system to an impulse input is, by definition, the impulse response, $h(t)$, and converting to the Laplace domain:

$$Out(s)_{\text{Impulse}} = TF(s) = \mathscr{L}\, h(t) \tag{7.7}$$

Taking the inverse Laplace transform of Equation 7.7 gives:

$$h(t) = \mathscr{L}^{-1}\, TF(s); \tag{7.8}$$

So the impulse response, $h(t)$, is the inverse Laplace transform of the transfer function.

If we assume sinusoidal steady-state conditions, then Equations 7.7 and 7.8 can be transposed into the frequency domain by letting $s = j\omega$. In the frequency domain, the Laplace transform is replaced by the Fourier transform. So the transfer function of an impulse in the frequency domain is the Fourier transform of the impulse function:

$$TF(\omega) = \text{FT}\,[h(t)] \tag{7.9}$$

Again taking the inverse Fourier transform of both sides of Equation 7.9, the impulse response in the time domain is given as:

$$h(t) = \text{FT}^{-1}[TF(\omega)] \tag{7.10}$$

So the impulse response can also be found directly from the frequency domain transfer function by taking the inverse Fourier transform of the transfer function.

Alternatively, if the system is physically available, as is often the case in the real world, the impulse response can be determined empirically by monitoring the response of the system to an impulse. Of course, a mathematically correct impulse function which is infinitely short but still has an area of 1.0 is impossible to produce in the real world, but an appropriately short pulse will serve just as well. Just what constitutes an appropriately short pulse depends on the dynamics of the system, and an example of how to determine if a pulse is a suitable surrogate for an impulse function is given in Example 7.4.

7.2.2 MATLAB Implementation

The convolution integral can, in principle, be evaluated analytically using standard calculus operations, but it quickly becomes tedious for complicated inputs or impulse response functions. However, the discrete form of the convolution integral (Equation 7.3) is easy to implement on a computer. In fact, the major application of convolution is in digital signal processing, where it is frequently used to apply filtering to signals. Example 7.3 explores just such an application.

It is not difficult to write a program to implement either version of Equation 7.3, but, of course, MATLAB has a routine to compute convolution, the `conv` routine:

```
y = conv(x,h,'option')
```

where `x` and `h` are vectors containing the waveforms to be convolved, and `y` is the output signal. If the option is `full`, or unspecified, all possible products of `h` and `x` are determined and zero padding is used as necessary. The length of the output waveform is equal to the length of `x` plus the length of `h` minus 1. In other words, the output, `y`, is longer than the input, `x`. Under the option `same` all possible combinations are still evaluated, but only the central portion of the result that is the same size as `x` is returned. If the option `valid` is specified, no padding is used, so the computation is performed only for time shifts where `x` and `h` completely overlap. Thus the number of points returned is the length of `x` minus the length of `h` plus 1 (assuming that `h` is the shorter vector). The use of this routine is shown in the next two examples.

An alternative routine, `filter`, can be used to implement convolution, and this routine always produces the same number of output points as in `x`. When used for convolution, the calling structure is:

```
y = filter(h,1,x);
```

where the variables are the same as is described above and the second variable is always 1.0. In the examples below, the `conv` route is used with the `same` option, but the reader is encouraged to experiment using `filter` where appropriate in the problems.

EXAMPLE 7.3

Find the output of the EEG signal shown in the upper trace of Figure 7.11 after you process it with a system having the first-order transfer function given in Equation 6.25 of Chapter 6 and repeated here:

$$TF(s) = \frac{1}{s + \omega_1} = \frac{(1/\tau)}{s + (1/\tau)} \tag{7.11}$$

where $\tau = 1/\omega_1 = 1$ sec. Plot the spectrum of this system first, determine the impulse function, then use convolution to find the output. Finally plot the spectrum of the output signal. Also plot the input (i.e., EEG signal) and output signals in the time domain.

Solution: We can do the entire problem in MATLAB, but in this example we determine the system's spectrum and impulse response analytically, then program the impulse function in MATLAB and use `conv` to determine the output. Using the analytical method first, we get the frequency domain transfer function by substituting $s = j\omega$. This leads to the transfer function:

$$TF(\omega) = \frac{1}{j\omega + \omega_1} = \frac{1}{1 + j\omega}$$

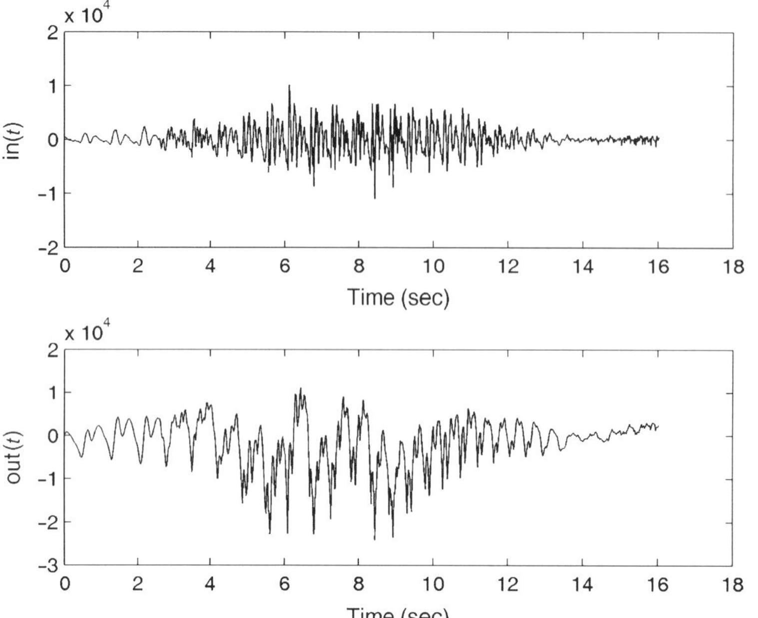

FIGURE 7.11 An EEG signal before (upper) and after (lower) filtering by an first-order process. Convolution is used to determine the output in the time domain. The process reduces the high-frequency components in the EEG signal.

This is the equation of a first-order system (Equation 5.34). Application of the Bode plot rules for plotting the magnitude transfer function leads to the plot shown in Figure 5.16 and repeated in Figure 7.12. The phase plot of this transfer function is shown in Figure 5.17.

The impulse response can be obtained by taking the inverse Laplace transform of the transfer function (Equation 7.10). From the Laplace transform table in Appendix B, entry 3 matches the transfer function of Equation 7.11.

$$\frac{1}{s+\alpha} \Rightarrow e^{-\alpha t}$$

Since $\alpha = 1.0$, the impulse response is:

$$h(t) = e^{-t}$$

This impulse response function is implemented in the code below, and the output EEG signal is determined using convolution; both input and output signals and their respective spectra are plotted.

```
% Example 7.3 Find the output of the EEG signal after filtering by a
% first-order system. Use convolution in conjunction with the impulse response
%
fs = 50;                  % Sample frequency (from EEG signal)
t1 = 0:1/fs:5;            % Time vector for impulse response
h = exp(-t1);             % Define h(t)
load eeg_data;            % Load EEG data
%
```

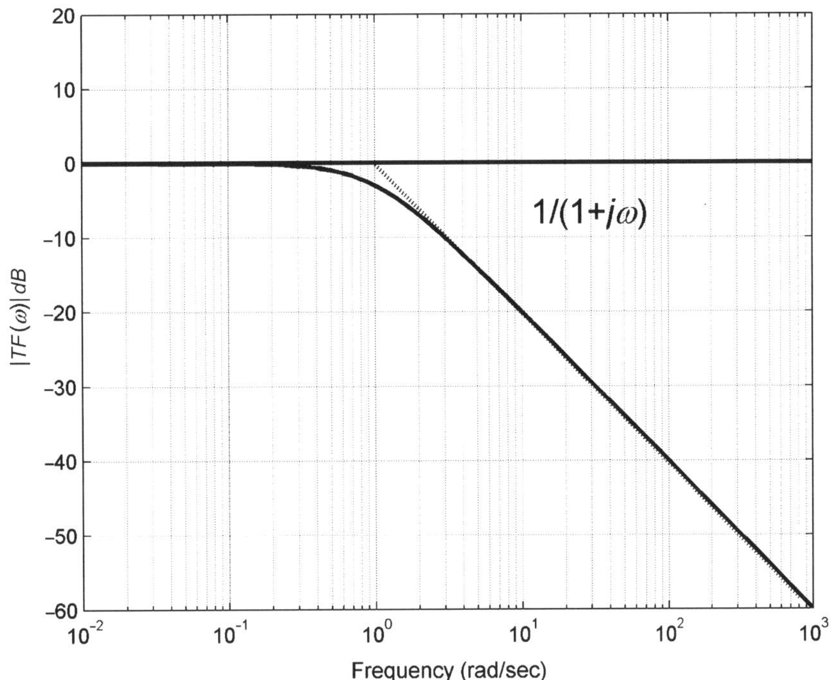

FIGURE 7.12 Magnitude spectrum of the first-order system having the transfer function given in Equation 7.11.

```
out = conv(eeg,h,'same');  % Perform convolution, use same length option.
t = (1:length(eeg))/fs;    % Construct time vector for plotting
  ......plot input and output signals, label and title......
%
Vin = abs(fft(eeg));       % Determine Fourier transform of EEG
Vout = abs(fft(out));      % and Vout
  ......plot frequency spectra, non-redundant points, label and title......
```

The output time response is shown in the lower trace of Figure 7.11, and the spectra of the input and output signals are shown in Figure 7.13. The reduction in frequency components above approximately 5 Hz is evident.

As mentioned in Chapter 6, the ideal impulse function is a pulse that is infinitely tall and infinitely narrow. Of course, such an idealization is not compatible with the real world, and what is actually used is a pulse that is narrow with respect to the response time of the system being "impulsed." A good rule of thumb is that the impulse function should be at least 10 times shorter than the fastest time constant in the system. Unfortunately, you rarely know the system's time constants, but in many cases an empirical approach can be used. If the pulse being used is more or less an impulse, decreasing its width should not change the output dynamics. In other words, if a pulse is already so

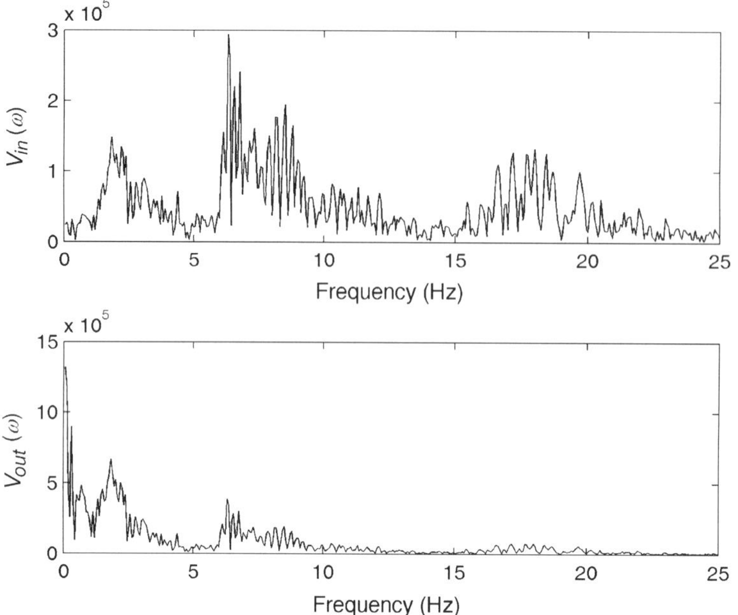

FIGURE 7.13 Frequency characteristics of the original EEG signal (upper plot) before and after processing by a first-order process. The reduction of the higher frequencies by this process is evident when comparing the two curves.

short that it behaves like an impulse, decreasing its width further will not make it any more of an impulse. If the system is physically available along with a pulse generator, finding the right pulse width is just a matter of decreasing the pulse width until further decreases no longer change the general shape of the response. The amplitude decreases as pulse width decreases, since there is less energy in the pulse, but the general shape of the response stays the same. If the system is not available, but you have the transfer function, you can use convolution to simulate the response of the system to stimuli having a range of pulse widths. This approach is explored in Example 7.4.

EXAMPLE 7.4

Find the maximum pulse width that can still be considered an impulse of a particular system. In this example, convolution is used to find the response of a system to pulse inputs having a range of widths. The system under exploration has a transfer function of:

$$TF(s) = \frac{1}{s^2 + 25s + 70} \tag{7.12}$$

Solution: Find the system's impulse response by applying the inverse Laplace transform to the transfer function in Equation 7.12. We could then use convolution to find the output of the system to any input, but in this case we are only interested in pulse inputs. When a longer and a shorter pulse produce essentially the same response, either pulse is effectively an impulse with respect to this system. To facilitate identification of pulses that produce essentially the same output, plot pulse responses in pairs.

To find the inverse Laplace transform of the transfer function of Equation 7.12, we first need to evaluate the damping coefficient to determine the proper Laplace transform equation:

$$2\delta\omega_n = 25; \quad \delta = \frac{25}{2\omega_n} = \frac{25}{2\sqrt{70}} = 1.49$$

Thus the system is overdamped and can be factored, or Laplace transform equation 11 (from Appendix B) can be used directly.

$$\left(\frac{b\beta - b\alpha + c}{2\beta}\right)e^{-(\alpha-\beta)t} + \left(\frac{b\beta + b\alpha - c}{2\beta}\right)e^{-(\alpha+\beta)t} \Leftrightarrow \frac{bs + c}{s^2 + 2\alpha s + \alpha^2 - \beta^2}$$

where: $b = 0$; $c = 1$; $2\alpha = 25$; $\alpha = 12.5$; and $\alpha^2 - \beta^2 = 70$,

$$\beta = \sqrt{\alpha^2 - 70} = \sqrt{156.25 - 70} = 9.29$$

Hence, the impulse response becomes:

$$h(t) = \left(\frac{1}{2(9.29)}\right)e^{-(12.5-9.29)t} + \left(\frac{-1}{2(9.29)}\right)e^{-(12.5+9.29)t} = 0.054(e^{-3.21t} - e^{-21.79t}) \tag{7.13}$$

This impulse response is explored with a range of pulse widths, and the shape of the responses is plotted. To aid shape comparison, the responses are plotted normalized to the same maximum amplitude. Also some care has to be taken to insure that the impulse response, $h(t)$, is appropriately represented in MATLAB, that is, the sampling time is adequate and the time vector used to generate $h(t)$ is long enough to produce the entire response. This can be done by trial and error, but note that the slowest time constant is in the first exponential: $\tau_1 = 1/3.21 = 0.31$ sec. Based on this, 1.5 sec (i.e., approximately 5 time constants) is adequate to represent the impulse response. Since the fastest time constant is in the second exponential, $\tau_2 = 1/21.79 = 0.046$ sec, we experiment with pulses around 0.01 and shorter. Given these numbers, a sampling frequency of 2 kHz will work, although a higher sampling frequency can be used if there is any doubt.

```
% Example 7.4
% Example to simulate a second-order transfer function
%
PW = [50 25 25 10 10 5 5 2.5]*2;             % Desired pulse widths in msec
fs = 2000;                                   % Sample frequency
t = 0:1/fs:1.5;                              % Generate a time vector from 0 to 1.5 sec.
h = .054*(exp(-3.21*t) - exp(-21.79*t));     % Construct the impulse response
  ......plot and labels; new figure......
for k = 1:4                                  % Construct four plots
  subplot(2,2,k); hold on;                   % Plot responses two by two
  for k1 = 1:2
    pw_index = 2*(k-1) + k1;                     % Get pulse index
    pulse = ones(1,PW(pw_index));            % Set pulse width from pulse width vector
    y = conv(h,pulse);                           % Construct impulse response
    y = y/max(y);                                % Normalize peak to 1.0
  ......plot, reduce horizontal axis to 0.2 sec, label......
  end
end
```

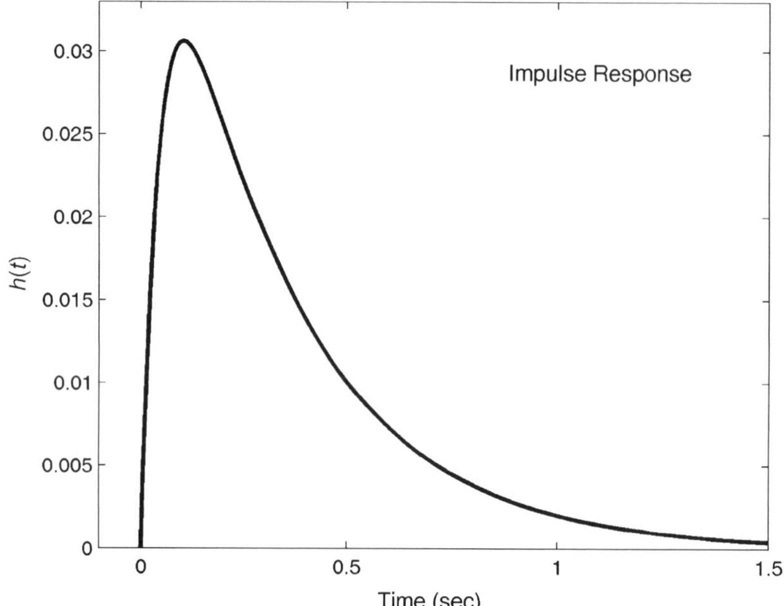

FIGURE 7.14 Impulse response defined in Equation 7.13 and used in Example 7.4. The 1.5-sec time vector used to generate this signal appears to be long enough to include most of the response.

Analysis: After specifying the sampling frequency, the desired pulse widths in msec are placed in a vector as they will be plotted: two at a time. These pulse widths are multiplied by two to give pulse width in samples ($f_s = 2$ kHz).

A time vector is then generated from 0 and 1.5 sec in increments of the sample interval. The impulse response h is constructed using this time vector based on Equation 7.13. The impulse response is plotted in Figure 7.14 (above) and the 1.5-sec data length appears to be adequate to represent most of the response.

A double loop is used to plot the 8 responses as pairs in 4 graphs. In each iteration, the pulse input is generated as an array of 1.0s having a length as specified by the PW vector. The simulated output is generated by convolving this pulse with impulse response, and the paired outputs are plotted. Only the first 0.2 sec of the response is plotted to emphasize the initial dynamics where the greatest differences will occur.

Results: The results show that as pulses reduce in width, the differences in response dynamics decrease. There is very little difference between the responses generated by the 5- and 2.5-msec pulses, so that a 5-msec pulse (or less) is probably close enough to a true impulse for this system. This example is a realistic simulation of the kind of experiment one might do to determine the impulse response empirically, provided the physical system and pulse generator are available. This also assumes the system can be stimulated with pulse waveforms and the response can be monitored, which is not always the case with biological systems.

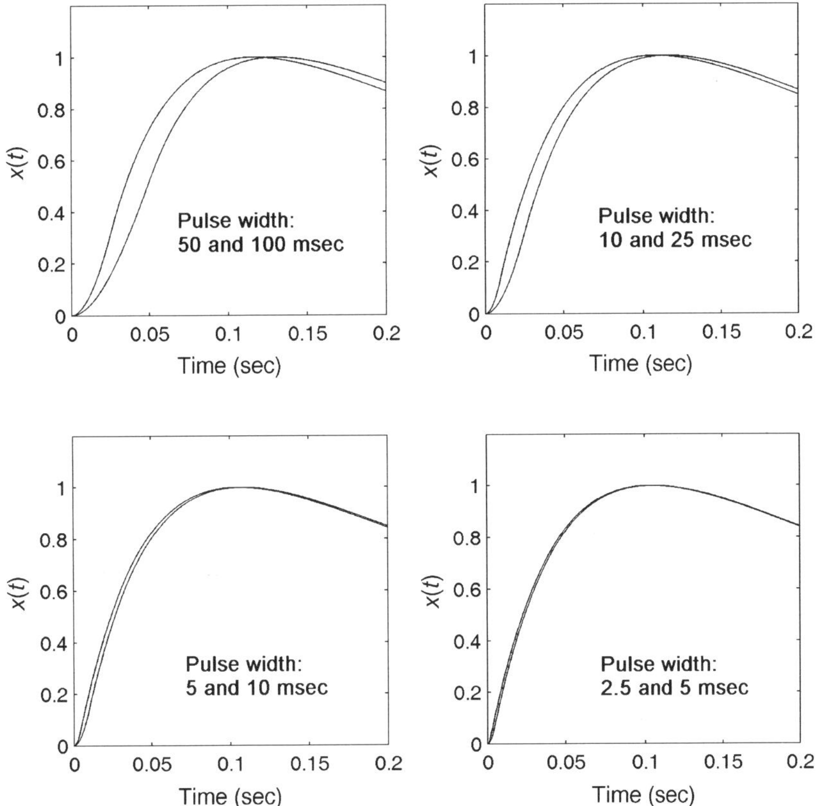

FIGURE 7.15 Pulse responses of the second-order system defined by the impulse response in Equation 7.13. Responses to two different pulse widths are plotted in each graph for comparison. The pulse width values are shown.

7.3 THE RELATIONSHIP BETWEEN CONVOLUTION AND FREQUENCY DOMAIN ANALYSIS

We now have several different ways of representing a system and for finding the output of a LTI system to any input:

1) **The Laplace transform approach.** Convert the signal, $x(t)$, to the Laplace domain, (i.e., a complex sinusoidal series) using the Laplace transform tables (see Appendix B), multiply by the Laplace transfer function, $TF(s)$, then convert the result back to the time domain using the Laplace transform tables in the reverse direction. This approach is used in Example 6.6.
2) **The frequency domain approach.** A special case of the Laplace domain approach for periodic signals: convert the signal, $x(t)$, to the frequency domain (i.e., a sinusoidal series) using the Fourier transform, multiply by the transfer function, $TF(f)$, then take

the inverse Fourier transform to get the output time response. This approach is used in Example 5.13.

3) **The convolution approach.** Convolve the input signal, $x(t)$, with the impulse response, $h(t)$, and get the time-domain output directly. This approach is used in Examples 7.1 through 7.4.

These three approaches are related. In Chapter 6 we learned that the Laplace domain and frequency domain approaches are really the same when we are dealing with signals that have a Fourier transform: We let $s = j\omega$ and the Laplace transfer function becomes the frequency transfer function. This is the easiest approach if the input signal has a Fourier transform, since the system response can be determined using MATLAB. If the signal does not have a Fourier transform, for example the step function, then you must use either Laplace transforms or convolution, at least for now. In the next section we explore the simulation approach, which works for any signal and any system, even many nonlinear systems.

The relationship between the frequency domain approach and convolution is given by Equations 7.9 and 7.10, restated here in words: The transfer function is the Fourier transform of the impulse response (Equation 7.9); and vice versa: the impulse response is the inverse Fourier transform of the transfer function (Equation 7.10). This allows us to flip easily between approaches 1) and 3) above. For example, if we had the impulse response of a system and wanted to find the output to a given signal, but for some reason we did not want to use convolution, we would use the transfer function method:

1) Convert the impulse response to a transfer function by taking its Fourier transform:

$$TF(\omega) = \mathrm{FT}[h(t)]$$

2) Convert the signal to the frequency domain using the Fourier transform:

$$In(\omega) = \mathrm{FT}[x(t)];$$

3) Multiply the two together: $Out(\omega) = TF(\omega) \cdot In(\omega)$
4) Take the inverse Fourier transform to get the time domain output: $y(t) = \mathrm{FT}^{-1}\,[Out(\omega)]$

In the next example, we try out this strategy and compare it to straightforward convolution using MATLAB. [*Top secret*: The four steps above may seem like a convoluted way to do convolution, but it is actually the way MATLAB does it in the `conv` routine. That is because the FFT algorithm and its inverse are so fast that it is faster to do convolution in the frequency domain and go back and forth than to execute all the time-domain multiplications and additions called for by Equation 7.3.] The reverse strategy is also possible: Convert the transfer function to the impulse response using the inverse Fourier transform, then apply convolution.

EXAMPLE 7.5

Given the system impulse response below, find the output of the system to a periodic sawtooth wave having a frequency of 400 Hz (Figure 7.16A). Use both convolution and frequency

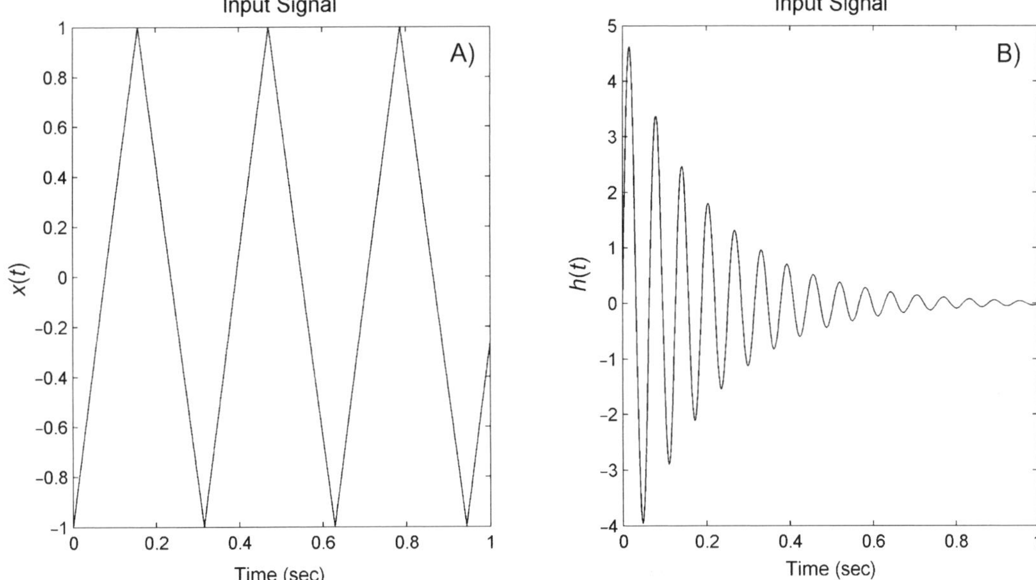

FIGURE 7.16 A) Sawtooth wave used as the input signal in Example 7.3. B) Impulse response of the system used in Example 7.3.

domain approaches. Determine and plot on a dB-semilog scale the magnitude transfer function of the system. Since the sawtooth has a frequency of 400 Hz, we use a sampling interval of 4 kHz.

$$h(t) = 5.0e^{-5t}\sin(100t)$$

Solution: First construct the impulse response function given above using a time vector of appropriate length (Figure 7.16B). Finding the response using convolution requires only the application of MATLAB's `conv`. To find the output using the frequency domain approach, convert the impulse response to the transfer function using the Fourier transform and the signal to the frequency domain also using the Fourier transform. Then multiply the two frequency domain functions together and take the inverse Fourier transform to get the time domain output signal.

```
% Example 7.5 System output found two different ways.
%
fs = 4000;                          % Sample frequency
T = 1;                              % Time in sec;
t = (0:1/fs:T);                     % Time vector
N = length(t);                      % Determine length of time vector
f = (1:N)*fs/(2*N);                 % Frequency vector
x = sawtooth(20*t,.5);              % Construct input signal
h = 5* exp(-5*t).*sin(100*t);       % Construct the impulse response


......Plot and label input signal......
% Calculate output using convolution.
```

```
y = conv(x,h);                              % Calculate output using convolution
figure;
subplot(2,1,1)
plot(t,y(1:N),'k');                         % Plot output from convolution
  ......label and title......
% Now calculate in the frequency domain.
TF = fft(h);                                % Get TF from impulse response
                                            % (Step 1)
X = fft(x);                                 % Take Fourier transform of input
                                            % (Step 2)
Y = X.*TF;                                  % Take product in the frequency
                                            % domain (Step 3)
y1 = ifft(Y);                               % Take the inverse FT to get y(t)
                                            % (Step 4)
subplot(2,1,2)
plot(t,y1,':k','LineWidth',2);              % Plot output from frequency
                                            % domain approach

  ......labels and title......
figure;
N_2 = round(N/2);                           % Plot transfer function,
                                            % only meaningful points
semilogx(f(1:N_2),20*log(abs(TF(1:N_2))));  % Plot in dB versus log frequency
  ......labels......
```

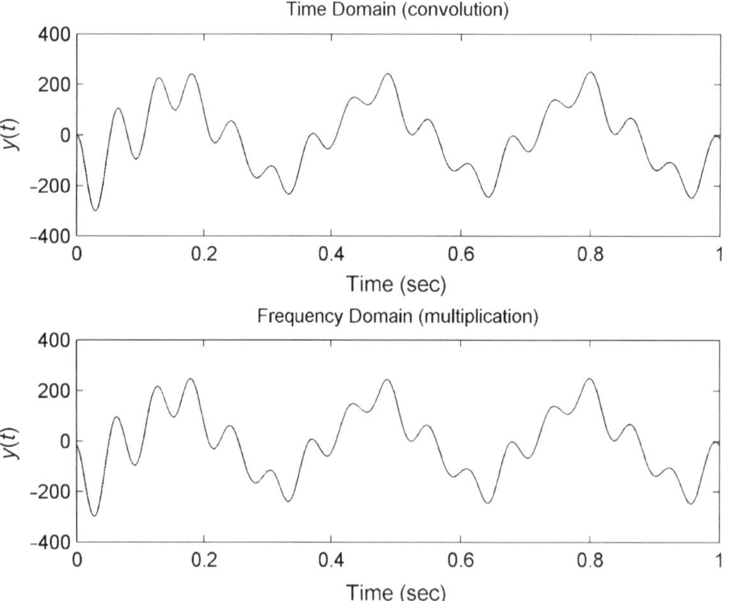

FIGURE 7.17 Output waveforms from the system defined by its impulse response in Example 7.5. Both the time domain (upper graph) and frequency domain (lower graph) produce the same result.

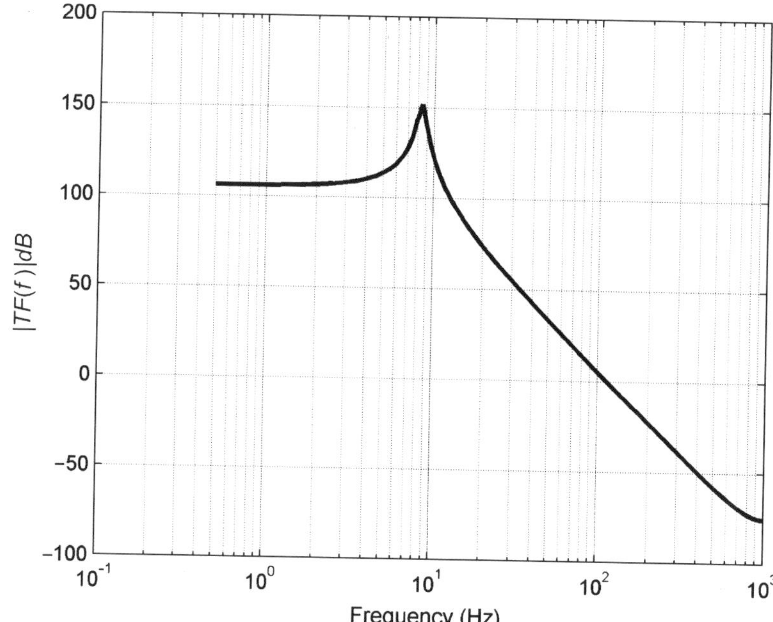

FIGURE 7.18 The magnitude transfer function of the system defined in Example 7.5 obtained by taking the Fourier transform of the system's impulse response.

Results: As shown in Figure 7.17, the output waveforms look like sinusoids riding on the input signal, and are the same using either time or frequency domain analysis. The Bode plot of this system's transfer function is found by taking the Fourier transform of the impulse response and is shown in Figure 7.18. As you might guess from the impulse response equation, the Bode plot is that of a second-order underdamped system.

7.4 CONVOLUTION IN THE FREQUENCY DOMAIN

Example 7.5 demonstrates that convolution in the time domain is the same as multiplication in the frequency domain. The reverse is also true: convolution in the frequency domain is the same as multiplication in the time domain. Using convolution in the frequency domain is not as useful, primarily because the multiplication of signals in the time domain is not all that common. Two important exceptions are sampling, where a continuous signal is multiplied by a series of pulses spaced T_s apart, and amplitude modulation (AM), where a signal is multiplied by a sinusoid or square wave.

There are also some uses for frequency domain convolution in signal processing theory. For example, in Chapter 4 we looked at the application of window functions to modify the spectra obtained when using the Fourier transform. A window function is a time function used to shorten a long data sequence so that it can fit in finite memory. If the data are simply truncated, it is like multiplying the longer data sequence by a rectangular window function, a function with a value of 1.0 over the time of the stored data and 0.0 everywhere else. Other functions can be used to shorten the data, and one of the more popular window functions, the Hamming window, is described in Chapter 4.

Window functions are applied by multiplying the time-domain window function by the time-domain data. In the frequency domain, this is the same as convolving the window's spectrum with the data's spectrum. So spectral convolutions can be used to demonstrate the influence of any given window function on the resulting spectrum. The next example illustrates the use of convolution to demonstrate the modification produced by two different window functions on a signal spectrum.

EXAMPLE 7.6

Determine the influence of the Hamming window and the Blackman–Harris window on the spectrum found as X in file spectrum1.mat on this book's accompanying CD. Plot the spectrum before and after application of the window function. Also plot the time plots of the two windows.

Solution: Take the Fourier transform of the two window functions. Recall the equations for the two windows are:

Hamming:

$$w[n] = 0.5 - 0.46 \cos\left(\frac{2\pi n}{N}\right) \tag{7.14}$$

Blackman:

$$w[n] = 0.41 + 0.5 \cos\left(\frac{2\pi n}{N}\right) + 0.08 \cos\left(\frac{4\pi n}{N}\right) \tag{7.15}$$

Convolve these with the spectrum and plot the original and convolved spectra superimposed. Also plot the window functions in the time domain. The hamming window function is generated using a routine in the Signal Processing Toolbox for variety although it could easily be generated using standard MATLAB. Since the window functions also scale the spectra slightly, the original and modified spectra are plotted normalized to their maximum values.

```
% Example 7.6 Demonstrate the influence of windowing on the spectrum
%
fs = 1000;                  % Sample frequency
load spectrum1;             % Load spectrum
N = length(X);              % Get spectrum length and
f = (1:N)*fs/(2*N);         % frequency vector
h_win = hamming(N);         % Get the Hamming window
H_win = abs(fft(h_win));    % Calculate the window spectrum
X_wind = conv(X,H_win);     % Convolve with the signal spectrum
subplot(2,1,1);             % Plot the original and modified spectra superimposed
 plot(f(1:N/2),X(1:N/2)/max(X),'k'); hold on;
 plot(f(1:N/2),X_wind(1:N/2)/max(X_wind),':k','LineWidth',2);
  ......labels......
%
win = .41-.5*cos(2*pi*(0:N-1)/N) + .08*cos(4*pi*(0:N-1)/N);
```

```
    % Get the Blackman–Harris window and repeat
  ......repeat code above......
figure;
plot(t,h_win,'k'); hold on;  % Plot the two window time functions
plot(t,bh_win,':k','LineWidth',2);
  ......labels......
```

Results: The original and modified *spectra* are shown for both windows in Figure 7.19. The original spectrum and the signal used to produce that spectrum were short ($N = 64$) to enhance the differences between the original and windowed data. Nonetheless, the modification produced by either window is slight, limited to a modest smoothing of the original spectrum. (This example is repeated in Problem 14 at the end of the chapter using different windows and longer data sets.) Time functions of the two windows are shown in Figure 7.20.

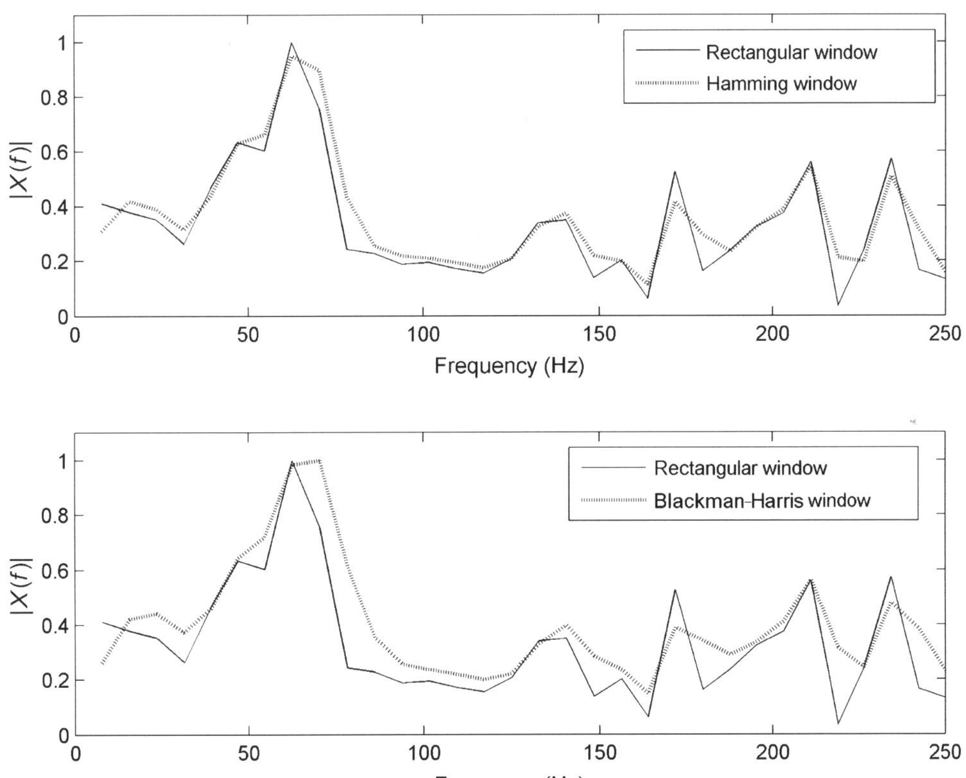

FIGURE 7.19 A) The spectra obtained with a rectangular window (solid curve) and with a Hamming window (dotted curve). B) The spectra obtained with a rectangular window (solid curve) and with a Blackman–Harris window (dotted curve). The windowed spectra are obtained by convolving the window's spectrum with the original spectrum in the frequency domain. This is equivalent to multiplying the window and original signal in the time domain.

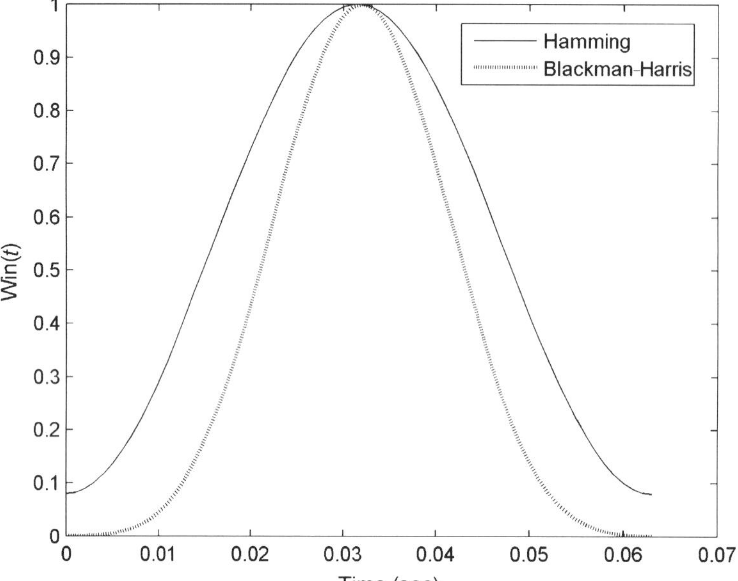

FIGURE 7.20 Time plots of the Hamming and Blackman–Harris window functions used in Example 7.6.

7.4.1 Sampling in the Frequency Domain

In Chapter 4, we find that data sampling produces additional frequencies—in fact an infinite number of additional frequencies. The sampling process is like multiplying a continuous time function by a series of impulse functions. Figure 7.21A shows a continuous signal, $x(t)$, being multiplied by a series of impulse functions spaced 50 msec apart. The result shown in Figure 7.21B is the same as a signal that is sampled at a frequency of $1/T_s = 1/0.05 = 20$ Hz. (A repeating series of pulses is referred to as a *pulse train*.) So the sampling process can be viewed as a time-domain multiplication of a continuous signal by a pulse train of impulses having a period equal to the sample interval T_s.

Since multiplication in the time domain is equivalent to convolution in the frequency domain, we should be able to determine the effect of sampling in a signal's spectrum using convolution. We just convolve the sampled signal's original spectrum with the spectrum of the pulse train. We get the former by applying the Fourier transform to the original signal. For the latter, the spectrum of a periodic series of impulse functions, we use the Fourier series analysis, since the functions are periodic. In Example 3.5, the complex form of the Fourier series equation is used to find the Fourier coefficients of a periodic series of ordinary pulses having a pulsewidth W:

$$C_m = \left(\frac{V_P W}{T}\right) \frac{\sin(2\pi m f_1 (W/2))}{2\pi m f_1 (W/2)} m = \pm 1, \pm 2, \pm 3, \ldots \tag{7.15}$$

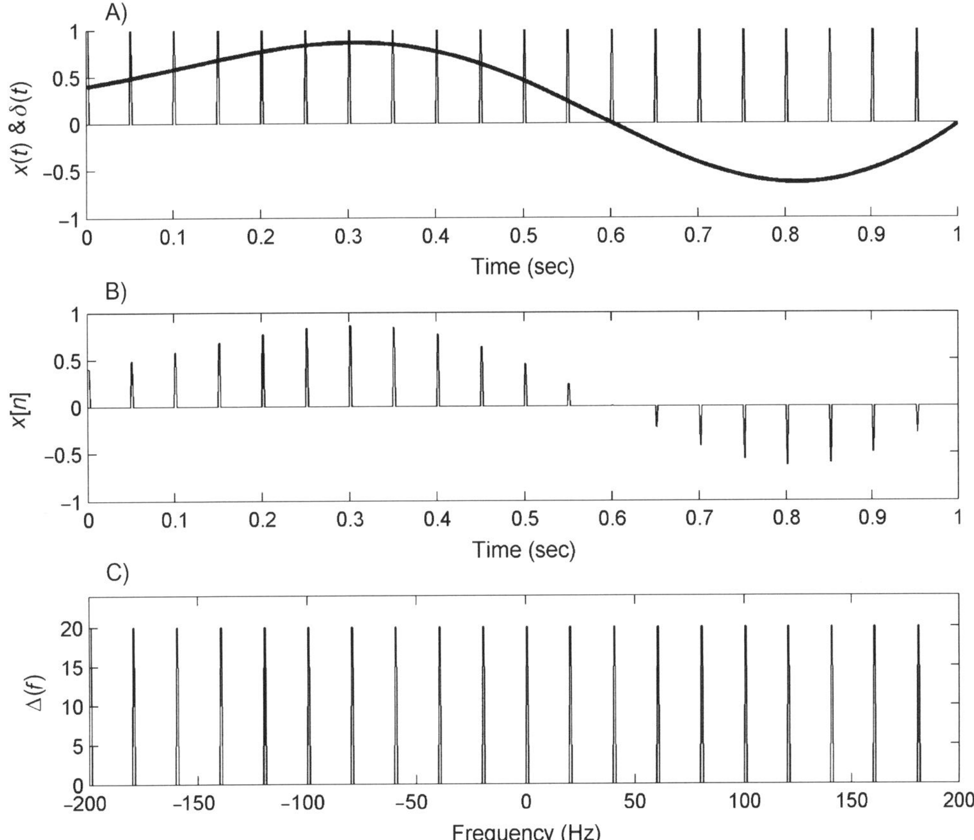

FIGURE 7.21 A) The sampling process can be viewed as the multiplication of a continuous signal by a periodic series of impulse functions. In this figure, the pulse train's period is 0.05 sec and is equivalent to the sample interval, T_s, or $1/f_s$. B) The sampled signal, $x[n]$, where $f_s = 1/0.05 = 20$ Hz. C) The frequency spectrum of a series of periodic impulse functions is itself a series of impulse functions.

For an impulse function, $W \to 0$, but the area, $V_P W$, remains equal to 1.0. For small angles, $\sin(x) \approx x$, so as $W \to 0$, Equation 7.15 becomes:

$$C_m = \lim_{W \to 0} \left| \left(\frac{V_P W}{T}\right) \frac{\sin(2\pi m f_1 (W/2))}{2\pi m f_1 (W/2)} = \left(\frac{V_P W}{T}\right) \frac{2\pi m f_1 (W/2)}{2\pi m f_1 (W/2)} = \frac{V_P W}{T} = \frac{1}{T} \right. \tag{7.16}$$

Since the period of the impulse train is $T = T_s$, the spectrum of this impulse train interval is:

$$C_m = \frac{1}{T} = \frac{1}{T_s} = f_s \tag{7.17}$$

So the spectrum of the impulse function is itself a pulse train at frequencies of mf_1, where $f_1 = 1/T = f_s$, as shown in Figure 7.21C. Since the complex version of the Fourier series equation is used, the resultant spectrum has both positive and negative frequency values. The mathematical expression for an impulse train's spectrum is:

$$\Delta(f) = \frac{1}{T}\sum_{k=-\infty}^{\infty} \delta(f_s - kf_s) = f_s \sum_{k=-\infty}^{\infty} \delta(f_s - kf_s) \tag{7.18}$$

If the frequency domain representation of a signal, $x(t)$, is $X(f)$ and the frequency spectrum of the impulse train is given by Equation 7.18, then the spectrum of the sampled signal is the convolution of the two. With the application of the discrete definition of convolution (Equation 7.3), the spectrum of the sample signal is given as:

$$X_{Samp}(f) = \sum_{n=-\infty}^{\infty} \left(f_s \sum_{k=-\infty}^{\infty} \delta(f_s - kf_s) \right) X(f - nf) \tag{7.19}$$

After we rearrange the sums and simplify by combining similar frequencies, Equation 7.19 becomes:

$$X_{Samp} = f_s \sum_{k=-\infty}^{\infty} X(f - kf_s) \tag{7.20}$$

This application of convolution in the frequency domain shows that the spectrum of a sampled signal is an infinite sum of shifted versions of the original spectrum. This is stated, but unproved, in Chapter 4.

7.5 SYSTEM SIMULATION AND SIMULINK

Simulation is an outgrowth of an early computation device known as an *analog computer*. In fact, simulation programs such as MATLAB's Simulink are basically digital computer representations of an analog computer. The basic idea behind simulation is similar to that used in convolution: calculate responses to a number of small time slices of the input signal. The main difference is that the response of each element is determined individually by calculating the element's response to a specific time slice of input. This element's response then becomes the input to any element(s) it feeds. This is illustrated for the two elements in Figure 7.22 showing the behavior over four time segments.

Simulation begins when the first time slice (Figure 7.22, shown in bold) is taken from the system input and sent to the first element. The time slice is the input signal's value at $t_s = 0$. The first element in Figure 7.22 is a scaling element that multiples the input by -0.5 and produces the bold time slice in $y_1(t_s)$. The time slice of both the input and the scalar output are passed to a summation element, which produces an algebraic sum as its output: the bold segment of $y_2(t_s)$. This progression continues through all other system elements (not shown) until it reaches the last element in the system. The response of this last element is the output of the system and is usually displayed or recorded. This whole process is then repeated for the second time slice, the one immediately following the bold slice. After the second time slice propagates through all the elements, the cycle is repeated

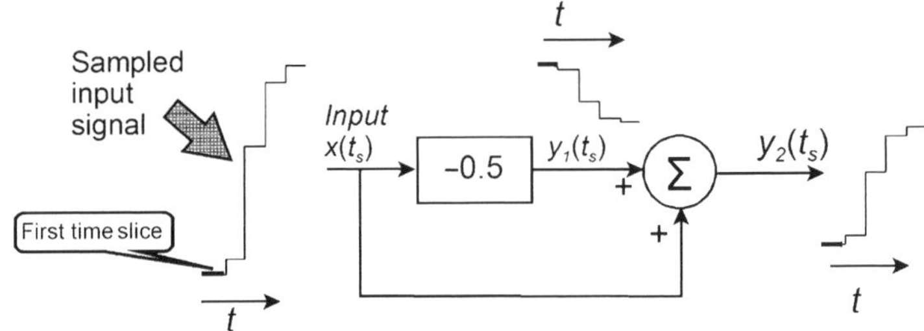

FIGURE 7.22 Two elements of a larger system showing hypothetical inputs and resultant outputs for four time slices. The first time slice is highlighted. Simulation software takes the input signal, $x(t_s)$, and applies the arithmetic operation of the first element, which in this case multiplies each time slice by -0.5. This produces the output $y_1(t_s)$, which is then fed to the summation element along with the original signal. The summation element applies a sum operation during each time slice to produce the output $y_2(t_s)$. Since this is just a portion of a larger system, this output would be sent to other elements.

for successive time slices until the input signal ends or some other stopping criterion is met. During simulation, the output signal and the response of each element evolves over time just as in a real system. In most simulators such as Simulink, it is possible to view the response of any element in the system as this evolution takes place.

All linear systems can be decomposed (at least in theory) into elements that perform addition or subtraction, scaling (i.e., gain elements), differentiation, and integration. In fact, it is possible to rearrange systems to eliminate derivative elements, but this is not necessary when using the simulation program Simulink. The simulation program must perform a number of tasks: 1) it must allow you to set up a model of the system you want to simulate; 2) generate the input waveform(s); 3) calculate and track the responses of the individual elements as the signals flow through the system; and 4) display and/or record the system's output. The program also has to decide on the size of the time slice and how long to continue the simulation. The latter is usually set by the user and the former can be modified by the user.

Step 1: Setting up the model. A systems model is a collection of elements that are interconnected in some specific way. In a systems model all interactions are explicitly indicated, usually by an arrow. Simulink, like most simulation programs, uses *interactive graphics* to set up the system configuration. Individual system elements are selected from a library of elements and the connections between them specified by dragging lines (representing signal influences) between the various elements. Lines can be dragged from the input of one element to the output of the other or vice versa, but the latter often works better. A step-by-step procedure is given in the next example.

Step 2: Generating input waveforms and setting up displays. In Simulink, the input signal usually comes from a waveform-generating element. Simulink has a wide range of waveform generators stored in a library group called "Sources." It is even possible to generate a waveform in a standard MATLAB program and pass it to your model as an input signal. Output and display elements are found in the library group labeled "Sinks."[1] These elements can be used to produce graphs and/or send the output to a file, the workspace, or a MATLAB program. Multiple output elements (i.e., sinks) can be connected to any element in your model.

Step 3: Calculating the responses of arithmetic elements such as addition, subtraction, or scaling (as in Figure 7.22). This is straightforward, but an integration element requires some special care. The output of an integrator could be calculated as the summation of all previous inputs but, like the trapezoidal rule for integration, this is only accurate if the time slice, also called the "step size," is very small. However, if the step size is very small, many time slices are required to complete the simulation and simulation time can become quite large. Moreover, computational errors due to round-off increase with both the number of elements and the number of steps required to complete the simulation. Simulink software goes to great lengths to optimize step size by using carefully designed integration algorithms. Unlike with signal sampling, there is no reason to keep step size constant, and Simulink often varies the size of the time slice to match the dynamic properties of system elements. In most simulations, including all of those covered here, the user does not need to be concerned with integration techniques and step size manipulation; Simulink sets these automatically. Users can alter these simulation parameters, including the algorithm used for integration, and sometimes these alterations are useful to create smoother response curves or to produce evenly sampled data for subsequent analyses.

To implement Step 3, the user only needs to specify the run time (or use the default) and click on the Start button. Simulink takes care of the rest.

7.5.1 Model Specification and Simulation

The examples below provide only an elementary introduction to Simulink's extensive library of elements, and only touch on the power of Simulink's simulation capabilities. They are intended to demonstrate the ability of Simulink to show the behavior of systems like those covered in the last few chapters. Detailed instructions can be found in MATLAB's Help assistance. Click on "Help" in the main MATLAB window, then select "Product Help" in the dropdown menu, then scroll down the list on the left side and select "Simulink." The main Simulink help page will open and a large variety of helpful topics will be available.

Example 7.7 provides a step-by-step demonstration of the use of Simulink to determine the response of a second-order system to a step input. Similar instructions can be found in MATLAB's help documentation following the links *Simulink, Getting Started, Creating a Simulink Model*, and ending up at *Creating a Simple Model*.

[1]Output and display devices are called "sinks" because the output signal flows into, and ends, at these elements just as water flows into a sink and disappears.

EXAMPLE 7.7

Use Simulink to determine the step response of the second-order system having the transfer function shown below:

$$TF(s) = \frac{(s/4)}{s^2 + 12.5s + 50} \tag{7.21}$$

Solution:

Step 1: The Simulink graphical interface is used to set up the systems model using mouse manipulations similar to those of other graphic programs such as PowerPoint. A model window is opened first, then elements are installed in the window and moved by dragging. They can also be duplicated using "copy" and "paste." To create a systems model, open Simulink by typing "simulink" in the MATLAB command window, or use the Simulink tab on the toolbar that contains the "open file" indicator. A new window containing the Simulink Library Browser opens with a right-hand frame that shows a number of icons representing different library groups of system elements (Figure 7.23). We use elements from the Commonly Used Blocks, Sources, and Continuous groups. Use the "File" dropdown menu and select "New"

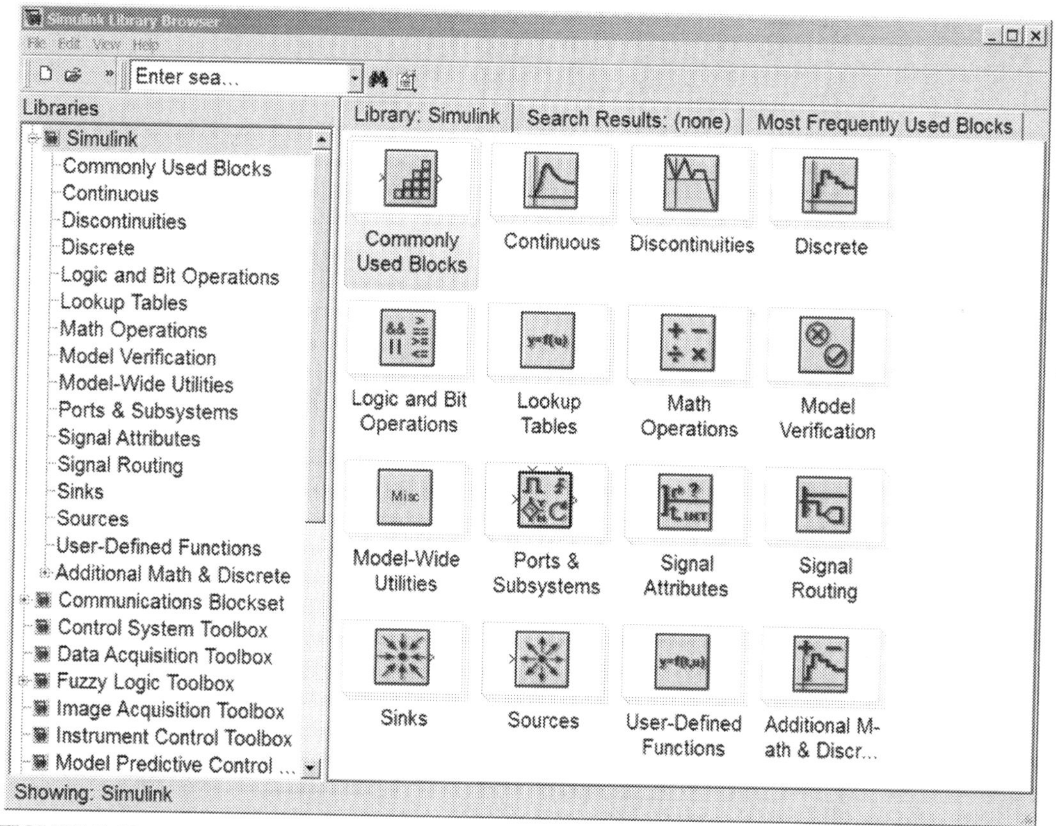

FIGURE 7.23 The main Simulink window showing the various library icons on the right side.

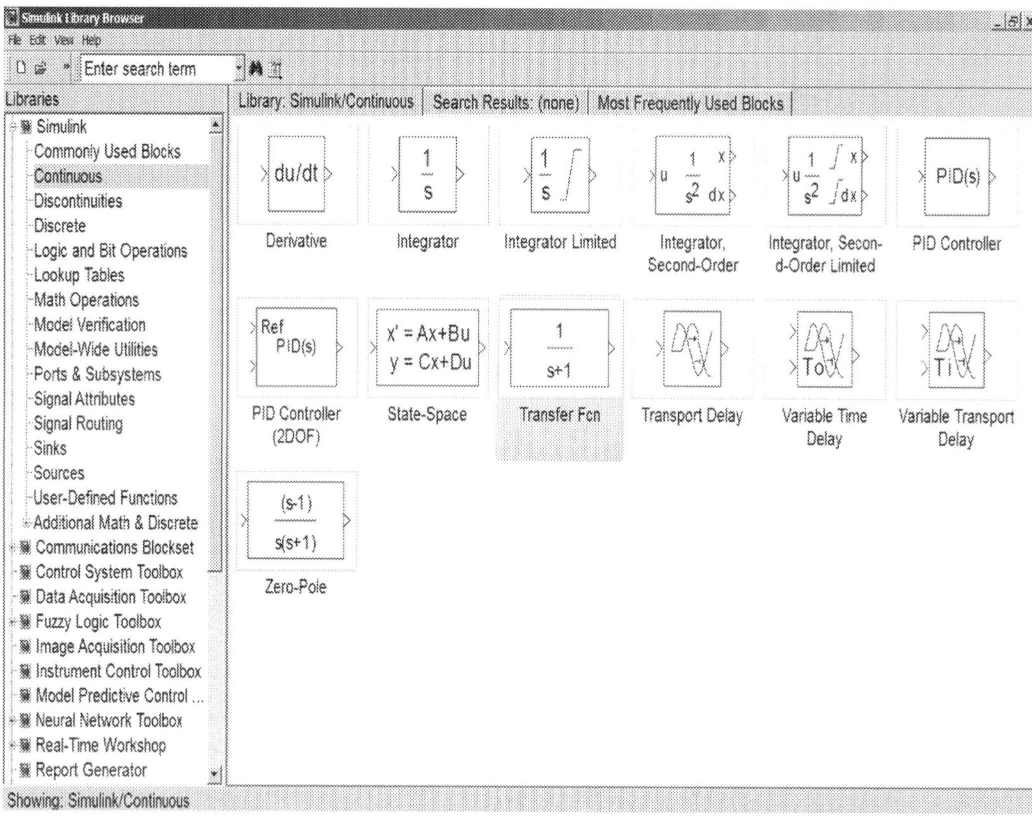

FIGURE 7.24 The window that opens for the Continuous library group. Several particularly useful elements are found in this group.

and then "Model." This opens a blank window with the title "untitled1." If you save this window as a model file (.mdl), the name you choose appears in the title. The model window can be saved, reopened, and resaved just as with any MATLAB program file.

The next operation in Step 1 is to populate the model window with the desired elements. Open one of the library groups by double-clicking, then drag the desired element into the model window. A duplicate of that element is created in the model window. Our model requires three elements: a step generator, the second-order element as given in Equation 7.21, and a display.

We begin with the second-order element, which can be found in the Continuous library, shown as shaded in Figure 7.23.[2] Double-click on the "Continuous" library icon (to the right of Commonly Used Blocks); this library contains some typical linear system elements

[2]Since linear systems deal with signals that are theoretically continuous in time, simulation of LTI systems is sometimes referred to as "continuous simulation" although the simulation approach is still based on discrete time slices. This continuous simulation terminology is the basis of the name MATLAB uses for this library.

including a differentiator, an integrator, and a first-order numerator and denominator element (termed *Zero-Pole*) along with other elements. The most versatile of these elements, and the one we use in this simulation, mimics a general transfer function and is labeled "Transfer Fcn" (the shaded icon in Figure 7.24). The icon shows a simple first-order transfer function, but the element can be modified to represent a transfer function of any complexity. To place the element in our model, drag it from library window to the model window.

After the Transfer Fcn block is dragged into the model window (Figure 7.25A), two parameters associated with that element must be set if the element is to match Equation 7.21. Element parameters are set by double-clicking on an element and filling in the necessary text boxes. Double-clicking on the Transfer Fcn element opens a window titled "Function Block Parameters: Transfer Fcn Second-order," as shown in Figure 7.25C. The text in this window succinctly describes the two parameters of interest: the Numerator coefficients and the Denominator coefficients. These define the numerator and denominator equations in terms of coefficients of the Laplace variable s. The coefficients are placed in a vector arranged in decreasing powers of s, just as in a standard Laplace transfer function. Hence, the last number in the vector is the constant (s^0); the next to last is the coefficient of s (s^1); the third to last is the coefficient of s^2; and so on. The numerator of Equation 7.21 is $0.25s + 0$, so the constant

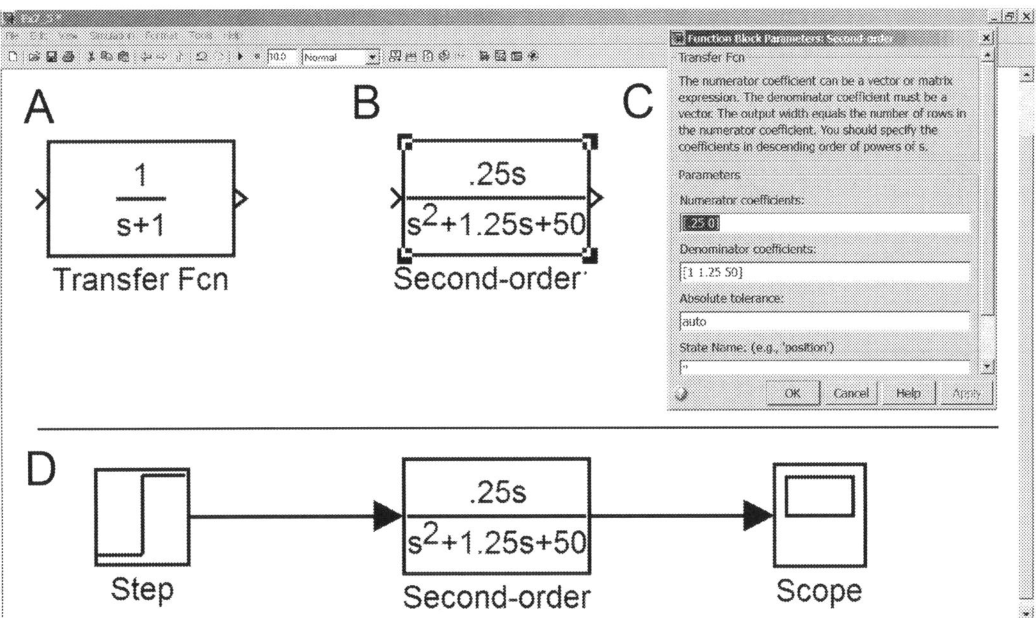

FIGURE 7.25 Model window in Example 7.7. A) The Transfer Fcn block has been dragged into the Model window. Input and output connection points can be seen on this block. B) The Transfer Fcn window has been modified to set up the block as a second-order element, and the name of the element has been changed to Second-order. C) The Transfer Fcn window has been opened and the appropriate numbers and vectors entered into the upper two text boxes to define the second-order element represented by Equation 7.21. D) An input signal element ("Step") and the display element ("Scope") have been added to the input and output of the second-order element. The input and output directions are indicated by arrows.

coefficient is zero, the coefficient of s is 0.25, and the upper box contains the vector [0.25, 0]. The coefficients of the denominator equation are, in decreasing powers of s: 1.0, 1.25, and 50. So the second text box should contain the vector [1, 1.25, 50] (Figure 7.25C). The lower two text boxes can be left with the defaults. When entering these values remember to include the brackets unless only a single number is entered. After you click "OK," the element appears in the center box, as shown in Figure 7.25B. The label below the element icon is changed when you double-click on the label and enter the desired name, in this case "Second-order."

Step 2: Two additional elements are needed to complete the model: the step stimulus and the display. The former can be found in the Sources library under the label "Step." After you drag this element into your model window, check its parameters by double-clicking on the element. This element has four parameters: Step time, Initial value, Final value, and Sample time. The first entry, Step time, is a scalar that defines, in seconds, when the step occurs. The default value is 1 second and can be left as is for this simulation. The Initial value default is zero and the Final value default is 1 which will produce a classic unit step signal, and these

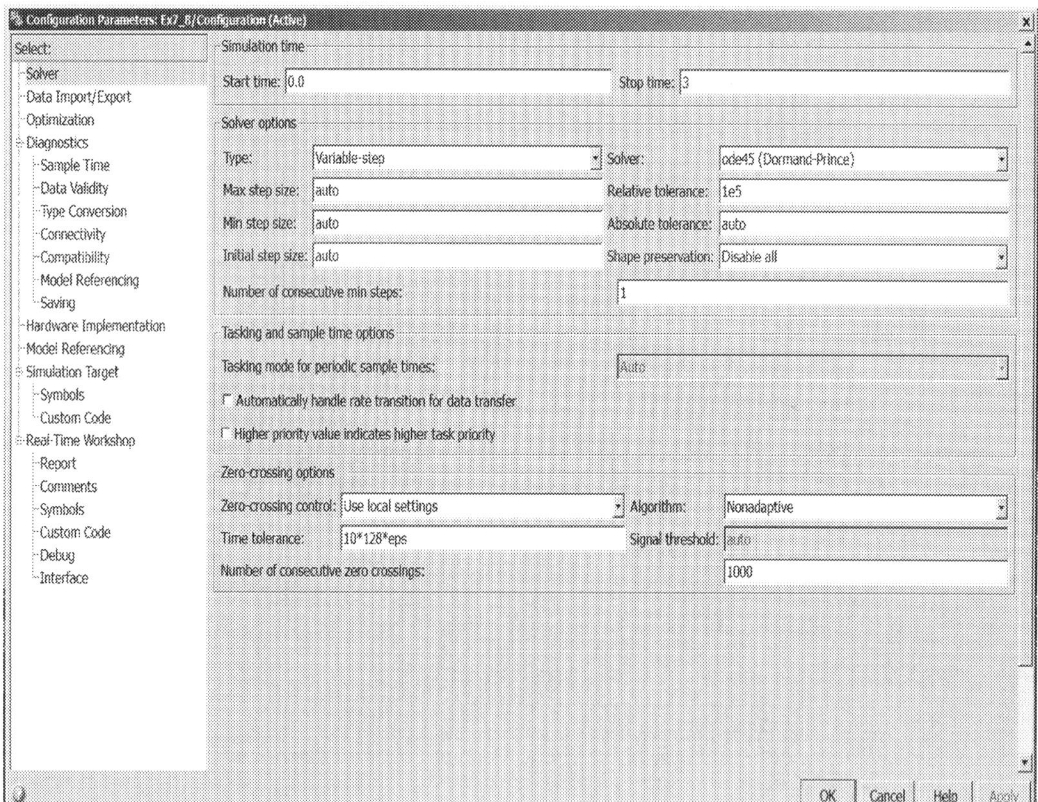

FIGURE 7.26 The Configuration Parameters window accessed through the Simulation pull-down menu. This window is used to set the simulation time (top two text boxes) and optional aspects of the simulation such as error tolerances.

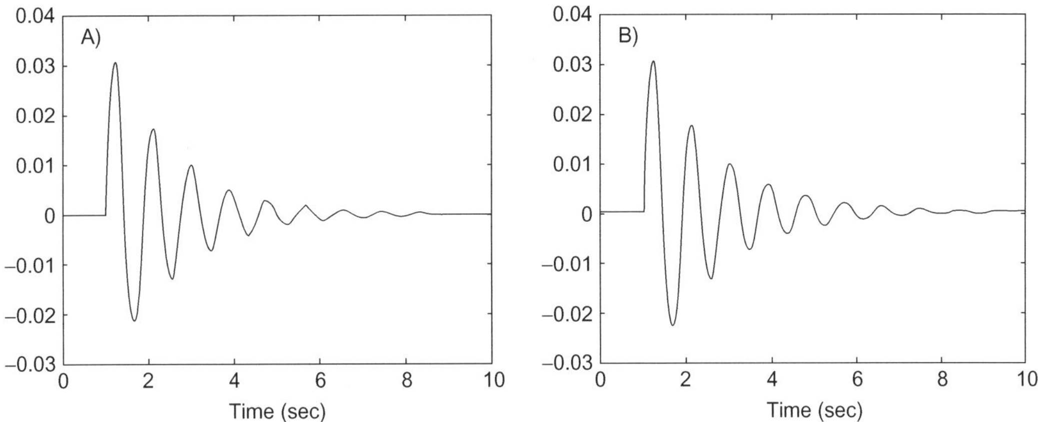

FIGURE 7.27 Time response of the second-order system given in Equation 7.21 to a unit step input. These graphs are similar to that produced by the Scope element of Simulink simulation program applied to the model of Figure 7.25. A) Response produced by the default error tolerance ($1e^{-3}$). B) Response produced by a lower error tolerance ($1e^{-8}$).

values can also be left at their default values. The Sample time parameter is not modified in standard use.

You also need to add a display to show the output. Output elements are found in the Sinks library, but the element we use is also found in the Commonly Used Blocks library. The Scope element provides a time display of its input signal similar to an oscilloscope. The default mode has only one display channel, but any number of these elements can be used throughout the model.

As is seen in the second-order element in Figure 7.25A, each element has an input terminal indicated by a ">" symbol, usually found on the left side of the element, and an output terminal indicated by an arrowhead symbol, usually on the right side of the element. After all the elements are placed in the window, connect them by dragging a line from the output of one element to the input of another. There are other methods for making connections, but dragging in the reverse direction is easy and straightforward. Figure 7.25D shows the complete model.

Step 3: The first entry on the pull-down Simulation menu is "Start" which, logically, starts the simulation. The simulation time is controlled by a text entry in the Configuration Parameters window which is accessed through the Simulation menu (Figure 7.26). The top two text entries set the total length of the simulation time in seconds. A 10-sec simulation time works well for this system (Figure 7.26). To visualize the output, double-click on the Scope icon which opens the scope display window. Click on the binoculars symbol to scale the y-axis automatically. A plot similar to the scope display is shown in Figure 7.27A.

The response shown in Figure 7.27A uses the default error tolerance, which results in a plot that is not very smooth. The error tolerance can be modified in the Configuration Parameters windows reached through the Simulation pull-down menu (Figure 7.26). Reducing the number found in the "Relative tolerance" text box from the default of 1e-3 shown in Figure 7.26 to 1e-8 produces a smoother response, as shown in Figure 7.27B.

7.5.2 Complex System Simulations

It is almost as simple to use Simulink to analyze the behavior of very complex systems. The next example applies the same methodology described in Example 7.7 to a more complicated system.

EXAMPLE 7.8

Simulate the step response of the system shown in Figure 7.28. Also simulate the response to a 200-msec pulse. Then determine the pulse width that represents an impulse input.

Solution: Construct the model shown in Figure 7.28, adding a step input and display. After simulating the step response, replace the step function element with a pulse generator element. Adjust the pulse width so that it appears as an impulse to the system using the strategy employed in Example 7.4.

From the Continuous library, we drag out three elements: two Transfer Fcn elements and an Integrator. From the Commonly Used Blocks library, we drag out two Sum elements (one will be changed to a subtraction element), a Gain element, and three display (i.e., Scope) elements. The Scope elements will allow us to monitor the output of three different elements. Then we go to the Sources library to pick up the Step element. The resulting model window appears as shown in Figure 7.29A, where the elements have been dragged into the window more or less randomly.

Next we modify the element parameters. Again, set all parameters by double-clicking on the block and filling in the associated parameter window. As in the last example, the two transfer function elements must be set with the coefficients of the numerator and denominator equations. The element in the upper path in Figure 7.28 has a transfer function of $5/(s+3)$, giving rise to a numerator term of 5 (this is a scalar so no brackets are needed) and a denominator term of [1, 3]. The transfer function in the lower branch, $1/(s^2+6s+40)$, has a numerator term of 1 and a denominator term of [1, 6, 40]. The gain term has one parameter that needs to be set: the Gain in its text box should be set to 50. The Sum element has a text box labeled "List of signs" which contains two + signs as defaults. One of these should be changed to a negative (−) sign to change the element to a subtractor. The element's icon will then show which input is negative.

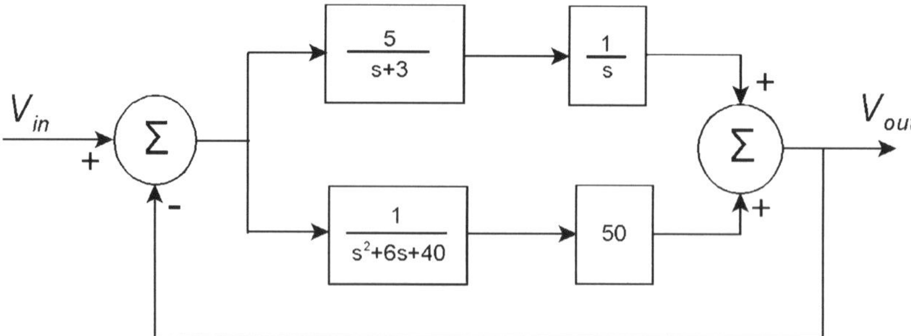

FIGURE 7.28 A more complicated system whose step and impulse responses are simulated in Example 7.8.

The Integrator block does not need to be changed. The Step element was modified so the step begins 100 msec after the start of the simulation.

The elements are then arranged in a configuration with the signals going from left to right (except for the feedback signal) and connected. Again, there are shortcuts to connecting these elements, but the technique of going for input back to output works well. The second connection does not have to be an element output terminal; it can be just a line which allows for "T" connections. Right angles are made by momentarily lifting up on the mouse button, as with other graphics programs. Scopes are connected to the signal lines and provide a display of those

A)

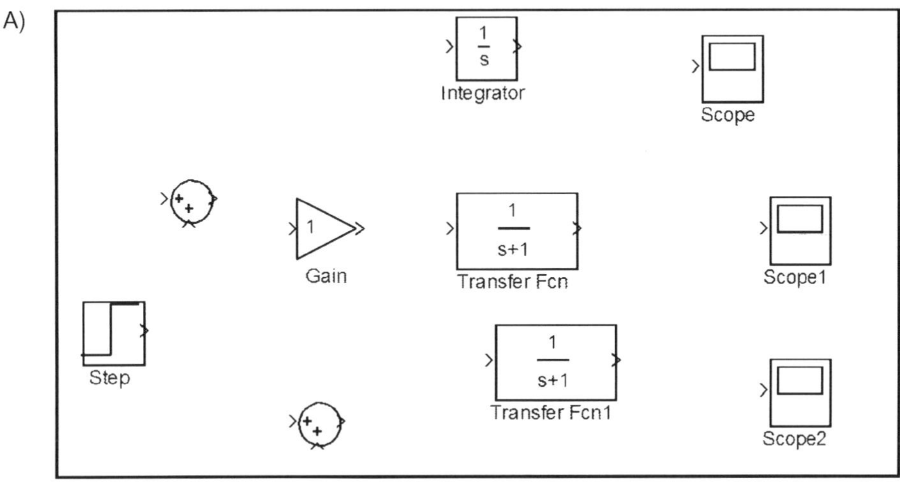

B)

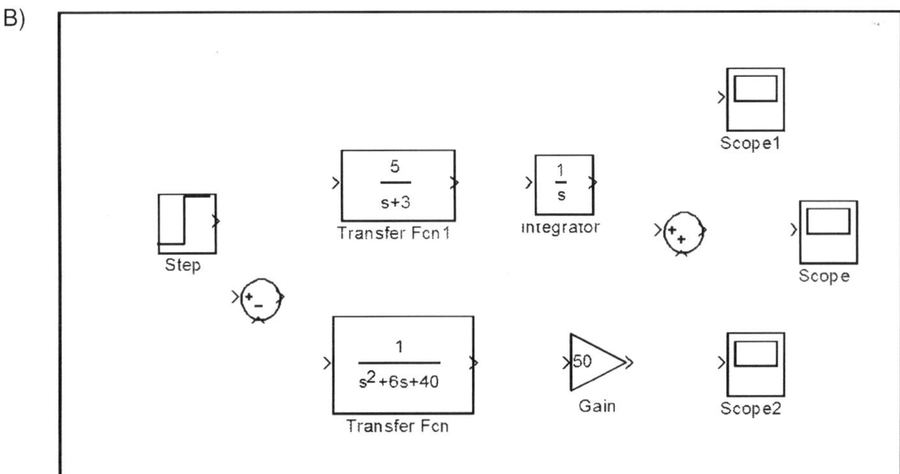

FIGURE 7.29 A) Collection of elements required to simulate the more complicated system shown in Figure 7.28. B) The elements are arranged to approximate the configuration of Figure 7.28.

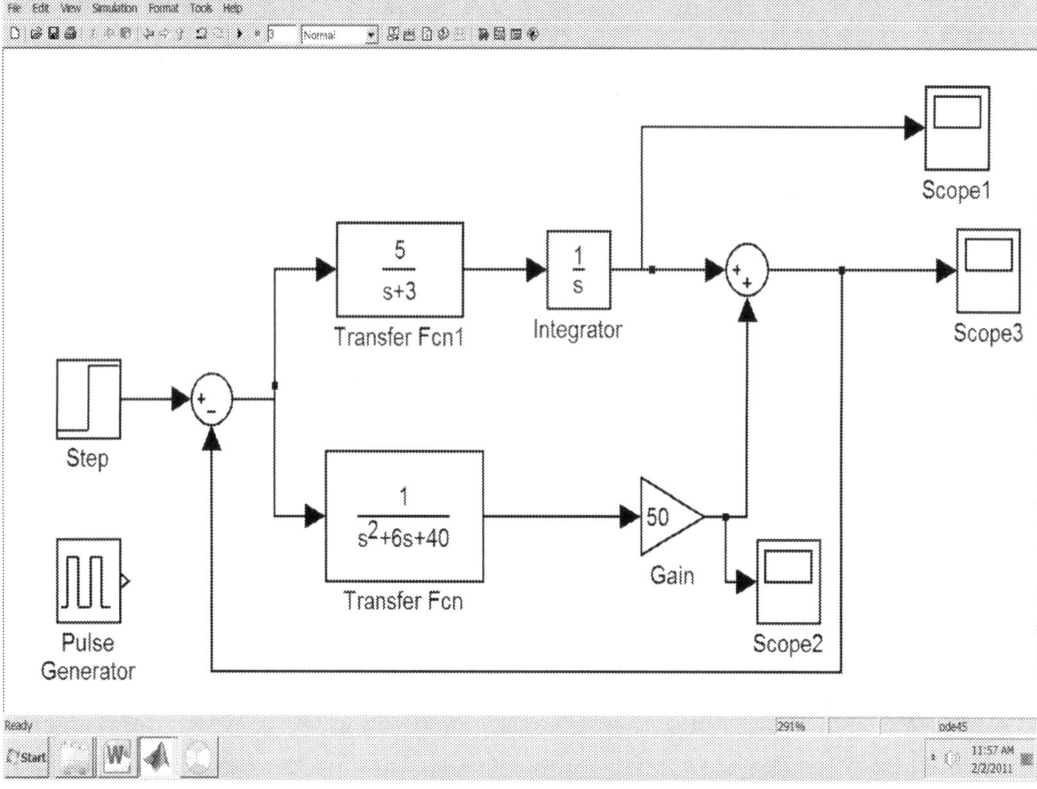

FIGURE 7.30 Final model configuration showing the interconnecting lines and appropriate signs on the subtractor, gain, and transfer functions. The step generator is shown as the input element with the pulse generator off to the side, unconnected. It will be swapped for the step generator in a second simulation. Note that the final model as drawn by Simulink closely follows the standard systems model block diagram, as in Figure 7.28.

signals. Three are used in this simulation: two are attached to the upper and lower pathways and one to the output. The resulting model window is shown in Figure 7.30.

After you set the simulation time to 10 sec and activate the Start tab, you see the unit step response on the output Scope display as in Figure 7.31 (right side). With Simulink, we can monitor any element in the system; in this simulation we use two other Scopes to monitor the upper and lower signal paths. The contributions of these two paths to the output signal are also shown in Figure 7.31 (lower two graphs on the right side). The upper pathway contributes a smooth exponential-like signal (actually it is a double exponential that continues to increase). The lower pathway contributes a transient component that serves to bring the response more quickly toward its final value, but also produces the inflection in the combined response.

To determine the pulse response of the system, replace the Step element with a Pulse Generator element, also found in the Sources library. Setting the "period" textbox of the Pulse

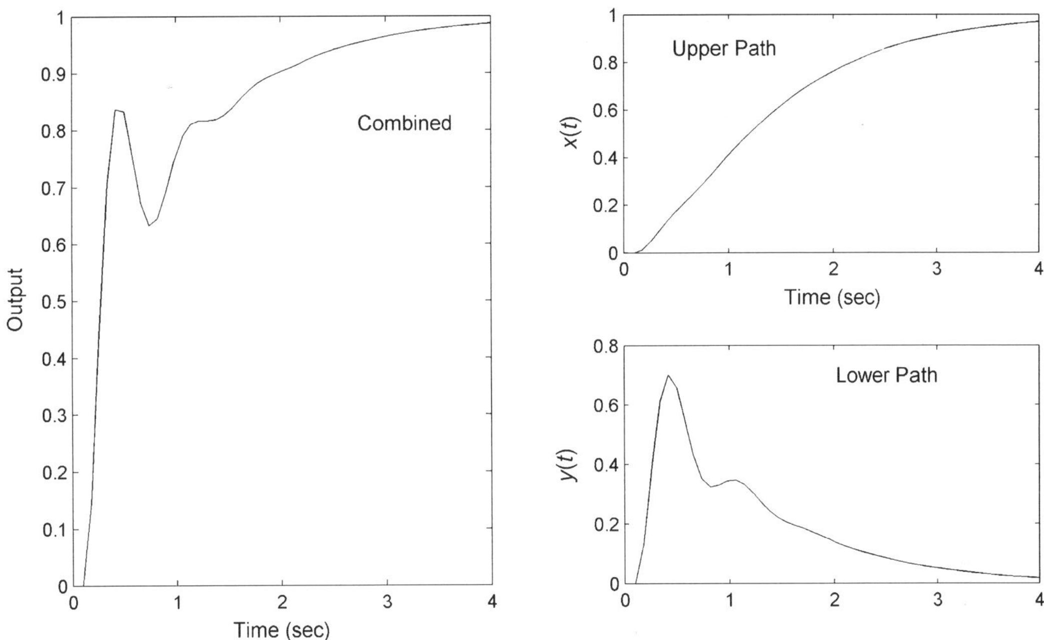

FIGURE 7.31 Simulink Scope displays showing the outputs of the upper (left) and lower (right) feedforward pathways in the model of Figure 7.28 to both step (right side) and pulse (left side) inputs.

Generator to 10 sec will ensure that the pulse occurs only once during the 10-sec simulation period. Setting the Pulse Width to 2% produces a 0.2-sec pulse, and the response is seen in Figure 7.32 (left upper plot). The contributions of the upper and lower pathways of the pulse response are also shown in Figure 7.32.

To estimate the pulse width that will approximate an impulse function, simulate the system while adjusting the duty cycle of the Pulse Generator element. Shorten the pulse to 0.1 sec (i.e., 1%), then to 0.05, 0.25, and finally 0.01 sec to produce the responses shown in Figure 7.32. These responses are normalized to the same amplitude to aid comparison. The responses produced by either the 10-msec or 25-msec pulse widths are very similar, so a pulse having a width of 25 msec or less can be taken as an impulse input with respect to this system. A unit impulse function should have an area of 1.0, so the amplitude of the 0.025-sec pulse should be set to $1/0.025 = 40$ while the amplitude of the 0.01-sec pulse should be 100.

7.6 BIOLOGICAL EXAMPLES

Numerous biomedical simulations use Simulink, many quite detailed involving more than 50 elements. For a comprehensive Simulink model of the cardiovascular system,

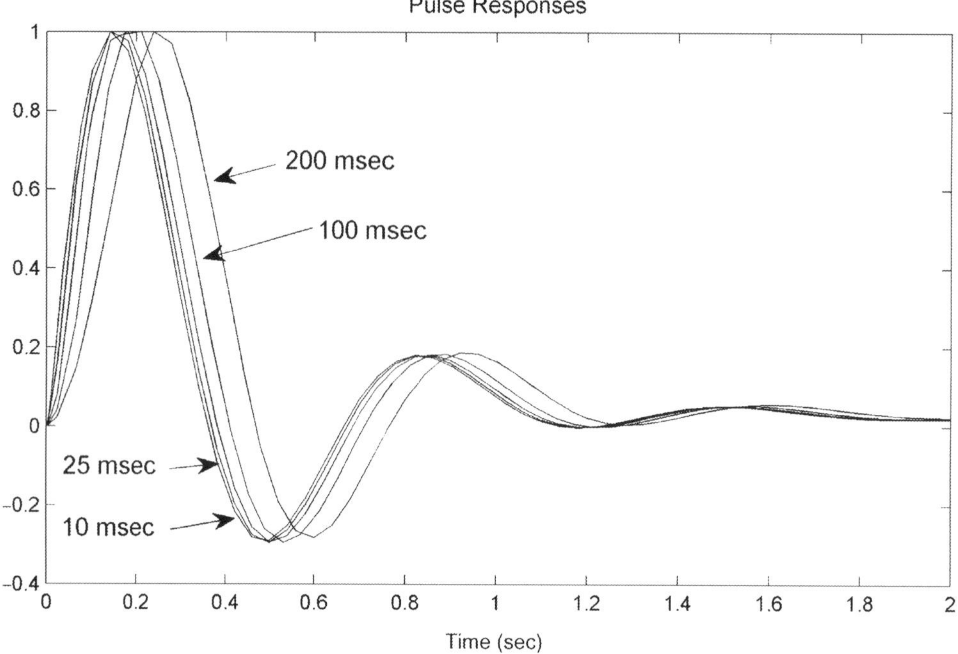

FIGURE 7.32 The response of the system shown in Figure 7.28 to pulses of varying widths. Pulses having widths of 10 or 25 msec produce nearly the same response, indicating that pulses less than 25 msec function as impulse inputs with respect to this system.

check out the model by Zhe Hu, downloadable from MATLAB Central.[3] This model, with over 100 elements, is far too large to be used here. Example 7.9 presents a realistic physiological model on a much smaller scale: a model of glucose and insulin balance in extracellular tissue.

EXAMPLE 7.9

Simulate a simplified version of the Stolwijk–Hardy model for glucose–insulin concentrations in extracellular space.

Solution: The Stolwijk–Hardy model consists of two integrators involved in a feedback system. The concentration of glucose or insulin in the extracellular fluid is determined by integration of the net flows of each in and out of the fluid divided by the effective volumes of the fluid for glucose and insulin. The Simulink version of the Stolwijk–Hardy model is shown in Figure 7.33. Concentrations of glucose and insulin are represented by the outputs of integrators labeled "Glucose" and "Insulin." The insulin integrator has a negative feedback signal modified by gain, "K5," and receives an input from the glucose concentration signal after it is modified by

[3]www.mathworks.com/matlabcentral/fileexchange/818-hsp

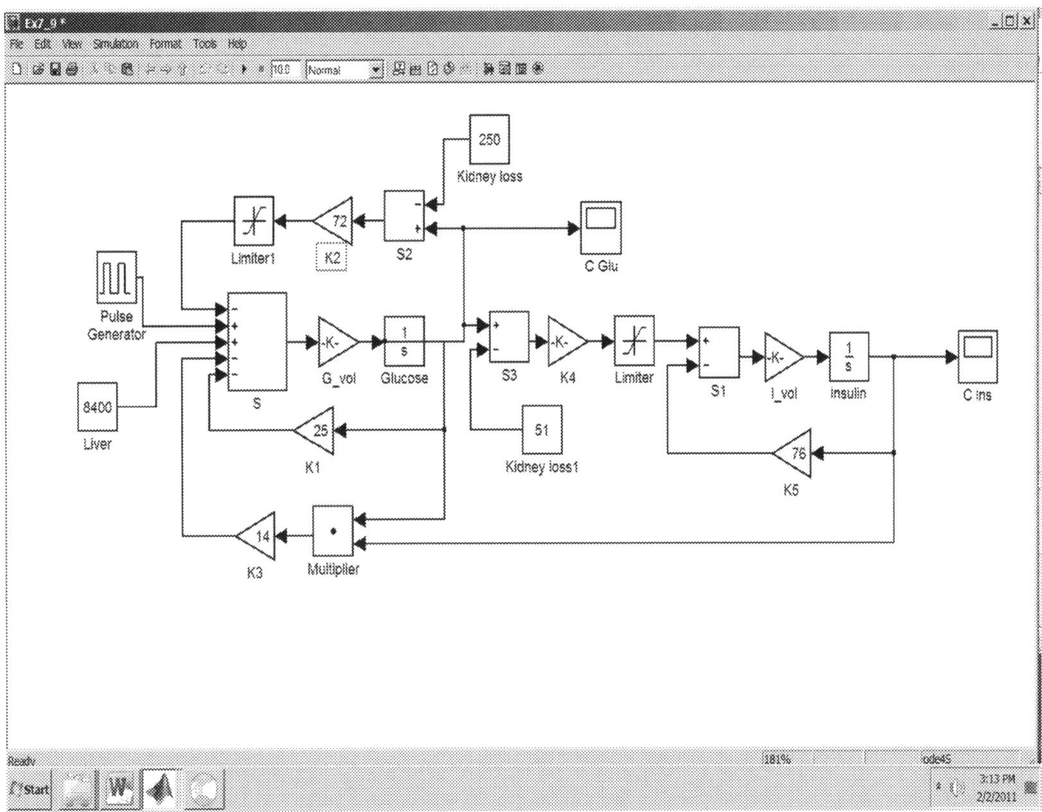

FIGURE 7.33 Simulink representation of the Stolwijk–Hardy model for glucose and insulin concentration in extracellular space.

glucose loss through the kidney represented by constant "Kidney loss 1." This signal also has an upper limit, or saturation, imposed by element "Limiter."

The glucose integrator receives two feedback signals: one directly through gain "K1" and the other modified by kidney loss and gain "K2." The kidney feedback signal also has a saturation implemented by "Limiter 1." This integrator also receives feedback from the insulin signal modified by its own output though the multiplier and gain "K4." The glucose integrator receives two input signals: a constant input from the liver and one representing glucose intake. The latter is implemented by a pulse generator set to produce a single half-hour pulse after a 1-hour delay. Both integrators are preceded by gain terms that reduce the signal in proportion to the effective volumes of the extracellular fluid for glucose and insulin, to produce signals that represent concentrations. Both integrators also have initial values incorporated into the integrator. Alternatively, a longer delay between the start of the simulation and the onset of the pulse can be used to allow the integrators to obtain appropriate initial values.

TABLE 7.1 Parameter Values for the Stolwijk–Hardy Model of Glucose and Insulin Concentrations

Parameter	Value	Description	Parameter	Value	Description
C_{G_init}	81 (% mg/mg)	Glucose integrator initial value	G_vol	1/150 (% ml)	Inverse effective volume of glucose extracellular
C_{I_init}	5.65 (% mg/mg)	Insulin integrator initial value	I_vol	1/150 (% ml)	Inverse effective volume of insulin extracellular
Liver	8400 (mg/hr)	Glucose inflow from liver	Limiter	10,000	Glucose saturation driving insulin
K1	25	Gain of glucose feedback	Limiter 1	10,000	Glucose saturation in feedback signal
K2	72	Gain of glucose feedback from kidney loss	Kidney loss	200 (mg/hr)	Renal spill in feedback signal
K3	14	Gain of insulin feedback to glucose	Kidney loss 1	51 (mg/hr)	Kidney loss in glucose feedback
K4	14	Gain of glucose drive to insulin	K5	76	Gain of insulin feedback

The limiters and multiplier elements make the equations for this system nonlinear so that they cannot be solved using the linear techniques presented in Chapters 5 and 6. Simulink, however, has no difficulty incorporating these and some other nonlinear elements into a model.

The various model parameters, from a description of the Stolwijk–Hardy model in Rideout (1991), are shown in Table 7.1.

The results of a 10-hour simulation of this model are shown in Figure 7.34. The influx of glucose, modeled as a half-hour pulse of glucose, results in a sharp rise in glucose concentration and a slightly slower rise of insulin. When the glucose input stops, glucose concentration immediately begins to fall and undershoots before returning to the baseline level several hours later. Insulin concentration falls much more slowly and shows only a slight undershoot, but does not return to baseline levels until about 7 hours after the stimulus.

Example 7.10 describes a model of the neuromuscular motor reflex. In this model, introduced by Khoo (2000), the mechanical properties of the limb and its muscles are modeled as a second-order system. The model addresses an experiment in which a sudden additional weight is used to exert torque on the forearm joint. The muscle spindles detect the stretching of arm muscles and generate a counterforce though the spinal reflex arc. The degree to which this counterforce compensates for the added load on the muscle can be determined by measuring the change in the angle of the elbow joint. The model consists of a representation of the forearm muscle mechanism in conjunction with the spindle feedback system. The input to the model is a mechanical torque applied by a step-like increase in the load on the arm; the output is forearm position measured at the elbow in degrees.

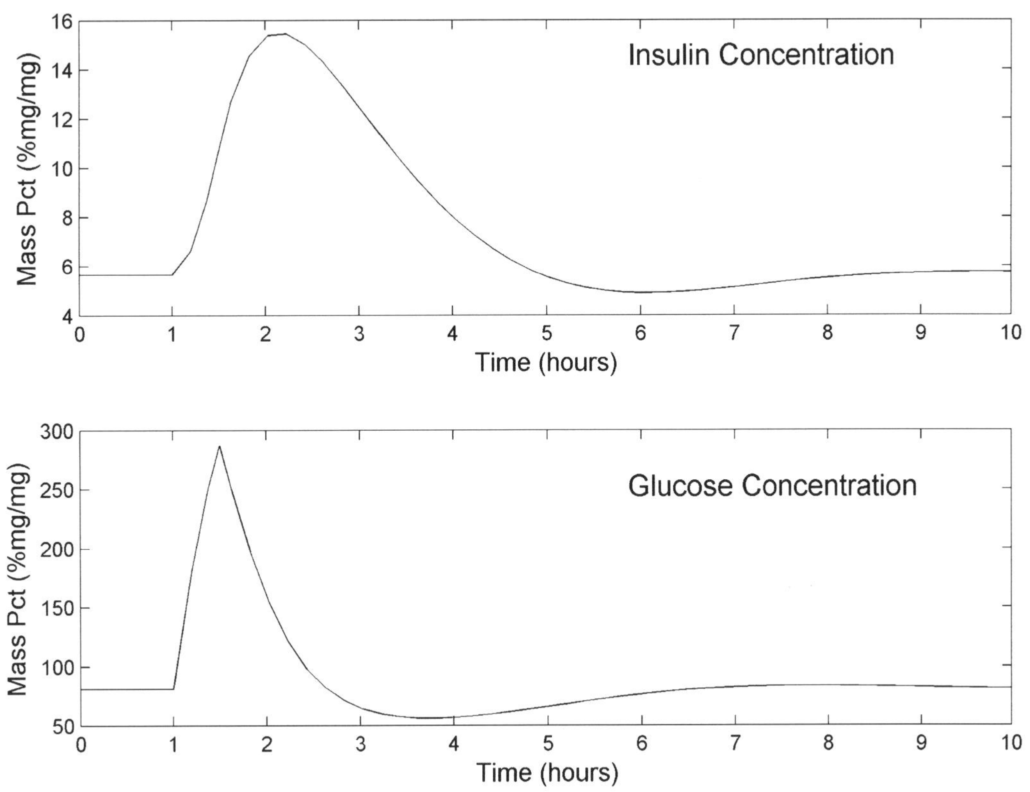

FIGURE 7.34 Simulated response of glucose and insulin concentration to a half-hour infusion of glucose. The 10-hour simulation shows that the glucose stimulus produces a sharp rise in glucose concentration and a slightly slower rise of insulin.

EXAMPLE 7.10

Model the neuromuscular reflex of the forearm to a suddenly applied load.

Solution: To generate a model of the forearm reflex, we need a representation of the mechanics of the forearm at the elbow joint and a representation of the stretch reflex. The transfer function of mechanical models is covered in Chapter 10, and we derive this function for the forearm joint from basic mechanical elements. For now, we accept that the transfer function of the forearm muscle mechanism is given by a second-order equation. Using typical limb parameters, the Laplace domain transfer function is:

$$TF(s) = \frac{V(s)}{M(s)} = \frac{1}{s^2 + 25s + 500} \tag{7.22}$$

where $V(s)$ is the velocity of the change in angle of the elbow joint and $M(s)$ is the applied torque. Since position is the integral of velocity, Equation 7.22 should be integrated to find position. In the Laplace domain, integration is achieved by dividing by s:

$$TF(s) = \frac{\theta(s)}{M(s)} = \frac{1}{s}\frac{V(s)}{M(s)} = \frac{250}{s(s^2+25s+500)} \tag{7.23}$$

The transfer function of Equation 7.23 represents the effector components in this system, which is sometimes referred to as the *plant*. The term is borrowed from early control systems work involving large chemical plants where a central controller was designed to manage numerous processes in the plant. Such control systems were said to consist of a *controller* and a *plant* in a feedback configuration.

Khoo (2000) shows that the muscle spindle system can be approximated by the transfer function:

$$TF(s) = \frac{M(s)}{\theta(s)} = \frac{s+60}{s+300} \tag{7.24}$$

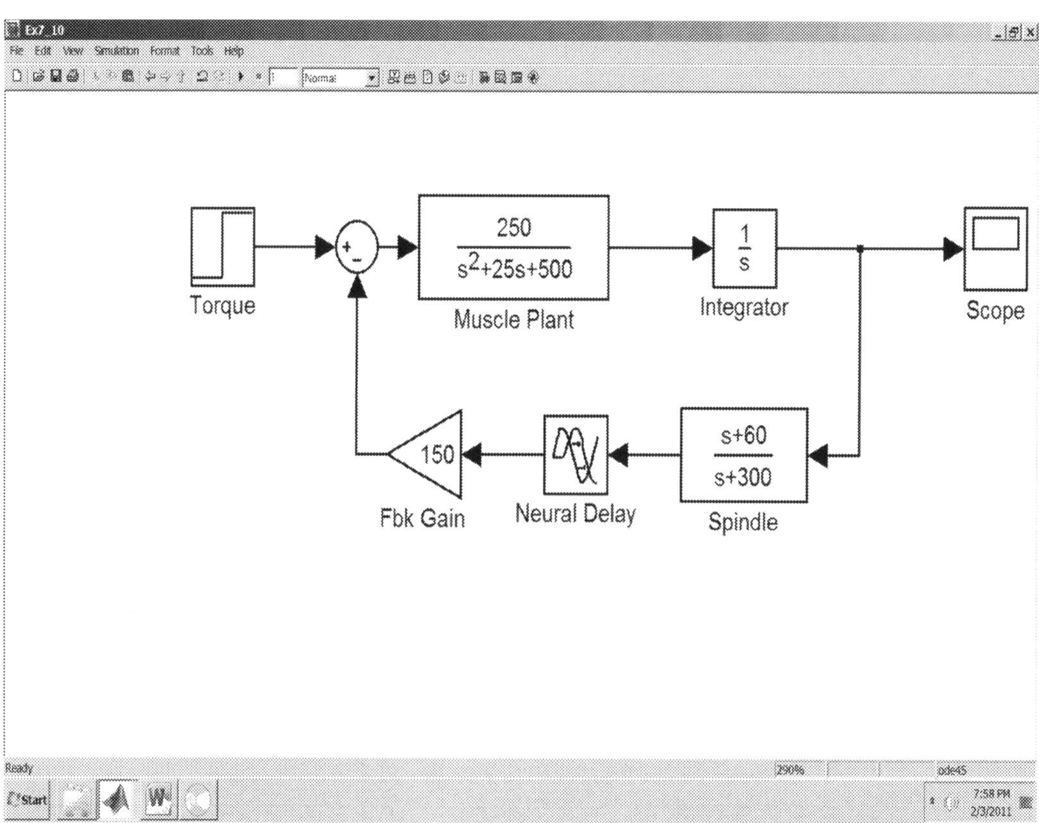

FIGURE 7.35 Model of the neuromuscular reflex system developed by Khoo (2000).

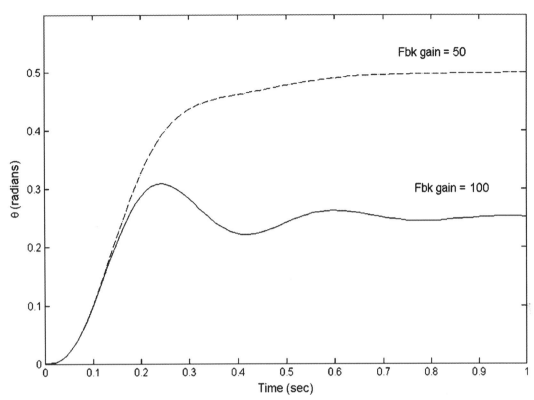

FIGURE 7.36 Model responses of the neuromuscular reflex to a suddenly added load. Simulated responses for two values of feedback gain are shown.

where $M(s)$ is the effective counter torque produced by the spindle reflex and $\theta(s)$ is the angle of the elbow joint. In addition, there is a gain term in the spindle reflex nominally set to 100 and a time delay (i.e., e^{-sT}) of 20 msec. Since the spindle system is a feedback system, the overall model is represented as shown in Figure 7.35.

Simulation of this model in response to a 5-kg step input is shown in Figure 7.36 for two values of feedback gain, 50 and 100. At the higher gain, the reflex response shows some oscillation, but the net displacement produced by the added weight is reduced. If the goal of the stretch reflex is to compensate, in part, for the change in load on a limb, the increase of feedback gain improves the performance of the reflex since it reduces the change produced by the added load. However, as seen in Figure 7.36, feedback systems with high loop gains tend to oscillate, a behavior that is exacerbated when there is a time delay in the loop. The effects of gain and delay on this oscillatory behavior are examined in Problems 17 and 18 at the end of this chapter, based on this model.

Example 7.11 is an extension of Example 7.10 and shows one way to integrate Simulink data with standard MATLAB routines.

EXAMPLE 7.11

Find the magnitude and phase spectrum of the neuromuscular reflex model used in Example 7.10. Use a feedback gain of 100.

Solution: There are several ways to do this, but the most straightforward is based on the fact that a system's spectrum is the Fourier transform of the impulse response (Equation 7.9). We will simulate an impulse response in Simulink then pass that data to a MATLAB program to calculate the Fourier transform. To generate the impulse response, we rely on the same approach used in Examples 7.4 and 7.8: Shorten the width of a pulse until the responses generated by a pair of short pulses are the same. At this point, shortening the pulse further will not change the response dynamics, even if the pulse were to be shortened all the way to a true impulse (which is not possible since true impulse functions are infinitely short). To find this pulse width, we replace the step generator in Figure 7.35 with a pulse generator and observe that pulse widths less than 2 msec produce the same response dynamics. The response to a 2-msec pulse with an amplitude of 5 kg is shown in Figure 7.37. For all practical purposes, this is the impulse response. The maximum amplitude of this response is quite small compared to that of the step response (Figure 7.36), because there is much less energy in the short pulse. (Recall that a true impulse has an amplitude of infinity.)

There are several ways to communicate between Simulink and standard MATLAB. Here we use the "To File" block in the Sinks library. This block takes an input and writes it to a `.mat` file. After you insert this block and connect it to the output of the integrator (the same connection as

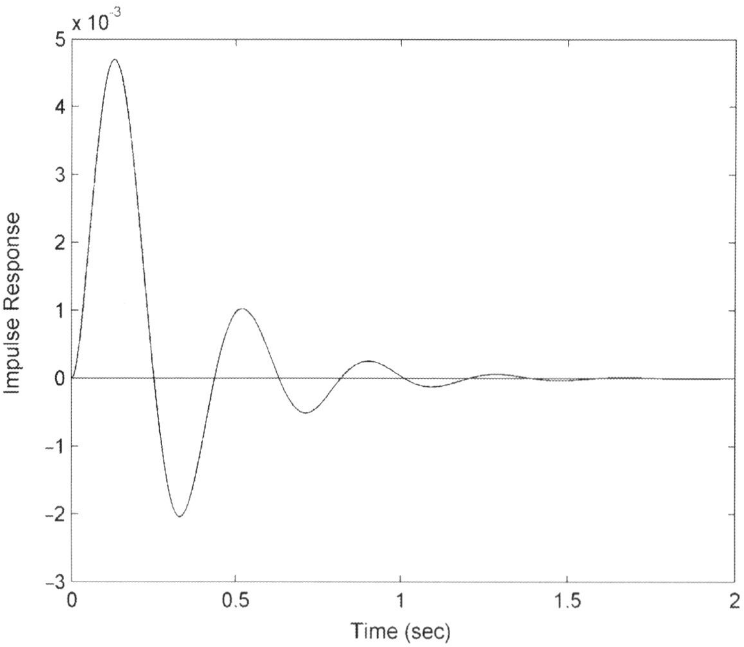

FIGURE 7.37 The impulse response of the model shown in Figure 7.35. A stimulus pulse that is equivalent to an impulse is found empirically by shortening the pulse width until there is no further change in response dynamics.

the scope in Figure 7.35), open the block in the usual manner (by double-clicking on the block) then enter the file name, `Ex7_11.mat` in the window labeled "File name:". In addition, set the "Save format" pull-down menu to "array." In this configuration, the element stores the time and the signal in a two-row matrix under the default name `ans` where time is in the first row and the signal is in the second.

One further modification is required if the simulated data are to be used in Fourier analysis. The default simulation routine does not use a fixed time step; the time step is adjusted depending on how fast model signals are changing. If the signals are changing slowly, the simulator uses longer time slices than when signals are changing rapidly. The Fourier transform (and most other signal analysis methods discussed in this text) require evenly spaced samples. To change the simulation approach, open the "Configuration parameters" window found in the "Simulation" pull-down menu. Under Solver options > Type, select "Fixed-step." This option allows you to set the step size. Set it to the convenient value of 0.001 sec, equivalent to a 1-kHz sampling rate ($f_s = 1/T_s$). After you run the simulation, use the MATLAB program below to calculate the magnitude and phase spectrum and plot responses.

```
% Ex 7_11. Determine the spectrum of a system from the impulse response.
%
load Ex7_11;                            % Load simulation results
t = ans(1,:);                           % Get time vector
x = ans(2,:);                           % Get model output
plot(t,x,'k'); hold on;                 % Plot simulation time response
plot([0 2],[0 0],'k');
  ......label......
%
N = length(x);                          % Get data length
fs = 1000;                              % Sampling frequency (set in simulation)
f = (1:N)*T;                            % Construct frequency vector for plotting
X = fft(x);                             % Take the Fourier transform of the impulse
N_2 = round(N/2);                       % Plot only valid spectral points
figure;
subplot(2,1,1);
Mag_dB = 20*log10(abs(X(1:N_2)));       % Convert magnitude to dB
semilogx(f(1:N_2),Mag_dB,'k');          % Plot magnitude spectrum
  ......labels, grid, axis limits......
subplot(2,1,2);
phase = unwrap(angle(X))*360/(2*pi);    % Convert phase to degrees
semilogx(f(1:N_2),phase(1:N_2),'k');    % and plot
  ......labels, grid, axis limits......
```

Results: The program loads the simulation output found in file `Ex7_11.mat` and puts the first row of `ans` (the default output variable name) in variable `t` and the second row in variable `x`. After plotting the time response, the program takes the Fourier transform of `x`, then calculates the magnitude in dB and phase in degrees. These two frequency components are plotted against log frequency in Hz. The resulting plot is shown in Figure 7.38. The magnitude curve shows a low-pass characteristic with a cutoff frequency of approximately 15 Hz and an attenuation of 60 dB/decade, as expected from a third-order system. Determining the spectral characteristics of a system through simulation is also explored in Problem 19 at the end of the chapter.

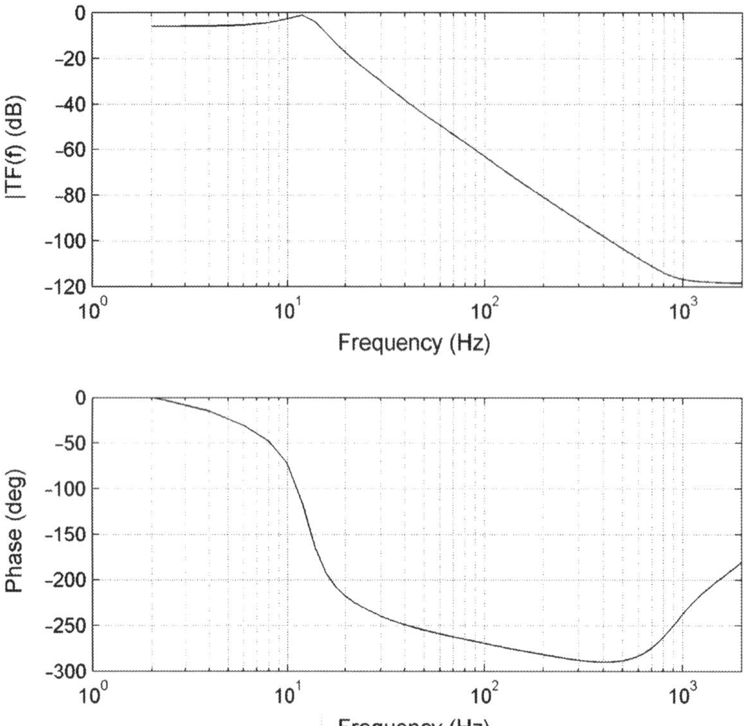

FIGURE 7.38 The spectral characteristics of the neuromuscular reflex system shown in Figure 7.35 as determined by taking the Fourier transform of the simulated impulse response.

7.7 SUMMARY

Convolution is a technique for determining the output of a system in response to any general input without leaving the time domain. The approach is based on the system's impulse response: The impulse response is used as a representation of the system's response to an infinitesimal segment of the input. If the system is linear and superposition holds, the impulse response from each input segment can be summed to produce the system's output. The convolution integral (Equations 7.1, 7.2, and 7.3) is basically a running correlation between the input signal and the impulse response. This integration can be cumbersome for complicated input signals or impulse responses, but is easy to program on a computer. Basic MATLAB provides two routines, `conv` and `filter`, to perform convolution. (The `conv` routine actually converts to the frequency domain, multiplies the two signals, then converts back to the time domain, and is faster than the `filter` routine.) Convolution is commonly used in signal processing to implement digital filtering, as is shown in Chapter 8.

Convolution requires knowledge of the system's impulse response. If the transfer function of the system is known or can be determined, then the impulse response can be obtained by taking the inverse of the Laplace domain transfer function (Equation 7.8). If the Fourier transform of the system is known, the impulse response is just the inverse

Fourier transform. Finally, if the system exists, then the impulse function can be determined directly by stimulating the system with an impulse.

Convolution performs the same function in the time domain as multiplication in the frequency domain. The reverse is also true: convolution in the frequency domain is like multiplication in the time domain. This fact is useful in explaining certain time-domain operations such as windowing and sampling.

Simulation is a powerful time-domain tool for exploring and designing complex systems. Simulation operates by automatically time slicing the input signal and determining the response of each element in turn until the final, output element is reached. This analysis is repeated time slice by time slice until a set time limit is reached. Since the response of each element is analyzed for each time slice, it is possible to examine the behavior of any element as well as the overall system behavior. Simulation also allows for incorporation of nonlinear elements such as the multiplier and limiter elements found in the model used in Example 7.9.

PROBLEMS

1. Use the basic convolution equation (Equation 7.1) to find the output of a system having an impulse response of $h(t) = 2e^{-10t}$ to an input that is a unit step function; that is, $x(t) = 1\ \mu(t)$. [*Hint*: If you reverse the step input, then for $t < 0$ there is no overlap and the output is 0. For $t > 0$, the integration limit is determined only by t, the leading edge of the reversed step function, so only a single integration, with limits from 0 to t, is required. See figure below.]

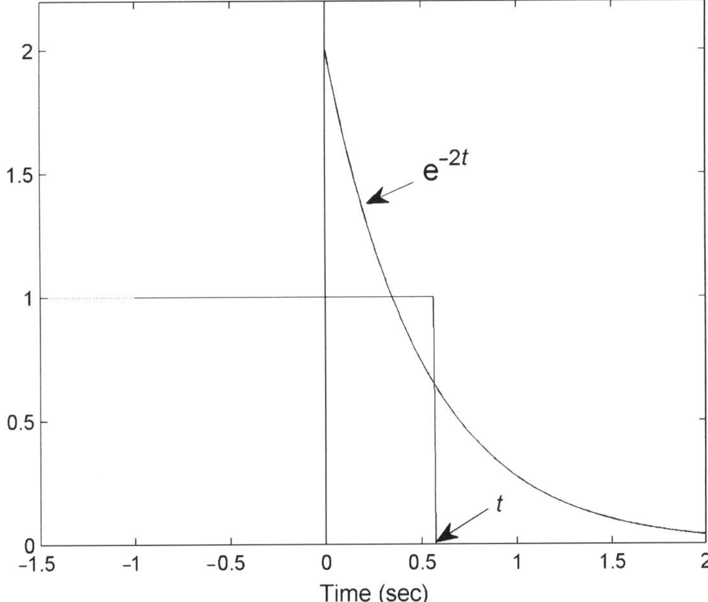

2. Use the basic convolution equation (Equations 7.1 and 7.2) to find the output of a system with an impulse response, $h(t) = 2(1 - t)$, to a 1-sec pulse having an amplitude of 1.

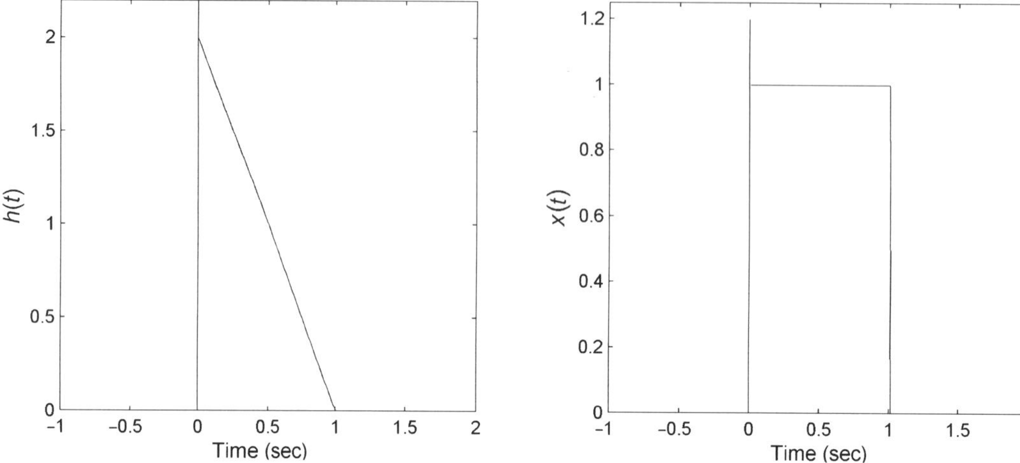

3. Use the basic convolution equation (Equations 7.1 and 7.2) to find the output of a system with an impulse response shown in the figure below to a step input with an amplitude of 5, that is, $x(t) = 5\,\mu(t)$.

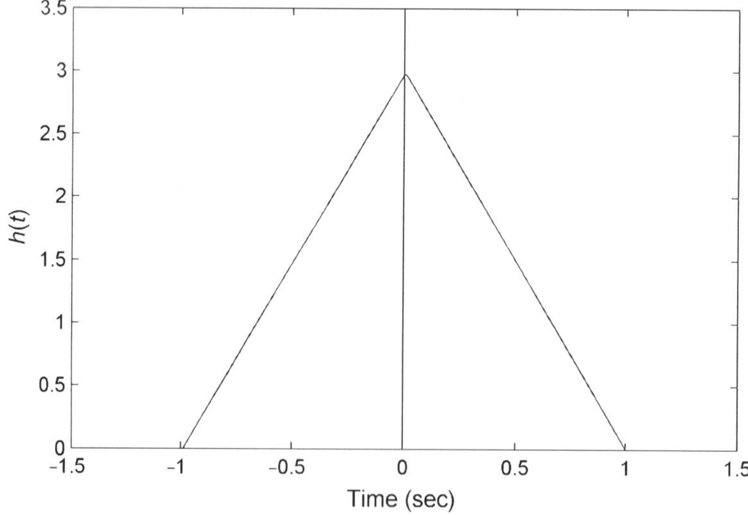

4. Find the convolution sum (Equation 7.3) for the discrete impulse response and discrete input signal shown in the figure below.

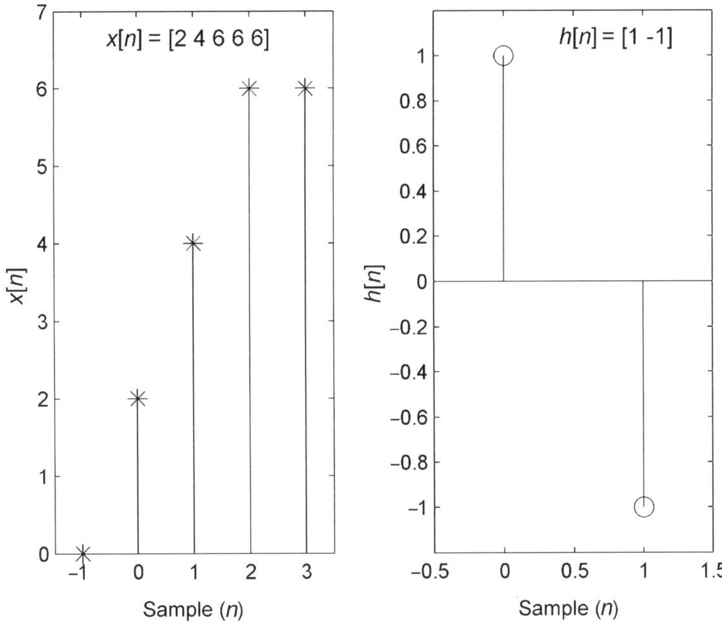

5. Find the convolution sum (Equation 7.3) for the discrete impulse response and discrete input signal shown in the figure below.

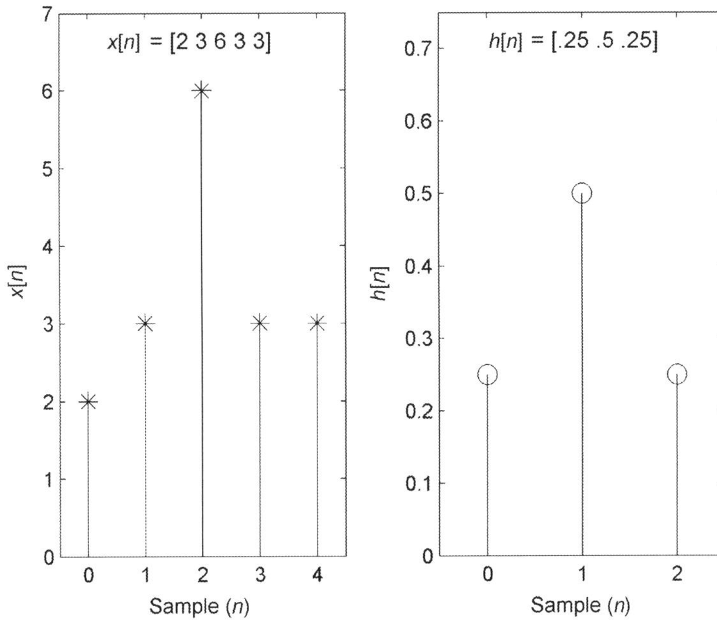

6. Use MATLAB and convolution to find the output of the second-order system used in Example 7.4 when the EEG signal is the input. The EEG signal is found as variable `eeg` in file `eeg_data.mat` on this book's accompanying CD and the impulse response is given in the example. Plot the second-order impulse response and the output of the convolution. Find the magnitude frequency spectrum of this system by taking the Fourier transform of the impulse function as given in the example. Plot only the nonredundant spectrum and correctly scale the frequency axis in Hz. Use the sampling frequency of the EEG signal which is 50 Hz. Note the strong attenuation of the system above about 5 Hz.

7. Systems known as filters are designed to alter the spectral characteristics of signals (discussed in Chapter 8). One such filter has an *impulse response* given by only four sequential data samples, as shown below. [*Note*: Filters often have short impulse responses.]

$$h[n] = \left[\frac{1+\sqrt{3}}{8} \quad \frac{1+\sqrt{3}}{8} \quad \frac{1-\sqrt{3}}{8} \quad \frac{1-\sqrt{3}}{8}\right]$$

a. Thus, the four samples are:

$$h[1] = \left[\frac{1+\sqrt{3}}{8}\right]; \quad h[2] = \left[\frac{1+\sqrt{3}}{8}\right]; \quad h[3] = \left[\frac{1-\sqrt{3}}{8}\right]; \quad h[4] = \left[\frac{1+\sqrt{3}}{8}\right]$$

Use convolution to apply this filter to a 1000-point *chirp* signal that increases from 0 to $f_s/4$ where $f_s = 1000$ Hz (i.e., 0–250 Hz). Plot the output of the convolution of the chirp signal with the four-point filter above. (Remember standard `conv` produces a data set which will be 3 points longer that the chirp signal, so either truncate the output of convolution to the length of the chirp signal or generate a longer time vector for plotting.) Since the chirp signal increases linearly with time, the time plot reflects the frequency characteristics of the system.

b. Determine and plot magnitude and phase frequency characteristics of this filter by taking the Fourier transform of the impulse response. Zero pad to 256 points (`fft(h,256)`) to improve the frequency plot. Plot the phase angle in degrees; you need to use the MATLAB `unwrap` function to get the correct phase plot.

This filter is known as a 4-coefficient *Daubechies filter*, and is one member of a family of filters used in Wavelet analysis.

8. Repeat Problem 2 using a simple 3-coefficient *moving average*-type filter. The impulse response for this filter is just three samples, each with a value of 1/3:

$$h[n] = \left[\frac{1}{3} \quad \frac{1}{3} \quad \frac{1}{3}\right]$$

Note that when this impulse response is applied to input data through convolution, it constructs an output that is an average of every three points in the input. Intuitively, this will reduce large point-to-point changes in the data. Such sharp changes are associated with high frequencies, so this should act as a low-pass filter. Plotting the

Fourier transfer in part B will confirm this. Finally, make note of the fact that even a very simple impulse response can have a significant impact on the data.

9. Repeat Problem 7 for one of the simplest of all filters having an impulse response consisting of only two samples each with a value of 1/2:

$$h[n] = \left[\frac{1}{2} \; \frac{1}{2}\right]$$

This filter is known as the *Haar filter* and while it is not a very strong filter, it is used in demonstrating some of the principles of Wavelet analysis.

10. Load the MATLAB data file `filter1.mat` and apply the impulse response variable `h` contained in that file using convolution with the EEG signal in `eeg_data.mat`. (Recall this file contains variable `eeg`.) Plot the EEG data before and after filtering, as in Example 7.3. What does this mystery filter do?
11. Load the MATLAB file `bp_filter` which contains the input response of a bandpass filter in variable `h`. Apply this filter to a chip signal used in Problems 7 through 9. Apply the filter in two ways:
    a. In the time domain using convolution.
    b. In the frequency domain using multiplication:
    Get the transfer function of the filter, and the frequency domain representation of the chirp, using the `fft`. Multiply the two frequency-domain functions (point-by-point), then take the inverse Fourier transform using MATLAB's `ifft` function to get the time-domain output. Compare the two time-domain responses. [*Hint*: To multiply the two frequency-domain functions together using point-by-point multiplication they must be at the same length, so when you take the `fft` of the filter impulse response, pad it to be the same length as the chirp signal. Also `ifft` produces a very small imaginary component in the output which will generate an error message in plotting, although the plot will still be correct. This error message can be eliminated by plotting only the real part of the output variable (e.g., `real(xout)`).]
12. Load the MATLAB data file `x_impulse.mat` containing the impulse response of an unknown system in variable `h`. Find the frequency characteristics of that unknown system in two ways: by probing the system with a series of sinusoids using convolution and by taking the Fourier transform of the impulse response, as in Example 7.5. When generating the sinusoids you will have to guess at how long to make them and what sampling rate to use. [*Suggestion*: A 1.0-second period and a sampling frequency of 2 kHz would appear to be a good place to start.]
13. The file `p7_13_data.mat` contains a waveform in variable `x` that serves as the input signal in this problem. This signal is the input to a system defined by the transfer function:

$$TF(\omega) = \frac{1000(s+10)}{s^2 + 10s + 400}$$

a. Convert the Laplace transfer function to the frequency domain ($s = j\omega$) and code it in MATLAB. Implement the transfer function using a 1000-point frequency vector ranging between 1 and 1000 Hz. Plot the transfer function.

Find the impulse response associated with this transfer function using the inverse Fourier transform, then find the system's output by convolving this impulse response with the input signal x. [*Note*: The inverse Fourier transform should be real but will contain imaginary components due to computational error, so take the real part before convolution.]

b. Confirm the result in the frequency domain. Take the Fourier transform of x, multiply by the transfer function, and take the inverse Fourier transform to get the output signal.

14. Use the code of Example 7.6 to evaluate the effect of a Blackman window function (Equation 7.15) on a long and a short spectrum obtained from the same signal ($N = 64$ and $N = 256$.) The spectra are found as variables `X_long` and `X_short` in file `spectrum2.mat`.

15. Use Simulink to find the step response of the system shown in Problem 10 of Chapter 6. Use values of 0.5 and 1 for k. [*Note*: To submit the MATLAB plot you can either print the scope screen or first store the data in a file using the "To File" block in the Sinks library as described in Example 7.11. Remember to set the "Save format" option of this block to "array" and select a filename. Alternatively you can store the output to your workspace using the "To Workspace" block. After you load the store files, you can make plots using all the plotting options provided in standard MATLAB.]

16. Use the approach of Examples 7.4 and 7.8 to find the maximum pulse width that can still be considered an impulse. Use Simulink to generate the pulse response of the system used in Problem 15 with the value of $k = 0.5$. To determine when responses are essentially the same dynamically, use the scope to measure carefully some dynamic characteristic of the response such as the time of the first peak. To get an accurate estimate of the time when the first peak occurs, use the magnifying feature on the scope. When two different pulse widths produce responses with the same peak time, the input pulse can be taken effectively as an impulse. Plot this impulse response.

17. Use Simulink and the neuromuscular reflex model developed by Khoo (2000) and presented in Example 7.10 to investigate the effect of changing feedback gain on final value. Set the feedback gain (Fbk gain in Figure 7.35) to values 10, 20, 40, 60, 100, 130, and 160. Measure the final error induced by the step-like addition of the 5-kg force: the same force used in Example 7.10. (Recall that the final value is a measure of the position error induced by the added force on the arm.) Lengthen the simulation time to 3 sec and use the scope output to find the value of the angle at the end of the simulation period. Plot the final angle (i.e., force-induced error) versus the feedback gain using standard MATLAB. Label the axes correctly.

18. Use Simulink and the neuromuscular reflex model developed by Khoo (2000) and presented in Example 7.10 to investigate the effect of changing feedback gain on response overshoot. Set the feedback gain to values: 40, 60, 100, 130, and 160. Use the scope to measure the response overshoot in percent: the maximum overshoot divided by the final angle. Measure the percent overshoot for three values of delay: 0, 20, and

30 msec. If there is no overshoot, record it as 0%. Plot percent overshoot versus feedback gain for the various delays using standard MATLAB. Label the axes correctly.

19. Use Simulink to simulate the fourth-order transfer function used in Example 6.11 and repeated here:

$$TF(s) = \frac{s^3 + 5s^2 + 3s + 10}{s^4 + 12s^3 + 20s^2 + 14s + 10}$$

This can be simulated with a single element (i.e., Transfer Fcn) but be sure to set the Solver options (Type in the Configuration Parameters window) to "Fixed-step," as described in Example 7.11. Save the output as a file using the "To File" block, then load the file into a standard MATLAB program and take the Fourier transform. Plot the magnitude in dB and phase in degrees against log frequency. Scale frequency to radians and limit the frequency axis to between 0.1 and 100 to facilitate comparison with Figure 6.15. Note that while the shapes are about the same, the magnitude curve is higher since the MATLAB `fft` routine does not scale the output.

20. The figure below shows a model of the vergence eye movement control system developed by Krishnan and Stark in the 1960s. (Vergence eye movements are those that move the eyes in opposite directions to view visual targets at different distances.) The lower path is called *derivative control* and functions to increase the speed of the response. Use simulation to find the step function when the "Gain" element in the derivative path has a gain of 0. The response will be relatively slow due to limiting dynamics of the eye muscles. Increase the value of the derivative "Gain" element until the response just begins to overshoot and note the greatly improved speed of the

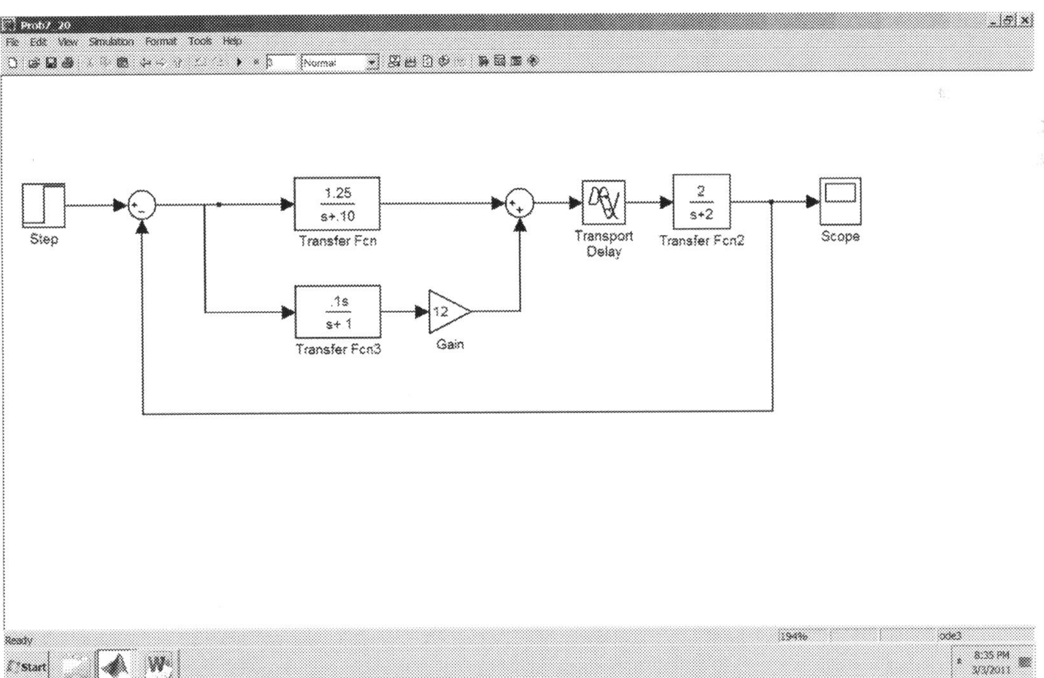

response. From the scope output, estimate the time constant for your optimal derivative gain and that for a derivative gain of 0. (i.e., no derivative control). Plot the responses to your optimal value of derivative gain and for no derivative control.

21. The model shown below is a two-compartment circulatory pharmacokinetic model developed by Richard Upton of the University of Adelaide in Australia. This model has interacting compartments and can become unstable if the rate constants become imbalanced. Simulate the model using Simulink for the following parameter values: k21 = k21a = 0.2; k12 = 0.5; Gain = 0.5; Gain1 = 0.1. Then reduce the parameter k21a to 0.15 and note that the model becomes unstable with an exponential increase in compartment concentrations. Find the lowest value of k21 for which the model is stable.

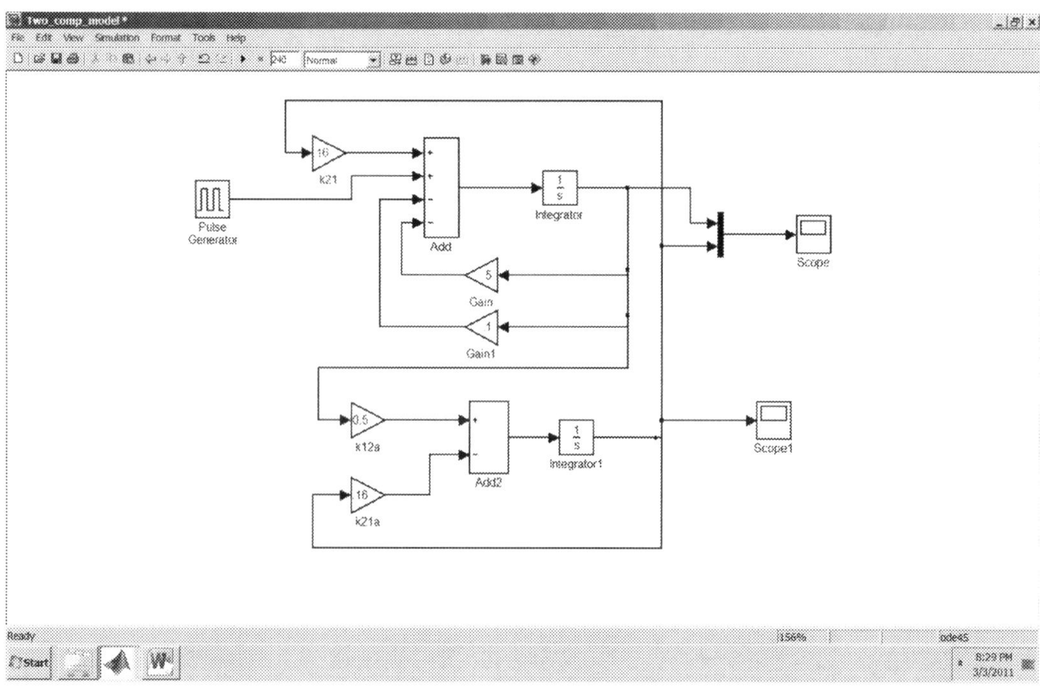

CHAPTER

8

Linear System Analysis: Applications

8.1 LINEAR FILTERS—INTRODUCTION

Filtering is a process that uses systems, usually LTI systems, to alter a signal in some desired manner. Linear filters are standard linear (LTI) systems, but they are called *filters* because their reason for being is to alter the spectral shape of the signal passing through them. Filters are often implemented in the time domain using convolution, although filtering is conceptualized in the frequency domain since the objective of filtering is to reshape a signal's spectrum. A frequent goal of filtering is to reduce noise. Most noise occurs over a wide range of frequencies (i.e., it is broadband), while most signals have a limited frequency range (i.e., they are narrowband). By reducing the energy at frequencies outside the signal range, the noise is reduced but not the signal, so the signal-to-noise ratio (SNR) is improved. In short, a filter is usually just an LTI system that reshapes a signal's spectrum in some well-defined, and we hope beneficial, manner.

Example 7.1 and some of the problems in Chapter 7 show how a signal's spectral properties can be altered substantially by a linear system. Example 7.1 uses convolution to implement the system, and this approach applies equally to filters. However, what we need to learn is how to construct the linear system that produces the desired change in a signal's spectrum. Construction of an appropriate system is termed *filter design* and is a principal topic of this chapter.

All linear filters are systems that act to produce some desired spectral modification. However, these systems differ in the way they achieve this spectral reshaping. Systems for filtering can be classified into two groups depending on their impulse response characteristics: those that have short well-defined impulse responses and those that produce impulse responses that theoretically go on forever. The former are termed *finite impulse response* (*FIR*) filters, while the latter are called *infinite impulse response* (*IIR*) filters. The advantages and disadvantages of each of the filter types are summarized in the next section. Both filter types are easy to implement in basic MATLAB; however, the design of IIR filters is best done using the MATLAB Signal Processing Toolbox and is not covered here.

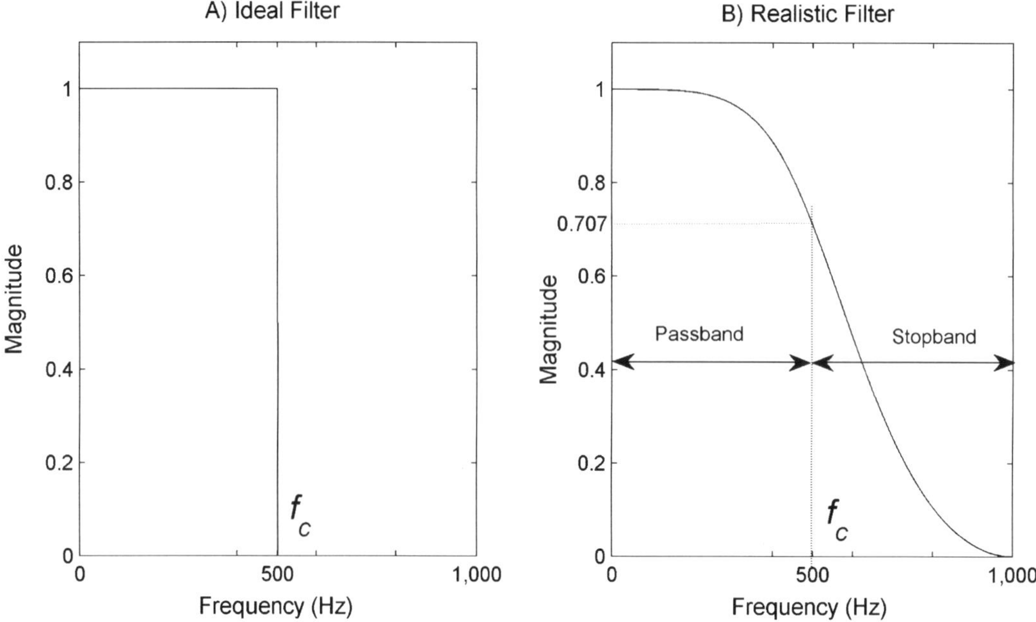

FIGURE 8.1 A) The magnitude spectrum of an "ideal" filter that has an infinitely steep attenuation at the cutoff frequency of 500 Hz. B) A realistic filter with the same cutoff frequency. As with signals, the cutoff frequency and the bandwidth are determined when the filter attenuation is 0.707 or −3 dB of the nominal gain of the filter. In filters, the nominal gain is known as the *passband gain*.

8.1.1 Filter Definitions

The function of filters is to change the spectral characteristics of a signal, and these spectral changes are set by the spectral characteristics of the filter. While you might think that a filter could have any imaginable magnitude and phase curve, in practice the variety of spectral characteristics of most filters is limited. Filter spectra can be defined by three basic properties: bandwidth, basic filter type, and attenuation slope.

8.1.1.1 Bandwidth

Bandwidth is introduced in Chapter 4 in terms of signals, but the same definitions apply to systems. The magnitude spectra of an ideal and a real filter are shown in Figure 8.1. Both filters have a cutoff frequency, f_c, of 500 Hz. In the case of the ideal filter, Figure 8.1A, the attenuation slope is infinite at 500 Hz so that all the frequencies below 500 Hz are "passed" through the filter and all those above 500 Hz are completely blocked. For the ideal filter, the cutoff frequency is obvious; it is where the infinite attenuation slope occurs. It is not possible to build a filter with an infinite attenuation slope, and the magnitude spectrum of a more realistic filter is shown in Figure 8.1B. Here signal attenuation occurs gradually as the frequency of the signal increases. The cutoff frequency is taken where the filter reduces the signal by 3 dB from the unattenuated level as defined in Section 4.5 of Chapter 4.

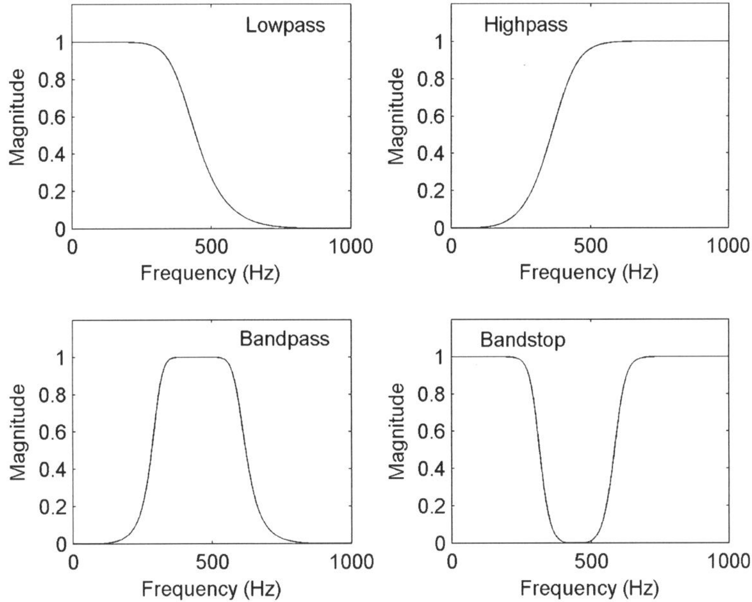

FIGURE 8.2 Magnitude spectral curves of the four basic filter types. A) Low-pass $f_c = 400$ Hz; B) high-pass $f_c = 400$ Hz; C) bandpass $f_c = 300$ and 600 Hz; and D) bandstop $f_c = 300$ and 600 Hz.

For a filter, the nominal gain region—the frequency region where the signal is not attenuated more than 3 dB—is termed the *passband,* and the gain of the filter in this region is called the *passband gain*. The passband gain of both filters in Figure 8.1 is 1.0 although the gain decreases toward the 3-dB point in the realistic filter, (Figure 8.1B). The frequency region where the signal is attenuated more than 3 dB is termed the *stopband*.

8.1.1.2 Filter Types

Filters are usually named according to the range of frequencies they do not suppress, in other words, the passband frequency range. Thus, a *lowpass* filter allows low frequencies to pass with minimum attenuation, while higher frequencies are attenuated. Conversely, a *high-pass* filter permits high frequencies to pass, but attenuates low frequencies. *Bandpass* filters allow a range of frequencies through, while frequencies above and below this range are attenuated. An exception to this terminology is the *bandstop* filter, which passes frequencies on either side of a range of attenuated frequencies: it is the inverse of the bandpass filter. The frequency characteristics of the four filter types are shown in Figure 8.2.

8.1.1.3 Filter Attenuation Slope—Filter Order

Filters are also defined by the sharpness with which they increase or decrease attenuation as frequency varies. The attenuation slope is sometimes referred to as the filter's *rolloff*. Spectral sharpness is specified in two ways: as an initial sharpness in the region where attenuation first begins and as a slope further along the attenuation curve.

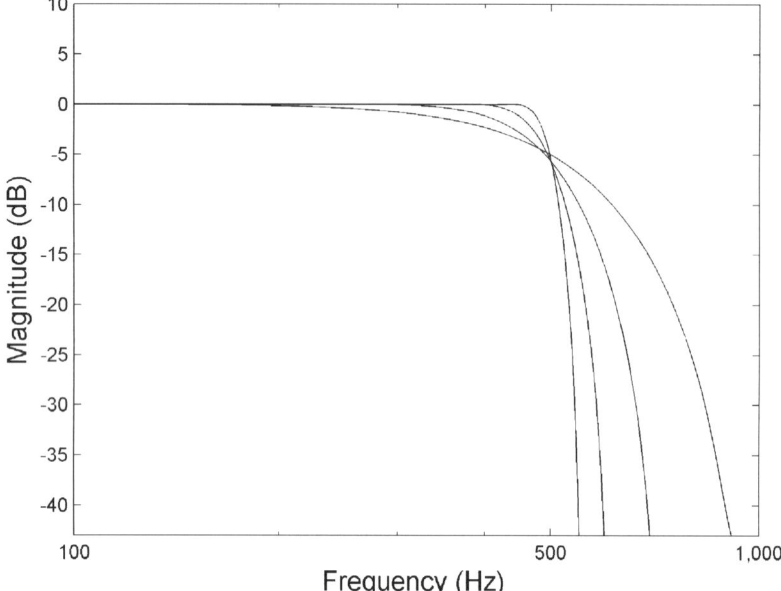

FIGURE 8.3 The magnitude spectral characteristics of four FIR filters that have the same cutoff frequency, $f_c = 500$ Hz, but increasing filter orders of 8, 16, 32, and 64. The highest filter order corresponds to the steepest attenuation slope.

For FIR filters the attenuation slope is directly related to the length of the impulse response, that is, the number of coefficients or samples that are used to represent the impulse response. In FIR filters, the filter order is defined as one less than the number of samples in the impulse response. Figure 8.3 shows the spectral characteristics of four FIR filters that have the same cutoff frequency but increasing filter orders, and the increase in attenuation slope is apparent.

The attenuation slope of IIR filters follows the pattern found for basic transfer functions analyzed in Chapter 5. The magnitude slope is determined by the order of the denominator polynomial: a first-order system has a slope of 20 dB/decade while a second-order system has a slope of 40 dB/decade. This generalizes to higher-order IIR filters so that an nth-order IIR filter has an attenuation slope of $20n$ dB/decade. In Chapter 10 we examine analog circuits and find that the order of a circuit's denominator polynomial is related to the complexity of the electrical circuit, specifically the number of energy storage elements in the circuit.

8.1.1.4 Filter Initial Sharpness

For both FIR and IIR filters, the attenuation slope increases with filter order, although for FIR filters it does not follow the neat 20 dB/decade per filter order slope of IIR filters. For IIR filters, it is also possible to increase the *initial* sharpness of the filter's attenuation characteristics without increasing the order of the filter, if you are willing to accept some unevenness, or *ripple*, in the passband. Figure 8.4 shows two low-pass, fourth-order IIR filters, differing in the initial sharpness of the attenuation. The one marked Butterworth has

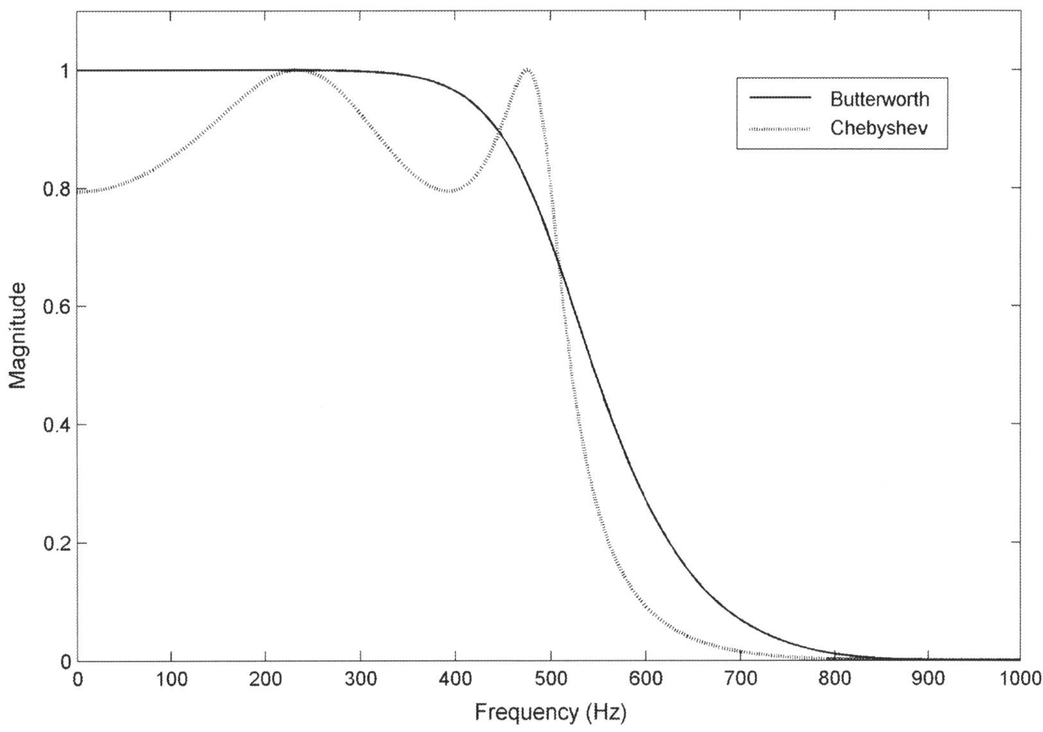

FIGURE 8.4 Two IIR filters having the same order (4-pole) and cutoff frequency, but differing in the sharpness of the initial slope. The filter marked "Chebyshev" has a much steeper initial slope or rolloff, but contains ripple in the passband. The final slope also looks steeper, but would not be if it were plotted on a log scale.

a smooth passband, but the initial attenuation is not as sharp as the one marked Chebyshev, which has a passband that contains ripples. This figure also suggests why the Butterworth filter is popular in biomedical engineering applications: it has the greatest initial sharpness while maintaining a smooth passband. For most biomedical engineering applications, a smooth passband is needed.

8.1.2 FIR versus IIR Filter Characteristics

There are several important behavioral differences between FIR and IIR filters aside from the characteristics of their impulse responses. FIR filters require more coefficients to achieve the same filter slope. Figure 8.5A compares the magnitude spectra of a 4th-order IIR filter with a 100th-order FIR filter having the same cutoff frequency of 50 Hz. They both have approximately the same slope, but the FIR filter requires 10 times the coefficients. (A 4th-order IIR filter has 10 coefficients and a 100th-order IIR filter has 101 coefficients.) In general, FIR filters do not perform as well for low cutoff frequencies as is the case here, and the discrepancy in filter coefficients is less if the cutoff frequency is

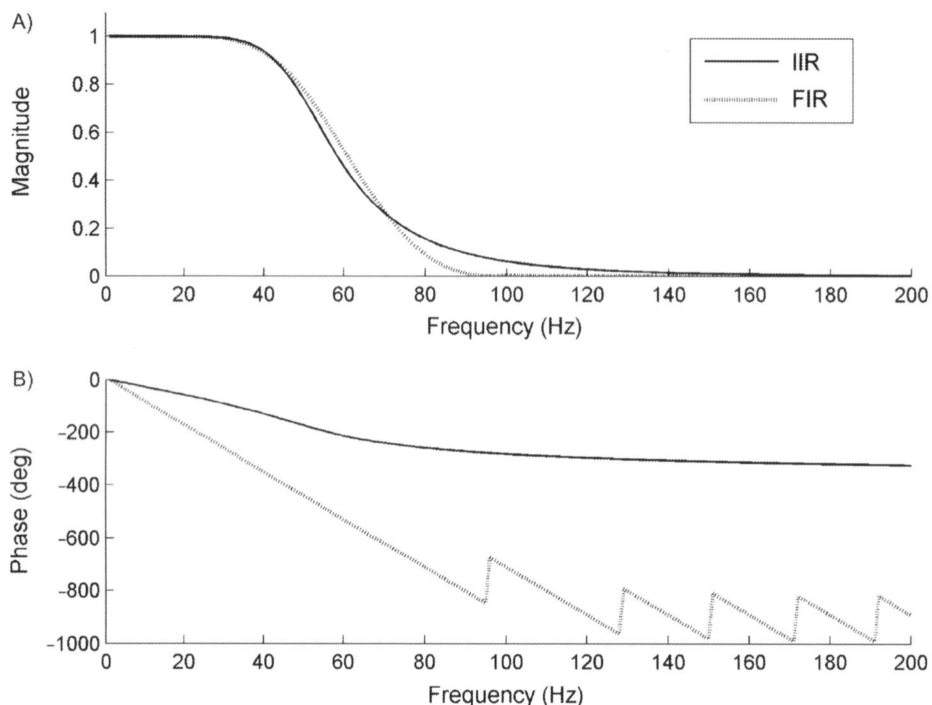

FIGURE 8.5 Comparison of the magnitude and phase spectra of a 100th-order FIR filter and an 4th-order IIR filter, both having a cutoff frequency, f_c, of 500 Hz. A) The magnitude spectra show approximately the same attenuation slope but the FIR filter requires 10 times the number of coefficients (an 4th-order IIR filter has 10 coefficients). B) The phase spectra show that the FIR filter has a greater phase change, but that change is linear within the passband. This turns out to be an important feature of FIR filters.

increased. The behavior is explored in the problem set. Another difference is the phase characteristics, as shown for the two filters in Figure 8.5B. The FIR has a larger change in phase, but the phase change is linear within the passband region of 0 to 50 Hz.

Figure 8.6 shows the application of the two filters to a fetal ECG signal that contains noise. Both low-pass filters do a good job of reducing the high-frequency noise, but the FIR filter consists of 10 times the number of coefficients, and the computer time to apply this filter is correspondingly longer. The filtered data are also shifted to the right in comparison to the original signal, more so for the FIR filtered data. This is because of the causal nature of the two filters and is further explored in Example 8.1 below.

The linear phase property of FIR filters shown in Figure 8.5B makes FIR filters much easier to design since we need only be concerned with the magnitude spectrum of the desired filter. FIR filters are also more stable than IIR filters. We show later that IIR filters are recursive in nature; this can produce unstable oscillatory outputs for certain combinations of filter coefficients. This is not a problem if the coefficients are fixed as in standard filters, but in adaptive filters where filter coefficients are modified on-the-fly, severe

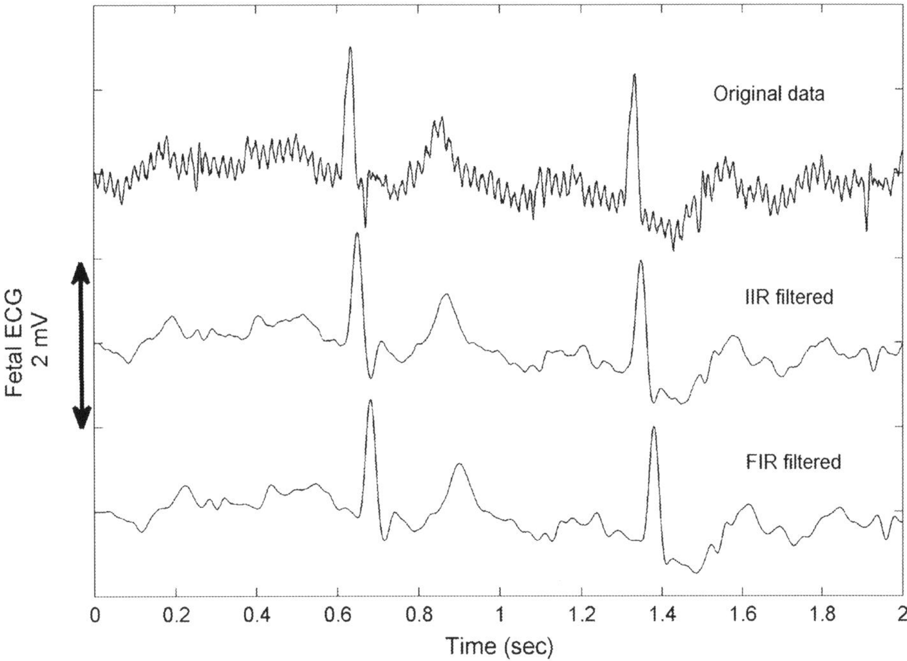

FIGURE 8.6 A fetal ECG signal recorded from the abdomen of the mother. The 60-Hz noise in the original is considerably reduced in both filtered signals, but the FIR filter requires more than 10 times the data samples and a proportionally longer computation time. Both filters induce a time shift because they are causal filters, as described below.

TABLE 8.1 FIR versus IIR Filters: Features and Applications

Filter Type	Features	Applications
FIR	Easy to design	Fixed, one-dimensional filters
	Stable	Adaptive filters
	Applicable to two-dimensional data (i.e., images)	Image filters
IIR	Requires fewer coefficients for the same attenuation slope, but can become unstable	Fixed, one-dimensional filters, particularly at low cutoff frequencies
	Particularly good for low cutoff frequencies	Real-time applications where speed is important
	Mimics analog (circuit) filters	

problems can occur. For this reason adaptive filters consist of FIR filters. IIR filters are also unsuitable for filtering two-dimensional data, so FIR filters are used for filtering images. A summary of the benefits and appropriate applications of the two filter types is given in Table 8.1.

8.2 FINITE IMPULSE RESPONSE (FIR) FILTERS

FIR filters have finite impulse responses that are usually fairly short, often consisting of less than 100 points. FIR filters are implemented using convolution, so in MATLAB either the `conv` or `filter` routine can be used.

The general equation for implementation of an FIR filter is the discrete form of a convolution equation presented in Equation 7.3 in Chapter 7 and repeated here:

$$y[n] = \sum_{k=1}^{K} b[k]x[n-k] \tag{8.1}$$

where $b[k]$ is the impulse response, also called the *filter coefficients* or *weighting function*, K is the filter length, $x[n]$ is the input, and $y[n]$ is the output. As we found in the last chapter, the spectrum of any system can be obtained by taking the Fourier transform of the impulse response, in this case the coefficients $b[k]$ (Equation 7.9 and repeated here for a discrete impulse response):

$$X[m] = \sum_{k=0}^{K-1} b[k]e^{-j2\pi mn/N} = FT\{\mathrm{b[k]}\} \tag{8.2}$$

A simple FIR filter is the *moving average*, where each sample in the filter output is the average of the last N points in the input. Such a filter, using only the current and past data, is a *causal* filter. As noted previously, all real-time systems must be causal since they do not have access to the future; they have no choice but to operate on current and past values of a signal. However, if the data are already stored in a computer, it is possible to use future signal values along with current and past values to compute an output signal; that is, future with respect to a given sample in the output signal. Filters (or systems) that use future values of a signal in their computation are *noncausal*.

The motivation for using future values in filter calculations is provided in Figure 8.7. The upper curve in Figure 8.7A is the response of the eyes to a target that jumps inward in a step-like manner. (The curve is actually the difference in the angle of the two eyes with respect to the straight-ahead position.) These eye movement data are corrupted by 60-Hz noise riding on top of the signal, a ubiquitous problem in the acquisition of biological signals. A simple 10-point moving average filter was applied to the noisy data and this filter does a good job of removing the noise as shown in the lower two curves. The 10-point moving average filter was applied in two ways: as a causal filter using only current and past values of the eye movement data (lower curve, Figure 8.7A) and as a noncausal filter using an approximately equal number of past and future values (middle curve, Figure 8.7A). The causal filter was implemented using MATLAB's `conv` routine and the noncausal filter using the `conv` routine with the option `'same'`:

```
y = conv(x,b,'same'); % Apply a noncausal filter.
```

where `x` is the noisy eye movement signal and `b` is an impulse response consisting of 10 equal value samples, `[1 1 1 1 1 1 1 1 1 1]/10`, to create a moving average. When the `same` option is invoked, the convolution algorithm returns only the center section of output, effectively shifting future values into the present.

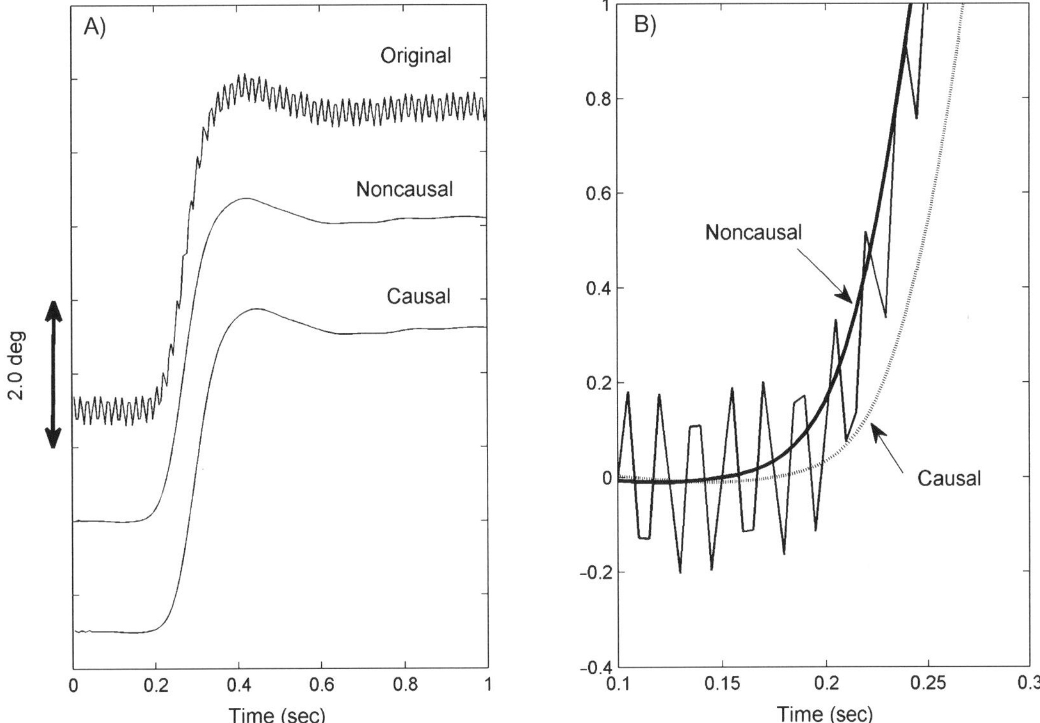

FIGURE 8.7 A) Eye movement data containing 60-Hz noise and the same response filtered with a 10-point moving average filter applied in a causal and noncausal mode. B) Detail of the initial response of the eye movement showing the causal and noncausal filtered data superimposed. The noncausal filter overlays the original data while the causal filter produces a time shift in the response because it operates only on past values.

Both filters do a good job of reducing the 60-cycle noise, but the causal filter has a slight time delay. In Figure 8.7B, the initial responses of the two filters are plotted superimposed over the original data and the delay in the causal filter is apparent. Eliminating the time shift inherent in causal filters is the primary motivation for using noncausal filters. In Figure 8.7, the time shift in the output is small, but can be much larger if filters with longer impulse responses are used or if the data are passed through multiple filters. However, in many applications a time shift does not matter, so a noncausal filter is adequate.

The noncausal implementation of a 3-point moving average filter is shown in Figure 8.8. This filter takes one past and one current sample from the input along with one future sample to construct the output sample. In Figure 8.8, a slightly heavier arrow indicates the future sample. This operation can be implemented in MATLAB using an impulse function of `[1/3 1/3 1/3]` and the `conv` routine with option `same`.

The frequency spectrum of any FIR filter can be determined by taking the Fourier transform of its impulse response. This applies to the 3-point moving average filter illustrated in Figure 8.8 or the 10-point moving average filter used in Figure 8.7 as well as any other

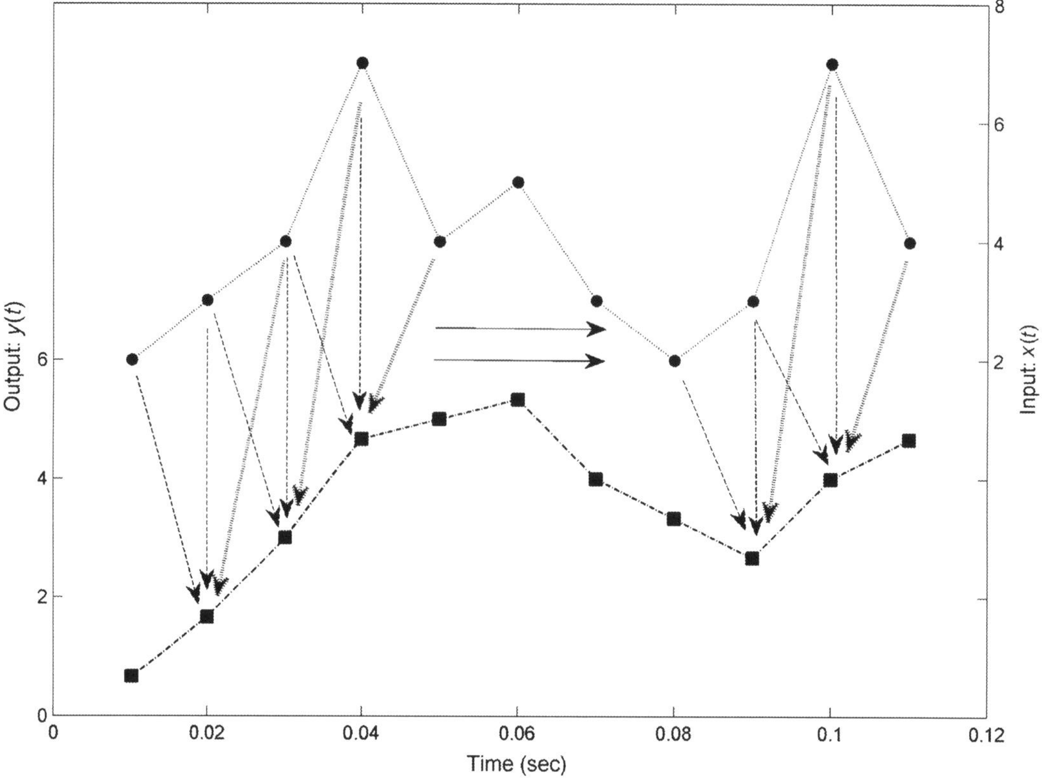

FIGURE 8.8. The procedure for constructing a 3-point moving average. The average of every 3 points from the input (upper trace) makes up the output point (lower trace). One point is taken from the past, one from the present, and one from the future (slightly darker arrow). This averaging process slides across the input signal to produce the output. Endpoints are usually treated by applying zero-padding to the input. This process is identical to convolving the input with an impulse response of [1/3 1/3 1/3] in conjunction with symmetrical convolution and produces a noncausal filter. The time axis is scaled assuming a sampling frequency of 100 Hz.

FIR filter. However, since the impulse responses of many FIR filters can be quite short (only 3 points for a 3-point moving average filter), it is appropriate to pad the response before taking the Fourier transform. The spectra of a 3-point and a 10-point moving average filter is determined in the next example.

EXAMPLE 8.1

Find the frequency characteristics of a 3-point moving average filter and a 10-point moving average filter. Assume a sampling frequency of 250 Hz.

Solution: Construct the two filter impulse responses, then take the Fourier transform using `fft`, but pad the data out to 256 points to get smooth plots. Plot the magnitude and phase of the complex Fourier transform provided by fft. Construct an accurate frequency vector based on the 250-Hz sampling rate. Plot only the valid data points (i.e., up to $f_s/2$).

```
%Example 8.1 Spectra of moving-average FIR filter
%
fs = 250;                    % Sampling frequency
% Construct moving average filters
b3 = [1 1 1]/3;
b10 = [1 1 1 1 1 1 1 1 1 1]/10;  % Construct filter impulse response
%
B3 = fft(b3,256);            % Find filter system spectrum
B3_mag = abs(B3);            % Compute magnitude spectrum
Phase3 = angle(B3);          % Compute phase angle
Phase3 = unwrap(Phase3);     % Remove 2 pi wrap around
Phase3 = Phase3*360/(2*pi);  % Phase angle in degrees
    ..........repeat for b10 and plot the magnitude and phase spectra..........
```

Results: Figure 8.8 shows the magnitude and phase spectra of the two filters assuming a sampling frequency of 250 Hz. Both are low-pass filters with a cutoff frequency of approximately 40 Hz for the 3-point moving average and approximately 16 Hz for the 10-point moving average filter. In addition to a lower cutoff frequency, the 10-point moving average filter has a much sharper rolloff. The phase characteristics decrease linearly from 0 to approximately 160 degrees in a repetitive manner.

The impulse response of the moving average filters used in Example 8.1 consists of equal value points; however, there is no reason to give the filter coefficients the same value. We can make up any impulse response we choose and quickly determine its spectral characteristics using the Fourier transform, as in Example 8.1. The problem is how do we make up an impulse response that does what we want it to do in the frequency domain? This inverse operation, going from a desired frequency response to an impulse response, $b[k]$, is known as *filter design*. We could use trial and error, and sometimes that is the best approach, but in the case of FIR filter design there are straightforward methods to get from the spectrum we want to the appropriate filter coefficients. Just as the frequency response of the filter is the Fourier transform of the filter coefficients (Equations 7.9 and 8.2), the desired impulse response is the inverse Fourier transform of the desired frequency response (Equation 7.10). This design strategy is illustrated in the next section.

8.2.1 FIR Filter Design and Implementation

To design a filter, we start with the desired frequency characteristic. Then all we need to do is take the inverse Fourier transform. We only need the magnitude spectrum because FIR filters have linear phase so the phase is uniquely determined by the magnitude spectrum. While there may be circumstances where some specialized frequency characteristic is needed, usually we just want to separate out a desired range of frequencies from everything else and we want that separation to be as sharp as possible. In other words, we usually prefer an ideal filter such as that shown in Figure 8.1A with a particular cutoff

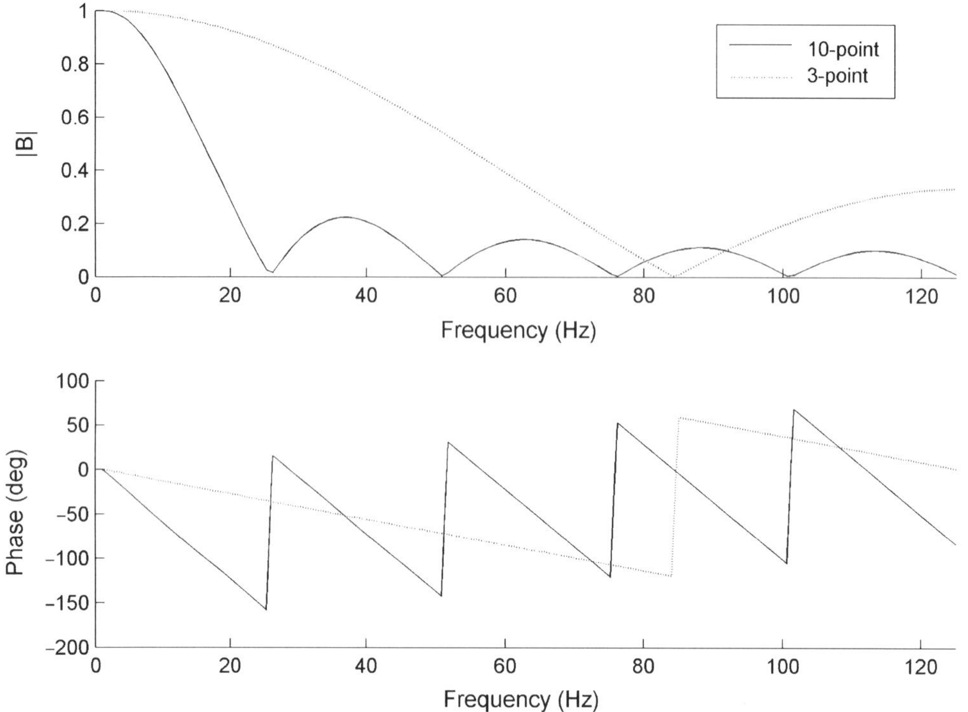

FIGURE 8.9 The magnitude (upper) and phase (lower) spectra of a 3-point and a 10-point moving average filter. Both filters show low-pass characteristics with a cutoff frequency, f_c, of approximately 40 Hz for the 3-point moving average and approximately 16 Hz for the 10-point moving average assuming the sampling frequency of 250 Hz. The phase characteristics change linearly and repetitively between 0 and approximately −160 degrees (Note the jumps in the phase curve are characteristic of the filter and are not due to transitions greater than 360 deg).

frequency. The ideal low-pass filter in Figure 8.1A has the frequency characteristics of a rectangular window and is sometimes referred to as a *rectangular window* filter.[1]

This spectrum of an ideal filter is a fairly simple function, so we ought to be able to find the inverse Fourier transform analytically from the defining equation given in Chapter 3 (Equation 3.29). (This is one of those cases where we are not totally dependent on MATLAB.) When using the complex form, we should must include the negative frequencies, so the desired filter's frequency characteristic is as shown in Figure 8.10. Frequency is shown in radians/sec to simplify the derivation of the rectangular window impulse response.

To derive the impulse response of a rectangular window filter, we apply the inverse discrete time Fourier transform equation (Equation 3.41) to the window function in Figure 8.9.

[1]This filter is sometimes called a *window filter*, but the term "*rectangular window filter*" is used in this text so as not to confuse such a filter with a *window function*, as described in Chapter 4. This can be particularly confusing since, as shown below, rectangular window filters use window functions!

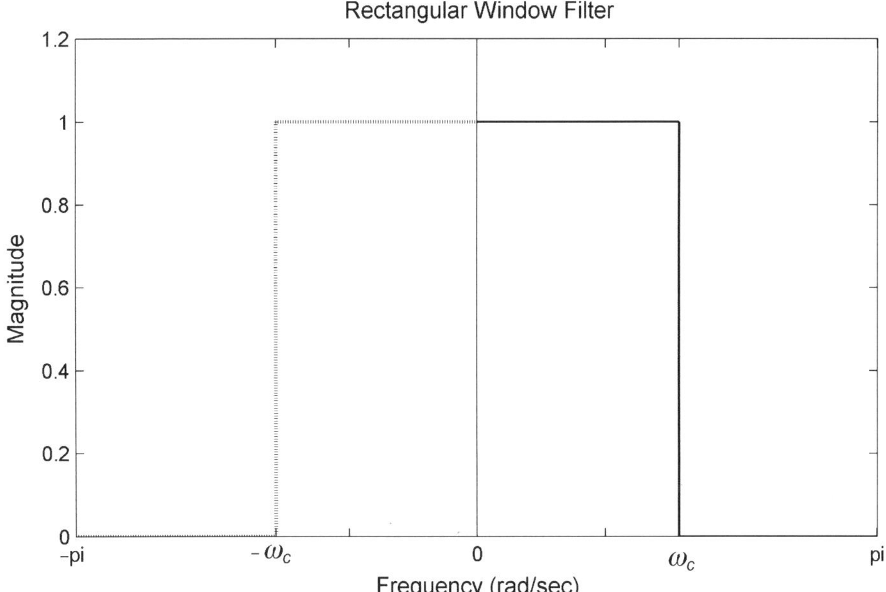

FIGURE 8.10. The magnitude spectrum of a rectangular window filter, which is an ideal filter. The negative frequency portion of this filter is also shown because it is needed in the computation of the complex inverse Fourier transform. The frequency axis is in radians per second and normalized to a maximum value of 1.0.

Since frequency is in radians, we should substitute $k\omega$ for $2\pi nf$ Equation 3.41 becomes:

$$b[k] = \frac{1}{2\pi}\int_{-\pi}^{\pi} B[\omega]e^{jk\omega}\,d\omega = \frac{1}{2\pi}\int_{-\omega_c}^{\omega_c} 1e^{jk\omega}\,d\omega \tag{8.3}$$

since the window function is 1.0 between $\pm\omega_c$ and zero elsewhere. Integrating and putting in the limits:

$$b[k] = \frac{1}{2\pi}\frac{e^{jk\omega}}{jk}\Bigg|_{-\omega_c}^{\omega_c} = \frac{1}{\pi k}\frac{e^{jk\omega_c} - e^{-jk\omega_c}}{2j} \tag{8.4}$$

The term $\frac{e^{jk\omega_c} - e^{-jk\omega_c}}{2j}$ is the exponential definition of the sine function and equals $\sin(k\omega)$. So the impulse response of a rectangular window is:

$$b[k] = \frac{\sin(k\omega_c)}{\pi k} = \frac{\sin(2\pi f_c k)}{\pi k} \tag{8.5}$$

The impulse response of a rectangular window filter has the general form of a *sinc* function: $\sin(x)/x$. The impulse response, $b[k]$ produced by Equation 8.5 is shown for 64 samples and two values of f_c in Figure 8.11. These cutoff frequencies are relative to the sampling frequency, as explained below.

From Figure 8.10 we can see that the cutoff frequency, ω_c, is relative to a maximum frequency of π. Converting frequency to Hz, the maximum frequency of sampled data is the

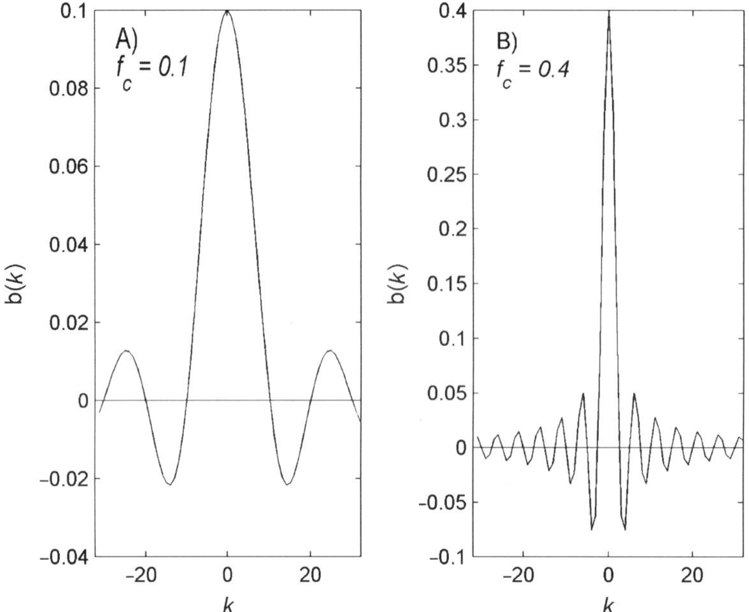

FIGURE 8.11 The impulse response of a rectangular window filter (Equation 8.5) for $L = 64$ coefficients. The cutoff frequencies are given relative to the sampling frequency, f_s, as is common for digital filter frequencies. A) Low-pass filter with a relative cutoff frequency of 0.1. B) Low-pass filter with a higher relative cutoff frequency of 0.4.

sampling frequency, f_s. Frequency specifications for digital filters are usually described relative to either the sampling frequency, f_s, or to half the sampling frequency $f_s/2$. (Recall that $f_s/2$ is also referred to as the Nyquist frequency.) In Equation 8.5, the cutoff frequency, f_c, is relative to the sampling frequency, f_s:

$$f_{\text{actual}} = f_c\, f_s \tag{8.6}$$

In the MATLAB filter design routines included in the Signal Processing Toolbox, the cutoff frequencies are relative to $f_s/2$, the Nyquist frequency: $f_{\text{actual}} = f_c(f_s/2)$.

The symmetrical impulse response shown in Figure 8.11 has both positive and negative values of k. Since MATLAB requires indexes to be positive, we need to shift the index, k, on the right side of Equation 8.5 half the total length of the filter. To be compatible with MATLAB, Equation 8.5 becomes:

$$b[k] = \frac{\sin[\omega_c(k - L/2)]}{\pi(k - L/2)} = \frac{\sin[2\pi f_c(k - L/2)]}{\pi(k - L/2)} \tag{8.7}$$

where f_c is the cutoff frequency relative to f_s and L is the length of the filter. If L is odd, then an adjustment is made so that $L/2$ is an integer, as shown in the next example.

When Equation 8.7 is implemented, a problem occurs when $k = L/2$ and the denominator goes to zero. The actual value of the function for $k = L/2$ can be obtained by applying the limits and noting that $\sin(x) \to x$ as x becomes small:

$$b[L/2] = \lim_{(k-L/2)\to 0} \left| \frac{\sin[2\pi f_c(k - L/2)]}{\pi(k - L/2)} \right. = \lim_{(k-L/2)\to 0} \left| \frac{2\pi f_c(k - L/2)}{\pi(k - L/2)} \right. = 2f_c \tag{8.8}$$

If frequency is in radians/sec, the value of $b[L/2]$ is:

$$b[L/2] = \lim_{(k-L/2)\to 0} \left| \frac{\sin[\omega_c(k-L/2)]}{\pi(k-L/2)} \right| = \lim_{(k-L/2)\to 0} \left| \frac{\omega_c(k-L/2)}{\pi(k-L/2)} \right| = \frac{\omega_c}{\pi} \tag{8.9}$$

Since we are designing *finite* impulse response filters with a finite number of filter coefficients (i.e., the length of $b[k]$ is finite), we need an impulse response that is also finite. Unfortunately the impulse response given in Equation 8.5 is infinite: it does not go to zero for finite values of filter length, L. In practice, we simply truncate the impulse response equation, Equation 8.5, to some finite value, L.

You might suspect that limiting a filter's impulse response by truncation has a negative impact on the filter's spectrum—we might not get the promised infinite attenuation at the cutoff frequency. In fact, truncation has two adverse effects: the filter no longer has an infinitely sharp cutoff, and oscillations are produced in the filter's spectrum. These adverse effects are demonstrated in the next example, where we show the spectrum of a rectangular window filter truncated to two different lengths.

EXAMPLE 8.2

Use the Fourier transform to find the magnitude spectrum of the rectangular window filter given in Equation 8.7 for two different lengths: $L = 17$ and $L = 65$. Use a cutoff frequency of 300 Hz assuming a sampling frequency of 1 kHz (i.e., a relative frequency of 0.3).

Solution: First generate the filter's impulse response, $b[k]$, by direct implementation of Equation 8.7. Note that the two filter lengths are both odd, so we reduce L by 1.0 to make them even, then shift by $k - L/2$. Also $b[L/2]$, where the denominator of Equation 8.7 is zero, should be set to $2f_c$ as given by Equation 8.8. After calculating the impulse response, find the spectrum by taking the Fourier transform of the response. Plot only the magnitude spectrum.

```
% Example 8.2 Generate two rectangular window impulse responses and
%  find their magnitude spectra.
%
N = 1024;                              % Number of samples for plotting
fs = 1000;                             % Sampling frequency
f = (1:N)*fs/N;                        % Frequency vector for plotting
fc = 300/fs;                           % Cutoff frequency (normalized to fs)
L = [17 67];                           % Filter lengths (filter order + 1)
for m = 1:2                            % Loop for the two filter lengths
  L1 = L(m)-1;                         % Make L1 even
  for k = 1:L(m
    n = k - L1/2;                      % n = k - L1/2 where L1 even
    if n == 0                          % Generate sin(n)/n function. Use Equation 8.7.
      b(k) = 2*fc;                     % Case where denominator is zero.
     else
      b(k) = sin(2*pi*fc*n)/(pi*n);    % Filter impulse response
    end
  end
  H = fft(b,N);                        % Calculate spectrum
```

```
    subplot(1,2,m);                        % Plot magnitude spectrum
    plot(f(1:N/2),abs(H(1:N/2)),'k');
    ..........labels and title..........
  end
```

The spectrum of this filter is shown in Figure 8.12 to have two artifacts associated with finite length. The oscillation in the magnitude spectrum is another example of a Gibbs artifact, first encountered in Chapter 3 and is due to the truncation of the infinite impulse response given by Equation 8.7. In addition, the slope is less steep when the filter's impulse response is shorter.

We can probably live with the less than ideal slope (we should never expect to get an ideal anything in the real world), but the oscillations in the spectrum are serious problems. Since Gibbs artifacts are due to truncation of an infinite function, we might reduce them if the function is tapered toward zero rather than abruptly truncated. Some of the tapering window functions described in Chapter 4 (Section 4.1.4) can be useful in eliminating the abrupt truncation. In Chapter 4, window functions such as the Hamming window are used to improve the spectra obtained from short data sets, and FIR impulse responses are usually short. There are many different window functions, but the two most popular and most useful for FIR impulse responses are the Hamming and the Blackman windows. The Hamming window equation is given in Equation 4.9 and repeated here:

$$w[n] = 0.5 - 0.46 \cos\left(\frac{2\pi n}{N}\right) \tag{8.10}$$

where N is the length of the window, which should be the same length as the data. The Blackman window is more complicated, but like the Hamming is easy to program in MATLAB[2]:

$$w[n] = 0.42 - 0.5 \cos\left(\frac{2\pi n}{N}\right) + 0.16 \cos\left(\frac{4\pi n}{N}\right) \tag{8.11}$$

Example 8.3 presents MATLAB code for the Blackman window and applies it to the filters of Example 8.2. Coding the Hamming window is part of Problems 5 and 6 at the end of this chapter.

EXAMPLE 8.3

Apply a Blackman window to the rectangular window filters used in Example 8.2. (Note that the word "window" is used in two completely different contexts in the last sentence. The first "window" is a time-domain function and the second is a frequency-domain function. Engineering terms are not always definitive.) Calculate and display the magnitude spectrum of the impulse functions after they have been windowed.

Solution: Write a MATLAB function, `blackman(N)`, to generate a Blackman window of length `N`. Apply it to the filter impulse responses using point-by-point multiplication (i.e., using the .*

[2]This equation is written assuming the popular value for constant α of 0.16.

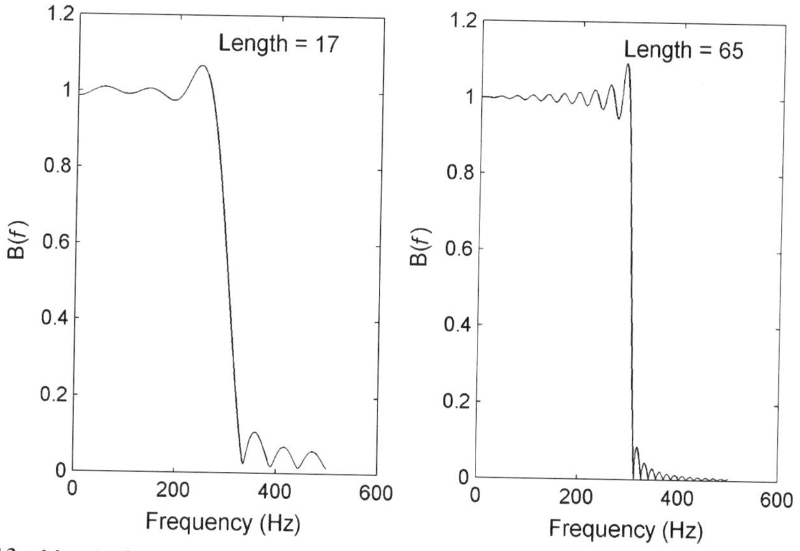

FIGURE 8.12 Magnitude spectrum of two FIR filters based on an impulse derived from Equation 8.7. The impulse responses are abruptly truncated at 17 and 65 coefficients. The low-pass cutoff frequency is 300 Hz for both filters with an assumed sample frequency of 1 kHz. The oscillations seen are Gibbs artifacts and are due to the abrupt truncation of what should be an infinite impulse response. Like the Gibbs artifacts seen in Chapter 3, they do not diminish with increasing filter length, but increase in frequency.

operator). Modify the last example applying the window to the filter's impulse response before taking the Fourier transform.

```
% Example 8.3 Apply the Blackman window to the rectangular window impulse responses
%  developed in Example 8.2.
%
N = 1024;                            % Number of points for plotting
.......... same code as in Example 8.2..........
if n == 0                            % Generate sin(n)/n function. Use Equation 8.7.
   b(k) = 2*fc;                      % Case where denominator is zero.
else
   b(k) = sin(2*pi*fc*n)/(pi*n);     % Filter impulse response
end
w = blackman(L);                     % Get Blackman window of length L
b = b.*w;                            % Apply window to impulse response
H = fft(b,N);                        % Calculate spectrum
..........same code as in Example 8.2, plot and label..........
%
function w = blackman(L)
% Function to calculate a Blackman window L samples long
%
n = (1:L);                           % Generate vector for window function
w = 0.42 - 0.5*cos(2*pi*n/(L-1)) + 0.08*cos(4*pi*n/(L-1));
```

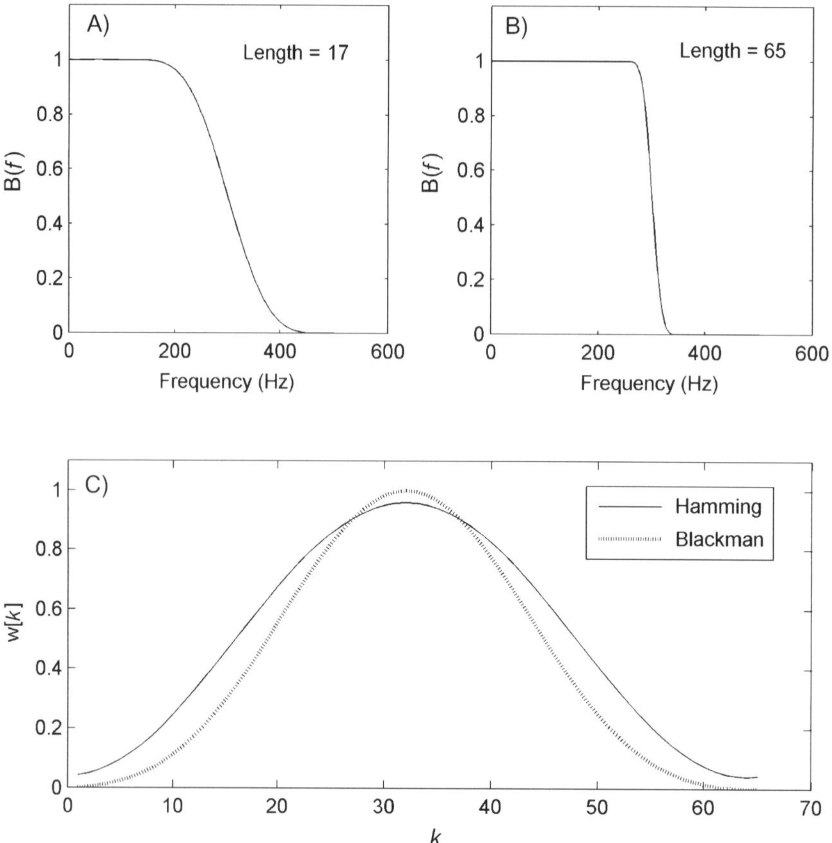

FIGURE 8.13 A) Magnitude spectrum produced by the 17-coefficient FIR filter in Figure 8.12 except a Blackman window is applied to the filter coefficients. The Gibbs oscillations seen in Figure 8.12 are no longer visible. B) Magnitude spectrum produced by the 65-coefficient FIR filter in Figure 8.12 also after application of the Blackman. C) Plot of the Blackman and Hamming window functions. Both have cosine-like appearances.

The Blackman window is easy to generate in MATLAB and, when applied to the impulse responses of Example 8.2, substantially reduces the oscillations, as shown in Figure 8.13. The filter rolloff is still not that of an ideal filter, but becomes steeper for a longer filter length. Of course, increasing the length of the filter increases the computation time required to apply the filter to a data set, a typical engineering compromise. Figure 8.13C shows a plot of the Blackman and Hamming windows. These popular windows are quite similar, having the appearance of cosine waves, not too surprising given the cosine function in both of their defining equations (Equations 8.10 and 8.11).

The next example applies a rectangular window low-pass filter to a signal of human respiration that was obtained from a respiratory monitor.

EXAMPLE 8.4

Apply a low-pass rectangular window filter to the 10-minute respiration signal shown in the top trace of Figure 8.14 and found as variable `resp` in file `Resp.mat`. The signal is sampled at 12.5 Hz. The low sampling frequency is used because the respiratory signal has a very low bandwidth. Use a cutoff frequency of 1.0 Hz and a filter length of 65. Use the Blackman window to truncate the filter's impulse response. Plot the original and filtered signal.

Solution: Reuse the code in Examples 8.2 and 8.3 and apply the filter to the respiratory signal using convolution. Again, MATLAB's `conv` routine is invoked using the option `same` to avoid a time shift in the output.

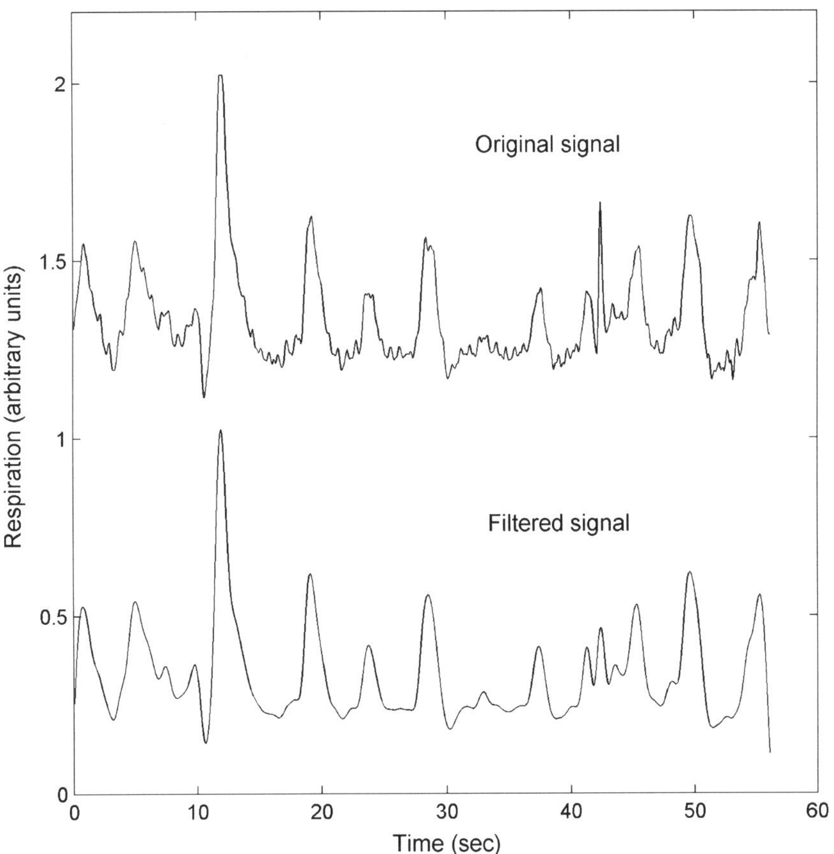

FIGURE 8.14 The output of a respiratory monitor is shown in the upper trace to have some higher frequency noise. In the lower trace, the signal has been filtered with an FIR rectangular low-pass filter with a cutoff frequency of 1.0 Hz and a filter length of 65. *Original data from PhysioNet, Goldberger et al., 2000.*

```
%Example 8.4 Application of a rectangular window low-pass filter to a respiratory
signal.
%
load Resp;                              % Get data
fs = 12.5;                              % Sampling frequency
N = length(resp);                       % Get data length
t = (1:N)/fs;                           % Time vector for plotting
L = 65;                                 % Filter lengths
fc = 1/fs;                              % Cutoff frequency: 1.0 Hz
plot(t,resp + 1,'k'); hold on;          % Plot original data (offset for clarity)
L1 = L -1;                              % Make L1 even
for k = 1:L                             % Generate sin(n)/n function symmetrically
  n = k - L1/2 ;                        % Same code as in previous examples
   if n == 0
     b(k) = 2*fc;                       % Case where denominator is zero.
   else
     b(k) = (sin(2*pi*fc*n))/(pi*n);    % Filter impulse response
   end
end
b = b.*blackman(L);                     % Apply Blackman window
%
y = conv(resp,b,'same');                % Apply filter
plot(t,y,'k');
```

Results: The filtered respiratory signal is shown in the lower trace of Figure 8.14. The filtered signal is smoother than the original signal as the noise riding on the original signal has been eliminated.

The FIR filter coefficients for high-pass, bandpass, and bandstop filters can also be derived by applying an inverse FT to rectangular spectra having the appropriate associated shape. These equations have the same general form as Equation 8.7 except they may include additional terms:

High-pass:

$$b[k] = \begin{cases} -\dfrac{\sin[2\pi f_c(k - L/2)]}{\pi(k - L/2)} & k \neq L/2 \\ 1 - 2f_c & k = L/2 \end{cases} \tag{8.12}$$

Bandpass:

$$b[k] = \begin{cases} \dfrac{\sin[2\pi f_h(k - L/2)]}{\pi(k - L/2)} - \dfrac{\sin[2\pi f_l(k - L/2)]}{\pi(k - L/2)} & k \neq L/2 \\ 2(f_h - f_l) & k = L/2 \end{cases} \tag{8.13}$$

Bandstop

$$b[k] = \begin{cases} \dfrac{\sin[2\pi f_l(k - L/2)]}{\pi(k - L/2)} - \dfrac{\sin[2\pi f_h(k - L/2)]}{\pi(k - L/2)} & k \neq L/2 \\ 1 - 2(f_h - f_l) & k = L/2 \end{cases} \tag{8.14}$$

The order, L, of high-pass and bandstop filters should always be even, so the number of coefficients in these filters will be odd ($L+1$). The next example applies a bandpass filter to the EEG introduced in Chapter 1.

EXAMPLE 8.5

Apply a bandpass filter to the `EEG` data in file `ECG.mat`. Use a lower cutoff frequency of 6 Hz and an upper cutoff frequency of 12 Hz. Use a Blackman window to truncate the filter's impulse to 129 coefficients. Plot the data before and after bandpass filtering. Also plot the spectrum of the original signal and superimpose the spectrum of the bandpass filter.

Solution: Construct the filter's impulse response using the bandpass equation (Equation 8.13). Note the special case where the denominator in the equation goes to zero (i.e., when $k = L/2$), the application of limits and the small angle approximation ($\sin(x) \to x$) gives a coefficient value of $b[L/2] = 2f_h - 2f_l$, as shown in Equation 8.13. Recall that the sampling frequency of the EEG signal is 100 Hz.

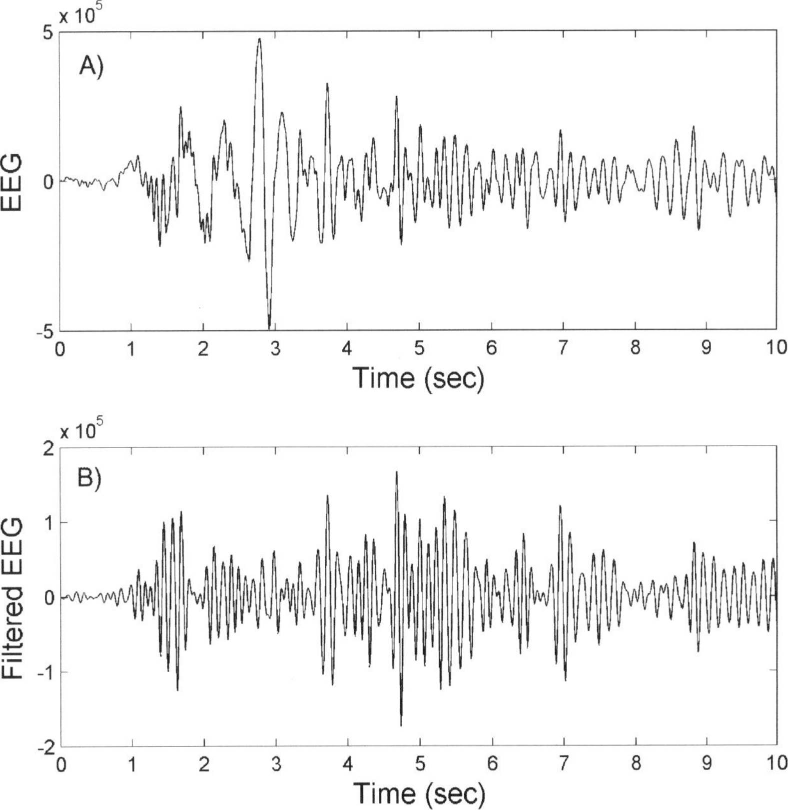

FIGURE 8.15 EEG signal before (A) and after (B) bandpass filtering between 6 and 12 Hz. A fairly regular oscillation is seen in the filtered signal.

After applying the filter and plotting the resulting signal, compute the signal spectrum using the Fourier transform and plot. The filter's spectrum is found by taking the Fourier transform of the impulse response ($b[k]$) and is plotted superimposed on the signal spectrum.

```
% Example 8.5 Apply a bandpass filter to the EEG data in file ECG.mat.
%
load EEG;                          % Get data
N = length(eeg);                   % Get data length
fs = 100;                          % Sample frequency
fh = 12/fs;                        % Set high-pass and
fl = 6/fs;                         % low-pass cutoff frequencies
L = 129;                           % Set number of weights
%
for k = 1:L                        % Generate bandpass filter impulse response
  L1 = L - 1;                      % Make L1 even
  n = k - L1/2;                    % Make symmetrical
  if n == 0
    b(k) = 2*fh - 2*fl;            % Case where denominator is zero
```

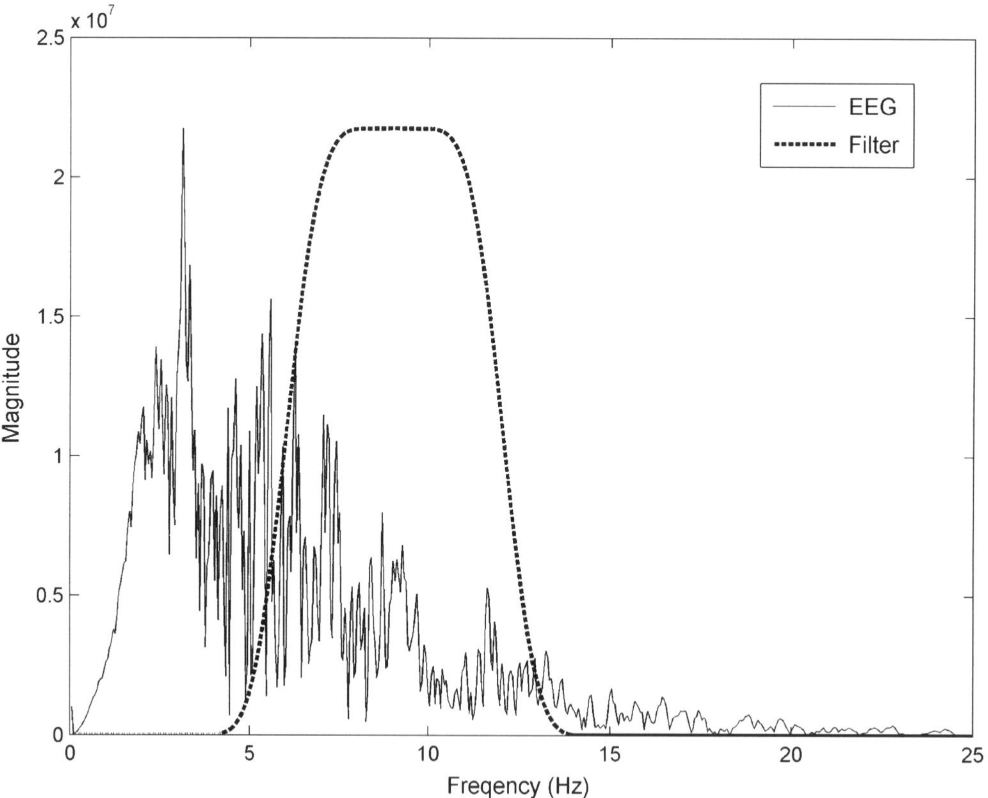

FIGURE 8.16 The magnitude spectrum of the EEG signal used in Example 8.5 along with the spectrum of a bandpass filter based in Equation 8.12. The bandpass filter range is designed to be between 6 and 12 Hz.

```
    else
      b(k) = sin(2*pi*fh*n)/...
        (pi*n) - sin(2*pi*fl*n)/(pi*n);% Filter impulse response
    end
  end
  b = b.* blackman(L);              % Apply Blackman window to filter coefficients;
  y = conv(eeg,b,'same');           % Filter the data using convolution
    ........plot eeg before and after filtering; plot eeg and filter spectrum........
```

Bandpass filtering the EEG signal between 6 and 12 Hz reveals a strong higher frequency oscillatory signal that is washed out by lower frequency components in the original signal, (Figure 8.15). This figure shows that filtering can significantly alter the appearance and interpretation of biomedical data. The bandpass filter spectrum shown in Figure 8.16 has the desired cutoff frequencies and, when compared to the EEG spectrum, is shown to reduce sharply the high- and low-frequency components of the EEG signal.

Implementation of other FIR filter types is found in the problems at the end of this chapter. A variety of FIR filters exist that use strategies other than the rectangular window to construct the filter coefficients. One FIR filter of particular interest is the filter used to construct the derivative of a waveform, since the derivative is often used in the analysis of biosignals. The next section explores a popular filter for this operation.

8.2.2 Derivative Filters—The Two-Point Central Difference Algorithm

The derivative is a common operation in signal processing and is particularly useful in analyzing certain physiological signals. Digital differentiation is defined as $dx[n]/dn$ and can be calculated directly from the slope of $x[n]$ by taking differences:

$$\frac{dx[n]}{dn} = \frac{\Delta x[n]}{T_s} = \frac{x[n+1] - x[n]}{T_s} \tag{8.15}$$

This equation can be implemented by MATLAB's `diff` routine. This routine uses no padding, so the output is one sample shorter than the input. As shown in Chapter 5 (Section 5.4.2), the frequency characteristic of the derivative operation increases linearly with frequency, so differentiation enhances higher frequency signal components. Since the higher frequencies frequently contain a greater percentage of noise, this operation tends to produce a noisy derivative curve. The upper curve of Figure 8.17A is a fairly clean physiological motor response, an eye movement similar to that used in Figure 8.7. The lower curve of Figure 8.17A is the velocity of the movement obtained by calculating the derivative using MATLAB's `diff` routine. Considering the relative smoothness of the original signal, the velocity curve obtained using Equation 8.15 is quite noisy.

In the context of FIR filters, Equation 8.15 is equivalent to a two-coefficient impulse function, $[+1, -1]/T_s$. (Note that since the positive and negative coefficients are reversed by convolution, so they are sequenced in reverse order in the impulse response.) A better approach to differentiation is to construct a filter that approximates the derivative at lower

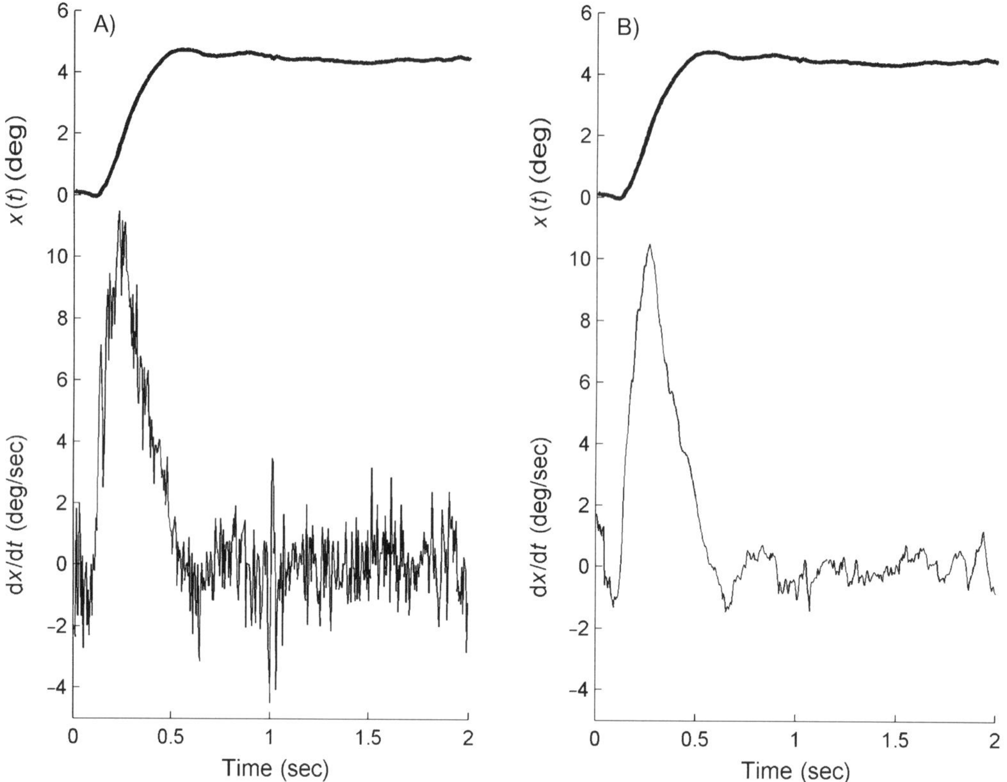

FIGURE 8.17 An eye movement response to a step change in target depth is shown in the upper trace, and its velocity (i.e., derivative) is shown in the lower trace. A) The derivative was calculated by taking the difference in adjacent points and scaling by the sample interval. The velocity signal is noisy even though the original signal is fairly smooth. B) The derivative was computed using the 2-point central difference algorithm with a skip factor of 4.

frequencies but attenuates at higher frequencies that are likely to be only noise. The *two-point central difference* algorithm achieves just such an effect, acting as a differentiator at lower frequencies and a low-pass filter at higher frequencies. Figure 8.17B shows the same response and derivative when this algorithm is used to estimate the derivative. The result is a much cleaner velocity signal that still captures the peak velocity of the response.

The two-point central difference algorithm still subtracts two points to get a slope but the two points are no longer adjacent; rather they may be spaced some distance apart. Putting this in FIR filter terms, the algorithm is based on an impulse function containing two coefficients of equal but opposite sign spaced L points apart. The equation defining this differentiator is:

$$\frac{dx[n]}{dn} = \frac{x(n+L) - x(n-L)}{2LT_s} \tag{8.16}$$

where L is now called the *skip factor* that defines the distance between the points used to calculate the slope, and T_s is the sample interval. The skip factor, L, influences the effective bandwidth of the filter as shown below. Implemented as a filter, Equation 8.16 leads to filter coefficients:

$$b[k] = \begin{cases} 1/2LT_s & k = -L \\ -1/2LT_s & k = +L \\ 0 & k \neq \pm L \end{cases} \tag{8.17}$$

Note that the $+L$ coefficient is negative and the $-L$ coefficient is positive since the convolution operation reverses the order of $b[k]$. As with all FIR filters, the frequency response of this filter algorithm can be determined by taking the Fourier transform of $b[k]$. Since this function is fairly simple, it is not difficult to take the Fourier transform analytically (trying to break from MATLAB whenever possible) as well as in the usual manner using MATLAB. Both methods are presented in the example below.

EXAMPLE 8.6

A) Determine the magnitude spectrum of the two-point central difference algorithm analytically, then B) use MATLAB to determine the spectrum and apply it to the EEG signal.

Analytical solution: Starting with the equation for the discrete Fourier transform (Equation 3.32) substituting k for n:

$$X[m] = \sum_{k=0}^{L-1} b[k]e^{-j2\pi mk/N}$$

Since $b[k]$ is nonzero only for $k = \pm L$, the Fourier transform, after the summation limits are adjusted for a symmetrical coefficient function with positive and negative n, becomes:

$$X(m) = \sum_{k=-L}^{L} b[k]e^{-j2\pi mk/N} = \frac{1}{2LT_s}e^{-j2\pi m(-L)/N} - \frac{1}{2LT_s}e^{-j2\pi mL/N}$$

$$X(m) = \frac{e^{-j2\pi m(-L)/N} - e^{-j2\pi nL/N}}{2LT_s} = \frac{-j\sin(2\pi mL/N)}{LT_s}$$

where L is the skip factor and N is the number of samples in the waveform. To put this equation in terms of frequency, note that $f = m/(N\,T_s)$, hence $m = f\,N\,T_s$. Substituting in $f\,N\,T_s$ for m and taking the magnitude of $X(m)$, now $X(f)$:

$$|X(f)| = \left| -j\frac{\sin(2\pi fLT_s)}{LT_s} \right| = \frac{|\sin(2\pi fLT_s)|}{LT_s}$$

This equation shows that the magnitude spectrum, $|X(f)|$, is a sine function that goes to zero at $f = 1/(LT_s) = f_s/L$. Figure 8.18 shows the frequency characteristics of the 2-point central difference algorithm for 2 different skip factors: $L = 2$ and $L = 6$.

MATLAB solution: Finding the spectrum using MATLAB is straightforward once the impulse response is constructed. An easy way to construct the impulse response is to use brackets and concatenate zeros between the two end values, essentially following Equation 8.17 directly.

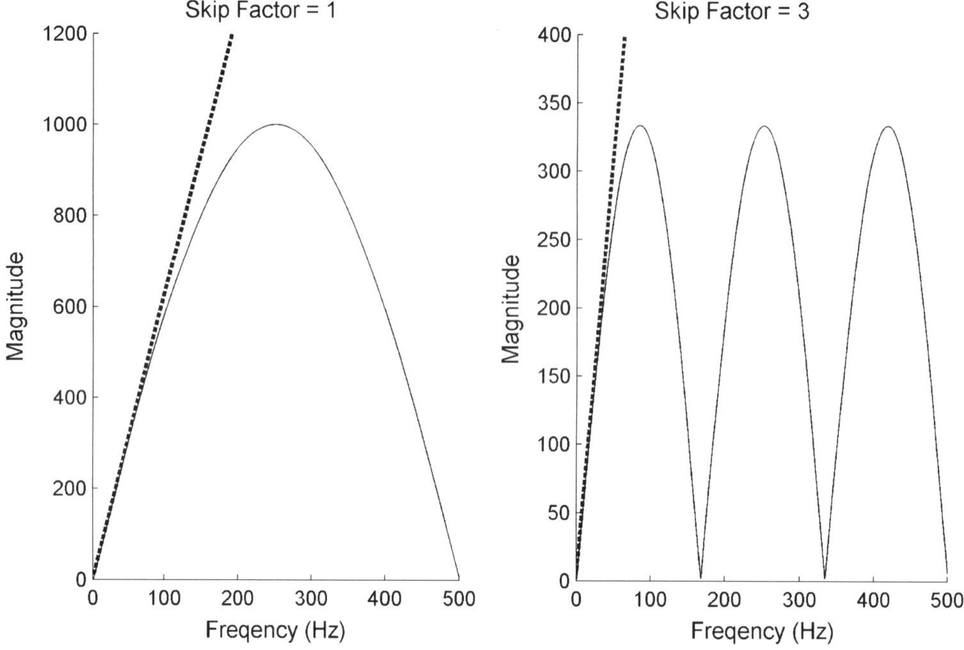

FIGURE 8.18 The frequency response of the 2-point central difference algorithm using 2 different skip factors: A) $L = 2$; B) $L = 6$. The dashed line shows the frequency characteristic of a simple differencing operation. The sample frequency is 1.0 kHz.

Again, the initial filter coefficient is positive and the final coefficient negative to account for the reversal produced by convolution.

```
% Example 8.6 Determine the frequency response of
% the two-point central difference algorithm used for differentiation.
%
Ts = .001;                                      % Assume a Ts of 1 msec. (i.e.,
                                                fs = 1 kHz)
N = 1000;                                       % Number of data points (time = I sec)
Ln = [1 3];                                     % Define two different skip factors
for m = 1:2                                     % Repeat for each skip factor
  L = Ln(m);                                    % Set skip factor
  b = [1/(2*L*Ts) zeros(1,2*L -1) -1/(2*L*Ts)];   % Filter impulse response
  H = abs(fft(b,N));                              % Calculate magnitude spectrum
  subplot(1,2,m);                               % Plot the result
........plot and label spectrum; plot straight line for comparison........
end
```

The results from this program and the analytical analysis is shown in Figure 8.18. A true derivative has a linear change with frequency: a line with a slope proportional to f as shown by the dashed lines in Figure 8.18. The 2-point central difference spectrum approximates a true derivative over the lower frequencies, but has the characteristic of a low-pass filter for higher frequencies.

Increasing the skip factor, L, has the effect of lowering the frequency range over which the filter acts like a derivative operator as well as lowering the low-pass filter range. Note that for skip factors >2, the response curve repeats above $f = 1/(LT_s)$. Usually the assumption is made that the signal does not contain frequencies in this range. If this is not true, then these higher frequencies can be removed by an additional low-pass filter as shown in Problem 9. It is also possible to combine the difference equation (Equation 8.15) with a low-pass filter and this is also explored in Problem 10.

8.2.3 Determining Cutoff Frequency and Skip Factor

Determining the appropriate cutoff frequency of a filter or the skip factor for the 2-point central difference algorithm can be somewhat of an art (meaning there is no definitive approach to a solution). If the frequency ranges of the signal and noise are known, setting cutoff frequencies is straightforward. In fact, there is an approach called *Wiener filtering* that leads to an optimal filter if these frequencies are accurately known. However this knowledge is not usually available in biomedical engineering applications. In most cases, filter parameters such as filter order and cutoff frequencies are set empirically based on the data. In one scenario, the signal bandwidth is progressively reduced until some desirable feature of the signal is lost or compromised. While it is not possible to establish definitive rules due to the task-dependent nature of filtering, the next example gives an idea of how these decisions are approached.

In Example 8.7, we return to the eye movement signal to evaluate several different skip factors to find the one that gives the best reduction of noise without reducing the accuracy of the velocity trace.

EXAMPLE 8.7

Use the 2-point central difference algorithm to compute velocity traces of the eye movement step response in file `eye.mat`. Use 4 different skip factors (1, 2, 5, and 10) to find the skip factor that best reduces noise without substantially reducing the peak velocity of the movement. (Peak velocity is being used here as a measure of the accuracy of the velocity trace.)

Solution: Load the file and use a loop to calculate and plot the velocity determined with the two-point central difference algorithm using the four different skip factors. Find the maximum value of the velocity trace for each derivative evaluation and display on the associated plot. The original eye movement, Figure 8.17 (upper traces), is in degrees so the velocity trace is in deg/sec.

```
% Example 8.7 To evaluate the two-point central difference algorithm using
% different derivative skip factors
%
load eye;                              % Get data
fs = 200;                              % Sampling frequency
Ts = 1/fs;                             % Calculate Ts
t = (1:length(eye_move))/fs;           % Time vector for plotting
L = [1 2 5 10];                        % Filter skip factors
```

```
for m=1:4                                              % Loop for different skip factors
  b=[1/(2*L(m)*Ts) zeros(1,2*L(m)-1) -1/(2*L(m)*Ts)];  % Construct filter
  der=conv(eye_move,b,'same');                         % Apply filter
  subplot(2,2,m);
  plot(t,der,'k');
  text(1,22,['L=',num2str(L(m))],'FontSize',12);
  text(1,18,['Peak=',num2str(max(der),2)],'FontSize',12);
  ........labels and axis........
end
```

The results of this program are shown in Figure 8.19. As the skip factor increases the noise decreases, and the velocity trace becomes quite smooth at a skip factor of 10. However, often we are concerned with measuring the maximum velocity of a response and the peak velocity also decreases with the higher skip factors. Examining the curve with the lowest skip factor ($L = 1$) suggests that the peak velocity shown, 23 degrees/sec, may be augmented a bit by noise. A peak velocity of 21 degrees/sec is found for both skip factors of 2 and 5. This peak velocity appears to be reflective of the true peak velocity. The peak found using a skip factor of 10 is clearly reduced by the derivative filter. Based on this empirical study, a skip factor of around 5 seems appropriate, but this is always a judgment call. In fact, judgment and intuition are all-too-frequently involved in signal processing and signal analysis tasks, a reflection that signal processing is still an art. Another example that requires empirical evaluation is found in Problem 8 at the end of the chapter.

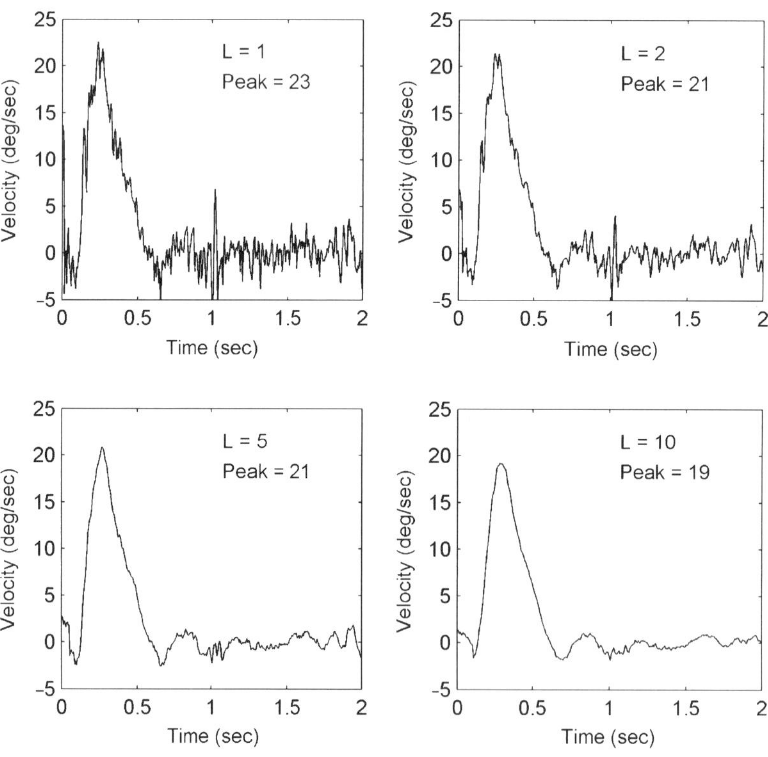

FIGURE 8.19 Velocity traces for the eye movement shown in Figure 8.17 (upper traces) calculated by the two-point central difference algorithm for different values of skip factor as shown. The peak velocities are shown in deg/sec.

8.3 TWO-DIMENSIONAL FILTERING—IMAGES

FIR filters can be extended to two dimensions and applied to images. A wide range of useful image filters can be designed using MATLAB's Image Processing Toolbox, but again a taste of image filtering can be had using standard MATLAB.

Two-dimensional filters have two-dimensional impulse responses. Usually these filters consist of square matrices with an odd number of rows and columns. Implementation of a two-dimensional filter can be achieved using two-dimensional convolution. The equation for two-dimensional convolution is a straightforward extension of the one-dimensional discrete convolution equation (Equation 7.3):

$$y(m,n) = \sum_{k_1=0}^{K_2-1} \sum_{k_2=0}^{K_1-1} x(k_1,k_2)b(m-k_1,n-k_2) \tag{8.18}$$

where K_1 and K_2 are the dimensions in pixels of the image, $b(k_1,k_2)$ are the filter coefficients, $x(k_1,k_2)$ is the original image, and $y(m,n)$ is the filtered image. In two-dimensional convolution, the impulse function, now a matrix, moves across the image as illustrated in Figure 8.20. The filtered image is obtained from the center pixel (black dot in Figure 8.20) where the value of this pixel is the two-dimensional summation of the product of the impulse response matrix (a 3×3 matrix in Figure 8.20) and the original image. As the impulse response matrix slides horizontally and vertically across the original image, the new filtered image is constructed.

Two-dimension convolution is implemented in MATLAB using:

```
y=conv2(x,b,'options');  % Two-dimensional convolution
```

where `y` is the output image, `x` is the input image, `b` is the impulse response matrix, and the `options` are the same as for one-dimensional convolution including the option `same`.

The next example applies a two-dimensional moving average filter to a noise-filled image of red blood cells.

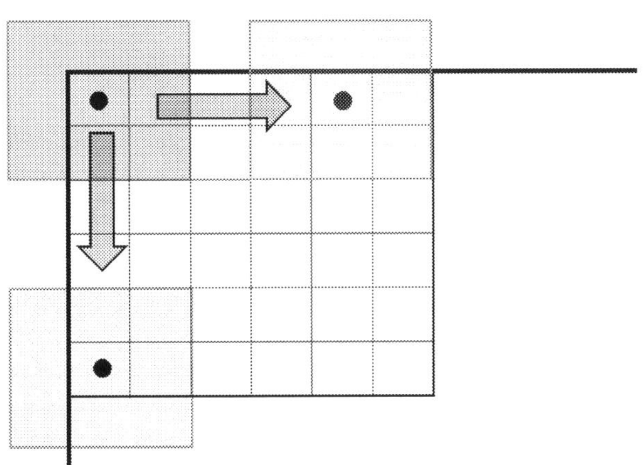

FIGURE 8.20. Illustration of two-dimensional convolution. The impulse response, which is now a matrix, moves across the image one pixel at a time. After one row is complete then the impulse response matrix moves across the next row, one pixel down. At each position, another pixel in the output image is generated as the sum of the product of each value in the impulse response matrix times the corresponding original image pixel. The edges are usually handled by zero-padding.

EXAMPLE 8.8

Load the image of blood cells found as variable `cells` in file `blood_cell_noisy.mat`. This image contains noise, black and white speckles known appropriately as *salt and pepper* noise. Filter the noise with a 5×5 moving average filter. Display the original and filtered images side by side.

Solution: Load the image and display using `pcolor` with the proper shading option as in previous examples. The matrix that generates a 5×5 moving average is:

$$b = \begin{vmatrix} 1/25 & 1/25 & 1/25 & 1/25 & 1/25 \\ 1/25 & 1/25 & 1/25 & 1/25 & 1/25 \\ 1.25 & 1/25 & 1/25 & 1/25 & 1/25 \\ 1/25 & 1/25 & 1/25 & 1/25 & 1/25 \\ 1/25 & 1/25 & 1/25 & 1/25 & 1/25 \end{vmatrix} \tag{8.19}$$

After generating this matrix, apply it to the image using `conv2` with the `same` option and display.

```
% Ex 8_8 Application of moving average filter to blood cell image with
% "salt and pepper" noise.
%
load blood_cell_noisy;            % Load blood cell image
subplot(1,2,1);
pcolor(cells);                    % Display image
shading interp;                   % Same as previous examples
colormap(bone);
axis('square');                   % Adjust image shape
```

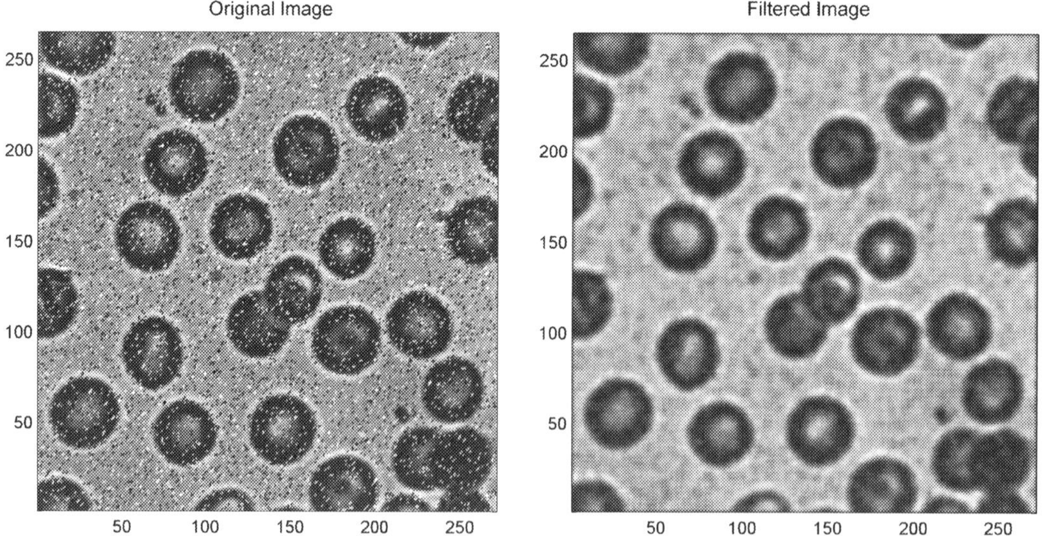

FIGURE 8.21 An image of red blood cells containing salt and pepper noise before and after filtering. The filter is a 5×5 moving average filter. Application of this filter using two-dimensional convolution substantially reduces the noise.

```
%
b = [ones(5,5)/25;                % Define moving average impulse response matrix
y = conv2(cells,b,'same');        % Filter image using convolution
subplot(1,2,2);
pcolor(y);                        % Display image as above
shading interp;
colormap(bone);
axis('square');
```

Results: The original and filtered images are shown in Figure 8.21. The noise in the original image is substantially reduced by the moving average filter.

It would be nice to know the frequency characteristics of the moving average filter used in Example 8.8. To determine the frequency characteristics of a two-dimensional filter we take the Fourier transform of the impulse response, but now we use a two-dimensional version of the Fourier transform. The two-dimensional discrete Fourier transform is defined as:

$$F(p,q) = \sum_{m=0}^{M-1}\sum_{n=0}^{N-1} b\,(m,n)\; e^{-j(2pm/M)}e^{-j(2qn/N)} \qquad (8.20)$$
$$p = 0,1,\ldots,M-1; \quad q = 0,1,\ldots,N-1$$

where the impulse response function, $b(m,n)$, is an $M \times N$ matrix. Of course the function can be any matrix, so this equation can also be used to obtain the discrete Fourier transform of an image. The two-dimensional Fourier transform is implemented in MATLAB using:

```
X = fft2(x,nrows,ncols);  % Two-dimensional Fourier transform
```

where `x` is the original matrix, `X` is the output, and `nrows` and `ncols` are optional arguments that are used to pad out the number of rows and columns in the input matrix. The output matrix is now in spatial frequency, cycles/distance. When displaying the two-dimensional Fourier transform, it is common to shift the zero frequency position to the center of the display and show the spectrum on either side. This improves visualization of the spectrum and is accomplished with routine `fftshift`:

```
Xshift = fftshift(X);  % Shift zero to center and reflect
```

where `X` is the original Fourier transform and `Xshift` is the shifted version. The spectrum can then be plotted using any of MATLAB's three-dimensional plotting routines such as `mesh` or `surf`. The next example provides an example of the use of these routines to determine the spatial frequency characteristics of the moving average filter.

EXAMPLE 8.9

Determine and plot the two-dimensional spectrum of the 5×5 moving average filter used in Example 8.8.

Solution: Generate the impulse response as in the last example, take the absolute value of the two-dimensional Fourier transform, shift, and then display using `mesh`.

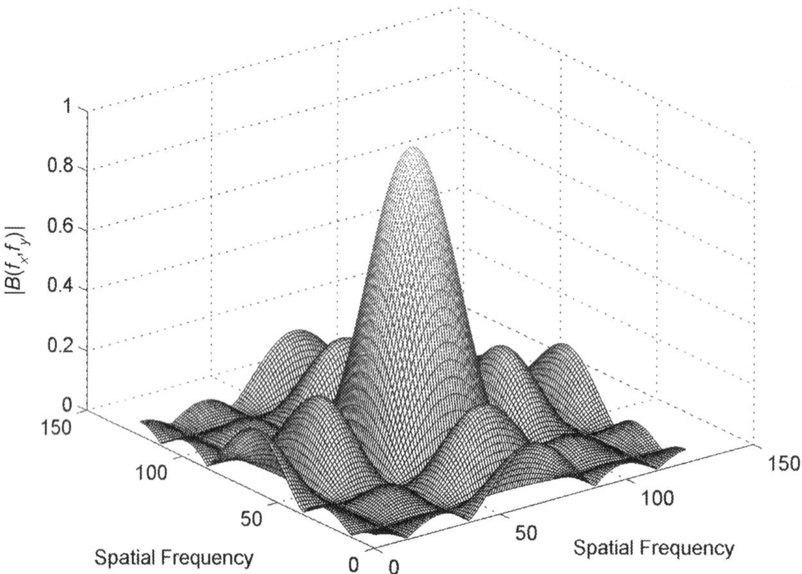

FIGURE 8.22 Spatial frequency characteristic of a 5 × 5 moving average filter. This filter has the general features of a two-dimensional low-pass filter reducing high spatial frequencies. If applied to an image 5 cm on a side, then the horizontal axes should be multiplied by 0.2 cycles/cm.

```
%EX 8.9 Spectrum of a 5×5 moving average filter
% Construct moving average filter
b = ones(5,5)/25;              % Filter impulse response matrix
B = abs(fft2(b,124,124));      % Determine 2D magnitude Fourier transform
B = fftshift(B);               % Shift before plotting
mesh(B);                       % Plot using mesh
```

Results: The plot generated by this program is shown in Figure 8.22. The spectrum is one of a two-dimensional low-pass filter that reduces higher spatial frequencies equally in both dimensions. The salt-and-pepper noise found in the image of blood cells has high spatial frequencies since it involves rapid changes in intensity values within 1 or 2 pixels. Since this low-pass filter significantly attenuates these frequencies, it is effective at reducing this type of noise. It is not possible to determine the spatial frequency scale of the horizontal axes without knowing image size, the equivalent of signal period in 2D data. Assuming the blood cell image is printed as 5 cm on a side, then the fundamental spatial frequency is $f_1 = 1/D = 1/5 = 0.2$ cycles/cm and the two axes should be multiplied by this scale factor.

Other more complicated filters can be designed, but most require support from software packages such as the MATLAB Image Processing Toolbox. An exception is the filter that does spatial differentiation. Use of the two-point central difference algorithm with a skip factor of 1 leads to an impulse response of b = [1 0 −1]. A two-dimensional equivalent might look like:

$$b = \begin{vmatrix} 1 & 0 & -1 \\ 2 & 0 & -2 \\ 1 & 0 & -1 \end{vmatrix} \tag{8.21}$$

The corner values are half the main horizontal value, because they are on the edges of the filter array and should logically be less. This filter is known as the *Sobel operator* and provides an estimate of the spatial derivative in the horizontal direction. The filter matrix can be transposed to operate in the vertical direction. The next example demonstrates the application of the Sobel operator.

EXAMPLE 8.10

Load the MR image of the brain in variable `brain` of file `Brain.mat`. Apply the Sobel filter to this image to generate a new image that enhances changes that go from dark to light (going left to right). Flip the Sobel operator left to right, then apply it again to the original image to generate another image that enhances change going in the opposite direction. Finally generate a fourth image that is white if either of the Sobel filtered images is above a threshold, and black otherwise. Adjust the threshold to highlight the vertical boundaries in the original image.

Solution: Load the image, generate the Sobel impulse response and apply as is, and flipped, to the original image using `conv2`. Use a double `for-loop` to check the value of each pixel in the two images and set the pixel in the new image to 1 if either pixel is above the threshold and below if otherwise. Adjust the threshold empirically to highlight the vertical boundaries.

```
% Example 8.10 Apply the Sobel operator to the image of bone
%
  load Brain;                                   % Load image
%
b = [1 0 -1; 2 0 -2; 1 0 -1];                   % Define Sobel filter
brain1 = conv2(brain,b,'same');                 % Apply Sobel filter
brain2 = conv2(brain,fliplr(b),'same');         % Apply Sobel filter flipped
  ........display brain, brain1, and brain2 images as in previous examples........
%
[M,N] = size(brain);                            % Find size of image
thresh = .4;                                    % Threshold. Determined
                                                % empirically.
brain3 = zeros(M,N);
for m = 1:M
  for n = 1:N
    if brain1(m,n) > thresh || brain2(m,n) > thresh % Test each pixel in each image
      brain3(m,n) = 1;                          % Set new image pixel to white
    else
      brain3(m,n) = 0;                          % else set to black
    end
  end
end
  ........display image brain4 and title........
```

Results: The images produced in this example are shown in Figure 8.23. The Sobel filtered images are largely gray since, except for the edges of the brain, the image intensities do not

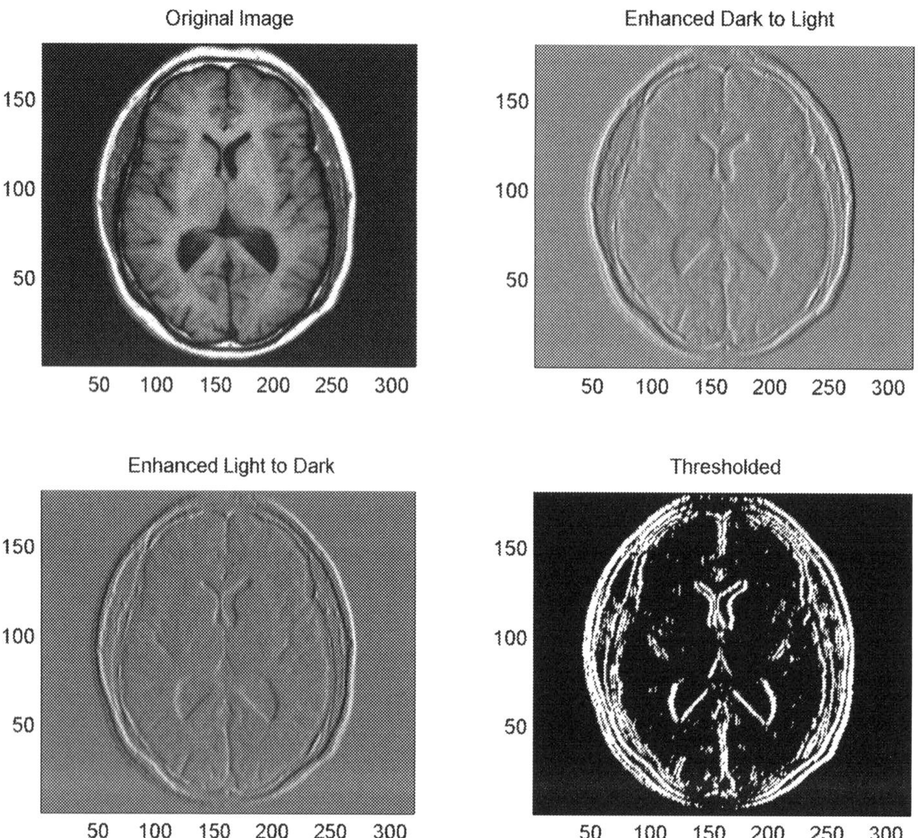

FIGURE 8.23 Original MR image of the brain along with two images filtered by the Sobel operator and a thresholded imaged constructed from the Sobel filtered images.

change abruptly. However, slightly lighter areas can be seen Problem 11 through 14 where more abrupt changes occur. If these two images are thresholded at the appropriate level, an image delineating the vertical boundaries is seen. This is a rough example of *edge detection* which is often an important step in separating out regions of interest in biomedical images. Identifying regions or tissue of interest in a biomedical image is known as *image segmentation*. Other examples of image filtering are given in the Problem 11 t.

8.4 FIR FILTER DESIGN USING MATLAB—THE SIGNAL PROCESSING TOOLBOX

The rectangular window equations given for one-dimensional filters facilitate the design of all the basic FIR filter types: low-pass, high-pass, bandpass, and bandstop. In some rare cases there may be a need for a filter with a more exotic spectrum and MATLAB offers considerable support for the design and evaluation of both FIR and IIR filters in the Signal Processing Toolbox.

Within the MATLAB environment, filter design and application occur in either one or two stages, each stage executed by separate but related routines. In the two-stage protocol, the user supplies information regarding the filter type and desired attenuation characteristics, but not the filter order. The first-stage routines determine the appropriate order as well as other parameters required by the second-stage routines. The second-stage routines then generate the filter coefficients, $b[k]$, based on the arguments produced by the first-stage routines, including the filter order. It is possible to bypass the first-stage routines if you already know, or can guess, the filter order. Only the second-stage routines are needed where the user supplies the filter order along with other filter specifications. Here we explore this later approach, since trial and error is often required to determine the filter order anyway.

If a filter task is particularly demanding, the Signal Processing Toolbox provides an interactive filter design package called `FDATool` (for Filter Design and Analysis Tool) that uses an easy graphical user interface (GUI) to design filters with highly specific or demanding spectral characteristics. Another Signal Processing Toolbox package, the `SPTool` (Signal Processing Tool), is useful for analyzing filters and generating spectra of both signals and filters. The MATLAB Help file contains detailed information on the use of these two packages.

Irrespective of the design process, the net result is a set of b coefficients, the filter's impulse response. The FIR filter represented by these coefficients is implemented as above using either MATLAB's `conv` or `filter` routine. Alternatively the Signal Processing Toolbox contains the routine `filtfilt` that, like `conv` with the `same` option, eliminates the time shift associated with standard filtering. A comparison between `filter` and `conv` with the `same` option is given in Problems 17 and 18.

One useful Signal Processing Toolbox routine determines the frequency response of a filter given the coefficients. We already know how to do this using the Fourier transform, but the MATLAB routine `freqz` also includes frequency scaling and plotting, making it quite convenient:

```
[H,f]=freqz(b,a,n,fs)
```

where `b` and `a` are the filter coefficients and `n` is optional and specifies the number of points in the desired frequency spectra. Only the `b` coefficients are associated with FIR filters so the value of `a` is set to 1.0. The input argument, `fs`, is also optional and specifies the sampling frequency. Both output arguments are also optional and are usually not given. If `freqz` is called without the output arguments, the magnitude and phase plots are produced. If the output arguments are specified, the output vector `H` is the complex frequency response of the filter (the same variable produced by `fft`) and `f` is a frequency vector useful in plotting. If `fs` is given, `f` is in Hz and ranges between 0 and $f_s/2$; otherwise `f` is in rad/sample and ranges between 0 and π.

The MATLAB Signal Processing Toolbox has a filter design routine, `fir1`, based on the rectangular window filters described above. The basic rectangular window filters provided by Equations 8.7, 8.12, 8.13, and 8.14 can be generated by `fir1`. The use of this routine is not explained here, but can easily be found in the associated MATLAB Help file. Like freqz, this routines does things we already know how to do; it is just more convenient.

A filter design algorithm that is more versatile than the rectangular window filters already described is fir2 which can generate a filter spectrum having an arbitrary shape. The command structure for fir2 is:

```
b=fir2(order,f,G)
```

where order is the filter order (i.e., the number of coefficients in the impulse response), f is a vector of normalized frequencies in ascending order, and G is the desired gain of the filter at the corresponding frequency in vector f. In other words, plot(f,G) shows the desired magnitude frequency curve. Clearly f and G must be the same length, but duplicate frequency points are allowed, corresponding to step changes in the frequency response. In addition, the first value of f must be 0.0 and the last value 1.0 (equal to $f_s/2$). As with all MATLAB filter design routines, frequencies are normalized *to* $f_s/2$ (not f_s): that is, an f = 1 is equivalent to the frequency of $f_s/2$. Some additional optional input arguments are mentioned in the MATLAB Help file. An example showing the flexibility of the fir2 design routine is given next.

EXAMPLE 8.11

Design a double bandpass filter that has one passband between 50 and 100 Hz and a second passband between 200 and 250 Hz. Use a filter order of 65. Apply this to a signal containing sinusoids at 75 and 225 Hz in 20 dB of noise. (Use sig_noise to generate the signal.) Plot the desired and actual filter magnitude spectrum and the magnitude spectrum of the signal before and after filtering.

Solution: We could construct a double bandpass filter by putting two bandpass filters in series, but we can also use fir2 to construct a single FIR filter having the desired filter characteristics. First we specify the desired frequency characteristics in terms of frequency and gain vectors. The frequency vector must begin with 0.0 and end with 1.0, but can have duplicate frequency entries to allow for step changes in gain. Then we construct the coefficients (i.e., impulse response) using fir2, and apply it using filter or conv. In this example we use filter for a little variety. Then the magnitude spectrum of the filtered and unfiltered waveform is determined using the standard Fourier transform.

```
% Ex 8.11 Design a double bandpass filter.
%
fs=1000;                        % Sample frequency
N=2000;                         % Number of points
L=65;                           % Filter order
fl1=50/(fs/2);                  % Define cutoff freqs: first peak low cutoff
fh1=100/(fs/2);                 % First peak high cutoff frEquation
fl2=200/(fs/2);                 % Second peak low frEquation cutoff
fh2=250/(fs/2);                 % Second peak high frEquation cutoff
%
x=sig_noise([75 225],-20,N);    % Generate noise waveform
%
```

```
% Design filter Construct frequency and gain vectors
f = [0 fl1 fl1 fh1 fh1 fl2 fl2 fh2 fh2 1];   % Frequency vector
G = [0 0 1 1 0 0 1 1 0 0];                   % Gain vector
subplot(2,1,1); hold on;                     % Set up to plot spectra
plot(f,G);                                   % Plot the desired frequency response
   ........labels........
b = fir2(L,f,G);                             % Construct filter
[H,f] = freqz(b,1,512,fs);                   % Calculate filter frequency response
subplot(2,1,2);
plot(f,abs(H));                              % Plot filter frequency response
........labels........
y = filter(b,1,x);                           % Apply filter
Xf = abs(fft(x));                            % Compute magnitude spectra of filtered
Yf = abs(fft(y));                            % and unfiltered data
........plot and label magnitude spectra of data........
```

Figure 8.24A is a plot of the desired magnitude spectrum obtained simply by plotting the gain vector against the frequency vector (`plot(f,A)`). Figure 8.24B shows the actual magnitude spectrum that is obtained by the `freqz` routine but could have been found using the Fourier transform. The effect of applying the filter to the noisy data is shown in magnitude spectra of

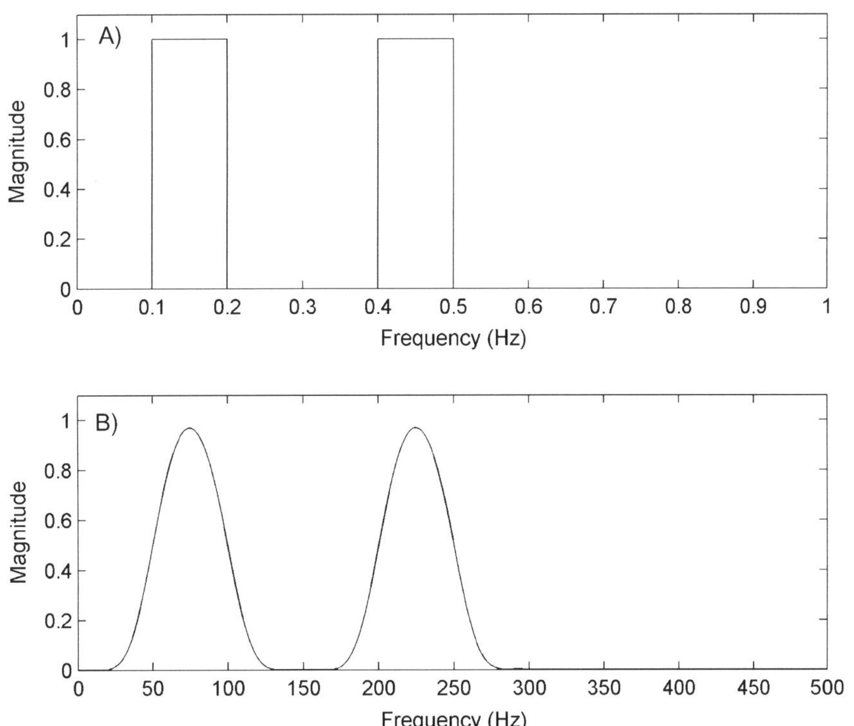

FIGURE 8.24 A) Desired magnitude response for the double bandpass filter in Example 8.10. B) Actual magnitude spectrum obtained from a 65th-order (i.e., 65 coefficients) FIR filter designed using MATLAB's `fir2` routine.

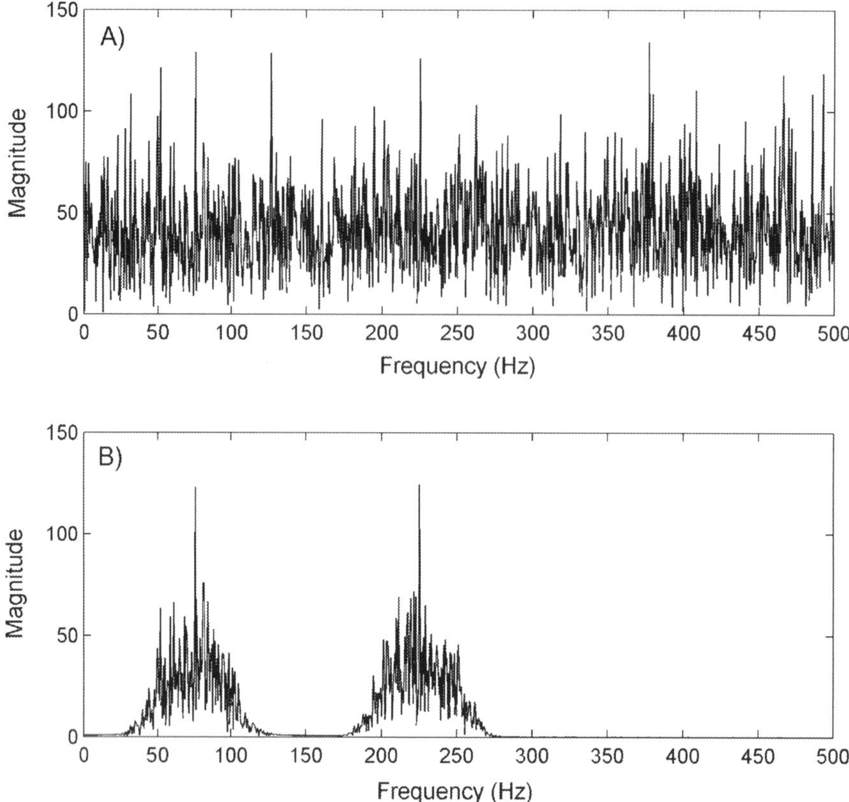

FIGURE 8.25 A) Magnitude spectrum of two sinusoids buried in a large amount of noise (SNR = −20 dB). The signal peaks at 75 and 225 Hz are visible, but a number of noise peaks have approximately the same amplitude. B) The spectrum of the signal and noise after filtering with a double bandpass filter having the spectral characteristic seen in Figure 8.24B. The signal peaks are no larger than in the unfiltered data, but are much easier to see as the surrounding noise has been greatly attenuated. Other interesting filters with unusual spectra are explored in the problems.

Figure 8.25. The spectrum of the unfiltered data, Figure 8.25A, shows the two peaks at 75 and 225 Hz, but many other noise peaks of nearly equal amplitude are seen. In the filtered data spectrum, Figure 8.25B, the peaks are not any larger but are clearly more evident as the energy outside the two passbands has been removed.

8.5 INFINITE IMPULSE RESPONSE FILTERS

To increase the attenuation slope of a filter, we can put two or more filters in series. As shown previously, the transfer function of two systems in series multiply to produce the overall transfer function (equation 5.9). In the frequency domain, the frequency spectra multiply, which leads to a steeper attenuation slope in the spectrum. Figure 8.26 shows the magnitude spectrum of a 12th-order FIR rectangular window filter (dashed line) and the spectrum of two such filters in series (dotted line). However, for the same computational effort you can use a single 24th-order filter and get an even steeper attenuation slope (Figure 8.26, solid line).

An intriguing alternative to adding filters in series is to filter the same data twice, once using an FIR filter, then feeding back the output of the FIR filter to another filter that also contributes to the output. This leads to the configuration shown in Figure 8.27, where the upper pathway is the standard FIR filter and the lower feedback pathway is another filter with its own impulse response, a[ℓ]. The problem with such an arrangement is apparent from Figure 8.27: the lower pathway receives its input from the filter's output but that input requires the lower pathway's output: the output, $y[n]$, is a function of itself. To avoid the *algebraic loop* (a mathematical term for vicious circle) that such a configuration produces, the lower feedback pathway only operates on *past values* of $y[n]$; that is, values that have already been determined. (In chicken and egg terms, the output, or rather *past output*, comes first!)

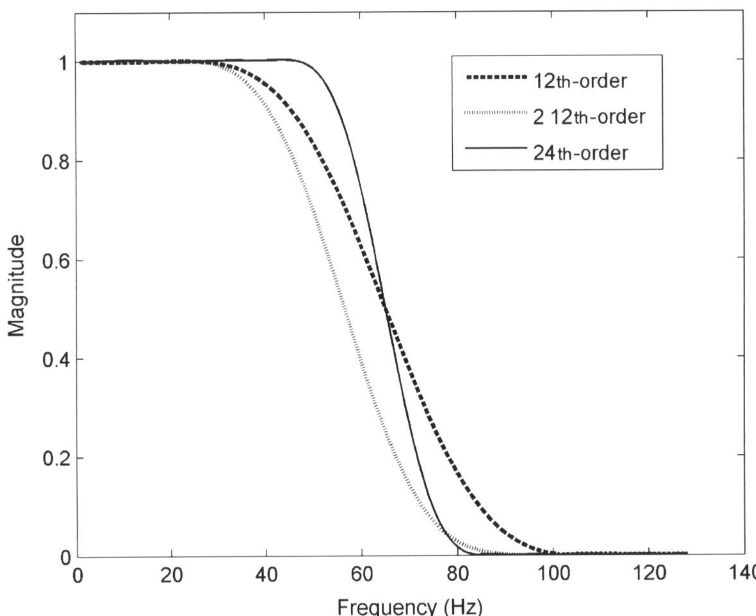

FIGURE 8.26 The magnitude spectrum of a 12th-order FIR filter (dashed line) and the spectrum obtained by passing a signal between two such filters in series (dotted line). The two series filters produce a curve that has twice the downward slope of the single filter. However, a single 24th-order filter produces an even steeper slope for the about the same computational effort (solid line).

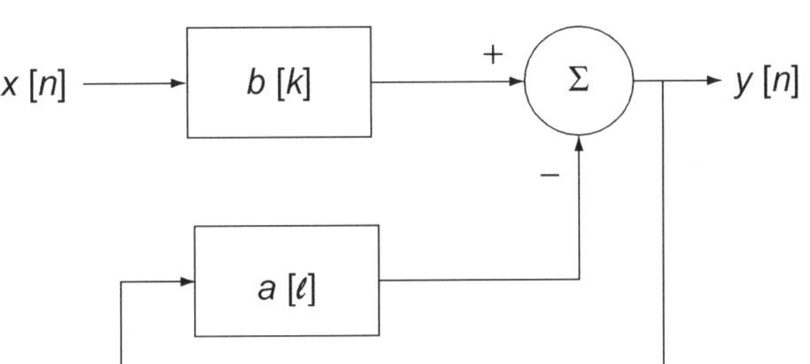

FIGURE 8.27 The configuration of an IIR filter containing two FIR filters. The upper filter plays the same role as in a standard FIR filter. The lower filter operates on the output, but only on past values of the output that have already been determined.

The equation for an IIR filter can be derived from the configuration given in Figure 8.27. The output is the convolution of the input with the upper FIR filter, minus the convolution of the lower FIR filter with the output, but only for past values of the output. This tacks on an additional term to the basic convolution equation given in Equation 8.1 to account for the feedback filter:

$$y[n] = \sum_{k=0}^{K} b[k]x[n-k] - \sum_{\ell=1}^{L} a[\ell]y[n-\ell] \tag{8.22}$$

where $b[k]$ are the upper filter coefficients, $a[\ell]$ are the lower filter coefficients, $x[n]$ is the input, and $y[n]$ the output. Note that while the b coefficients are summed over delays beginning at $k=0$, the a coefficients are summed over delays beginning at $\ell = 1.0$. Hence, the a coefficients are only summed over past values of $y[n]$. Since part of an IIR filter operates on past values of the output, these filters are also referred to as *recursive* filters and the a coefficients as the *recursive coefficients*. The primary advantages and disadvantages of IIR filters have been summarized in Table 8.1.

8.5.1 IIR Filter Implementation

IIR filters are applied to signals using Equation 8.22 where x is the signal and a and b are the coefficients that define the filter. While it would not be difficult to write a routine to implement this equation, it is unnecessary as the MATLAB `filter` routine works for IIR filters as well:

```
y = filter(b,a,x);     % IIR filter applied to signal x
```

where `b` and `a` are the same filter coefficients used in Equation 8.22, `x` is the input signal, and `y` is the output. (An FIR filter can be considered an IIR filter with $a=1$, so FIR filters are implemented using: `y = filter(b,1,x).`)

As mentioned above, the time shift of either FIR or IIR filters can be eliminated by using a noncausal implementation. Eliminating the time shift also reduces the filter's phase shift, and noncausal techniques can be used to produce zero-phase filters. Since noncausal filters use both future as well as past data samples (future only with respect to a given sample in the computer), they require the data to exist in computer memory. The Signal Processing Toolbox routine `filtfilt` employs noncausal methods to implement filters with no phase shift. The calling structure is exactly the same as MATLAB's `filter` routine:

```
y = filtfilt(b,a,x);     % Noncausal IIR filter applied to signal x
```

Problem 22 at the end of this chapter dramatically illustrates the difference between the use of `filter` and `filtfilt`.

8.5.2 Designing IIR Filters with MATLAB

The design of IIR filters is not as straightforward as that for FIR filters; however, the MATLAB Signal Processing Toolbox provides a number of advanced routines to assist in this process. IIR filters are similar to analog filters and originally were designed using tools developed for analog filters. In an analog filter there is a direct relationship between the number of independent energy storage elements in the system and the filter's rolloff slope: each energy storage element adds 20 dB/decade to the slope. As we find in Chapter 10, each energy storage device adds a $j\omega$ term to the denominator of the system transfer function, which accounts for its influence on the system's spectrum. In IIR digital filters, the first a coefficient, $a[0]$, always equals 1.0, but each additional a coefficient adds 20 dB/decade to the slope. So an eighth-order analog filter and an eighth-order IIR filter have the same slope. Since the downward slope of an IIR filter increases by 20 dB/decade for each filter order, determining the filter order needed for a given desired attenuation is straightforward.

IIR filter design under MATLAB follows the same procedures as FIR filter design, only the names of the routines are different. In the MATLAB Signal Processing Toolbox, the two-stage design process is supported for most of the IIR filter types. As with FIR design, a single-stage design process can be used if the filter order is known, and that is the approach that is covered here.

The Yule–Walker recursive filter is the IIR equivalent of the `fir2` FIR filter routine in that it allows for the specification of a general desired frequency response curve. The calling structure is also very similar to that of `fir2`.

```
[b,a]=yulewalk(order,f,G);
```

where `order` is the filter order, and `f` and `G` specify the desired frequency characteristic in the same manner as `fir2`: `G` is a vector of the desired filter gains at the frequencies specified in `f`. As in all MATLAB filter design routines, the frequencies in `f` are relative to $f_s/2$, and the first point in `f` must be 0 and the last point 1. Again, duplicate frequency points are allowed and correspond to steps in the frequency response.

EXAMPLE 8.12

Design the double bandpass filter that is used in Example 8.11. The first passband has a range of 50 to 100 Hz and a second passband ranges between 200 and 250 Hz. Use an IIR filter order of 12 and compare the results with the 65th-order FIR filter used in Example 8.11. Plot the frequency spectra of both filters superimposed for easy comparison.

Solution: Modify Example 8.11 to add the Yule–Walker filter, determine its spectrum using `freqz`, and plot this spectrum superimposed with the FIR filter spectrum. Remove the code relating to the signal as it is not needed in this example.

```
%Example 8.12 Design a double bandpass IIR filter and compare with a similar FIR filter
%
  ........same intial code as in Example 8.8........
% Design filter
f=[0 fl1 fl1 fh1 fh1 fl2 fl2 fh2 fh2 1];  % Construct desired
G=[0 0 1 1 0 0 1 1 0 0];                   % frequency characteristic
subplot(2,1,1);
plot(f,G); hold on;                        % Plot the desired magnitude spectrum
b1=fir2(L1,f,G);                           % Construct FIR filter
```

```
[H1,f1] = freqz(b1,1,512,fs);           % Calculate FIR filter spectrum
[b2 a2] = yulewalk(L2,f,G);             % Construct IIR filter
[H2 f2] = freqz(b2,a2,512,fs);          % Calculate IIR filter spectrum
........labels........
subplot(2,1,2);
plot(f1,abs(H1),'k'); hold on;          % Plot FIR filter magnitude spectrum
plot(f2,abs(H2),':k','LineWidth',2);    % Plot IIR filter magnitude spectrum
```

The spectral results from two filters are shown in Figure 8.28. The magnitude spectra of both filters look quite similar despite the fact that the IIR filter has far fewer coefficients. The FIR filter has 66 *b* coefficients while the IIR filter has 13 *b* coefficients and 13 *a* coefficients.

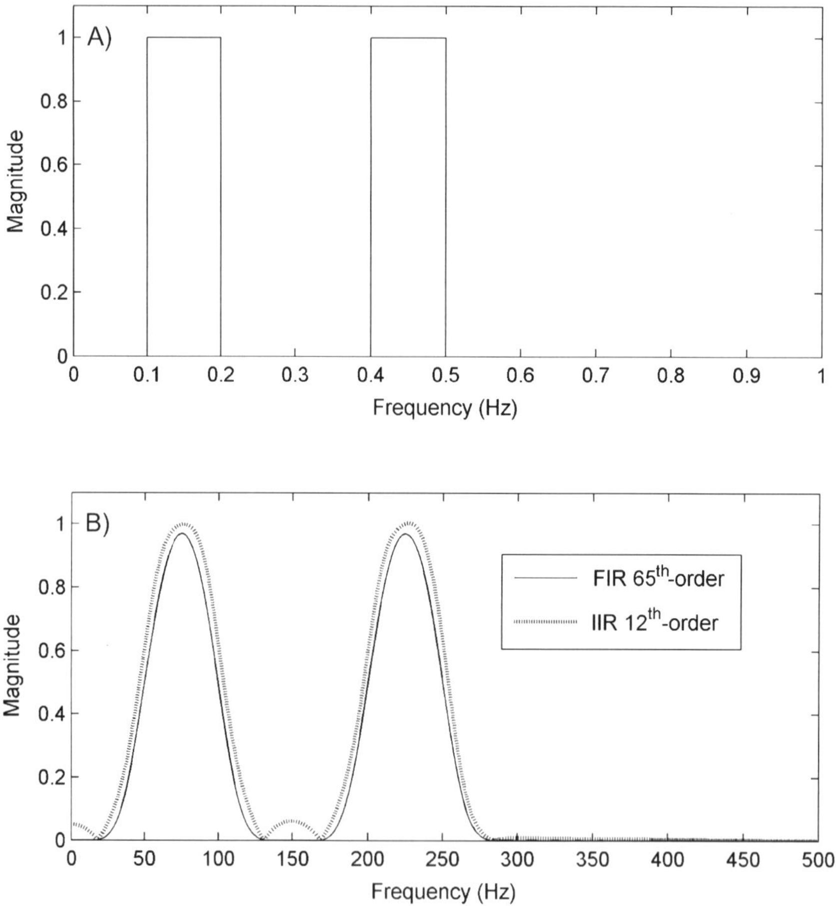

FIGURE 8.28 A) Desired magnitude response for the double bandpass filter in Examples 8.10 and 8.12. B) Actual magnitude spectrum obtained from a 65th-order FIR filter and a 12th-order IIR filter. The IIR filter has 26 coefficients while the FIR filter has 66 coefficients.

In an extension of Example 8.12, the FIR and IIR filters are applied to a random signal of 10,000 data points, and MATLAB's tic and toc are used to evaluate the time required for each filter operation.

```
% Example 8.12 (continued) Time required to apply the two filters in Ex. 8.12
% to a random array of 10,000 samples.
%
x = rand(1,10000);       % Generate random data
tic  % Start clock
y = filter(b2,a2,x);     % Filter data using the IIR filter
toc % Get IIR filter operation time
tic % Restart clock
y = filter(b1,1,x);      % Filter using the FIR filter
toc % Get FIR filter operation time
```

Surprisingly, despite the larger number of coefficients, the FIR filter is approximately 30% faster than the IIR filter. However, if conv is used to implement the FIR filter, then it takes three times as long.

As mentioned above, well-known analog filter types can be duplicated as IIR filters. Specifically, analog filters termed Butterworth, Chebyshev Type I and II, and Elliptic (or Cauer) designs can be implemented as IIR digital filters and are supported in the MATLAB Signal Processing Toolbox. Butterworth filters provide a frequency response that is maximally flat in the passband and monotonic overall. To achieve this characteristic, Butterworth filters sacrifice rolloff steepness; hence, the Butterworth filter has a less sharp initial attenuation characteristic than other filters. The Chebyshev Type I filter features a faster rolloff than Butterworth filters, but has ripple in the passband. The Chebyshev Type II filter has ripple only in the stopband, its passband is monotonic, but it does not roll off as sharply as Type I. The ripple produced by Chebyshev filters is termed *equi-ripple* since it is of constant amplitude across all frequencies. Finally, Elliptic filters have a steeper rolloff than any of the above, but have equi-ripple in both the passband and stopband. While the sharper initial rolloff is a desirable feature as it provides a more definitive boundary between passband and stopband, most biomedical engineering applications require a smooth passband making Butterworth the filter of choice.

The filter coefficients for a Butterworth IIR filter can be determined using the MATLAB routine:

```
[b,a] = butter(order,wn,'ftype'); % Design Butterworth filter
```

where order and wn are the order and cutoff frequencies, respectively. (Of course wn is relative to $f_s/2$.) If the 'ftype' argument is missing, a low-pass filter is designed if wn is scalar or a bandpass filter if wn is a two-element vector. In the latter case, wn = [w1 w2], where w1 is the low cutoff frequency and w2 is the high cutoff frequency. If a high-pass filter is desired, then wn should be a scalar and 'ftype' should be high. For a stopband filter, wn would be a two-element vector indicating the frequency ranges of the stopband and 'ftype' should be stop. The outputs of butter are the b and a coefficients.

While the Butterworth filter is the only IIR filter you are likely to use, the other filters are easily designed using the associated MATLAB routine. The Chebyshev Type I and II filters are designed with similar routines except an additional parameter is needed to specify the allowable ripple:

```
[b,a]=cheby1(order,rp,wn,'ftype'); % Design Chebyshev Type I
```

where the arguments are the same as in `butter` except for the additional argument, `rp`, which specifies the maximum desired passband ripple in dB. The Type II Chebyshev filter is designed using:

```
[b,a]=cheby2(n,rs, wn,'ftype'); % Design Chebyshev Type II
```

where again the arguments are the same, except `rs` specifies the stopband ripple again in dB but with respect to the passband gain. In other words, a value of 40 dB means that the ripple does not exceed 40 dB *below* the passband gain. In effect, this value specifies the minimum attenuation in the stopband.

The Elliptic filter includes both stopband and passband ripple values:

```
[b,a]=ellip(n,rp,rs,wn,'ftype'); % Design Elliptic filter
```

where the arguments presented are in the same manner as described above, with `rp` specifying the passband gain in dB and `rs` specifying the stopband ripple relative to the passband gain.

The example below uses these routines to compare the frequency response of the four IIR filters discussed above.

EXAMPLE 8.13

Plot the frequency response curves (in dB) obtained from an 8th-order low-pass filter using the Butterworth, Chebyshev Type I and II, and Elliptic filters. Use a cutoff frequency of 200 Hz and assume a sampling frequency of 2 kHz. For all filters, the ripple or maximum attenuation should be less than 3 dB in the passband, and the stopband attenuation should be at least 60 dB.

Solution: Use the design routines above to determine the *a* and *b* coefficients, use `freqz` to calculate the complex frequency spectrum, take the absolute of this spectrum and convert to dB (`20*log(abs(H))`), then plot using `semilogx`. Repeat this procedure for the four filters.

```
% Example 8.11 Frequency response of four IIR 8th-order low-pass filters
%
fs=2000;                    % Sampling filter
n=8;                        % Filter order
wn=200/1000;                % Filter cutoff frequency
rp=3;                       % Maximum passband ripple
rs=60;                      % Maximum stopband ripple
% Determine filter coefficients
[b,a]=butter(n,wn);         % Butterworth filter coefficients
[H,f]=freqz(b,a,256,fs);    % Calculate complex spectrum
H=20*log10(abs(H));         % Convert to magnitude in dB
subplot(2,2,1);
semilogx(f,H,'k');          % Plot spectrum in dB vs. log frequency
  ........labels and title........
%
[b,a]=cheby1(n,rp,wn);      % Chebyshev Type I filter coefficients
[H,f]=freqz(b,a,256,fs);    % Calculate complex spectrum
```

```
H=20*log10(abs(H));           % Convert to magnitude in dB
subplot(2,2,2);
semilogx(f,H,'k');            % Plot spectrum in dB vs. log frequency
........labels and title........
%
[b,a]=cheby2(n,rs,wn);        % Chebyshev Type II filter coefficients
[H,f]=freqz(b,a,256,fs);      % Calculate complex spectrum
H=20*log10(abs(H));   % Convert to magnitude in dB
subplot(2,2,3);
semilogx(f,H,'k');            % Plot spectrum in dB vs. log frequency
........labels and title........
%
[b,a]=ellip(n,rp,rs,wn);      % Elliptic filter coefficients
[H,f]=freqz(b,a,256,fs);      % Calculate complex spectrum
H=20*log10(abs(H));   % Convert to magnitude in dB
subplot(2,2,4);
semilogx(f,H,'k');            % Plot spectrum in dB vs. log frequency
........labels and title........
```

The spectra of the four filters are shown in Figure 8.29. As described above, the Butterworth is the only filter that has smooth frequency characteristics in both the passband and stopband; it

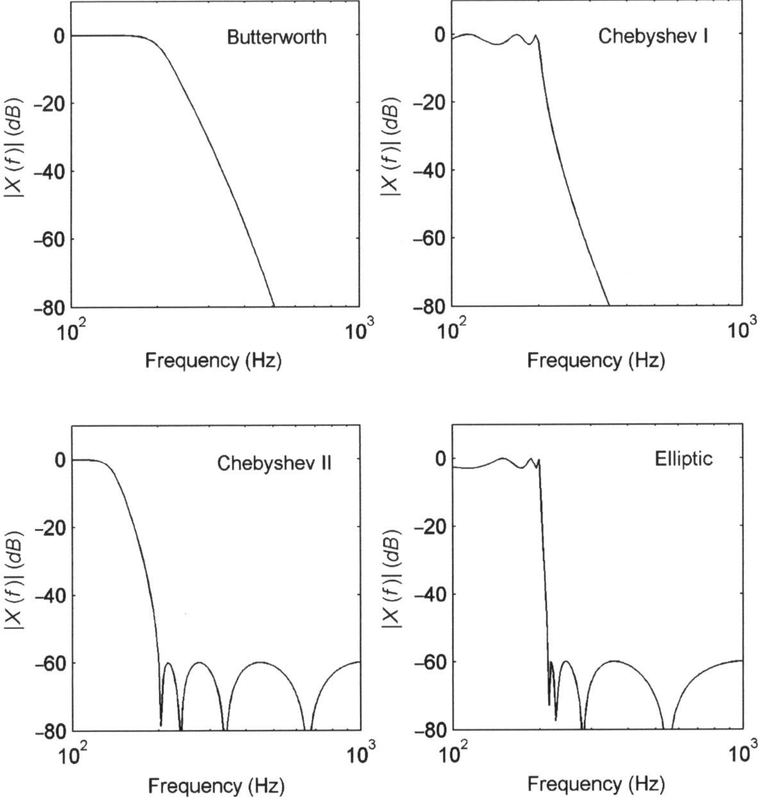

FIGURE 8.29 The spectral characteristics of four different 8th-order IIR low-pass filters with a cutoff frequency of 200 Hz. The assumed sampling frequency is 2 kHz. The spectral characteristics show that it is possible to increase significantly the initial sharpness of an IIR filter if various types of ripple in the spectral curve can be tolerated.

is this feature that makes it popular in biomedical signal processing both in its analog and digital incarnations. The Chebyshev Type II filter also has a smooth passband and a slightly steeper initial slope than the Butterworth, but it does have ripple in the stopband which can be problematic in some situations. The Chebyshev Type I has an even sharper initial slope, but also has ripple in the passband which limits its usefulness. The sharpest initial slope is provided by the Elliptic filter, but ripple is found in both the passband and stopband. Reducing the ripple of these last three filters is possible in the design process, but this also reduces the filter's initial sharpness, another illustration of the compromises that continually arise in engineering.

Other types of filters exist in both FIR and IIR forms; however, the filters described above are the most common and are the most useful in biomedical engineering.

8.6 THE DIGITAL TRANSFER FUNCTION AND THE *z*-TRANSFORM

Frequency analysis techniques such as the Fourier transform, the Bode plot, and the transfer function are the most popular and useful tools for analyzing systems or signals. These frequency methods are applicable as long as the waveforms are either periodic, can be assumed to be periodic, or are aperiodic and finite. The Fourier transform is used to convert signals into the frequency domain, and the transfer function is used to analyze any general system's response to these signals. The Fourier transform can be applied analytically, as was done in Example 3.2 (if the signal is not too complicated), but usually it is implemented in the digital domain using a computer. The Bode plot is a system analysis tool that provides a frequency or spectral plot of a system's transfer function. It too can be implemented either analytically, using graphical tricks (Examples 5.8. 5.10, 5.11, and 5.12), or using a computer (Examples 5.7, and 5.12).

Frequency methods fall apart when confronted with so-called transient signals: functions that change and never return to their original levels such as step functions, or in systems with initial conditions. In such cases the Laplace transform comes to the rescue, but Laplace transform methods are strictly analytical. The Laplace transform is difficult to evaluate numerically on a computer; transformation tables and analytical calculations are necessary. Most of us would agree that it is far easier to use the Fourier transform in MATLAB than do paper and pencil analysis of a system using Laplace transforms. To analyze systems with transient signal inputs or nonzero initial conditions, we are best off turning to simulation. Simulation can be used to analyze any system with any signals, which explains its popularity, but it usually does not provide as much insight into the system.

We have most of the tools we need to analyze linear signals or systems, but in a few cases it is useful to have a digital version of the Laplace transform. This allows us to construct a strictly digital transfer function. To develop a digital transform, a transform especially suited to numerical series, we return to the approach that was used to develop the Laplace transform: extending the Fourier transform. In this approach, the complex frequency s (where $s = \sigma + j\omega$) is inserted into the Fourier transform in place of the standard frequency term $j\omega$. This leads to the Laplace transform:

$$X(\sigma, \omega) = \int_0^{\infty} x(t)e^{-\sigma t}e^{-j\omega t}\,dt = \int_0^{\infty} x(t)e^{-st}\,dt \tag{8.23}$$

The addition of the $e^{-\sigma t}$ term in the Laplace transform insures convergence for a wide range of input functions providing σ takes on the proper value.

The *z-transform* is a modification of the Laplace transform that is more suited to the digital domain. It is much easier to make the transformation between the *discrete time domain* and the z-transform domain. Since we need a discrete equation, the first step in modifying the Laplace transform is to convert it to discrete form by replacing the integral with a summation and substituting sample number n for the time variable t. (Actually $t = nT_s$, but we assume that T_s is normalized to 1.0 for this development.) These modifications give:

$$X(\sigma, \omega) = \sum_{n=-\infty}^{\infty} x[n]e^{-\sigma n}e^{-j\omega n} \tag{8.24}$$

Next, define a new variable:

$$r = e^{\sigma} \tag{8.25}$$

Note that $\sigma = \ln(r)$ if we take the natural logarithm of both sides of Equation 8.21. Substituting r for e^{σ} into Equation 8.20 gives:

$$X(r, \omega) = \sum_{n=-\infty}^{\infty} x[n]r^{-n}e^{-j\omega n} \tag{8.26}$$

Equation 8.26 is a valid equation for the z-transform, but in the usual format the equation is simplified by defining another new variable, a complex variable. This is essentially the idea used in the Laplace transform where the complex variable s is introduced to represent $\sigma + j\omega$. The new variable, z, is defined as:

$$z = re^{j\omega} = |z|e^{j\omega} \tag{8.27}$$

where r is now just the magnitude of the new complex variable z. Substituting z into Equation 8.26 gives:

$$X(z) = Z[x[n]] = \sum_{n=-\infty}^{\infty} x[n]z^{-n} \tag{8.28}$$

This is the defining equation for the z-transform, notated as $Z\,[x[n]]$. In any real application, the limit of the summation in Equation 8.28 is finite, usually the length of $x[n]$. As with the Laplace transform, the z-transform is based around a complex variable: in this case the arbitrary complex variable z which equals $|z|e^{jw}$. Like the analogous variable s in the Laplace transform, z is termed the *complex frequency*. As with the Laplace variable s, it is also possible to substitute $e^{j\omega}$ for z to perform a strictly sinusoidal analysis.[3] This is a very useful property as it allows us to easily determine the frequency characteristic of a z-transform transfer function.

While the substitutions made to convert the Laplace transform to the z-transform may seem arbitrary, they greatly simplify conversion between time and frequency in the digital domain. Unlike the Laplace transform, which requires a table and often considerable algebra, the conversion between discrete complex frequency representation and the discrete time function is easy. For example, it is easy to convert from the z-transform transfer function of an IIR filter (the complex frequency domain) to its equivalent filter coefficients (the *a*s and *b*s, the discrete time representation) and vice versa. Moreover, knowing the z-transform of a filter, we can easily find the spectral characteristics of the filter: substitute $e^{j\omega}$ for z and apply standard techniques like the Fourier transform.

In Equation 8.28 the x and z are *both* functions of the discrete time variable n. This means that every data sample in the sequence $x[n]$ is associated with a unique power of z. Recall that for a discrete variable $x[n] = [x_1, x_2, x_3, \ldots x_N]$, the value of n defines a sample's position in the sequence. The higher the value of n in the z-transform, the further along in the time sequence a given data sample is located. So in the z-transform, the value of n indicates a time shift for the associated input waveform, $x[n]$. This time-shifting property of z^{-n} is formally stated as:

$$Z[x[n-k]] = z^{-k} Z[x[n]] \tag{8.29}$$

As is shown below, this time-shifting property of the z-transform makes it easy to implement the z-transform conversion from a data sequence to the z-transform representation and vice versa.

8.6.1 The Digital Transfer Function

One of the most useful applications of the z-transform lies in its ability to define the digital equivalent of a transfer function. By analogy to linear system analysis, the digital transfer function is defined as:

$$H(z) = \frac{Y(z)}{X(z)} = \frac{Z[y[n]]}{Z[x[n]]} \tag{8.30}$$

where $X(z)$ is the z-transform of the input signal, $x[n]$, and $Y(z)$ is the z-transform of the system's output, $y[n]$. As an example, a simple linear system known as a *unit delay* is shown in Figure 8.30. In this system, the time-shifting characteristic of z is used to define processes where the output is the same as the input but shifted (or delayed) by one data sample. The z-transfer function for this process is $H(z) = z^{-1}$.

[3]If $|z|$ is set to 1, then from Equation 8.27, $z = e^{j\omega}$. This is called evaluating z on the unit circle. See Smith (1997) or Bruce (2001) for a clear and detailed discussion of the properties of z and the z-transform.

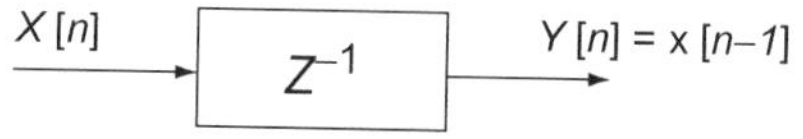

FIGURE 8.30 A unit delay is a linear system that shifts the input by one data sample. Other powers of z can be used to provide larger shifts. The transfer function of this system is z^{-1}.

Most transfer functions are more complicated than that of Figure 8.30 and can include polynomials of z in both the numerator and denominator, just as continuous time domain transfer functions contain polynomials of s in those places:

$$H(z) = \frac{b[0] + b[1]z^{-1} + b[2]z^{-2} + \cdots + b[K]z^{-K}}{1 + a[1]z^{-1} + a[2]z^{-2} + \cdots a[L]z^{-L}} \tag{8.31}$$

where the $b[k]$s are constant coefficients of the numerator and the $a[\ell]$s are coefficients of the denominator.[4] While $H(z)$ has a structure similar to the Laplace domain transfer function $H(s)$, there is no simple relationship between them.[5] For example, unlike analog systems, the order of the numerator, K, need not be less than, or equal to, the order of the denominator, L, for stability. In fact, systems that have a denominator order of 1 (i.e., such as FIR filters) are more stable that those having higher-order denominators.

Note that just as in the Laplace transform, a linear system is completely defined by the denominator, a, and numerator, b, coefficients. Equation 8.31 can be more succinctly written as:

$$H(z) = \frac{\sum_{k=0}^{K} b[k]z^{-k}}{\sum_{\ell=0}^{L} a[\ell]z^{-\ell}} \tag{8.32}$$

From the digital transfer function, $H(z)$, it is possible to determine the output given any input:

$$Y(z) = X(x)H(z) \quad y(n) = Y^{-1}(z) \tag{8.33}$$

where $Y^{-1}(z)$ is the inverse z-transform. For many systems, it is possible to accomplish the same thing by applying Laplace transforms to the continuous time version of the system, but Laplace transforms are more cumbersome to handle analytically and, since they represent continuous operations, cannot be evaluated numerically.

While the input-output relationship can be determined from Equation 8.31 or 8.32, it is more common to use the difference equation, which is analogous to the time domain

[4]This equation can be found in a number of different formats. A few authors reverse the role of the a and b coefficients with as being the numerator coefficients and bs being the denominator coefficients. More commonly, the denominator coefficients are written with negative signs changing the nominal sign of the a coefficient values. Finally, it is common to start the coefficient series with $a[0]$ and $b[0]$ so the coefficient index is the same as the power of z. This is the notation used here, which departs from that used by MATLAB where each coefficient in the series is indexed beginning with a 1.

[5]Nonetheless, the z-transfer function borrows from the Laplace *terminology*, so the term *pole* is sometimes used for denominator coefficients and the term *zero* for numerator coefficients.

equation and can be obtained from Equation 8.32 by applying the time-shift interpretation to the term z^{-n}:

$$y[n] = \sum_{k=0}^{K-1} b[k]x[n-k] - \sum_{\ell=1}^{L-1} a[\ell]y[n-\ell] \tag{8.34}$$

This equation assumes that $a[0] = 1$, as specified in Equation 8.31. Equation 8.34 is the same as Equation 8.22 introduced to describe the IIR filter. Hence any system represented by the z-transform transfer function can be realized using the IIR filter equation and the MATLAB IIR filter routines (i.e., `filter` or `filtfilt`).

Just as in the Laplace domain, if the signals are limited to sinusoids we can substitute $e^{j\omega}$ for z. All we need to determine a spectrum are sinusoids. With this substitution, Equation 8.32 becomes:

$$H(m) = \frac{\sum_{k=0}^{K-1} b[k]e^{-j\omega k}}{\sum_{\ell=0}^{L-1} a[\ell]e^{-j\omega \ell}} = \frac{\sum_{k=0}^{K-1} b[k]e^{-2\pi mn/N}}{\sum_{\ell=0}^{L-1} a[\ell]e^{-2\pi mn/N}} = \frac{fft\{\mathrm{b[k]}\}}{fft\{\mathrm{a}[\ell]\}} \tag{8.35}$$

where N is the length of $b[k]$ (which can be zero-padded) and *fft* indicates the Fourier transform. As with all Fourier transforms, the actual frequency can be obtained from the variable m by multiplying by f_s/N or, equivalently, $1/(NT_s)$. Equation 8.35 shows us how to move from the time domain, the a and b coefficients, to the z domain, then to the frequency domain, for any linear system. This equation is easily implemented in MATLAB and is the equation used in the routine `freqz`.

8.6.2 MATLAB Implementation

Some of the MATLAB functions used in filter design and application can also be used in digital transfer function analysis. The MATLAB routine `filter` has been used in several previous examples to implement FIR and IIR filters, but since it is based on Equation 8.34, it can be used to implement any LTI system given the z-transform transfer function. From Equation 8.34 we can see why the filter routine also works for FIR filters: if all the a coefficients are zero (except $a[0]$ which equals 1), then Equation 8.34 becomes identical to the convolution equation.

EXAMPLE 8.14

Find and plot the frequency spectrum (magnitude and phase) of a system having the digital transfer function shown below. Also plot the step response of this system. Assume a sampling rate of 1000 Hz.

$$H(z) = \frac{0.2 + 0.5z^{-1}}{1 - 0.2z^{-1} + 0.8z^{-2}} \tag{8.36}$$

Solution: First determine the a and b coefficients from the digital transfer function. This can be done by inspecting $H(z)$: $b = [0.2, 0.5]$ and $a = [1.0, 0.2, 0.8]$. Next find $H(f)$ using Equation 8.35 and noting that $f = mf_s/N$. To find the step response, just treat the system like a filter since there is no difference between a system and a filter. (We use the word "filter" when the intent is to adjust the spectral content of a signal, while "system" is the more general term.) So to find the step response, we generate a step signal and apply Equation 8.34 to that signal via MATLAB's `filter` routine.

```
% Example 8.14 Plot the Frequency characteristics and step response of a linear
% system.
%
fs = 1000;                          % Sampling frequency
N = 512;                            % Number of points (arbitrary)
%
% Define a and b coefficients based digital transfer function
a = [1 .2 .8];                      % Denominator coefficients
b = [.2 .5];                        % Numerator coefficients
%
H = fft(b,N)./fft(a,N);             % Compute H(m). Equation 8.35.
Hm = 20*log10(abs(H));              % Get magnitude in dB
Theta = (angle(H))*360/(2*pi);      % and phase in degrees
f = (1:N/2) *fs/N;                  % Frequency vector for plotting
  ........plot magnitude and phase transfer function with grid and labels........
%
x = [0, ones(1,N1)];                % Generate a unit step
y = filter(b,a,x);                  % Calculate output using Equation 8.34
........plot first 60 points for clarity and label........
```

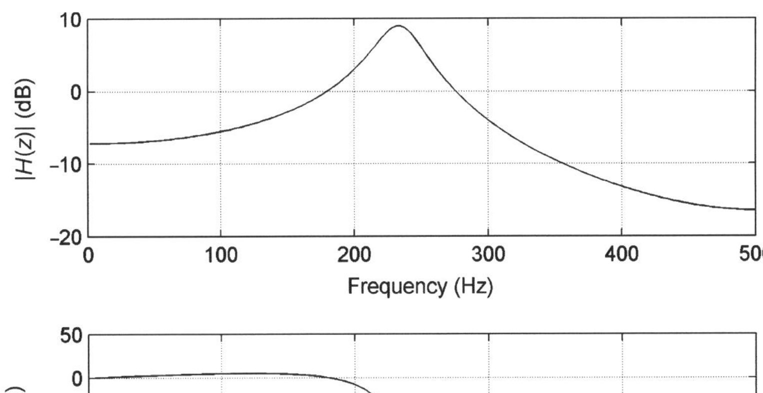

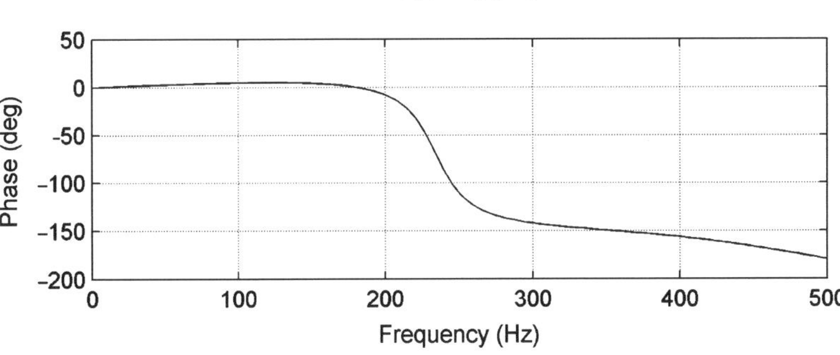

FIGURE 8.31 Plot of the frequency characteristic (magnitude and phase) of the digital transfer function given in Equation 8.36. Using our Bode plot skills, we can determine that the spectrum is one of an underdamped system with a resonant frequency of approximately 230 Hz.

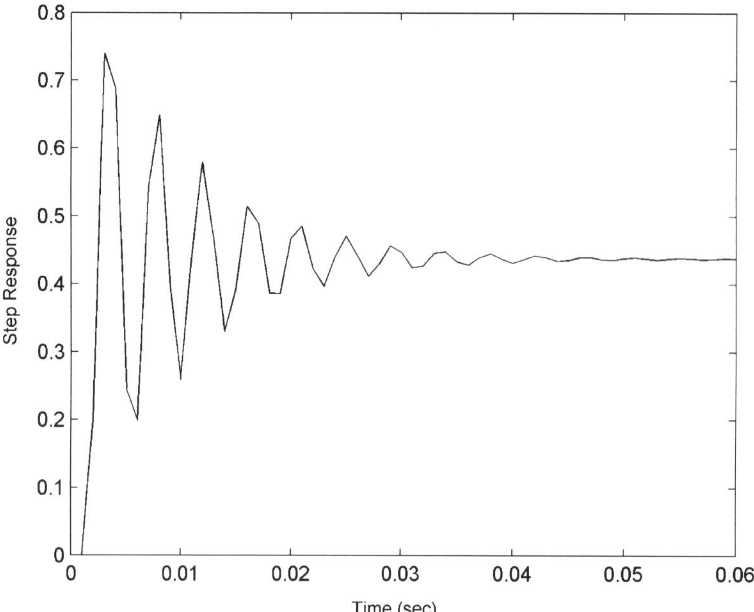

FIGURE 8.32 Step response of the system represented by the digital transfer function given in Equation 8.36. The decaying oscillatory response is characteristic of an underdamped system as expected from the Bode plot in Figure 8.31.

The spectrum of the system defined by Equation 8.36 is given in Figure 8.31. The spectrum has a hump at around 230 Hz, indicating that the system is underdamped with a resonant frequency of approximately 230 Hz. The step response given in Figure 8.32 is clearly that of an underdamped system showing a decaying oscillation. We can estimate the time difference between oscillatory peaks as roughly 0.004 sec which is equivalent to 250 Hz.

This example demonstrates how easy it is to determine the time and frequency behavior of a system defined by a z-transform transfer function. The approach used here can readily be extended to any transfer function of any complexity. Other examples of the use of the z-transform are given in Problems 27 and 28 at the end of the chapter.

8.7 SUMMARY

Filters are used to shape the spectrum of a signal, often to eliminate frequency ranges that include noise or enhance frequencies of interest. Filters come in two basic versions: finite impulse response (FIR) and infinite impulse response (IIR). The former are just systems such as studied previously with impulse response functions specifically designed to appropriately alter the input signal's spectrum. They are implemented using standard convolution. They have linear phase characteristics, but their spectra do not cut off as sharply as IIR filters of the same complexity. They can be applied to images where the impulse response becomes a two-dimensional matrix. FIR filters can be designed based on the inverse Fourier transform as in rectangular window filters, but MATLAB routines also exist that simplify design.

IIR filters consist of two impulse response functions, one of which is applied to the input signal thought standard convolution and the other to a delayed version of the output also using convolution. They can produce greater attenuation slopes; some types can produce very sharp initial cutoffs at the expense of ripple in the bandpass region. An IIR filter known as the Butterworth filter produces the sharpest initial cutoff without bandpass ripple and is the most commonly used in biomedical applications. The design of IIR filters is more complicated than FIR filters and is best achieved using MATLAB routines.

A special digital transfer function based on the z-transform can be used to analyze IIR filters or any other LTI system within the digital domain. The spectral characteristics of z-domain transfer functions can easily be determined using the Fourier transform. Unlike the Laplace and frequency domain transfer functions, there is no easy way to convert between z-domain transfer functions and transfer functions in the other two domains.

PROBLEMS

1. Use MATLAB to find the frequency response of a 3-point moving average filter (i.e., [1 1 1]/3) and a 10-point moving average filter. Use appropriate zero-padding to improve the spectra. Plot both the magnitude and phase spectra assuming a sample frequency of 500 Hz.
2. Use `sig_noise` to generate a 20-Hz sine wave in 5 dB of noise (i.e., SNR = −5 dB) and apply the two filters from Problem 1 using the MATLAB `filter` routine. Plot the time characteristics of the two outputs. Use a data length (`N`) of 200 in `sig_noise` and remember that `sig_noise` assumes a sample frequency of 1 kHz.
3. Find the magnitude spectrum of an FIR filter with a weighting function of `b = [.2 .2 .2 .2 .2]` in two ways: a) apply the `fft` with padding to the filter coefficients as in Problem 1 and plot the magnitude spectrum of the result; b) pass white noise through the filter using `conv` and plot the magnitude spectra of the output. Since white noise has, theoretically, a flat spectrum, the spectrum of the filter's output to white noise should be the same as the spectrum of the filter. In the second method, use a 20,000-point noise array; that is, `y = conv (b, randn(20000,1))`. Use the Welch averaging method described in Section 4.3 to smooth the spectrum. For the Welch method, use a suggested segment size of 128 points and a 50% segment overlap. Since the `welch` routine produces the power spectrum, you need to take the square root to get the magnitude spectrum for comparison with the FT method. The two methods use different scaling, so the vertical axes are slightly different. Assume a sampling frequency of 200 Hz for plotting the spectra.
4. Use `sig_noise` to construct a 512-point array consisting of two closely spaced sinusoids of 200 and 230 Hz with SNR of −8 dB and −12 dB, respectively. Plot the magnitude spectrum using the FFT. Generate a 25-coefficient rectangular window bandpass filter using the approach in Example 8.5. Set the low cutoff frequency to 180 Hz and the

high cutoff frequency to 250 Hz. Apply a Blackman window to the filter coefficients, then filter the data using MATLAB's `filter` routine. Plot the magnitude spectra before and after filtering.

5. Write a program to construct the coefficients of a 15th-order low-pass rectangular window filter (i.e., 16 coefficients). Assume $f_s = 1000$ and make the cutoff frequency 200 Hz. Apply the appropriate Hamming (Equation 8.10) and Blackman (Equation 8.11) windows to the filter coefficients. Find and plot the spectra of the filter without a window and with the two windows. Plot the spectra superimposed to aid comparison and pad the windows to $N = 256$ when determining the spectra. As always, do not plot redundant points. You can use the `blackman` routine for the Blackman window, but you need to write your own code for the Hamming window.
6. Comparison of Blackman and Hamming windows. Generate the filter coefficients of a 128th-order rectangular window filter. Apply the Blackman and Hamming windows to the coefficients. Apply these filters to an impulse function using the MATLAB filter routine. The impulse input should consist of a 1 followed by 255 zeros. The impulse response will look nearly identical, so take the Fourier transform of each the two responses and plot the magnitude and phase. Use the MATLAB `unwrap` routine on the phase data before plotting. Is there any difference in the spectra of the impulses produced by the two filters? Look carefully.
7. Load file ECG_9`.mat` which contains 9 sec of ECG data in variable `x`. These ECG data have been sampled at 250 Hz. The data have a low-frequency signal superimposed over the ECG signal, possibly due to respiration artifact. Filter the data with a high-pass filter having 65 coefficients and a cutoff frequency of 8 Hz. Plot the spectrum of the filter to confirm the correct type and cutoff frequency. Also plot the filtered and unfiltered ECG data. [*Hint*: You can modify a section of the code in Example 8.2 to generate the high-pass filter.]
8. ECG data are often used to determine the heart rate by measuring the time interval between the peaks that occur in each cycle known as the *R wave*. To determine the position of this peak accurately, ECG data are first prefiltered with a bandpass filter that enhances the QRS complex. Load file `ECG_noise.mat` which contains 10 sec of noisy ECG data in variable `ecg`. Filter the data with a 64-coefficient FIR bandpass filter to best enhance the R-wave peaks. Determine the low and high cutoff frequencies empirically, but they will be in the range or 2 to 30 Hz. The sampling frequency is 250 Hz.
9. Load the variable `x` found in `Prob8_9_data.mat`. Take the derivative of these data using the two-point central difference algorithm with a skip factor of 6, implemented using the `filter` routine. Now add an additional low-pass filter using a rectangular window filter with 65 coefficients and a cutoff frequency 25 Hz. Use the `conv` routine with option `same` to eliminate the added delay that is induced by the `filter` routine. Plot the original time data, the result of the two-point central difference algorithm, and the low-pass filtered derivative data. Also plot the spectrum of the two-point central difference algorithm, the low-pass filter, and the combined spectrum. The combined spectrum can be obtained by simply multiplying the two-point central difference spectrum point-by-point with the low-pass filter spectrum.
10. Load the variable `x` found in `Prob8_10_data.mat`. These data were sampled at 250 Hz. Use these data to compare the two-point central difference algorithm (Equation 8.15)

with a simple differencer (i.e., MATLAB's diff routine) in combination with a low-pass filter. Use a skip factor of 8 for the two-point central difference algorithm and to filter the differencer output use a 67th-order rectangular window lowpass filter (Equation 8.7) with a cutoff frequency of 20 Hz. Use `diff` to produce the differencer output then divide by T_s and filter this output with the low-pass filter. Note that the derivative taken this way is smooth, but has low-frequency noise.

11. The two-dimensional filter below is used to sharpen an image.

$$b = \begin{vmatrix} -1 & -4 & -1 \\ -4 & 26 & -4 \\ -1 & -4 & -1 \end{vmatrix}$$

This filter works somewhat like the human visual system, in which receptive fields respond to central activity but are inhibited by surrounding activity ("lateral inhibition"). (Note that the positive central number is surrounded by negative numbers.) Construct this sharpening filter impulse response matrix and apply it to the brain image given as variable `brain` in file `Brain.mat`. After you apply this filter increase the contrast of the output image by limiting the display range using MATLAB's `caxis (caxis([0 1]))`. Display the original and filtered image after contrast enhancement. Note the greater sharpness of the filtered image.

12. Plot the spectra of the sharpening filter used in Problem 11 and the spectra of the Sobel filter given in Equation 8.21. Pad the Fourier transform to improve the apparent resolution of the spectra. Assume that both filters are applied to an image that is 10×10 cm and provide the correct horizontal axes. Since you are plotting the magnitude spectrum, both peaks of the Sobel filter spectrum are positive. Note the high-pass characteristics of the sharpening filter.

13. Load the noisy image of the cell found as variable `cell_noise` in file `Cell_noise.mat`. Filter this image with a 3×3 and 7×7 moving average filter. Plot the original image and plot the two filtered images side by side for easy comparison. Also determine the magnitude spectra of the two filters and plot them side by side in a separate figure. How do the two spectra compare?

14. Extend Example 8.10 to include enhancement of the other two horizontal boundaries, top to bottom and bottom to top, in addition to the two vertical boundaries found in Example 8.10. All boundaries between light and dark should be highlighted in the thresholded image. You may want to change the threshold slightly to improve the image.

15. Use `sig_noise` to construct a 512-point array consisting of two widely separated sinusoids: 150 and 350 Hz, both with SNR of -14 dB. Use MATLAB's `fir2` to design a 65-coefficient FIR filter having a spectrum with a double bandpass. The bandwidth of the two bandpass regions should be 20 Hz centered about the two peaks. Plot the filter's spectrum superimposed on the desired spectrum. [*Hint*: Referring to the calling structure of `fir2`, plot `m` versus `f` after scaling the frequency vector, `f`, appropriately.] Also plot the spectrum before and after filtering.

16. The file `Prob8_16_data.mat` contains an ECG signal in variable `x` that was sampled at 250 Hz and has been corrupted by 60-Hz noise. The 60-Hz noise is at a high frequency compared to the ECG signal, so it appears as a fuzzy line superimposed on the signal. Construct a 127-coefficient FIR rectangular bandstop filter with a center frequency of 60 Hz and a bandwidth of 10 Hz, and apply it to the noisy signal. Implement the filter using either `filter` or `conv` with the `same` option, but note the time shift if you use the former. Plot the signal before and after filtering and also plot the filter spectrum to insure the filter is correct. [*Hint*: You can modify a portion of the code in Example 8.5 to implement the bandstop filter.]

17. Comparison of causal and noncausal filter FIR filter implementation. Generate the filter coefficients of a 64th-order rectangular window filter. Apply a Blackman window to the filter. Then apply the filter to the sawtooth wave, `x`, in file `sawth.mat`. This waveform was sampled at $f_s = 1000$ Hz. Implement the filter in two ways. Use the causal routine `filter` and noncausal routine `conv` with the option `same`. (Without this option, `conv` is like `filter` except it produces extra points.) Plot the two waveforms along with the original superimposed for comparison. Note the differences.

18. Given the advantage of a noncausal filter with regard to the time shift shown in Problem 17, why not use noncausal filter routinely? This problem shows the downsides of noncausal FIR filtering. Generate the filter coefficients of a 33rd-order rectangular window filter with a cutoff frequency of 100 Hz, assuming $f_s = 1$ kHz. Use a Blackman window on the truncated filter coefficients. Generate an impulse function consisting of a 1 followed by 255 zeros. Now apply the filter to the impulse function in two ways: using the MATLAB `filter` routine, and the `conv` routine with the `same` option. The latter generates a noncausal filter since it performs symmetrical convolution. Plot the two time responses separately limiting the x-axis to 0 to 0.05 sec to better visualize the responses. Then take the Fourier transform of each output and plot the magnitude and phase. Use the MATLAB `unwrap` routine on the phase data before plotting. Note the strange spectrum produced by the noncausal filter (i.e., `conv` with the `same` option.). This is because the noncausal filter has truncated the initial portion of the impulse response. To confirm this, rerun the program using an impulse that is delayed by 10 sample intervals (i.e., `impulse=[zeros(1,10) 1 zeros(1,245)];`). Note that the magnitude spectra of the two filters are now the same. The phase spectrum of the noncausal filter shows reduced phase shift with frequency, as would be expected. This problem demonstrates that noncausal filters can create artifacts with the initial portion of an input signal because of the way it compensates for the time shift produced by causal filters.

19. Differentiate the variable `x` in file `Prob8_10_data.mat` using the two-point central difference operator with a "skip factor" of 10. Construct another differentiator using a 16th-order least-square IIR filter implemented in MATLAB's `yulewalk` routine. The filter should perform a modified differentiator operation by having a spectrum that has a constant upward slope until some frequency f_c, then has a rapid attenuation to zero. Adjust f_c to minimize noise and still maintain derivative peaks. (A relative frequency around 0.1 is a good place to start.) In order to maintain the proper slope, the desired gain at the 0.0 Hz should be 0.0 and the gain at $f = f_c$ should be $f_c f_s \pi/2$. Plot the original data and derivative for each method. The derivative should be scaled for reasonable viewing. Also plot the new filter's magnitude spectrum for the value of f_c you selected.

20. Compare the step response of an 8th-order Butterworth filter and a 64th-order rectangular window filter, both having a cutoff frequency of 0.2 f_s. Assume a sampling frequency of 2000 Hz for plotting. (Recall that MATLAB routines normalize frequencies to $f_s/2$ while the rectangular window filters are normalized to f_s.) Use a Blackman window with the FIR filter and implement both filters using MATLAB's `filter` routine. Use a step of 256 samples but offset the step by 20 samples for better visualization (i.e., the step change should occur at the 20th sample). Plot the time responses of both filters. Also plot the spectra of both filters. Use Equation 8.2 to find the spectrum of the FIR filter and Equation 8.35 to find the spectrum of the IIR filter
21. Repeat Problem 16 above, but use an eighth-order Butterworth bandstop filter to remove the 60-Hz noise. Implement the filter using either `filter` or `filtfilt`, but note the time shift if you use the former. Plot the signal before and after filtering; also plot the filter spectrum to insure the filter is correct.
22. This problem demonstrates a comparison of a causal and a noncausal IIR filter implementation. Load file `Resp_noise1.mat` containing a noisy respiration signal in variable `resp_noise1`. Assume a sample frequency of 125 Hz. Construct a 14th-order Butterworth filter with a cutoff frequency of 0.15 $f_s/2$. Filter that signal using both `filter` and `filtfilt` and plot the original and both filtered signals. Plot the signals offset on the same graph to allow for easy comparison. Also plot the noise-free signal found as `resp` in file `Resp_noise1.mat` below the other signals. Note how the original signal compares with the two filtered signals in terms of the restoration of features in the original signal and the time shift.
23. This problem also illustrates the downsides of noncausal filtering. This problem is similar to Problem 18 except that the filter is an IIR filter and the MATLAB routine `filtfilt` is used to implement the noncausal filter. Generate the filter coefficients of an eighth-order Butterworth filter with a cutoff frequency of 100 Hz assuming $f_s = 1$ kHz. Generate an impulse function consisting of a 1 followed by 255 zeros. Now apply the filter to the impulse function using both the MATLAB `filter` routine and the `filtfilt` routine. The latter generates a noncausal filter. Plot the two time responses separately, limiting the x-axis to 0 to 0.05 sec better to visualize the responses. Then take the Fourier transform of each output and plot the magnitude and phase. Use the MATLAB `unwrap` routine on the phase data before plotting. Note the differences in the magnitude spectra. The noncausal filter (i.e., `filtfilt`) has ripple in the passband. Again, this is because the noncausal filter has truncated the initial portion of the impulse response. To confirm this, rerun the program using an impulse that is delayed by 20 sample intervals (i.e., `impulse = [zeros(1,20) 1 zeros(1,235)];`). Note that the magnitude spectra of the two filters are now similar. The phase spectrum of the noncausal filter shows reduced phase shift with frequency, as would be expected. However the time response of the noncausal filter shows a small response that precedes the stimulus.
24. Load the data file `ensemble_data.mat`. Filter the average with a 12th-order Butterworth filter. Select a cutoff frequency that removes most of the noise, but does not unduly distort the response dynamics. Implement the Butterworth filter using `filter` and plot the data before and after filtering. Implement the same filter using `filtfilt` and plot

the resultant filter data. Compare the two implementations of the Butterworth filter. For this signal, where the interesting part is not near the edges, the noncausal filter is appropriate. Use MATLAB's `text` command to display the cutoff frequency on the plot containing the filtered data.

25. FIR–IIR filter comparison. Construct a 12th-order Butterworth high-pass filter with a cutoff frequency of 80 Hz assuming $f_s = 300$ Hz. Construct an FIR high-pass filter using Equation 8.12 having the same cutoff frequency. Apply a Blackman window to the FIR filter. Plot the spectra of both filters and adjust the order of the FIR filter to approximately match the slope of the IIR filter. Use Equation 8.35 to find the spectrum of the IIR filter. Count the number of a and b coefficients in the IIR filter and compare them with the length of the FIR filter.

26. Find the power spectrum of an LTI system four ways: 1) use white noise as the input and take the Fourier transform of the output; 2) use white noise as the input and take the Fourier transform of the autocorrelation function of the output; 3) use white noise as the input and take the Fourier transform of the crosscorrelation function of the output and the input; and 4) apply Equation 8.35 to the a and b coefficients. The third approach works even if the input is not white noise. As a sample LTI system use a 4th-order Butterworth bandpass filter with cutoff frequencies of 150 and 300 Hz. For the first three methods use a random input array of 20,000 points. Use `axcor` to calculate the auto- and crosscorrelation and `welch` to calculate the power spectrum. For `welch` use a window of 128 points and a 50% overlap. [Due to the number of samples, this program may take 30 sec or more to run, so you may want to debug it with fewer samples initially.]

27. Find the spectrum (magnitude and phase) of the system represented by the z-transform:

$$H(z) = \frac{0.06 - 0.24z^{-1} + 0.37z^{-2} - 0.24z^{-3} + 0.06z^{-4}}{1 - 1.18z^{-1} + 1.61z^{-2} - 0.93z^{-3} + 0.78z^{-4}}$$

(Note you may have to use `unwrap` to get the correct phase spectrum.) Also find the step response of this system over a time period of 0.5 sec assuming a sample frequency of 2000 Hz.

28. Write the z-transform equation for a 4th-order Butterworth high-pass filter with a relative cutoff frequency of 0.3. [*Hint*: Get the coefficients from MATLAB's `butter` routine.]

PART III

CIRCUITS

CHAPTER

9

Circuit Elements and Circuit Variables

9.1 CIRCUITS AND ANALOG SYSTEMS

Chapters 9 and 10 provide the tools for analyzing systems containing analog elements. Such systems are called *circuits*, *networks*,[1] or *analog systems*. Analyzing a circuit usually means finding the mathematical representation of one or more signals in the circuit or finding the transfer function. Once the transfer function is determined, the circuit can be viewed with any of the tools already discussed, in either the time or frequency domain (Bode plots, convolution analysis, or Fourier analysis).

The primary application of the tools covered in these two chapters is the analysis of electrical and electronic circuits. Biomedical instrumentation often contains analog circuitry that provides an interface between devices that monitor physiological variables (i.e., biotransducers) and the computer that processes the data. Figure 9.1 shows an example of an electronic circuit that is used to amplify low-level biological signals such as the ECG. The analysis of this electronic circuit is discussed in Chapter 12.

The analysis of analog representations of physiological systems is another biomedical application of circuit analysis. Figure 9.2 shows what looks like an electric circuit but is actually an analog model of the cardiovascular system. A discussion of the difference between analog models and system models is given in Chapter 1. While analog models provide a closer link to the processes they represent, they have lost ground in recent years to systems models due in large part to the power and simplicity of programs such as MATLAB's Simulink described in Chapter 7.

Analysis of an electric circuit or analog model generally calls for the solution of one or more differential equations. Even the solution of a simple circuit defined by a first-order differential equation can become tedious if the signals are complicated. Fortunately, both

[1]The terms *network* and *circuit* are completely interchangeable. Here they are both used more or less randomly.

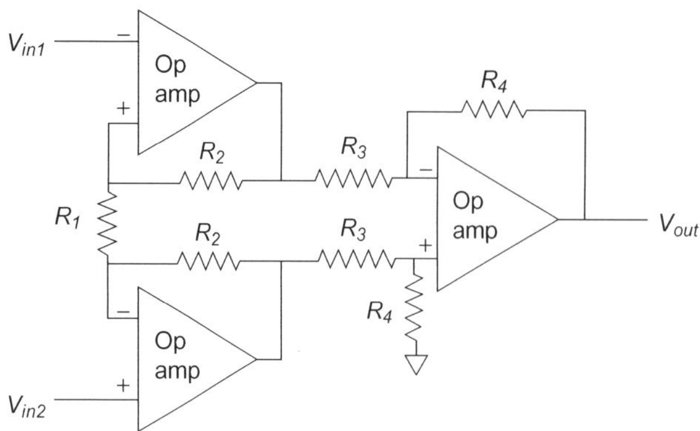

FIGURE 9.1 An electronic circuit that amplifies low-level signals such as the ECG.

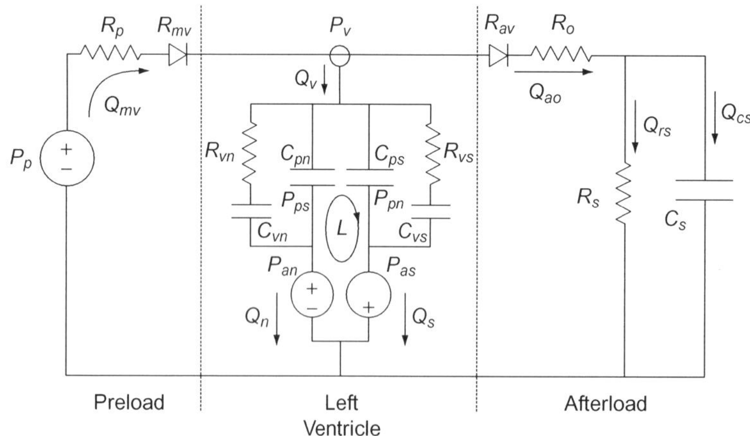

FIGURE 9.2 Analog model that represents elements of the cardiovascular system as electrical elements. *(Model courtesy of Prof. J. K. Li, Rutgers University.)*

the phasor techniques covered in Chapter 5 and the Laplace techniques covered in Chapter 6 can be applied directly to circuits to simplify the analysis. Of course, both these approaches require that the underlying processes be linear, and this can present severe constraints when circuits are used to model physiological systems, as most living systems are nonlinear. (This is another reason why system modeling analyzed using Simulink has become popular.) However, electrical elements, themselves, are very near linear and electric circuits are as close to LTI systems as can be found in the real world, so the application of linear analysis techniques is justified.

Choosing between phasor and Laplace analysis for a specific circuit follows the same rules used for systems. The Laplace transform is the most general approach, since it applies to circuits and systems exposed to transient or step-like inputs, or to circuits with nonzero initial conditions. The phasor approach is easier to use, but applies only to circuits that have sinusoidal signals or periodic signals that can be decomposed into sinusoids. No matter which approach is selected, the basic rules for analyzing the circuit are the same.

9.2 SYSTEM VARIABLES

Electric or electronic circuits consist of an arrangement of elements that define relationships between voltages and currents. In electric circuits, such arrangements may be used to generate, detect, and/or process signals. In analog models, elements are used to represent a biological mechanism with analogous electrical or mechanical elements. In the cardiovascular model shown in Figure 9.2, electrical elements are used to represent mechanical elements in the cardiovascular system, and the electrical signals in the analog model represent pressures and flows. Nonetheless, both the analogous cardiovascular mechanical elements and the circuit electric elements have the same mathematical representations, and the tools for analyzing these representations are the same. The approaches developed in the next three chapters are of value to biomedical engineers both for the design of medical instruments and the analysis of physiological systems.

In both electric circuits and mechanical systems, only two variables are needed to describe a system's behavior. An element, be it electrical or mechanical, can be viewed as simply the defining relationship between these two variables. While these variables are different for electrical and mechanical systems, they have much in common: one variable is associated with potential energy and the other variable is related to kinetic energy or movement. The potential energy variable may be viewed as the cause of an action, while the kinetic energy variable is the effect. To emphasize the strong relationship between mechanical and electrical systems, the potential and kinetic energy variables for both systems are treated together.

9.2.1 Electrical and Mechanical Variables

For electric circuits or electric analog models, the major variables are voltage and current. Voltage, the potential energy variable, is sometimes called *potential*. Voltage applied to a circuit element causes current to flow through that element. Voltage is the push behind current. It is defined as the potential energy with respect to an electric charge:

$$v = \frac{dE}{dq}\left(\frac{\text{joules}}{\text{coul}}\right)\text{volts} \tag{9.1}$$

where v is voltage, E is the energy of an electric field, and q is charge. (A slightly different typeface is used to represent the electrical variable voltage, v, and the mechanical variable velocity, υ, but noting the type of system the variable is used in, mechanical or electrical, is the best way to minimize confusion.)

The kinetic energy or flow variable that results from voltage is current. Current is the flow of charge:

$$i = \frac{dq}{dt}\left(\frac{\text{coul}}{\text{sec}}\right)\text{amps} \tag{9.2}$$

where i is current and q is charge.

Power is defined as energy per unit time:

$$P = \frac{dE}{dt} \tag{9.3}$$

To relate power to the variables v and i, note that from Equation 9.1:

$$dE = vdq \tag{9.4}$$

Substituting this definition of energy into Equation 9.3 gives:

$$P = \frac{dE}{dt} = v\left(\frac{dq}{dt}\right) = vi \text{ watts} \tag{9.5}$$

since dq/dt is just the definition of current, i. Hence, power is the product of the potential and kinetic energy variables.

In mechanical systems, the two variables are force, which is related to potential energy, and velocity, which is related to kinetic energy. The relationship between force and energy can be derived from the basic definition of energy in mechanics:

$$E = Work = \int F dx = Fx \text{ joules} \quad (\text{if } F \text{ constant over } x) \tag{9.6}$$

where F is force in dynes and x is distance in cm.

The kinetic energy variable, the flow variable that results from force, is velocity:

$$v = \frac{dx}{dt} \text{ cm/sec} \tag{9.7}$$

Velocity is directly related to the kinetic energy of a mass by the well-known equation:

$$E = \frac{1}{2} mv^2 \tag{9.8}$$

Power in mechanical systems is again the product of the potential and kinetic energy variables. From the definition of power given by Equation 9.3:

$$P = \frac{dE}{dt} = F\left(\frac{dx}{dt}\right) = Fv \text{ watts} \tag{9.9}$$

where P is power, F is force, x is distance, and v is velocity.

Table 9.1 summarizes the variables used to describe the behavior of mechanical and electrical systems. Because the tools that are developed in this chapter are more often used to analyze circuits, they are introduced in this context. Section 9.7 later in this chapter shows that these tools apply equally well to certain mechanical systems. The mechanical

TABLE 9.1 Major Variables in Mechanical and Electrical Systems

Domain	Potential Energy Variable (units)	Kinetic Energy Variable (units)
Mechanical	Force, F (joules/cm = dyne)	Velocity, v (cm/sec)
Electrical	Voltage, v (joules/coulomb = volts)	Current, i (coulomb/sec = amps)

analysis described later in this chapter is sometimes applied to mechanical systems, but for biomedical engineers the main application is in mechanical analog models of physiological processes.

9.2.2 Voltage and Current Definitions

As mentioned above, "analyzing" a circuit usually means finding the voltages (or currents) in the circuit and/or finding its transfer function. Determination of voltage is also required for the transfer function since the transfer function of a circuit is usually given as the ratio of output to input voltages ($TF = V_{out}/V_{in}$).

Voltage is always a relative variable: it is the difference between the voltages at two points. In fact the proper term for voltage is *potential difference* (abbreviated p.d.), but this term is rarely used by engineers. Subscripts are sometimes used to indicate the points from which the potential difference is measured. For example, in Equation 9.10 the notation v_{ba} means "the voltage at point b with respect to point a."

$$v_{ba} = v_b - v_a \tag{9.10}$$

The positive voltage point, point b in Equation 9.10, is indicated by a plus sign when drawn on a circuit diagram, as shown in Figure 9.3. Given the position of the plus sign in Figure 9.3, it is common to say that there is a voltage drop from point b to point a (i.e., left to right), or a voltage rise from a to b. By this convention, it is logical that v_{ab} should be the negative of $v_{ba} : v_{ab} = -v_{ba}$. Voltage always has a direction, or *polarity*, usually indicated by a " + " sign to show the side assumed to have a greater voltage (Figure 9.3).

A source of considerable confusion is that the " + " sign only indicates the point that is assumed to have a more positive value for the purpose of analysis or discussion. Sometimes the selection of the plus position is an arbitrary decision. It could be that the voltage polarity is in actuality the opposite of what was originally assigned. As an example, suppose there is actually a rise in voltage from b to a in Figure 9.3 even though we assumed that b was the more positive as indicated by the " + " sign. When this occurs, we do not change our original polarity assignment. We merely state that v_{ba} has a negative value. So a negative voltage does not imply negative potential energy, it is just that the actual polarity is the reverse of what was assumed.

By convention, but with some justification, as is shown later, the voltage of the earth is assumed to be at 0.0 volts, so voltages are often measured with respect to the voltage of the earth, or some common point referred to as *ground*. A common ground point is indicated by either of the two symbols at the bottom of the simple circuit shown in Figure 9.4. Some conventions use the symbol on the right side to mean a ground that is actually connected to the earth while the symbol on the left side indicates a common reference point not necessarily connected to the earth, but still assumed to be 0.0 volts. However, this usage is not standardized and the only assumption that can be made with certainty is that

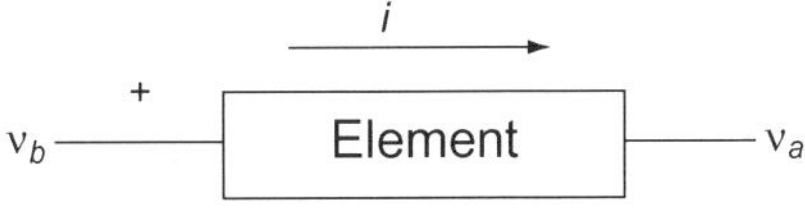

FIGURE 9.3 A generic electric circuit element demonstrating how voltage and current directions are specified. The straight lines on either side indicate wires connected to the element.

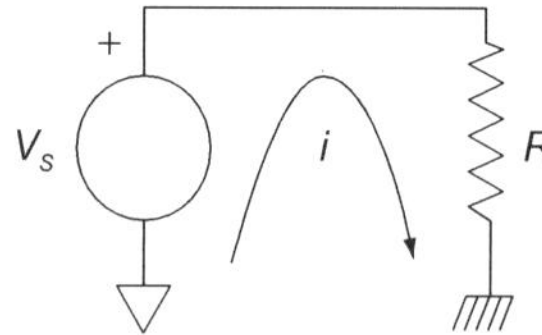

FIGURE 9.4 A simple circuit consisting of a voltage source, V_S, and a resistor R. Two different symbols for grounding points are shown. Sometimes the symbol on the right is taken to mean an actual connection to the earth, but this is not standardized.

both symbols represent a common grounding point which may or may not be connected to the earth, but is assumed to be 0.0 volts with respect to the rest of the circuit. Hence when a voltage is given with only one subscript, v_a, it is understood that this is voltage at a with respect to ground or a common reference point.

Current is a flow, so it must have a direction. This direction, or rather the assumed direction, is indicated by an arrow, as in Figures 9.3 and 9.4. By convention, the direction of the arrow indicates the direction of assumed positive charge flow. In electric or electronic circuits, charge is usually carried by electrons that have a negative value for charge. So the particles carrying the charge are actually flowing in the opposite direction of positive charge flow. Nonetheless, the convention of defining positive charge flow was established by Benjamin Franklin before the existence of electrons was known and has not been modified because it really does not matter which direction is taken as positive as long as we are consistent.

As with voltage, it may turn out that positive charge flow is actually in the direction opposite to that indicated by the arrow. Again, we do not change the direction of the arrow, rather we assign the current a negative value. So a negative value of current does not mean that some strange positrons or antiparticles are flowing but only that the actual current direction is opposite to our assumed direction.

This approach to voltage polarity and current direction may seem confusing, but it is actually quite liberating. It means that we can make our polarity and direction assignments without worrying about reality, that is, without concern for the voltage polarity or current direction that actually exists in a given circuit. We can make these assignments more or less arbitrarily (there are some rules that must be followed, as described below) and, after the circuit is analyzed, allow the positive or negative values to indicate the actual direction or polarity.

9.3 ELECTRICAL ELEMENTS

The elements as described below are idealizations: true elements only approximate the characteristics described. However, actual electrical elements come quite close to these idealizations, so their shortcomings can usually be ignored, at least with respect to that famous engineering phrase "for all practical purposes."

Electrical elements are defined by the mathematical relationship they impose between the voltage and current. They are divided into two categories based on their energy generation characteristics: active elements can supply energy to a circuit; passive elements do not. In fact, active elements do not always supply energy; in some cases they actually absorb energy, but at least they have the ability to supply energy. For example, a battery is an active element that can supply energy, but when it is being charged it is absorbing energy. Passive elements are further divided into two categories: those that use up, or dissipate, energy and those that store energy.

9.3.1 Passive Electrical Elements

9.3.1.1 Energy Users—Resistors

The resistor is the only element in the first group of passive elements: elements that use up energy. Resistors dissipate energy as heat. The defining equation for a resistor, the basic voltage–current relationship, is the classic Ohm's law:

$$v_R = Ri_R \text{ volts} \tag{9.11}$$

where R is the resistance in volts/amp, better known as ohms (Ω), i is the current in amps (A), and v is the voltage in volts (V). The resistance value of a resistor is a consequence of a property of the material from which it is made known as resistivity, ρ. The resistance value is determined by this resistivity and the geometry and is given by:

$$R = \rho \frac{l}{A} \Omega \tag{9.12}$$

where ρ is the resistivity of the resistor material, l is the length of the material, and A is the area of the material. Table 9.2 shows the resistivity, ρ, of several materials commonly used in electric components.

The power that is dissipated by a resistor can be determined by combining Equation 9.5, the basic power equation for electrical elements, and the resistor defining equation, Equation 9.11:

$$P = vi = v\left(\frac{v}{R}\right) = \frac{v^2}{R} \text{ watts} \tag{9.13}$$

or

$$P = vi = (iR)i = i^2R \text{ watts}$$

TABLE 9.2 Resistivity

Conductors	ρ (ohm-meters)	Insulators	ρ (ohm-meters)
Aluminum	2.74×10^{-8}	Glass	$10^{10} - 10^{14}$
Nickel	7.04×10^{-8}	Lucite	$>10^{13}$
Copper	1.70×10^{-8}	Mica	$10^{11} - 10^{15}$
Silver	1.61×10^{-8}	Quartz	75×10^{16}
Tungsten	5.33×10^{-8}	Teflon	$>10^{13}$

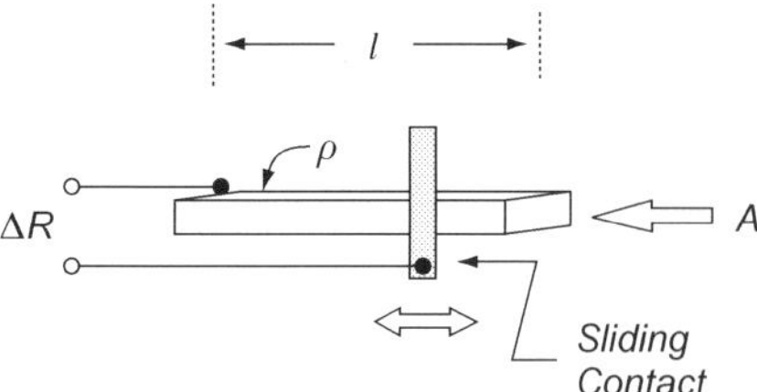

FIGURE 9.5 A variable resistor made by changing the effective length, Δl, of the resistive material.

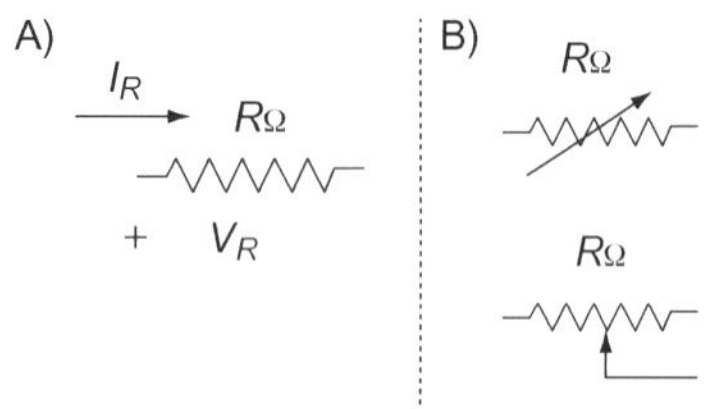

FIGURE 9.6 A) The symbol for a resistor along with its polarity conventions. For a resistor, as with all passive elements, the current direction must be such that it flows from positive to negative. In other words, current flows into the positive side of the element. B) Two symbols that denote a variable resistor or potentiometer.

The voltage–current relationship expressed by Equation 9.11 can also be stated in terms of the current:

$$i = \frac{1}{R}v = Gv \text{ amps} \tag{9.14}$$

The inverse of resistance, R, is termed the *conductance*, G, and has the units of mhos (ohms spelled backwards, a rare example of engineering humor) and is symbolized by the upside down omega, ℧ (even more amusing).

Equation 9.12 can be exploited to make a device that varies in resistance, usually by varying the length l, as shown in Figure 9.5. Such a device is termed a "potentiometer" or "pot" for short. The two symbols that are used to indicate a variable resistor are shown in Figure 9.6B.

By convention, power is positive when it is being lost or dissipated. Hence resistors must always have a positive value for power. In fact, one way to define a resistor is to say that it is a device for which $P > 0$ for all t. For P to be positive, the voltage and current must have the same orientation; that is, the current direction must point in the direction of the voltage drop. This polarity restriction is indicated in Figure 9.6A along with the symbol that is used for a resistor in electric circuit diagrams or *schematics*. Stated another way, the voltage and current polarities must be set so that current flows into the positive side of the resistor. Either the voltage direction (+ side), or the current direction can be chosen arbitrarily, but not both. Once either the voltage polarity or the current direction is selected, the other is fixed by power considerations. The same will be true for other passive elements, but not for source elements. This is because source elements can, and usually do, supply energy, so their associated power is usually negative.

EXAMPLE 9.1

Determine the resistance of 100 ft of #14 AWG copper wire.

Solution: A wire of size #14 AWG (American Wire Gauge also know as B. & S. gauge) has a diameter of 0.064 inches (see Appendix D). The value of ρ for copper is 1.70×10^{-8} Ω-m (Table 9.2). Convert all units to the *cgs* (centimeters, grams, dynes) system, then apply Equation 9.12. To insure proper unit conversion, we carry the dimensions into the equation and make sure they cancel to give us the desired dimension. This approach is sometimes known as dimensional analysis and can be very helpful where there are a lot of unit conversions.

$$l = 100\ \text{ft}\left(\frac{12\text{in}}{1\ \text{ft}}\right)\left(\frac{2.54\ \text{cm}}{\text{in}}\right) = 3048\ \text{cm}$$

$$A = \pi r^2 = \pi\left(\frac{d\ \text{cm}}{2}\right)^2 = \pi\left(\frac{2.54\ \text{cm}}{\text{in}}\frac{0.064\ \text{in}}{2}\right)^2 = 0.0208\ \text{cm}^2$$

$$R = \rho\left(\frac{l}{A}\right) = 1.7\times 10^{-8}\ \Omega\text{m}\left(\frac{100\ \text{cm}}{1\ \text{m}}\right)\left(\frac{3048\ \text{cm}}{0.0208\ \text{cm}^2}\right) = 0.2491\ \Omega$$

9.3.1.2 Energy Storage Devices—Inductors and Capacitors

Energy storage devices can be divided into two classes: *inertial* elements and *capacitive* elements. The corresponding electrical elements are the inductor and capacitor, respectively, and the voltage–current equations for these elements involve differential or integral equations. Current flowing into an inductor carries energy that is stored in a magnetic field. The voltage across an inductor is the result of a self-induced electromotive force that opposes that voltage and is proportional to the time derivative of the current:

$$v_L = L\frac{di_L}{dt} \tag{9.15}$$

where L is the constant of proportionality termed the *inductance* measured in henrys (h). (The henry is actually weber turns per amp, or volts per amp/sec, and is named for the American physicist Joseph Henry, 1797–1878.) An inductor is simply a coil of wire that utilizes mutual flux coupling (i.e., mutual inductance) between the wire coils. The inductance is related to the magnetic flux, Φ, carried by the inductor and by the geometry of the coil and the number of loops, or "turns," N:

$$L = \frac{N\Phi}{i}\text{henrys} \tag{9.16}$$

This equation is not as useful as the corresponding resistor equation (Equation 9.12) for determining the value of a coil's inductance because the actual flux in the coil depends on the shape of the coil. In practice, coil inductance is determined empirically.

The energy stored can be determined from the equation for power (Equation 9.5) and the voltage–current relationship of the inductor (Equation 9.15):

$$P = vi = Li\left(\frac{di}{dt}\right) = \frac{dE}{dt} \quad \text{solving for } i$$

$$dE = Pdt = Li\left(\frac{di}{dt}\right)dt = Li\,di$$

The total energy stored as current increases from 0 to a value I given by:

$$E = \int dE = L\int_0^I i\,di = \frac{1}{2}LI^2 \tag{9.17}$$

The similarity between the equation for kinetic energy of a mass (Equation 9.8) and the energy in an inductor (Equation 9.17) explains why an inductor is considered an inertial element. It behaves as if the energy is stored as kinetic energy associated with a mass of moving electrons, and that is a good way to conceptualize the behavior of an inductor even though the energy is actually stored in an electromagnetic field. Inductors follow the same polarity rules as resistors. Figure 9.7 shows the symbol for an inductor, a schematic representation of a coil, with the appropriate current–voltage directions.

If the current through an inductor is constant, so-called *direct current* (DC), then there will be no energy stored in the inductor and the voltage across the inductor will be zero, regardless of the amount of steady current flowing through the inductor. (The term "DC" stands for direct current, but it has been modified to mean "constant value" so that it can be applied to either current or voltage: as in "DC current" or "DC voltage"). The condition when voltage across an element is zero for any current is also known as a *short circuit*. Hence an inductor appears as a short circuit to a DC current, a feature that can be used to solve certain electrical problems encountered later in this book.

The *v-i* relationship of an inductor can also be expressed in terms of current. Solving Equation 9.15 for *i*:

$$\begin{aligned} &v_L = L\frac{di_L}{dt}; \quad di_L = \frac{1}{L}v_L dt; \quad \int di_L = \int \frac{1}{L}v_L dt \\ &i_L = \frac{1}{L}\int v_L dt \end{aligned} \tag{9.18}$$

The integral of any function is continuous, even if that function contains a discontinuity, as long as that discontinuity is finite. A continuous function is one that does not change instantaneously, that is, for a continuous function, $f(t)$:

$$f(t-) = f(t+) \text{ for any } t \tag{9.19}$$

Since the current through an inductor is the integral of the voltage across the inductor (Equation 9.18), the current will be continuous in real situations because any voltage discontinuities that occur will surely be finite. Thus, the current though an inductor can change slowly or rapidly (depending on the current), but it can never change

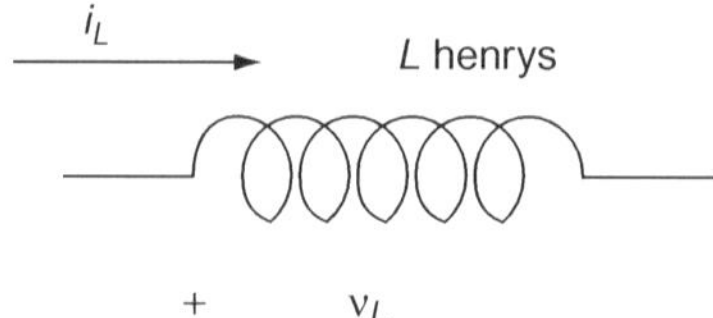

FIGURE 9.7 Symbol for an inductor showing the polarity conventions for this passive element.

instantaneously; i.e., in a discontinuous or step-like manner. In mathematical terms this is stated as:

$$i_L(t-) = i_L(t+) \tag{9.20}$$

The integral relationship between current and voltage in an inductor insures that current passing through an inductor is always continuous. Indeed, one of the popular usages of an inductor is to reduce current spikes (i.e., discontinuities) by passing the current through an inductor. The integration property tends to choke off the current spikes, so an inductor used in this manner is sometimes called a *choke*.

A capacitor also stores energy, in this case in an electromagnetic field created by oppositely charged plates. (Capacitors are nicknamed "caps" and engineers frequently use that term. Curiously, no such nicknames exist for resistors or inductors, although in some applications inductors are called "chokes" as noted above.) In the case of a capacitor, the energy stored is proportional to the charge on the capacitor, and charge is related to the time integral of current flowing through the capacitor. This gives rise to voltage–current relationships that are the reverse of the relationships for an inductor:

$$v_C = \frac{1}{C}\int i_C dt \tag{9.21}$$

or solving for i_C:

$$i_C = C\frac{dv}{dt} \tag{9.22}$$

where C, the *capacitance*, is the constant of proportionality and is given in units of *farads* which are coulombs/volt. (The farad is named after Michael Faraday, an English chemist and physicist who, in 1831, discovered electromagnetic induction.) The inverse relationship between the voltage–current equations of inductors and capacitors is an example of *duality*, a property that occurs often in electric circuits. The symbol for a capacitor is two short parallel lines reflecting the parallel plates of a typical capacitor (Figure 9.8).

Capacitance is well-named because it describes the ability (or capacity) of a capacitor to store (or release) charge without much change in voltage. Stated mathematically, this relationship between charge and voltage is:

$$C = \frac{q}{v} \text{ farads} \tag{9.23}$$

where q is charge in coulombs and v is volts. A large capacitor can take on (or give up) charge, q, with only a small change in its voltage, but in a small capacitor only a little charge flowing in (or out) results in a relatively large change in voltage. The largest capacitor readily available to us, the earth, is considered to be an infinite capacitor: its voltage

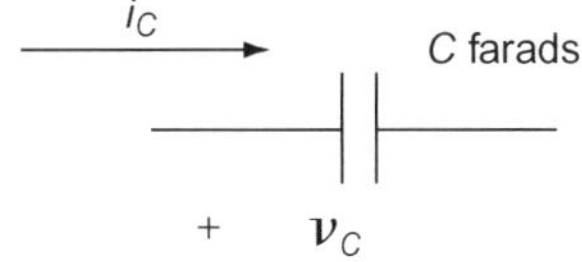

FIGURE 9.8 The symbol for a capacitor showing the polarity conventions.

remains constant (for all practical purposes) no matter how much current flows into or out of it. Consider a large lake: taking out or pouring in a cup, or even a bucket, of water will have little effect on the height of the lake. This is why the earth is a popular ground point or reference voltage; it is always at the same voltage and we just all agree that this voltage is 0.0 volts.

Most capacitors are constructed from two approximately parallel plates. Sometimes the plates are rolled into a circular tube to reduce space. The capacitance for such a parallel plate capacitor is given as:

$$C = \frac{q}{v} = \varepsilon \frac{A}{d} \tag{9.24}$$

where A is the area of the plates, d is the distance separating the plates, and ε is a property of the material separating the plates, termed the *dielectric constant*. Although Equation 9.24 is only an approximation for a real capacitor, it does correctly indicate that capacitance can be increased either by increasing the plate area, A, or by decreasing the plate separation, d. However, if the plates are closely spaced and a high voltage is applied, the dielectric material may break down, allowing current to flow directly across it, sometimes leading to dramatic failure involving smoke or even a tiny explosion. The device, if it survives, is now a resistor (or short circuit), not a capacitor. So capacitors that must work at high voltages need more distance between their plates. However, a greater separation between the plates, d, means that the area, A, must now be increased to get the same capacitance so the physical volume of the capacitor will be larger. There are special dielectrics that can sustain higher voltages with smaller plate separations. These can be used, but they are more expensive. Referring again to Equation 9.24, given a fixed plate separation, d, capacitors having larger capacitance values require more plate area, again leading to larger physical volume. Hence for a given dielectric material, the size of a capacitor is related to the product of its capacitance and voltage rating, at least for larger capacitance-voltage sizes.

EXAMPLE 9.2

Calculate the dimensions of a 1-farad capacitor. Assume a plate separation of 1.0 mm with air between the plates.

Solution: Use Equation 9.24 and the dielectric constant for a vacuum. The dielectric constant for a vacuum is $\varepsilon_0 = 8.85 \times 10^{-12}$ coul^2/newton m^2 and is also used for air.

$$A = \frac{Cd}{\varepsilon_0} = \frac{1\ \text{coul}/v\ (10^{-3}\ \text{m})}{8.85 \times 10^{-12}\ \text{col}^2/\text{newton m}^2} = \frac{\dfrac{1\ \text{coul}}{\text{newton}\ m/\text{coul}}}{8.85 \times 10^{-12}\text{col}^2/\text{newton m}^2} = 1.13 \times 10^8\ \text{m}^2$$

This is an area of about 6.5 miles on a side! This large size is related to the units of farads, which are very large for practical capacitors. Typical capacitors are in microfarads (1 μf $= 10^{-6}$ f) or picofarads (1 pf $= 10^{-12}$ f), giving rise to much smaller plate sizes. An example calculating the dimensions of a practical capacitor is given in Problem 5 at the end of this chapter.

The energy stored in a capacitor can be determined using modified versions of Equations 9.4 and 9.23:

$$v = \frac{q}{C}$$

and from Equation 9.4:

$$dE = vdq = \frac{q}{C}dq$$

Hence, for a capacitor holding a specific charge, Q:

$$E = \int dE = \frac{1}{C}\int_0^Q qdq = \frac{1}{2}\frac{Q^2}{C}$$

Substituting $V = \frac{Q}{C}$:

$$E = \frac{1}{2}CV^2 \text{ joules} \tag{9.25}$$

Capacitors in parallel essentially increase the effective size of the capacitor plates, so when two or more capacitors are connected in parallel their values add. If they are connected in series, their values add as reciprocals. Such series and/or parallel combinations are discussed at length in Chapter 11.

While inductors will not allow an instantaneous change in current, capacitors will not allow an instantaneous change in voltage. Since the voltage across a capacitor is the integral of the current, capacitor voltage is continuous based on the same arguments used for inductor current. Thus, for a capacitor:

$$v_C(t-) = v_C(t+) \tag{9.26}$$

It is possible to change the voltage across a capacitor either slowly or rapidly depending on the current, but never instantaneously. For this reason, capacitors are frequently used to reduce voltage spikes just as inductors are sometimes used to reduce current spikes. The fact that the behavior of voltage across a capacitor is similar to the behavior of current through an inductor is another example of duality. Again, this behavior is useful in the solution of certain types of problems encountered later.

Capacitors and inductors have reciprocal responses to situations where voltages and currents are constant, that is, DC conditions. Since the current through a capacitor is proportional to the derivative of voltage (Equation 9.22), if the voltage across a capacitor is constant, the capacitor current will be zero irrespective of the value of the voltage (the derivative of a constant is zero). An *open circuit* is defined as an element having zero current for any voltage; hence, capacitors appear as open circuits to DC current. For this reason, capacitors are said to "block DC" and are sometimes used for exactly that purpose.

The continuity and DC properties of inductors and capacitors are summarized in Table 9.3. A general summary of passive and active electrical elements is presented in Table 9.4.

TABLE 9.3 Energy Storage and Response to Discontinuous and DC Variables in Inductors and Capacitors

Element	Energy Stored	Continuity Property	DC Property
Inductor	$E = \frac{1}{2}LI^2$	Current continuous $i_L(t-) = i_L(t+)$	If i_c = constant (DC current) $v_L = 0$ (short circuit)
Capacitor	$E = \frac{1}{2}CV^2$	Voltage continuous $v_C(t-) = v_C(t+)$	If v_C = constant (DC voltage) $i_C = 0$ (open circuit)

TABLE 9.4 Electrical Elements—Basic Properties

Element	Units	Equation $v(t) = f\,[i(t)]$	Symbol
Resistor (R)	Ω (volts/amp)	$v(t) = Ri(t)$	
Inductor (L)	h = henry (weber turns/amp)	$v(t) = L\, di/dt$	
Capacitor (C)	f = farad (coulombs/volt)	$v(t) = 1/C \int i dt$	
Voltage source (V_S)	v = volts (joules/coulomb)	$v(t) = V_S(t)$	+ V_S
Current source (I_S)	a = amperes (coulombs/sec)	$i(t) = I_S(t)$	I_S

9.3.1.3 Electrical Elements—Reality Check

The equations given above for passive electrical elements are idealizations of the actual elements. In fact, real electrical elements do have fairly linear voltage–current characteristics. However, all real electric elements will contain some resistance, inductance, and capacitance. These undesirable characteristics are termed *parasitic elements*. For example, a real resistor will have some inductor- and capacitor-like characteristics, although these generally will be small and can be ignored except at very high frequencies. (Resistors made by winding resistance wire around a core, so-called "wire-wound resistors," have a large inductance, as might be expected for this coil-like configuration. However, these are rarely used nowadays.) Similarly, real capacitors also closely approximate ideal capacitors except for some parasitic resistance. This parasitic element appears as a large resistance in parallel with the capacitor (Figure 9.9), leading to a small *leakage* current through the capacitor. Low-leakage capacitors can be obtained at additional cost with parallel resistances exceeding 9^{12}–9^{14} Ω, resulting in very small leakage currents. Inductors are constructed by winding a wire into coil configuration. Since all wire contains some resistance, and a considerable amount of wire may be used in an inductor, real inductors generally include a fair amount of series resistance. This resistance can be reduced by using wire with a larger diameter (as suggested by Equation 9.12), but this results in increased physical size.

In most electrical applications, the errors introduced by real elements can be ignored. It is only under extreme conditions, involving high-frequency signals or the need for very high resistances, that these parasitic contributions need be taken into account. While the inductor is the least ideal of the three passive elements, it is also the least used in conventional electronic circuitry, so its shortcomings are not that consequential.

9.3.2 Electrical Elements—Active Elements or Sources

Active elements can supply energy to a system, and in the electrical domain they come in two flavors: voltage sources and current sources. These two devices are somewhat self-explanatory. Voltage sources supply a specific voltage which may be constant or time-varying but is always defined by the element. In the ideal case, the voltage is independent of the current going through the source: a voltage source is concerned only about maintaining its specified voltage; it does not care about what the current is doing. The voltage polarity is part of the voltage source definition and must be included with the symbol for a voltage source, as shown in Figure 9.10. The current through the source can be in either direction (again, the source does not care). If the current is flowing into the positive end of the source, the source is being "charged" and is removing energy from the circuit. If current flows out of the positive side, then the source is supplying energy. The voltage–current equation for a voltage source is $v = V_{Source}$; not only does the voltage

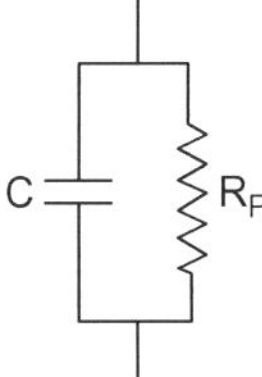

FIGURE 9.9 The schematic of a real capacitor showing the parasitic resistance that leads to a leakage of current through the capacitor.

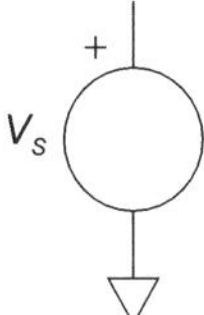

FIGURE 9.10 Schematic representation of a voltage source, V_S. This element specifies only the voltage, including the direction or polarity. The current value and direction are unspecified and depend on the rest of the circuit. Voltage sources are often used with one side connected to ground as shown.

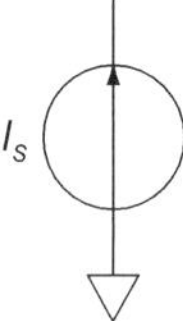

FIGURE 9.11 Schematic representation of a current source, I_S. This element specifies only the value and direction of the current: the voltage value and polarity are unspecified and are whatever is needed to produce the specified current.

source not care about the current going through it, it has no direct control over that current (except indirectly with regard to the rest of the circuit). The energy supplied or taken up by the source is still given by Equation 9.5: $P = vi$. The voltage source in Figure 9.10 is shown as "grounded"; that is, one side is connected to ground. Voltage sources are commonly used in this manner and many commercial power supplies have this grounding built in. Voltage sources that are not grounded are sometimes referred to as *floating* sources. A battery can be used as a floating voltage source since neither its positive nor negative terminal is necessarily connected to ground.

An ideal current source supplies a specified current which again can be fixed or time varying. It cares only about the current running through it. Current sources are less intuitive since current is thought of as an effect (of voltage and circuit elements), not as a cause. One way to think about a current source is that it is really a voltage source whose output voltage is somehow automatically adjusted to produce the desired current. A current source manipulates the cause, voltage, to produce a desired effect, current. However, the current source does not directly control the voltage across it: it will be whatever it has to be to produce the desired current. Figure 9.11 shows the symbol used to represent an ideal current source. Current direction is part of the current source specification and is indicated by an arrow (see Figure 9.11). Since a current source does not care about voltage (except indirectly) it does not specify a voltage polarity. The actual voltage and voltage polarity is whatever it has to be to produce the desired current.

Again, these are idealizations; real current and voltage sources usually fall short. Real voltage sources do care about the current they have to produce, at least if it gets too large, and their voltages will drop off if the current requirement becomes too high. Similarly, real current sources do care about the voltage across them, and their current output will decrease if the voltage required to produce the desired current gets too large. More realistic representations for voltage and current sources are given in Chapter 11 under the topics of Thévenin and Norton equivalent circuits.

Table 9.4 summarizes the various electrical elements giving the associated units, the defining equation, and the symbol used to represent that element in a circuit diagram.

9.3.3 The Fluid Analogy

One of the reasons analog modeling is popular is that it parallels human intuitive reasoning. To understand a complex notion, we often describe something that is similar but

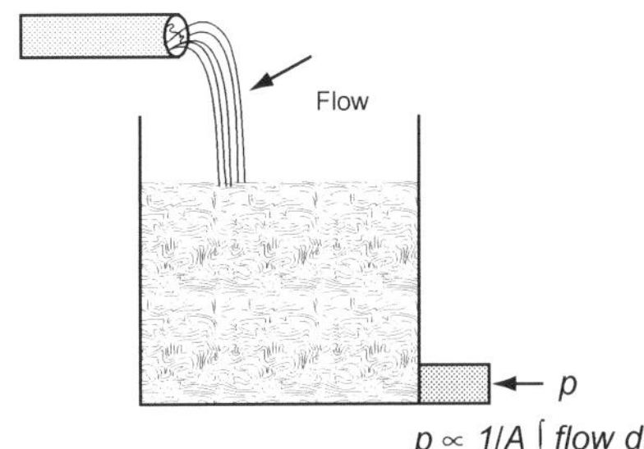

FIGURE 9.12 Water analogy of a capacitor. Water pressure at the bottom is analogous to voltage across a capacitor, and water flow is analogous to the flow of charge, current. The amount of water contained in the vessel is analogous to the charge on a capacitor.

easier to comprehend. Some intuitive insight into the characteristics of electrical elements can be made using an analogy based on the flow of a fluid such as water. In this analogy, the flow-volume of the water is analogous to the flow of charge in an electric circuit (i.e., current), and the pressure behind that flow is analogous to voltage. A resistor is a constriction, or pipe, inserted into the path of water flow. As with a resistor, the flow through this pipe is linearly dependent on the pressure (voltage) and inversely related to the resistance offered by the constricting pipe. The equivalent of Ohm's law is: flow = pressure/resistance ($i = v/R$). Also as with a resistor, the resistance to flow generated by the constriction increases linearly with its length and decrease with its cross-sectional area, as in Equation 9.12: pipe resistance = constant * length/area ($R = \rho \frac{l}{A}$).

The fluid analogy to a capacitor is a container with a given cross-sectional area. The pressure at the bottom of the container is analogous to the voltage across the capacitor, and the water flowing into or out of the container is analogous to current. As with a capacitor, the height of the water is proportional to the pressure at the bottom, and is linearly related to the integral of water flow and inversely related to the area of the container (see Figure 9.12). A container with a large area (i.e., a large capacity) is analogous to a large capacitor: it has the ability to accept larger amounts of water (charge) with little change in bottom pressure (voltage). Conversely, a vessel with a small area fills quickly, so the change in pressure at the bottom changes dramatically with only minor changes in the amount of water in the vessel. Just as in a capacitor, it is impossible to change the height of the water, and therefore the pressure at the bottom, instantaneously unless you had an infinite flow of water. With a high flow, you could change the height and related bottom pressure quickly, but not instantaneously. If the flow is outward, water will continue to flow until the vessel is empty. This is analogous to fully discharging a capacitor. In fact, even the time course of the outward flow (an exponential) would parallel that of a discharging capacitor. Also, for the pressure at the bottom of the vessel to remain constant, the flow into or out of the vessel must be zero, just as the current has to be zero for constant capacitor voltage.

A water container, like a capacitor, stores energy. In a container, the energy is stored as the potential energy of the contained water. Using a dam as an example, the amount of energy stored is proportional to the amount of water contained behind the dam and the pressure squared. A dam, or any other real container, has a limited height, and if the inflow of water continues for too long it overflows. This is analogous to exceeding the voltage rating of the capacitor, when the inflow of charge causes the voltage to rise until some type of failure occurrs. It is possible to increase the overflow value of a container by increasing its height, but this leads to an increase in physical size just as in the capacitor. A real container might also leak, in which case water stored in the container is lost, either rapidly or slowly depending on the size of the leak. This is analogous to the leakage of current that exists in all real capacitors. Even if there is no explicit current outflow, eventually all charge on the capacitor is lost due to leakage, and the capacitor's voltage reduced to zero.

In the fluid analogy, the element analogous to an inductor is a large pipe with negligible resistance to flow, but in which any change in flow requires some pressure to overcome the inertia of the fluid. This parallel with inertial properties of a fluid demonstrates why an inductor is sometimes referred to as an "inertial element." For water traveling in this large pipe, the change in flow velocity (the time derivative of flow) is proportional to the pressure applied. The proportionality constant is related to the mass of the water. Hence the relationship between pressure and flow in such an element is:

$$p = k \text{ flow velocity} = k\frac{d(\text{flow})}{dt} \tag{9.27}$$

which is analogous to Equation 9.15, the defining equation for an inductor. Energy is stored in this pipe as kinetic energy of the moving water.

The greater the applied pressure, the faster the water changes velocity but, just as with an inductor, it is not possible to change the flow of a mass of water instantaneously using finite pressures. Also as with an inductor, it is difficult to construct a pipe holding a substantial mass of water without some associated resistance to flow analogous to the "parasitic" resistance found in an inductor.

In the fluid analogy, a current source is an ideal, constant-flow pump. It generates whatever pressure is required to maintain its specified flow. A voltage source is similar to a very large capacity vessel, such as a dam. It supplies the same pressure stream no matter how much water flows out of it, or even if water flows into it, or if there is no flow at all.

9.4 PHASOR ANALYSIS

If a system or element is being driven by sinusoidal signals, or signals that can be converted to sinusoids via the Fourier series or Fourier transform, then phasor analysis techniques can be used to analyze the system's response (Chapter 5, Section 5.2). The defining equations for an inductor (Equation 9.15) and capacitor (Equation 9.21) contain the calculus operations of differentiation and integration respectively. In Chapter 5, we found that when the system variables were in phasor notation, differentiation was replaced by multiplying by $j\omega$ (Equation 5.18), and integration was replaced by dividing by $j\omega$ (Equation 5.20). If we use phasors, the calculus operation in the defining equations for an inductor and capacitor become algebraic operations, multiplication and division by $j\omega$.

Since phasor analysis is an important component of circuit analysis, a quick review of Section 5.2 might be worthwhile. In those cases where the variables are not sinusoidal, or cannot be reduced to sinusoids, we again turn to the Laplace domain to convert calculus operations to algebraic operations. The Laplace domain representation of electrical elements is covered in Section 9.6.

9.4.1 Phasor Representation—Electrical Elements

It is not difficult to convert the voltage–current equation for a resistor from time to phasor domain, nor is it particularly consequential since the time-domain equation (Equation 9.11) is already represents an algebraic relationship. Accordingly, the conversion is only a matter of restating the voltage and current variables in phasor notation:

$$V(\omega) = RI(\omega) \tag{9.28}$$

Rearranged as a voltage–current ratio:

$$R = \frac{V(\omega)}{I(\omega)} \tag{9.29}$$

Since the voltage–current relationship of a resistor is algebraic anyway, nothing is gained by converting it to the phasor domain. However, converting the voltage–current equation of an inductor or capacitor to phasor notation does make a considerable difference, as the differential or integral relationships become algebraic. For an inductor, the voltage–current equation in the time domain is given in Equation 9.15 and repeated here:

$$v_L = L\frac{di(t)}{dt} \tag{9.30}$$

But in phasor notation, the derivative operation becomes multiplication by $j\omega$:

$$\frac{di(t)}{dt} \Leftrightarrow j\omega I(\omega)$$

so the voltage–current operation for an inductor becomes:

$$V_L(\omega) = Lj\omega\, I(\omega) = j\omega\, LI(\omega) \tag{9.31}$$

Since the calculus operation has been removed it is now possible to rearrange Equation 9.31 to obtain a voltage-to-current ratio similar to that for the resistor (Equation 9.29):

$$\frac{V(\omega)}{I(\omega)} = j\omega L\ \Omega \tag{9.32}$$

The ability to express the voltage–current relationship as a ratio is part of the power of the phasor domain method. Thus the term $j\omega L$ becomes something like the equivalent resistance of an inductor. This resistor-like ratio is termed *impedance*, represented by the letter Z, and has the units of ohms (volts/amp), the same as for a resistor:

$$Z_L(\omega) = \frac{V_L(\omega)}{I_L(\omega)} = j\omega L\ \ \Omega \tag{9.33}$$

Impedance, the ratio of voltage to current, is not defined for inductors or capacitors in the time domain since the voltage–current relationships for these elements contain integrals or differentials and it is not possible to determine a V/I ratio. In general, the impedance is a function of frequency except for resistors, although often the impedance is written simply as Z with the frequency term understood. Since impedance is a generalization of the concept of resistance (it is the V/I ratio for any passive element), the term is often used in discussion of any V/I relationships, even if only resistances are involved. The concept of impedance extends to mechanical and thermal systems as well: for a system element, impedance is the ratio of the two system variables. For example, in mechanical systems, the impedance of an element would be the ratio of force to velocity:

$$Z(\omega) = \frac{F(\omega)}{v(\omega)} \tag{9.34}$$

To apply phasor analysis and the concept of impedance to the capacitor, we start with the basic voltage–current equation for a capacitor (Equation 9.21), repeated here:

$$v_C(t) = \frac{1}{C}\int i_c(t)dt \tag{9.35}$$

Noting that integration becomes the operation of dividing by $j\omega$ in the phasor domain:

$$\int i(t)dt \Leftrightarrow \frac{I(\omega)}{j\omega}$$

so the phasor voltage–current equation for a capacitor becomes:

$$V(\omega) = \frac{I(\omega)}{j\omega C} \tag{9.36}$$

The capacitor impedance is:

$$Z_C(\omega) = \frac{V_C(\omega)}{I_C(\omega)} = \frac{1}{j\omega C} = \frac{-j}{\omega C}\ \Omega \tag{9.37}$$

Active elements producing sinusoidal voltages or currents can also be represented in the phasor domain by returning to the original phasor description of the sinusoid given as Equation 5.16 in Chapter 5:

$$A\cos(\omega t + \theta) \Leftrightarrow Ae^{j\theta} \tag{9.38}$$

Using this equation, the phasor representation for a voltage source becomes:

$$V_s(t) = V_s\cos(\omega t + \theta) \Leftrightarrow V_s e^{j\theta} \equiv V_s \angle\theta \tag{9.39}$$

and for a current source:

$$I_s(t) = I_s\cos(\omega t + \theta) \Leftrightarrow I_s e^{j\theta} \equiv I_s \angle\theta \tag{9.40}$$

(Recall that capital letters are used for phasor variables.)

Usually the frequency, ω, is not explicitly included in the phasor expression of a current or voltage, but it is often shown in its symbolic representation on the circuit diagram as shown in Figure 9.13. Here the frequency is defined in the definition of the voltage source as $\omega = 2$. An elementary use of phasors is demonstrated in the example below.

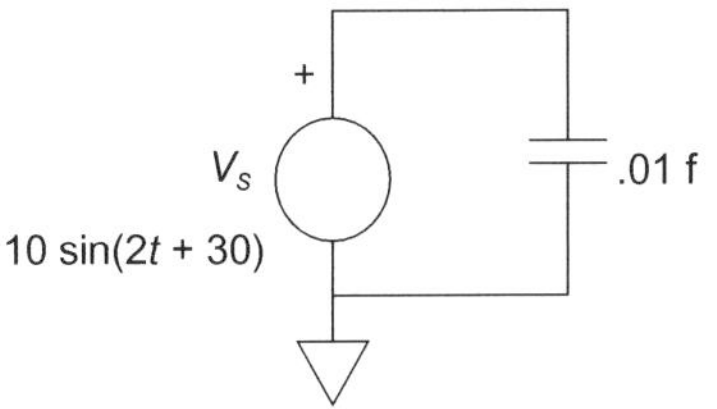

FIGURE 9.13 A simple circuit consisting of a voltage source and the capacitor used in Example 9.3. The two elements are represented by their voltage–current relationships in the phasor domain.

EXAMPLE 9.3

Find the current through the capacitor in the circuit of Figure 9.13.

Solution: Since the voltage across the capacitor is known (it is the voltage of the voltage source, V_s), the current through the capacitor can be determined by a phasor extension of Ohm's Law: $V(\omega) = I(\omega)\ Z(\omega)$:

$$\text{Starting with } V_c = I_c Z_c; \quad \text{solving for } I_c\text{:} \quad I_c = \frac{V_c}{Z_c}$$

The voltage across the capacitor is the same as the voltage source, V_S:

$$V_c = V_S = 10 \sin(2t + 30) \Rightarrow V_c(\omega) = V_S(\omega) = 10\angle{-60}$$

Next, we find the phasor notation of the capacitor:

$$Z_c = \frac{1}{j\omega C} = \frac{1}{j2(.01)} = -j50\ \Omega = 50\angle{-90}\ \Omega$$

Then solving for I_C using Ohm's law:

$$I_C(\omega) = \frac{V_c(\omega)}{Z_c(\omega)} = \frac{10\angle{-60}}{50\angle{-90}}$$

Recall that to divide two complex numbers, convert them to polar notation, divide the magnitudes, and subtract the denominator angle from the numerator angle (see Appendix E for details). In the equation above, the complex numbers are already in polar form.

$$I_C(\omega) = \frac{10\angle{-60}}{50\angle{-90}} = 0.2\angle 30 \text{ amps}$$

Converting back to the time domain from the phasor domain:

$$i_C(t) = 0.2\cos(2t + 30) \text{ amps}$$

The solution to the problem requires only algebra, although it is complex algebra.

The phasor approach can be used whenever the voltages and currents are sinusoids, or can be decomposed to sinusoids. In the latter case, the circuit needs to be solved separately for each sinusoidal frequency applying the law of superposition. Like Fourier series analysis, this can become quite tedious if manual calculations are involved, but such analyses are not that difficult using MATLAB. An example of solving for a voltage in a simple circuit over a large number of frequencies using MATLAB is shown below.

EXAMPLE 9.4

The circuit shown in Figure 9.14 is one of the earliest models of the cardiovascular system. This model, the most basic of several known as *windkessel* models, represents the cardiovascular load on the left heart. The voltage source, $v(t)$, represents the pressure in the aorta and the current, $i(t)$, the flow of blood into the periphery. The resistor represents the resistance to flow of the arterial system and the capacitor represents the compliance of the arteries. Of course the heart output is pulsatile, not sinusoidal, so the analysis of this model is better suited to Laplace methods, but a sinusoidal analysis can be used to find the frequency characteristics of the load. Assume that the voltage source is a sinusoid that varies in frequency between 0.01 and 10 Hz in 0.01-Hz intervals. Find the frequency impedance characteristics, $Z(f)$, of the peripheral vessels in the windkessel model. Plot the log of the magnitude impedance against log frequency as used in constructing Bode plots.

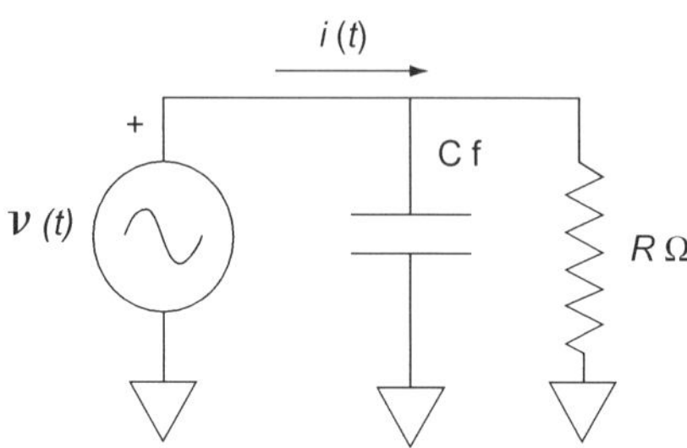

FIGURE 9.14 The most basic form of the windkessel model representation of the cardiovascular load on the heart. The voltage $v(t)$ represents the pressure generated in the aorta, the current $i(t)$ is the blood flowing into the periphery, R is the resistance, and C the compliance of the peripheral vessels.

Solution: The peripheral impedance is $Z(f) = V(f)/I(f)$ where $V(f)$ is the phasor representation of the aortic pressure and $I(f)$ is the phasor notation for the blood flow. (Since this example describes frequency ranges in Hz instead of radians, we use f instead of ω.) One approach is to apply a $V(f)$ having a known value and solving for the current, $I(f)$. The impedance then is just the ratio. Since it does not matter what value of input pressure we use, we might as well use $v(t) = \cos(2\pi ft)$ which leads to a phasor representation of $V(f) = 1.0$.

First convert the elements into phasor notation. The resistor does not change, the voltage source becomes 1.0, and the capacitor becomes $1/j\omega C = 1/j2\pi fC$. Note that since the voltage represents pressure, it is in mmHg and the current is in ml/sec. Typical values for R and C are: $R = 1.05$ mmHg-s/ml (pressure/flow) and $C = 1.1$ ml/mmHG. In the next chapter, an algorithmic method for solving any circuit problem is presented, but for now note that the total current flow is the current though the resistor plus the current through the capacitor. By Ohm's law and the equivalent for a capacitor (Equation 9.37):

$$V(f) = I_R(f)R; \quad I_R(f) = \frac{V(f)}{R} \quad V(f) = \frac{I_C(f)}{j2\pi fC}; \quad I_C(f) = V(f)(j2\pi fC)$$

$$I(f) = I_R(f) + I_C(f) = \frac{V(f)}{R} + V(f)(j2\pi fC) = V(f)\left(\frac{1}{R} + j2\pi fC\right)$$

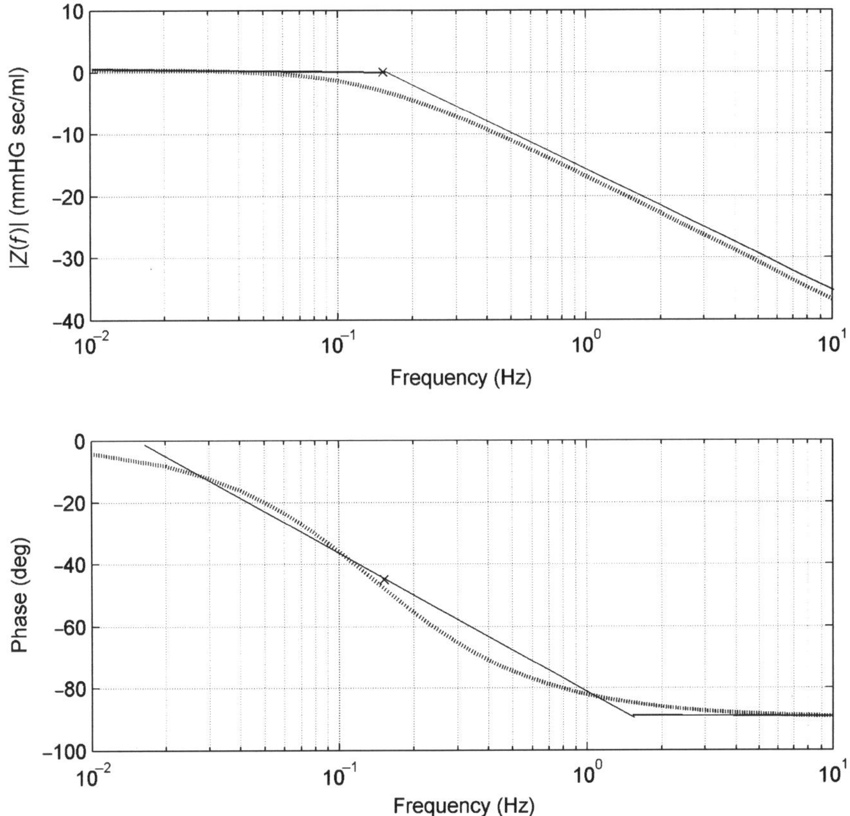

FIGURE 9.15 The impedance of the cardiovascular system as a function of frequency as determined from a simple 2-element windkessel model. The impedance decreases above 0.152 Hz due to the compliance of the arteries (capacitance in the model).

Substituting in appropriate values for R and C:

$$I(f) = V(f)\left(\frac{1}{1.05} + j2\pi f 1.1\right) = V(f)(0.95 + j6.9f)$$

$$Z(f) = \frac{V(f)}{I(f)} = \frac{1.0}{0.95 + j6.9f} = \frac{1.05}{1 + j6.55f}$$

From our Bode plot analysis, this equation has the form of a first-order system with a cutoff frequency of $1/6.55 = 0.153$ Hz. The Bode plot for the magnitude and phase of $Z(f)$ is shown in Figure 9.15.

A more complicated version of the windkessel model is analyzed using a more realistic pulse-like signal in Chapter 10.

9.5 LAPLACE DOMAIN—ELECTRICAL ELEMENTS

If the voltages and currents are not sinusoidal, periodic, or aperiodic signals, then they must be represented in the Laplace domain. The Laplace transform cannot be used in situations where $t<0$ since the transform equation (Equation 6.4) diverges for negative time. In the Laplace domain, all signals must begin at $t=0$ with the initial, and perhaps only, transition taking place at this time. If we are only interested in the behavior of the system for $t>0$, then circuit analysis in the Laplace domain is a straightforward extension of phasor analysis as described in the next section. However, as is described in Chapter 6, there is a way to take into account activity in a circuit for $t<0$ by summarizing that activity as an initial condition at $t=0$. The way in which nonzero initial conditions alter the Laplace transform of inductors and capacitors is described in Section 9.5.2.

9.5.1 Zero Initial Conditions

For elements that do not have an initial condition, the Laplace representation is straightforward: differentiation and integration are represented by the Laplace operators, s and $1/s$ respectively. In the Laplace domain, the defining equation for an inductor becomes:

$$V(s) = sLI(s) \tag{9.41}$$

and the Laplace impedance for an inductor is:

$$Z_L(s) = \frac{V(s)}{I(s)} = sL\ \Omega \tag{9.42}$$

Impedance in the Laplace domain, like impedance in the phasor domain, is given in ohms.

For a capacitor, the defining equation is:

$$V(s) = \frac{1}{Cs} I(s) \tag{9.43}$$

and the impedance of a capacitor is:

$$Z_C(s) = \frac{v(s)}{i(s)} = \frac{1}{Cs}\ \Omega \tag{9.44}$$

The defining equations and the impedance relationships are summarized for electrical elements in Table 9.5 for the time, phasor, and Laplace domain representations.

TABLE 9.5 V–I Relationships and Impedances for Electrical Elements

		Phasor Domain		Laplace Domain	
Element	**v–i Time Domain**	**V–I Equation**	**$Z(\omega)$**	**V–I Equation**	**$Z(s)$**
Resistor	$v = Ri$	$V(\omega) = RI(\omega)$	$R\ \Omega$	$V(s) = RI(s)$	$R\ \Omega$
Inductor	$v = L\ di/dt$	$V(\omega) = j\omega LI(\omega)$	$j\omega L\ \Omega$	$V(s) = sLI(s)$	$sL\ \Omega$
Capacitor	$v = (1/C)\int i dt$	$V(\omega) = I(\omega)/j\omega C$	$1/j\omega C\Omega$	$V(s) = 1/sCI(s)$	$1/s\ C\Omega$

9.5.2 Nonzero Initial Conditions

Circuit activity that occurs for $t<0$ can be included in a Laplace analysis by determining the initial voltages and/or currents that are produced during negative time. Essentially we are taking all of the past history of an element and boiling it down to a single voltage or current at $t=0$. Only the Laplace transforms of passive energy storage units is altered by nonzero initial conditions, specifically nonzero initial voltages on capacitors and nonzero initial currents in inductors.

In an inductor, energy is stored in the form of current flow, so the salient initial condition for an inductor is the current at $t=0$. This can be obtained based on the past history of the inductor's voltage:

$$I_L(0) = \frac{1}{L}\int_{-\infty}^{0} v_L dt \tag{9.45}$$

An inductor treats current flowing through it the same way a capacitor treats voltage: the current flowing through an inductor is continuous and cannot be changed instantaneously. So no matter what change occurs at $t=0$, the current just before the $t=0$ transition, $I_L(0-)$, will be the same after the transition, $I_L(0+)$.

For an inductor with nonzero initial conditions, consider the equation that defines the derivative operation in the Laplace domain (Equation 6.5), repeated here:

$$\mathscr{L}\frac{dx(t)}{dt} = sX(s) - x(0-) \tag{9.46}$$

Start with the defining equation for an inductor ($v(t) = L\frac{di(t)}{dt}$) and apply Equation 9.46 for the derivative operation, substituting $i(t)$ for $x(t)$ and noting that the Laplace transform of a constant is just the constant:

$$\begin{gathered}\mathscr{L}v(t) = \mathscr{L}\left[L\frac{di(t)}{dt}\right] = L\mathscr{L}\left[\frac{di(t)}{dt}\right] \\ V(s) = L(sI(s) - i(0)) = sLI(s) - Li(0)\end{gathered} \tag{9.47}$$

So the initial current, $i(0)$, becomes part of the voltage–current relationship of the inductor. Often these values are known, but in some cases they must be calculated from the past history of the system.

As it stands, it is not possible to determine an impedance term for an inductor with an initial current from Equation 9.47 since you cannot get a single ratio $V(s)/I(s)$. However, there is a clever way to deal with the initial condition term, the $Li(0)$ term, and still retain the concept of impedance. Regarding the two terms in the v–i equation for an inductor (Equation 9.47), the first term is the impedance, sL, and the second term is a constant, $Li(0)$. The constant term can be viewed as a constant voltage source with a value of $Li(0)$. It is a strange voltage source, being dependent on an initial current and the inductance, but is still clearly a constant voltage source in the Laplace domain. (To verify this, just set the first term to zero, then $V(s) = Li(0)$ where both L and $i(0)$ are constants.) Thus the symbol for an inductor in the Laplace domain is actually a combination of two elements: a

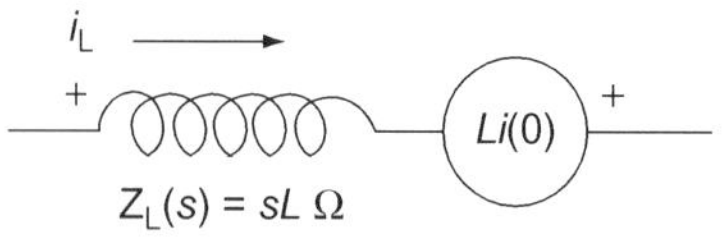

i_L + $Li(0)$ + $Z_L(s) = sL\ \Omega$

FIGURE 9.16 The Laplace domain representation of an inductor with a nonzero initial current. The inductor becomes two elements in this representation: a Laplace domain inductor having an impedance of sL, and a voltage source with a value of $Li(0)$ where $i(0)$ is the initial current. Note the polarity of the voltage source, which is based on the negative sign in Equation 9.47.

Laplace impedance representing the inductor in series with the peculiar voltage source representing the initial current condition (Figure 9.16).

For a capacitor, the entire $t<0$ history can be summarized as a single voltage at $t=0$ using the basic equation that defines the voltage–current relationship of a capacitor:

$$V_C(0) = \frac{1}{C}\int_{-\infty}^{0} i_C(t)dt \tag{9.48}$$

This equation has the same form as the right-hand term of the Laplace transform of the integration operation (Equation 6.7), repeated here:

$$\mathscr{L}\left[\int_0^T x(t)dt\right] = \frac{1}{s}X(s) + \frac{1}{s}\int_{-\infty}^{0} x(t)dt \tag{9.49}$$

Because it is not possible to change the voltage across a capacitor instantaneously (Table 9.3), the voltage on the capacitor when t is on the negative side of zero ($t=0-$) will be the same as when t is on the positive side of zero ($t=0+$). Even if there is a change in stimulus conditions at $t=0$, the voltage on the capacitor will be the same as it was just before that change: $V_C(0+) = V_C(0-)$.

From the basic defining equation for a capacitor (Equation 9.21: $v = 1/C\int idt$) using Equation 9.49 for the Laplace transform of the integral operation substituting in $i(t)$ for $x(t)$:

$$\mathscr{L}v(t) = \mathscr{L}\left[\frac{1}{C}\int idt\right] = \frac{1}{sC}I(s) + \frac{1}{sC}\int_{-\infty}^{0} i(t)dt \tag{9.50}$$

The second term is the voltage on a capacitor at $t=0$ (Equation 9.48), divided by s, so the second term is $V_C(0)/s$ and Equation 9.50 becomes:

$$V(s) = \frac{1}{sC}I(s) + \frac{V_C(0)}{s} \tag{9.51}$$

The Laplace domain representation of a capacitor having an initial voltage (Equation 9.51), can also be interpreted as a capacitance impedance, sC, in series with a voltage source. In this case, the voltage source is $V_C(0)/s$. This leads to the combined Laplace elements shown in Figure 9.17. The polarity of the voltage source in this case has the same polarity as the voltage on the capacitor.

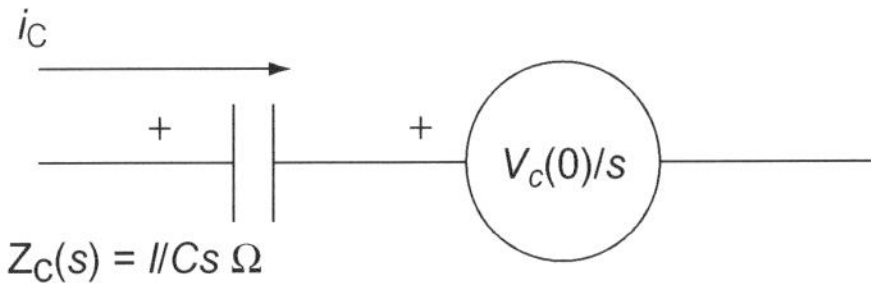

FIGURE 9.17 Laplace domain representation for a capacitor with an initial voltage. The element now consists of two components: an impedance element, $1/Cs$, and a voltage source element representing the initial condition, $V_C(0)/s$. The polarity of the voltage element is in the same direction as polarity of the impedance element, as given by Equation 9.51.

EXAMPLE 9.5

A 0.1-f capacitor with an initial voltage of 10 volts is connected to a 100-Ω resistor at $t=0$. Find the value of the resistor voltage for $t \geq 0$.

Solution: Configure the circuit using Laplace notation, as shown in Figure 9.18.

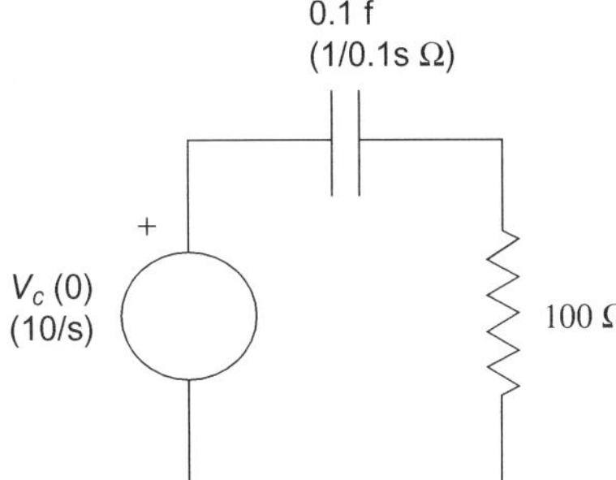

FIGURE 9.18 The network used in Example 9.5 shown in Laplace notation. Note that the 0.1-f capacitor is also shown with its impedance value of $1/sC = 1/0.1s$ Ω. The ohm symbol indicates that the capacitance has been converted to impedance.

From Figure 9.18 note that the voltage source representing the initial condition, $V_C(0)$, must be equal to the voltage across the resistor and capacitor. The voltage source is due to the initial condition and is a function of s as given in Equation 9.51. This voltage equals the voltage across the two passive elements which is: the current though the resistor and capacitor times their impedances:

$$V_C = Z_C I(s) + RI(s) = \frac{1}{sC} I(s) + RI(s)$$

$$\frac{10}{s} = \frac{1}{0.1s} I(s) + 100I(s) = I(s)\left(\frac{10}{s} + 100\right)$$

Solving for current, $I(s)$:

$$I(s) = \frac{10}{s(10/s + 100)} = \frac{10}{100s + 10} = \frac{0.1}{s + 0.1}$$

To find the voltage across the resistor, again apply Ohm's law:

$$V_R(s) = RI(s) = 100\frac{.1}{s + 0.1} + \frac{10}{s + 0.1}$$

Convert to the time domain using entry #3 of the Laplace transform table in Appendix B to give the required answer.

$$v_R(t) = 10e^{-0.1t} \text{ volts}$$

9.6 SUMMARY—ELECTRICAL ELEMENTS

In the time domain, impedance, the ratio of voltage to current for a component, can only be defined for a resistor (i.e., Ohm's law). In the phasor and Laplace domains, impedance can be defined for inductors and capacitors as well. Using phasors or Laplace analysis, it is possible to treat these so-called "reactive elements" (inductors and capacitors) as if they were resistors, at least from a mathematical standpoint. This allows us to generalize Ohm's law, $V = IZ$, to include inductors and capacitors and to treat the mathematical equations using algebra. The application of such an extended form of Ohm's law is shown in Examples 9.3, 9.4, and 9.5.

In the Chapter 10, rules are introduced that capitalize on the generalized form of Ohm's law. These rules lead to a step-by-step process to analyze any network of sources and passive elements no matter how complicated. However, the first step in circuit analysis is always the same as that shown here: convert the electrical elements into their phasor or Laplace representations. This can include the generation of additional elements if the energy storage elements have initial conditions. In Chapter 12, the analysis of circuits containing electronic elements are presented.

9.7 MECHANICAL ELEMENTS

The mechanical properties of many materials often vary across and through the material so that analysis requires an involved mathematical approach known as *continuum mechanics*. However, if only the overall behavior of an element or collection of elements is needed, then the properties of each element can be grouped together and a *lumped-parameter analysis* can be performed. An intermediate approach facilitated by high-speed computers is to apply lumped-parameter analysis to small segments of the material, then to compute how each of these segments interacts with its neighbors. This approach is often used in biomechanics and is known as *finite element analysis*.

Lumped-parameter mechanical analysis is similar to that used for electrical elements, and most of the mathematical techniques described above and elsewhere in this text can be applied to this type of mechanical analysis. In lumped-parameter mechanical analysis, the major variables are force and velocity, and the mechanical element has a well-defined relationship between these variables, a relationship that is similar to the voltage–current relationship defined by electrical elements. In mechanical systems, the flow-like variable analogous to current is velocity, while the potential energy variable analogous to voltage is force. Hence, mechanical elements define a relationship between force and velocity. Mechanical elements can be either active or passive and, with electrical elements, passive elements can either dissipate or store energy.

9.7.1 Passive Mechanical Elements

Dynamic friction is the only mechanical element that dissipates energy and, as with the resistor, that energy is converted to heat. The force–velocity relationship for a friction element is also similar to that for a resistor: the force generated by the friction element is proportional to its velocity:

$$F = k_f v \tag{9.52}$$

where k_f is the constant proportionality and is termed simply *friction*, F is force, and v is velocity.

In the cgs (centimeters, grams, dynes) metric system used in this text, the unit of force is dynes and the unit of velocity is cm/sec, so the units for friction are dyne/cm/sec (force in dynes divided by velocity in cm/sec) or dyne-sec/cm. The other commonly used measurement system is the "mks" (meters, kilograms, seconds) system preferable for systems having larger forces and velocities. Conversion between the two is straightforward (see Appendix D).

The equation for the power lost as heat in a friction element is analogous to that of a resistor:

$$P = Fv \tag{9.53}$$

The symbol for such a friction element is termed a *dashpot*, and is shown in Figure 9.19. Friction is often a parasitic element, but can also be a device specifically constructed to produce friction. Devices designed specifically to produce friction are sometimes made using a piston that moves through a fluid (or air); for example, shock absorbers on a car or some door-closing mechanisms. This construction approach, a moving piston, forms the basis for the schematic representation of the friction element that is shown in Figure 9.19.

As with passive electrical elements, passive mechanical elements have a specified directional relationship between force and velocity: specifically, the direction of positive force is opposite to that of positive velocity. Again, the direction of one of the variables can be chosen arbitrarily, after which the direction of the other variable is determined. These conventions are illustrated in Figure 9.19.

In addition to elements specifically designed to produce friction (such as shock absorbers), friction occurs as a parasitic component of other elements, just as resistance is often found in other electrical elements such as inductors. For example, a mass sliding on a surface exhibits some friction no matter how smooth the surface. Irrespective of whether friction arises from a dashpot element specifically designed to create friction or is associated with another element, it is usually represented by the dashpot schematic shown in Figure 9.19.

There are two mechanical elements that store energy just as there are two energy-storing electrical elements. The inertial-type element corresponding to inductance is, not surprisingly, inertia associated with mass. It is termed simply *mass*, and is represented by

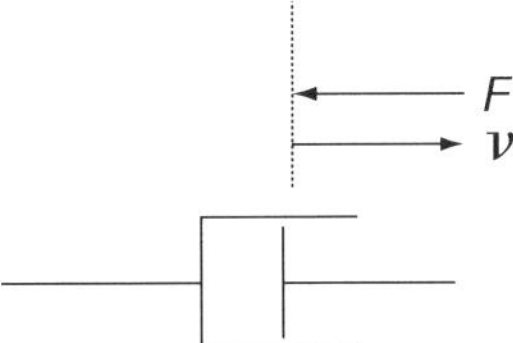

FIGURE 9.19 The schematic representation of a friction element showing the convention for the direction of force and velocity. This element is also referred to as a dashpot.

the letter m. The equation for the force–velocity relationship associated with mass is a version of Newton's law:

$$F = ma = m\frac{dv}{dt} \tag{9.54}$$

The mass element is schematically represented as a rectangle, again with force and velocity in opposite directions, as shown in Figure 9.20.

A mass element stores energy as kinetic energy following the well-known equation given in Equation 9.8 and repeated here:

$$E = \frac{1}{2}mv^2 \tag{9.55}$$

The parallel between the inertial electrical element and its analogous mechanical element, mass, includes the continuity relationship imposed on the flow variable. Just as current moving through an inductor must be continuous and cannot be changed instantaneously, moving objects tend to continue moving (to paraphrase Newton), so the velocity of a mass cannot be changed instantaneously without applying infinite force. Hence, the velocity of a mass is continuous, so that $v_m(0-) = v_m(0+)$. It is possible to change the force on a mass instantaneously, just as it is possible to change the voltage applied to an inductor instantaneously, but not the velocity.

The mechanical energy storage element that is analogous to a capacitor is a spring or elasticity, and it has a force–velocity equation that is in the same form as the equation of a capacitor:

$$F = k_e x(t) = k_e \int v dt \tag{9.56}$$

where k_e is the *spring constant* in dynes/cm. A related term frequently used is the compliance, C_k, which is just the inverse of the spring constant ($C_k = 1/k_e$), and its use makes the equation of spring and capacitor even more similar in form:

$$F = \frac{1}{C_k}x(t) = \frac{1}{C_k}\int v dt \tag{9.57}$$

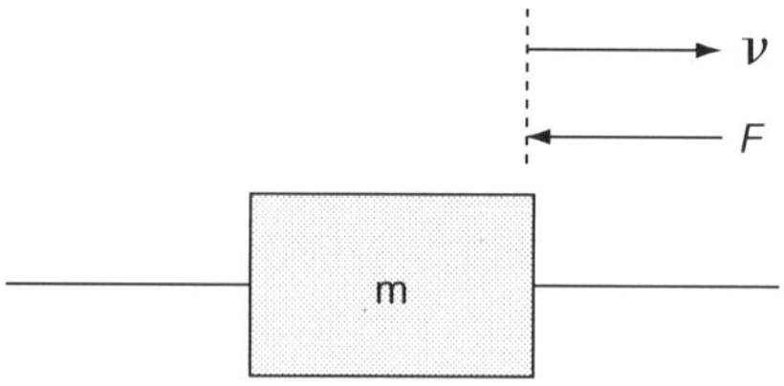

FIGURE 9.20 The schematic representation of a mass element which features the same direction conventions as the friction element.

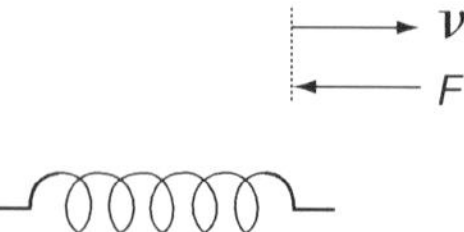

FIGURE 9.21 The symbol for a spring showing the direction conventions for this passive element. Schematically, the spring looks like an inductor—an unfortunate coincidence because it acts like a capacitor.

While a spring is analogous to a capacitor, the symbol used for a spring is similar to that used for an inductor, as shown in Figure 9.21. Springs look schematically like inductors, but they act like capacitors; no analogy is perfect.

As with a capacitor, a spring stores energy as potential energy. A spring that is stretched or compressed generates a force that can do work if it is allowed to move through distance. The work or energy stored in a spring is:

$$E = \int F dx = \int k_e x dx = \frac{1}{2} k_e x^2 \tag{9.58}$$

Displacement, x, is analogous to charge, q, in the electrical domain, so the equation for energy stored in a spring is analogous to the equation for energy stored in a capacitor found in the derivation of Equation 9.25 and repeated here:

$$E = \frac{Q^2}{2C} \tag{9.59}$$

As with a capacitor, it is impossible to change the force on a spring instantaneously using finite velocities. This is because force is proportional to length ($F_e = k_e x$), and the length of a spring cannot change instantaneously. Using high velocities, it is possible to change spring force quickly but not instantaneously; hence, for a spring, force is continuous: $F_e(0-) = F_e(0+)$.

Since passive mechanical elements have defining equations similar to those of electrical elements, the same analysis techniques such as phasor and Laplace analysis can be applied. Moreover, the rules for analytically describing combinations of elements (i.e., mechanical systems) are similar to those for describing electrical circuits. Table 9.6 is analogous to Table 9.3 and shows the energy, continuity, and DC properties of mass and elasticity.

9.7.2 Elasticity

Elasticity relates to the spring constant of a spring, but because it is such an important component in biomechanics, a few additional definitions related to the spring constant and compliance are presented here. Elasticity is most often distributed through or within a material and is defined by the relationship between *stress* and *strain*. Stress is a normalized force, one that is normalized by the cross-sectional area:

$$Stress = \frac{\Delta F}{A} \tag{9.60}$$

Strain is a normalized stretching or elongation. Strain is the change in length with respect to the rest length, which is the length the material would assume if no force were applied:

$$Strain = \frac{\Delta \ell}{\ell} \tag{9.61}$$

TABLE 9.6 Energy Storage and Response to Discontinuous and DC Variables in Mass and Elasticity

Element	Energy Stored	Continuity Property	DC Property
Mass	$E = 1/2LI^2$	Velocity continuous $v_m(0-) = v_m(0+)$	If v_m = constant (DC velocity) $F_m = 0$
Elastic Element	$E = 1/2k_e x^2$	Force continuous $F_e(0-) = F_e(0+)$	If F_e = constant (DC force) $v_m = 0$

TABLE 9.7 Young's Modulus of Selected Materials

Material	**Y_M (dynes/cm^2)**
Steel (drawn)	19.22×10^{10}
Copper (wire)	10.12×10^{10}
Aluminum (rolled)	$6.8\text{–}7.0 \times 10^{10}$
Nickel	$20.01\text{–}21.38 \times 10^{10}$
Constantan	$14.51\text{–}14.89 \times 10^{10}$
Silver (drawn)	7.75×10^{10}
Tungsten (drawn)	35.5×10^{10}

The ratio of stress to strain is a normalized measure of the ability of a material to stretch and is given as an elastic coefficient termed *Young's Modulus*:

$$Y_M = \frac{\text{Stress}}{\text{Strain}} = \frac{\Delta F/A}{\Delta \ell/\ell} \tag{9.62}$$

If a material is stretched by a load or weight produced by a mass, m, then the equation for Young's modulus can be written as:

$$Y_M = \frac{mg/\pi r^2}{\Delta \ell/\ell} \tag{9.63}$$

where g is the gravitational constant, 980.665 cm/sec^2. Values for Young's modulus for a wide range of materials can be found in traditional references such as the *Handbook of Chemistry and Physics* (Haynes, 2011). Some values for typical materials are shown in Table 9.7. Example 9.6 and 9.7 illustrate applications of Young's modulus and the related equations given above.

EXAMPLE 9.6

A 10-lb weight is suspended by a #12 (AWG) copper wire 10 inches long. How much does the wire stretch?

Solution: To find the new length of the wire use Equation 9.54 and solve for $\Delta\ell$. First convert all constants to cgs units:

$$m = 10\ \text{lb} = 10\ \text{lb}\left(\frac{1\ \text{kg}}{2.205\ \text{lb}}\right)\left(\frac{1000\ \text{gm}}{1\ \text{kg}}\right) = 4535\ \text{gm}$$

$$l = 10\ \text{in}\left(\frac{2.54\ \text{cm}}{1\ \text{in}}\right) = 25.4\ \text{cm}$$

To find the diameter of 12-gauge (AWG) wire, use Table 5 in Appendix D

$$d = 0.081 \text{ in}$$

from Table 5 in Appendix D

$$r = \frac{d}{2} = \left(\frac{0.081 \text{ in}}{2}\right)\left(\frac{2.54 \text{ cm}}{1 \text{ in}}\right) = 0.103 \text{ cm}$$

Then solve for Δl, using the value of Y_M for copper from Table 9.7:

$$Y_M = \frac{mg/A}{\Delta l/l} = \frac{mg/\pi r^2}{\Delta l/l}; \quad \Delta l = \frac{mgl}{\pi r^2 Y_M} = \frac{(4535)(980.6)(25.4)}{\pi(0.103)^2(10.12 \times 10^{10})} = 0.0335 \text{ cm}$$

EXAMPLE 9.7

Find the elastic coefficient of a steel bar with a diameter of 0.5 mm and length of 0.5 m.

Solution: From Equation 9.49, $F = k_e x$, so $k_e = F/x$ where in this case $x = \Delta\ell$. After rearranging Equation 9.62, k_e is found in terms of Young's modulus:

$$k_e = \frac{F}{x} = \frac{F}{\Delta\ell} = Y_M \frac{A}{\ell}$$

Use the dimensions given and the material in Equation 9.62 to find Young's modulus:

$$Y_M = \frac{\Delta F/A}{\Delta\ell/\ell} = \frac{F\ell}{\Delta\ell A}; \quad F = \frac{Y_M A \Delta\ell}{\ell}; \quad k_e = \frac{F}{\Delta\ell} = \frac{Y_M A}{\ell} = \frac{Y_M \pi(d/2)^2}{\ell}$$

$$k_e = \frac{19.22 \times 10^{10}\pi(.05/2)^2}{50} = \frac{3.77 \times 10^8}{50} = 7.55 \times 10^6 \text{dynes/cm}^2$$

9.7.3 Mechanical Sources

Sources supply mechanical energy in the form of force or velocity or displacement. Displacement is another word for "a change in position" and is the integral of velocity: $x = \int v\, dt$. As mentioned earlier, displacement is analogous to charge in the electrical domain since $q = \int i dt$, and current is analogous to velocity. While sources of constant force or constant velocity do occasionally occur in mechanical systems, most sources of mechanical energy are much less ideal than their electrical counterparts. Sometimes a mechanical source can look like either a velocity (or displacement) generator or a force generator, depending on the characteristics of the load, that is, the mechanical properties of the elements connected to the source. For example for a muscle contracting under a light, constant load is undergoing *isotonic contraction* because the force (i.e., "tonus") opposing the contraction is constant (i.e., "iso"). Under this condition, the muscle acts like a velocity generator, although the velocity being generated is not constant throughout the contraction. However, if the muscle's endpoints are not allowed to move, the muscle is

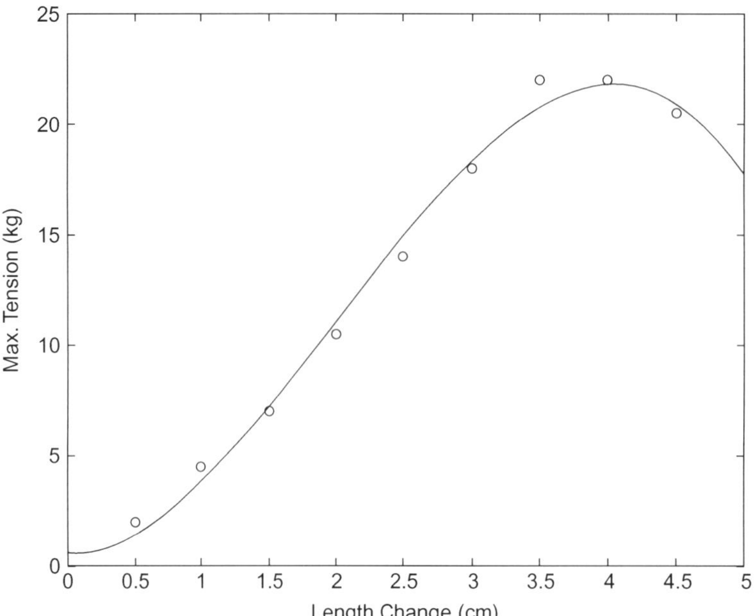

FIGURE 9.22 The length–tension relationship of skeletal muscle: the relationship between the maximum force a muscle can produce depends strongly on its length. An ideal force generator produces the same force irrespective of its length, or its velocity for that matter. (This curve is based on early measurements made on the human triceps muscle.)

involved in an *isometric contraction* because the muscle's length (i.e., "metric") is constant (again "iso"). Under this condition, the muscle acts like a force generator

In fact, a muscle is neither an ideal force generator nor an ideal velocity generator. An ideal force generator puts out the same force no matter what the velocity; however, the maximum force developed by a muscle depends strongly on its initial length. Figure 9.22 shows the classic *length–tension curve* for skeletal muscle and shows how maximum force depends on position with respect to rest length. (The rest length is the position the muscle assumes where there is no force applied to the muscle.) When it operates as a velocity generator under constant load, muscle is again far from ideal as the velocity generated is highly dependent on the load. As shown by another classic physiological relationship, *force–velocity curve*, as the force resisting the contraction of a muscle is increased its velocity decreases, Figure 9.23. Velocity can even reverse if the opposing force becomes great enough. Of course, electrical sources are not ideal either, but they are generally more nearly ideal than mechanical sources. The characteristics of real sources, mechanical and electrical, are explored in Chapter 11.

With these practical considerations in mind, a force generator is usually represented by a circle, or simply an F with a directional arrow (Figure 9.24).

A mass placed in a gravitational field looks like a force generator. The force developed on a mass by gravity is in addition to its inertial force described by Equation 9.54. The forces developed by the inertial properties of a mass, or *inertial mass*, and its gravitational properties, *gravitational mass*, need not necessarily be coupled if they are the result of separate physical mechanisms. However, careful experiments have shown them to be linked

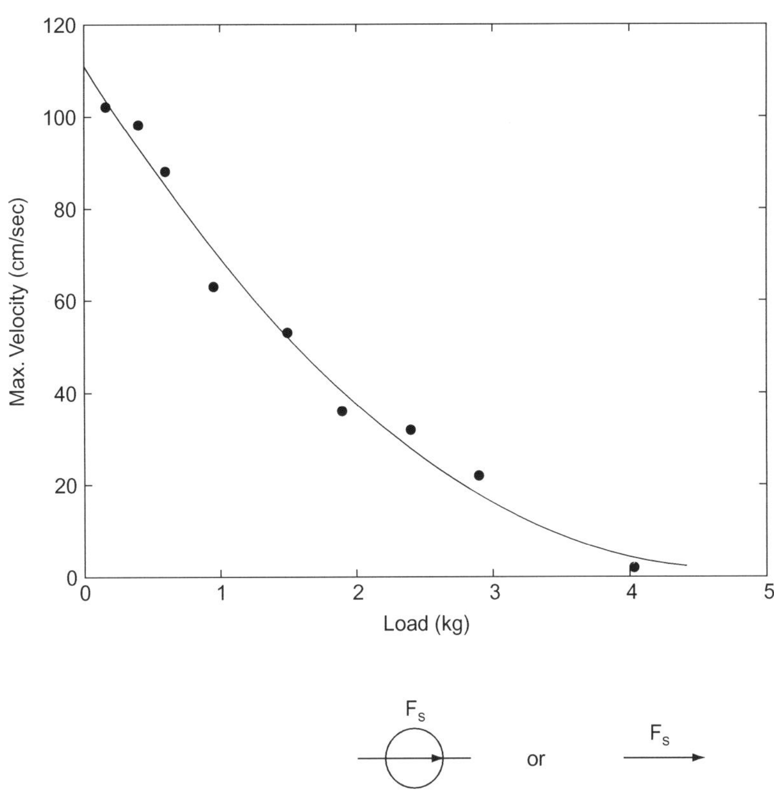

FIGURE 9.23 As a velocity generator, muscle is hardly ideal. As the load increases, the maximum velocity does not stay constant, as would be expected of an ideal source, but decreases with increasing force and can even reverse direction if the force becomes too high. This is known as the force–velocity characteristic of muscle. (This curve is based on early measurements made on the human pectoralis major muscle.)

F_S or F_S

FIGURE 9.24 Two schematic representations of an ideal force generator showing direction of force.

down to very high resolutions, indicating that they are related to the same underlying physics. The force is proportional to the value of the mass and the gravitational constant:

$$F = mg \tag{9.64}$$

where m is the mass in grams and g is the gravitational constant in cm/sec^2. Note that a gm-cm/sec^2 equals a dyne of force. The value of g at sea level is 980.665 cm/sec^2. Provided the mass does not change significantly in altitude (which would vary g), then the force produced by a mass due to gravity is nearly ideal: the force produced is even independent of velocity if it is in a vacuum and there is no friction due to wind resistance.

In some mechanical systems that include mass, the force due to gravity must be considered, while in others it is cancelled by some sort of support structure. In Figure 9.25, the system on the left side has a mass supported by a surface (either a frictionless surface or with the friction incorporated in k_f) and exerts only the inertial force defined in Equation 9.54. In the system on the right-hand side, the mass is under the influence of gravity and produces both an inertial force that is a function of velocity (Equation 9.54) and a gravitational force that is constant and defined by Equation 9.64. This additional force would be represented as a force generator acting in the downward direction with a force of mg.

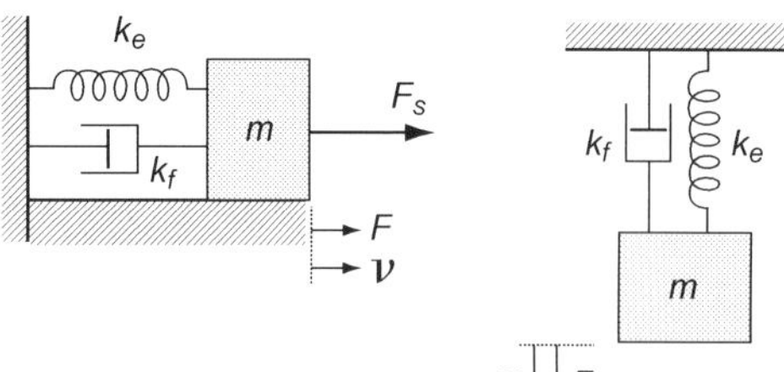

FIGURE 9.25 Two mechanical systems containing mass, m. In the left-hand system, the mass is supported by a surface so the only force involved with this element is the inertial force. In the right-hand system, gravity is acting on the mass so that it produces two forces: a constant force due to gravity (mg) and its inertial force.

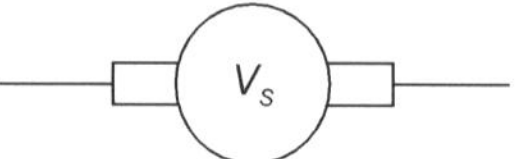

FIGURE 9.26 Symbol used to represent a motor. A motor can be either a force (torque) or velocity (rpm) generator.

TABLE 9.8 Mechanical Elements

Element (units)	Equation $F(t) = f[v(t)]$	Phasor Equation Laplace Equation	Impedance $Z(\omega)$ $Z(s)$	Symbol
Friction (k_f) (dyne-sec/cm)	$F(t) = k_f v(t)$	$F(\omega) = k_f\, v(\omega)$	k_f	
		$F(s) = k_f\, v(s)$	k_f	
Mass (m) (gms)	$F(t) = m\, dv/dt$	$F(\omega) = j\omega m\, v(\omega)$	$j\omega m$	m
		$F(s) = sm\, v(s)$	sm	
Elasticity (k_r) (spring) (dynes/cm)	$F(t) = k_e \int v\, dt$	$F(\omega) = (k_e/jw)\, v(\omega)$	$k_e/j\omega$	
		$F(\omega) = (k_e/s)v(\omega)$	k_e/s	
Force generator (F_S)	$F(t) = F_S(t)$	$F(\omega) = F_S(\omega)$	—	F_S
		$F(s) = F_S(s)$		
Velocity or displacement generator (V_S or X_S)	$v(t) = V_S(t)$	$v(\omega) = V_S(\omega)$	—	V_S
	$x(t) = X_S(t)$	$v(s) = V_S(s)$		

A velocity or displacement generator is represented as in Figure 9.24, but with the letter V_S if it is a velocity generator or X_S if it is a displacement generator. Motors can be viewed as either sources of torque (i.e., force generators) or sources of velocity. The schematic representation of a motor is shown in Figure 9.26.

9.7.4 Phasor Analysis of Mechanical Systems—Mechanical Impedance

Phasor analysis of mechanical systems is the same as for electrical systems; only the names, and associated letters, change. The mechanical elements, their differential and integral equations, and their phasor representations are summarized in Table 9.8; the electrical elements are summarized in Table 9.4. Impedance is defined for mechanical elements as:

$$Z(\omega) = \frac{F(\omega)}{v(\omega)} \tag{9.65}$$

Mechanical impedance has the units of dyne-cm/sec.

The application of phasors to mechanical systems is given in Example 9.8. More complicated systems are presented in Chapter 10.

EXAMPLE 9.8

Find the velocity of the mass in the mechanical system shown in Figure 9.27. The force, F_S, is 5 cos (9t) dynes and the mass is 5 gm. The mass is supported by a frictionless surface.

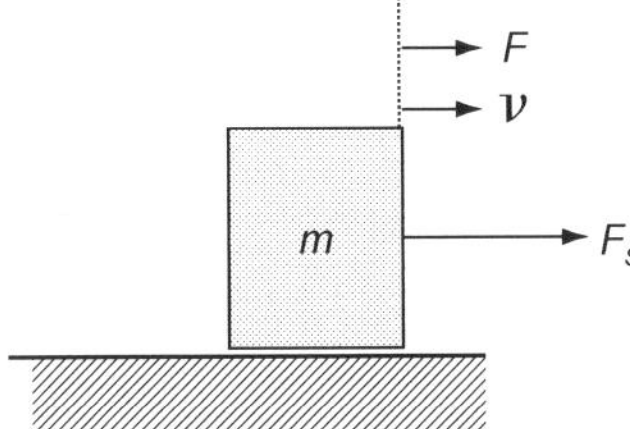

FIGURE 9.27 Mechanical system consisting of a mass with a force applied, used in Example 9.8.

Solution: Convert the force to a phasor and apply the appropriate phasor equation from Table 9.8. Solve for $v(\omega)$.

Converting the force to phasor notation:

$$5 \cos(9t) \Leftrightarrow 5\angle 0 \text{ dynes}$$

$$F(\omega) = j\omega m v(\omega); \quad v(\omega) = \frac{F(\omega)}{j\omega m}$$

$$v(\omega) = \frac{5\angle 0}{j9(5)} = \frac{5\angle 0}{45\angle 90} = 0.11\angle -90 \text{ cm/sec}$$

Converting back to the time domain (if desired):

$$v(t) = 0.11 \cos(9t - 90) = 0.11 \sin(9t) \text{ cm/sec}$$

9.7.5 Laplace Domain Representations of Mechanical Elements with Nonzero Initial Conditions

The analogy between electrical and mechanical elements holds for initial conditions as well. The two energy storage mechanical elements can have initial conditions that need to be taken into account in the analysis. A mass can have an initial velocity that clearly produces a force, and a spring can have a nonzero rest length that also produces a force.

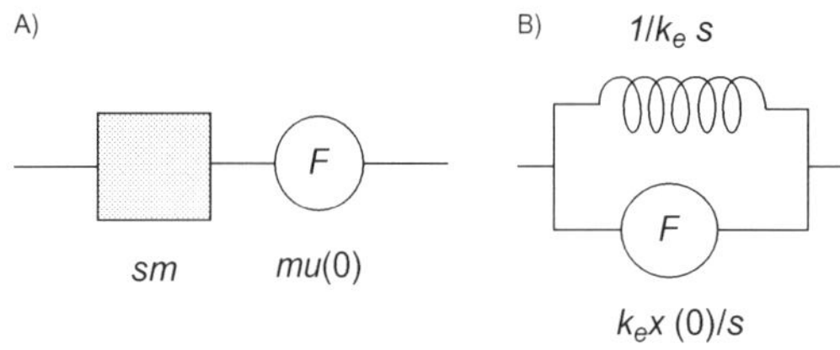

FIGURE 9.28 The Laplace representation of mechanical energy storage elements with nonzero initial conditions. A) A mass with a nonzero initial velocity is represented in the Laplace domain as a mass impedance plus a series force generator having a value of $mv(0)$. B) A spring having a nonzero rest length is represented in the Laplace domain as a spring impedance with a parallel force generator of $k_e x(0)/s$, which is also the initial force divided by s.

For a mass, an initial velocity produces a force that is equal to the mass times the initial velocity. Take the Laplace transform of an equation defining the force–velocity relationship of a mass (Newton's law; Equation 9.54):

$$\mathscr{L}F(t) = \mathscr{L}\left[m\frac{dv(t)}{dt}\right] = m\mathscr{L}\left[\frac{dv(t)}{dt}\right]$$

Applying the Laplace transform equation for the derivative operation, Equation 9.46, the force–velocity relationship of a mass with initial conditions becomes:

$$F(s) = m(sV(s) - v(0)) = smV(s) - mv(0) \tag{9.66}$$

The Laplace representation of a mass with an initial velocity consists of two elements: an impedance term related to the Laplace velocity and a force generator (Figure 9.28A). The Laplace elements in Figure 9.28A should be used to represent the mass when solving mechanical systems that contain mass with an initial velocity.

The same approach can be used to determine the Laplace representation of a spring with an initial nonzero rest length. Applying the Laplace transform to both sides of the equation defining the force–velocity relationship of a spring (Equation 9.57):

$$\mathscr{L}F(t) = \mathscr{L}\left[k_e\int v dt\right] = k_e\mathscr{L}\left[\int v dt\right]$$

Applying the Laplace transform for an integral operation:

$$F(s) = \frac{k_e}{s}\left(V(s) + \int_{-\infty}^{0} v(t)dt\right) = \frac{k_e}{s}V(s) + \frac{k_e}{s}\int_{-\infty}^{0} v(t)dt$$

The second term is just the initial displacement ($x(0) = \int_{-\infty}^{0} vdt$) times the spring constant divided by s, so the force–velocity equation becomes:

$$F(s) = \frac{k_e}{s}V(s) + \frac{k_e x(0)}{s}$$

Note that $k_e x(0)$ is also the initial force on the spring. Thus the Laplace representation of a spring having a nonzero rest length again includes two elements: the spring impedance and a parallel force generator having a value of $k_e x(0)/s$ (Figure 9.28B).

An example of a mechanical system having energy storage elements with nonzero initial conditions is given next; there are more examples in Chapter 10.

EXAMPLE 9.9

Find velocity of the mass in Figure 9.29A for $t \geq 0$. Assume the mass is 5 gm, the spring constant is 0.4 dynes/cm, and the spring is initially stretched 2 cm beyond its rest length.

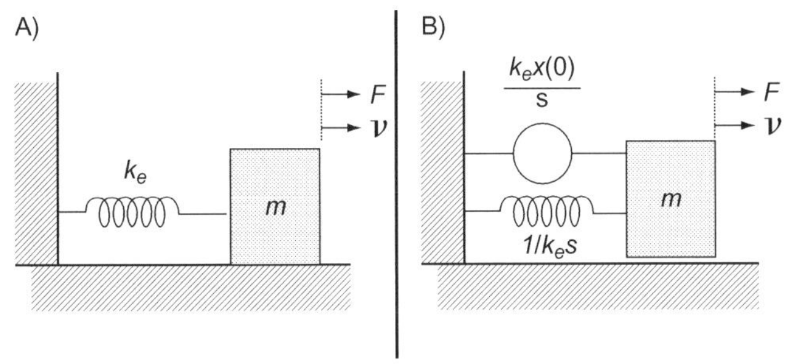

FIGURE 9.29 A mechanical system composed of a mass and a spring, used in Example 9.10. The mass is 5 gm, the spring constant is 0.4 dynes/cm, and the spring is stretched 2 cm beyond its rest length.

Solution: Replace the time variables by their Laplace equivalents and add a force generator to account for the initial condition on the spring (Figure 9.29B). The force generator will have a value of:

$$F_s(0) = \frac{k_e x(0)}{s} = \frac{0.4(2)}{s} = \frac{0.8}{s} \text{ dynes}$$

This initial force is directly applied to both the mass and the spring, so

$$Fs(0) = F(m(s)) + F(k(s)); \quad \frac{0.8}{s} = msV(s) + \frac{k_e}{s}V(s) = V(s)\left(5s + \frac{0.4}{s}\right)$$

Solving for $V(s)$:

$$V(s) = \frac{0.8/s}{5s + (0.4/s)} = \frac{0.8}{5s^2 + 0.4} = \frac{0.16}{s^2 + 0.08}$$

This matches entry #6 in the Laplace transform table of Appendix B where

$$\beta = \sqrt{0.08} = 0.283$$

$$V(s) = 0.565\frac{0.283}{s^2 + 0.283^2} \Leftrightarrow v(t) = 0.565 \sin(0.283t) \text{ dynes/sec}$$

Chapter 10 explores solutions to more complicated mechanical systems using an (algorithmic-like approach that parallels an approach developed for electric circuits.

9.8 SUMMARY

The most complicated electrical and mechanical systems are constructed from a small set of basic elements. These elements fall into two general categories: active elements that usually supply energy to the system and passive elements that either dissipate or store energy. In electrical systems, the passive elements are described and defined by the relationship they enforce between voltage and current. In mechanical systems, the defining relationships are between force and velocity. These relationships are linear and involve only scaling, differentiation, or integration. Active electrical elements supply either a specific well-defined voltage and are logically termed *voltage sources*, or a well-defined current, *current sources*. Active mechanical elements are categorized as either sources of force or sources of velocity (or displacement), although mechanical sources are generally far from ideal. All of these elements are defined as idealizations. Many practical elements, particularly electrical elements, approach these idealizations; some of the major deviations have been described.

These basic elements are combined to construct electrical and mechanical systems. Since some of the passive elements involve calculus operations, differential equations are required to describe most electrical and mechanical systems. Using phasor or Laplace analysis, it is possible to represent these elements so that only algebra is needed for their solution.

PROBLEMS

1. A resistor is constructed of thin copper wire wound into a coil (a "wire-wound" resistor). The wire has a diameter of 1 mm.
   a. How long is the wire required to be in order to make a resistor of 12 Ω?
   b. If this resistor is connected to a 5-volt source, how much power will it dissipate as heat?

2. **a.** A length of size #12 copper wire has a resistance of 0.05 Ω. It is replaced by #16 (AWG) wire. What is the resistance of this new wire?
   b. Assuming both wires carry 2 amps of current, what is the power lost in the two wires?

3. The figure below shows the current passing through a 2 h inductor.
   a. What is the voltage drop across the inductor?
   b. What is the energy stored in the inductor after 2 seconds?

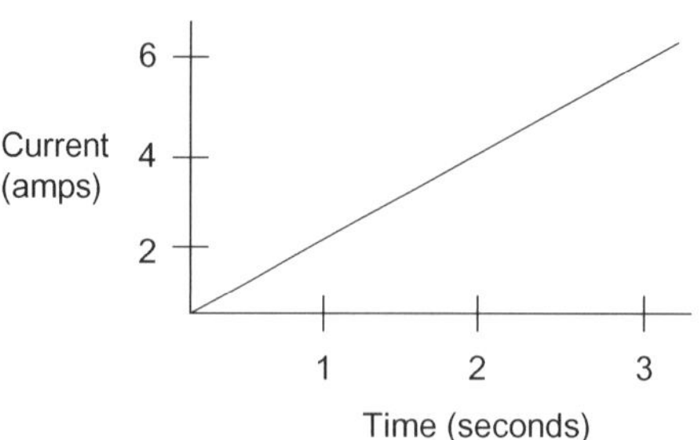

4. The voltage drop across a 10 h inductor is measured as 10 cos(20*t*) volts. What is the current through the inductor?

5. A parallel plate capacitor has a value of 1 μf (10^{-6} f). The separation between the two plates is 0.2 mm. What is the area of the plates?
6. The current waveform shown in Problem 3 passes through a 0.1 f capacitor.
   a. What is the equation for the voltage across the capacitor?
   b. What is the charge, q, contained in the capacitor after 2 seconds (assuming it was unchanged or $t = 0$)?
7. A current of 1 amp has been flowing through a 1 f capacitor for 1 second.
   a. What is the voltage across the capacitor, and what is the total energy stored in the capacitor?
   b. Repeat for a 90 f capacitor.
8. In the circuit below, find the value of the current, $i(t)$, through the inductor using the phasor extension of Ohm's law.

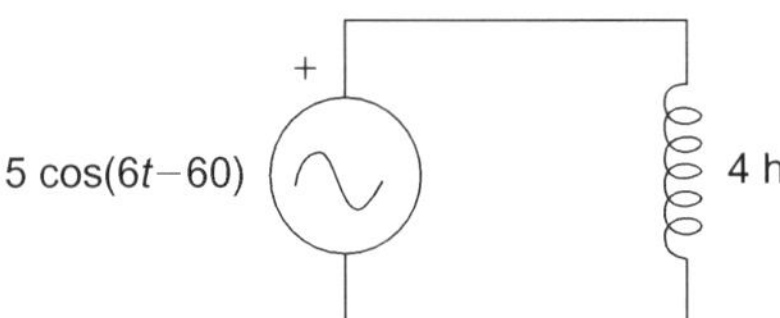

9. The sinusoidal source in Problem 8 is replaced by a source that generates a step function at $t = 0$. Find the current, $i(t)$, through the inductor for $t \geq 0$. [*Hint*: Use the same approach as in Problem 8, but replace the source and inductor by their Laplace representations.]
10. In Problem 9, assume the inductor has an initial current of 2 amps going toward ground (i.e., from top to bottom in the Figure for Problem 8). Find the current, $i(t)$, through the inductor for $t \geq 0$. [*Hint*: This problem is similar to Problem 9 except you now have two voltage sources in series. These two sources can be combined into one by algebraic summation (be sure to get the signs right).]
11. The 5 h inductor shown below has an initial current of 0.5 amps in the direction shown. Find the voltage across the 50 Ω resistor. [*Hint*: This is the same as Example 9.5 except an inductor is used instead of the capacitor. The approach is the same.]

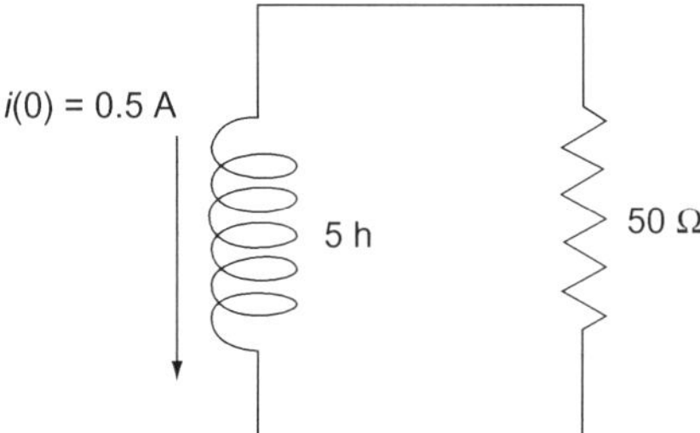

12. In the circuit shown below, the voltage source is a sinusoid that varies in frequency between 10 and 10,000 Hz in 10-Hz intervals. Use MATLAB to find the voltage across the resistor for each of the frequencies and plot this voltage as a function of frequency. The plot should be in dB versus log frequency. The resulting curve should look

familiar. [*Hint*: The same current flows thought all three elements. Convert the elements to the phasor representation, write the extended Ohm's law equation, solve for $I(f)$, and solve for $V_R(f)$. In MATLAB, define a frequency vector, f, that ranges between 10 and 10,000 in steps of 10 Hz, and use the vector to find $V_R(f)$. Since the code that solves for $V_R(f)$ will contain a vector, you need to use the ./ divide operator to perform the division.]

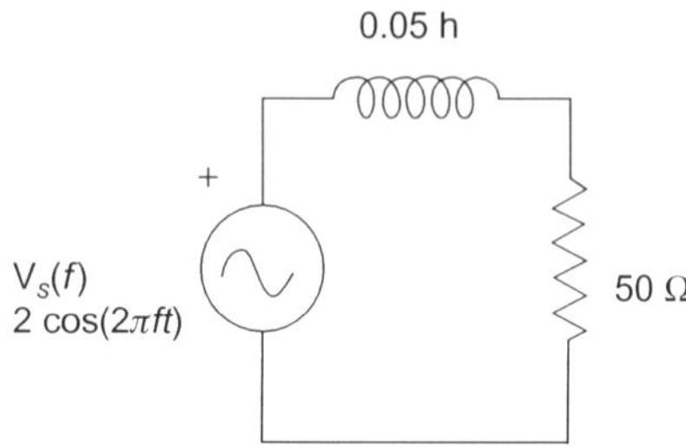

13. In a physiological preparation, the left heart of a frog is replaced by a sinusoidal pump that has a pressure output of $v(t) = \cos(2\pi t)$ mmHg ($f = 1$ Hz). Assuming that the windkessel model of Figure 9.14 accurately represents the aorta and vasculature system of the frog, what is the resulting blood flow? If the pump frequency is increased to 4 Hz, what is the blood flow?
14. Use MATLAB to find cardiac pressure ($v(t)$ in Figure 9.14) of the windkessel model used in Example 9.4 with a flow waveform that gives a more realistic waveform of blood flow. In particular, the aortic flow ($i(t)$ in Figure 9.14) should be a periodic function having a period T and defined by:

$$i(t) = \begin{cases} I_0 \sin^2\left(\dfrac{\pi t}{T_1}\right) & 0 \le t < T_1 \\ 0 & T_1 \le t < T \end{cases}$$

This function will have the appearance of a sharp half-rectified sine wave. Assume $T_1 = 0.3$ sec, the period, T, is 1 sec, and $I_o = 500$ ml/sec. Plot both the pressure and velocity waveforms. [*Hint*: Since $i(t)$ is periodic it can be decomposed into a series of sinusoids using the Fourier transform, multiplied by $V(f)/I(f)$, and the inverse Fourier transform taken to get the pressure wave, $v(t)$. Use a sampling frequency of 1 kHz to define a 1-sec time vector and also use it to define $i(t)$. Set values of $i(t)$ above 0.33 sec in the 1.0 sec period to zero as defined in the equation above. Due to round-off errors, you need to take the real part of $v(t)$ before plotting.]
15. A constant force of 9 dynes is applied to a 5-gm mass. The force is initially applied at $t = 0$ when the mass is at rest.
    a. At what value of t does the speed of the mass equal 6 dyne/sec?
    b. What is the energy stored in the mass after 2 seconds?

16. A force of 9 cos(6t + 30) dynes is applied to a spring having a spring constant of 20 dynes/cm.
    a. What is the equation for the velocity of the spring?
    b. What is the instantaneous energy stored in the spring at $t = 2.0$ sec?
17. A 90-foot length of silver wire having a diameter of 0.02 inches is stretched by 0.5 inches. What is the tension (stretching force) on the wire?
18. Use MATLAB to find the velocity of the mass in Example 9.8 for the 5-dyne cosine source where frequency varies from 1 to 40 rad/sec in increments of 1 rad/sec. Plot the velocity in dB as a function of log frequency in radians. [*Hint*: The approach is analogous to that of Problem 12. Generate a frequency vector between 1 and 40 using the MATLAB command `w=1:40`, then solve for a velocity vector, `vel`, by dividing the source value, 5, by `j*5*w`. Since `w` is a vector, you will need to do point-by-point division using the `./` command. Plot the magnitude of the velocity vector.]

CHAPTER

10

Analysis of Analog Circuits and Models

10.1 CONSERVATION LAWS—KIRCHHOFF'S VOLTAGE LAW

In this chapter we learn how to analyze systems composed of analog elements such as those covered in the last chapter. When electrical elements are involved, the analysis of such systems is known as *network analysis* or *circuit analysis*. (Recall that the words network and circuit are interchangeable.) Usually such systems are standard electric circuits, but occasionally they are analog representations of physiological mechanisms such as the windkessel model introduced in Chapter 9. We may only want to find a specific voltage or current in the circuit, but the tools developed here also enable us to construct the transfer function of the system in either frequency or Laplace notation. Once we have the transfer function, we can apply any of the techniques developed in previous chapters, such as Bode plots or the inverse Laplace transform, to analyze the behavior of the system.

Having defined the players (i.e., the elements) in Chapter 9, we now need to find the rules of the game: the rules that describe the interactions between elements. For both mechanical and electrical elements, the rules are based on two conservation laws: 1) conservation of energy and 2) conservation of mass or charge. For electrical elements, these rules are termed *Kirchhoff's voltage law* for conservation of energy and *Kirchhoff's current law* for conservation of charge. With these two rules and our extension of Ohm's law we can analyze any network.

Kirchhoff's voltage law (KVL) is based on conservation of energy: The total energy in a closed system must be zero. Since voltage is a measure of potential energy, the law implies that voltage that increases or decreases around a closed loop must sum to zero. Simply stated, what goes up must come down (in voltage):

$$\sum_{\text{Loop}} v = 0 \tag{10.1}$$

All elements that do anything useful must have current flowing through them. Otherwise, as shown in the last chapter, the voltage across the element will be zero, and for all practical purposes the element does nothing and might as well be ignored. However, current can only flow in a closed circuit (also known as a loop); if there is no loop there will be no current because it has no place to go. So, by straightforward deductive logic, all elements that do anything must be connected in some kind of loop. That means that KVL is applicable to all elements of significance in the circuit, and this law will allow us to write an equation for all electrical elements connected in a loop. Figure 10.1 illustrates KVL. It also gives an example of a "useless" element that is connected to the loop but is not itself a loop. Since no current can flow though this element, the voltage across the element, V_4, must be zero. The voltage on one side of this element is the same as the voltage on the other side, so it might as well not be there.

More complicated circuits may contain a number of loops, and some elements may be involved in more than one loop, but KVL still applies. The analysis of a circuit with any number of loops is a straightforward extension of the analysis for a single loop.

While all circuits can be analyzed using only KVL, in some situations the analysis is simplified by using the other conservation law that is based on the conservation of charge. This law is known as Kirchhoff's current law (KCL) and states that the sum of currents into a connection point (otherwise known as a *node*) must be zero:

$$\sum_{\text{Node}} i = 0 \tag{10.2}$$

In other words, what goes in must come out (with respect to charge at a connection point). For example, consider the three currents going into the connection point or node in Figure 10.2. According to KCL the three currents must sum to zero: $i_1 + i_2 + i_3 = 0$. Of course we know that one, or maybe two, of the currents is actually flowing out of the node, but this just means that one (or two) of the current values will be negative.

In the analysis of some electronic circuits, both KVL and KCL are applied, but to analyze the networks covered in this chapter only one of the two rules is required. As mentioned above, network analysis involves the determination of the network's transfer function, the output to a specific input, and/or a specific voltage or current in the network. While either rule can be used to analyze the networks in this chapter, the rule of choice is usually the one that results in the least number of equations that need to be solved. This is determined by the configuration of the network as described below.

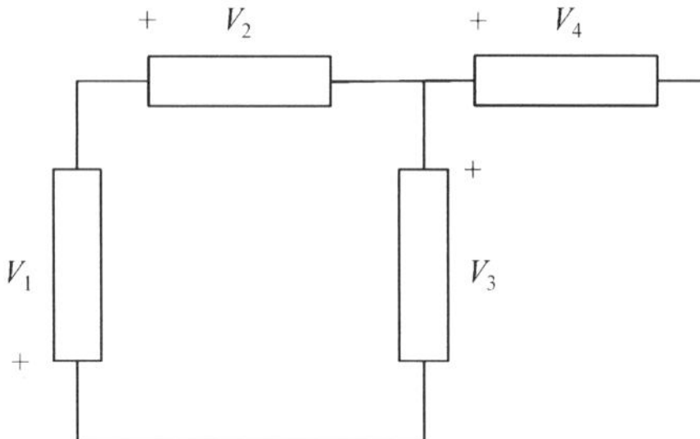

FIGURE 10.1 Illustration of Kirchhoff's voltage law (KVL). Three generic elements are arranged in a loop. The three voltages must sum to zero: $V_1 + V_2 + V_3 = 0$. This circuit also includes a nonfunctional element that is not part of a loop. Since the element with voltage V_4 is not in a loop, there is no current through it and therefore no voltage across it. So the voltages on either side of this element are the same and the element acts like a plain wire: it does nothing, and the circuit behaves no differently than if it were not there.

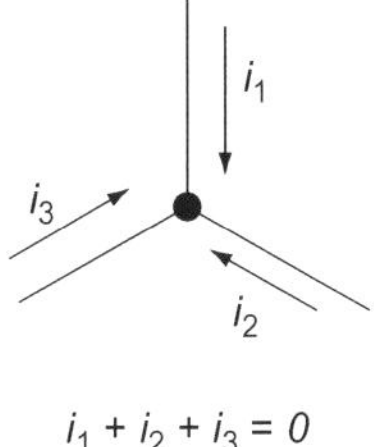

FIGURE 10.2 Illustration of Kirchhoff's current law (KCL). The sum of the three currents flowing into the connection point, or node, must be zero. In reality one (or two) of these currents will actually be flowing out of the node and such outward flowing currents will have negative values.

Mesh analysis is the term given to the circuit analysis approach that uses KVL. This terminology makes more sense when you understand that the word "mesh" is technical jargon for a circuit loop. Mesh analysis using KVL generates one equation for each loop in the circuit. *Nodal analysis* is the term used for the circuit analysis approach that uses KCL, and it generates one equation for each node in the circuit minus one node, since one node can be assumed to be at 0.0 volts (i.e., grounded) and so does not require an equation. The equations generated by either method must be solved simultaneously to find the voltages and currents in the circuit or the transfer function. The method that leads to the fewest number of simultaneous equations is an obvious choice, particularly if we must find the solutions manually. However, if we use MATLAB to solve the circuit equations, it matters little how many equations we need to solve, so we could stick with just one approach, usually KVL. Nonetheless, some electronic networks require use of both rules, so we learn to use them both.

10.1.1 Mesh Analysis—Single Loops

Mesh analysis employs KVL (Equation 10.1) to generate the equations that lead to the circuit currents and voltages. In mesh analysis you write equations based on voltages in the loop but solve for loop currents. Once you have the loop currents, you can go back and find any of the voltages in the loop by applying the basic voltage/current definitions given in Chapter 9. (As you might guess, in nodal analysis it is the opposite: You write an equation or equations based on currents, but end up solving for node voltages.) Mesh analysis can be done using an algorithm that provides a step-by-step procedure that can be applied to any non-electronic circuit. (Electronic circuits may require use of both KVL and KCL and are covered in Chapter 12.) An example of this mesh analysis algorithm is given in Example 10.1.

EXAMPLE 10.1

Find the voltage across the capacitor in the network of Figure 10.3. Note that the source is sinusoidal, so phasor analysis can be used in this example.

Solution: You may already have solved a network similar to this in Problem 12 of Chapter 9 using the extended version of Ohm's law. However, the algorithmic approach presented here can be applied to much more complicated networks.

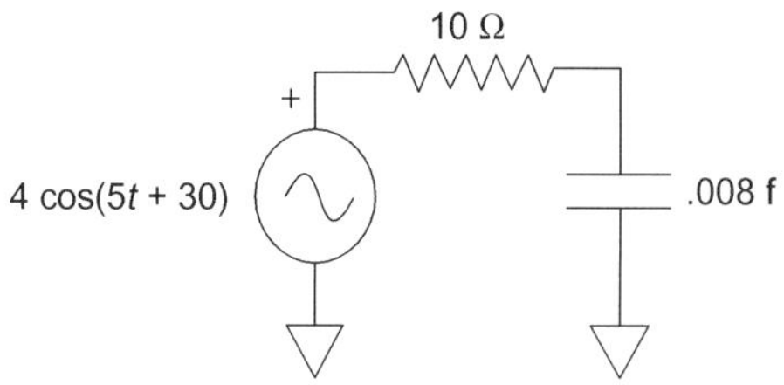

10 Ω
+
4 cos(5*t* + 30)
.008 f

FIGURE 10.3 Single mesh (i.e., loop) circuit used in Example 10.1.

The circuit has one mesh (i.e., loop) and three nodes. Nodal analysis requires the simultaneous solution of two equations, while mesh analysis requires the solution of only one equation, making it the obvious choice. The only trick in mesh analysis is to keep accurate tabs on the direction, or *polarity*, of the voltage changes: up or down corresponding to voltage increases or decreases. Actually, this is not too difficult when using the algorithmic approach described here. The steps are presented in considerable detail for this example, but are directly applicable to a wide range of network problems.

Step 1. Apply a network transformation so that all elements are represented by their phasor or Laplace domain notation. These conversions are summarized in Chapter 9 (Table 9.5), and can be used to get the phasor or Laplace representations of the various elements. Since we are dealing in this example with a sinusoidal stimulus, we use the former. In the transformed circuit, the sources are represented by phasor variables such as $V_S\angle\theta$, while the passive elements are given their respective phasor impedances: R Ω, $j\omega L$ Ω, or $1/j\omega C$ Ω. Putting the Ω symbol on the figure indicates to us that the elements have been transformed into phasor or Laplace notation. Sometimes voltage sources use RMS values in the phasor domain, but peak values will be used in this text as is more common. It really does not matter as long as you are consistent and know which units are being used.

Step 2. In mesh analysis, this step consists of defining the mesh current (or currents if more than one loop is involved). This current goes completely around the loop in either a clockwise or counterclockwise direction, theoretically your choice. The loop in this example is closed by the two grounds, which are at the same voltage and therefore essentially connected. To be consistent, we always assume in this text that the current travels clockwise around the mesh. Of course, the current might really be traveling in the opposite direction, but this just means that the value we find for this current will be negative. Defining the current direction defines the voltage polarities for the passive elements, since current must flow into the positive side of a passive element. Remember, a voltage source does not care about current and comes with its polarity already assigned. After completion of these two steps, the circuit looks as shown in Figure 10.4.

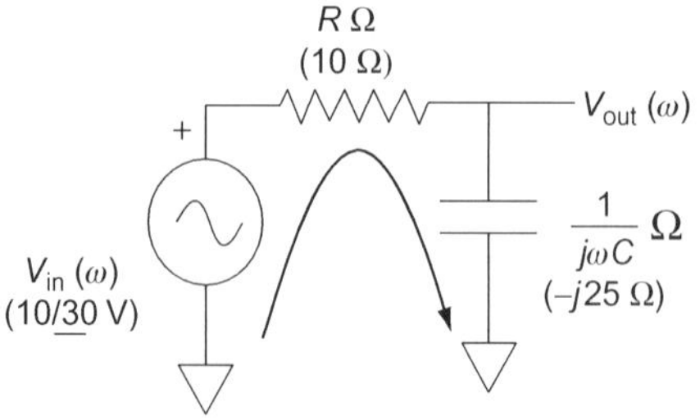

FIGURE 10.4 The circuit in Figure 10.3 after steps 1 and 2. The elements and source have been converted to their phasor representation (indicated by the Ω symbol) and the mesh current, $I(\omega)$, has been assigned including the direction, which is always assumed to be clockwise in this text. If the actual current flow is counterclockwise, $I(\omega)$ will have a negative value.

Step 3. Apply KVL. We simply go around the mesh summing the voltages, but it is an algebraic summation. Assign positive values if there is an increase in voltage and negative values if there is a decrease in voltage. Start at the lower left corner (below the source) and mentally proceed around the loop in a clockwise direction. Traversing the source leads to a voltage rise, so this entry is positive; the next two components have a voltage drop (from + to −), so their entries are negative:

$$V_S - V_R - V_C = 0$$

Normally we would substitute in the actual numbers for these voltages, but for now we will write the equation in terms of R and C. Substituting in $V_R = RI(\omega)$, and:

$$Vc = \frac{1}{j\omega C} I(\omega)$$

The KVL equation becomes:

$$V_S - RI(\omega) - \frac{1}{j\omega C} I(\omega) = 0$$

$$V_S - I(\omega)\left(R + \frac{1}{j\omega C}\right) = 0$$

In fact, to develop this mesh equation we could have started anywhere in the loop and gone in either the clockwise or counterclockwise direction, but again for consistency, and since it really does not matter, we always go in clockwise and begin at the lower left corner of the circuit.

Step 4. Solve for the current. Put the source(s) on one side and the terms for the passive elements on the other. Then solve for $I(\omega)$.

$$V_S(\omega) = (R + Z_C)I(\omega)$$
$$I(\omega) = \frac{V_S}{R + Z_C}$$

Step 5. Solve for any voltages of interest. In this problem, we want the voltage across the capacitor. From the equation that defines a capacitor: $V_C(\omega) = Z_C(\omega)I(\omega)$. Substituting our solution for $I(\omega)$ above into this equation:

$$V_C = Z_C(\omega)I(\omega) = \frac{Z_C(\omega)V_S}{R + Z_C(\omega)}$$

Sometimes you need to leave the solution in this form, with variables for your element values, for example, when you do not know the specific values, or when several different values may be used in the circuit, or when the problem will be solved on the computer. In this case we have specific values for our elements, so we can substitute in the values for R, Z_C, and V_S and solve:

$$V_C = \frac{(-j25)4\angle 30}{10 - j25} = \frac{(25\angle -90)4\angle 30}{27\angle -68} = \frac{100\angle -60}{27\angle -68} = 3.7\angle 8 \text{ volts}$$

The next example applies this 5-step algorithm to a slightly more complicated single loop circuit.

EXAMPLE 10.2

Example of the general solution of an *RLC* circuit. Find the general solution for V_{out} in the circuit below. The arrows on either side of V_{out} indicate that this output voltage is the voltage across the capacitor. As is often the case, the ground voltage points are not shown as connected, but a connection between the two points can be assumed since they are both at 0.0 volts.

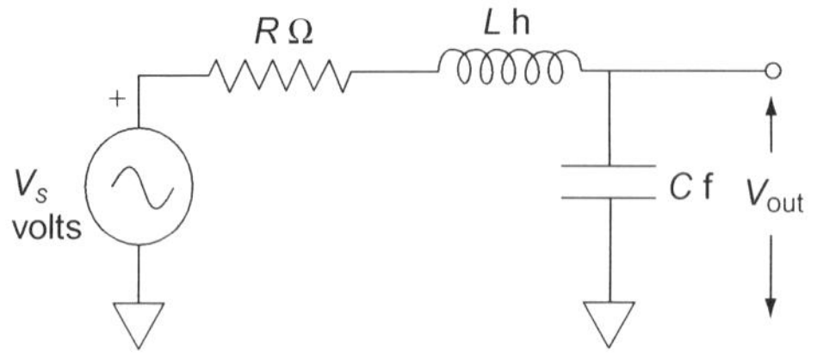

FIGURE 10.5 A circuit consisting of resistor, inductor, capacitor, and source. This can also be looked at as an input-output system where V_S is the input and V_{out} is the output.

Steps 1 and 2. These lead to the circuit shown in Figure 10.6. Passive elements are shown with units in ohms to help remind us that we are now in the phasor domain.

Step 3. Now write the basic equation going around the loop:

$$V_S(\omega) - RI(\omega) - j\omega LI(\omega) - \frac{1}{j\omega C}I(\omega) = 0$$

$$V_S(\omega) - I(\omega)\left(R + j\omega L + \frac{1}{j\omega C}\right) = 0$$

Step 4. Solve for $I(\omega)$:

$$V_S(\omega) = I(\omega)\left(R + j\omega L + \frac{1}{j\omega C}\right)$$

To clean things up a bit, clear the fraction in the denominator and rearrange the right-hand side into real and imaginary parts. (Recall that $j^2 = -1$.)

$$I(\omega) = \frac{V_S(\omega)j\omega C}{R(j\omega C) + j\omega L(j\omega C) + 1} = \frac{V_S(\omega)j\omega C}{1 - \omega^2 LC + j\omega RC} \text{ amps}$$

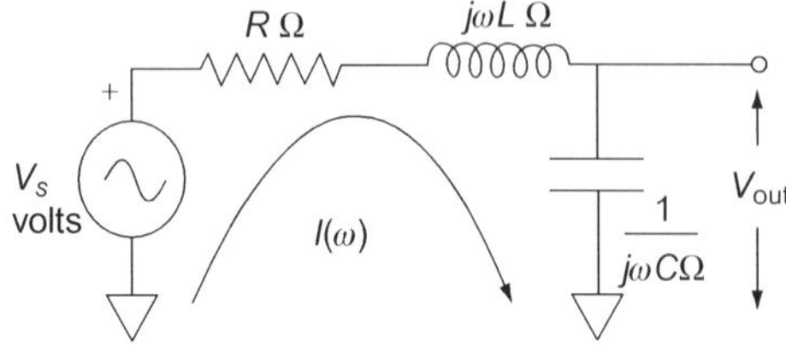

FIGURE 10.6 The circuit shown in Figure 10.5 after converting the elements into phasor notation and defining the loop current. In this case the specific values for the elements are not known so they must remain as variables *R*, *L*, and *C*. However, we can substitute in phasor representations $j\omega L$ for $Z_L(\omega)$ and $1/j\omega C$ for $Z_C(\omega)$.

Step 5. Now find the desired voltage, the output voltage, V_{out}. Use the same strategy used in the last example: multiply $I(\omega)$ by the capacitance impedance, $1/j\omega C$:

$$V_{out}(\omega) = \frac{V_S(\omega)j\omega C}{1 - \omega^2 LC + j\omega RC}\left(\frac{1}{j\omega C}\right) = \frac{V_S(\omega)}{1 - \omega^2 LC + j\omega RC} \text{ volts} \tag{10.3}$$

To find a specific value for V_{out} it is necessary to put in numerical values for R, L, and C as well as for the input voltage, V_S, which would define the frequency ω. However, using some of the tools developed in previous chapters much can be learned from the general form of the equation. For example, Equation 10.3 has the form of a second-order equation in phasor notation. By equating coefficients, we could even determine values for ω_n and δ without solving the equation, as will be done in the next example.

As indicated above, the network in this example can be also viewed as a linear process or input-output system where V_S is the input and V_{out} is the output. When thought of in these terms, the network is usually drawn as in Figure 10.7.

If the network is thought of as an input-output system, it is possible to represent it by a transfer function: $TF(\omega) = V_{out}(\omega)/V_S(\omega)$. To repeat the caveat stated in the Laplace transform chapter, Chapter 6, the term "transfer function" should only be used for a function that is written in terms of the Laplace variables. However, the concept is so powerful that it is used to describe almost any input-output relationship, even qualitative relationships. To find the transfer function for the system shown in Figure 10.7, simply divide both sides of Equation 10.3 $V_S(\omega)$.

$$\frac{V_{out}(\omega)}{V_S(\omega)} = \frac{1}{1 - \omega^2 LC + j\omega RC} \tag{10.4}$$

This is clearly the transfer function of a second-order system (compare to Equation 5.45). Since this transfer function is in the phasor domain, it is limited to sinusoid and periodic functions. However this circuit could just as easily be analyzed using Laplace domain variables as illustrated in the next example.

EXAMPLE 10.3

Find the Laplace domain transfer function for the system/network shown in Figure 10.8.

Solution: The analysis of this network in the Laplace domain follows the same steps for the phasor domain analysis, except that in step 1 elements are represented by their Laplace equations: R, sL, and $1/sC$.

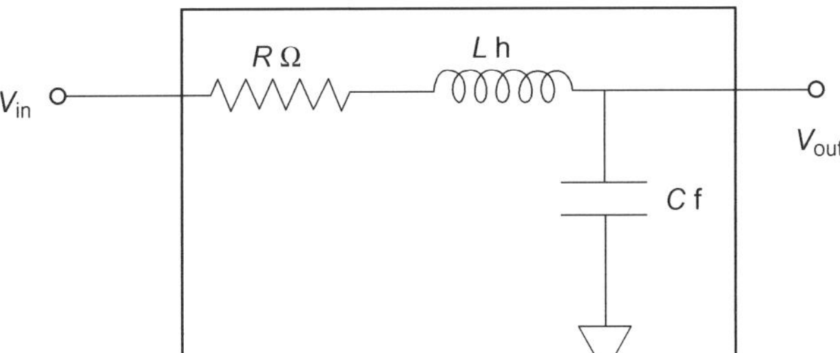

FIGURE 10.7 The network used in Example 10.2 viewed as a linear process or *input-output system*. The input and output voltages are assumed to be referenced to ground but these grounds are not explicitly shown in this diagram.

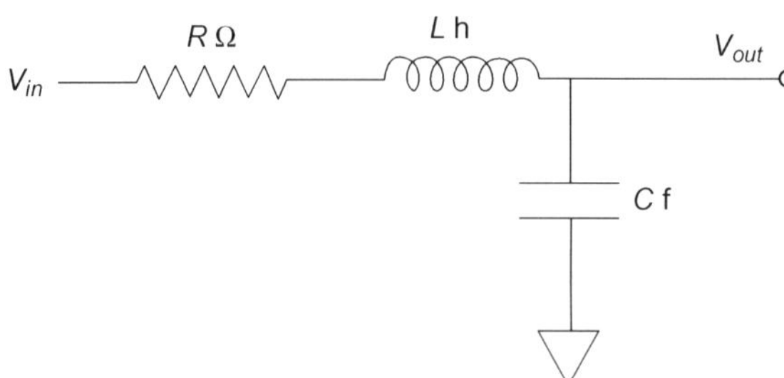

FIGURE 10.8 The three-element, single loop network used in Example 10.3.

Step 1. The elements' values are replaced by the Laplace domain equivalents. These modified values are in Ω to show they are impedances.

Step 2. Currents are assigned, again clockwise, but in Laplace notation: $I(s)$. Note that the current actually does go around in a loop though an implicit (i.r., not shown) ground and source element, $V_{in}(s)$. After application of these two modified steps the circuit appears as shown in Figure 10.8.

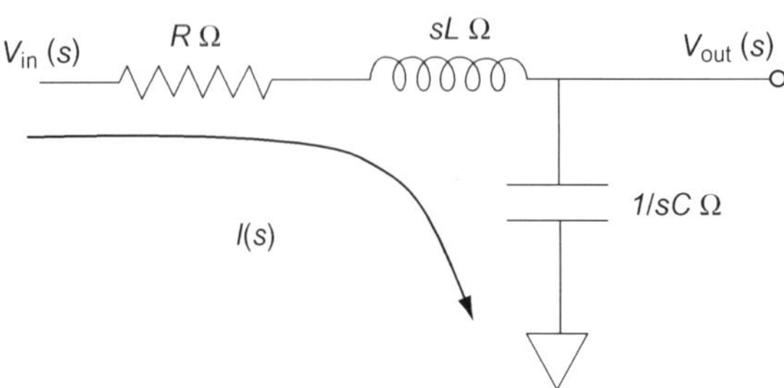

FIGURE 10.9 The circuit shown in Figure 10.8 after converting the elements into Laplace notation and defining the loop current also in Laplace format.

Step 3. Writing the loop equation:

$$V_{\text{in}}(s) - RI(s) - sLI(s) - \frac{I(s)}{sC} = 0$$

$$V_{\text{in}}(s) - I(s)\left(R + sL + \frac{1}{sC}\right) = 0; \quad V_{\text{in}}(s) = I(s)\left(R + sL + \frac{1}{sC}\right)$$

Step 4. Solving for $I(s)$:

$$I(s) = \frac{V_{\text{in}}(s)}{R + sL + (1/sC)}$$

Step 5. Solving for $V_{\text{out}}(s) = I(s)/sC$:

$$V_{\text{out}}(s) = \frac{V_{\text{in}}(s)sC}{R + sL + (1/sC)}\left(\frac{1}{sC}\right) = \frac{V_{\text{in}}(s)}{sRC + s^2CL + 1}$$

Rearranging into standard Laplace format where the highest power of s (in this case the s^2) has a coefficient of 1.0.

$$V_{\text{out}}(s) = \frac{(V_{\text{in}}(s)/CL)}{s^2 + (R/L)s + (1/LC)}$$

To find the transfer function, just divide by $V_{in}(s)$:

$$TF(s) = \frac{V_{out}(s)}{V_{in}(s)} = \frac{(1/CL)}{s^2 + (R/L)s + (1/LC)} \tag{10.5}$$

This is the Laplace transform transfer function of a second-order system having the same format as Equation 6.31, repeated here.

$$TF(s) = \frac{1}{s^2 + 2\delta\omega_n s + \omega_n^2} \tag{10.6}$$

Equating coefficients to solve for ω_n in terms of R, L, and C:

$$\omega_n^2 = (1/LC); \quad \omega_n = \sqrt{\frac{1}{LC}} \tag{10.7}$$

Solving for δ:

$$2\delta\omega_n = \frac{R}{L}; \quad \delta = \frac{R}{2L\omega_n}$$

Substituting in for ω_n

$$\delta = \frac{R}{2L\left(\frac{1}{\sqrt{LC}}\right)} = \frac{R\sqrt{LC}}{2L} = \frac{R}{2}\sqrt{\frac{C}{L}} \tag{10.8}$$

The values of ω_n and δ are of particular importance in resonant systems as described in the section on resonance below. That section includes examples of other analyses using Laplace notation.

There are two implicit assumptions in the transfer function given in Equation 10.5 embedded in the analysis: 1) the input, $V_s(s)$, is an *ideal* voltage source; that is, $V_s(s)$ will produce whatever current is necessary to maintain its prescribed voltage; and 2) nothing is connected to the output: "nothing" meaning that no current flows out of the output terminals. Assumption 2 is also described by stating that $V_{out}(s)$ is connected to an *ideal load*. (An ideal load is one where $Z_L \rightarrow \infty$ which is really no load at all.) In a real circuit application it is possible that one or both of these assumptions are violated, but in many real circuits the conditions are close enough to these idealizations that for all practical purposes the assumptions are valid.

10.1.2 Mesh Analysis—Multiple Loops

Any single-loop circuit can be solved using the "5-step process,"; and a surprising number of useful circuits consist of only a single loop. Nonetheless, it is not difficult to extend the approach to contend with two or more loops, although the complex arithmetic can become tedious for three loops or more. This is not really a problem since MATLAB can handle the necessary math for a large number of loops without breaking a sweat. The example below uses the 5-step approach to solve a two-loop network and indicates how larger networks can be solved.

EXAMPLE 10.4

An extension of the windkessel model used in Example 9.4 is shown in Figure 10.10. In this version, an additional resistor has been added to account for the resistance of the aorta so the model is now a two-mesh circuit. In this example, we find the relationship between $v(t)$, which represents the pressure in the left heart, and the current $i(t)$, which represents the blood flow in the aorta. This is done using phasor notation in conjunction with an extension of the approach developed for single loop circuits.

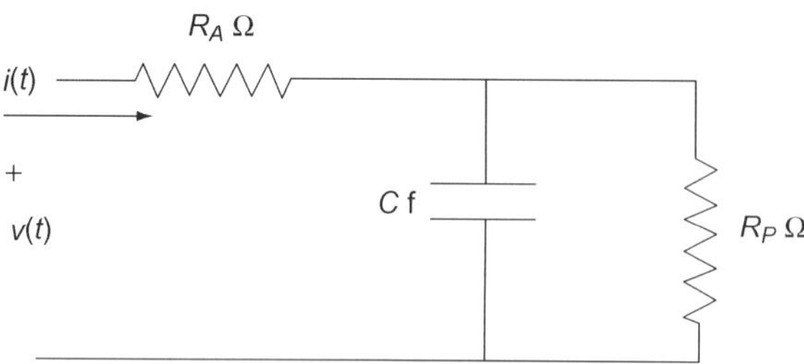

FIGURE 10.10 The three-element windkessel model analyzed in Example 10.4. The values for the components are $R_A = 0.033$ mmHg/ml/sec, $C = 1.75$ ml/mmHG, $R_p = 0.79$ mmHg/ml/sec. In this figure all the connections are explicitly shown: no implicit connections through grounds are depicted; however, the voltage source, $v(t)$, is not explicitly shown.

Solution: To find the phasor relationship between v_{in} and i_{in}, follow the same 5-step plan, but with modifications to step 2. Since this problem is solved manually, we substitute in the actual component values to simplify the complex algebra.

Step 1. Represent all elements by their equivalent phasor or Laplace representations. This step is always the same in any analysis. Note that the capacitor impedance in phasor notation becomes:

$$\frac{1}{j\omega C} = \frac{1}{j1.75\omega} = \frac{0.57}{j\omega}$$

Step 2. Define the mesh currents. This step is essentially the same as for single loop circuits. The only trick is that there are two mesh currents and they are defined as going around each loop separately: they are limited to their respective loop. These currents must be defined as going in the same direction, either clockwise or counterclockwise. Here, for consistency, we always define mesh current in the clockwise direction. Of course real currents are not limited to an individual loop in such an organized fashion. Mesh currents are artificial constructs that aid in solving multi-loop problems. Nonetheless, the two mesh currents do account for all of the currents in the circuit. For example, the current through the capacitor would be the difference between the two mesh currents, $I_1(\omega) - I_2(\omega)$. This is not a problem as long as this difference current is used when solving for the voltage across elements common to both loops (the capacitor in this case).

Steps 1 and 2 lead to the circuit below.

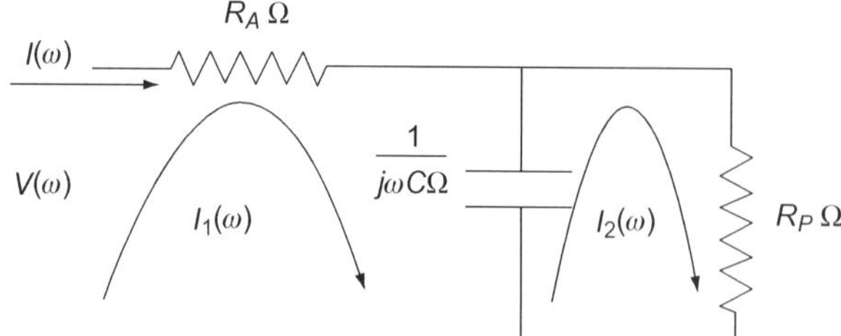

FIGURE 10.11 The three-element windkessel model of Figure 10.10 after the elements are represented by their phasor equivalents and the mesh currents are defined. By definition, the mesh currents go in the clockwise direction and are limited to a single loop.

Step 3. Apply KVL around each loop, keeping in mind that the voltage drop (or rise) across the element(s) shared by both meshes will be due to two currents, and since the currents are flowing in opposite directions their voltage contributions will have opposite signs: $I_1(\omega)$ produces the usual voltage drop, but $I_2(\omega)$ produces a voltage rise since it flows into the bottom of the resistor (again, mentally going clockwise around the loop). Each loop gives rise to an equation.

Mesh 1: KVL following standard procedure, beginning in the lower left-hand corner.

$$V(\omega) - 0.79I_1(\omega) - \frac{0.57}{j\omega}I_1(\omega) + \frac{0.57}{j\omega}I_2(\omega) = 0$$

The capacitor produces a voltage term due to both I_1 and I_2; the term produced by I_2 is positive because mesh current I_2 produces a voltage rise when going around the loop clockwise.

Mesh 2: KVL using the same procedure:

$$\frac{0.57}{j\omega}I_1(\omega) - \frac{0.57}{j\omega}I_2(\omega) - 0.033I_2(\omega) = 0$$

Now the capacitor contributes a voltage rise from current I_1 as we mentally go around the second loop clockwise, in addition to a voltage drop from I_2.

Step 4. Solve for the current(s). Rearranging the two equations, placing current on the right side and sources on the left, and separating the coefficients of the two current variables, gives us two equations to be solved simultaneously. Pay particular attention to keeping the signs straight.

$$V(\omega) = \left(0.033 + \frac{0.57}{j\omega}\right)I_1(\omega) - \frac{0.57}{j\omega}I_2(\omega)$$
$$0 = -\frac{0.57}{j\omega}I_1(\omega) + \left(0.79 + \frac{0.57}{j\omega}\right)I_2(\omega)$$

With only two equations, it is possible to solve for the currents using substitution, avoiding matrix methods. However, with more than two meshes, matrix methods are easier and lend themselves to computer solutions. (The solution of a three-mesh circuit using MATLAB is given below.) The spacing used in the above equations facilitates transforming them into matrix notation:

$$\begin{vmatrix} V(\omega) \\ 0 \end{vmatrix} = \begin{vmatrix} 0.033 + \dfrac{0.57}{j\omega} & -\dfrac{0.57}{j\omega} \\ -\dfrac{0.57}{j\omega} & 0.79 + \dfrac{0.57}{j\omega} \end{vmatrix} \begin{vmatrix} I_1(\omega) \\ I_2(\omega) \end{vmatrix} \qquad (10.9)$$

Solve for $I(\omega) \equiv I_1(\omega)$, which is also using the method of determinants (Appendix G):

$$I_1(\omega) = \frac{\begin{vmatrix} V(\omega) & -\dfrac{0.57}{j\omega} \\ 0 & 0.79 + \dfrac{0.57}{j\omega} \end{vmatrix}}{\begin{vmatrix} 0.033 + \dfrac{0.57}{j\omega} & -\dfrac{0.57}{j\omega} \\ -\dfrac{0.57}{j\omega} & 0.79 + \dfrac{0.57}{j\omega} \end{vmatrix}} = \frac{V(\omega)\left(0.79 + \dfrac{0.57}{j\omega}\right)}{0.026 + \dfrac{0.02}{j\omega} + \dfrac{0.45}{j\omega} + \left(\dfrac{0.57}{j\omega}\right)^2 - \left(\dfrac{0.57}{j\omega}\right)^2}$$

$$I_1(\omega) = \frac{V(\omega)\left(0.79 + \dfrac{0.57}{j\omega}\right)}{0.026 + \dfrac{0.47}{j\omega}} = \frac{V(\omega)(0.57 + j0.79\omega)}{0.47 + j0.026\omega}$$

$$I_1(\omega) = \frac{0.57}{0.47}\frac{V(\omega)(1 + j1.39\omega)}{1 + j0.055\omega} = \frac{1.21V(\omega)(1 + j1.39\omega)}{1 + j0.055\omega}$$

Even this relatively simple two-mesh circuit involved considerable complex arithmetic, but we find that it is easy to solve these problems in MATLAB.

Step 5. In this case we want to find the cardiac pressure as a function of blood flow. In this model, pressure is analogous to the input voltage, $V(\omega)$, and flow to the current, $I(\omega)$. To find $V(\omega)$ as a function of $I(\omega)$, we just invert the equation and factor out the $V(\omega)$:

$$\frac{V(\omega)}{I(\omega)} = \frac{0.83(1 + j0.055\omega)}{1 + j1.39\omega} \tag{10.10}$$

The relationship between cardiac blood flow, $i(t)$, in this model and cardiac pressure, $v(t)$, here, can be used to determine the pressure given the flow or vice versa. It is only necessary to have a quantitative description of the flow or the pressure. The pressure or flow does not need to be sinusoidal (which would not be very realistic); as long as it is periodic it can be decomposed into sinusoids using the Fourier transform. This approach is illustrated in the next example.

EXAMPLE 10.5

Use the three-element windkessel model as described by Equation 10.10 to find the cardiac pressure given the blood flow. Use a realistic pulsatile waveform for blood flow as given in Equation 10.11. (This waveform is also used in Problem 14 of Chapter 9.) It might seem more logical to have cardiac pressure as the input and blood flow as the output, following the causal relationship, but in certain experimental situations we might measure blood flow and want to use the model to determine the cardiac pressure that produced the measured flow.

$$i(t) = \begin{cases} I_0 \sin^2\left(\dfrac{\pi t}{T_1}\right) & 0 \le t < T_1 \\ 0 & T_1 \le t < T \end{cases} \tag{10.11}$$

Solution: To find the pressure we use the standard frequency domain approach: convert $i(t)$ to the frequency domain using the Fourier transform, multiplying $I(\omega)$ by Equation 10.10 to get $V(\omega)$, then take the inverse Fourier transform of $V(\omega)$ to get $v(t)$, the pressure. It is good to remember that in this approach you are solving for $V(\omega)$ a large number of times, each at a different frequency (i.e., the frequencies determined by the Fourier transform of $i(t)$). The fact that the overall answer can be obtained as just the sum of all the individual solutions is possible because superposition holds for this linear system.

Solutions are obtained using MATLAB and the resulting pressure waveform plotted for one cardiac cycle. We assume the period of the heartbeat, T, is 1 sec and T_1 in Equation 10.11 is 0.3 sec; in other words, blood flow occurs during approximately a third of the cardiac cycle (the *systolic* time period). We use a sampling frequency of 1000 Hz to construct the blood flow waveform given in Equation 10.11.

```
% Ex 10.4 Part 2 Plot the cardiac pressure from the windkessel model
%
fs = 1000;                          % Sampling frequency
t = (1:1000)/fs;                    % Time vector, 1 sec long
T1 = .3;                            % Period of blood flow in sec
it = round(T1*fs);                  % Index of blood flow
w = 2*pi*(1:fs);                    % Frequency vector in radians
i = 500*(sin(pi*t/T1)).^2;          % Define blood flow waveform, Equation 10.11
i(it:1000) = 0;                     % Zero period when no blood flow
subplot(2,1,1);
plot(t,i,'k');                      % Plot blood flow waveform
  .......labels and title.......
```

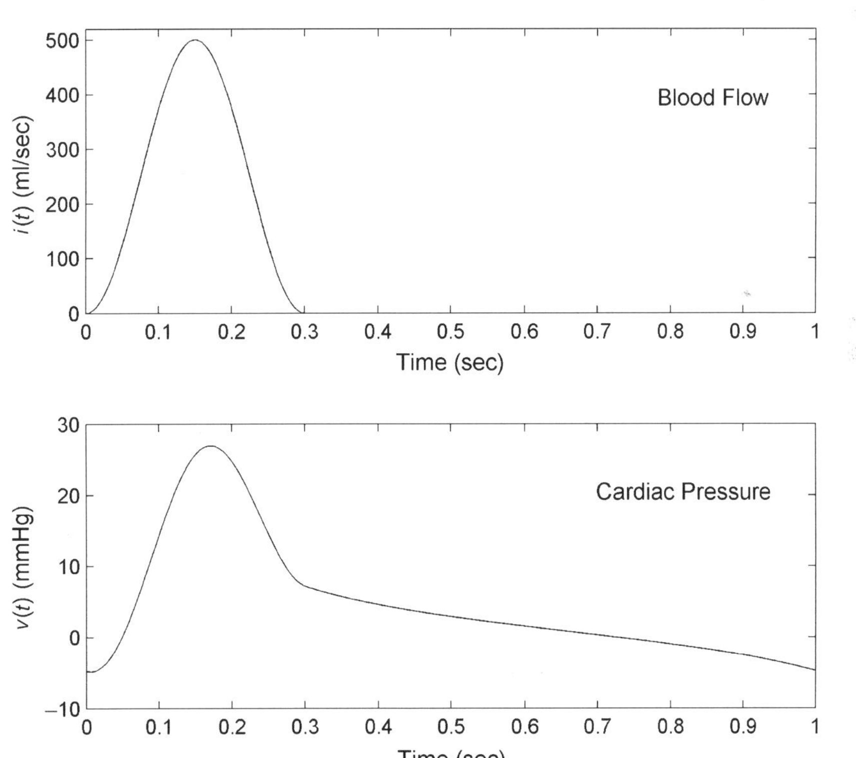

FIGURE 10.12 Pressure in the left heart (lower curve) as a function of time as determined from the three-element windkessel model. The upper curve shows the blood flow used to find the pressure curve.

```
I = fft(i);                                  % Take Fourier transform
ZTF = 0.82 *(1 + j*0.055*w)./(1 + j*1.39*w); % Pressure/flow relationship
V = TF.*I;                                   % Find V (in phasor)
v = ifft(V);                                 % Find v(t), Inverse FT
subplot(2,1,2);
plot(t,real(v),'k');                         % Plot pressure waveform
```

Results: The plots produced by this program are shown in Figure 10.12. As cardiac flow increases the pressure increases, but when cardiac flow goes to zero, the pressure does not drop to zero, but decreases gradually though the rest of the cardiac cycle. A more complicated 4-element windkessel model is analyzed in Problem 10.

10.1.2.1 Shortcut Method for Multimesh Circuits

There is a shortcut that makes it possible to write the matrix equation directly from inspection of the circuit. Regard the matrix equation, Equation 10.9, derived for the circuit in Figure 10.11 and repeated here:

$$\underbrace{\begin{vmatrix} V(\omega) \\ \\ 0 \end{vmatrix}}_{V} = \underbrace{\begin{vmatrix} 0.033 + \dfrac{0.57}{j\omega} & -\dfrac{0.57}{j\omega} \\ -\dfrac{0.57}{j\omega} & 0.79 + \dfrac{0.57}{j\omega} \end{vmatrix}}_{Z} \underbrace{\begin{vmatrix} I_i(\omega) \\ \\ I_2(\omega) \end{vmatrix}}_{I}$$

This equation has the general form of $V = ZI$ as indicated by the brackets. The left-hand side contains only the sources; the right-hand side contains a matrix of impedances multiplied by a vector containing the mesh currents. The source vector contains the source in Mesh 1 in the upper position and the source in Mesh 2 in the lower position which, in Example 10.4, is zero because there are no sources in Mesh 2. A similar arrangement holds for the current vector that features the two mesh currents. The impedance matrix also relates topographically to the circuit: the upper left entry is the sum of impedances in Mesh 1, the lower right is the sum of impedances in Mesh 2, and the off-diagonals (upper right and lower left) contain the negative of the sum of impedances common to both loops. In this circuit there is only one element common to both meshes, the capacitor, so the off-diagonals contain the negative of this element, but other circuits could have several elements common to the two meshes. Putting this verbal description into mathematical form:

$$\begin{vmatrix} \Sigma V_S \text{ Mesh1} \\ \Sigma V_S \text{ Mesh2} \end{vmatrix} = \begin{vmatrix} \Sigma Z \text{ Mesh1} & -\Sigma Z \text{ Mesh1\&2} \\ -\Sigma Z \text{ Mesh1\&2} & \Sigma Z \text{ Mesh2} \end{vmatrix} \begin{vmatrix} I_1 \\ I_2 \end{vmatrix} \tag{10.12}$$

In this equation the impedance sums, ΣZ, are additions; there is no need to worry about signs except if the impedance itself is negative, as may be for a capacitor. However, the summation of voltage sources, the ΣV_S, still requires some care, since the summations must take the voltage source signs into consideration. For example, the source in Mesh 1 (ΣV_S Mesh1) in Example 10.4 has a positive sign because it represents a voltage rise when going around the loop clockwise.

This shortcut rule also applies to circuits that use phasor or Laplace representations. The approach can easily be extended to circuits having any number of meshes, although the subsequent calculations become tedious for three or more meshes unless computer assistance is used. The extension to three meshes is given in the example below, but the solution is determined using MATLAB.

10.1.3 Mesh Analysis—MATLAB Implementation

EXAMPLE 10.6

Solve for V_{out} in the three-mesh network in Figure 10.13. This circuit uses realistic values for *R*, *L*, and *C*.

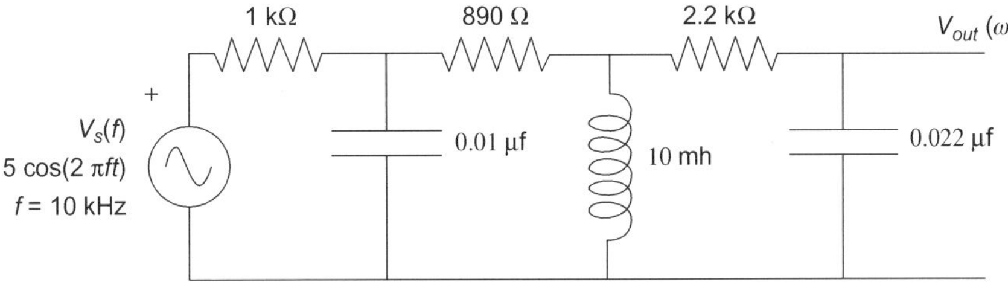

FIGURE 10.13 Three-mesh circuit used in Example 10.6. The values for *R*, *L*, and *C* used in this network are closer to those found in real circuits.

Solution: Follow the steps used in the previous examples, but use the shortcut method in step 3. In step 4, solve for the currents using MATLAB, and in step 5 solve for V_{out} using MATLAB.

Steps 1 and 2. The figure shows both the original circuit and the circuit after steps 1 and 2 with the elements represented in phasor representation and the phasor currents defined.

The impedances for *L* and C are determined as:

$$Z_L = j\omega L = j2\pi fL = j2\pi 10^4(10 \times 10^{-3}) = j628\Omega$$

$$Z_{C2} = \frac{1}{j\omega C_2} = \frac{1}{j2\pi fC_2} = \frac{1}{j2\pi 10^4(0.022 \times 10^{-6})} = \frac{1}{j1.38 \times 10^{-3}} = -j723\Omega$$

$$\text{Similarly: } Z_{C2} = \frac{1}{j2\pi fC_2} = \frac{1}{j2\pi 10^4(0.01 \times 10^{-6})} = j1592\Omega$$

Step 3. The matrix equation for step 3 is a modification of Equation 10.11 where the voltage and current vectors have three elements each and the impedance matrix is extended to a 3×3 matrix written as:

$$\begin{vmatrix} \Sigma V_S \text{ Mesh1} \\ \Sigma V_S \text{ Mesh2} \\ \Sigma V_S \text{ Mesh3} \end{vmatrix} = \begin{vmatrix} \Sigma Z \text{ Mesh1} & -\Sigma Z \text{ Mesh1\&2} & -\Sigma Z \text{ Mesh1\&3} \\ -\Sigma Z \text{ Mesh1\&2} & \Sigma Z \text{ Mesh2} & -\Sigma Z \text{ Mesh2\&3} \\ -\Sigma Z \text{ Mesh1\&3} & -\Sigma Z \text{ Mesh2\&3} & \Sigma Z \text{ Mesh3} \end{vmatrix} \begin{vmatrix} I_1 \\ I_2 \\ I_3 \end{vmatrix} \tag{10.13}$$

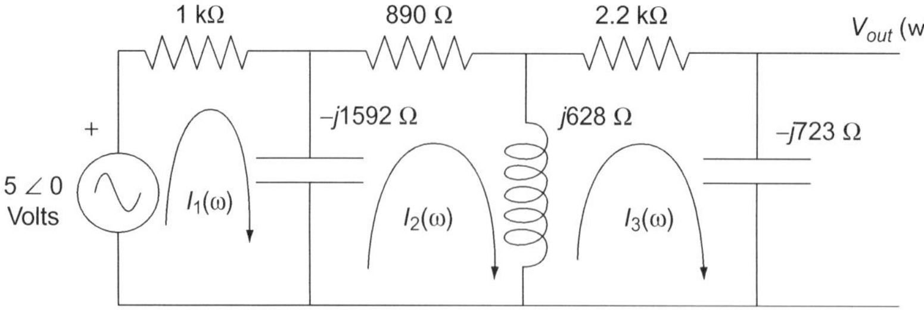

FIGURE 10.14 The network shown in Figure 10.13 after the circuit elements have been converted to phasor representation.

where: ΣZ Mesh1, ΣZ Mesh2, and ΣZ Mesh3 are the sum of impedances in the three meshes; ΣZ Mesh1 and ΣZ Mesh2 is the sum of impedances common to Meshes 1 and 2; ΣZ Mesh1 and ΣZ Mesh3 is the sum of impedances common to Meshes 1 and 3; and ΣZ Mesh2 and ΣZ Mesh3 is the sum of impedances common to Meshes 2 and 3. Note that all the off-diagonals are negative sums (although they could be positive if they contain capacitors) and the impedance matrix has symmetry about the diagonal. Such symmetry is often found in matrix algebra, and a matrix with this symmetry is termed a *Toplitz matrix*. In this particular network there are no impedances common to Meshes 1 and 3, but there are circuits in which all three meshes have at least one element in common.

Applying Equation 10.13 to the network in Figure 10.14 gives rise to the matrix equation for this network:

$$\begin{vmatrix} 5 \\ 0 \\ 0 \end{vmatrix} = \begin{vmatrix} 1000 - j1592 & j1592 & 0 \\ j1592 & 890 - j964 & -j628 \\ 0 & -j628 & 2200 - j95 \end{vmatrix} \begin{vmatrix} I_1(\omega) \\ I_2(\omega) \\ I_3(\omega) \end{vmatrix}$$

Note that the $-j964$ in the middle term is the algebraic sum of $j628 - j1592$ and the $-j95$ in the lower right is the sum of $j628 - j723$.

Steps 4 and 5. Solve for the currents, in this case $I_3(\omega)$, then the desired voltage, which is $-j723\ I_3(\omega)$. The MATLAB program does both. First it defines the voltage vector and impedance matrix then solves the matrix equation as presented below.

```
% Example 10.6 Solution of a 3-mesh network
%
V = [10 0 0]';              % Note the use of the transpose symbol, '
% Define the impedance matrix
Z = [1000 -1i*1592, 1i*1592, 0; 1i*1592, 890 -1i*964, -1i*628; 0, -1i*628, 2200 -1i*95];
I = Z\V                     % Solve for the currents
Vout = I(3)*(-723*1i)  % Output the requested voltage
Vmag = abs(Vout)            % also as magnitude and phase
Vphase = angle(Vout)*306/(2*pi) % Output angle in deg.
```

Results: The output of this program for the 3-mesh currents is:

```
I = 0.0042 + 0.0007i
0.0037 - 0.0029i
0.0008 + 0.0011i
```

The output for the voltages is:

```
Vout = 0.7908 - 0.5659i
Vmag = 0.9724 Vphase = -30.2476
```

MATLAB accepts either `i`, `1i`, or `j` to represent a complex number, but outputs only using `i`. (In newer versions of MATLAB, the symbol `1i` is recommended to represent an imaginary number instead of either `i` or `j` for improved speed and stability.)

The time-domain output is determined directly from the phasor output given above. Recall that that $\omega = 10^4$ rad/sec:

$$V_{out}(t) = 0.97\cos\,(2\pi 10^4 t - 30.2)\text{ volts}$$

Analysis: In the MATLAB program, the voltage vector is written as a row vector, but it should be a column vector as in Equation 10.13, so the MATLAB transpose operator (single quote) is used. It could just as easily been entered as a column vector directly (`V = [10; 0; 0]`); the transposed row vector was the whim of the programmer. The second line defines the impedance matrix using standard MATLAB notation. The third line solves for the three currents by matrix inversion, implementing the equation: $I = Z^{-1}V$ using the backslash (\) operator. The fourth line multiplies the third mesh current, $I_3(\omega)$, by the capacitor impedance to get V_{out}. The next two lines convert V_{out} into polar form.

If MATLAB can solve one matrix equation, it can solve it many times over. By making the input a sinusoid of varying frequency, we can generate the Bode plot of any network. The next example treats a network as an input-output system where the source voltage is the input and the voltage across one of the elements is the output. MATLAB is called upon to solve the problem many times over at different sinusoidal frequencies and for different element values.

EXAMPLE 10.7

Plot the magnitude Bode plot in Hz of the transfer function of the network in Figure 10.15 with input and output as shown. Plot the Bode plot for three values of the capacitance: 0.01 μf, 0.1 μf, and 1.0 μf. The three capacitors all have the same values, and the three resistors are all 2.0 kΩ.

Solution: We could first find the phasor domain transfer function and plot that function for the various capacitance values. However, that would require some tedious algebra, particularly

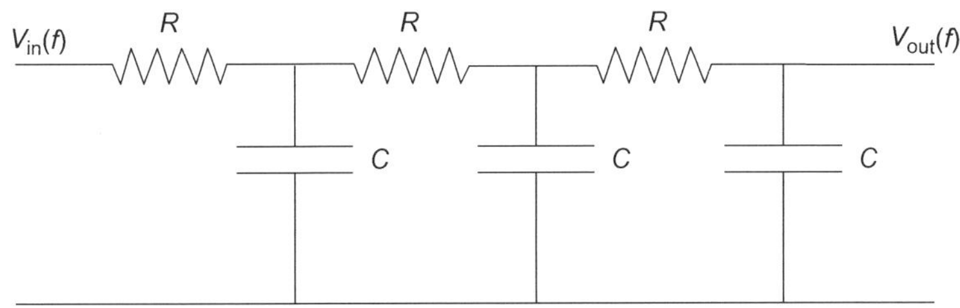

FIGURE 10.15 Three-mesh circuit used in Example 10.6. This circuit is viewed as an input-output system with input $V_{in}(\omega)$ and output $V_{out}(\omega)$.

as we have to leave the capacitance values as variables. It is easier on us to use MATLAB to determine and plot the output voltage to a range of inputs at different frequencies.

Follow the 5-step procedure to generate the KVL matrix and solve for the third mesh current using MATLAB. Then find V_{out}, the voltage across the right-hand capacitor. Assume V_{in} is an ideal sinusoidal source varying in frequency. Select a frequency range that includes any interesting changes in the Bode plot. This might require trial and error, so to begin try a large frequency range of around 10 Hz and 100 kHz. To keep computational time down, use 20-Hz intervals. Use one loop to vary the frequency and an outside loop to change the capacitor values.

Steps 1 and 2. At this point you can implement these steps by inspection. The capacitors become $(1/j\omega C)$ and the mesh currents are $I_1(\omega)$, $I_2(\omega)$, and $I_3(\omega)$. We do not need to redraw the network.

Step 3. Again the KVL equation for this circuit can be done by inspection.

$$\begin{vmatrix} V_{\text{in}}(s) \\ 0 \\ 0 \end{vmatrix} = \begin{vmatrix} R + \dfrac{1}{j\omega C} & -\dfrac{1}{j\omega C} & 0 \\ -\dfrac{1}{j\omega C} & R + \dfrac{2}{j\omega C} & -\dfrac{1}{j\omega C} \\ 0 & -\dfrac{1}{j\omega C} & R + \dfrac{2}{j\omega C} \end{vmatrix} \begin{vmatrix} I_1(\omega) \\ I_2(\omega) \\ I_3(\omega) \end{vmatrix}$$

Steps 4 and 5. These steps are implemented in the MATLAB program below noting that

$$V_{\text{out}}(\omega) = \frac{1}{j\omega C} I_3(\omega)$$

The MATLAB implementation of these steps is shown below. The matrix equation and its solutions are placed in a loop that solves the equation for 4000 values of frequency ranging from 10 Hz to 80 kHz. In this example, frequency is requested in Hz. The desired voltage V_{out} is determined from mesh current I_3. A Bode plot is constructed by plotting $20 \log|V_{out}|$ against log frequency since $V_{in} = 1$ so $TF(f) = V_{out}(f)$. This loop is placed within another loop that repeats the 4000 solutions for the three values of capacitance.

```
R = 2000;                                   % Resistor values
C = [1 0.1 0.01]*1e - 6;                    % Capacitance values
V = [1; 0; 0];                              % Mesh voltages: Vin = 1
for i = 1:3                                 % Capacitance value loop
  for k = 1:4000                            % Frequency loop
    f(k) = 10 + (k -1)*20;                  % Determine frequency: 10-80 kHz
    w = 2*pi*f(k);                          % Calculate frequency in radians
    Xc = 1/(j*w*C(i));                      % Calculate capacitance impedance
    Z = [R + Xc, -Xc, 0;...
    -Xc, R + 2*Xc, -Xc; 0, -Xc, R + 2*Xc];  % Mesh equation
    I = Z\V;                                % Solve for currents
    Vout(k) = Xc*I(3);                      % Solve for Vout
  end
  semilogx(f,20*log10(abs(Vout)),'k','LineWidth',2); hold on;% Bode plot
end
```

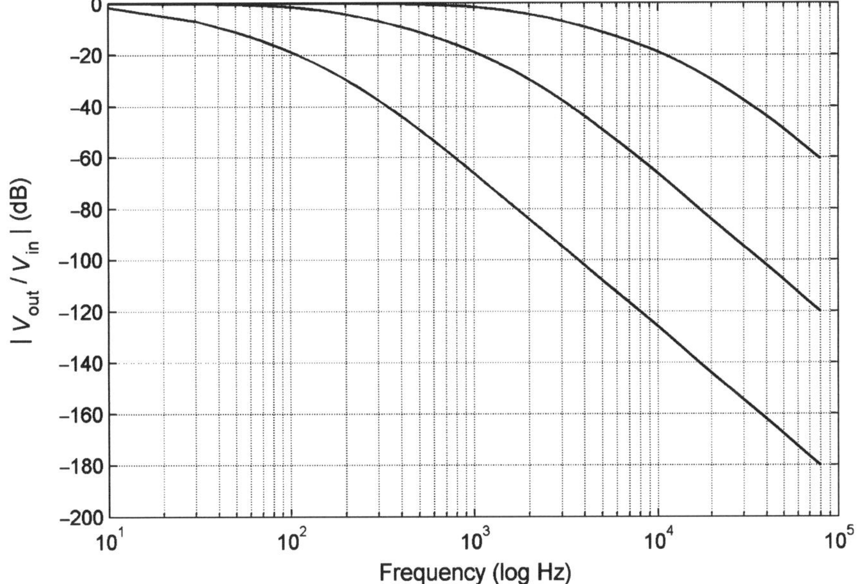

FIGURE 10.16 Magnitude plot of the transfer function of the circuit in Figure 10.15 obtained by solving the network mesh equation over a wide range of frequencies and for three values of *C*.

The Bode plots generated by this program are shown in Figure 10.15. The plots of the three curves look like low-pass filters with downward slopes of 60 dB/decade, each having a different cutoff frequency (i.e., −3 dB attenuation points). Based just on the Bode plot slope, the network of Figure 10.15 looks like a third-order low-pass filter with a cutoff frequency that varies with the value of *C*. As we can see from the circuit diagram, Figure 10.14, that is exactly what it is. In fact, the cutoff frequency is approximately $\frac{1}{2\pi RC}$, but the initial slope is not as sharp as that of a Butterworth filter. Usually filters above first order are constructed using active elements (see Chapter 12), but the circuit of Figure 10.14 is easy to construct and is occasionally used as a "quick and dirty" third-order filter in real-world electronics.

10.2 CONSERVATION LAWS—KIRCHHOFF'S CURRENT LAW: NODAL ANALYSIS

Kirchhoff's current law can also be used to analyze circuits. This law, based on the conservation of charge, was given in Equation 10.2 and is repeated here:

$$\sum_{\text{Node}} i = 0 \tag{10.14}$$

KCL is best suited to analyzing circuits with many loops but only a few connection points. Figure 10.17 shows the Hodgkin–Huxley model for a nerve membrane. The three voltage–resistor combinations represent the potassium membrane channel, the sodium

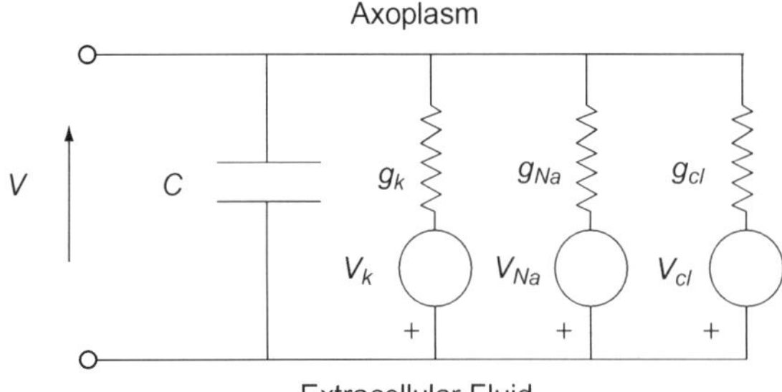

FIGURE 10.17 Model of nerve membrane developed by Hodgkin and Huxley. The three voltage–resistor combinations represent ion channels in the membrane that sustain the resting voltage and mediate an action potential. The equation describing this model is best developed using KCL and nodal analysis.

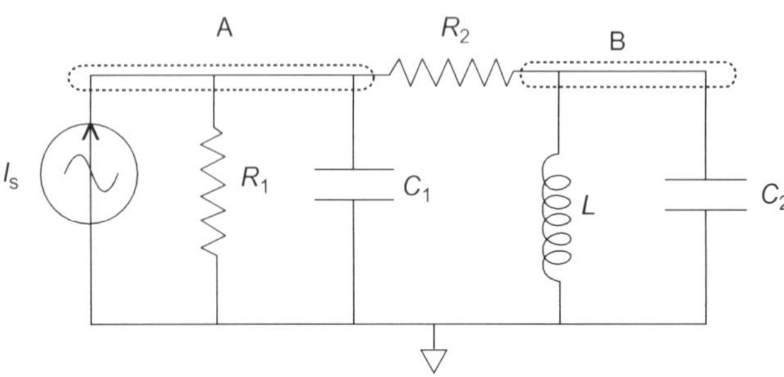

FIGURE 10.18 A circuit consisting of four meshes but only two nodes. The nodes are the connection points labeled A and B. Nodes include all the connections that are at the same voltage, as indicated by the dashed lines. The ground point (the line across the bottom) is not considered an independent node since its voltage is, by definition, fixed at zero. Nodal analysis works with currents using KVL and is easiest if the sources are current sources.

membrane channel, and the chloride membrane channel, while C is the membrane capacitance. Analyzing this circuit requires four mesh equations, but only one nodal equation. In this model, most of the components are nonlinear, at least during an action potential, so the model cannot be solved analytically as is done with our linear processes. Nonetheless, the defining equation(s) would be generated using nodal analysis and can be solved using computer simulation.

Another example of a circuit appropriate for nodal analysis is shown in Figure 10.18. This circuit has four meshes, and mesh analysis would give rise to four simultaneous equations. This same circuit only has two nodes (marked A and B, again ground points do not count) and would require solving only two nodal equations. If MATLAB is used, then solving a four-equation problem is really not all that much harder than solving a two-equation problem; it is just a matter of adding a few more entries into the voltage vector and impedance matrix. However, when circuits are used as models representing physiological processes, as in Figure 10.17, the more concise description given by nodal equations is of great value.

The circuit in Figure 10.18 contains a current source, not a voltage source as we have seen in previous examples. This is because nodal analysis is an application of a current law, so it is easier to implement if the sources are current sources. A similar statement could be made about mesh analysis: mesh analysis involves voltage summation and it is easier to implement if all sources are voltage sources. The need to have only current sources may seem like a drawback to the application of nodal analysis, but we see in Chapter 11 that it is easy to convert voltage sources to equivalent current sources and vice versa, so this requirement is not really a handicap. In this chapter, nodal analysis examples use current sources as the technique can be applied equally well to voltage sources after a simple conversion.

Analyzing circuits using nodal analysis follows the same 5-step procedure as that used in mesh analysis. In fact, steps 4 and 5 are the same. Step 1 could also be the same, but often elements are converted to $1/Z$, instead of simply Z. The inverse impedance, $Y = 1/Z$, is termed the *admittance*. In step 2 the node voltages are assigned rather than the mesh currents and in step 3 the equations are generated using KCL.

The equations developed from KCL have a sort of inverse symmetry with those of mesh analysis. In mesh analysis, we are writing matrix equations of the form:

$$\boldsymbol{v} = \boldsymbol{Z}\boldsymbol{i} \tag{10.15}$$

where $\boldsymbol{v}$ is a voltage vector, $\boldsymbol{i}$ is a current vector, and $\boldsymbol{Z}$ is an impedance matrix (Equation 10.12 and 10.13). In nodal analysis we are writing matrix equations in the form:

$$\boldsymbol{i} = \boldsymbol{Y}\boldsymbol{v} \tag{10.16}$$

where $\boldsymbol{Y}$ is a matrix, termed the *admittance matrix* containing the inverse impedances. The terms $\boldsymbol{v}$ and $\boldsymbol{i}$ are vectors as in Equation 10.15.

EXAMPLE 10.8

Find the voltage V_A in the circuit in Figure 10.19.

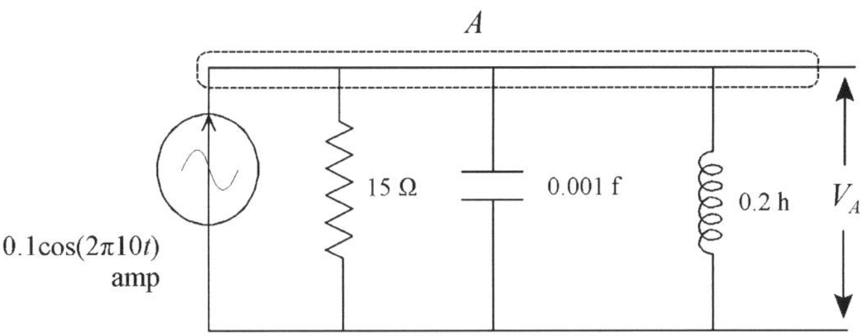

FIGURE 10.19 Circuit used in Example 10.8 where KCL is used to solve for the voltage, V_A.

Solution: This circuit requires two mesh equations (after conversion of the current source to an equivalent voltage source as explained in Chapter 11), but only one nodal equation. Moreover, it conveniently contains a current source, making the nodal analysis even easier. There are four currents flowing into or out of the single node at the top of the circuit labeled A. The current in the current source branch is $0.1\cos(2\pi 10t)$ and the current in each of the other three branches is equal to the voltage, V_A, divided by the impedance of the branch; i.e., $I(\omega) = V_A(\omega)/Z_{Branch}$. By KCL, these four currents will sum to zero.

After applying steps 1 and 2, the network becomes as shown in Figure 10.20. If we define $V_A(\omega)$ as a positive voltage, then the current though the passive elements flows downward as shown due to the voltage-current polarity rule for passive elements. The frequency in radians is $\omega = 2\pi f = 62.8$ rad/sec.

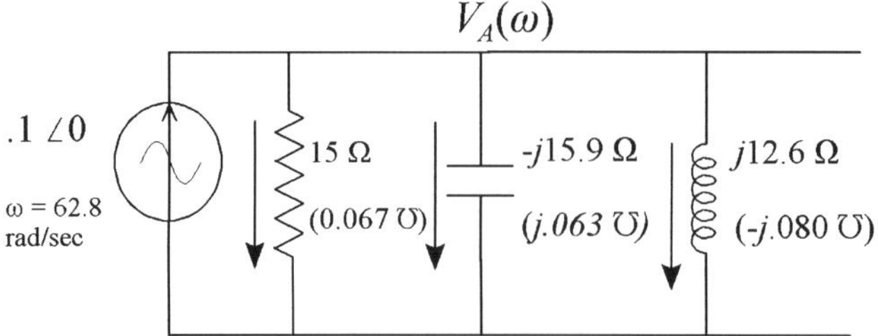

FIGURE 10.20 The circuit shown in Figure 10.19 with the elements represented in phasor notation. This is the same operation that is applied in either KVL or KCL approaches.

Step 3. This applies the fact that the four currents sum to zero. As with mesh analysis, care has to be taken that the signs are correct. The current source flows into node A, but the other three currents flow out.

$$i_S(\omega) - i_R(\omega) - i_C(\omega) - i_L(\omega) = 0 \qquad (KCL)$$

$$I_S - \frac{V_A(\omega)}{R} - \frac{V_A(\omega)}{(1/j\omega C)} - \frac{V_A(\omega)}{j\omega L} = 0$$

$$0.1 + \frac{V_A(\omega)}{15} + \frac{V_A(\omega)}{-j15.9} + \frac{V_A(\omega)}{j12.6} = 0$$

Now we can solve this single equation for $V_A(\omega)$. The equation is easier written in terms of admittances: $Y = 1/Z$. The values of the admittances are shown in parentheses in the circuit above. Using admittances:

$$I_S + Y_R V_A(\omega) + Y_C V_A(\omega) + Y_L V_A(\omega) = 0$$

$$I_S + V_A(\omega)\left(\frac{1}{R} + j\omega C + \frac{1}{j\omega L}\right) = 0$$

$$0.1 + V_A(\omega)(0.067 + j0.063 - j0.08) = 0$$

$$V_A(\omega) = \frac{0.1}{0.067 + j0.063 - j0.08} = \frac{0.1}{0.067 - j0.017}$$

$$V_A(\omega) = \frac{0.1}{0.069\angle -14} = 1.5\angle 14 \text{ volts}$$

Moving to multinodal systems, we go directly to the shortcut, matrix equation. If we apply KCL to circuits with multiple nodes, we find that the equations fall into a pattern similar to that of mesh analysis, except that they have the form of Equation 10.16: $i = Yv$. The admittance matrix consists of the summed admittances (i.e., $1/Z$'s) that are common to each node along the diagonal and the summed admittances between nodes on the off-diagonals. This general format is shown here for a three-node circuit:

$$\begin{vmatrix} \Sigma I_1 \\ \Sigma I_2 \\ \Sigma I_3 \end{vmatrix} = \begin{vmatrix} \Sigma Y\ \text{Node1} & -\Sigma Y\ \text{Node1\&2} & -\Sigma Y\ \text{Node\&3} \\ -\Sigma Z\ \text{Mesh1\&2} & \Sigma Y\ \text{Node2} & -\Sigma Y\ \text{Node2\&3} \\ -\Sigma Y\ \text{Node1\&3} & -\Sigma Y\ \text{Node2\&3} & \Sigma Y\ \text{Node3} \end{vmatrix} \begin{vmatrix} V_1 \\ V_2 \\ V_3 \end{vmatrix} \qquad (10.17)$$

The application of Equation 10.17 is straightforward and follows the same pattern as in mesh analysis. An example of nodal analysis to the two-node circuit is given in Example 10.9.

EXAMPLE 10.9

Find the voltage, v_2 in the circuit shown in Figure 10.21. This circuit is similar to the one shown in Figure 10.19 except an additional component has been added so the network now has two nodes.

Solution: Apply nodal analysis to this two-node circuit. Follow the step-by-step procedure outlined above, but in step 3 write the matrix equation directly as given in Equation 10.9. Implement step 4 to solve for V_B using MATLAB.

Step 1. Convert all the elements to phasor admittances. Note that $\omega = 20$ rad/sec.

Step 2. Assign nodal voltages. This has already been done in the circuit. The circuit after modification by these two steps is shown in Figure 10.22.

Step 3. Generate the matrix equations directly following a reduced (i.e., two node) version of Equation 10.15. Note that inductors now have $-j$ values while capacitors have $+j$ values. Also note that the two nodes share two components, so the shared admittance will be the sum of the admittances from each component:

$$\sum Y_{\text{node1,2}} = .004 - j.007$$

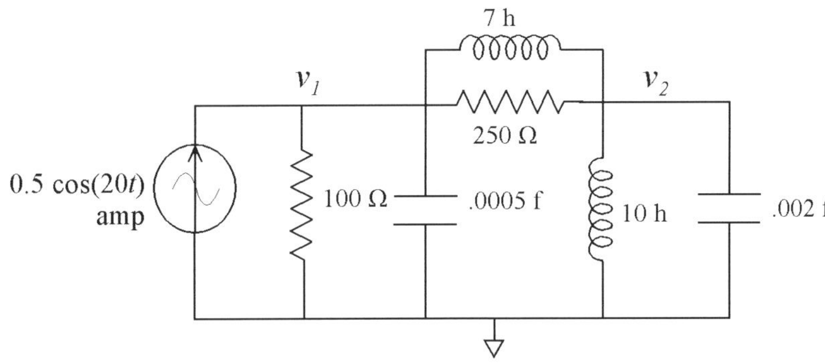

FIGURE 10.21 Circuit used in Example 10.9.

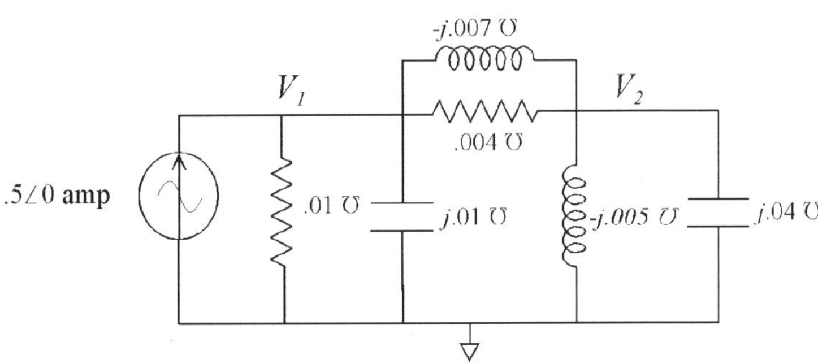

FIGURE 10.22 The circuit shown in Figure 10.21 after all the elements are converted to phasor notation. Since this is a KCL approach, the passive elements are converted to phasor admittances. Nodal voltages have also been assigned.

Hence the KCL circuit equation becomes:

$$\begin{vmatrix} 0.5 \\ 0 \end{vmatrix} = \begin{vmatrix} 0.01 + 0.004 + j0.01 - j0.007 & -0.004 + j0.007 \\ -0.004 + j0.007 & 0.004 - j0.005 - j0.007 + j0.04 \end{vmatrix} \begin{vmatrix} V_1 \\ V_2 \end{vmatrix}$$

$$\begin{vmatrix} 0.5 \\ 0 \end{vmatrix} = \begin{vmatrix} 0.014 + 0.003 & -0.004 + j0.007 \\ -0.004 + j.007 & 0.004 - j0.028 \end{vmatrix} \begin{vmatrix} V_1 \\ V_2 \end{vmatrix}$$

Step 4. This matrix equation can easily be solved using MATLAB as illustrated by the code below.

```
% Example 10.9 Solution of two-node matrix equation
%
I = [.5; 0];                          % Assign current vector
Y11 = 0.01 + .004 + 1i*.01 -1i*.007;  % Assign admittances
Y12 = .004 -1i*.007;
Y22 = 0.004 -1i*.005 -1i*.007 + 1i*.04;
Y = [Y11 - Y12; - Y12 Y22];           % Admittance matrix
V = Y\I;                              % Solve for voltages
Magnitude = abs(V(2))                 % Magnitude and phase of V2
Phase = angle(V(2))*360/(2*pi)
```

The output gives the magnitude and phase of V_2 as:

```
Mag = 8.7628; Phase - 149.6324
```

In the time domain:

$v_2(t) = 8.76 \cos(20t - 149)$ volts

This approach can be extended to three-node or even higher node circuits without a great deal of additional difficulty. A three-node problem is given in Problem 14 at the end of the chapter. Nodal analysis applies equally well to networks represented in Laplace notation. The basic 5-step approach can also be applied to the analysis of lumped-parameter mechanical systems as described in the next section.

10.3 CONSERVATION LAWS—NEWTON'S LAW: MECHANICAL SYSTEMS

The analysis of lumped-parameter mechanical systems also uses a conservation law, one based on conservation of energy. In mechanical systems, force is potential energy, since force acting though distance produces work. The mechanical version of KVL (the electrical law based on conservation of energy) states that the sum of the forces around any one connection point must be zero. This is a form of the classic law associated with Newton:

$$\sum_{Point} F = 0 \tag{10.18}$$

In this application, a connection point includes all connections between mechanical elements that are at the same velocity. Figure 10.23 shows one version of the linear model for skeletal muscle. Skeletal muscle has two different elastic elements: a parallel elastic element, C_p, and a series elastic element, C_s. [*Note*: the symbol C indicates they are given as compliances, $1/k_e$.] The force generator, F_o, represents the active contractile element, and k_f represents viscous damping inherent in the muscle tissue. The muscle model has two

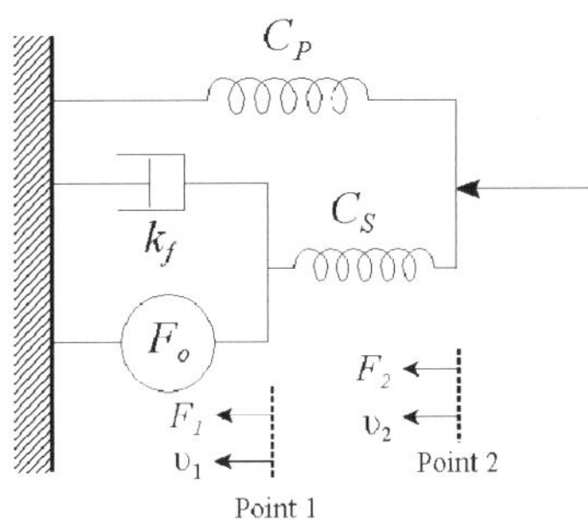

FIGURE 10.23 Linear mechanical model of skeletal muscle. F_o is the force produced by the active contractile element, C_p and C_s are known as the parallel and series elasticity, and k_f is viscous damping associated with the tissue.

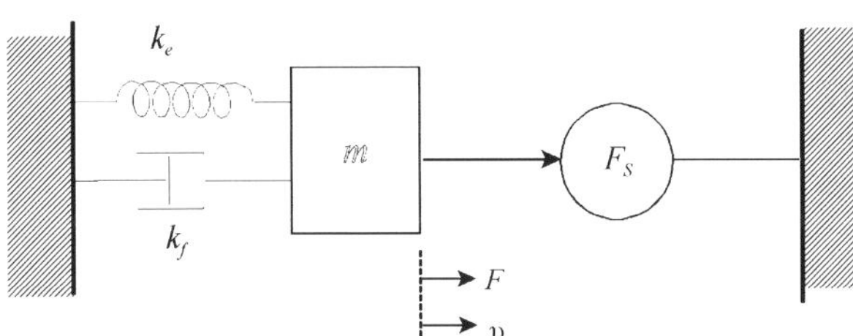

FIGURE 10.24 A mechanical system with one independent velocity v used in Example 10.10. The connections on either side of the mass are at the same point from an analysis point of view.

connection points that may have different velocities, labeled Point 1 and Point 2. The positive force is defined inward, reflecting the fact that muscles can only generate contractile force. This is the reason they are so often found in agonist–antagonist pairs.

Since this system has two different velocities, its analysis would require the simultaneous solution of equations. The system is the mechanical equivalent of a two-mesh electrical circuit. The two equations would be written around Points 1 and 2: the sum of forces around each point must be zero. The graphic on the left represents a zero-velocity point or a solid wall, the analog of a ground point in an electrical system.

The muscle model will be analyzed in Example 10.11, but for a first example of the application of Newton's Law (Equation 10.18) we turn to a less complicated, one-equation system with a single velocity point as shown in Figure 10.24.

EXAMPLE 10.10

Find the velocity and displacement of the mass in the mechanical system shown in Figure 10.24. The force, $F_S(t) = 5\cos(2t + 30)$ dynes. The following parameters also apply:

$k_f = 10$ sec/cm; $k_e = 8$ dynes/cm; $m = 3$ gm

Solution: In this example, the units are all comparable and are in the cgs metric system, the standard for this text. (In Example 10.11, conversion of units is required.) To analyze this system, we follow the same five-step plan developed for electric circuits.

Step 1. Convert variables to phasor notation and represent the passive elements by their phasor impedances. Since $\omega = 2$:

$$Z_f = k_f = 5; \quad Z_e = \frac{k_e}{j\omega} = \frac{8}{j2} = -j4; \quad Z_m = j\omega\, m = j2(3) = j6$$

Step 2. Assign variable directions. In mechanical systems, we use the convention of assigning the force and velocity in the same direction, but the direction (right or left) is arbitrary (in this example it is to the right). This is analogous to the assigning currents as counterclockwise and keeping track of voltage polarities by going in the same direction. By assigning force and velocity in the same direction, the polarity of passive elements is always negative just as in electric circuits, so the equations look similar.

Step 3. Apply Newton's Law about the force-velocity point (next to the mass):

$$\sum F = 0; \qquad F_S(\omega) - k_f v(\omega) - \frac{k_e}{j\omega} v(\omega) - j\omega m v(\omega) = 0$$

$$5\angle 30 - v(\omega)(5 - j4 + j6) = 0$$

The first three steps follow a path parallel to that followed in KVL analysis while the last two steps are essentially identical: solve for the velocity (analogous to current), then any other variable of interest such as a force or, in this case, a displacement.

Step 4. Solve for the phasor velocity:

$$v(\omega) = \frac{5\angle 30}{5 - j4 + j6} = \frac{5\angle 30}{5 + j2} = \frac{5\angle 30}{5.39\angle 21.8} = 0.93\angle 8.2 \text{ cm/sec}$$

Step 5. Solve for displacement. Since $x(t) = \int v dt$; integration in the phasor domain is division by $j\omega$, so $x(\omega) = v(\omega)/j\omega$:

$$x(\omega) = \frac{v(\omega)}{j\omega} = \frac{0.93\angle 8.2}{j2} = \frac{0.93\angle 8.2}{2\angle 90} = 0.465\angle -81.8 \text{ cm}$$

Both $v(\omega)$ and $x(\omega)$ can be converted to the time domain if desired:

$$v(t) = 0.93 \cos(2t + 8.2) \text{ cm/sec}$$
$$x(t) = 0.4610 \cos(2t - 81.8) \text{ cm}$$

The next example is more complicated in two ways: 1) the component values are not given and must be carried though the algebra, and 2) there are two summation points in the problem so two equations are required that must be solved simultaneously.

EXAMPLE 10.11

Find the force out of the skeletal muscle model in Figure 10.23.

Solution: After converting to the phasor domain, write Newton's Law (Equation 10.18) around Points 1 and 2. Solve for $v_2(\omega)$, using the equation:

$$F_2(\omega) = v_2(\omega) Z_{C_P}(\omega) = v_2(\omega)\frac{1}{j\omega C_P}$$

In this solution the algebra is a bit tedious because the various parameters remain as variables, but the procedure is otherwise straightforward.

Steps 1 and 2. The force and velocity assignments are given in Figure 10.23. The various components become:

$$F_o \to F_o(\omega);\quad k_f \to k_f;\quad C_P \to \frac{1}{j\omega C_P}\quad C_S \to \frac{1}{j\omega C_S}$$

The equation around Point 1 is:

$$F_o - k_f v_1(\omega) - \frac{1}{j\omega C_s}(v_1(\omega) - v_2(\omega)) = 0$$

Note that the force generated by the elastic element C_s depends on the difference in velocities between Points 1 and 2. The force that is generated by $v_1(\omega) - v_2(\omega)$ is in the opposite direction of F_1 which accounts for the negative sign in front of this term.

Step 3. The equation around Point 2 is:

$$0 - \frac{1}{j\omega C_p} v_2(\omega) - \frac{1}{j\omega C_s}(v_2(\omega) - v_1(\omega)) = 0$$

Rearranging to separate coefficients of $v_1(\omega)$ and $v_2(\omega)$:

$$F_o = \left(k_f + \frac{1}{j\omega C_s}\right) v_1(\omega) - \frac{1}{j\omega C_s} v_2(\omega)$$

$$0 = \frac{1}{j\omega C_s} v_1(\omega) + \left(\frac{1}{j\omega C_s} + \frac{1}{j\omega C_p}\right) v_2(\omega)$$

Step 4. Solving $v_2(\omega)$:

$$v2(\omega) = \frac{\begin{vmatrix} F_o & -\dfrac{1}{j\omega C_s} \\ 0 & \dfrac{1}{j\omega C_s} + \dfrac{1}{j\omega C_p} \end{vmatrix}}{\begin{vmatrix} kf + \dfrac{1}{j\omega C_s} & -\dfrac{1}{j\omega C_s} \\ -\dfrac{1}{j\omega C_s} & \dfrac{1}{j\omega C_s} + \dfrac{1}{j\omega C_p} \end{vmatrix}}$$

$$= \frac{F_o\left(\dfrac{1}{j\omega C_s} + \dfrac{1}{j\omega C_p}\right)}{\dfrac{k_f}{j\omega C_s} + \dfrac{k_f}{j\omega C_p} + \left(\dfrac{1}{j\omega C_s}\right)^2 + \dfrac{1}{(j\omega)^2 C_p C_s} - \left(\dfrac{1}{j\omega C_s}\right)^2}$$

Canceling the two terms, noting that $j^2 = -1$, and multiplying through by $(j\omega)^2$:

$$v_2(\omega) = \frac{j\omega F_o\left(\frac{1}{C_s} + \frac{1}{C_p}\right)}{\frac{1}{C_pC_s} + j\omega\left(\frac{k_f}{C_s} + \frac{k_f}{C_p}\right)} = \frac{j\omega F_o(C_s + C_p)}{1 + j\omega(k_fC_s + k_fC_p)}$$

Step 5. Now to find the force at the output, multiply $v_2(\omega)$ by the impedance of the parallel elastic element, C_P.

$$F(\omega) = v_2(\omega)\left(\frac{1}{j\omega C_p}\right) = \frac{j\omega F_o(C_s + C_p)}{1 + j\omega(k_fC_s + k_fC_p)}\left(\frac{1}{j\omega C_p}\right) = \frac{F_o((C_s + C_p)/C_p)}{1 + j\omega(k_fC_s + k_fC_p)}$$

In the next chapter we find that this equation, if viewed as a transfer function

$$\frac{F(\omega)}{v_2(\omega)}$$

has the same general properties as a low-pass filter. While real muscle can be fairly well described by the elements in Figure 10.23, the equation does not take into account the component nonlinearities. Nonetheless, this linear analysis provides a starting point for more complicated models and analyses.

The final mechanical example solves a two-equation system that includes a mass. This somewhat involved example also demonstrates how to convert various measurement units to the cgs system.

EXAMPLE 10.12

Find the velocity of the mass in the system shown in Figure 10.25. Assume that $F_S(t) = 0.001 \cos(20t)$ newtons and that the following parameter assignments apply:

$m = 1.0$ oz; $k_{f1} = 200$ dyne-sec/cm
$k_{e1} = 6000$ dyne/sec; $k_{e2} = 0.05$ lbs/in

Solution: We first need to convert F_S, m, and k_{e2} to cgs units. Converting F_S is relatively straightforward since it is in MKS metric units. Using the conversion factors in Appendix D: 1 newton $= 10^5$ dynes. Hence $F_S = 100 \sin(20t)$ dynes. To convert m from English units (oz) to cgs

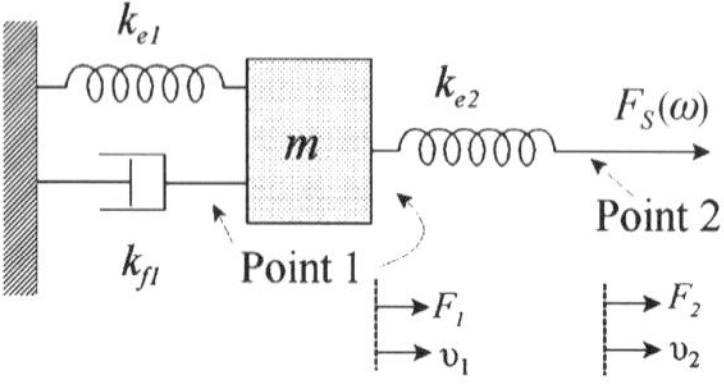

FIGURE 10.25 The mechanical system analyzed in Example 10.12. The system includes two independent force and velocity points as in the last example, so two equations need to be solved simultaneously.

metric units, use the conversion factors in Appendix D in conjunction with dimensional analysis:

$$m = 1.0\ \text{oz}\frac{1\ \text{lb}}{16\ \text{oz}}\frac{1\ \text{kg}}{2.2046\ \text{lb}}\frac{1000\ \text{gm}}{1\ \text{kg}} = 28.4\ \text{gm}$$

In the English system the lb is actually a measure of mass, but is often used as a measure of force, as is the case here. The assumption is that a lb weight is equal to the force produced by a lb accelerated by gravity (i.e., $F = mg$). So it is not surprising that the cgs equivalent of 1 lb is 4103.109 gm,[1] a measure of mass; however, the cgs equivalent of a 1-lb weight is 4.448 N, a measure of force. The same usage is sometimes found in the cgs system where grams are given to represent a weight as well as a mass. In cgs, the gravitational constant, g, is equal to 980.6610 cm/sec^2, so a *gm weight* is 980.6610 dynes. (Similarly a kg weight is 9.807 newtons.) Applying the conversion factors from lbs weight to dynes and inches to cm:

$$k_{e2} = 0.05\frac{\text{lb}}{\text{in}}\frac{4.448\ \text{N}}{\text{lb wt}}\frac{10^5\ \text{dynes}}{1\ N}\frac{1\text{in}}{2.54\ \text{cm}} = 8756\ \text{dynes/cm}$$

This is a lot of work to go through before we even get to step 1, but real-world problems often involve messy situations like inconsistent units.

Step 1. Following the conversion to cgs units, the determination of the phasor impedances is straightforward. With $\omega = 20$ rad/sec:

$$Z_{f1} = k_{f1} = 200\ \text{dynes/cm}; \qquad Z_m = j\omega m = j20(28.4) = 568\ \text{dynes/cm}$$

$$Z_{e1} = \frac{k_{e1}}{j\omega} = \frac{6000}{j20} = -j300\ \text{dynes/cm}; \quad Z_{e2} = \frac{k_{e2}}{j\omega} = \frac{8756}{j20} = -j438\ \text{dynes/cm}$$

Step 2. The force and velocity directions have already been assigned with F_1 and F_2 as positive to the right. After the first two steps, the phasor representation of the system is as shown in Figure 10.26.

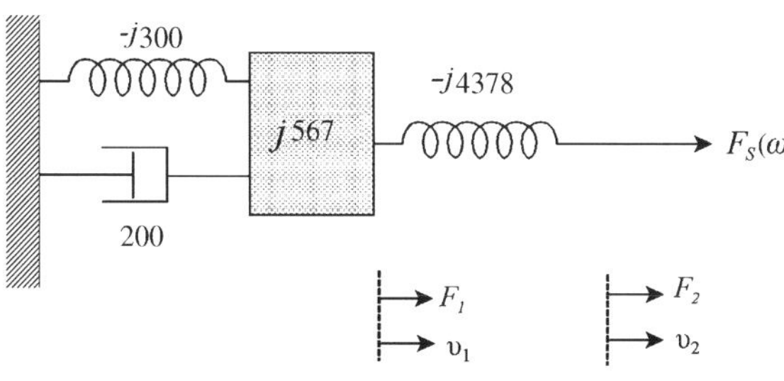

FIGURE 10.26 The mechanical system of Figure 10.25 after the elements have been given consistent cgs units and transformed into phasor notation.

[1]To make matters even more confusing there are two units of mass in the English system termed pounds and abbreviated lbs. The most commonly used pound is termed the commercial or "avoirdupois pound," while a less commonly used measure is the "troy" or "apothecary pound." To convert: 1 troy lb = 0.822857 avoirdupois lb. In this text only avoirdupois pounds are used, but conversions for both can be found in Appendix D.

In writing the equations about the two points, we must take into account the fact that the spring on the right side of the mass has nonzero velocities on both sides. Therefore, the net velocity across these two elements is $v_2 - v_1$. Thus, the force across the right-hand spring is

$$\frac{k_{e2}}{j\omega}(v_2 - v_1)$$

With this in mind, the equation around the input point becomes:

$$F_S - \frac{k_{e2}}{j\omega}(v_2 - v_1) = F_S - (-j438)(v_2 - v_1) = 0$$

With respect to the point between the mass and right spring, the force exerted by that spring has the same magnitude, but is opposite in direction (i.e., positive with respect to F_1). Thus the equation for the force around that point is:

$$\frac{k_{e2}}{j\omega}(v_2 - v_1) - \left(j\omega m - k_f - \frac{k_{e1}}{j\omega}\right)v_1 = 0$$
$$-j438(v_2 - v_1) - j568v_1 + 200v_1 - (-j300)v_1 = 0$$

Rearranging the two equations as coefficients of v_1 and v_2:

$$\begin{aligned} 0 &= (200 + j568 - j300 + j438)v_1(\omega) && -(-j438)v_2(\omega) \\ 100 &= \qquad -(-j438)v_1(\omega) && -j438v_2(\omega) \end{aligned}$$

and in matrix notation:

$$\begin{vmatrix} 0 \\ 100 \end{vmatrix} = \begin{vmatrix} 200 - j170 & j438 \\ j438 & -j438 \end{vmatrix}\begin{vmatrix} v_1(\omega) \\ v_2(\omega) \end{vmatrix}$$

Solving for v_2, the velocity, this time manually:

$$v_2(\omega) = \frac{\begin{vmatrix} 200 - j170 & 0 \\ j438 & 100 \end{vmatrix}}{\begin{vmatrix} 200 - j170 & j438 \\ j438 & -j438 \end{vmatrix}} = \frac{20000 - j17000}{-j85600 - 74460 + 191844}$$

$$v_1(\omega) = \frac{20000 - j17000}{117384 - j87600} = \frac{26249\angle -40}{146470\angle -36.7} = 0.18\angle\ -3.3\ \text{cm/sec}$$

Of course, this solution can be obtained easily using MATLAB.

10.4 RESONANCE

Resonance is a frequency-dependent behavior characterized by a sharp increase (or decrease) in some system variables over a limited range of frequencies. It can occur in almost any feedback system including mechanical and electrical systems as well as

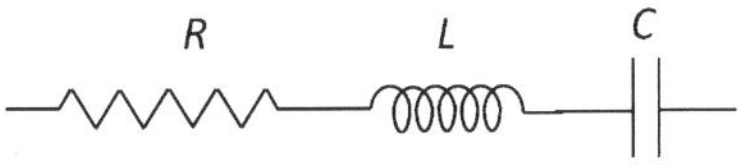

R L C

FIGURE 10.27 A series RLC circuit used to illustrate resonance.

chemical and molecular systems. Often it is beneficial and exploited as in: proton resonance used in magnetic resonance imaging (MRI), optical resonance used in spectroscopy to identify molecular systems, or electrical resonance used to isolate specific frequencies or generate sinusoidal signals. However, it can also be undesirable, particularly in mechanical systems. For example, shock absorbers are friction elements placed on cars to increase damping and reduce the resonant properties of automotive suspension systems. Resonance is discussed in terms of mechanical and electrical systems since these are the systems we have been studying, but the basic concepts are applicable to other systems.

10.4.1 Resonant Frequency

In electrical and mechanical systems resonance occurs when the impedance of an inertial-type element (mass or inductor) equals and cancels the impedance of a capacitive-type element. Consider the impedance of a series RLC circuit, as shown in Figure 10.27.

In the frequency (phasor) domain, the impedance of this series combination is:

$$Z(\omega) = R + j\omega L + \frac{1}{j\omega C} = R + j\left(\omega L - \frac{1}{\omega C}\right)\Omega \tag{10.19}$$

At some value of ω, the capacitor's impedance will be equal to the inductor's impedance and the two impedances will cancel. This will leave only the resistor to contribute to the total impedance. To determine the frequency at which this cancellation takes place, simply set the impedances equal and solve for frequency:

$$\begin{aligned} &\omega_o L = \frac{1}{\omega_o C}; \quad \omega_o L(\omega_o C) = 1; \quad \omega_o^2 = \frac{1}{LC} \\ &\omega_o = \frac{1}{\sqrt{LC}} \text{ rad/sec} \end{aligned} \tag{10.20}$$

where ω_o is the *resonant frequency*. Note that this is the same equation as for the undamped natural frequency, ω_n, in a second-order representation of the RLC circuit (see Equation 10.7). If we plot the magnitude of the impedance in Equation 10.20, we get a curve that reached a minimum value of R at $\omega = \omega_o$ with the curve increasing on either side (Figure 10.28). The sharpness of the curve relates to the bandwidth of the resonant system, as discussed in the next system.

10.4.2 Resonant Bandwidth, Q

When a system approaches the resonant frequency, the system variables (voltage-current or force-velocity) will increase (or decrease) to a maximum (or minimum). The

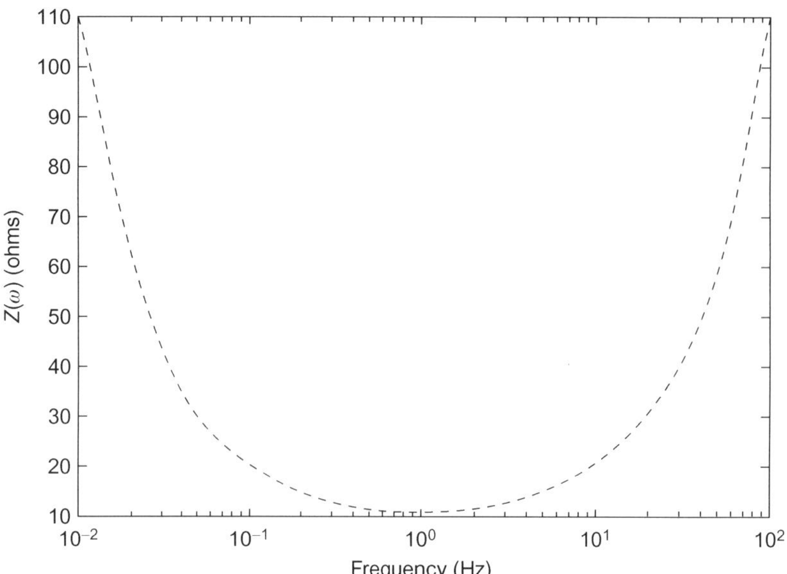

FIGURE 10.28 The impedance of series RLC network as a function of frequency. $R = 10\ \Omega$, $L = 1$ h, and $C = 1$ f. The impedance reaches a minimum of $10\ \Omega$, the impedance of the resistor, at $\omega = 1/\sqrt{LC}$ rad/sec.

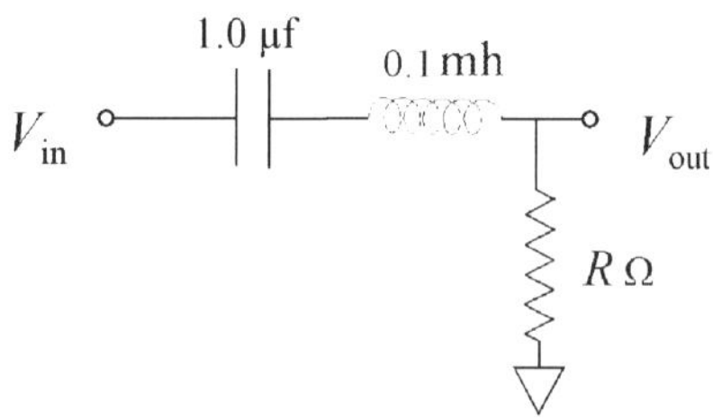

FIGURE 10.29 Input-output system consisting of an RLC network with realistic values for L and C. The transfer function and Bode plot of this system is determined in Example 10.13 for four different values of R.

sharpness of that curve depends on the energy dissipation element (resistance or friction). Figure 10.29 shows an RLC circuit configured as an input-output system. In the next example we find and plot the Bode plot of the transfer function of this system for different values of R and show how the sharpness of the resonant peak depends on R: as R increases the sharpness decreases.

EXAMPLE 10.13

Plot the Bode plot of the transfer function of the network shown in Figure 10.29 for 4 values of resistance: 0.1, 1.0, 10, and 100 Ω. Plot the Bode plot in radians per sec (for variety).

Combining steps 1 (conversion), 2 (current assignment), and 3 (KVL), the equation for the circuit current becomes:

$$V_{\text{in}}(\omega) = I(\omega)\left(R + j\omega L + \frac{1}{j\omega C}\right)$$

Steps 4 and 5. Since we will be using MATLAB to plot the Bode plot, we can leave the L and C as variables. Solving for $I(\omega)$, then finding $V_{out}(\omega)$:

$$I(\omega) = \frac{V_{in}(\omega)}{R + j\omega L + (1/j\omega C)}$$

$$V_{out}(\omega) = I(\omega)R = \frac{RV_{in}(\omega)}{R + j\omega L + (1/j\omega C)}$$

The transfer function becomes:

$$TF(\omega) = \frac{V_{out}(\omega)}{V_{in}(\omega)} = \frac{R}{R + j\omega L + (1/j\omega C)} \tag{10.21}$$

Rearranging into the standard Bode plot format with the lowest coefficient of ω equal to 1 (in this case the constant):

$$TF(\omega) = \frac{V_{out}(\omega)}{V_{in}(\omega)} = \frac{j\omega RC}{j\omega RC + j^2\omega^2 LC + 1} = \frac{j\omega RC}{j\omega RC - \omega^2 LC + 1}$$

$$TF(\omega) = \frac{V_{out}(\omega)}{V_{in}(\omega)} = \frac{j\omega RC}{1 - \omega^2 LC + j\omega RC} \tag{10.22}$$

Note that since we will be solving the transfer function equation on the computer it would have been just as easy to use Equation 10.21 directly; the extra work of putting it into standard Bode plot form was just an exercise.

The resonant frequency of the transfer function is

$$\omega_o = \frac{1}{\sqrt{LC}} = 1 \times 10^5 \text{ rad/sec}$$

and the Bode plot should include a couple of orders of frequency above and below this frequency, say, from 10^3 to 10^7 rad per sec. Solving Equation 10.22 over this frequency range leads to the MATLAB program shown below.

```
% Example 10.13 Bode plot of the TF of an RLC circuit with 4 different values of R.
%
R=[.1 1 10 100 100];                          % Resistance values
L=10^-4;                                      % Inductance value
C= 10^-6;                                     % Capacitance value
w=(1000:1000:10000000);                       % Frequency range
for k=1:length(R)
TF=(j*w*R(k)*C)./(1-w.^2* L*C+j*w*R(k)*C);    % TF equation
TF = 20*log10(abs(TF));                       % Convert to dB
semilogx(w,TF,'k');                           % Plot as Bode plot
  ....... text labels......
end
  ....... labels and axis.......
```

This program produces the graph shown in Figure 10.30.

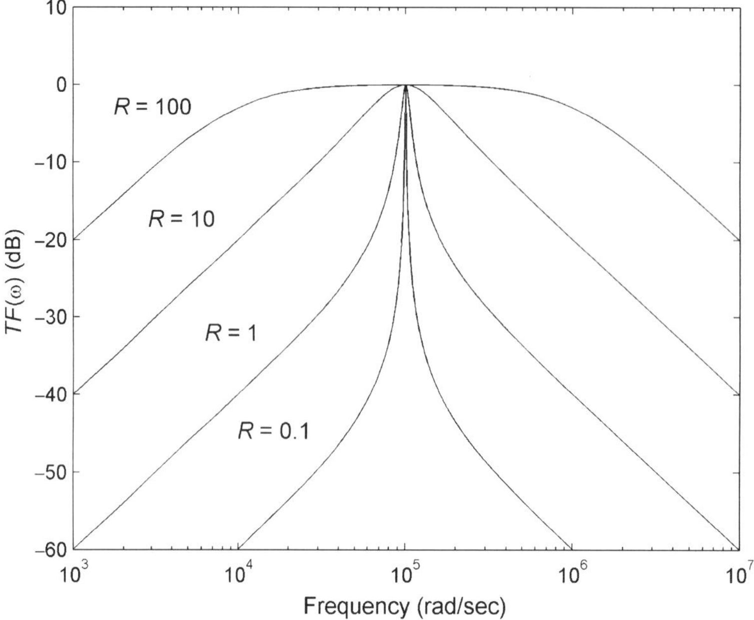

FIGURE 10.30 Magnitude frequency characteristics of the transfer function of the RLC circuit shown in Figure 10.29 with four different resistor values. The resonant frequency, 100,000 rad/sec, depends only on L and C, but sharpness of the curve depends strongly on R.

Figure 10.29 shows that the transfer function peak occurs at the same frequency for all values of R. This is expected since the resonant frequency is a function of only L and C, Equation 10.20. Specifically, the peak occurs at:

$$\omega_o = \frac{1}{\sqrt{LC}} = \frac{1}{\sqrt{10^{-4}10^{-6}}} = 100{,}000 \text{ rad/sec}$$

Peak sharpness increases as R decreases. In the configuration shown in Figure 10.28, the circuit is a bandpass filter; however, if the resistor and capacitor are interchanged, the circuit becomes a low-pass filter as used in previous examples. Nonetheless, it still has a resonant peak at the same frequency and the sharpness of the peak varies in the same manner. This is demonstrated in Problem 23.

The sharpness of the frequency curve around the resonant frequency is a very important property of a resonant system. This characteristic is often described by a number known as Q which is defined as the resonant frequency divided by the bandwidth:

$$Q = \frac{\omega_o}{BW} \tag{10.23}$$

where ω_o is the resonant frequency and BW is the bandwidth, the difference between the high-frequency cutoff and the low-frequency cutoff (−3 dB points). This definition is also referred to as *selectivity* as Q has an alternate, classical definition described below.

Classically, Q is defined in terms of energy storage and energy loss to describe how close energy storage elements such as inductors and capacitors approach the ideal. Ideally such elements should only store energy, but real elements also dissipate energy due to parasitic resistance. In this context, Q is defined as the energy stored over the energy lost in one cycle:

$$Q = 2\pi \frac{\text{Energy stored at resonance}}{\text{Energy dissipated at resonance}} \tag{10.24}$$

To calculate Q for an inductor, assume an inductor having inductance L h also has a parasitic series resistance of $R\ \Omega$. Since inductors are simply coils of wire, often considerable lengths of wire, inductors inevitably have some resistance due to the wire. (Inductors are the least ideal of the passive electrical elements.) The energy lost in this resistance over one sinusoidal cycle is equal to the power integrated over the cycle, and power is just vi, or for a resistor, Ri^2. Assuming the current through the resistor is $i_R(t) = I\sin(\omega_o t)$, the energy lost over one cycle becomes:

$$E_{lost} = \int_{Cyc} vi dt = \int_0^{2\pi} R(I\sin\omega_o t)^2 dt = \frac{2\pi RI^2}{2\omega_o}\text{joules}$$

The energy stored in an inductor is also the integral of vi over one cycle. The current through the inductor is the same as through the resistor, $i_L(t) = I\sin(\omega_o t)$, and from the definition of an inductor (Equation 9.15), the voltage is L times the derivative of the current:

$$v_L(t) = L\frac{di}{dt} = \omega_o LI\cos(\omega_o t)$$

$$E_{stored} = \int_{Cyc} vi dt = \int_0^{2\pi} \omega_o LI^2 \cos(\omega_o t)\sin(\omega_o t) dt = \frac{\omega_o LI^2}{2\omega_o}$$

Plugging the two energies into Equation 10.24, the Q of an inductor becomes:

$$Q_L = \frac{2\pi\left(\frac{\omega_o LI^2}{2\omega_o}\right)}{\left(\frac{2\pi RI^2}{2\omega_o}\right)} = \frac{\omega_o L}{R} \tag{10.25}$$

Similarly, it is possible to derive the Q of a capacitor having a capacitance value C and a parasitic resistance R as:

$$Q_C = \frac{1}{\omega_o RC} \tag{10.26}$$

However, in a circuit that includes both an inductor and a capacitor, most of the parasitic resistance is from the inductor, and any contribution from the capacitor is usually ignored so Equation 10.25 can be used to find the Q.

Based on these definitions, it is possible to derive Equation 10.23 from the definition of bandwidth. Returning to the RLC circuit in Figure 10.28, the transfer function of this circuit as given in Equation 10.21 is:

$$TF(\omega) = \frac{R}{R + j\omega L + (1/j\omega C)} = \frac{R}{Z(\omega)} = \frac{1}{(Z(\omega)/R)}$$

where $Z(\omega)$ is just the series R, L, C impedance.

$Z(\omega)/R$ can also be written as:

$$\frac{Z}{R} = \frac{R + j\omega L - \frac{j}{\omega C}}{R} = 1 + j\left(\frac{\omega L}{R} - \frac{1}{\omega C}\right)$$

multiplying both imaginary terms by $\frac{\omega_o}{\omega_o}$ and rearranging:

$$\frac{Z}{R} = 1 + j\left(\frac{\omega}{\omega_o}\left(\frac{\omega_o L}{R}\right) - \frac{\omega_o}{\omega}\left(\frac{1}{\omega_o C}\right)\right)$$

This allows us to substitute in the definition of Q into the equation for Z/R:

$$\frac{Z}{R} = 1 + j\left(\frac{\omega Q}{\omega_o} - \frac{\omega_o Q}{\omega}\right) = 1 + jQ\left(\frac{\omega}{\omega_o} - \frac{\omega_o}{\omega}\right) \tag{10.27}$$

At the cutoff frequencies, $|TF(\omega_{H,L})| = 0.707\ |TF(\omega = \omega_o)|$. However, at the resonant frequency, $Z(\omega = \omega_o) = R$ and $TF(\omega = \omega_o)$. Hence, at the cutoff frequencies $|TF(\omega_{H,L})| = 0.707$. So to find the bandwidth (which is just the difference between the cutoff frequencies), set

$$\left|\frac{Z}{R}\right| = \frac{1}{.707} = 1.414$$

and solve for ω_H and ω_L. Setting the magnitude of Equation 10.27 to 1.414, we get:

$$\left|1 + jQ\left(\frac{\omega}{\omega_o} - \frac{\omega_o}{\omega}\right)\right| = \sqrt{2}; \qquad (\text{For } |1 + jB| = \sqrt{2}, \quad B = \pm 1)$$

$$Q\left(\frac{\omega}{\omega_o} - \frac{\omega_o}{\omega}\right) = \pm 1$$

There are two solutions to the equation for $+1$ and -1:

$$\omega_L = \omega_0\left(1 - \frac{1}{2Q}\right); \qquad \omega_H = \omega_0\left(1 + \frac{1}{2Q}\right)$$

$$BW = \omega_H - \omega_L = \frac{\omega_0}{Q}; \qquad Q = \frac{\omega_0}{BW}$$

We can also relate Q to the standard coefficients of a second-order underdamped equation. Again referring to the RLC circuit in Figure 10.29, the transfer function can be easily determined in terms of the Laplace variable s. To make the solution more general, we continue using variables L and C to represent the inductance and capacitance:

Steps 1, 2, and 3:

$$V_{\text{in}}(s) = I(s)\left(R + sL + \frac{1}{sC}\right)$$

Steps 4 and 5:

$$I(s) = \frac{V_{\text{in}}(s)}{R + sL + (1/sc)}; \qquad V_{\text{out}}(s) = I(s)R = \frac{RV_{\text{in}}(s)}{R + sL + (1/sC)}$$

Solving for the transfer function, $V_{\text{out}}(s)/V_{in}(s)$:

$$TF(s) = \frac{V_{\text{out}}(s)}{V_{\text{in}}(s)} = \frac{R}{R + sL + (1/sC)}$$

Putting the transfer function equation into standard Laplace domain format where the highest power of s has a coefficient of 1:

$$TF(s) = \frac{V_{\text{out}}(s)}{V_{\text{in}}(s)} = \frac{sR}{sR + s^2L + (1/C)} = \frac{s(R/L)}{s(R/L) + s^2 + (1/LC)}$$

Rearranging and comparing with the standard form of a second-order equation analyzed in Chapter 6 (Equation 6.31):

$$TF(s) = \frac{(R/L)s}{s^2 + (R/L)s + (1/LC)} = (RCs)\frac{(1/LC)}{s^2 + (R/L)s + (1/LC)} = (RCs)\frac{\omega_n^2}{s^2 + 2\delta\omega_n s + \omega_n^2}$$

where $\omega_n \equiv \omega_o$. Equating coefficients:

$$2\delta\omega_o = \frac{R}{L}; \qquad \delta = \frac{R}{2\omega_o L}; \qquad \text{but } Q = \frac{\omega_o L}{R} \qquad \text{so}$$
$$\delta = \frac{1}{2Q}; \qquad Q = \frac{1}{2\delta} \tag{10.28}$$

The relationship between Q and δ is amazingly simple.

The characteristics of resonance and the various definitions and relationships described apply equally to mechanical systems. In fact, most mechanical systems exhibit some resonant behavior and often that behavior is detrimental to the system's performance (consider a car with bad shock absorbers). An investigation of the resonant properties of a mechanical system is given in the next example.

EXAMPLE 10.14

Find the Q of the mechanical system shown in Figure 10.31. The system coefficients are: $k_f = 6$ dyne sec/cm; m = 8 gms; $k_e = 10$ dynes/cm.

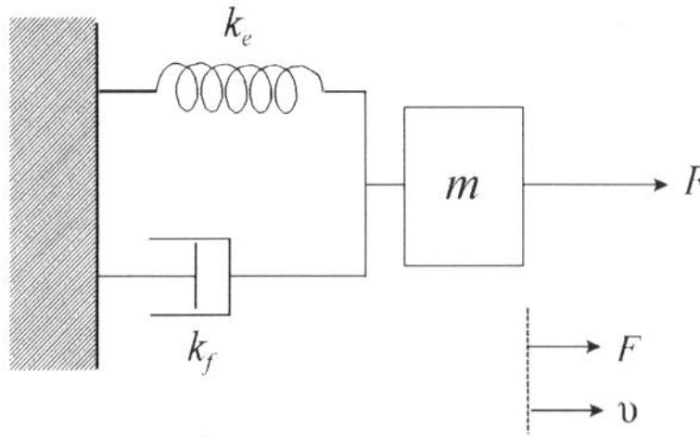

FIGURE 10.31 Mechanical system containing friction, mass, and a spring used in Example 10.13. This system is analogous to an RLC circuit.

Solution: Q can be determined directly from δ in the transfer function using Equation 10.28. So we need to find the transfer function $\upsilon(s)/F_s(s)$. Applying the standard analysis based on Equation 10.18:

$$F_S(s) - \upsilon(s)\left(k_f + ms + \frac{k_e}{s}\right) = 0; \qquad \upsilon(s) = \frac{F_S(s)}{k_f + ms + (k_e/s)}$$

$$\frac{\upsilon(s)}{F_S(s)} = \frac{(s/m)}{s^2 + (k_f/m)s + (k_e/m)} = \frac{(s/m)}{s^2 + 2\delta 2\omega_n s + \omega_n^2}$$

Equating coefficients:

$$\omega_n = \sqrt{\frac{k_e}{m}}; \qquad 2\delta\omega_n = \frac{k_f}{m}; \qquad \delta = \frac{k_f}{2m\omega_n} = \frac{k_f}{2\sqrt{k_e m}} = \frac{6}{2\sqrt{10(8)}} = 0.335$$

$$Q = \frac{1}{2\delta} = 1.49$$

The next example uses MATLAB to explore the ways in which the Bode plot and impulse response of a generic second-order system vary for different values of Q.

EXAMPLE 10.15

Plot the frequency characteristics and impulse response of a second-order system in which $Q = 1$, 10, and 100. Use MATLAB and assume a resonant frequency, ω_n, of 1000 rad/sec.

Solution: Begin with the standard second-order Laplace transfer function equation, but substitute Q for δ. Convert this equation to the time domain by taking the inverse Laplace transform to get the impulse response. (Recall, the inverse Laplace transform of the transfer function is the impulse response.) Then convert the transfer function equation to the phasor domain to get the Bode plot equation. Use MATLAB to plot both the Bode plot and the time response.

The standard Laplace second-order equation is:

$$TF(s) = \frac{\omega_n^2}{s^2 + 2\delta\omega_n s + \omega_n^2}$$

Substituting in $\delta = \frac{1}{2Q}$ and $\omega_n = 1000$

$$TF(s) = \frac{\omega_n^2}{s^2 + \frac{\omega_n}{Q}s + \omega_n^2} = \frac{10^6}{s^2 + \frac{1000}{Q}s + 10^6}$$

To find the impulse response, take the inverse Laplace transform of the transfer function. Since $Q \geq 1$, δ will be ≤ 0.5, so the system is underdamped for all values of Q and entry 15 in the Laplace transform table can be used:

$$x(t) = \frac{\omega_n}{\sqrt{1-\delta^2}}\left[e^{-\delta\omega_n t}\sin\left(\omega_n\sqrt{1-\delta^2}t\right)\right] = \frac{10^3}{\sqrt{1-\frac{1}{4Q^2}}}\left[e^{-1000t/2Q}\sin\left(1000\sqrt{1-\frac{1}{4Q^2}}t\right)\right]$$

To find the frequency response, convert to phasor and plot for the requested values of Q:

$$TF(\omega) = \frac{10^6}{(j\omega)^2 + j\omega\frac{1000}{Q} + 10^6} = \frac{1}{1-\left(\frac{\omega}{10^3}\right)^2 + \frac{j\omega}{Q10^3}}$$

Plotting of the frequency and time domain equations is done in the program below.

```
% Example 10.15
%
wn = 1000;                                      % Define resonant frequency
w = (100:10:10000);                             % Define a frequency vector for
                                                Bode plot
t = (10^-5:10^-5:.2);                           % Define a time vector
Q = [1 10 100];                                 % Define Q's
for k = 1:length(Q)                             % Calculate and plot the freq. plots
  TF = 1./(1 - (w/1000).^2 + j*w/(Q(k)*1000));  % Frequency equation
  TF = 20*log10(abs(TF));                       % Convert to dB
  semilogx(w,TF,'k');                           % Bode plot
end
  .......labels and axis.......
%
% Now construct the impulse responses
figure;
for k = 1:3                                     % Calculate and plot the time
                                                % response
   d = sqrt(1 - 1/(4*Q(k)^2));                  % Define square root of 1 - δ²
   x = (wn/d)*(exp(-wn*t/(2*Q(k)))).*sin (wn*d*t);
   subplot(3,1,k);                              % Plot separately for clarity
   plot(t,x,'k');
   ylabel('x(t)');
end
xlabel('Time (sec)');
```

This program generates the plots shown in Figure 10.32 and 10.33. Figure 10.32 reiterates the message in Figure 10.30, that high-Q systems can have very sharp resonance peaks.

In the time domain, sharp resonance peaks correspond to sustained oscillations, a sinusoid at the resonance frequency that diminishes very slowly (Figure 10.33).

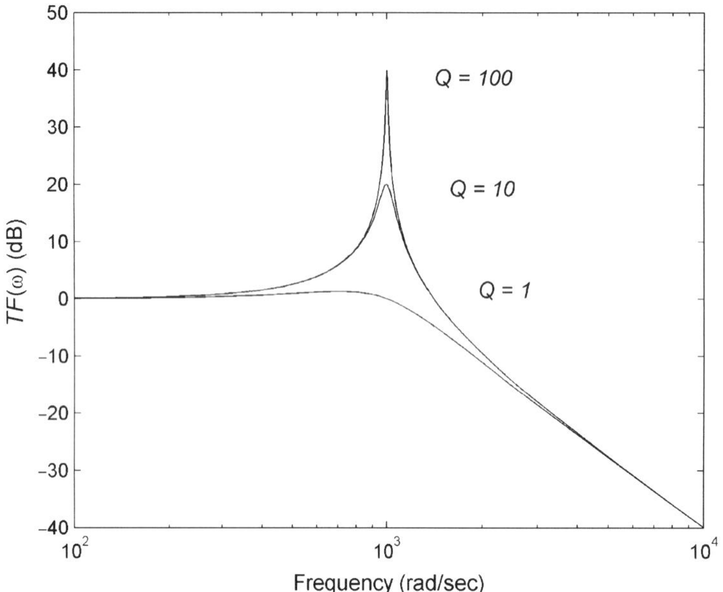

FIGURE 10.32 Bode plot of a second-order system with a Q of 1, 10, and 100.

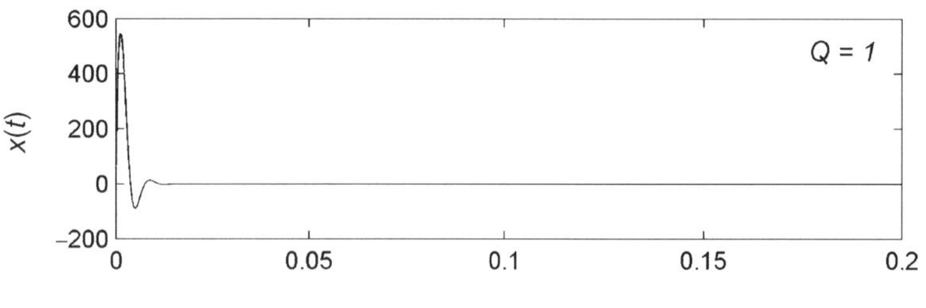

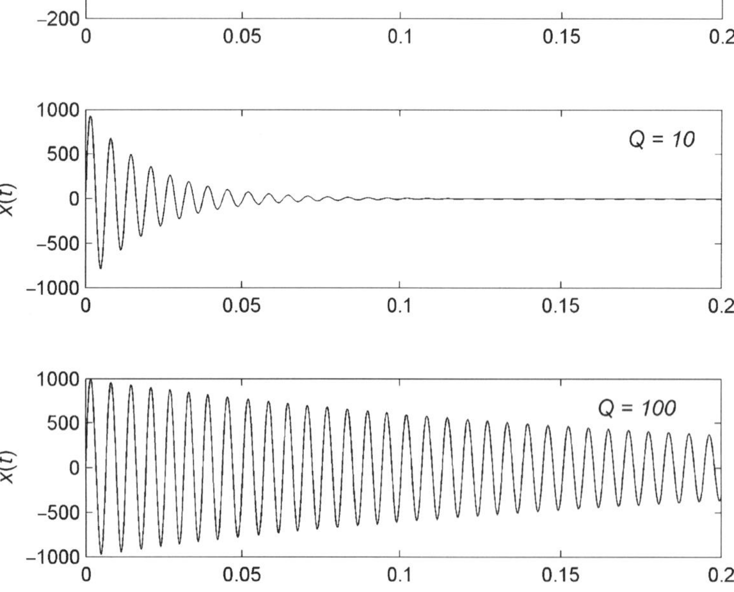

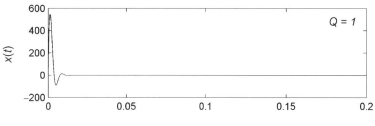

FIGURE 10.33 Time plot of the impulse response of a second-order system with a Q of 1, 10, and 100.

A common example of a high-Q mechanical system is a large bell. Tone continues to sound long after it is struck. Striking a bell is like applying an impulse. In fact, sustained oscillation is a characteristic of high-Q systems and is sometimes referred to as "ringing" even in electrical systems.

If a high-Q circuit is used as a filter, it can be quite selective with regard to the frequencies it allows to pass though. This is demonstrated in the final example.

EXAMPLE 10.16

Generate a 1.0-sec chirp waveform that continually increases in frequency from 1 Hz to 1 kHz. (Recall, a chirp waveform increases linearly in frequency with time.) Assume a sampling frequency of 5 kHz.

Modify the RLC circuit of Figure 10.29 so that the resonant frequency, f_o, is 100 Hz and the circuit has a Q of 10. Assume the capacitor value is the same as that shown in Figure 10.29 (i.e., 1 μf), and adjust L to get the desired resonant frequency. Simulate the operation of passing the chirp waveform through the filter and plot the output. Repeat for two other values of resonant frequency: $f_o = 250$ and 500 Hz.

To simulate the operation the circuit performs on the signal, find the circuit's impulse response and convolve that response with the input to get the simulated output.

Solution: First we are requested to design the network to have the desired resonant frequency and Q. The resonant frequency is entirely determined by L and C. Since we already have a value for C, we can determine the value of L from Equation 10.20.

$$\omega_0 = 2\pi f_0 = \frac{1}{\sqrt{LC}}; \quad f = \frac{1}{2\pi\sqrt{LC}}$$

Squaring both sides:

$$f_0^2 = \frac{1}{4\pi 2LC}$$

Solving for L:

$$L = \frac{1}{4\pi^2 f_0^2 C} = \frac{1}{4\pi^2(10^4)(10^{-6})} = 2.53\text{ h} \quad \text{when } f_0 = 100\text{ Hz}$$

For the other two resonant frequencies:

$$L = \frac{1}{4\pi^2 f_0^2 C} = \frac{1}{4\pi^2(250^2)(10^{-6})} = 0.41\text{ h} \qquad f_0 = 250\text{ Hz}$$

$$L = \frac{1}{4\pi^2(500^2)(10^{-6})} = 0.1\text{ h} \qquad f_0 = 500\text{ Hz}$$

To find the resistor values required to give the desired Q, we can use Equation 10.28:

$$Q = \frac{\omega_0 L}{R} = \frac{2\pi f_0 L}{R}; \qquad R = \frac{2\pi f_0 L}{Q}$$

$$R = \frac{2\pi(100)(2.53)}{10} = 158.9\ \Omega \qquad f_0 = 100\text{ Hz}$$

TABLE 10.1 Resonant Frequencies and Respective Component Values of the Circuit Given in Figure 10.29

Resonant Frequency (Hz)	R (Ω)	L (h)	C (μf)
100	158.9	2.53	1.0
250	64.4	0.41	1.0
500	31.4	0.10	1.0

Solving for the other two values of R, the component values for the three resonant frequencies are summarized in Table 10.1.

Now that we have the circuit designed we need to find the impulse response. We can get this from the inverse of the Laplace transfer function. In Example 10.13 we find the transfer function of the network in Figure 10.28. Substituting in s for $j\omega$ in Equation 10.21 we get:

$$TF(s) = \frac{R}{R + Ls + (1/CS)} = \frac{RCs}{RCs + CLs^2 + 1} = \frac{(R/L)s}{s^2 + (R/L)s + (1/CL)} \quad (10.29)$$

To find the inverse Laplace transform, we note that since $Q = 10$ in all cases, $\delta = 1/2Q = .05$, so the system is underdamped and we can use entry 13 of the Laplace transform table (Appendix B) where:

$$b = \frac{R}{L} \qquad \alpha = \frac{R}{2L} \qquad c = 0$$

$$\alpha^2 + \beta^2 = \frac{1}{LC} \quad \beta^2 = \frac{1}{LC} - \alpha^2 = \frac{1}{LC} - \alpha^2 \quad \beta = \sqrt{\frac{1}{LC}} - \alpha$$

Taking the inverse Laplace transform, the impulse response becomes:

$$\delta(t) = e^{-\alpha t}\left[\left(\frac{-(R/L)\alpha}{\beta}\right)\sin \beta t + \frac{R}{L}\cos \beta t\right] \quad (10.30)$$

where α and β are defined above (it is easier to program if we leave these as variables and substitute in their specific values in the MATLAB routine). We could combine this equation into a single sinusoid with phase, but since we are going to program it in MATLAB, further reduction is not necessary.

To construct the chirp signal, we define a time vector, t, that goes from 0 to 1 in steps of T_s where $T_s = 1/f_s = 1/5000$. This time vector is then used to construct a frequency vector of the same length that ranges from 1 to 1000. The chirp signal is then constructed as the sine of the product of f and t (i.e., `sin(pi*f.*t)`) so that the frequency increases linearly as t increases. [*Note*: use π, not 2π to make the final frequency 1000 Hz.]

We then program the impulse response using the equation above and the various values of R, L, and C. Convolving the input signal with the impulse response produces the output signal. The solution of Equation 10.30 is performed three times in a loop for the various values of R, L, and C and plotted.

```
% Example 10.16
%
clear all; close all;
% Sampling and component definitions
fs = 5000;                          % Sample frequency
Ts = 1/fs;                          % Sample interval
C = 0.000001;                       % Capacitor value (fixed)
R = [158.9 64.4 31.4];              % Resistor values
L = [2.53 0.41 0.1];                % Inductor values
f_0 = [100 250 500];                % For plot labels
t = (0:Ts:1);                       % Define time vector (0 - 5 sec; Ts = .0002)
f = t*1000;                         % Frequency goes from 1 to 1000 Hz
xin = in(pi*f.*t);                  % Construct "chirp"
%
for k = 1:3                         % Calculate impulse response
   alpha = R(k)/(2*L(k));           %Define alpha and beta, then the impulse
                                    response
   beta = sqrt(1/(L(k)*C)) - alpha;
   h = exp(-alpha*t).*((-R(k)*alpha)/(L(k)*beta)*...
   sin (beta*t) + (R(k)/L(k))*cos (beta*t));
   xout = conv(h,xin);              % Filter vin
   xout = xout(1:length(t));        % Remove extra points
   subplot(3,1,k);                  % Plot separately
   plot(t,xout,'k');
   ylabel('xout(t)','FontSize',14);
   title(['f_0 = ',num2str(f_0(k))],'FontSize',14);
end
%
figure;                             % Plot spectrum of chirp
XIN = abs(fft(xin));                % FFT
f1 = (1:length(t))*fs/length(t);    % Construct frequency vector for plotting
plot(f1(1:2500),XIN(1:2500),'k');   % Plot only valid points
  .......labels......
```

Analysis: The initial portion of the program defines the sampling parameters, the component values, and the chirp. (If you have a sound system on your computer, try typing `sound(xin)` in MATLAB after executing this code. The resulting chirp sound is both startling and amusing.)

A loop is used to evaluate the impulse response for the three component configurations and find the circuit's output to the chirp input using convolution. This output is then plotted and is shown in Figure 10.34. The RLC circuit functions as a bandpass filter with a center frequency that is equal to the resonant frequency. Since the chirp signal's frequency depends on time, signal will appear in the output only when chirp's frequency corresponds to the circuit's resonant frequency. For example, the chirp signal's frequency approaches 500 Hz at 0.5 sec, so if the resonant frequency of the filter is 500 Hz, only the signal around 0.5 sec passes through the filter to the output (see Figure 10.34, lower curve). This selective ability of a high-Q filter is well demonstrated in this example. As the Q increases, the selectivity also increases. This is demonstrated in Problem 26.

The last section of the program plots the frequency spectrum of the chirp signal using the standard `fft` command. The spectrum is seen in Figure 10.34 to be relatively flat up to 200 Hz as expected.

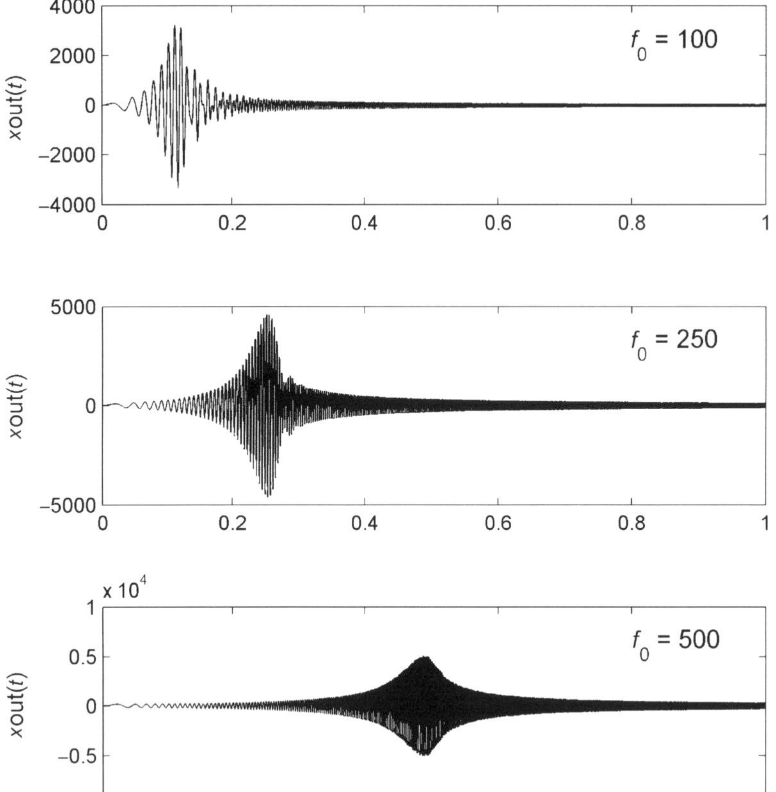

FIGURE 10.34 The output of the RLC filter to the chirp signal. The filter's ability to select a specific frequency range is shown for three different resonant frequencies.

10.5 SUMMARY

Conservation laws are invoked to generate an orderly set of descriptive equations from any collection of mechanical or electrical elements. In electric circuits, the law of conservation of energy leads directly to Kirchhoff's voltage law (KVL) which states that the voltages around the loop must sum to zero. Combining this rule with the phasor representation of network elements leads to an analysis technique known as mesh analysis. In mesh analysis, equations are constructed as an extension of Ohm's law: $v = Zi$. These equations are solved for the mesh currents, i, which then can be used to determine any voltage in the system. The law of conservation of charge leads to Kirchhoff's current law (KCL) which can also be used to find the voltages and currents in any network. Application of KCL leads to an equation of the form $i = v/Z$ that is solved for the node voltages, v. From the node voltages any current in the circuit can be found. This analysis, termed nodal analysis, leads to fewer equations in networks that contain many loops, but only a few nodes.

The conservation law active in mechanical systems is Newton's law, which states that the forces on any element must sum to zero. Again using phasor representation of

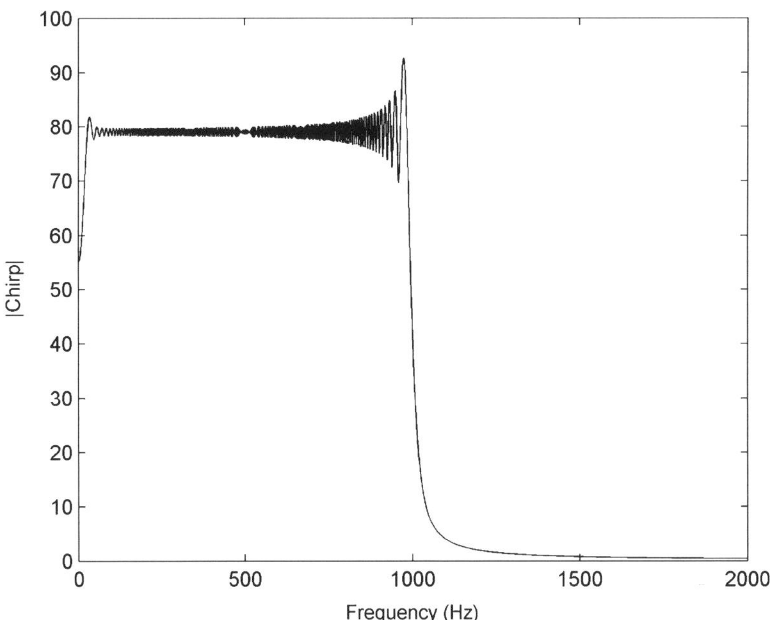

FIGURE 10.35 Frequency spectrum of the chirp signal used in Example 10.15. The spectrum is fairly flat between 0 and 1000 Hz.

mechanical elements, this law can be applied to generate equations of the form $F = Zv$. These equations are solved for velocities, and these velocities can be used to determine all of the forces in the system.

These conservation laws, and the analysis procedures they lead to, allow us to develop equations for even very complex electrical or mechanical systems. Large systems may generate equations that are difficult to solve analytically, but MATLAB makes short work out of even the most challenging equations.

Resonance is a phenomenon commonly found in nature. In electrical and mechanical systems, it occurs when two different energy storage devices have equal (but oppositely signed) impedances. During resonance, energy is passed back and forth between the two energy storage devices. For example, in an oscillating mechanical system, the moving mass stretches the spring transferring the kinetic energy of the mass to potential energy in the spring. Once the spring is fully compressed (or extended), the energy is passed back to the mass as the spring recoils. Without friction this process would continue forever, but friction removes energy from the system so the oscillation gradually decays. Electrical RLC circuits behave in exactly the same way, passing energy between the inductor and capacitor while the resistor removes energy from the system. (Recall both friction and resistance remove energy in the form of heat.)

The quantity Q is a measure of the ratio of energy storage to energy dissipation of a system. The Q of a system is inversely proportional to the damping factor, δ. In the frequency domain, higher Q corresponds to a sharper resonance peak. In the time domain, higher Q systems have longer impulse responses; they continue "ringing" long after the impulse because energy is only slowly removed from the system. Such systems are useful in frequency selection: with their sharp resonant peaks they are able to select out a narrow range of frequencies from a broadband signal. Standard systems analysis techniques such as Bode plots and the transfer function can be used to describe the behaviors of these systems.

PROBLEMS

1. In the circuit below, the voltage across element 1 is 2 cos(2t + 60). What is the voltage, V_2, across element 2?

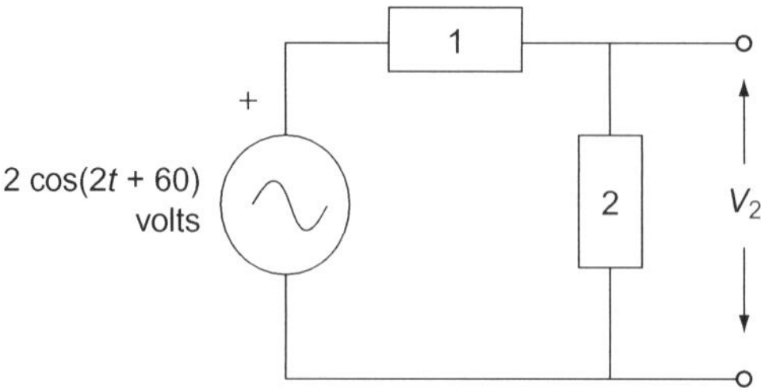

2. Find the voltage across the 50-Ω resistor.

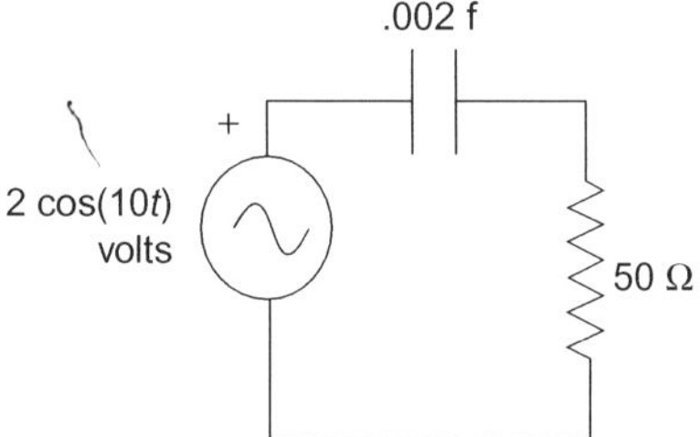

3. The loop current in the circuit below is 0.2 cos(10t − 103). What is element 2 (i.e., is it an R, L, or C), and what is its value?

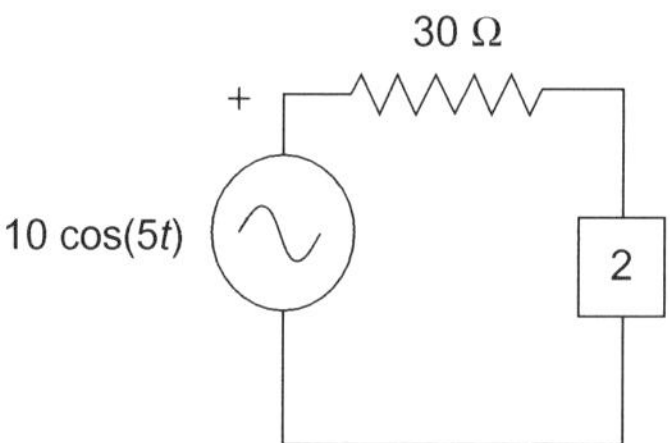

4. Find the voltage across the inductor, v_L, for $\omega = 10$ and 20 rad/sec.

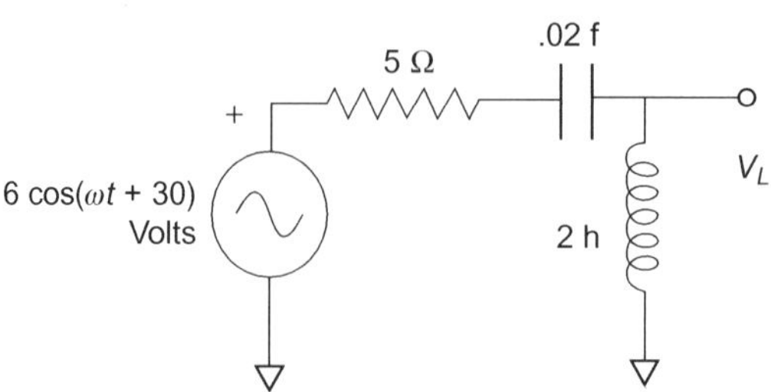

5. Find the voltage across the 5-h inductor in the network below.

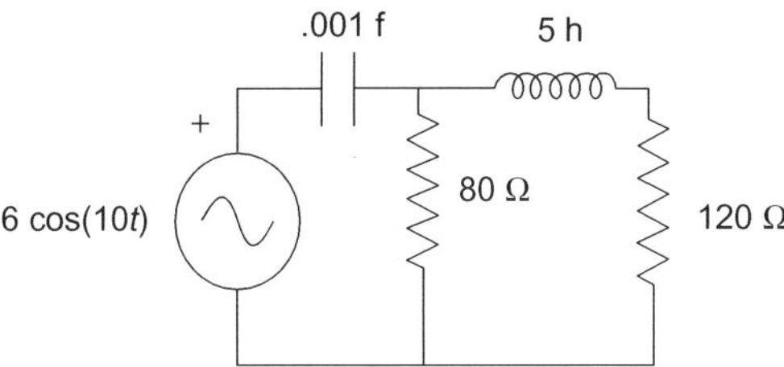

6. What is the voltage in the center 80-Ω resistor in the network given in Problem 5? [*Note*: the total current through the resistor is $i_R = i_1 - i_2$.]
7. Find the current through the 2-h inductor in the circuit below. Is the voltage source, V_2, supplying energy or storing energy?

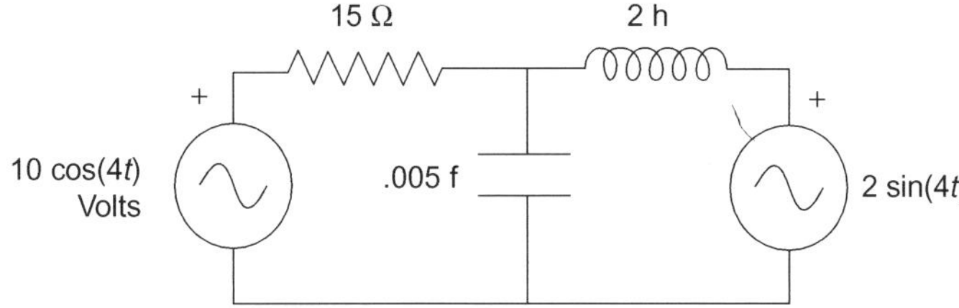

8. Find V_1.

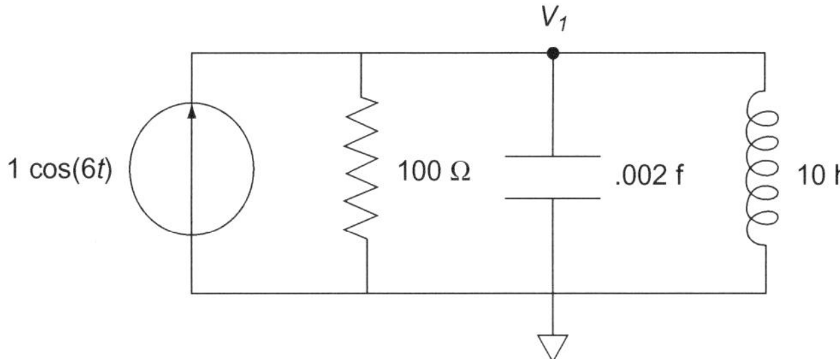

9. Find the voltage across the 10-Ω resistor.

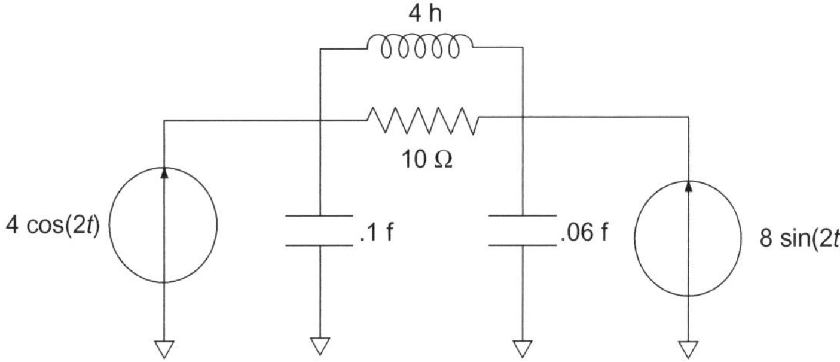

10. The circuit shown below is a four-element windkessel model. The element values in this model of the cardiovascular system are: $R_A = 0.05$ mmHg/ml/sec; $C =$ 1.75 ml/mmHG; $R_p = 0.79$ mmHg/ml/sec; and $L = 0.005$ mmHG/ml. Write the three-mesh equation and use the frequency domain approach of Example 10.5 to find the pressure in the left heart associated with the blood flow given by Equation 10.11.

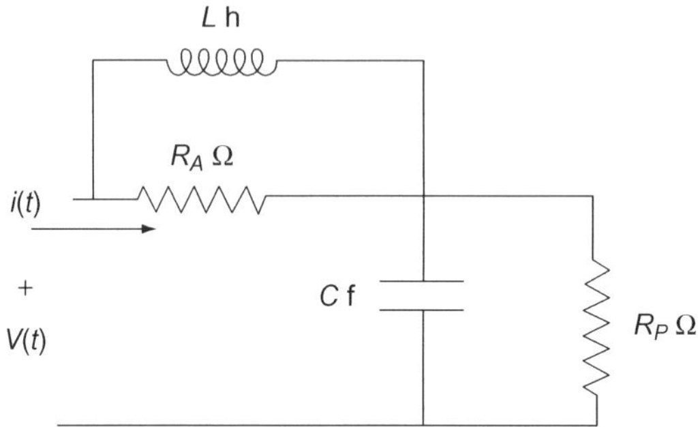

11. Given the mechanical system shown where: $k_f = 8$ dyne sec/cm; $k_e = 12$ dynes/cm; $m = 2$ gms; $F_s(t) = 10\cos(2t)$ dynes, find the length of the spring when $t = 0.10$ sec. [*Hint*: solve for $x(\omega)$ where $x(\omega) = v(\omega/j\omega)$. Then convert to time domain, find $x(t)$, and solve for $t = 0.10$ sec.]

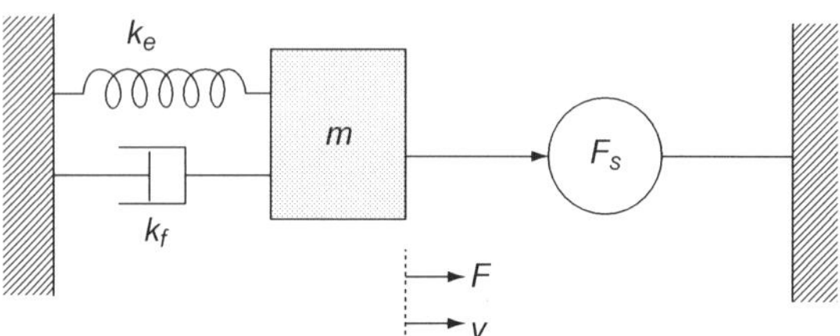

12. In the mechanical system of Problem 11, what must be the value of k_f to limit the maximum velocity to ±1 cm/sec?
13. Find V_{out} using MATLAB. [*Hint*: Example 10.6 provides a framework for the code.]

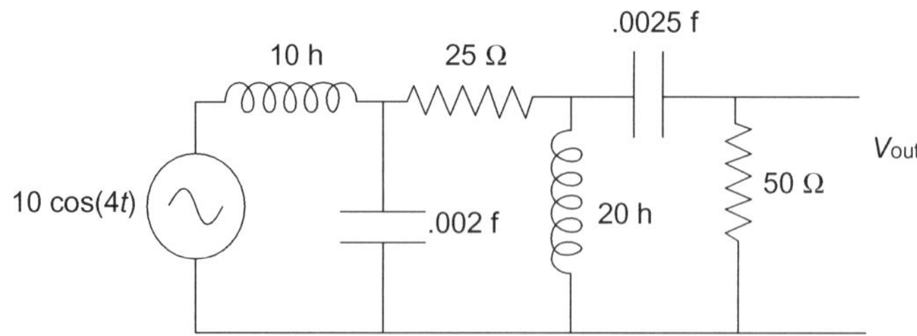

14. Find V_{out} using MATLAB and nodal analysis. Remember: $10\sin(10t) = 10\cos(10t - 90)$.

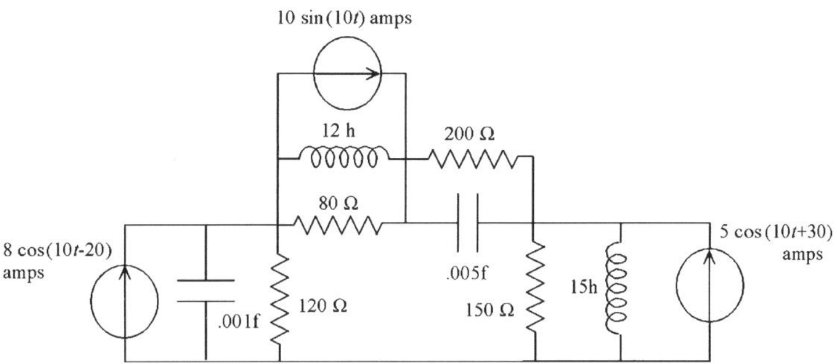

15. Find v_{out} for the circuit below for frequencies ranging between 1 and 1000 rad/sec. Plot the magnitude of v_{out} as a function of frequency. [*Hint*: write the MATLAB impedance matrix in terms of a variable frequency, set up ω as a vector, then solve for v_{out}. The vector ω should be a vector ranging between 1 and 1000 rad/sec in increments of 1 rad/sec. Then plot abs($v_{out}(\omega)$).]

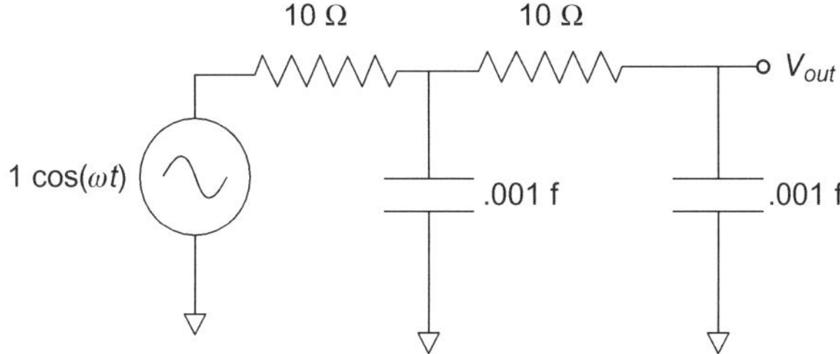

16. Repeat Problem 15 for the mechanical system shown in Problem 11. Use the parameter values: $k_f = 10$ dyne sec/cm; $k_e = 1100$ dyne/cm; $m = 1.10$ gms; [?] $F_S(t) = \cos(\omega t)$.
17. Find and plot $V_{out}(\omega)$ over a range of frequencies from 0.1 to 100 rad/sec.

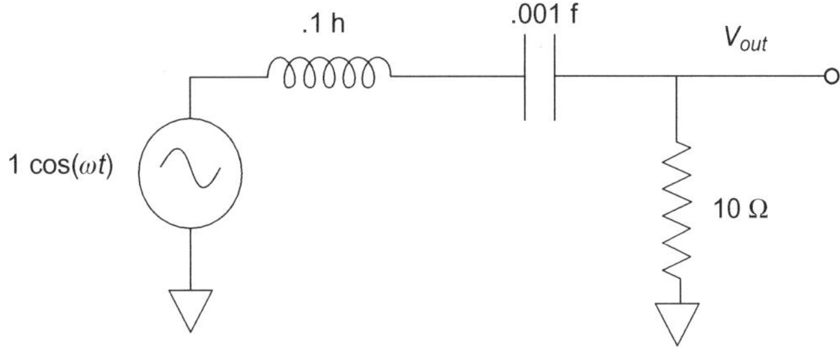

18. Use the graphical (i.e., manual) technique developed in Chapter 5 to plot the magnitude Bode plot for the RLC circuit shown in Problem 4. Assume that $L = 1$ h, $C = 0.0001$ f, and $R = 10\ \Omega$.
19. Use MATLAB to plot the magnitude transfer function of the circuit in Problem 4 for values of $R = 1\ \Omega$, $10\ \Omega$, $70\ \Omega$, and $500\ \Omega$. Assume that $L = 1$ h and $C = .0001$ f. How does the resistance affect the shape of the frequency plot?
20. Use MATLAB to plot the Bode plot (magnitude and phase) for the transfer function for the mechanical system in Problem 11 for values of $k_e = 2$ dynes/cm, m = 3 gm, and

$k_f = 1, 10, 70$, and 500 dynes sec/cm. How does the friction element affect the shape of the frequency plot?

21. Use MATLAB to plot the magnitude Bode plot of the transfer function of the circuit below. Note the more realistic units for the elements. What type of filter is it? Plot between 100 and 10^6 rad/sec, but plot in Hz.

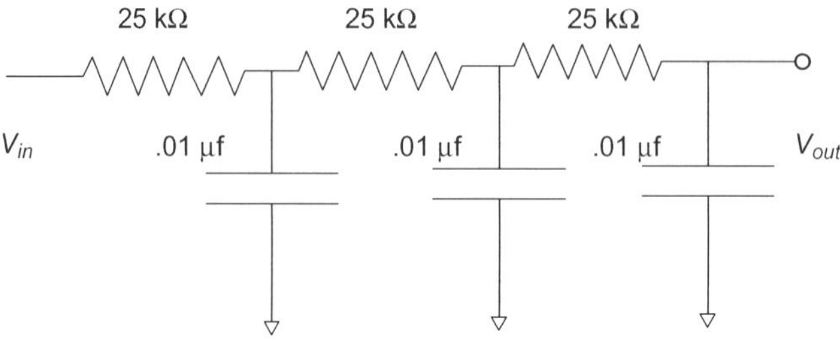

22. Reverse the resistors and capacitors for the network of Problem 21 and plot the magnitude Bode plot. How does it compare with that of the original network?

23. Plot the Bode plot (magnitude and phase) for the transfer function of the RLC circuit below. Reverse the position of the resistor and capacitor and replot. That is, put the resistor across the top and connect the capacitor to ground. How does this change the frequency characteristics of the filter?

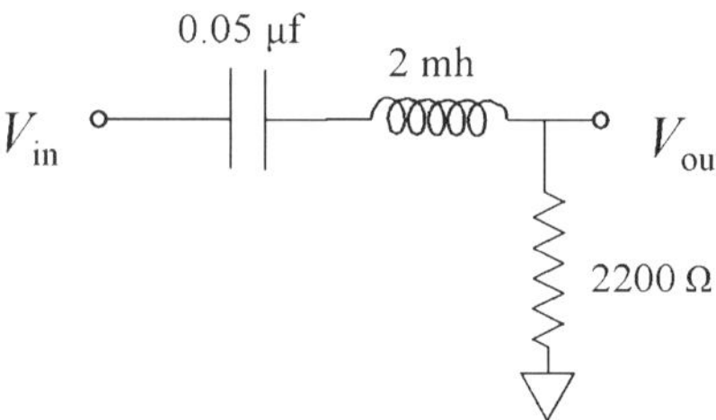

24. Plot the Bode plot of the network used in Example 10.16 from the transfer function. Note that the transfer function equation was derived in Example 10.13 and is given by Equation 10.21. Plot the Bode plot for the three configurations of component values given in Table 10.1 over a frequency range of 1 to 10,000 Hz.

25. Plot the Bode plot of the network used in Example 10.16 as in Problem 24, but use the time-domain approach and the impulse response. Plot the Bode plot for the three configurations of component values of a frequency range of 1 to 1000 Hz. Plot in dB versus log frequency and take care to construct the correct frequency vector. [*Hint*: The code in Example 10.15 already determines the impulse response; you just need to add the code that determines the frequency response (using the Fourier transform) and generates the frequency plots (Equation 10.30).]

26. Modify the code of Example 10.16 so that the simulated circuit has a resonant frequency of 400 Hz and plots the circuit's response to the chirp for three different values of *Q*: 2, 10, and 20. Note the effect of increasing and decreasing the filter's selectivity (i.e., Q).

CHAPTER

11

Circuit Reduction: Simplifications

11.1 SYSTEM SIMPLIFICATIONS—PASSIVE NETWORK REDUCTION

Sometimes it is desirable to simplify a circuit or system by combining elements. Many quite complex circuits can be reduced to just a few elements using an approach known as *network reduction*. Such simplifications can be valuable: They can be used to simplify analysis and provide a summary-like representation of a complicated system. New insights can be had about the properties of a network after it has been simplified. Network reduction can be particularly useful when two networks or systems are to be connected together, as the reduced forms highlight how the interconnection affects each network and the passage of information between them. Finally, the principles of network reduction are useful in understanding the behavior of real sources.

It is necessary to understand reduction principles in order to analyze biomedical measurements. Making a measurement on a biological system can be viewed as connecting the biological and measurement systems together. All measurements require drawing some energy from the system being measured. How much energy is taken depends on the match between the biological and measurement systems. This match can be quantified in terms of a generalized concept of impedance: the match between the output impedance of the biological system and the input impedance of the measurement system

Before we can reduce complex networks or systems, we must first learn to reduce simple configurations of elements such as series and parallel combinations. Network reduction is based on a few simple rules for combining series and/or parallel elements. The approach is straightforward, although implementation can become tedious for large networks (that is when we turn to MATLAB). After we introduce the reduction rules for networks consisting only of passive elements, we will expand our guidelines to include networks with sources.

11.1.1 Series Electrical Elements

Electrical elements are said to be in *series* when they are interconnected and no other elements share that interconnection (Figure 11.1A). While series elements are often drawn in line with one another, they can be drawn in any configuration and still be in series as long as they follow the "connected with no other" rule. The three elements in Figure 11.1B are also in series as long as no other elements are connected at their interconnections. Simple application of KVL demonstrates that when elements are in series their impedances add. The voltage across three series elements in Figure 11.1B is:

$$v_{total} = v_1 + v_2 + v_3 = (Z_1 + Z_2 + Z_3)i$$

The total voltage can also be written as:

$$v_{total} = Z_{eq}\, i; \quad \text{where: } Z_{eq} = Z_1 + Z_2 + Z_3$$

So series elements can be represented by a single element that is the sum of the individual elements as in Equation 11.1.

$$Z_{eq} = Z_1 + Z_2 + Z_3 + \ldots + Z_N \tag{11.1}$$

If the series elements are all resistors or all inductors, they can be represented by a single resistor or inductor that is the sum of the individual elements:

$$R_{eq} = R_1 + R_2 + R_3 + \ldots \tag{11.2}$$

$$L_{eq} = L_1 + L_2 + L_3 + \ldots \tag{11.3}$$

If the elements are all capacitors, then their reciprocals add, since the impedance of a capacitor is a function of $1/C$ (i.e., $1/j\omega C$ or $1/sC$):

$$1/C_{eq} = 1/C_1 + 1/C_2 + 1/C_3 + \ldots \tag{11.4}$$

If the string of elements includes different element types, then the individual impedances can be added using complex arithmetic to determine a single equivalent impedance. In general, this equivalent impedance is complex, as shown in the example below.

A)

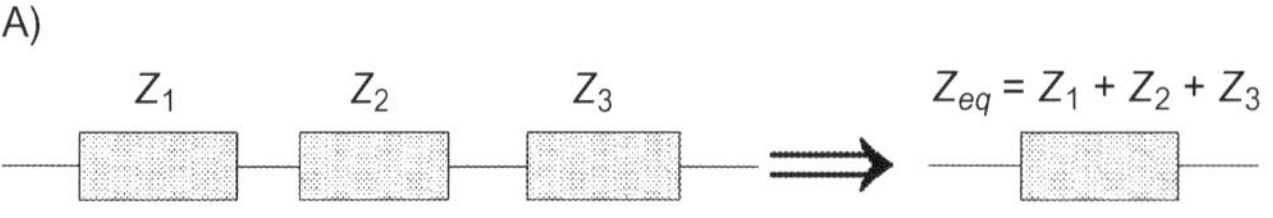

B)

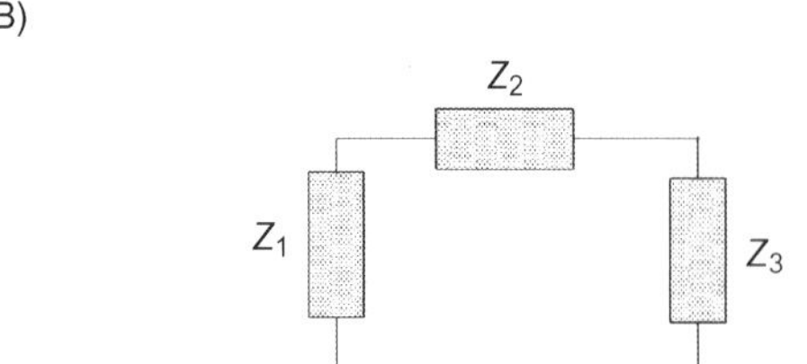

FIGURE 11.1 A) Three elements in series, Z_1, Z_2, and Z_3, can be converted into a single equivalent element, Z_{eq}, that is the sum of the three individual elements. Elements are in series when they share one node and no other elements share this node. B) Series elements are often drawn in line, but these elements are also in series as long as nothing else is connected between the elements.

EXAMPLE 11.1

Series element combination. Find the equivalent single impedance, Z_{eq}, of the series combination in Figure 11.2.

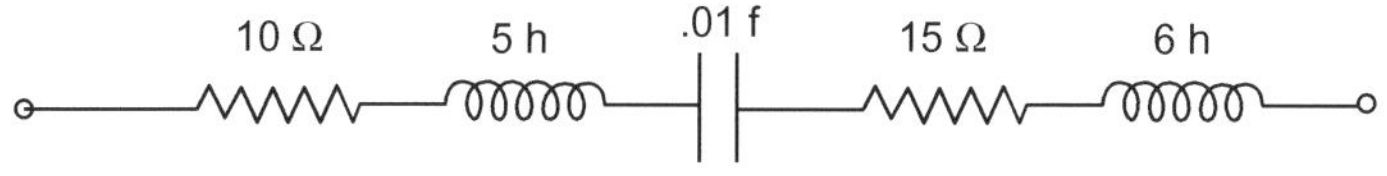

FIGURE 11.2 A combination of series elements that can be mathematically combined into a single element. Usually this element is complex (i.e., contains a real and an imaginary part).

Solution: First combine the two resistors into a single 25 Ω resistor, then combine the two inductors into a single 11 h inductor, then add the three impedances (R_{eq}, $j\omega L_{eq}$, and $1/j\omega C$). Alternatively, convert each element to its equivalent phasor impedance, then add these impedances.

$$Z_{eq} = 10 + j\omega 5 + \frac{1}{j\omega .01} + 15 + j\omega 6 = 25 + j\omega 11 + \frac{100}{j\omega}$$

If a specific frequency is given, for example $\omega = 2.0$ rad/sec, then Z_{eq} can be evaluated as a single complex number.

$$Z_{eq}(\omega = 2) = 25 + j(11(2) - j100/2) = 25 - j28\ \Omega = 37.5\angle{-48}\ \Omega$$

EXAMPLE 11.2

Find the equivalent capacitance for the three capacitors in series in Figure 11.3.

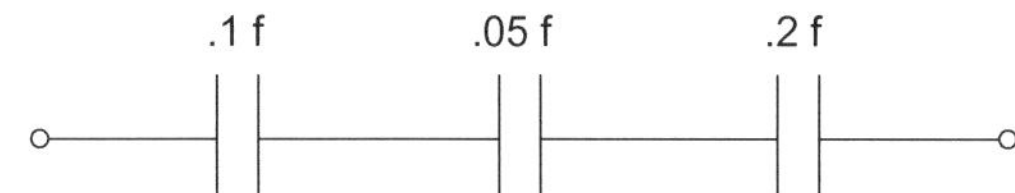

FIGURE 11.3 Three series capacitors. They can be combined into a single capacitor by adding reciprocals, as shown in Example 11.2.

Solution: Since all the elements are capacitors, they add as reciprocals:

$$1/C_{eq} = 1/C_1 + 1/C_2 + 1/C_3 = 1/0.1 + 1/0.05 + 1/0.2 = 10 + 20 + 5 = 35$$

$$C_{eq} = 1/35 = 0.029\ \text{f}$$

11.1.2 Parallel Elements

Elements are in parallel when they share both connection points, as shown in Figure 11.4. For parallel electrical elements, it does not matter if other elements share these mutual connection points, as long as both ends of the elements are connected.

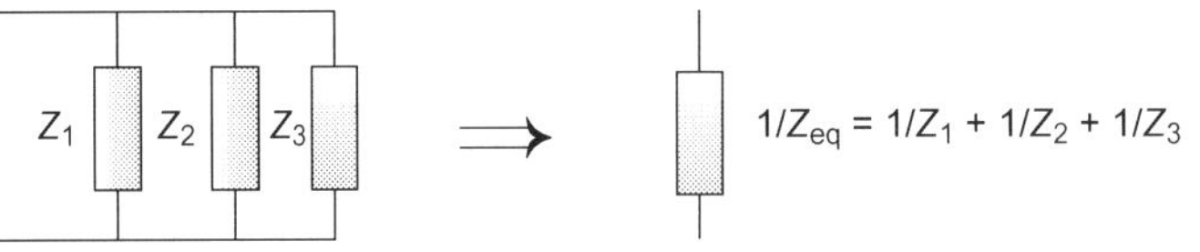

FIGURE 11.4 Parallel elements share both end-connection points. Three elements in parallel, Z_1, Z_2, and Z_3, can be converted into a single equivalent impedance, Z_{eq}, that is the reciprocal of the sum of the reciprocals of the three individual impedances.

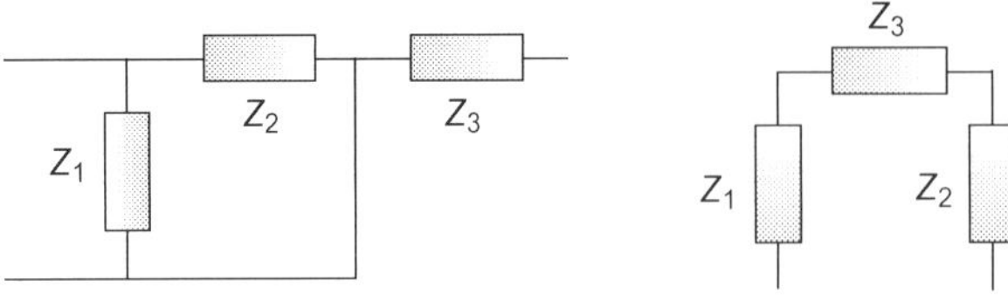

FIGURE 11.5 Left circuit: The elements Z_1 and Z_2 are connected at both ends and are therefore electrically in parallel even though they are not drawn parallel. Right circuit: Although they are drawn parallel, elements Z_1 and Z_2 are not electrically parallel because they are not connected at either end.

When looking at electrical schematics, it is important to keep in mind the definition of parallel and series elements because series elements may not necessarily be drawn in line with one another (Figure 11.1B) and parallel elements may not necessarily be drawn as geometrically parallel. For example, the two elements, Z_1 and Z_2, on the left side of Figure 11.5 are in parallel because they connect at both ends, even though they are not drawn in parallel geometrically. Conversely, elements Z_1 and Z_2 are drawn parallel on the right side of Figure 11.5, but they are not in parallel electrically because they are not connected at either end.

As can be shown by KCL, parallel elements combine as the reciprocal of the sum of the reciprocals of each impedance. With application of KCL to the upper node of the three parallel elements in Figure 11.4, the total current flowing through the three impedances is: $i_{total} = i_1 + i_2 + i_3$. Substituting in v/Z for the currents through the impedances, the total current is:

$$i_{total} = v/Z_1 + v/Z_2 + v/Z_3 = v(1/Z_1 + 1/Z_2 + 1/Z_3)$$

This equation, restated in terms of an equivalent impedance, becomes:

$$i_{total} = v/Z_{eq} \quad \text{where} \quad 1/Z_{eq} = 1/Z_1 + 1/Z_2 + 1/Z_3$$

Hence:

$$1/Z_{eq} = 1/Z_1 + 1/Z_2 + 1/Z_3 + \cdots \tag{11.5}$$

Equation 11.5 also holds for the value of parallel resistor and inductors:

$$1/R_{eq} = 1/R_1 + 1/R_2 + 1/R_3 + \cdots \tag{11.6}$$

$$1/L_{eq} = 1/L_1 + 1/L_2 + 1/L_3 + \cdots \tag{11.7}$$

Parallel capacitors, however, simply add.

$$C_{eq} = C_1 + C_2 + C_3 + \cdots \tag{11.8}$$

Hence if the three capacitors in Example 11.2 are in parallel, their equivalent capacitance is simply the addition of the three values:

$$C_{eq} = 0.1 + 0.05 + 0.2 = 0.35 \text{ f}$$

EXAMPLE 11.3

Find the equivalent single impedance for the parallel RLC combination in Figure 11.6.

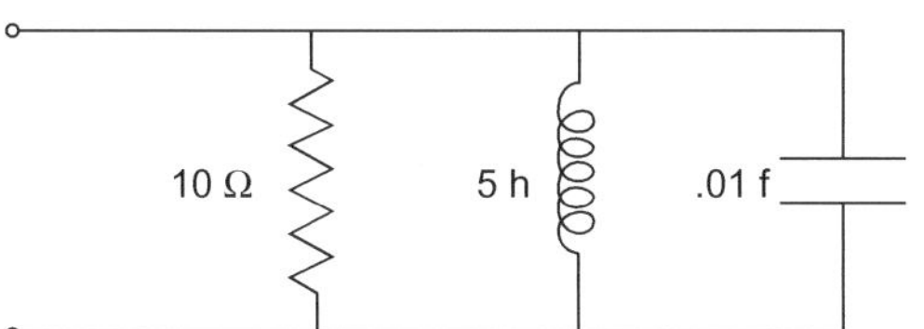

FIGURE 11.6 Three parallel elements can be combined into a single element, as shown in Example 11.3.

Solution: First take the reciprocals of the impedances:

$$1/R = 1/10 = 0.1\ \mho;\ 1/j\omega L = 1/j5\omega\ \mho;\ j\omega C = j0.01\ \omega\ \mho$$

Then add, and invert:

$$\frac{1}{Z_{eq}} = Yeq = 0.1 + \frac{1}{j\omega 5} + j\omega 0.01 = 0.1 + j\left(0.01\omega - \frac{1}{5\omega}\right)$$

$$Zeq = \frac{1}{Yeq} = \frac{1}{0.1 + j\left(0.01\omega - \frac{1}{5\omega}\right)}$$

Once a value of frequency, ω, is given, this equation can be solved for a specific impedance value. Alternatively, we can solve for Z_{eq} over a range of frequencies using MATLAB. This is given as an exercise in Problem 8.

EXAMPLE 11.4

Find the equivalent resistance of the parallel combination of three resistors: 10 Ω, 15 Ω, and 20 Ω.

Solution: Calculate reciprocals, add them and invert:

$$1/R_{eq} = 1/R_1 + 1/R_2 + 1/R_3 = 1/10 + 1/15 + 1/20 = 0.1 + 0.0667 + 0.05 = 0.217\ \mho$$

$$R_{eq} = 1/0.217 = 4.61\ \Omega$$

Note that the equivalent resistance of a parallel combination of resistors is always less than the smallest resistor in the group. The same is true of inductors, but the opposite is true of parallel capacitors; the parallel combination of capacitors is always larger than the largest capacitor in the group.

11.1.2.1 Combining Two Parallel Impedances

Combining parallel elements via Equation 11.5 is mildly irritating, with its double inversions. Most parallel element combinations involve only two elements, so it is useful to have an equation that directly states the equivalent impedance of two parallel elements without the inversion. Starting with Equation 11.5 for two elements:

$$\frac{1}{Z_{eq}} = \frac{1}{Z_1} + \frac{1}{Z_2} = \frac{Z_2}{Z_2 Z_1} + \frac{Z_1}{Z_1 Z_2} = \frac{Z_1 + Z_2}{Z_1 Z_2}$$
$$\text{Inverting:} \qquad Z_{eq} = \frac{Z_1 Z_2}{Z_1 + Z_2} \tag{11.9}$$

Hence the equivalent impedance, Z_{eq}, of two parallel elements equals the product of the two impedances divided by their sum.

11.2 NETWORK REDUCTION—PASSIVE NETWORKS

The rules for combining series and parallel elements can be applied to networks that include a number of elements. Even very involved configurations of passive elements can usually be reduced to a single element. Obviously, it is easier to grasp the significance of a single element than a confusing combination of many elements.

11.2.1 Network Reduction—Successive Series–Parallel Combinations

In the last section, we see that it is possible to combine a number of series or parallel combinations. Even when most of the elements are not in either series or parallel configurations, it is possible to combine them into a single impedance using the techniques of network reduction. In this section, the networks being reduced consist only of passive elements, but in Section 11.3 we learn how to expand network reduction to include networks with sources.

In the network in Figure 11.7, most of the elements are neither in series nor in parallel. It is important to realize that the elements across the top of this network—inductor, resistor, and inductor—are not in series because their connection points are shared by other elements, capacitors in this case. To be in series, the elements must share one connection point and must be the only elements to share that point. If we could somehow eliminate the two capacitors (we cannot), then these three elements would be in series. Nor are any

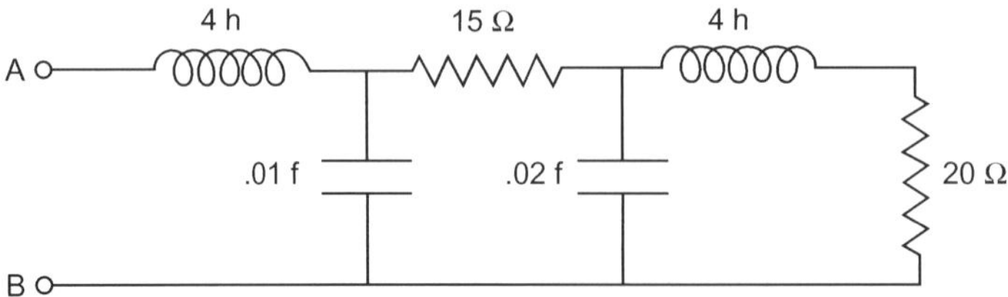

FIGURE 11.7 A network containing *R*, *L*, and *C*s where most of the elements are neither in series or in parallel. Nonetheless, this network can be reduced to a single equivalent impedance (with respect to nodes A and B), as is shown in Example 11.5.

of the elements in parallel, since no elements share both connection points. If they did, they would be in parallel even if other elements share these connection points. However, there are two elements in series: the 4 h inductor and the 20 Ω resistor on the right-hand side of the network. We could combine these two elements using Equation 11.1. After combining these two elements, we now find that the new element is in parallel with the 0.02 f capacitor. We can then combine these two parallel elements using the parallel rule given in Equation 11.9. While the reduction process becomes difficult to follow at this point without actually going through it, the single element newly combined from the parallel elements is now in series with the 15 Ω resistor. Most reductions of passive networks proceed in this fashion: Find a series or parallel combination to start with, combine them, and then look to see if the new combination produces a new series or parallel combination. Then just continue down the line

The example below uses the approach based on sequential series–parallel combinations to reduce the network in Figure 11.7 to a single equivalent impedance.

EXAMPLE 11.5

In a network reduction using sequential series–parallel combinations, find the equivalent impedance between the nodes A and B (Figure 11.7). Find the impedance at only one frequency, $\omega = 5.0$ rad/sec. Using network reduction, we can find the equivalent impedance leaving frequency as a variable (i.e., $Z(\omega)$), but this will make the algebra more difficult. With a specific frequency, we are able to use complex arithmetic instead of complex algebra.

Solution: Convert all elements to their equivalent impedances at $\omega = 5.0$ rad/sec (to simplify subsequent calculations). Then begin the reduction by combining the two series elements on the right-hand side together. As a first step, first convert the elements to their phasor impedances (at $\omega = 5$ rad/sec). Then combine the two series elements using Equation 11.1, leading to the network shown in Figure 11.8 on the right-hand side.

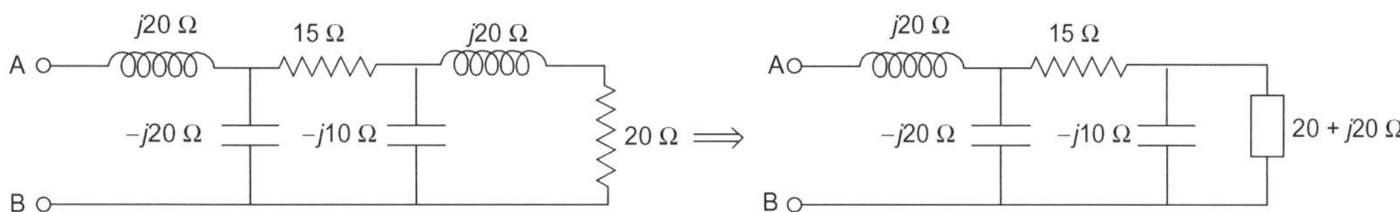

FIGURE 11.8 The network on the left is a phasor representation of the network in Figure 11.7; a partial reduction of this network is shown on the right.

This combination puts two elements in parallel: the newly formed impedance and the $-j10\ \Omega$ capacitor. These two parallel elements can be combined using Equation 11.9:

$$Z_{eq} = \frac{Z_1 Z_2}{Z_1 + Z_2} = \frac{(20 + j20)(-j10)}{20 + j20 - j10} = \frac{200 - j200}{20 + j10} = \frac{282.8\angle -45}{22.36\angle 26.6}$$

$$Z_{eq} = 12.65\angle -71.6\ \Omega = 4 - j12\ \Omega$$

This leaves a new series combination that can be combined as shown below. In the third step of network reduction, the newly formed element from the parallel combination ($4 - j12\ \Omega$) is now in series with the $15\ \Omega$ resistor, and this equivalent series element will be in parallel with the 0.01 f capacitor (Figure 11.9).

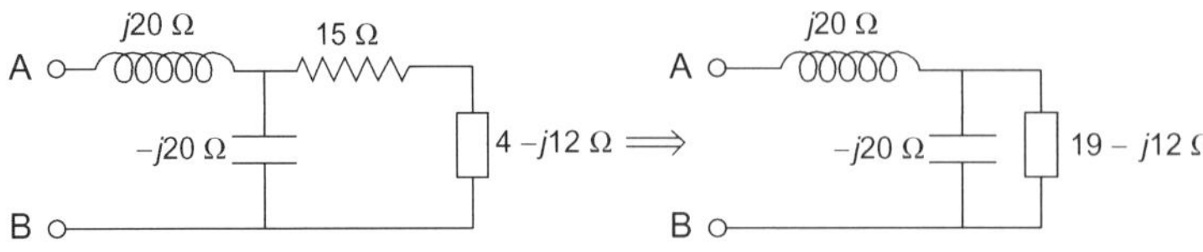

FIGURE 11.9 The next step in the reduction of the network is shown in Figure 11.7. The newly combined elements are now in series with the $15\ \Omega$ resistor, and the element formed from that combination is in parallel with the $-j20\ \Omega$ (0.01 f) capacitor.

Combining these two parallel elements:

$$Z_{eq} = \frac{-j20(19 - j12)}{-j20 + 19 - j12} = \frac{-240 - j380}{19 - j32} = \frac{449.4\angle 238}{37.2\angle -59}$$

$$Z_{eq} = 12.07\angle 297\ \Omega = 5.49 - j10.76\ \Omega$$

This leads to the final series combination and, after combination, a single equivalent impedance with a value of:

$$Z_{eq} = 5.49 + j9.24\ \Omega = 10.75\angle 59\Omega$$

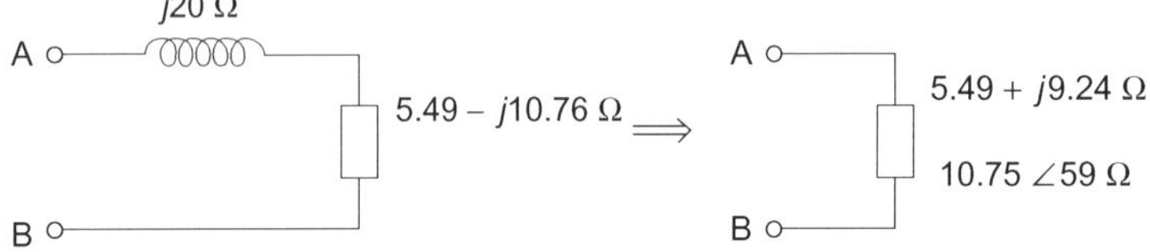

FIGURE 11.10 A single element represents the effective impedance between nodes A and B in the network of Figure 11.7.

Network reduction is always done from the point of view of two nodes, such as nodes A and B in this example. In principle, any two nodes can be selected for analysis and the equivalent impedance can be determined between these nodes. Generally the nodes selected have some special significance, for example, the nodes that make up the input or output of the circuit. Network reduction usually follows the format of this example: sequential combinations of series elements, then parallel elements, then series elements, and so on. In a few networks there are no elements either in series or in parallel to start with, and an alternative method described in the next section must be used. This method works for all networks and any combination of two nodes, but it is usually more computationally intensive. On the other hand, it lends itself to a computer solution using MATLAB, which ends up being less computationally intensive.

11.2.1.1 Resonance Revisited

In Chapter 10 we found that resonance occurs in electrical and mechanical systems when the impedance of the two types of energy storage devices, inertial and capacitive, are equal and cancel. This leads to a minimum in total impedance: zero if there are no dissipative elements present. For example, in a series RLC circuit the total impedance is $R + j\omega L + 1/j\omega C$ and resonance occurs when the impedance of the inductor and capacitor are the same. The total impedance is just that of the resistor. This occurs when $\omega = \sqrt{1/LC}$ (Eq. 10.7). If the two energy storage elements are in parallel, then an *anti-resonance* occurs where the net impedance goes to infinity, since by Equation 11.9:

$$Z = \frac{Z_1 Z_2}{Z_1 + Z_2} = \frac{j\omega L(1/j\omega C)}{j\omega L + 1/j\omega C} \rightarrow \infty \quad \text{when} \quad j\omega L = \frac{1}{j\omega C}$$

An example of the impedance of a 5.0 μf capacitor in series with a 50 mh inductor is shown as a function of log frequency in Figure 11.11A. The sharp anti-resonance peak at 2000 rad/sec is evident.

A network can exhibit both resonance and antiresonance, at different frequencies, of course. In the network shown in Figure 11.12, the parallel inductor and capacitor produce an antiresonance at 2000 rad/sec, but at a higher frequency the combination becomes capacitive and that parallel combination will resonate with the series inductor. This is shown in the next example.

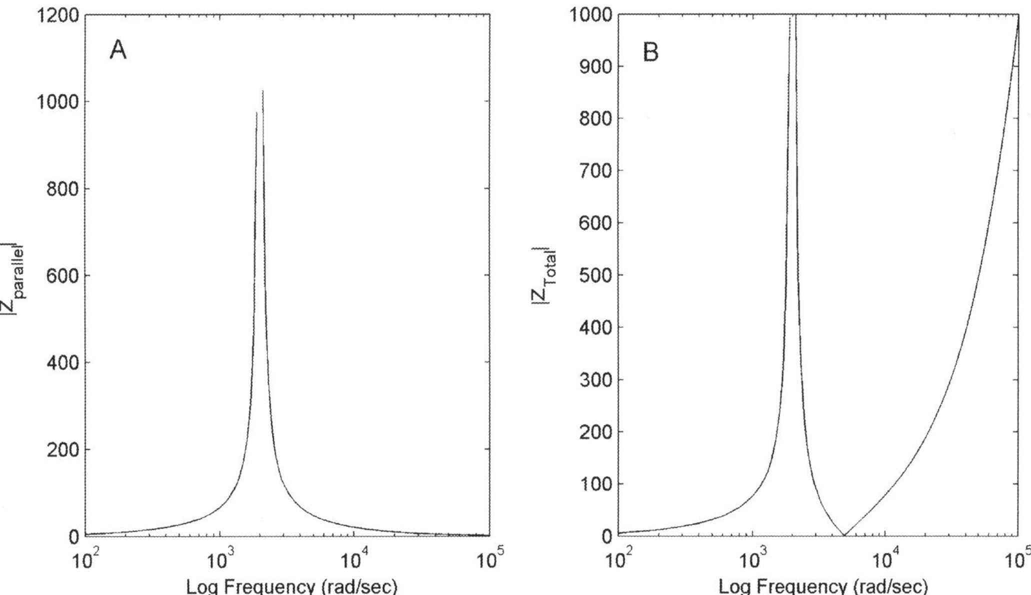

FIGURE 11.11 A) The impedance of a parallel LC combination. An antiresonance peak is seen at 2000 rad/sec when the impedances of the two elements are equal. B) An example of both resonance and antiresonance phenomena from the network shown in Figure 11.12. This graph is constructed in Example 11.6 using MATLAB.

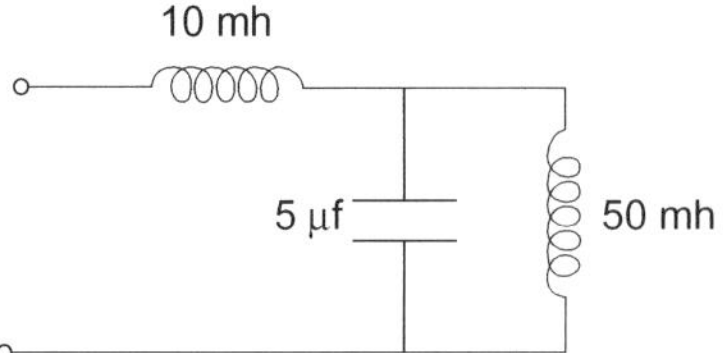

FIGURE 11.12 A circuit that exhibits both resonance and antiresonance. This circuit is analyzed in Example 11.6.

EXAMPLE 11.6

Use MATLAB to find the net impedance of the circuit in Figure 11.12 and plot the impedance as a function of log frequency. Also plot the impedance characteristic of the parallel LC separately.

Solution: After defining the component values, calculate the impedance of each component. Use the parallel element equation, Equation 11.9, to find the impedance of the parallel LC, then add in the impedance of the series inductor to find the total impedance. The program is shown below.

```
% Ex 11.6 Example to show resonance and antiresonance
%
w = (100:100:1000000);            % Define frequency vector
C = 05e-6;                        % Assign component values
L1 = 0.05;
L2 = .01;
ZL1 = j*w*L1;                     % Calculate component impedances
ZL2 = j*w*L2;
ZC = 1./(j*w*C);
Zp = (ZL1 .* ZC)./(ZL1 + ZC);     % Calculate parallel impedance
Z = ZL2 + Zp;                     % Calculate total impedance
 ......plot using semilogx and label.......
```

Results: The resulting plots from this program are shown in Figure 11.11. Figure 11.11B plots the total impedance and shows both the antiresonance peak, when $Z_{Eq} \rightarrow \infty\ \Omega$ at 2000 rad/sec, and the resonance peak, when $Z_{Eq} \rightarrow 0\ \Omega$ at 5000 rad/sec.

11.2.2 Network Reduction—Voltage-Current Method

The other way to find the equivalent impedance of a network follows the approach that is used given an actual network in a laboratory setting. Suppose you are asked to determine the impedance between two nodes of a network, what is known as a classical "two-terminal" problem. Perhaps the actual network is inaccessible to you and all you have available are two nodes, as shown in Figure 11.13.

The actual network could contain any number of nodes, but in this scenario only two are accessible for measurement. Even without "opening the box," it is possible to determine the equivalent impedance between these two terminals. What you do in this situation (or, at least, what I would do) is to apply a known voltage to the two terminals,

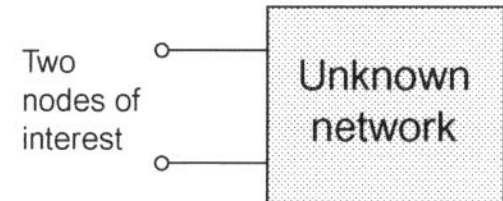

FIGURE 11.13 An unknown network that has only two terminals available for use or for measurement. This is a classic "two-terminal" device.

measure the resulting current, and calculate the equivalent impedance Z_{eq}, using Ohm's law:

$$Z_{eq} = \frac{V_{known}}{I_{measured}} \tag{11.10}$$

Of course, V_{known} has to be a sinusoidal source of known amplitude, phase, and frequency unless you know, *a priori,* that the network is purely resistive, in which case a DC source suffices. Moreover, you are limited to determining Z_{eq} at only one frequency, the frequency of the voltage source, but most laboratory sinusoidal sources offer a range of selectable frequencies so the impedance can be determined over a range of frequencies. If the unknown network contains sources, a slightly different strategy developed in the next section can be used instead.

This same approach can be applied to a network that exists only on paper, such as the network in the last example. Using the tools that we have acquired thus far, we simply connect a source of our choosing to the network and solve for the current into the network. The source can be anything. Moreover, this approach can be applied to simplify any network, even one that does not have any series or parallel elements. In the next example, this method is applied to the network in Example 11.5 and, in a subsequent example, to a more challenging network.

EXAMPLE 11.7

Passive network reduction using the source-current method. Find the equivalent impedance between nodes A and B in the network of Figure 11.7 for a frequency of $\omega = 5$ rad/sec.

Solution: Apply a known source across nodes A and B and solve for the resulting current using mesh analysis. Theoretically we can choose any sinusoidal source as long as it has a frequency of 5 rad/sec; here we choose something simple: a 1 volt source with 0.0 degree phase: $v(t) = \cos(5t) \rightarrow V(\omega) = 1\angle 0$. The desired impedance Z_{ab} will then be: $Z_{ab} = V(\omega)/I_1(\omega) = 1/I_1(\omega)$. Mesh analysis can be used to solve for $I_1(\omega)$. The three-mesh network is shown using phasor notation in Figure 11.14.

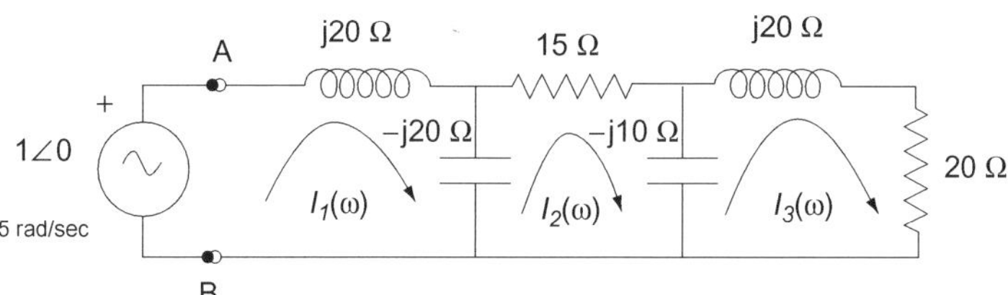

FIGURE 11.14 The network shown in Figure 11.7 with a voltage source attached to nodes A and B and represented in the phasor domain.

To find the input current, we analyze this as a straightforward three-mesh problem solving for I_1. However, several alternatives are also possible: using network reduction to convert the last three elements to a single element and solving as a two-mesh problem; converting the voltage source to an equivalent current source (as shown later in this chapter) and solving as a two-node problem; or combining these approaches and solving as a one-node problem. Here we solve the three-mesh problem directly, but use MATLAB to reduce the computational burden. Applying standard mesh analysis to the network above, the basic matrix equation can be written as:

$$\begin{vmatrix} 1.0 \\ 0 \\ 0 \end{vmatrix} = \begin{vmatrix} j20 - j20 & +j20 & 0 \\ +j20 & 15 - j20 - j10 & +j10 \\ 0 & +j10 & 20 + j20 - j10 \end{vmatrix} \begin{vmatrix} i_1 \\ i_2 \\ i_3 \end{vmatrix}$$

Simplifying by complex addition:

$$\begin{vmatrix} 1.0 \\ 0 \\ 0 \end{vmatrix} = \begin{vmatrix} 0 & j20 & 0 \\ j20 & 15 - j30 & j10 \\ 0 & j10 & 20 + j10 \end{vmatrix} \begin{vmatrix} i_1 \\ i_2 \\ i_3 \end{vmatrix}$$

Solving for I_1, then Z_{eq} using MATLAB:

```
% Example 11.7 Find the equivalent impedance of a network by applying a source and
solving for the resultant current.
%
% First assign values for v and Z
v = [1 0 0];
Z = [0 20j 0; 20j 15 - 30j 10j; 0 10j 20 + 10j];
%
i = Z\v                                      % Solve for currents
Zeq = 1/i(1)                                 % Solve for Zeq
Zeq = [abs(Zeq) angle(Zeq)*360/(2*pi)]       % Output in polar form
```

The output from this program is: $Z_{eq} = 5.49 + 9.24i\ \Omega$ or $10.75\angle 59\ \Omega$, which is the same as that found in the previous example using network reduction. Again, this approach could be used to reduce any passive network of any complexity to an equivalent impedance between any two terminals. The latter portion of this chapter shows how to reduce, and think about, networks that also contain sources. Before proceeding to the next section, Example 11.8 shows how to reduce a passive network when the two terminals of interest are in more complicated positions (for example, separated by more than a single element). We also solve for the impedance over a range of frequencies.

EXAMPLE 11.8

Find the equivalent impedance of the circuit shown in Figure 11.15 between terminals A and B. Use MATLAB to find Z_{eq} over a range of frequencies and plot Z_{eq} as a function of log frequency. Use a frequency range of 0.01 to 100 rad/sec. (This frequency range was established by trial and error in MATLAB.)

Solution: Apply a source between terminals A and B and solve for the current flowing out of the source. In this problem, the source has a fixed amplitude of 1.0 volts with a phase of 0

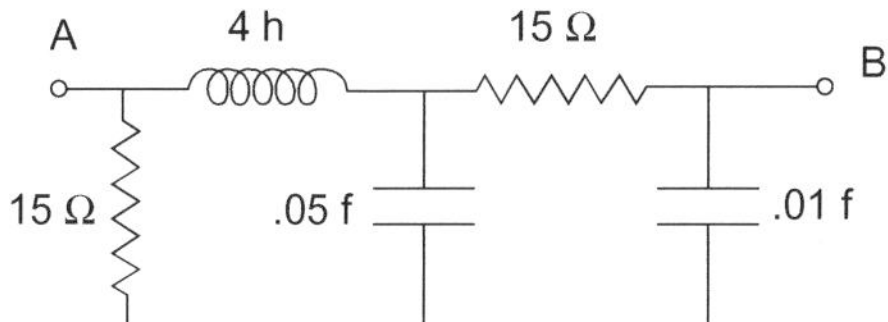

FIGURE 11.15 Network used in Example 11.8. The goal is to find the equivalent impedance between nodes A and B over a frequency range of 0.01 to 100 rad/sec.

degrees, but the frequency is variable so that Z_{eq} can be determined over the specified range of frequencies. Before you apply the source, it is helpful to rearrange the network so that the meshes can be more readily identified. Sometimes this topographical reconfiguration can be the most difficult part of the problem, especially to individuals who are spatially challenged.

In this network, simply rotating the network by 90 degrees makes the mesh arrangement evident, as shown in Figure 11.16.

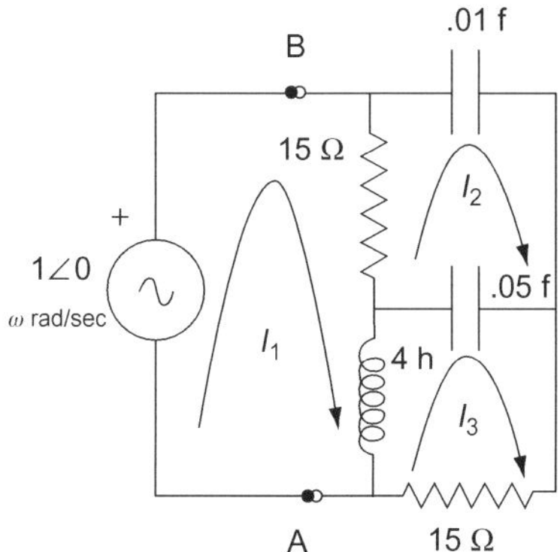

FIGURE 11.16 The network shown in Figure 11.15 after rotating by 90 degrees to make the three-mesh topology more evident.

This network cannot be reduced by simple series–parallel combination strategy as employed previously. While more complicated transformations can be used,[1] we can just as easily apply the voltage–current method. Again the problem can be solved using standard network analysis. Apply a voltage source and assign mesh currents as shown in Figure 11.16. The total current out of the source flows into both the 15 Ω resistor and the 0.01 f capacitor (the source is actually across both branches). Hence the current into node B, the current we are looking for, is actually the sum of $i_1 + i_2$.

After you convert all elements to phasor notation, write the matrix equation as:

$$\begin{vmatrix} 1 \\ 0 \\ 0 \end{vmatrix} = \begin{vmatrix} 15 + j4\omega & -15 & -j4\omega \\ -15 & 15 - \dfrac{j100}{\omega} - \dfrac{j20}{\omega} & \dfrac{j20}{\omega} \\ -j4\omega & \dfrac{j20}{\omega} & 15 + j4\omega - \dfrac{j20}{\omega} \end{vmatrix} \begin{vmatrix} I_1 \\ I_2 \\ I_3 \end{vmatrix}$$

[1]A slightly more complicated transformation exists that allows configurations such as this to be reduced by the reduction method, specifically a transformation known as the *Π (pi) to T* transformation. However, the voltage-current approach applies to any network and lends itself well to computer analysis.

Note that the frequency, ω, must remain a variable in this equation because we need to find the value of Z_{eq} over a range of frequencies. MATLAB is used to find the values of Z_{eq} over the desired range of frequencies and also plots these impedance values.

```
% Example 11.8 Find and plot the values of an equivalent impedance
% between .01 and 100 rad/sec
%
% Define frequency, use .01 rad/sec increments
w=.01:.01:100;              % Define frequency vector
v=[1; 0; 0];                % Define voltage vector
%
% Loop over all frequencies, solving for Zeq
for k=1:length(w)
% Define impedance vector (Use continuation statement)
  Z [15+4j*w(k), −15, −4j*w(k); −15, 15 −120j/w(k), 20j/w(k);...
  −4j*w(k), 20j/w(k), 15+4j*w(k) − 20j/w(k)];
  i = Z\v;                  % Solve for current
  Zeq(k)=1/i(1) + i(2));    % Solve for Zeq
end
  .......plot and label magnitude and phase........
```

The graph produced by this program is shown in Figure 11.17. Both the magnitude and phase of Z_{eq} are functions of frequency and both change sharply in value at around 3 rad/sec.

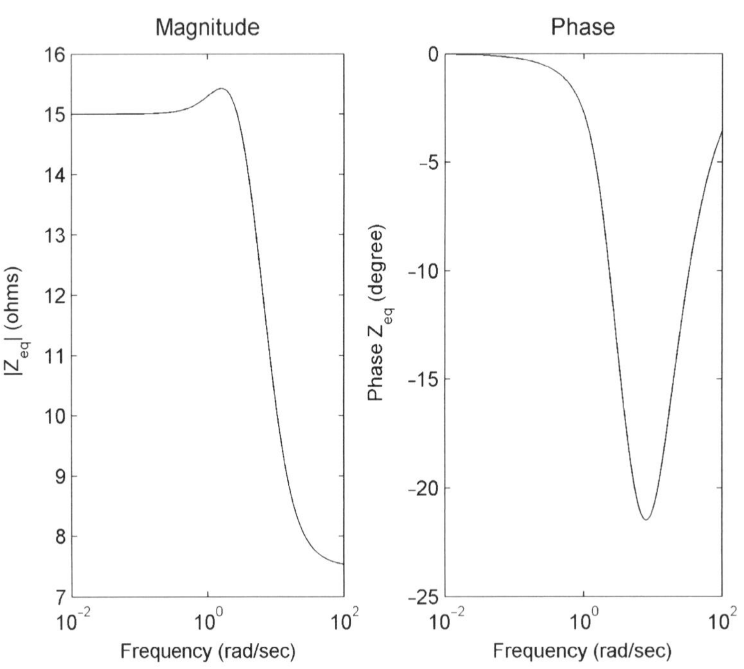

FIGURE 11.17 The value of Z_{eq} as a function of frequency for the network in Example 11.8. The impedance is plotted in terms of magnitude and phase. These plots are obtained from the MATLAB code used in Example 11.8.

It is easy to find the maximum and minimum of MATLAB variables using the MATLAB `max` or `min` functions. Apply these functions to the `Zeq`:

```
and [max_Zeq_abs, abs_freq]=max(abs(Zeq));
[max_Zeq_phase, phase_freq]=min(angle(Zeq)*360/(2*pi));
%
% Now display the maximum and minimum values including frequency
disp([max_Zeq_abs, w(abs_freq); max_Zeq_phase, w(phase_freq)])
```

This produces the values:

$|Z_{eq}|_{max} = 15.43\ \Omega \qquad f_{max} = 1.6\ \text{rad/sec}$
$\angle Z_{eq\_max} = -21.48\ \text{deg} \qquad f_{max} = 8.15\ \text{rad/sec}$

The `max` and `min` functions give the maximum or minimum values of a vector and the second argument gives the index where these maximum and minimum values occur. To convert the index to the appropriate frequency, just take `w(index)`, as in the code above.

The remainder of this chapter examines the characteristics of sources, both real and ideal, and develops methods for reducing networks that contain sources. Many of the principles used here in passive networks are also used with more general networks.

11.3 IDEAL AND REAL SOURCES

Before we develop methods to reduce networks that contain sources, it is helpful to revisit the properties of ideal sources described in Chapter 9 (Section 9.3.2) and examine how ideal and real sources differ. In this discussion, only constant output sources (i.e., DC sources) are considered, but the arguments presented generalize with only minor modifications to time varying sources, as shown in the next section.

11.3.1 The Voltage-Current or v-i Plot

Essentially an ideal source can supply any amount of energy that is required by whatever is connected to that source. Discussions of real and ideal sources often use plots of voltage against current (or force against velocity) which provide a visual representation of the source characteristics. Such "*v-i*" plots are particularly effective at demonstrating the equivalent resistance of an element. Consider the *v-i* plot of pure resistors as shown in Figure 11.18. The plots of five different resistors are shown: 0, 10, 100, 1000, and $\infty\ \Omega$. The voltage-current relationship for a resistor is given by Ohm's law, $v = Ri$. Comparing Ohm's law with the equation for a straight line, $y = mx + b$, where m is the slope of the line and b is the intercept, shows that the voltage-current relationship for a resistor plot is a straight line with a slope equal to the value of the resistance and an intercept of 0.0.

The reverse argument says that an element that plots as a straight line on a *v-i* plot is either a resistor or contains a resistance, and the slope of the line indicates the value of the resistance. The steeper the slope, the greater the resistance: a vertical line with a slope of

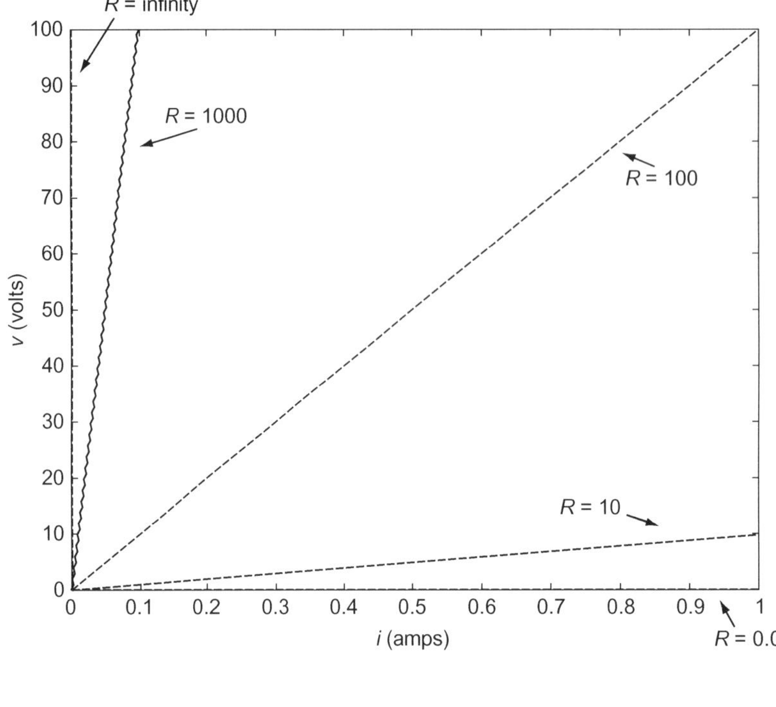

FIGURE 11.18 A v-i plot, showing voltage against current, for resistors having five different values between 0.0 Ω and infinity Ω. This shows that the v-i plot of a resistor is a straight line passing through the origin and having a slope equal to the resistance.

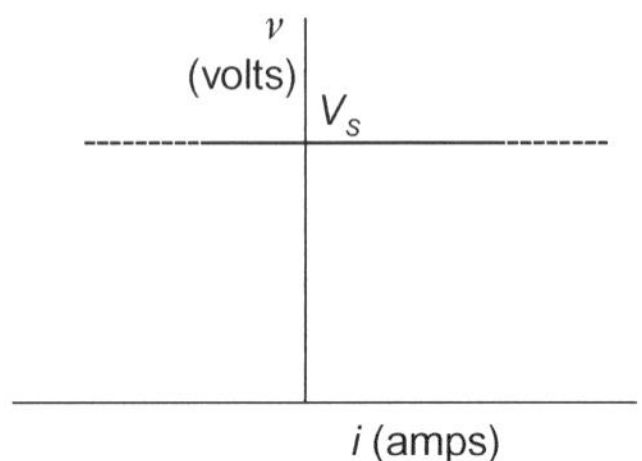

FIGURE 11.19 A v-i plot of an ideal voltage source. This plot shows that the resistor-like properties of a voltage source have zero value.

infinity indicates the presence of an infinite resistance while a horizontal line with a zero slope indicates the presence of a 0.0 Ω resistance.

The v-i plot of an ideal DC voltage source follows directly from the definition: a source of voltage that is constant irrespective of the current through it. For a time-varying source, such as a sinusoidal source, the voltage varies as a function of time, but not as a function of the current. An ideal voltage source cares naught about the current through it. Hence, the v-i plot of an ideal DC voltage source, V_S, is a horizontal line intersecting the vertical (voltage) axis at $v = V_S$, as shown in Figure 11.19. If the voltage source were time-varying, the v-i plot would look essentially the same except that the vertical position of the horizontal line would vary in time.

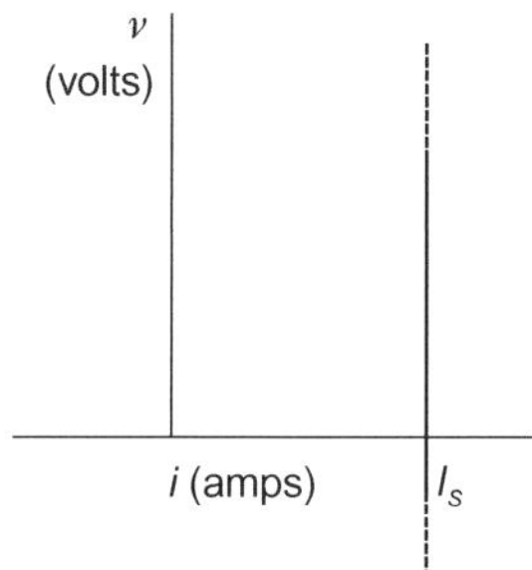

FIGURE 11.20 A *v-i* plot of an ideal current source. This plot shows that the resistor-like properties of a current source have an infinite value.

The *v-i* plot of a voltage source with its horizontal line demonstrates that the resistive component of an ideal voltage source is zero. In other words, the equivalent resistance of an ideal voltage source is 0.0 Ω. With regard to the *v-i* plot, an ideal source looks like a resistor of 0.0 Ω with an offset of V_S. It may seem strange to talk about the equivalent resistance of a voltage source, especially since it has a resistance of 0.0 Ω, but it turns out to be a very useful concept. The *equivalent resistance* of a source is its resistance ignoring other electrical properties. It is especially useful in describing real sources where the equivalent resistance is no longer zero. The concept of equivalent resistance, or more generally equivalent impedance, is important in network reduction and has a strong impact on transducer analysis and design.

Resistive elements having zero or infinite resistance have special significance and have their own terminology. Resistances of zero will produce no voltage drop regardless of the current running through them. As mentioned in Chapter 9, an element that has zero voltage for any current is termed a *short circuit* since current flows freely through such elements. Although it may seem contradictory, a voltage source is actually a short circuit device, but with a voltage offset.

Resistors with infinite resistance have the opposite voltage-current relationship: They allow no current irrespective of the voltage (assuming that it is finite). Devices that have zero current irrespective of the voltage are termed *open circuits* since they do not provide a path for current. These elements have infinite resistance. As described below, current sources also have nonintuitive equivalent resistance: They are open circuit devices but with a current offset.

The *v-i* plot of an ideal current source is evident from its definition: an element that produces a specified current irrespective of the voltage across it. This leads to the *v-i* plot shown in Figure 11.20 of a vertical line that intersects the current axis at $i = I_S$. By the arguments above, the equivalent resistance of such an ideal current source is infinite.

The concepts of ideal voltage and ideal current sources are somewhat counterintuitive. An ideal voltage source has the resistive properties of a short circuit, but also somehow maintains a nonzero voltage. The trick is to understand that an ideal voltage source is a short circuit with respect to current, but not with respect to voltage. A similar apparent contradiction applies to current sources: They are open circuits with respect to voltage, yet produce a specified current. Understanding these apparent contradictions is critical to understanding real and ideal sources.

In this section, voltage and current sources have been described in terms of fixed values (i.e., DC sources), but the basic arguments do not change if V_S or I_S are time-varying. This generalization also holds for the concepts presented next.

11.3.2 Real Voltage Sources—The Thévenin Source

Unlike ideal sources, real voltage sources are not immune to the current flowing through them, nor are real current sources immune to the voltage falling across them. In real voltage sources, the source voltage drops as more current is drawn from the source. This gives rise to a *v-i* plot such as the one in Figure 11.21, where the line is no longer horizontal but decreases with increasing current. The decrease indicates the presence of an internal, nonzero, resistance having a value equal to the negative slope of the line. The slope is negative because the voltage drop across the internal resistance is opposite in sign to V_s and hence subtracts from the value of V_s.

Real sources, then, are simply ideal sources with some nonzero resistance, so they can be represented as an ideal source in series with a resistor, as shown in Figure 11.22. This configuration is known as a Thévenin source, named after the engineer who developed the network reduction theory described below. Finding values for V_T and R_T given a physical (and therefore "real") source is fairly straightforward. The value of the internal ideal

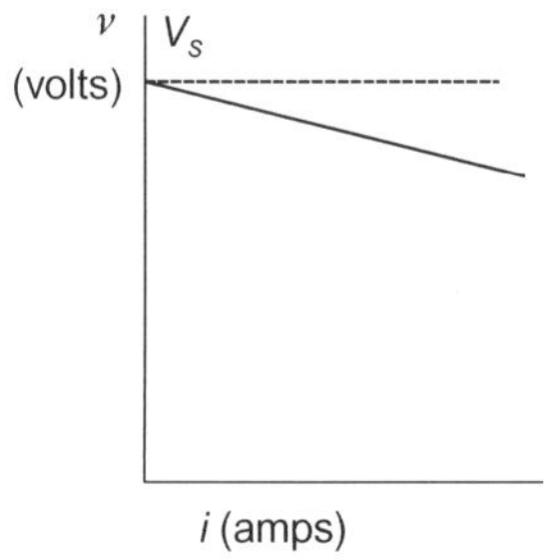

FIGURE 11.21 The *v-i* plot of a real voltage source (solid line). The nonzero slope of this line shows that the source contains an internal resistor. The slope indicates the value of this internal resistance.

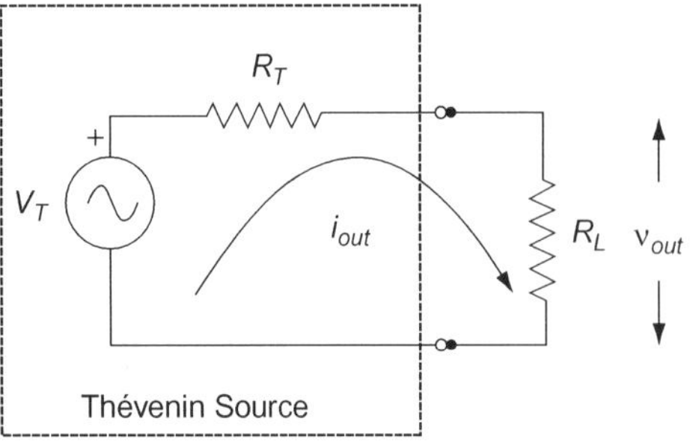

FIGURE 11.22 Representation of a real source known as a Thévenin circuit. The Thévenin circuit is inside the dashed rectangle and is shown with a load resistor, R_L. The load resistor is used in Example 11.8 to draw current from the source. In a high-quality voltage source, R_T is small.

source, V_T, is the voltage that would be measured at the output, v_{out}, if no current is flowing through the circuit; that is, if the output is an open circuit. For this reason, V_T is equivalent to the *open circuit voltage*, denoted v_{oc}. To find the value of the internal resistance, R_T, we need to draw current from the circuit and measure how much the output voltage decreases. A resistor placed across the output of the Thévenin source will do the trick. This resister, R_L, is often referred to as a *load resistor* or just the *load* because it makes the source do work by drawing current from the source, specifically; $P = v_{out}^2/R_L$. The smaller the load resistor, the more current that will be drawn for the source, and the more power the source must supply.

Assuming a current i_{out} is being drawn from the source and the voltage measured is v_{out}, then the difference in voltage between the no-current voltage and current conditions is: $v_D = V_T - v_{out}$. Since the voltage difference v_D is due entirely to R_T, the value of R_T can be determined as:

$$R_T = \frac{v_D}{i_{out}} = \frac{(V_T - v_{out})}{i_{out}} \tag{11.11}$$

The maximum current the source is capable of producing occurs when the source is connected to a short circuit (i.e., $R_L \to 0\ \Omega$), and the current that will flow, the *short-circuit current*, i_{sc}, is:

$$i_{sc} = \frac{V_T}{R_T} \tag{11.12}$$

Remembering that V_T is the voltage that would be measured under open circuit conditions and defining the open circuit voltage as v_{oc}, we write:

$$\begin{aligned} V_T &= v_{oc} = R_T i_{sc} \\ R_T &= \frac{v_{oc}}{i_{sc}} \end{aligned} \tag{11.13}$$

In words, the internal resistance is equal to the open circuit voltage (v_{oc}) divided by the short-circuit current (i_{sc}). Using Equation 11.13 is a viable method for determining R_T in theoretical problems, but is not practical in real situations with real sources because shorting a real source may draw excessive current and damage the source. It is safer to draw only a small amount of current out of a real source by placing a large resistor across the source, not a short-circuit. In this case, Equation 11.11 is used to find R_T since v_{out} and i_{out} can be measured and V_T can be determined by a measurement of open-circuit voltage. Example 11.9 takes this approach to determine the internal resistance of a voltage source.

EXAMPLE 11.9

In the laboratory, the voltage of a real source is measured using a *voltmeter* that draws negligible current from the source; hence the voltage recorded can be taken as the open-circuit voltage. (Most high-quality voltmeters require very little current to make their measurement of voltage.) The voltage measured at the output terminals of the circuit in Figure 12.22 is 9.0 volts when there is no load resistor, R_L, across the output. When a resistor, R_L, is placed across the output terminals a current of $i_{out} = 5$ mA (5×10^{-3} A) flows from the source. Assume this current is measured using an ideal current measurement device, although it could also be calculated from v_{out} if the value of R_L is known: $i_{out} = v_{out}/R_L$. Under this load condition, the output voltage, v_{out}, falls to 8.6 volts. What is the internal resistance of the source, R_T? What is the resistance of the load, R_T, that produced this current?

Solution: When there is no load resistor (i.e., $R_L = \infty$), then $i_{out} = 0$ and $v_{out} = V_T = 9$ volts. When the load resistor is attached to the output, $i_{out} = 5$ mA, and $v_{out} = 8.6$ volts. Applying Equation 11.11:

$$R_T = \frac{v_D}{i_{out}} = \frac{(V_T - v_{out})}{i_{out}} = \frac{(9.0 - 8.6)}{0.005} = \frac{0.4}{0.005} = 80\ \Omega$$

To find the load resistor, R_L, use Ohm's law:

$$R_L = \frac{v_{out}}{i_{out}} = \frac{8.6}{0.005} = 1720\ \Omega$$

In summary, a real voltage source can be represented by an ideal source with a series resistance. In the examples shown here the sources are DC and the series element a pure resistor. In the more general case, the Thévenin source could generate sinusoids or other waveforms (but would still be ideal), and the series element might be a Thévenin impedance, Z_T. In the case where the source is sinusoidal or periodic, the equations above still hold, but require phasor analysis for their solution.

11.3.3 Real Current Sources—The Norton Source

Current sources are difficult to grasp intuitively because we think of current as an effect of voltage: Voltage pushes current through a circuit. Current sources can be viewed as sources that adjust their voltage as needed to produce the desired current. For an ideal current source, the larger the load resistor, the more work they have to do since they must generate a larger voltage for a given current. Current sources prefer small load resistors, the opposite of voltage sources. For a real current source, as the load resistor increases the voltage requirement increases. Eventually, the voltage required exceeds the voltage limit of the source and the output current will fall off. This is reflected by the *v-i* plot circuit in Figure 11.23, and can also be represented by an internal resistor.

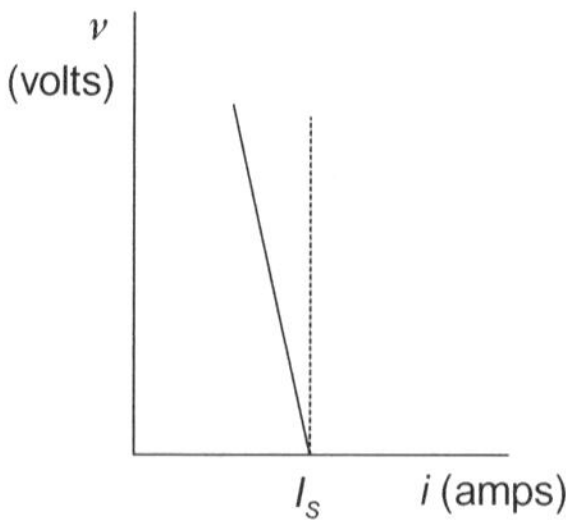

FIGURE 11.23 The *v-i* plot of a real current source (solid line). The less than infinite slope of this line shows that the current source cannot keep up with the voltage required when higher voltages are needed to produce the desired current. The resultant reduction in current output with increased voltage requirements can be represented by an internal resistor, and the slope is indicative of its value.

For current sources, the negative slope of the line in the *v*-*i* plot is equal to the internal resistance just as for voltage sources. The circuit diagram of a real current source is shown in Figure 11.24 to be an ideal current source in parallel with an internal resistance. This configuration is often referred to as a *Norton equivalent* circuit. Inspection of this circuit shows how this circuit represents the falloff in output current that occurs when higher voltages are needed to push the current though a large resistor. As the voltage at the source output increases, more current flows though the internal resistor, R_N, and less comes out of the source. If there were no internal resistor, all of the current would have to flow out of the source irrespective of the output voltage, as expected from an ideal current source.

As shown in Figure 11.24, when the output is a short circuit (i.e., $R_L = 0.0\ \Omega$), then the current that flows through R_N is zero, so all the current flows though the output. (Note that analogous to the open-circuit condition for a voltage source, the short-circuit condition produces the least work for a current source—in fact no work at all.) By KCL:

$$\begin{aligned} I_N - i_{R_N} - i_{out} &= 0; \\ i_{out} = i_{sc} &= I_N \end{aligned} \tag{11.14}$$

Hence I_N equals the short-circuit current, i_{sc}. When R_L is not a short circuit, some of the current flows through R_N and i_{out} decreases. Essentially, the internal resistor "steals" current from the current source when the output voltage is anything other than zero. Applying KCL to the upper node of the Norton circuit (Figure 11.24), paying attention to the current directions:

$$\begin{aligned} I_N - i_{R_N} - i_{out} &= 0 \\ I_N - \frac{v_{out}}{R_N} - i_{out} &= 0 \end{aligned} \tag{11.15}$$

Solving for R_N:

$$\frac{v_{out}}{R_N} = I_N - i_{out}; \qquad R_N = \frac{v_{out}}{(I_N - i_{out})}$$

Since

$$I_N = isc: \qquad R_N = \frac{v_{out}}{(i_{sc} - i_{out})} \tag{11.16}$$

When the output of the Norton circuit is an open circuit (i.e., $R_L \rightarrow \infty$), all the current flows through the internal resistor, R_N. Hence:

$$v_{oc} = I_N R_N \tag{11.17}$$

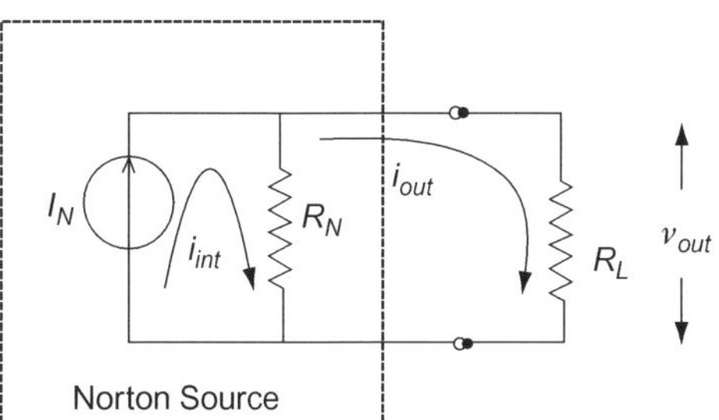

FIGURE 11.24 A circuit diagram of a real current source (inside the dashed rectangle) that is connected to a resistor load. The current source circuit is referred to as a Norton circuit. When a large external load is presented to this circuit, more of the current flows though the internal resistor, R_N. In a high-quality current source, R_N is very large.

Combining this equation with Equation 11.14, we can solve for R_N in terms of the open-circuit voltage, v_{oc}, and the short-circuit current, i_{sc}:

$$v_{oc} = I_N R_N = i_{sc} R_N$$
$$R_N = \frac{v_{oc}}{i_{sc}} \tag{11.18}$$

This relationship is the same as for the Thévenin circuit as given in Equation 11.13 if we equate R_T and R_N.

EXAMPLE 11.10

A real current source produces a current of 500 mA under short-circuit conditions and a current of 490 mA when the short is replaced by a 20 Ω resistor. Find the internal resistance.

Solution: Find v_{out} when the load is 20 Ω, then apply Equation 11.16 to find R_N.

$$v_{out} = R_L i_{out} = 20(.49) = 9.8 \text{ volts}$$
$$R_N = \frac{v_{out}}{i_{sc} - i_{out}} = \frac{9.8}{0.5 - 0.49} = \frac{9.8}{0.01} = 980\ \Omega$$

The Thévenin and Norton circuits have been presented in term of sources, but they can also be used to represent entire networks as well as mechanical and other nonelectrical systems. These representations can be especially helpful when two systems are being connected. Imagine you are connecting two systems and you want to know how the interconnection will affect the behavior of the overall system. If you represent the system serving as a source as a Thévenin or Norton source and also determine the effective input impedance of the system serving as load, you are able to calculate the loss of signal due to the interconnection. The same could be stated for a biological measurement where the biological system is the source and the measurement system is the load. You may not have much control over the nature of the source, but as a biomedical engineer you have some control over the effective impedance of the load (i.e., your biotransducer). These concepts are explored further in Section 11.5.

11.3.4 Thévenin and Norton Circuit Conversion

It is easy to convert between the Thévenin and Norton equivalent circuits, that is, to determine a Norton circuit that has the same voltage-current relationship as a given Thévenin circuit and vice versa. Such conversions allow you to apply KVL to systems with current sources (by converting them to an equivalent voltage source) or to use KCL in systems with voltage sources (by converting them to an equivalent current). Consider the voltage-current relationship shown in the *v-i* plot of Figure 11.25. Since the curve is a straight line, it is uniquely determined by any two points. The horizontal and vertical intercepts, v_{oc} and i_{sc}, are particularly convenient points to use.

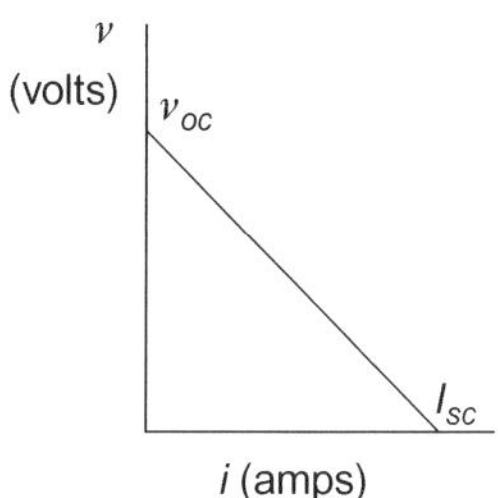

FIGURE 11.25 The v-i plot of the output of a device to be represented as either a Thévenin or Norton circuit. The voltage-current relationship plots as a straight line and can be uniquely represented by two points: v_{oc} and i_{sc}, for example.

The equations for equivalence are easy to derive based on relationships given in Equations 11.13 and 11.14:

For a Thévenin circuit:

$$v_{ocT} = V_T$$

and from Equation 11.12:

$$R_T = \frac{v_{ocT}}{i_{scT}}; \quad i_{scT} = \frac{v_{ocT}}{R_T}$$

For a Norton circuit:

$$I_N = i_{scN}$$

and from Equation 11.18:

$$R_N = \frac{v_{ocN}}{i_{scN}}; \quad v_{ocN} = R_N i_{scN}$$

For a Norton circuit to have the same v-i relationship as a Thévenin, $i_{scN} = i_{scT}$ and $v_{ocN} = v_{ocT}$. Equating these terms in the equations above:

$$I_N = i_{scN} = i_{scT} = \frac{v_{ocT}}{R_T};$$

but

$$v_{ocT} = V_T$$
$$I_N = \frac{V_T}{R_T} \tag{11.19}$$
$$R_N = \frac{v_{ocN}}{i_{scN}} = \frac{v_{ocT}}{i_{scT}};$$

but

$$\frac{v_{ocT}}{i_{scT}} = R_T$$
$$R_N = R_T \tag{11.20}$$

Equations 11.19 and 11.20 show us how to design a Norton circuit using components V_N and R_N that is equivalent to a Thevenin circuit with components V_T and R_T. To go the other way and convert from a Norton to an equivalent Thévenin:

$$V_T = v_{ocT} = v_{ocN};$$

but

$$v_{ocN} = I_N R_N$$
$$V_T = I_N R_N$$
$$R_T = \frac{v_{ocT}}{i_{scT}} = \frac{v_{ocN}}{i_{scN}}; \tag{11.21}$$

but

$$\frac{v_{ocN}}{i_{scN}} = R_N$$
$$R_T = R_N \tag{11.22}$$

These four equations allow for easy conversion between Thévenin and Norton circuits. Note that the internal resistance, R_N or R_T, is the same for either configuration. This is reasonable, since the internal resistance defines the slope of the *v-i* curve, so to achieve the same *v-i* relationship you need the same slope curve and hence the same resistor.

The ability to represent any linear *v-i* relationship by either a Thévenin or a Norton circuit implies that it is impossible to determine whether a real source is in fact a current or voltage source based solely on external measurement of voltage and current. If the *v-i* relationship of a source is more-or-less a vertical line, as in Figure 11.23, indicating a large internal resistance, we might guess that the source is probably a current source. In fact, one simple technique for constructing a crude current source in practice is to place a voltage source in series with a large resistor. Alternatively, if the *v-i* relationship is approximately horizontal as in Figure 11.21, a nonideal voltage source would be a better guess. However, if the *v-i* curve is neither particularly vertical nor horizontal, as in Figure 11.25, it is anyone's guess as to whether it is a current or voltage source and either would be an equally appropriate representation unless other information is available.

Conversion between Thévenin and Norton circuits can be used in order to use nodal analysis in circuits that contain voltage sources or to use mesh analysis in circuits that contain current sources. This application of Thévenin–Norton conversion is shown in the example below.

EXAMPLE 11.11

Find the voltage, V_A, in the circuit shown in Figure 11.26 using nodal analysis.

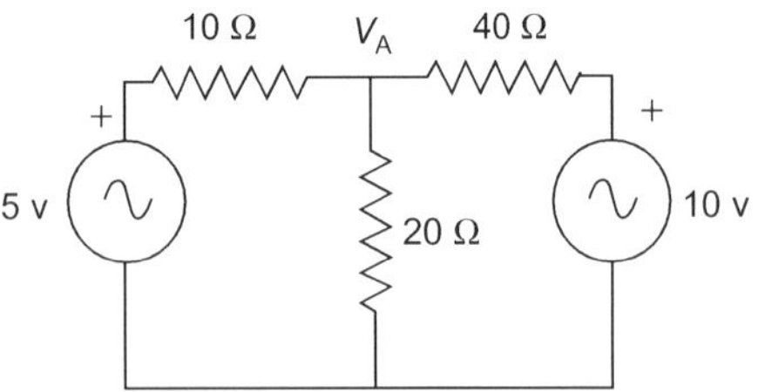

FIGURE 11.26 A two-mesh network containing voltage sources. If the Thévenin circuits on either side are converted to their Norton equivalents, this circuit becomes a single-node circuit and can be evaluated using a single nodal equation.

Solution: This circuit can be viewed as containing two Thévenin circuits: a 5-volt source and 10-Ω resistor, and a 10-volt source and 40-Ω resistor. After converting these two Thévenin circuits to equivalent Norton circuits using Equations 11.19 and 11.20, we apply standard nodal analysis. Figure 11.27 shows the circuit after Thévenin–Norton conversion.

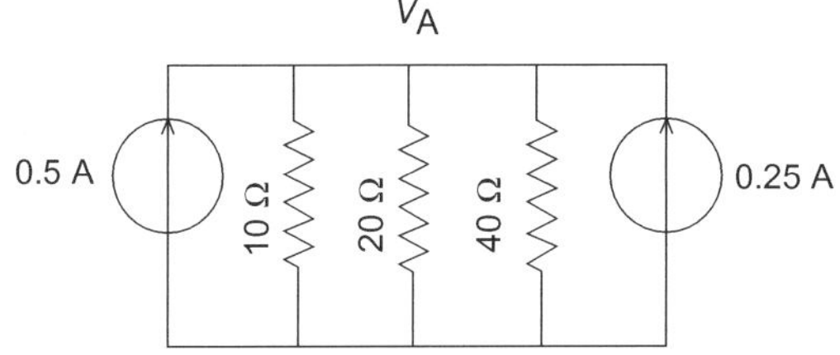

FIGURE 11.27 The network in Figure 11.26 after the two Thévenin circuits have been converted to their Norton equivalents. This is now a single-node network.

Writing the KCL equation around node A.

$$i_{S1} + i_{S2} + i_{R1} + i_{R1} + i_{R3} = 0$$

$$0.5 + 0.25 - \frac{V_A}{R_1} - \frac{V_A}{R_2} - \frac{V_A}{R_3} = 0$$

$$0.5 + 0.25 - V_A\left(\frac{1}{10} + \frac{1}{20} + \frac{1}{40}\right) = 0$$

$$V_A = \frac{0.5 + 0.25}{\frac{1}{10} + \frac{1}{20} + \frac{1}{40}} = \frac{.75}{0.1 + 0.05 + 0.025} = \frac{0.75}{0.175} = 4.29 \text{ volts}$$

The Thévenin and Norton circuits and their inter-conversions are useful for network reduction of circuits that contain sources, as shown in the next section. These concepts also apply to mechanical systems, with appropriate modifications, as illustrated in Section 11.6.

11.4 THÉVENIN AND NORTON THEOREMS—NETWORK REDUCTION WITH SOURCES

The Thévenin theorem states that any network of passive elements and sources can be reduced to a single voltage source and series impedance. Such a reduced network would look like a Thévenin circuit such as that shown in Figure 12.22, except that the internal resistance, R_T, would be replaced by a generalized impedance, $Z\angle\theta$. The Norton theorem makes the same claim for Norton circuits, which is reasonable since Thévenin circuits can easily be converted to Norton circuits via Equations 11.19 and 11.20.

There are a few constraints on these theorems. The elements in the network being reduced must be linear, and if there are multiple sources in the network they must be at the same frequency. As has been done in the past, the techniques for network reduction will be developed using phasor representation and hence will be limited to networks with sinusoidal sources. However, the approach is the same in the Laplace domain.

There are two approaches to finding the Thévenin or Norton equivalent of a general network. One is based on solving for the open-circuit voltage, v_{oc}, and the short-circuit current, i_{sc}. The other method evaluates only the open-circuit voltage, v_{oc}, then determines R_T (or R_N) through network reduction. During network reduction, a source is replaced by its equivalent resistance, that is, short circuits ($R = 0$) substitute for voltage sources, and open circuits ($R \rightarrow \infty$) substitute for current sources. Both network reduction methods are straightforward, but the open-circuit voltage/short-circuit current approach can be implemented on a computer. These approaches are demonstrated in the next two examples.

EXAMPLE 11.12

Find the Thévenin equivalent of the circuit in Figure 11.28 using both the v_{oc}-i_{sc} method and the v_{oc}-network reduction technique.

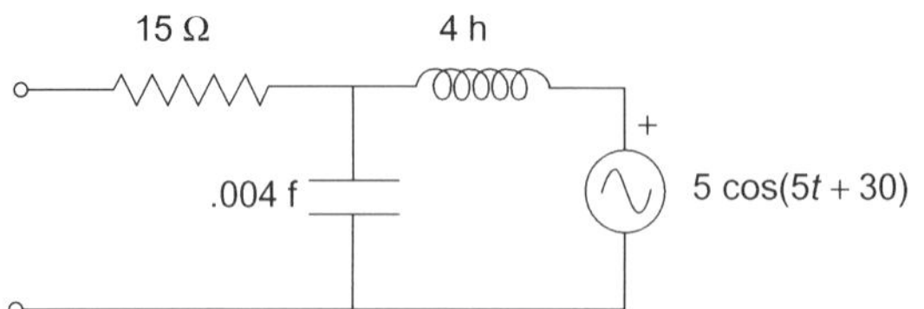

FIGURE 11.28 A network that will be reduced to a single impedance and source using two different strategies.

Solution, v_{oc}-reduction method: First find the open-circuit voltage, v_{oc}, using standard network analysis. Convert all network elements to their phasor representation as shown in Figure 11.28.

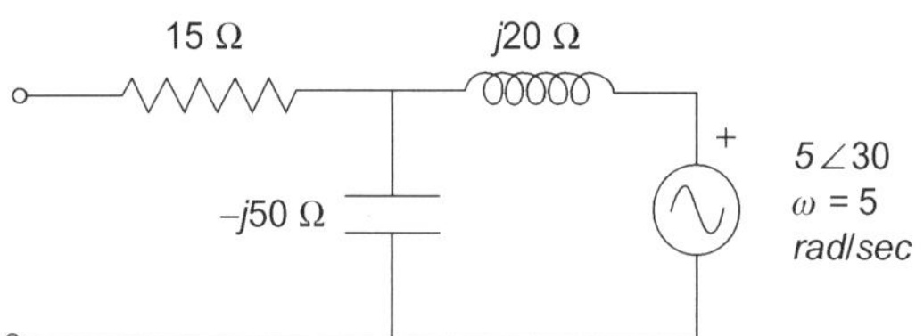

FIGURE 11.29 The network of Figure 11.28 after conversion to the phasor domain.

Since in the open-circuit case no current flows through the 15 Ω resistor, there is no voltage drop across this resistor, so the open-circuit voltage is the same as the voltage across the capacitor. The resistor is essentially not there with respect to open-circuit voltage. The resistor is not totally useless: It does play a role in determining the equivalent impedance and is involved in the calculation of short-circuit current. The open-circuit voltage, the voltage across the capacitor, can be determined by writing the mesh equation around the loop consisting of the capacitor, inductor, and source. Using the usual directional conventions, defining the mesh current as clockwise and going around the loop in a clockwise direction, note that the voltage source is negative since there is a voltage drop going around in the clockwise direction.

$$-5\angle 30 - i(-j50 + j20) = 0$$

$$i = \frac{-5\angle 30}{-j50 + j20} = \frac{-5\angle 30}{-j30} = \frac{-5\angle 30}{30\angle -90} = -0.167\angle 120 \text{ amps}$$

$$v_{oc} = iZ_C = -0.167\angle 120(-j50) = -0.167\angle 120(50\angle -90)$$

$$v_{oc} = -8.35\angle 30 \text{ volts}$$

Note that the magnitude of the Thévenin equivalent voltage is actually larger than the source voltage. This is the result of a partial resonance between the inductor and capacitor. Also the voltage is given as negative based on the clockwise direction of current flow. With respect to the upper terminal the voltage will be positive.

Next, find the equivalent impedance by reduction. To reduce the network, essentially turn off the sources and apply network reduction techniques to what is left. Turning off a source does not mean you remove it from the circuit; to turn off a source, you replace it by its equivalent resistance. For an ideal voltage source $R_T \to 0\ \Omega$: so the equivalent resistance of an ideal voltage source is $0\ \Omega$: that is, a short circuit. To turn off a voltage source, you replace it by a short circuit. After you replace the source by a short circuit, you are left with the network shown in Figure 11.30 on the left-hand side. Series–parallel reduction techniques will work for this network.

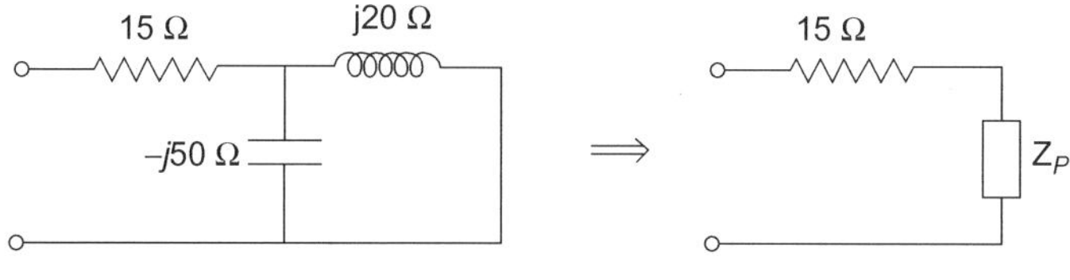

FIGURE 11.30 *Left*: The network given in Figure 11.29 after effectively "turning off" the voltage source. The voltage source is replaced by its equivalent resistance which, since it is an ideal source, is $0.0\ \Omega$. *Right*: The network after combining the two parallel impedances into a single impedance, Z_P.

After replacing the voltage source by a short circuit, we are left with a parallel combination of inductor and capacitor. This combines to a single impedance, Z_P:

$$Z_P = \frac{Z_C Z_L}{Z_C + Z_L} = \frac{-j50(j20)}{-j50 + j20} = \frac{1000}{-j30} = \frac{1000}{30\angle -90}$$

$$Z_P = 33.33\angle 90\ \Omega$$

We are then left with the series combination of Z_P and the $15\ \Omega$ resistor:

$$Z_T = 15 + 33.33\angle 90 = 15 + j33.33 = 36.55\angle 65.8\ \Omega$$

Hence, the original circuit can be equivalently represented by the Thévenin circuit shown in Figure 11.31

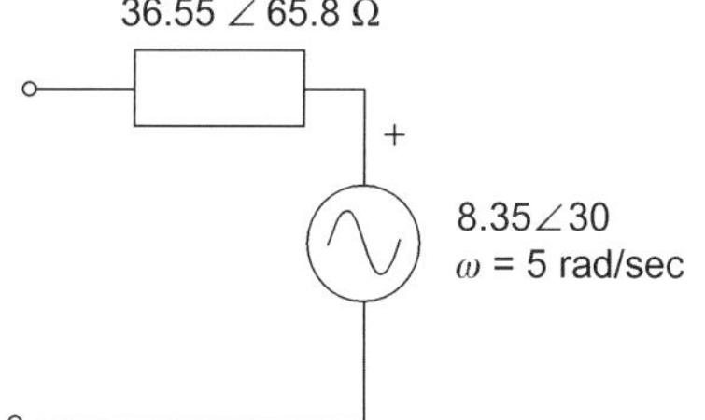

FIGURE 11.31 The Thévenin equivalent of the network given in Figure 11.28 determined by the v_{oc}-reduction method. The sign on the voltage source is positive with respect to the upper terminals.

Solution, v_{oc}-i_{sc}: In this method, we solve for the open-circuit voltage and short-circuit current. We have already found the open-circuit voltage above, so it is only necessary to find the short-circuit current. After shorting out the output and converting to phasor notation, the circuit is shown in Figure 11.32.

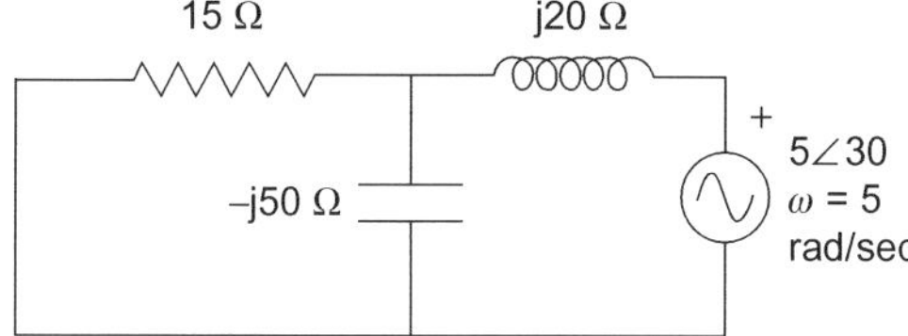

FIGURE 11.32 The circuit shown in Figure 11.28 after shorting the output terminals and converting to phasor notation.

If we solve this using mesh analysis, it is a two-mesh circuit, but if we convert the inductor–source combination to a Norton equivalent it becomes a single-node equation. To implement the conversion, use Equations 11.19 and 11.20:

$$I_N = \frac{V_T}{R_T} = \frac{5\angle 30}{j20} = \frac{5\angle 30}{20\angle 90} = 0.25\angle -60 \text{ volts;} \quad R_N = R_T = j20\ \Omega$$

The new network is shown in Figure 11.33.

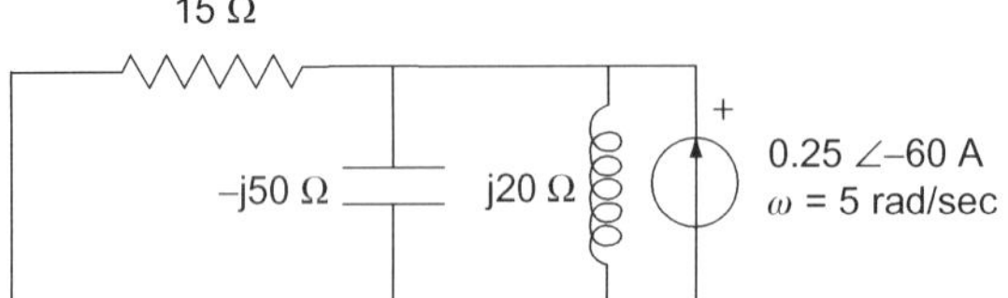

FIGURE 11.33 The circuit of Figure 11.32 after the voltage source and series impedance are converted to a Norton equivalent source. This makes this a one-node problem that can be solved from a single equation.

$$.25\angle -60 - v\left(\frac{1}{15} + \frac{1}{-j50} + \frac{1}{j20}\right) = 0$$

$$v = \frac{.25\angle -60}{\frac{1}{15} + \frac{1}{-j50} + \frac{1}{j20}} = \frac{.25\angle -60}{0.0667 + j0.02 - j0.05} = \frac{.25\angle -60}{0.0667 - j0.03}$$

$$v = \frac{.25\angle -60}{0.073\angle -24.2} = 3.42\angle -35.8 \text{ volts}$$

$$i_{sc} = \frac{v}{15} = \frac{3.42\angle -35.8}{15} = 0.228\angle -35.8 \text{ amps}$$

Now solve for R_T:

$$R_T = \frac{v_{oc}}{i_{sc}} = \frac{8.35\angle 30}{0.228\angle -35.8} = 36.6\angle 65.8\ \Omega$$

This is the same value for R_T found using the v_{oc}-reduction method above. More complicated networks can be reduced using MATLAB as shown in the example below.

EXAMPLE 11.13

Find the Norton equivalent of the circuit of Figure 11.34 with the aid of MATLAB.

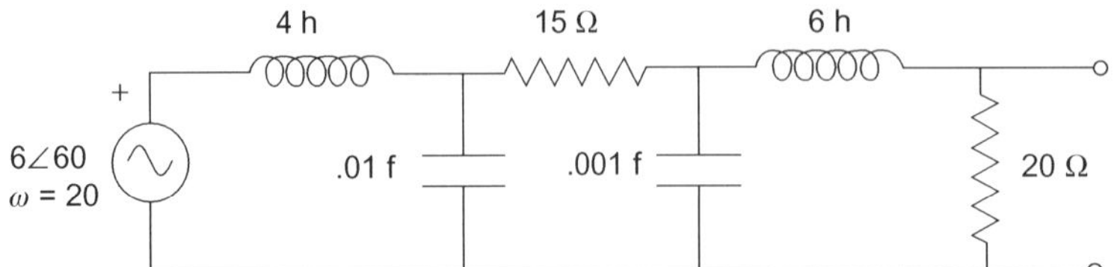

FIGURE 11.34 A complicated network that is reduced to a Norton equivalent in Example 11.13. To ease the mathematical burden, MATLAB is used to solve the equation.

Solution: The open-circuit voltage and short-circuit current can be solved directly using mesh analysis in conjunction with MATLAB. In fact, the mesh equations in both cases (solving for v_{oc} or i_{sc}) are similar. The only difference is that when solving for the short-circuit current, the 20-Ω resistor will be short circuited and not appear in the impedance matrix.

First convert to phasor notation, then encode the network directly into MATLAB. Since we are using MATLAB and the computational load is reduced, we keep ω as a variable in case we want to find the Norton equivalent for other frequencies.

After converting to phasor notation and assigning the mesh current, the circuit is as shown in Figure 11.35. Note that the open-circuit voltage, v_{oc}, is the voltage across the 20 Ω resistor and the short-circuit current, i_{sc}, is just i_3 when the resistor is shorted.

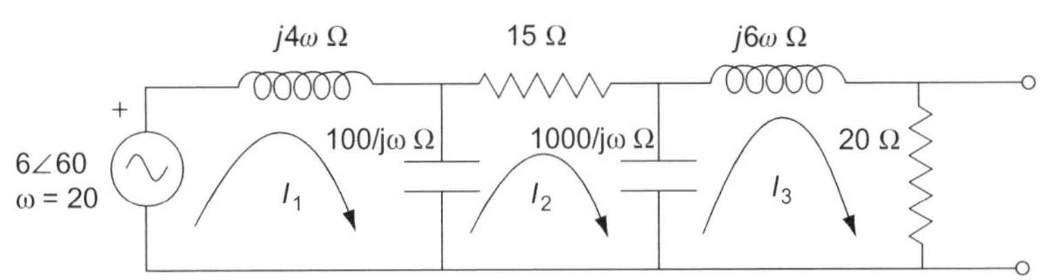

FIGURE 11.35 The circuit of Figure 11.34 after conversion to phasor notation. Since mesh analysis will be used, the mesh currents are shown.

Writing the KVL equations for the open-circuit condition:

$$\begin{vmatrix} 6\angle 60 \\ 0 \\ 0 \end{vmatrix} = \begin{vmatrix} j4\omega + \dfrac{100}{j\omega} & -\dfrac{100}{j\omega} & 0 \\ -\dfrac{100}{j\omega} & 15 + \dfrac{100+1000}{j\omega} & -\dfrac{1000}{j\omega} \\ 0 & -\dfrac{1000}{j\omega} & 20 + j6\omega + \dfrac{1000}{j\omega} \end{vmatrix} \begin{vmatrix} i_1 \\ i_2 \\ i_3 \end{vmatrix}$$

where: $v_{oc} = 20\ i_3$.

The mesh equation for the short-circuit condition is quite similar:

$$\begin{vmatrix} 6\angle 60 \\ 0 \\ 0 \end{vmatrix} = \begin{vmatrix} j4\omega + \dfrac{100}{j\omega} & -\dfrac{100}{j\omega} & 0 \\ -\dfrac{100}{j\omega} & 15 + \dfrac{100+1000}{j\omega} & -\dfrac{1000}{j\omega} \\ 0 & -\dfrac{1000}{j\omega} & j6\omega + \dfrac{1000}{j\omega} \end{vmatrix} \begin{vmatrix} i_1 \\ i_2 \\ i_3 \end{vmatrix}$$

where: $i_{sc} = i_3$.

The program to solve these equations and find I_N and R_N is shown below:

```
% Example 11.13 To find the Norton equivalent of a three-mesh circuit
%
w = 20;                              % Define frequency
theta = 60*2*pi/360;
```

```
VS = 6*cos(theta) + 6*sin(theta)*j;  % Define Vs as rectangular
v = [VS 0 0]';                       % Define voltage vector
%
% Define open-circuit impedance matrix
Zoc = [4j*w + 100/(j*w), -100/(j*w), 0; -100/(j*w), 15 + 1100/(j*w),...
   -1000/(j*w); 0, -1000/(j*w), 20 + 6j*w + 1000/(j*w)];
ioc = Zoc\v;                         % Solve for currents
voc = 20*ioc(3);                     % and open-circuit voltage
%
% Define short-circuit impedance matrix and solve for short-circuit current
Zsc = [4j*w + 100/(j*w), -100/(j*w), 0; -100/(j*w), 15 + 1100/(j*w),...
   -1000/(j*w); 0, - 1000/(1j*w), 6j*w + 1000/(j*w)];
i = Zsc\v                            % Solve for currents
isc = i(3);                          % Find isc
Zn = voc/isc;                        % Solve for ZN (Equation 11.18)
%
% Output magnitude and phase of IN and RN
IN_mag = abs(isc)
IN_phase = angle(isc)*360/(2*pi)
ZN_mag = abs(Zn)
ZN_phase = angle(Zn)*360/(2*pi)
```

This program produces the following outputs:

$I_N = 0.0031\angle 20.6$ amp

$Z_N = 19.36\angle 9.8\ \Omega$

It is easy to modify this program to find, and plot, the Norton element values over a range of frequencies. This exercise is given in Problem 14.

11.5 MEASUREMENT LOADING

We now have the tools to analyze the situation when two systems are connected together. For biomedical engineers not involved in electronic design, this situation most frequently occurs when making a measurement, so we analyze the problem in that context. However, the approach followed here applies to any situation when two processes are connected together.

11.5.1 Ideal and Real Measurement Devices

One of the important tasks of biomedical engineers is to make measurements, usually on a living system. Any measurement requires withdrawing some energy from the system of interest and that, in turn, alters the state of the system and the value of the measurement. This alteration is referred to here as "measurement loading." The word "load" is applied to any device that is attached to a system of interest, and "loading" is the influence the attached device has on the system. So measurement loading is the influence a measurement device has on the system being measured. This well-known phenomenon extends down to the smallest systems and has a significant impact on fundamental principles in particle physics. The concepts developed above can be used to analyze the effect of a measurement device on a given system. In fact, such an evaluation of measurement loading is one of the major applications of Thévenin and Norton circuit models.

Just as there are ideal and real sources, there are ideal and real measurement devices or, equivalently, ideal and real loads. An ideal voltage source supplies a given voltage at any current, and an ideal current source supplies a given current at any voltage. Since power is the product of voltage times current ($P = vi$), an ideal source can supply any amount of energy or power required: infinite if need be. Ideal measurement devices (or ideal loads) have the opposite characteristics: They can make a measurement without drawing any energy or power from the system being measured. Of course, we know from basic physics that such an idealization is impossible, but some measurement devices can provide nearly ideal measurements, at least for all practical purposes.

The goal in practical situations is to be able to make a measurement without significantly altering the system being measured. The ability to attain this goal depends on the characteristics of the source as well as the load; a given device might have little effect on one system providing a reliable measurement, yet significantly alter another system giving a measurement that does not reflect the underlying conditions. It is not just a matter of how much energy a measurement device requires (i.e., load impedance), but how much energy the system being measured can supply without significant change (i.e., source impedance).

Just as ideal voltage and current sources have quite different properties, ideal measurement devices for voltage and current differ significantly. A device that measures voltage is a *voltmeter*. An ideal voltmeter draws no power from the circuit being measured. Since $P = vi$, and v cannot be zero (that is what is being measured), an ideal voltmeter must draw no current while making the measurement. The current is zero for any voltage only if the equivalent resistance of the voltmeter is infinite. An ideal voltmeter is effectively an open circuit, and the v-i plot is a vertical line. Practical voltmeters do not have infinite resistances, but they do have very large impedances, of the order of hundreds of megohms (1 megohm = 1 MΩ = 10^6 Ω), and can be considered ideal for all but the most challenging conditions. The characteristics of ideal sources and loads are summarized in Table 11.1.

Current measuring devices are called *ammeters*. An ideal ammeter also needs no power from the circuit to make its measurement. Again, since $P = vi$, and i cannot be zero in an ammeter, voltage must be zero if no energy is to be drawn from the system being

TABLE 11.1 Electrical Characteristics of Ideal Sources and Loads

	Sources		Measurement Devices (Loads)	
Characteristics	**Voltage**	**Current**	**Voltage**	**Current**
Impedance	0.0 Ω	∞ Ω	∞ Ω	0.0 Ω
Voltage	V_S	up to ∞ v	$V_{measured}$	0.0 v
Current	up to ∞ A	I_S	0.0 ∞	$I_{measured}$

measured. This means that an ideal ammeter is effectively a short circuit having an equivalent resistance of 0.0 Ω, Table 11.1. The *v-i* plot of an ideal ammeter would be a horizontal line. Practical ammeters are generally not even close to ideal, having resistances approaching a tenth of an ohm or more. However, current measurements are rarely made in practice because that involves breaking a circuit connection to make the measurement unless special "clip-on" ammeters are used.

An illustration of the error caused by a less-than-ideal ammeter is given in the example below.

EXAMPLE 11.14

A practical ammeter having an internal resistance of 2 Ω is used to measure the short-circuit current of the three-mesh network used in Example 11.13. How large is the error, that is, how much does the measurement differ from the true short-circuit current?

Solution: As with all issues of measurement loading, it is easiest to use the Thévenin or Norton representation of the system being loaded. The Norton equivalent of the three-mesh circuit is determined in Example 11.13 and is shown in Figure 11.36 loaded by the ammeter. From the Norton circuit, we know the true short-circuit current is: $i_{sc} = I_N = 0.0031\angle 20.6$ amp. The measured short-circuit current can be determined by applying nodal analysis to the circuit.

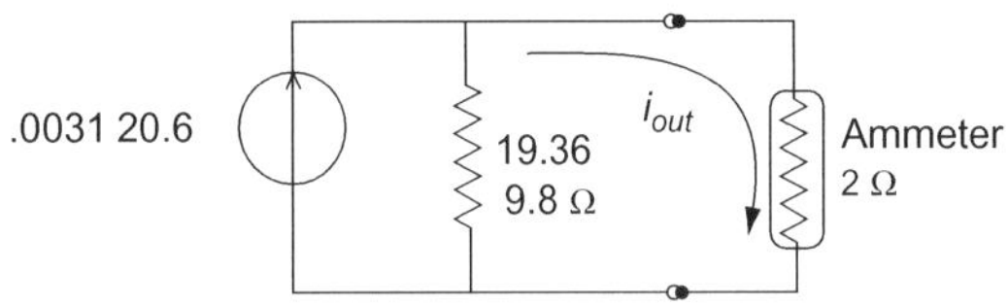

FIGURE 11.36 The output current of the Norton equivalent of the circuit analyzed in Example 11.13, measured by a less-than-ideal ammeter. Rather than having an internal resistance of 0.0 Ω, the ammeter has an internal resistance of 2 Ω. The effect of this nonzero internal resistance of the ammeter is determined in Example 11.14.

Applying KCL:

$$.0031\angle 20.6 - v\left(\frac{1}{19.36\angle 9.8} + \frac{1}{2}\right) = 0$$

$$v = \frac{.0031\angle 20.6}{\dfrac{1}{19.36\angle 9.8} + \dfrac{1}{2}} = \frac{.0031\angle 20.6}{.052\angle -9.8 + 0.5} = \frac{.0031\angle 20.6}{.051 - j.009 + 0.5} = \frac{.0031\angle 20.6}{0.55\angle -0.9}$$

$$v = 0.0056\angle 21.5 \text{ volts}$$

$$i_{sc\_measured} = \frac{v}{R} = \frac{.0056\angle 21.5}{2} = .0028\angle 21.5 \text{ amps}$$

The measured short-circuit current is slightly less than the actual short-circuit current, which is equal to I_N. The error is:

$$Error = \frac{(.0031 - .0028)}{.0031}100 = 9.7\%$$

The difference in the measured current and the actual current is due to some current flowing through the internal resistor. Since the external load is much (an order of magnitude) less than the internal resistor, the current taking the internal pathway is small. So the smaller the external impedance is with respect to the internal impedance, the closer the measurement to the ideal. In this case, the load resistor is approximately one-tenth the value of the internal resistor, leading to an error of approximately 10%. For some measurements, this may be sufficiently accurate. Usually a ratio of one to 100 (i.e., two orders of magnitude) between source and load impedances is adequate for the type of accuracy required in biomedical engineering measurements.

The same rule of thumb can be used for voltage measurements, except now the load resistance should be much greater than the internal resistance (or impedance). In voltage measurements, if the load impedance is 100 or more times the source impedance, the loading usually can be considered negligible and the measurement sufficiently accurate.

These general rules also apply whenever one network is attached to another. If voltages carry the signal (usually the case), the influence of the second network on the first can be ignored if the effective input impedance of the second network is much greater than the effective output impedance of the first network, Figure 11.37.

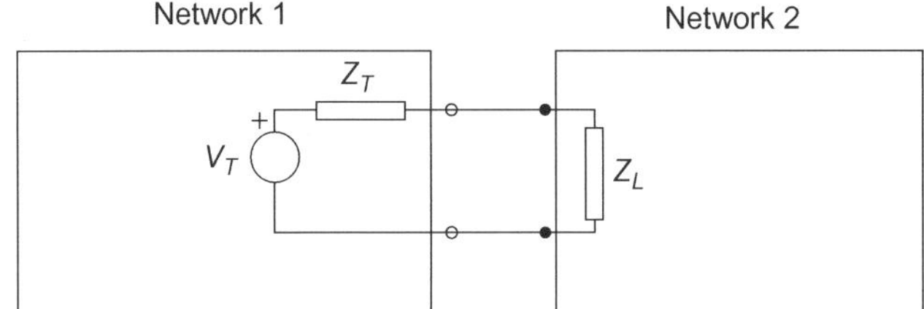

FIGURE 11.37 The input of Network 2 is connected to the output of Network 1. If the equivalent input impedance of Network 2, Z_L, is much greater than the output impedance of Network 1, Z_T, the output of Network 1 will not be significantly altered by the connection to Network 2.

This means that if in Figure 11.37 $Z_L >> Z_T$, the transfer functions derived for each of the two networks independently can be taken as unchanged when the two are interconnected. This comes from the two basic assumptions used in the derivations of the transfer function: that the input is an ideal source and that the output is an ideal load ($Z_L \rightarrow \infty$). If $Z_L >> Z_T$ (say, 100 times), these assumptions are reasonably met and the original transfer functions are valid. Then the transfer function of the combined network can be taken as $TF_1(\omega)*TF_2(\omega)$. Of course the input to Network 1 and the load on Network 2 could still present problems that would depend on the relative input and output impedances connected to the combined network. An analysis of voltage loading is explored in several problems.

If the signal is carried as a current, then the opposite would be true. The output impedance of Network 1 should be much greater than the input impedance of Network 2; that is, $Z_L << Z_T$. In this situation, the current loading of Network 2 can be considered negligible with respect to Network 1. Signals are rarely viewed in terms of currents except for the output of certain transducers, particularly those that respond to light, and in these cases the signal is converted to a voltage by a special amplifier circuit. (See Chapter 12, Section 12.7.4.)

What if these conditions are not met; for example, if Z_L, were approximately equal to Z_T? In this case, you have two choices: Calculate the transfer function of the two-network combination, or estimate the error that will occur due to the interconnection, as in the last example. Sometimes you actually want to increase the load on a system, purposely making Z_L close to the value of Z_T. The motivation for such a strategy is explained in the next section.

11.5.2 Maximum Power Transfer

The goal in most measurement applications is to extract minimum energy from the system. This is also the usual goal when one system is connected to another. Assuming voltage signals, the load impedance should be much greater than the internal impedance of the source, or vice versa if the signal is carried by current. This results in the maximum voltage (or current), and minimum energy, out of the source. But what if the goal is to extract maximum energy from the system? To determine the conditions for maximum power out of the system, we consider the Thévenin circuit with its load resistor in Figure 11.38.

To address this question, we assume that R_T is part of a source and cannot be adjusted; that is, it is an internal (and therefore inaccessible) property of the source from which we are trying to extract the maximum power. If only R_L can be adjusted, the question becomes, what is the value of R_L that will extract maximum power from the system? The power out of the system is: $P = v_{out}\, i_{out}$ or $P = R_L\, i_{out}^2$. To find the value of the load resistor R_L that will deliver the maximum power to itself, we use the standard calculus trick for maximizing a function: Solve for P in terms of R_L, then take dP/dR_L and set it to zero:

$$P = R_L i^2; \quad i = \frac{V_T}{(R_L + R_T)};$$

hence,

$$P = \frac{R_L V_T^2}{(R_L + R_T)^2}$$

Solving for $\frac{dP}{dR_L}$ by parts:

$$\frac{dP}{dR_L} = \frac{V_T^2(R_L + R_T)^2 - 2V_T^2 R_L(R_L + R_T)}{(R_L + R_T)^4} = 0$$

$$V_T^2(R_L + R_T)^2 = 2V_T^2 R_L(R_L + R_T)$$

$$R_L + R_T = 2R_L;$$

$$R_L = R_T \tag{11.23}$$

So for maximum power out of the system, R_L should equal R_T (or, more generally, $Z_L = Z_T$), a condition known as *impedance matching*. Since the power in a resistor is proportional to the resistance (i^2R) and the two resistors are equal, the power transferred to R_L will be half the total power (the other half is dissipated by R_T).

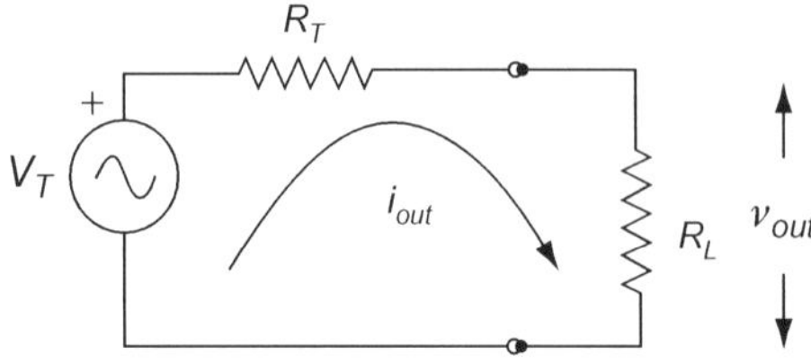

FIGURE 11.38 A Thévenin circuit is shown with a load resistor, R_L. For minimum power out of the system, R_L should be much greater than R_T. That way output current approaches 0.0, as does power. In this section we seek to determine the value of R_L that will extract maximum power from the Thévenin source.

Equation 11.23 is known as the *maximum power transfer theorem*. Using this theorem, it is possible to find the value of load resistance that extracts maximum power from any network. Just convert the network to a Thévenin equivalent and set R_L equal to R_T. Recall that the maximum power transfer theorem applies when R_T is fixed and R_L is varied. If R_T can be adjusted, then just by inspection of Figure 11.38 we can see that maximum power will be extracted from the circuit when $R_T = 0$. When sinusoidal or other signals are involved, Equation 11.23 extends directly to impedances; however the reactive part of the load impedance should have the opposite sign as the source impedance. Mathematically, the source and load impedances should be complex conjugates.

11.6 MECHANICAL SYSTEMS

All of the concepts described in this chapter are applicable to mechanical. The concepts of equivalent impedances and impedance matching are often used in mechanical systems, particularly in acoustic applications. Of particular value are the concepts of real and ideal sources and real and ideal loads or measurement devices. As mentioned in Chapter 9, an ideal force generator produces a specific force irrespective of the velocity, just as an ideal voltage source produces the required voltage at any current. An ideal force generator will generate the same force at 0.0 velocity, or 10 mph, or 10,000 mph, or even at the speed of light (clearly impossible), if necessary. The force produced by a real force generator will decrease with velocity. This can be expressed in a force–velocity plot (analogous to the *v-i* plot) as shown in Figure 11.39A. An ideal velocity (or displacement) generator will produce a specific velocity against any force, be it 1 oz or 1 ton, but the velocity produced by a real velocity generator will decrease as the force against it is increased (Figure 11.39B). Again there can be an ambiguity differentiating between real force and velocity generators. The device producing the force-velocity curve shown in Figure 11.39C could be interpreted as a nonideal force generator or a nonideal velocity generator: It is not possible to determine its true nature from the force-velocity plot.

Ideal measurement devices follow the same guiding principle in mechanical systems: They should extract no energy from the system being measured. For a force-measuring device, a *force transducer*, velocity must be zero and position a constant. A constant

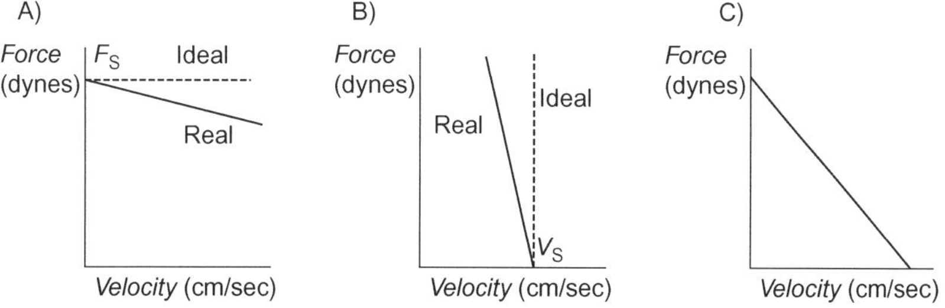

FIGURE 11.39 A) The force-velocity plot of an ideal (dashed line) and a real (solid line) force generator. The force produced by a real generator decreases the faster it must move to generate that force. B) The force-velocity plot of an ideal (dashed line) and a real (solid line) velocity generator. The velocity produced by a real generator decreases as the strength of the force that opposes its movement increases. C) The force-velocity plot of a generator that may be interpreted as either a nonideal force or velocity generator.

position condition is also known as an *isometric* condition. So an ideal force transducer requires no movement to make its measurement—it appears as a solid, immobile wall to the system being measured. Since mechanical impedance is defined as F/v in Equation 9.65, if v is zero for all F then the mechanical impedance must be infinite. For an ideal velocity transducer, the force required to make a measurement is zero. If F is zero for all v, then the mechanical impedance must be zero. The characteristics of ideal mechanical sources and loads are given in Table 11.2 in a fashion analogous to the electrical characteristics in Table 11.1.

The concept of equivalent impedances and sources is useful in determining the alteration produced by a nonideal measurement device or load. The analog of Thévenin and Norton equivalent circuits can also be constructed for mechanical systems. The mechanical analog of a Thévenin circuit is a force generator with a parallel impedance (the configuration is reversed in a mechanical system), as shown in Figure 11.40A, while the mechanical analog of a Norton circuit is a velocity (or displacement) generator in series with the equivalent impedance, shown in Figure 11.40B. To find the values for either of the two equivalent mechanical systems in Figure 11.40, we use the analog of the v_{oc}-i_{sc} method, but we call it the F_{iso}-$v_{no\ load}$ method: find isometric force, the force when velocity is zero (position is constant), and the unloaded velocity, the velocity when no force is applied to the system.

Most practical lumped-parameter mechanical systems are not so complicated and do not usually require reduction to a Thévenin- or Norton-like equivalent circuit. Nonetheless, all of the concepts regarding source and load impedances developed previously apply to mechanical systems. If the output of the mechanical system is a force, the minimum load is one with a large equivalent mechanical impedance, a load that tends to produce a large opposing force and allows very little movement. A solid wall is an example of high mechanical impedance. Conversely, if the output of the mechanical system is a velocity, the minimum load is produced by a system having very little opposition to

TABLE 11.2 Mechanical Characteristics of Ideal Sources and Loads

	Sources		Measurement Devices (Loads)	
Characteristics	**Force**	**Velocity**	**Force**	**Velocity**
Impedance	0.0 dynes/cm	∞ dynes/cm	∞ dynes/cm	0.0 dynes/cm
Force	F_S	up to ∞ dynes	$F_{measured}$	0.0 dynes
Velocity	up to ∞ cm/sec	V_S	0.0 cm/sec	$V_{measured}$

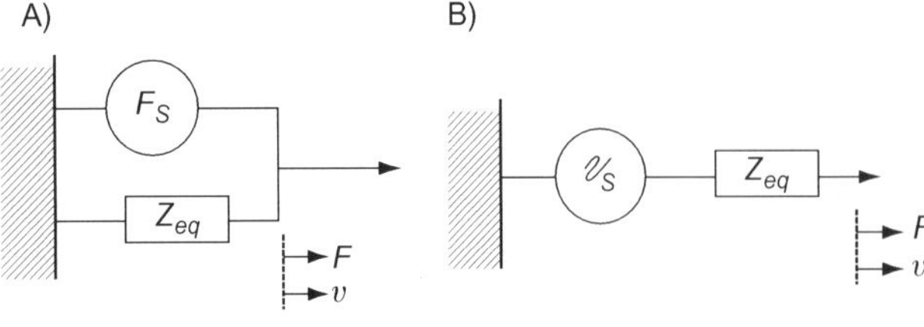

FIGURE 11.40 A) A mechanical analog of a Thévenin equivalent circuit consisting of an ideal force generator with a parallel mechanical impedance. B) The mechanical analog of a Norton circuit consisting of an ideal velocity generator with series mechanical impedance. These configurations can be used to determine the effect of loading by a measurement device or by another mechanical system.

movement, thus exhibiting small mechanical impedance. Air, or better yet a vacuum, is an example of a low impedance mechanical system. Finally, if the goal is to transfer maximum power from the source to the load, their mechanical impedances should be complex congugates. These principles are explored in the examples below.

EXAMPLE 11.15

The mechanical elements of a real force generator are shown on the left side of Figure 11.41 while the mechanical elements of a real force transducer are shown on the right-hand side.

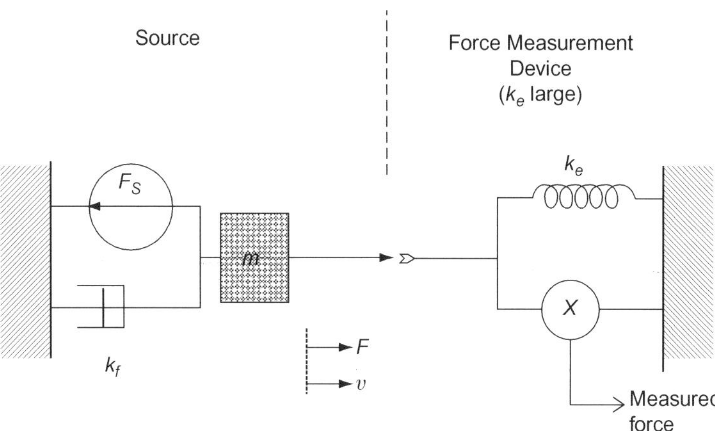

FIGURE 11.41 A real force generator on the left is connected to a real force transducer on the right. These combined systems are used to examine the effect of loading on a real force generator in Example 11.15.

The left side shows a real force generator consisting of an ideal force generator (F_S) in parallel with friction and mass. The right side is a real force transducer consisting of a displacement transducer (marked X) with a parallel spring. This transducer actually measures the displacement of the spring which is proportional to force ($F = k_e\, X$).

Find the force that would be measured by an ideal force transducer, that is, the force at the interface that would be produced by the generator if the velocity were zero. The force transducer actually measures displacement of the spring (the usual construction of a force transducer). What is the force that would be measured by this nonideal transducer? How could this transducer be improved to make a measurement with less error? The system parameters are:

$$k_f = 20 \text{ dyne sec/cm};\quad \text{m} = 5 \text{ gms};\quad k_e = 2400 \text{ dynes/cm};\quad F_S = 10\sin(4t)$$

Solution: To find the force measured by an ideal force transducer, write the equations for the force generator and set velocity to zero as would be the case for an ideal force transducer. Note that F_S is negative based on its defined direction (the arrow pointing to the left).

$$F_{measured} = -F_S - v(j\omega m + k_f)$$

if

$$v = 0$$
$$F_{measured} = -F_S$$

As expected, the force measured by an ideal force transducer is just the force of the ideal component of the source. If the measurement device or load is ideal, then is does not matter what the internal impedance is, the output of the ideal component in the source is measured at the output. However, when the measurement transducer is attached to a less-than-ideal transducer, the velocity is no longer zero and some of the force is distributed to the internal impedance.

To determine the force out of the force generator under the nonideal load, write the equations for the combined system. Since there is only one independent velocity in the combined system, only a single equation will be necessary, but we need to pay attention to the signs.

$$-F_S - v\left(j\omega m + k_f + \frac{k_e}{j\omega}\right) = 0$$

$$-10 - v\left(j4(5) + 20 + \frac{2400}{j4}\right) = -10 - v(j20 + 20 - j600) = 0$$

$$v = \frac{-10}{20 - j580} = \frac{-10}{580.3\angle -88} = -17.2 \times 10^{-3}\angle 88 \text{ cm/sec}$$

$$F_{measured} = vZ_{k_e} = -17.2 \times 10^{-3}\angle 88(-j600) = -10.3\angle -2 \text{ dynes}$$

Hence the measured force is 3% larger than the actual force. The measured force is larger because of a very small resonance between the mass in the force generator and the elastic element in the transducer. Note that the elastic element is very stiff ($k_e = 2400$ dynes/cm) in order to make the transducer a good force transducer: The stiffer the elastic element, the closer the velocity will be to zero. To improve the measurement further, this element could be made even stiffer. For example, if the elasticity, k_e, is increased to 9600 dynes/cm, the force measured becomes $10.08\angle -0.5$, reducing the error to less than a percent. The determination of improvement created by different transducer loads is presented in Problem 18.

The next example addresses the measurement of a velocity generator.

EXAMPLE 11.16

The mechanical elements of a real velocity generator are shown on the left side of Figure 11.42. The left side shows a real velocity generator consisting of an ideal velocity generator (V_S) in series with a friction element and an elastic element. The right side is a real velocity transducer consisting of an ideal velocity transducer (marked X, its output is proportional to v_2) with a parallel spring. Assume V_S is a sinusoid with $\omega = 4$ rad/sec and $k_f = 20$ dyne sec/cm; $k_{e1} = 20$ dynes/cm; $k_{e2} = 2$ dynes/cm.

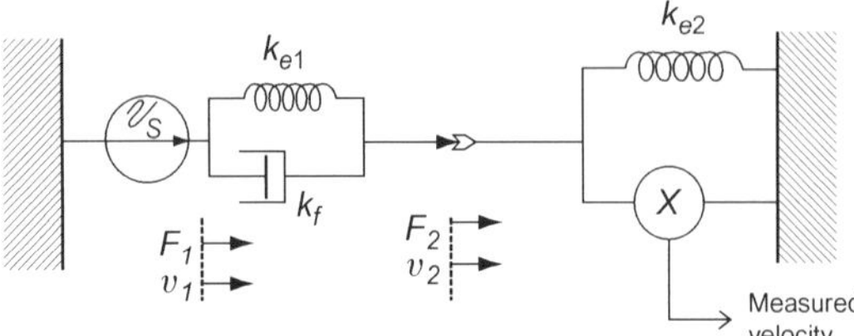

FIGURE 11.42 A realistic velocity generator is connected to a realistic velocity transducer. This configuration is used in Example 11.16 to examine the difference between the measured velocity and that produced by the ideal source.

Find the velocity that is measured by the velocity transducer on the right side of the figure. Note that this transducer is the same as the force transducer except the elasticity is very much lower. Improvement in the accuracy of this transducer is given in Problem 19.

Solution: The solution proceeds in exactly the same manner as in the previous problem except that now there are two velocity points. However, since velocity v_1 is equal to V_S, it is not an independent variable so only one equation needs to be solved.

Writing the equation for the sum of forces equation about the point labeled F_2 and v_2:

$$-\left(\frac{k_{e1}}{j\omega}+k_f\right)(v_2-v_1)-\left(\frac{k_{e2}}{j\omega}\right)v_2=0$$

Substituting $v_1 = V_S$ and values for k_{e1}, k_{e2} and k_f:

$$-\left(\frac{20}{j4}+20\right)(v_2-V_S)-\left(\frac{2}{j4}\right)v_2=(j5-20)(v_2-V_S)+j0.5(v_2)=0$$

$$(-20+j5.5)v_2-(-20+j5)V_S=0$$

Solving for v_2

$$v_2=\frac{(-20+j5)V_S}{-20+j5.5}=\frac{(20.6\angle 165.9)V_S}{20.7\angle 164.6}=(0.995\angle 1.3)V_S\ \text{cm/sec}$$

The measured value of v_2 is very close to that of V_S. That is the value of v_2 that would have been measured by an ideal velocity transducer, one that produced no resistance to movement. The measurement error is small because the impedance of the transducer, while not zero, is still much less than that of the source (i.e., $k_{e2} = 0.1\ k_{e1}$). The measurement elasticity provides the only resistance to movement in the transducer, so if it is increased, the error increases and if it is reduced, the error is reduced. The influence of transducer impedance is demonstrated in several of the problems.

In some measurement situations, it is better to match the impedance of the transducer with that of the source. Matching mechanical impedances is particularly important in ultrasound imaging. Ultrasound imaging uses a high-frequency (1 MHz and up) pressure pulse wave that is introduced into the body and reflects off of various internal surfaces. The time-of-flight for the return signal is used to estimate the depth of a given surface. Using a scanning technique, a large number of individual pulses are directed into the body at different directions and a two-dimensional image is constructed. Because the return signals can be very small, it is important that the maximum energy be obtained from the return signals. The following example illustrates the advantage of matching acoustic (i.e., mechanical) impedances in ultrasound imaging.

EXAMPLE 11.17

An ultrasound transducer that uses a barium titanate piezoelectric device is applied to the skin. The transducer is round with a diameter of 2.5 cm. Use an acoustic impedance of 24.58×10^6 kg-sec/m^2 for barium titanate and an acoustic impedance of 1.63×10^6 kg-sec/m^2 for the skin. What is the maximum percent power that can be transferred into the skin with and without impedance matching?

Solution: The interface between skin and transducer can be represented as two of mechanical impedances as shown in Figure 11.43. The idea is to transfer maximum power (Fv) to Z_2. We know from the maximum power transfer theory, that maximum power will be transferred to Z_2 when $Z_1 = Z_2$ (Equation 11.23). The problem also requires us to find how much power is transferred in both the matched and unmatched conditions.

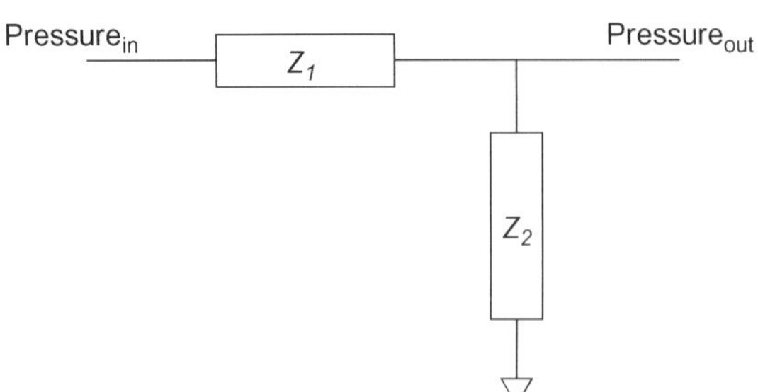

FIGURE 11.43 A model of the piezoelectric-skin interface in terms of acoustic (i.e., mechanical) impedance. Z_1 is the impedance of the transducer and Z_2 the impedance of the skin. In Example 11.16 the power transferred to Z_2 is determined for both the impedance-matched and -unmatched cases.

The transducer is applied directly to the skin, so the area of the transducer and the area of the skin are the same and we can substitute pressures for forces since the areas cancel. In an electrical circuit, the two impedances act as a voltage divider, and in a mechanical circuit they act as a force or pressure divider:

$$\frac{F_{Z_2}}{F_{in}} = \frac{p_{Z_2}a}{p_{in}a} = \frac{p_{Z_2}}{p_{in}} = \frac{Z_2}{Z_1 + Z_2}$$

where F are forces and p are pressures. The power into the skin is given by $P_{Z2} = p_{Z2}\, v_{Z2}$ where p is the pressure across Z_2 and v is the velocity of Z_2.

$$\frac{P_{Z_2}}{P_{in}} = \frac{p_{Z_2}v}{p_{in}v} = \frac{p_{Z_2}}{p_{in}} = \frac{Z_2}{Z_1 + Z_2} \tag{11.24}$$

Note that the impedances are given above in MKS units (meters, kilograms, seconds), not the CGS units (centimeters, grams, seconds) used throughout this text. However, since they are used in a ratio, the units cancel so there is no need to convert them to CGS in this example. In addition, these impedances are primarily dissipative (resistor-like) so we do not need complex notation.

When the impedances are matched, $Z_1 = Z_2$, then half the power is transferred to Z_2 as given by the maximum power transfer theorem. Under the unmatched conditions, the power ratio can be calculated as:

$$\frac{P_{Z_2}}{P_{in}} = \frac{Z_2}{Z_1 + Z_2} = \frac{1.63}{24.58 + 1.63} = 0.062$$

So the power ratio in the unmatched case is 6.2%, as opposed to 50% in the matched case. This approximately 8-fold gain in power transferred into the body shows the importance of matching impedances to improve power transfer. Since we have no control over the impedance of the skin, we must adjust the impedance of the ultrasound transducer. In fact, special coatings are applied to the active side of the transducer to match tissue impedance. In addition, a gel having the same impedance is used to improve coupling and insure impedance matching between the transducer and skin.

11.7 MULTIPLE SOURCES—REVISITED

In Chapter 5 we apply multiple sinusoids at different frequencies to a single input to find the frequency characteristics of a system. Superposition allows us to compute the transfer function for a range of frequencies with the assurance that if these frequencies are applied to the system, the responses are just the summation of the responses to individual frequencies. But what if the sources have both different frequencies and different locations? (If the sources are at the same frequency, but different locations, we have no problem, as standard analysis techniques apply; see Problems 7, 9 and 14 in Chapter 9.)

Even if the sources have different locations and different frequencies, superposition still can be used to analyze the network. We can solve the problem for each source separately knowing that the total solution will be the algebraic summation of all the partial solutions. We simply "turn off" all sources but one, solve the problem using standard techniques, and repeat until the problem has been solved for all sources. Then we add all the partial solutions for a final solution. As stated previously, turning off a source does not necessarily mean removing it from the system; it is replaced by its equivalent impedance. Hence, voltage sources are replaced by short circuits and current sources by open circuits (so current sources are actually removed from the circuit). Similarly, force sources become straight connections and velocity sources essentially disappear. The example below uses superposition in conjunction with source equivalent impedance to solve a circuit problem with two sources having different frequencies.

EXAMPLE 11.18

Find the voltage across the 30-Ω resistor in the circuit of Figure 11.44.

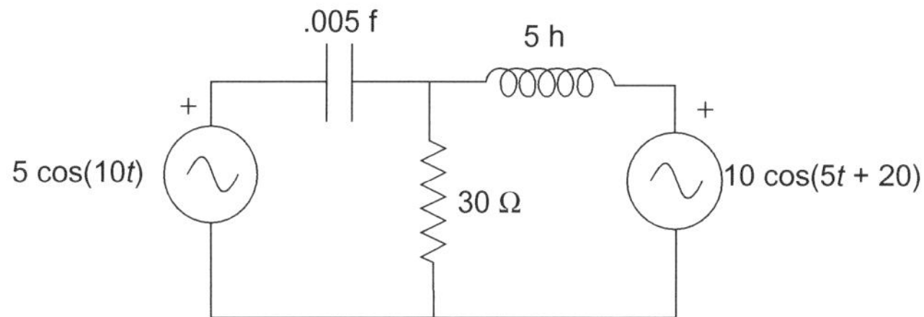

FIGURE 11.44 A two-mesh circuit that has both sources in different locations; each source has a different frequency. This type of problem can be solved using superposition.

Solution: First "turn off" the right-hand source by replacing it with a short circuit (its internal resistance), solve for the currents through the 30-Ω resistor, and then solve for the voltage across it. Then "turn off" the left-hand source and repeat the process. Note that the impedances of the inductor and capacitor will be different since the frequency is different. Add the two partial solutions to get the final voltage across the resistor.

Turning off the right-hand source and converting to the phasor domain results in the circuit in Figure 11.45.

Applying KVL leads to a solution for the voltage across the center resistor:

$$\begin{vmatrix} 5 \\ 0 \end{vmatrix} = \begin{vmatrix} 30 - j20 & -30 \\ -30 & 30 + j50 \end{vmatrix} \begin{vmatrix} I_1(\omega) \\ I_2\omega) \end{vmatrix}$$

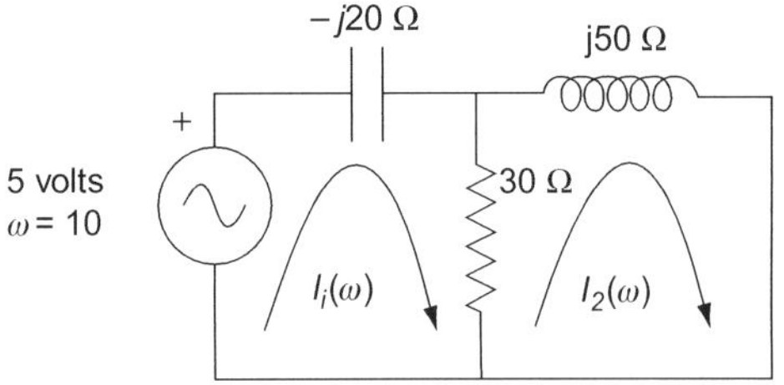

FIGURE 11.45 The circuit of Figure 11.44 with the right-hand source turned off. Since this source is a voltage source, turning it off means replacing it with a short, the equivalent impedance of an ideal voltage source.

Solving:

$I_1(\omega = 10) = 0.217\angle 17$; $I_2(\omega = 10) = 0.112\angle -42$; $V_R(\omega = 10) = (I_1(\omega) - I_2(\omega))R = 5.57\angle 48$ volts.

Now, turning off the left-hand source and converting to the phasor domain leads to the circuit in Figure 11.46. Note that the impedances are different since the new source has a different frequency.

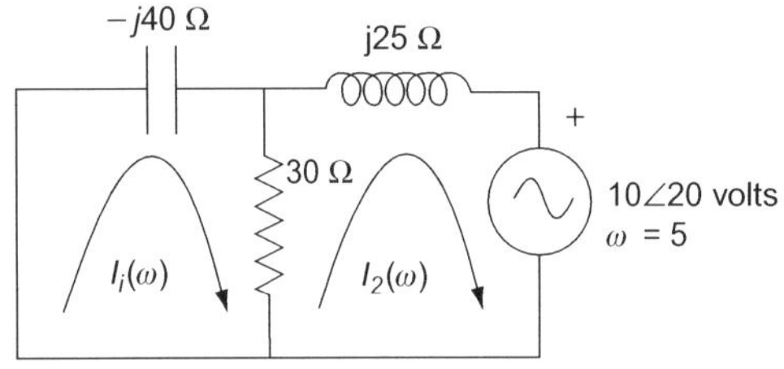

FIGURE 11.46 The circuit of Figure 11.44 with the left-hand source turned off.

Again apply the standard analysis.

$$\begin{vmatrix} 0 \\ -10\angle 20 \end{vmatrix} = \begin{vmatrix} 30 - j40 & -30 \\ -30 & 30 + j25 \end{vmatrix} \begin{vmatrix} I_1(\omega) \\ I_2(\omega) \end{vmatrix}$$

Solving:

$I_1(\omega = 5) = 0.274\angle -135$; $I_2(\omega = 5) = 0.456\angle 171$; $V_R(\omega = 5) = (I_1(\omega) - I_2(\omega))R = 10.94\angle -45$ volts.

The total solution is just the sum of the two partial solutions: $v_R(t) = 5.57\cos(10t + 48) + 10.94\cos(5t - 45)$ volts.

This approach extends directly to any number of sources. It applies equally well to current sources as shown in Problem 21.

11.8 SUMMARY

Even very complicated circuits can be reduced using the rules of network reduction. These rules allow networks containing one or more sources (at the same frequency) and any number of passive elements to be reduced to a single source and a single impedance. This single source–impedance combination could be either a voltage source in series with the impedance, called a Thévenin source, or a current source in parallel with the impedance, a Norton source. Conversion between the two representations is straightforward.

One of the major applications of network reduction is to evaluate the performance of the system when circuits are combined together. The transfer function of each isolated network is determined based on the assumption that the circuit is driven by an ideal source and connected to an ideal load. This can be taken as true if the impedance of the source driving the network is much much less than the network's input impedance and the impedance of the load is much much greater than the network's output impedance. Network reduction techniques provide a method for determining these input and output impedances.

The ratio of input to output impedance is particularly important when making physiological measurements. Often the goal is to make the input impedance of the measurement device as high as possible to load the process being measured as little as possible; that is, to draw minimum energy from the process. Sometimes it is desirable to transfer a maximum amount of energy between the process being measured and the measurement system. This is often true if the measurement device must also inject energy into the process in order to make its measurement. In such situations, an impedance matching strategy is used where the input impedance of the measuring device is adjusted to equal the complex congugate of output impedance of the process being measured. By matching impedances, maximum power is transferred between the process and the measurement device.

All of the network reduction tools apply equally to mechanical systems. Indeed, one of the major applications of impedance matching in biomedical engineering is in ultrasound imaging, where the ultrasound transducer acoustic impedance must be matched with the acoustic impedance of the tissue.

This chapter concludes with the analysis of networks containing multiple sources at different frequencies. To solve these problems, the effect of each source on the network must be determined separately and the solutions for each source added together. Superposition is an underlying assumption of this "summation of partial solutions" approach. When solving for the influence of each source on the network, the other sources are turned off by replacing them with their equivalent impedances: Voltage sources are replaced by short circuits and current sources are replaced by open circuits.

PROBLEMS

1. Find the combined values of the elements below.

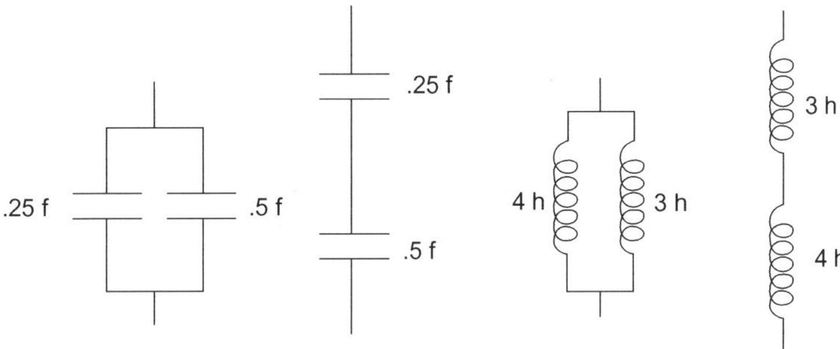

2. Find the value of R so the resistor combination equals 10 Ω. [*Hint*: Use Equation 11.9.]

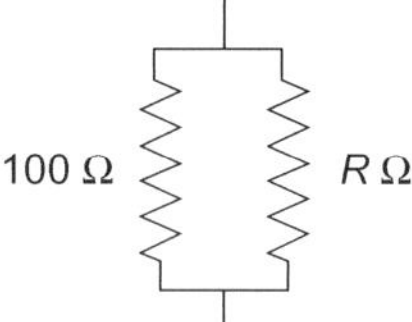

3. A two-terminal element has an impedance of $100\angle -30$ Ω at $\omega = 2$ rad/sec. The element consists of two components in series. What are they: two resistors? two capacitors? resistor capacitor? resistor inductor? or two inductors? What are their values?
4. An impedance, Z, has a value of $60\angle 25$ Ω at $\omega = 10$ rad/sec. What type of element should be added in series to make the combination look like a pure resistance at this frequency? What is the value of the added element at $\omega = 10$ rad/sec? What is the value of the combined element at $\omega = 10$ rad/sec?
5. Use network reduction to find the equivalent impedance of the network below between terminals A and B.

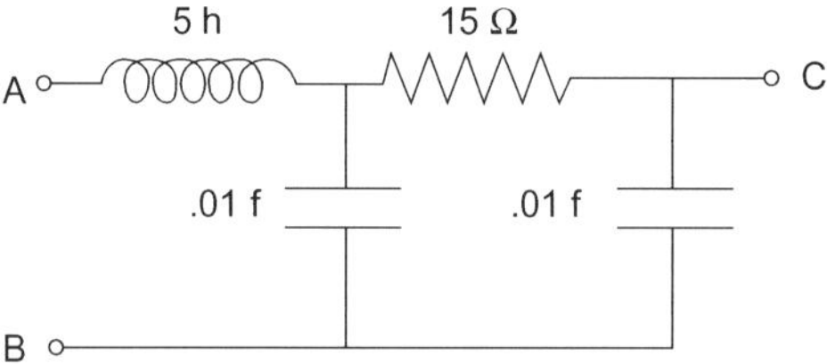

6. Use network reduction to find the equivalent impedance of the network in Problem 5 between the terminals A and C.
7. Find the equivalent impedance of the network below between terminals A and B.

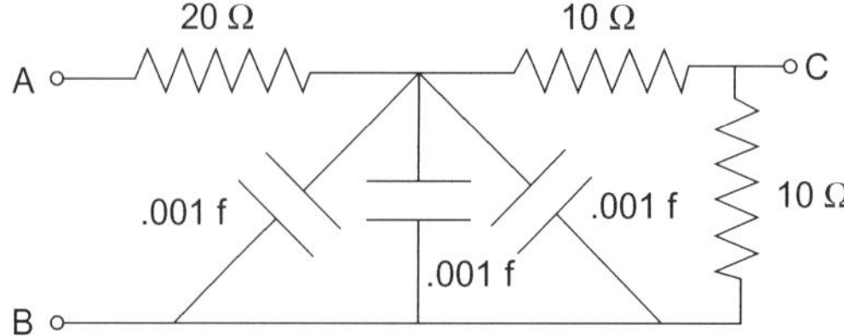

8. Plot the magnitude and phase of the impedance, Z_{eq}, in Example 11.3 over a range of frequencies from $\omega = 0.01$ to 1000 rad/sec. Plot both magnitude and phase in log frequency and plot the phase in degrees.
9. The following v-i characteristics were measured on a two-terminal device. Model the device by a Thévenin circuit and a Norton circuit.

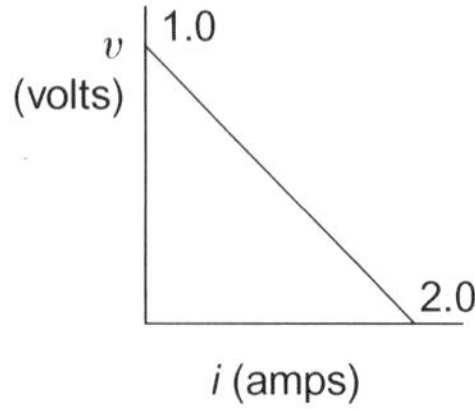

10. Plot the v-i characteristics of this network at two different frequencies: $\omega = 2$ and $\omega = 200$ rad/sec. Note that the v-i plot is a plot of the voltage magnitude against current magnitude. [*Hint*: Find $|V_T|$ at the two frequencies.]

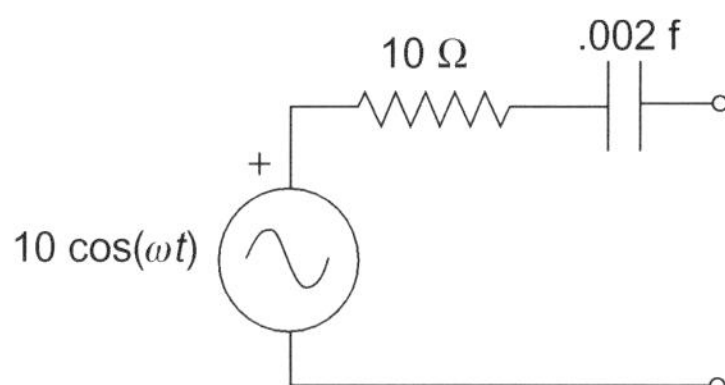

11. Two different resistors are placed across a two-terminal source known to contain an ideal voltage source in series with a resistor (i.e., a battery). The voltages obtained using the two load resistors are: when $R = 1000\ \Omega$, $V = 8.5$ volts; and when $R = 100\ \Omega$, $V = 8.2$ volts. Find the load resistor, R_L, that will extract the maximum power from this source. What is the power extracted?

12. The following magnitude v-i plot was found for a two-terminal device at the two frequencies shown. Model the device as a Thévenin equivalent. [*Hint*: To find R_T, use Equation 11.11 generalized for impedances.]

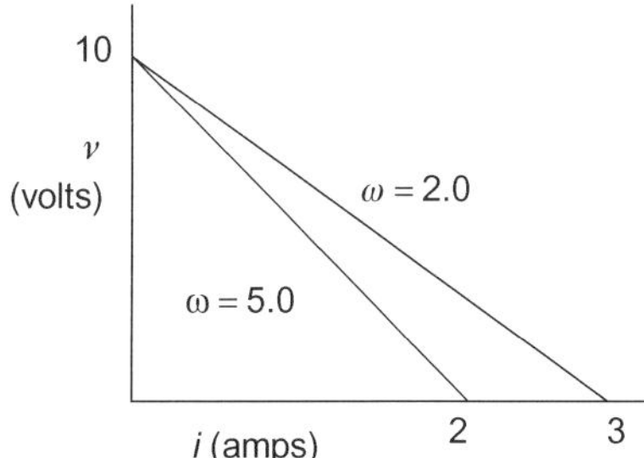

13. Find the Thévenin equivalent circuit of the network below. [*Hint*: Modify the code in Example 11.13.]

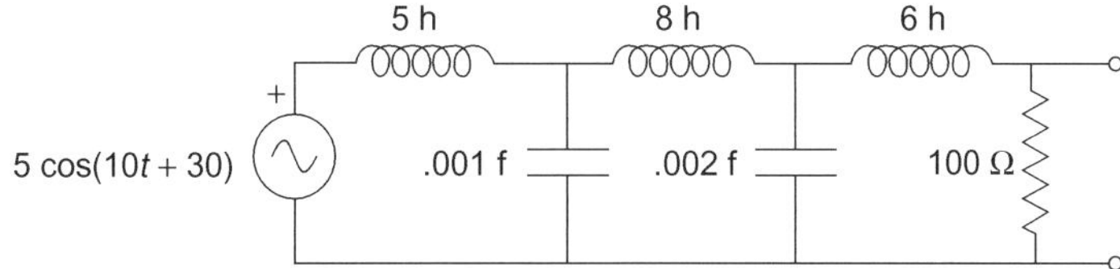

14. For the circuit in Problem 13, further modify the code in Example 11.13 to determine and plot the Norton current source and Norton impedance as a function of frequency. Plot both magnitude and phase of both variables for a range of frequency between 0.1 and 100 Hz.

15. The voltage of a nonideal voltage source shown below is measured with two voltmeters. One has an internal resistance of 10 MΩ while the other, a cheapie, has an internal resistance of only 100 KΩ. What voltages will be read by the two voltmeters?

How do they compare with the true ideal source voltage? (Assume the voltmeters read peak-to-peak voltage, although voltmeters usually read in rms.) *Note*: The components in the Thévenin source have values commonly encountered in real circuits.

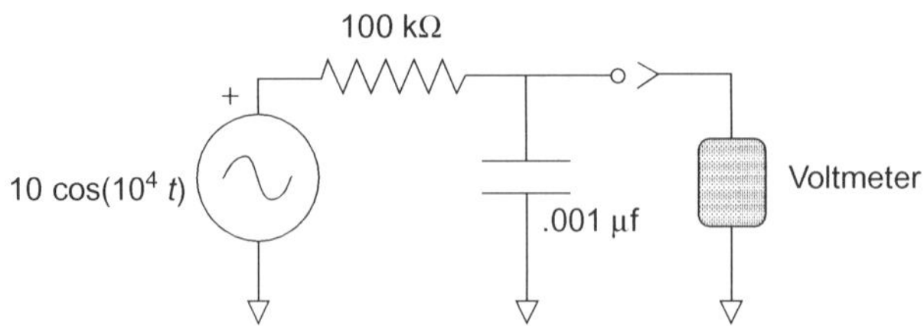

16. For the network shown below, what should the value of Z_L be to extract maximum power from the network? [*Hint*: Use an approach similar to that in Example 11.13.]
If the voltage source is increased to 15 cos(20t), what should the value of Z_L be to extract maximum power from the network?

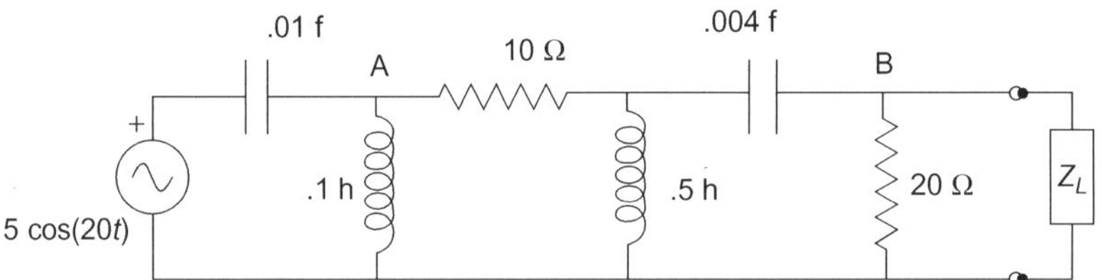

17. In the network shown in Problem 16, what is the impedance between nodes A and B, assuming Z_L is not attached? Use the approach shown in Example 11.8. Apply a hypothetical 1-volt source between these two points and use MATLAB-aided mesh analysis to find the current. Also since you only want the impedance, you need to remove the influence of the 5-volt source by turning it off; that is, replacing it by a short circuit. [*Note*: This can be reduced to a three-mesh problem, but even as a four-mesh problem it requires only a few lines of MATLAB code to solve.]

18. For the mechanical system shown in Example 11.15, find the difference between the measured force $F_{measured}$, and F_S (the ideal source) if the friction in the source, k_f, is increased from 20 dyne-sec/cm to 60 dyne cm/sec. Find the difference between $F_{measured}$, and F_S with the increased friction in the source if the elasticity of the transducer (k_e) is also increased by a factor of 3.

19. For the mechanical system shown in Example 11.16, find the difference between the measured velocity, v_2, and V_S if the elastic coefficient of the transducer, k_{e2} is doubled, quadrupled, or halved.

20. For the mechanical system shown in Example 11.16, find the difference between the measured velocity, v_2, and V_S if the velocity transducer (i.e., right side of system) contains a mass of 4 gm.

21. Find the voltage across the 0.01 f capacitor in the circuit below. Note that the two sources are at different frequencies.

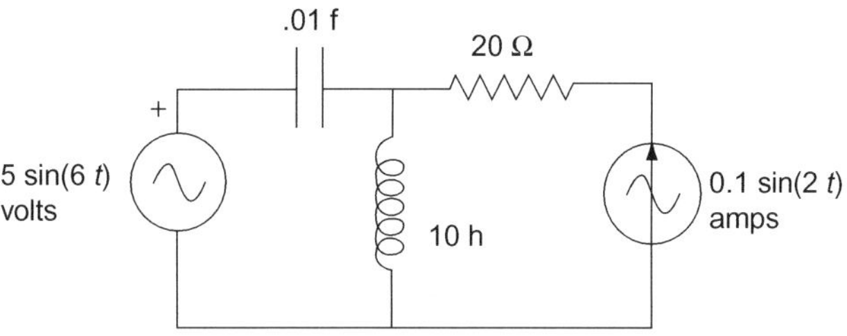

CHAPTER

12

Basic Analog Electronics: Operational Amplifiers

Electronic circuits come in two basic varieties: analog and digital. Digital circuits feature electronic components that produce only two voltage levels, one high and the other low. This limits the signals they can handle to two values, 0 and 1. To transmit information, these high/low values, termed *bits*, are combined in groups to make up a binary number. Groupings of 8 are common; an 8-bit binary number is called a *byte*. Bytes can be combined to make larger binary numbers or can be used to encode alphanumeric characters, most often in a coding scheme known as ASCII code. Digital circuits form the basis of all modern computers and microprocessors. While nearly all medical instruments contain one or more small computers (i.e., microprocessors) along with related digital circuitry, bioengineers are not usually concerned with their design. They may be called upon to develop some, or all, of the software but not the actual electronics, except perhaps for some basic interface circuits.

Analog circuit elements support a continuous range of voltages, and the information they carry is usually encoded as a time-varying continuous signal similar to those used throughout this text. All of the circuits described thus far are analog circuits. Analog circuitry is a necessary part of most medical instrumentation because the devices that perform physiological measurements, the transducers or "biotransducers," usually produce analog electric signals. This includes devices that measure movement, pressure, bioelectric activity, sound and ultrasound, light, and other forms of electromagnetic energy. Bioelectric signals such as the EEG, ECG, and EMG are analog signals. Before these analog signals can be processed by a digital computer, some type of manipulation is usually required while the signals are still in the analog domain. This analog signal processing may only consist of increasing the amplitude of the signal, but can also include filtering and other basic signal-processing operations. Unlike digital circuitry, the design of analog circuits is often the responsibility of the biomedical engineer. After analog signal processing, the signal is usually converted to a digital signal using an analog-to-digital converter. The components of a typical biomedical instrument are summarized in Figure 12.1.

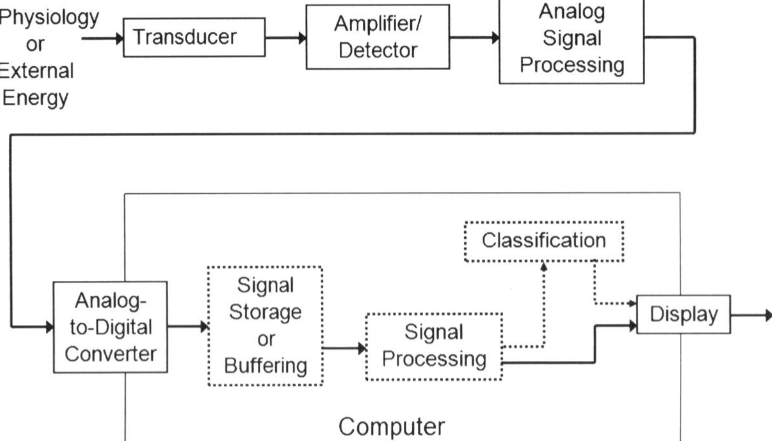

FIGURE 12.1 Basic elements of a typical biomedical instrument.

This chapter discusses the design and construction of some basic analog circuits such as amplifiers and filters. The design of biotransducers, often the most important biomedical instrument, can be found elsewhere (Northrop, 2004) along with the development of signal processing software (Semmlow, 2009).

12.1 THE AMPLIFIER

Increasing the amplitude or gain of an analog signal is termed *amplification* and is achieved using an electronic device known as an *amplifier*. The properties of an amplifier are commonly introduced using a simplification called the *ideal amplifier*. In this pedagogical scenario, the properties of a real amplifier are described as deviations from the idealization. In many practical situations, real amplifiers closely approximate the idealization; the limitations of real amplifiers and their deviations from the ideal become important only in more challenging applications. Nevertheless, the biomedical engineer involved in circuit design must know these limitations to understand when a typical amplifier circuit is being challenged.

An ideal amplifier is characterized by three features:

1. It has a well-defined gain at all frequencies (or at least over a specific range of frequencies);
2. Its output is an ideal source ($Z_{out} = 0.0\ \Omega$);
3. Its input is an ideal load ($Z_{in} \to \infty\ \Omega$).

As a systems element, an ideal amplifier is simply a gain term having ideal input and output properties, features 2 and 3 above. The schematic and system representation of an ideal amplifier is shown in Figure 12.2.

The transfer function of this amplifier would be:

$$\frac{V_{out}(s)}{V_{in}(s)} = \frac{V_{out}(\omega)}{V_{in}(\omega)} = G \tag{12.1}$$

where G would be a constant, or a function of frequency.

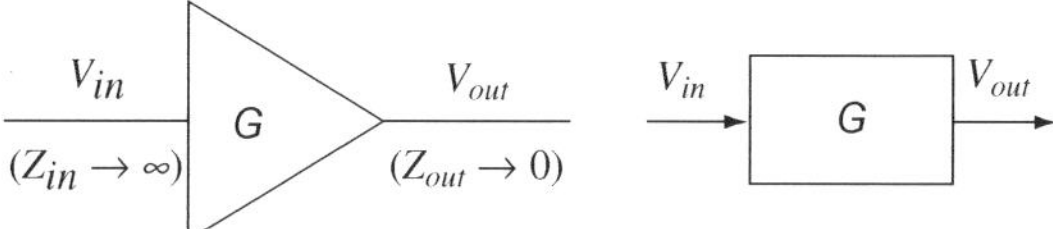

FIGURE 12.2 Schematic (*left*) and block diagram (*right*) of an ideal amplifier with gain of G.

Many amplifiers have a differential input configuration; that is, there are two separate inputs and the output is the gain constant times the difference between the two inputs. Stated mathematically:

$$V_{out} = G(V_{in2} - V_{in1}) \tag{12.2}$$

The schematic for such a *differential amplifier* is shown in Figure 12.3. Note that one of the inputs is labeled " + " and the other " − " to indicate how the difference is taken: The minus terminal subtracts its voltage from the plus terminal. (It is common to draw the negative input above the positive input.) The " + " terminal is also known as the *noninverting input*, while the " − " terminal is referred to as the *inverting input*.

Some transducers, and a few amplifiers, produce a differential signal: actually two signals that move in opposite directions with respect to the information they represent. For such differential signals, a differential amplifier is ideal, since it takes advantage of both input signals. Moreover, the subtraction tends to cancel any signal that is common to both inputs, and noise is often common to both inputs. However, most of the time only a single signal is available. In these cases, a differential amplifier may still be used, but one of the inputs is set to zero by grounding. If the positive input is grounded and the signal is sent into the negative input (Figure 12.4, left side), then the output will be the inverse of the input:

$$V_{out} = G(-V_{in}) \tag{12.3}$$

In this case the amplifier may be called an *inverting amplifier* for obvious reasons. If the opposite strategy is used and the signal is sent to the positive input while the negative input is grounded as in Figure 12.4 (right side), the output will have the same direction as the input. This amplifier is termed a *noninverting amplifier* (sort of a double negative).

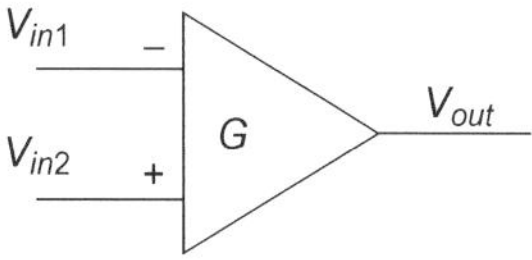

FIGURE 12.3 An amplifier with a *differential input* configuration. The output of this amplifier is G times $V_{in2} - V_{in1}$; i.e., $V_{out} = G(V_{in2} - V_{in1})$.

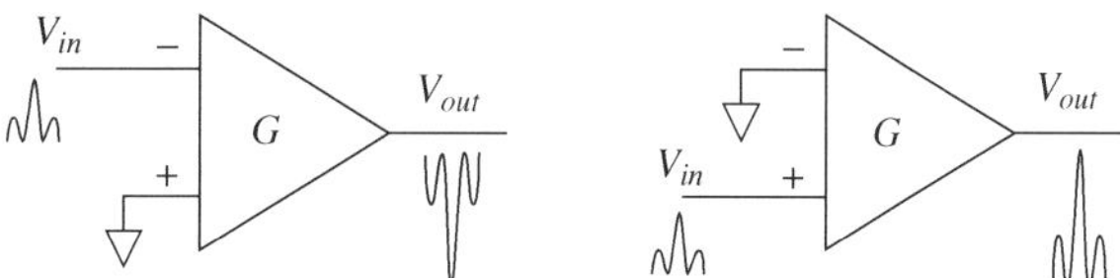

FIGURE 12.4 A differential amplifier configured as an inverting (*left*) and noninverting (*right*) amplifier.

12.2 THE OPERATIONAL AMPLIFIER

The *operational amplifier*, or *op amp*, is a basic building block for a wide variety of analog circuits. One of its first uses was to perform mathematical operations, such as addition and integration, in analog computers, hence the name "operational" amplifier. Although the functions provided by analog computers are now performed by digital computers, the op amp remains a valuable, perhaps the most valuable, tool in analog circuit design.

In its idealized form, the op amp has the same properties as the ideal amplifier described above except for one curious departure: It has infinite gain. Thus an ideal op amp has infinite input impedance (an ideal load), zero output impedance (an ideal source), and a gain, A_v, of infinity. (The symbols A_v or A_{VOL} are commonly used to represent the gain of an operational amplifier.) Obviously an amplifier with a gain of infinity is of limited value, so an op amp is rarely used alone, but is usually used in conjunction with other elements that reduce its gain to a finite level.

Negative feedback can be used to limit the gain. Consider the feedback system in Figure 12.5. Assume that the upper path (i.e., the feedfoward path) represents the op amp with its gain of A_v and lower path is the feedback gain, β.

The gain of the overall system can be found from the basic feedback equation introduced and derived in Example 5.1. When we insert A_V and β into the feedback equation, Equation 5.10, the overall system gain, G, becomes:

$$G = \frac{A_V}{1 + A_V\beta} \tag{12.4}$$

Now if we let the feedforward gain, A_V, the gain of the operational amplifier, go to infinity:

$$G = \lim_{A_V \to \infty} \left| \frac{A_V}{1 + A_V\beta} \right. = \lim_{A_V \to \infty} \left| \frac{A_V}{A_V\beta} \right. = \frac{1}{\beta} \tag{12.5}$$

The overall gain expressed in dB becomes:

$$G_{dB} = 20 \log G = 20 \log \left(\frac{1}{\beta}\right) = -20 \log \beta \tag{12.6}$$

If $\beta < 1$, then the gain, G, is >1 and if $\beta = 1$, $G = 1$. A $\beta > 1$ leads to a gain <1 and a reduction in signal amplitude. If a reduction in gain is desired, it is easier to use a passive voltage divider, a series resistor pair with one resistor to ground. So in real operational amplifier circuits, the feedback gain, β, is ≤ 1 to achieve a gain ≥ 1. This is fortunate, since all we need to produce a feedback gain <1 is a voltage divider network: One end of a pair of series resistors is connected to the output, the other end to ground, and the reduced

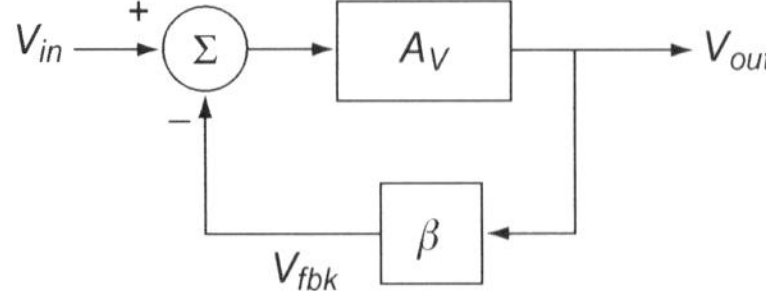

FIGURE 12.5 A basic feedback control system used to illustrate the use of feedback to set a finite gain in a system that has infinite feedforward gain, A_V.

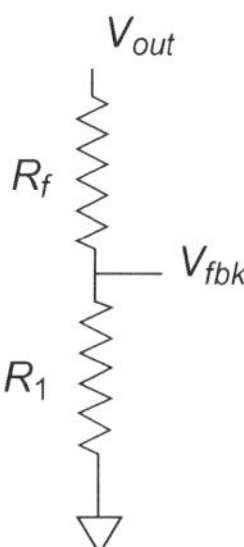

FIGURE 12.6 A voltage divider network that can be used to feed back a portion of the output signal to the inverting or negative input of an op amp.

feedback signal is taken from the intersection of the two resistors (Figure 12.6). A feedback gain of $\beta = 1$ is even easier; just feed the entire output back to the input.

The approach of beginning with an amplifier that has infinite gain, then reducing that gain to a finite level with the addition of feedback, seems needlessly convoluted. Why not design the amplifier to have a finite fixed gain to begin with? The answer is summarized in two words: flexibility and stability. If feedback is used to set the gain of an op amp circuit, then only one basic amplifier needs to be produced: one with an infinite (or just very high) gain. Any desired gain can be achieved by adjusting a simple feedback network. More importantly, the feedback network is almost always implemented using passive components, resistors, and sometimes capacitors. Passive components are always more stable than transistor-based devices; that is, they are more immune to fluctuations due to changes in temperature, humidity, age, and other environmental factors. Passive elements can also be more easily manufactured to tighter tolerances than active elements. For example, it is easy to buy resistors that have a 1% error in their values, while most common transistors have variations in gain of a factor of 100% or more. Finally, back in the flexibility category, a wide variety of different feedback configurations can be used, allowing one operational amplifier chip to perform many different signal processing operations. Some of these different functions are explored in Section 12.7 at the end of this chapter.

12.3 THE NONINVERTING AMPLIFIER

Negative feedback can be achieved using a simple voltage divider to feed a reduced version of the output back to the inverting (i.e., negative) input of an op amp. Consider the voltage divider network in Figure 12.6. The feedback voltage can be found by the simple application of KVL. Assuming that V_{out} is an ideal source (which it is because it is the output of an ideal amplifier):

$$V_{out} - i(R_f + R_1) = 0; \quad i = \frac{V_{out}}{R_f + R_1}:$$

$$V_{fbk} = iR_1 = V_{out}\left(\frac{R_1}{R_1 + R_f}\right) \tag{12.7}$$

For the system diagram in Figure 12.5, we see that $\beta = V_{fbk}/V_{out}$. Using Equation 12.7 we can solve for β in terms of the voltage divider network:

$$\beta = \frac{V_{fbk}}{V_{out}} = \frac{R_1}{R_1 + R_f} \tag{12.8}$$

The transfer function of an operational amplifier circuit that uses feedback to set the gain is just $1/\beta$, as given in Equation 12.5:

$$G \equiv \frac{V_{out}}{V_{in}} = \frac{1}{\beta} = \frac{1}{\frac{R_1}{R_1 + R_f}} = \frac{R_1 + R_f}{R_1} \tag{12.9}$$

An op amp circuit using this feedback network is shown in Figure 12.7. The gain of this amplifier is given by Equation 12.9. Since the input signal is fed to the positive side of the op amp, this circuit is termed a *noninverting amplifier*.

The transfer function for the circuit in Figure 12.7 can also be found through circuit analysis, but a couple of helpful rules are needed.

1. Since the input resistance of the op amp approaches infinity, no current flows into, or out of, either of the op amp's input terminals. Realistically, the input resistance of an op amp is so high that any current that does flow into the op amp's input is negligible.
2. Since the gain of the op amp approaches infinity, the only way the output can be finite is if the input is zero; that is, the difference between the plus input and the minus input must be zero. Stated yet another way, the voltage on the plus input terminal must be the same as the voltage on the minus input terminal of the op amp and vice versa. Realistically, the very high gain means that for reasonable output voltages, the voltage difference between the two input terminals is very small and can be ignored.

In practical op amps the gain is large (up to 10^6) but not infinite, so the voltage difference in a practical op amp circuit might be a few millivolts, but this small difference can generally be ignored. Similarly, the input resistance, while not infinite, is quite large: values of r_{in} (resistances internal to the op amp are denoted in lower case) are usually greater than $10^{12}\ \Omega$, so any input current will be very small and can be disregarded (especially since the input voltage must be zero or at least very small by rule 2, above). We use these two rules to solve the transfer function of a noninverting amplifier in the following example.

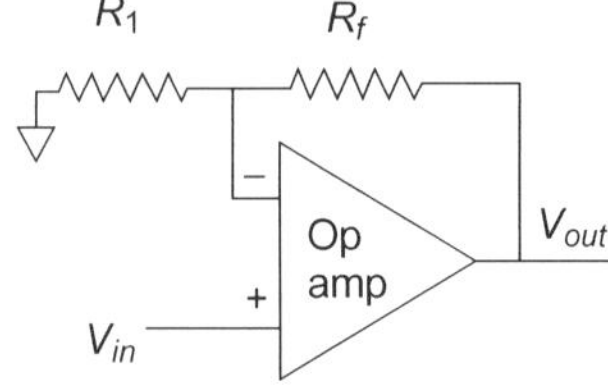

FIGURE 12.7 Noninverting op amp circuit. The gain of this op amp circuit is determined solely by the two resistors as given by Equation 12.9.

EXAMPLE 12.1

Find the transfer function of the noninverting op amp circuit in Figure 12.7 using network analysis.

Solution: First, by rule 2 (above) the voltage between the two resistors, the voltage at the negative terminal, must equal V_{in} since V_{in} is applied to the lower terminal and the voltage difference between the upper and lower terminals is zero.

Define the three currents in and out of the node between the two resistors and apply KCL to that node as shown in Figure 12.8:

By KCL: $-i_1 - i_{in} + i_f = 0$, but $i_{in} = 0$ according to rule 1.

Hence if: $-i_1 - i_{in} + i_f = 0$; then: $i_1 = i_f$.

Applying Ohm's law to substitute the voltages for the currents and noting that the voltage at the node must equal V_{in} by rule 2:

$$i_1 = \frac{V_{in}}{R_1} \quad \text{and} \quad i_f = \frac{V_{out} - V_{in}}{R_f}$$

$$\text{Then, since } i_1 = i_f\text{:} \quad \frac{V_{in}}{R_1} = \frac{V_{out} - V_{in}}{R_f}$$

Solving for V_{out}:

$$V_{out} - V_{in} = V_{in}\left(\frac{R_f}{R_1}\right); \quad V_{out} = V_{in}\left(1 + \frac{R_f}{R_1}\right) = V_{in}\left(\frac{R_f + R_1}{R_1}\right)$$

and the transfer function becomes:

$$\frac{V_{out}}{V_{in}} = \frac{R_f + R_1}{R_1}$$

This is the same transfer function found using the feedback equation (Equation 12.9).

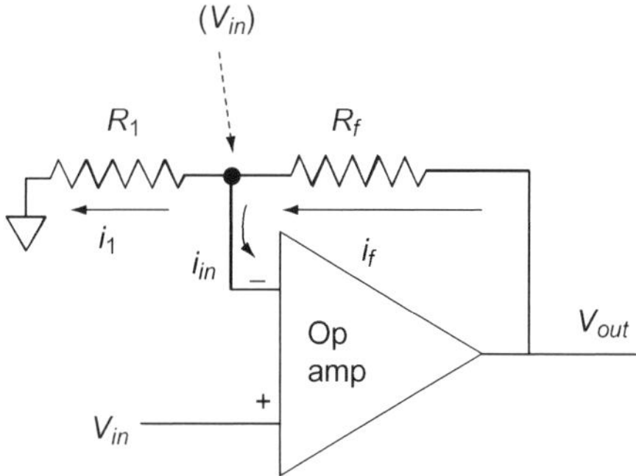

FIGURE 12.8 A noninverting operational amplifier circuit.

12.4 THE INVERTING AMPLIFIER

To construct an amplifier circuit that inverts the input signal, the ground and signal inputs of the noninverting amplifier are reversed, as shown in Figure 12.9.

The transfer function of the inverting amplifier circuit is a little different than that of the noninverting amplifier, but can be easily found using the same approach (and tricks) used in Example 12.1.

EXAMPLE 12.2

Find the transfer function, or gain, of the inverting amplifier circuit shown in Figure 12.9.

Solution: Define the currents and apply KCL to the node between the two resistors, as shown in Figure 12.10. In this circuit the node between the two resistors must be at 0 volts by rule 2; that is, since the plus side is grounded and the difference between the two terminals is 0, the minus side is effectively grounded. In fact, the inverting input terminal in this configuration is sometimes referred to as a *virtual ground* (see Figure 12.10).

As in Example 12.1, we apply KCL to the inverting terminal and find that: $i_1 = i_f$, and by Ohm's Law:

$$i_1 = \frac{0 - V_{in}}{R_1}; \quad i_f = \frac{V_{out} - 0}{R_f}; \quad \frac{-V_{in}}{R_1} = \frac{V_{out}}{R_f}$$

Again solving for V_{out}/V_{in}:

$$\frac{V_{out}}{V_{in}} = -\frac{R_f}{R_1} \tag{12.10}$$

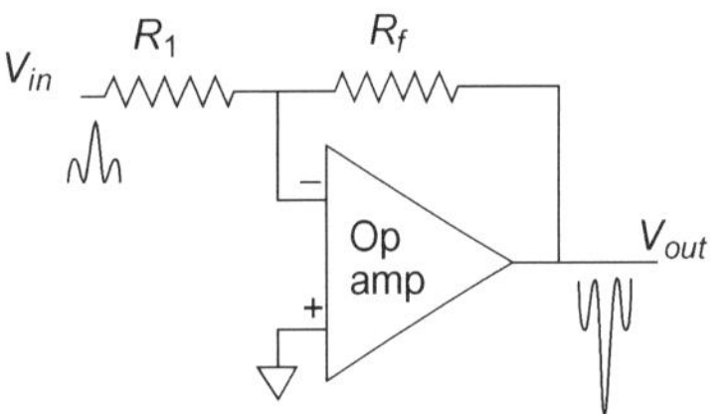

FIGURE 12.9 The op amp circuit used to construct an inverting amplifier.

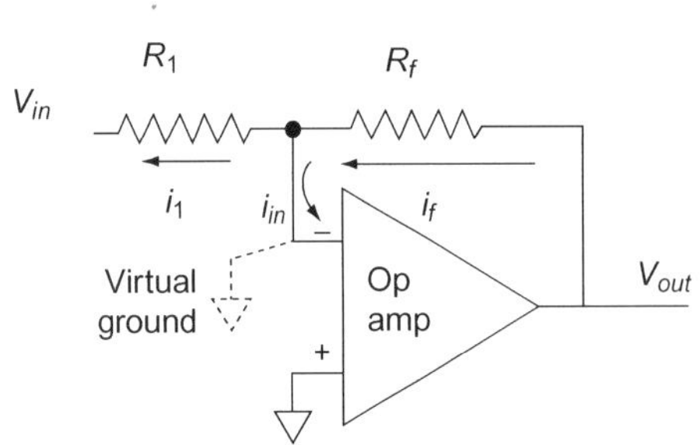

FIGURE 12.10 Inverting op amp configuration showing currents assigned to the upper node. The upper node in this configuration must be at 0 volts by rule 2 (see Section 12.3), and is referred to as virtual ground.

The negative sign demonstrates that this is an inverting amplifier: The output is the negative, or inverse, of the input. The output is also larger than the input by a factor of R_f/R_1. If $R_1 > R_f$, the output will actually be reduced. The point is that the circuit designer has control over the gain of this circuit simply by varying the values of R_1 and/or R_f. If one of the resistors is variable (i.e., a potentiometer) then the amplifier will have a variable gain. This design feature holds for both inverting and noninverting amplifier circuits although the gain equations are different (Equation 12.9 vs. Equation 12.10).

EXAMPLE 12.3

Design an inverting amplifier circuit with a variable gain between 10 and 100. Assume you have a variable 1 MΩ potentiometer; that is, a resistor that can be varied between 0 and 1 MΩ. Also assume you have available a wide range of fixed resistors.

Solution: The amplifier circuit will have the general configuration of Figure 12.9. It is possible to put the variable resistance as part of either R_{in} or R_f, but let us assume that the potentiometer is part of R_f along with a fixed input resistance. This results in the circuit shown in Figure 12.11.

Since the variable resistor is in the feedback path, we vary feedback resistance, R_f, to get the desired gain variation. Assume that the variable resistor is 0 Ω when the gain is 10 and 1 MΩ when the gain is 100. We can write two gain equations based on Equation 12.10 for the gain limits and solve for our two unknowns, R_1 and R_2.

For $G = 10$ the variable resistor is 0 Ω:

$$\frac{R_f}{R_1} = 10; \quad \frac{R_2 + 0}{R_1} = 10; \quad R_2 = 10R_1$$

For $G = 100$ the variable resistor is 10^6 Ω:

$$\frac{R_2 + 10^6}{R_1} = 100; \quad \text{Substituting for } R_2; \quad 10R_1 + 10^6 = 100R_1$$

$$90R_1 = 10^6; \quad R_1 = 11.1\ \text{k}\Omega; \text{ and } \quad R_2 = 10R_1 = 111\ \text{k}\Omega$$

The final circuit is shown in Figure 12.12.

The equation for the gain of noninverting and inverting op amp circuits can be extended to include feedback networks that contain capacitors and inductors. To modify these equations to

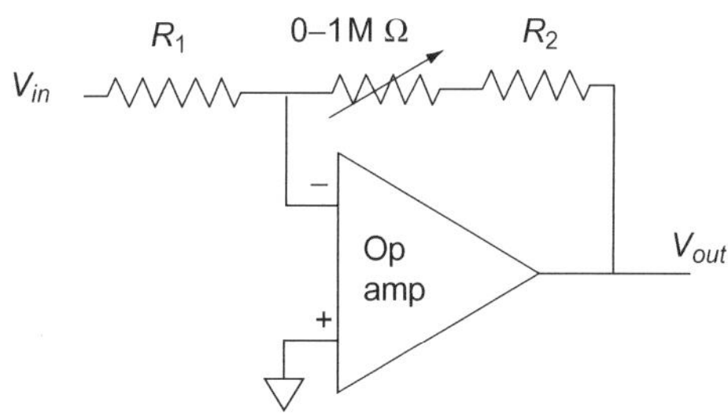

FIGURE 12.11 An inverting op amp circuit with a variable resistor in the feedback path to allow for variable gain. In Example 12.3, the variable and fixed resistors are adjusted to provide an amplifier gain from 10 to 100.

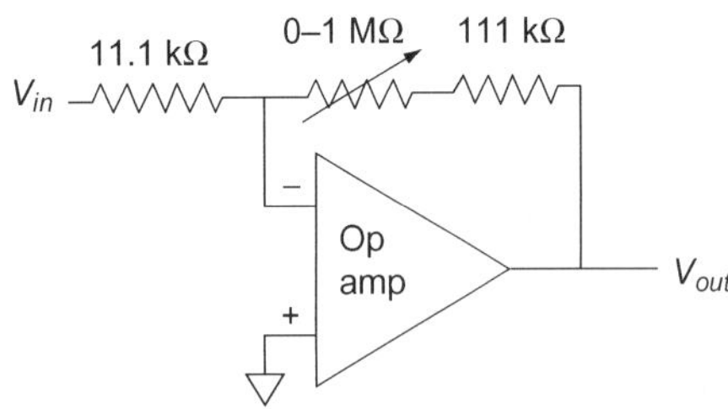

FIGURE 12.12 An inverting amplifier circuit with a gain between 10 and 100.

include components other than resistors, simply substitute impedances for resistors. So the gain equation for a noninverting op amp circuit becomes:

$$\frac{V_{out}}{V_{in}} = \frac{Z_f + Z_1}{Z_1} \tag{12.11}$$

and the equation for an inverting op amp circuit becomes:

$$\frac{V_{out}}{V_{in}} = -\frac{Z_f}{Z_1} \tag{12.12}$$

12.5 PRACTICAL OP AMPS

Practical op amps, the kind that you buy from electronics supply houses, differ in a number of ways from the idealizations used above. In many applications, perhaps most applications, the limitations inherent in real devices can be ignored. The problem is that any bioengineer designing analog circuitry must know when the limitations are important and when they are not, and to do this he or she must understand the characteristics of real devices. Only the topics that involve the type of circuits the bioengineer is likely to encounter are covered here. Several excellent references can be found to extend these concepts to other circuits (Horowitz and Hill, 1989).

Deviations of real op amps from the ideal can be classified into three general areas: deviations in input characteristics, deviations in output characteristics, and deviations in transfer characteristics. Each of these areas is discussed in turn, beginning with the area likely to be of utmost concern to biomedical engineers: transfer characteristics.

12.5.1 Limitations in Transfer Characteristics of Real Op Amps

The most important limitations in the transfer characteristics of real op amps are bandwidth limitations and stability. In addition, real op amps have large, but not infinite, gain. Bandwidth limitations occur because an op amp's magnitude gain is not only finite, but decreases with increasing frequency. The lack of stability results in unwanted oscillations and is due to the op amp's increased phase shifts at higher frequencies.

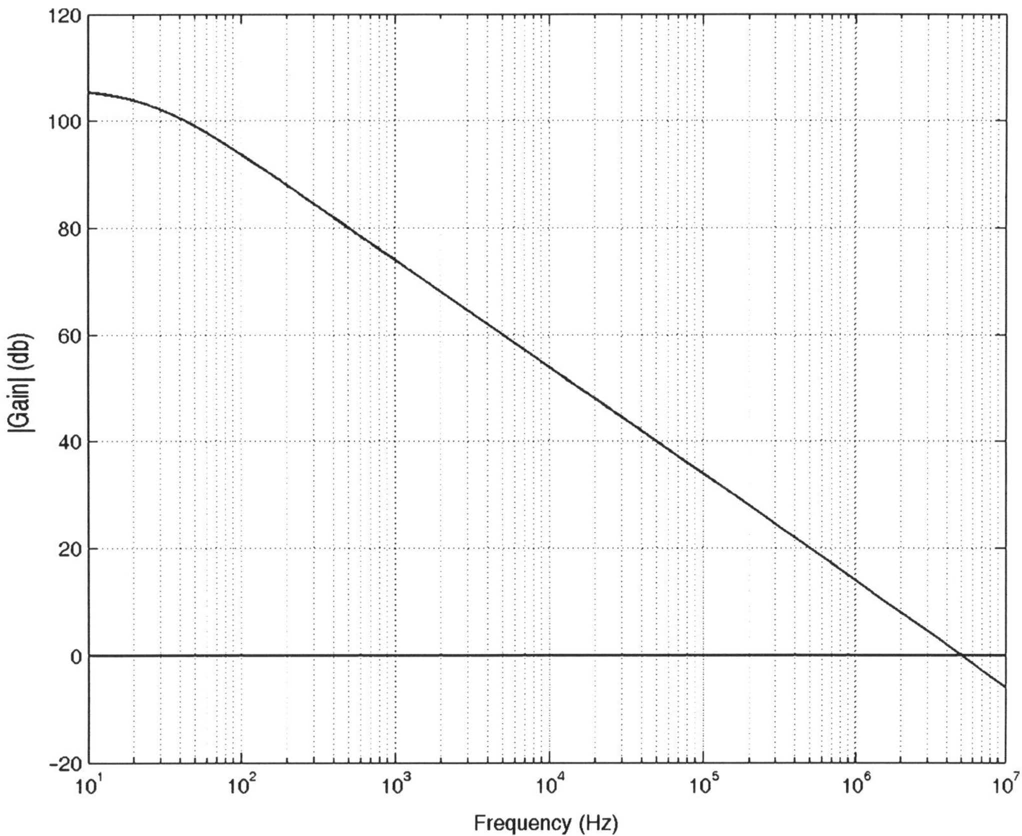

FIGURE 12.13 The open-loop magnitude gain characteristics of a popular operational amplifier, the LF356.

12.5.1.1 Bandwidth

The magnitude frequency characteristic of a popular op amp, the LF356, is shown in Figure 12.9. Not surprisingly, even at low frequencies the gain of this operational amplifier is less than infinity. This in itself would not be a cause for much concern as the gain is still quite high: approximately 106 dB or a gain of 199,530. The problem is that this gain is also a function of frequency, so that at higher frequencies the gain becomes quite small. In fact there is a frequency above which the gain is less than one. Thus at higher frequencies the transfer function equations no longer hold since they were based on the assumption that op amp gain is infinite. Since the bandwidth of a real op amp is limited, the bandwidth of an amplifier using such an op amp must also be limited. Essentially, the gain of an op amp circuit is limited by either the bandwidth limitations of the op amp or the feedback, whichever is lower. An easy technique for determining the bandwidth of an op amp circuit is to plot the gain produced by the feedback circuit superimposed on the bandwidth curve of the real op amp. The former is referred to as the *closed-loop gain* since it includes the feedback, while the latter is termed the *open-loop gain* since it is the gain of the operational amplifier without a feedback loop.

To determine the transfer function of a real op-amp circuit, we start with the fact that the theoretical closed-loop transfer function is $1/\beta$ (Equation 12.5). The real transfer function gain is either this value or the op amp's open-loop gain, whichever is lower. (The gain in an op amp circuit can never be greater than what the op amp is capable of producing.) So to get the real gain, we plot theoretical gain, $1/\beta$, superimposed on the open-loop curve. The real gain is simply the lower of the two curves. If the feedback network consists only of resistors, β will be constant for all frequencies, so $1/\beta$ will plot as a straight line on the frequency curve. (Although real resistors have some small inductance and capacitance, the effect of these "parasitic elements" can be ignored except at very high frequencies.)

For example, assume that 1/100 of the signal is fed back to the inverting terminal of a real op amp. Then feedback gain is: $\beta = 0.01 = -40$ dB and $1/\beta = 100 = 40$ dB.

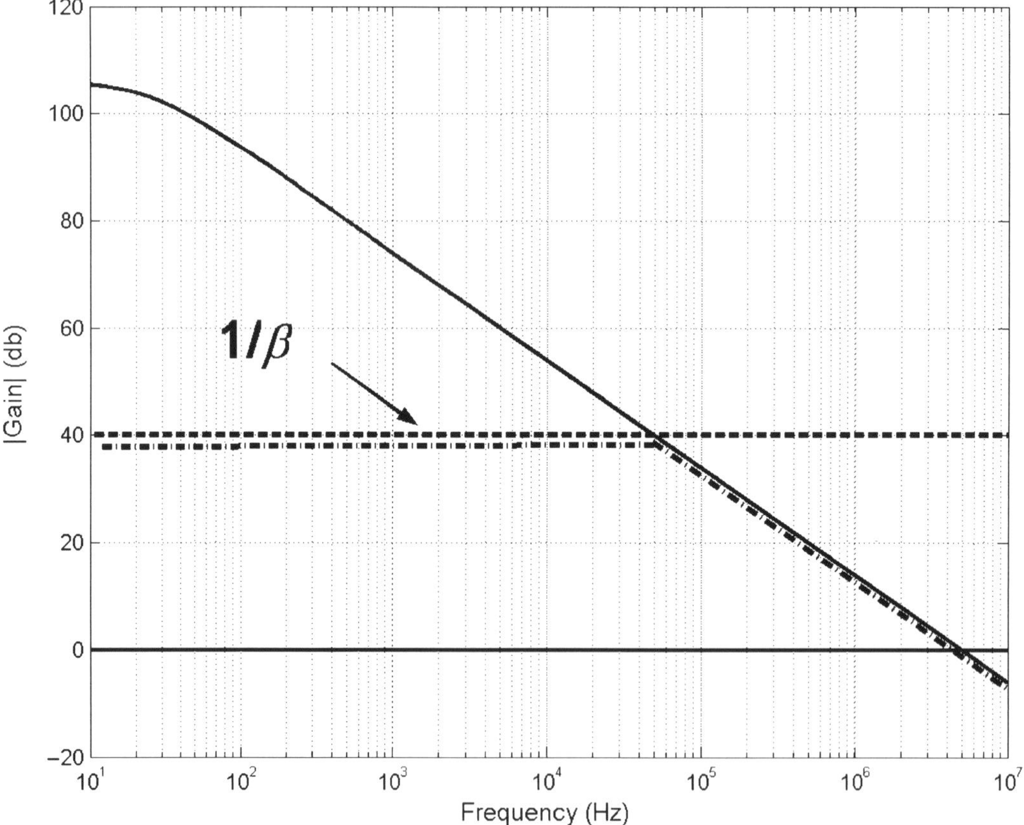

FIGURE 12.14 The open-loop magnitude characteristics of the LF356 (solid line) as shown in Figure 12.13, with a plot of $1/\beta = 40$ dB superimposed (dashed line). The overall gain follows the dashed line until it intersects the solid open-loop curve, then follows the open-loop curve downward (dash-dot line).

Figure 12.14 shows the open-loop gain characteristics of a typical op amp (LF356) with the plot of $1/\beta$ superimposed (dashed line). The overall gain will follow the dashed line until it intersects the op amp's open-loop curve (solid line), where it will follow that curve since this is less than $1/\beta$. Hence, the amplifier's magnitude frequency characteristic will follow the heavy dash-dot lines seen in Figure 12.14. Given this particular op amp and this value of β, the bandwidth of the amplifier circuit as determined by the -3 dB point is approximately 50 kHz. The feedback gain, β, is the same for both inverting and noninverting op amp circuits, so this approach for determining amplifier bandwidth is the same in both configurations. The value $1/\beta$ is sometimes referred to as the *noise gain* because it is also the gain factor for input noise and errors, again irrespective of the specific configuration.

EXAMPLE 12.4

Find the bandwidth of the inverting amplifier circuit in Figure 12.15.

Solution: First determine the feedback gain, β, then plot the inverse ($1/\beta$) superimposed on the open-loop gain curve obtained from the op amp's spec sheets.

$$\frac{1}{\beta} = \frac{R_f + R_{in}}{R_{in}} = \frac{20+1}{1} \approx 20 = 26\text{ dB}$$

From the superimposed plot in Figure 12.16, we see that the desired gain curve (i.e., $1/\beta$) intersects the op amp's actual gain curve at approximately 200 kHz. Above 200 kHz the circuit gain drops off at 20 dB per decade so this intersection is used to define the circuit bandwidth, although the actual -3 dB point will be at a slightly higher frequency.

As shown in the open-loop gain curve of Figures 12.13, 12.14, and 12.16, a typical op amp has very high values of gain at low frequencies, but also has a low cutoff frequency. Above this cutoff frequency, the gain curve rolls off at 20 dB/decade as in a first-order low-pass filter. For example, the LF356 has a low-frequency gain of around 106 dB but the open loop bandwidth is only around 25 Hz. The addition of feedback dramatically lowers the overall gain, but just as dramatically increases the bandwidth. Indeed, the idea of using negative feedback in amplifier circuits was first introduced to improve bandwidth by trading reduced gain for increased cutoff frequency.

If the feedback gain is a multiple of 10, as is often the case, the bandwidth can be determined without actually plotting the two curves. The bandwidth can be determined

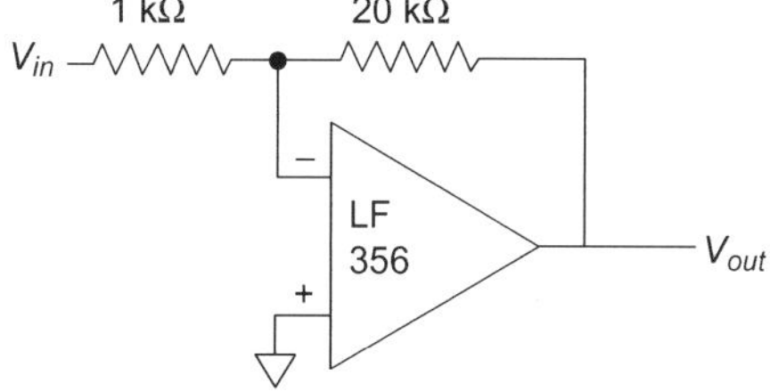

FIGURE 12.15 The bandwidth of this inverting amplifier circuit is determined in Example 12.4.

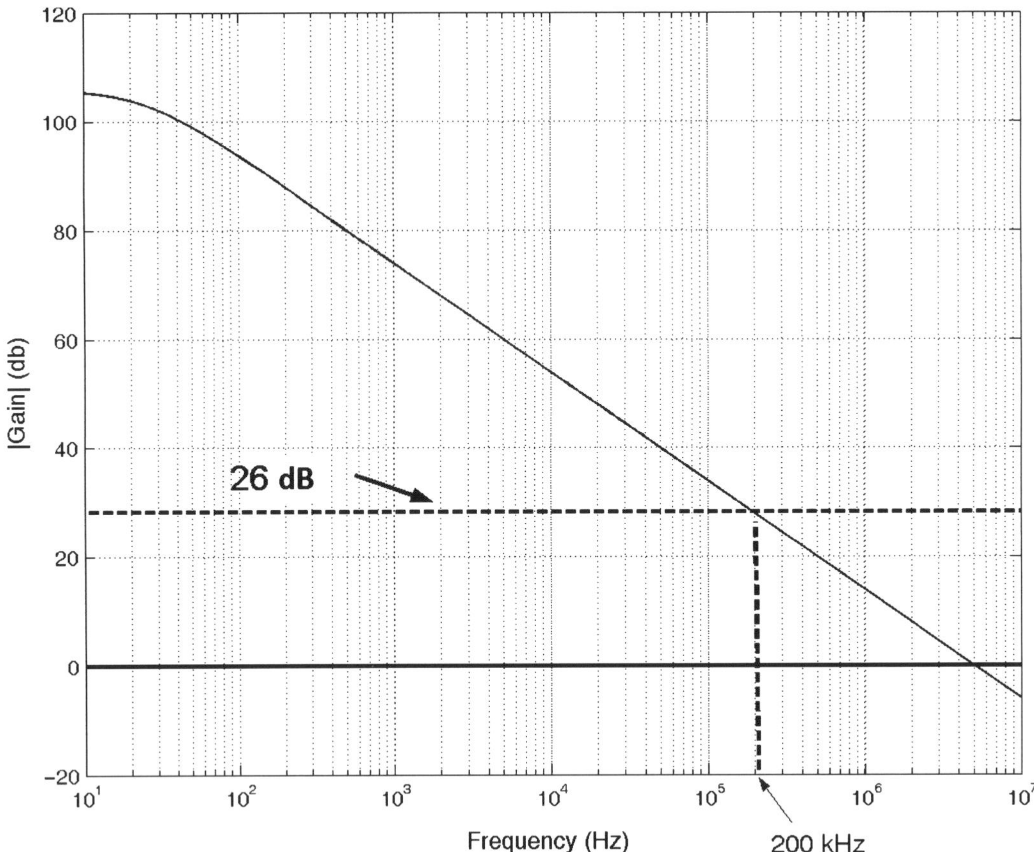

FIGURE 12.16 The open-loop gain curve of the LF356 with the $1/\beta$ line superimposed. This line intersects the op amp's open-loop gain curve at around 200 kHz. This intersection defines the bandwidth of the amplifier circuit.

directly from the amplifier gain and the frequency at which the open-loop curve intersects 0 dB. This frequency, where op amp gain falls to 1.0 (i.e., 0 dB), is termed the *gain bandwidth product* (GBP). The GBP is given in the op amp specifications; for example, the GBP of the LF356 is 5.0 MHz. To use the GBP to find the bandwidth of a feedback amplifier, start with the GBP frequency and knock off one decade in bandwidth for every 20 dB in amplifier gain. Essentially you are sliding up the 20 dB/decade slope starting at 0 dB, the GBP. For example, if the amplifier has a gain of 1 (0 dB), the bandwidth equals the GBP, 5 MHz in the case of the LF356. For every 20 dB (or factor of 10) above 0 dB, the bandwidth is reduced one decade. So an amplifier circuit with a gain of 10 (20 dB) would have bandwidth of 500 kHz and an amplifier circuit with a gain of 100 (40 dB) would have a bandwidth of 50 kHz, assuming these circuits used the LF356. If the gain is 1000 (60 dB), the bandwidth would be reduced to 5 kHz. If the $1/\beta$ gain is not a power of 10, logarithmic interpolation would be required and it is easiest to use the plotting technique.

EXAMPLE 12.5

Using an LF356 op amp, what is the maximum amplifier gain (i.e., closed-loop gain) that can be obtained with a bandwidth of 100 kHz?

Solution: From the open-loop curve given in the Figure 12.13, the open-loop gain at 100 kHz is approximately 30 dB. Hence this is the maximum close-looped gain that will reach the desired cutoff frequency. Designing the appropriate feedback network to attain this gain (and bandwidth) is straightforward using Equation 12.9 (noninverting) or Equation 12.10 (inverting).

12.5.1.2 Stability

Most op amp circuits require negative feedback, which is why the feedback voltage, V_{fbk}, is fed to the inverting (i.e., minus) input of the op amp. Except in special situations, positive feedback is to be avoided. Positive feedback creates a vicious circle: the feedback signal enhances the feedforward signal which enhances the feedback signal which enhances.... A number of things can happen to a positive feedback network, most of them bad. The two most likely outcomes are that the output oscillates, sometimes between the maximum and minimum possible voltage levels, or the output is driven into saturation and locked at the maximum or minimum level. When the word stability is used in context with an op amp circuit it means the absence of oscillation or other deleterious effects associated with positive feedback. The oscillation produced by positive feedback is a sustained repetitive waveform which could be a sinusoid, but it may also be more complicated.

Positive feedback oscillation occurs in a feedback circuit where the overall gain or *loop-gain* (i.e., the gain of the feedback and feedforward circuits) is greater than or equal to 1 and has a phase shift of 360 degrees:

$$\text{Loop gain (for oscillation)} \equiv \text{feedforward gain} \times \text{feedback gain} \geq 1.0\angle 360 \text{ deg} \qquad (12.13)$$

When this condition occurs, any small signal will feed back positively and grow to produce a sustained oscillation. Sometimes this amp oscillation will ride on top of the signal, sometimes it will overwhelm the signal, but in either case it is unacceptable.

Since the feedback signal is sent to the inverting input of the op amp, positive feedback should not occur. The noninverting input induces a phase shift of 180 degrees, so the feedback signal is negative since a sinusoid shifted by 180 degrees is the inverse of an unsifted sinusoid: $\sin(2\pi ft) = -\sin(2\pi ft + 180)$. However, if the op amp induces an additional phase shift of 180 degrees, then the negative feedback becomes positive feedback since the total phase shift is 360 degrees. If the loop gain happens to be ≥ 1 when this occurs, the conditions of Equation 12.13 are met and the circuit will oscillate. The base frequency of that oscillation will be equal to the frequency where the total phase shift becomes 360 degrees; that is, the frequency where the op amp contributes a phase shift of an additional 180 degrees. Hence, oscillation is a result of phase shifts induced by op amps at high frequencies.

A rigorous analysis of stability is beyond the scope of this text. However, since stability is so often a problem in op amp circuits, some discussion is warranted. Since the inverting input always contributes a 180-degree phase shift (to make the feedback negative), to ensure stability we must make sure that everything else in the feedback loop contributes a phase shift that is less than 180 degrees. If β is a constant, then any additional phase shift must come from the op amp, so all we need are op amps that never approach a phase shift of 180 degrees, at least for loop gains greater than or equal to 1.0. Unfortunately, all op amps will reach a phase shift of 180 degrees if we go high enough in frequency, so the only alternative strategy is to ensure that when the op amp reaches a phase shift of 180 degrees, the overall loop gain is $<$1.0.

The overall loop gain is just $A_V(\omega)\beta(\omega)$. Putting the condition for stability in terms of the gain symbols we have used thus far, the condition for stability is:

$$\text{Loop gain}_{\text{stability}} = A_V(\omega)\beta(\omega) < 1.0\angle 360 \text{ deg} \tag{12.14}$$

where A_V is the gain of the op amp at a specific frequency and β is the feedback gain. Alternatively, if we want to build an oscillator, the condition for oscillation is:

$$\text{Loop gain}_{\text{oscillation}} = A_V(\omega)\beta(\omega) \geq 1.0\angle 360 \text{ deg} \tag{12.15}$$

With respect to β, the worst case for stability occurs when β is the largest. In most op amp circuits β is $<$1, but can sometimes be as large as 1. As mentioned, a feedback gain of $\beta = 1$ corresponds to the lowest op amp gain: a gain of 1 ($V_{out} = V_{in}$). While it is somewhat counterintuitive, this means that stability is more likely to be a problem in low gain amplifier circuits where β is large. Stability is even more likely to be a problem when the gain is 1.0, since $\beta = 1$ in this case.

If the op amp gain, A_V, is less than 1.0 when its phase shift hits 180 degrees, and β is at most 1.0, then $Av\beta$ will be $<$1, and the conditions for stability are met (Equation 12.14). Stated in terms of phase, the op amp's phase shift should be less than 180 degrees for all frequencies where its gain is ≥ 1. In fact, most op amps have a maximum phase shift that is less than 120 degrees to be on the safe side for gains greater than, or equal to, 1.0. Such op amps are said to be *unity gain stable* because they will not oscillate even when $\beta = 1$ and the amplifier has unity gain. However, to achieve this criterion requires some compromise on the part of the op amp manufacturer, usually some form of phase compensation that reduces the gain bandwidth product. In many op amp applications where gain will be high and β correspondingly low, unity gain stability is overkill and the unity gain compensation results in a needless reduction of performance.

Op amp manufacturers have come up with two strategies to overcome performance loss due to unity gain stability: produce different versions of the same basic op amp, one that has higher bandwidth but requires a minimum gain while another that is unity gain stable but with a lower bandwidth; or produce a single version but have the user supply the compensation (usually as an external capacitor) to suit the application requirements. The former has become more popular since it does not require additional circuit components. The LF356 is an example of the former strategy. The LF356 has a GBP of 5 MHz and is unity gain stable while its "sister" chip, the LF357, requires a $1/\beta$ gain of 5 or more, but has a GBP of 20 MHz.

Unfortunately, using an op amp with unity gain stability still does not guarantee a stable circuit because stability problems can still occur if the feedback network introduces phase shifts. A feedback network containing only resistors might be considered safe, but

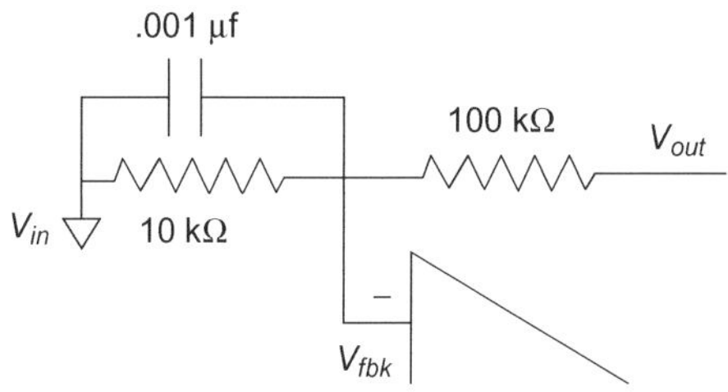

FIGURE 12.17 A feedback network that includes some capacitance either due to the intentional addition of a capacitor or due to parasitic elements. This feedback network will introduce an additional phase shift into the loop. This shift could be beneficial if it moves the overall loop phase shift away from 360 degrees or harmful if it moves it closer to 360 degrees.

parasitic elements, small inductances, and capacitances can create an undesirable phase shift. Consider the feedback circuit in Figure 12.17 in which a capacitor is placed in parallel with one of the resistors. With a capacitor in the feedback network, β becomes a function of frequency and will introduce an additional phase shift in the loop. Will this additional capacitance present a problem with regard to stability?

To answer this question, we need to find the phase shift of the network at the frequency when the loop gain is 1. The loop gain will be 1 when:

$$A_V\beta = 1; \quad A_V = \frac{1}{\beta} \tag{12.16}$$

Hence the loop gain is 1 when $1/\beta$ equals the op amp gain A_V. On the plot of A_V and $1/\beta$, such as in Figure 12.16, this is when the two curves intersect. In the circuit presented in Figure 12.17, the $1/\beta$ curve will not be a straight line, but can easily be found using the phasor circuit techniques.

For the voltage divider:

$$V_{fbk} = \frac{Z_1}{Z_1 + Z_f} V_{out};$$

Solving for β:

$$\frac{V_{fbk}}{V_{out}} = \beta = \frac{Z_1}{Z_1 + Z_f}$$

where:

$$Z_1 = \frac{R_1\left(\frac{1}{j\omega C}\right)}{R_1 + \left(\frac{1}{j\omega C}\right)} = \frac{10^4\left(\frac{10^9}{j\omega}\right)}{10^4 + \left(\frac{10^9}{j\omega}\right)} = \frac{10^4}{1 + \frac{j\omega}{10^5}}$$

$$Z_f = 10^5\ \Omega$$

Substituting in Z_1 and Z_f and solving for β:

$$\beta = \frac{\frac{10^4}{1 + \frac{j\omega}{10^5}}}{10^5 + \frac{10^4}{1 + \frac{j\omega}{10^5}}} = \frac{10^4}{10^5 + j\omega + 10^4} = \frac{0.09}{1 + \frac{j\omega}{1.1 \times 10^5}} \tag{12.17}$$

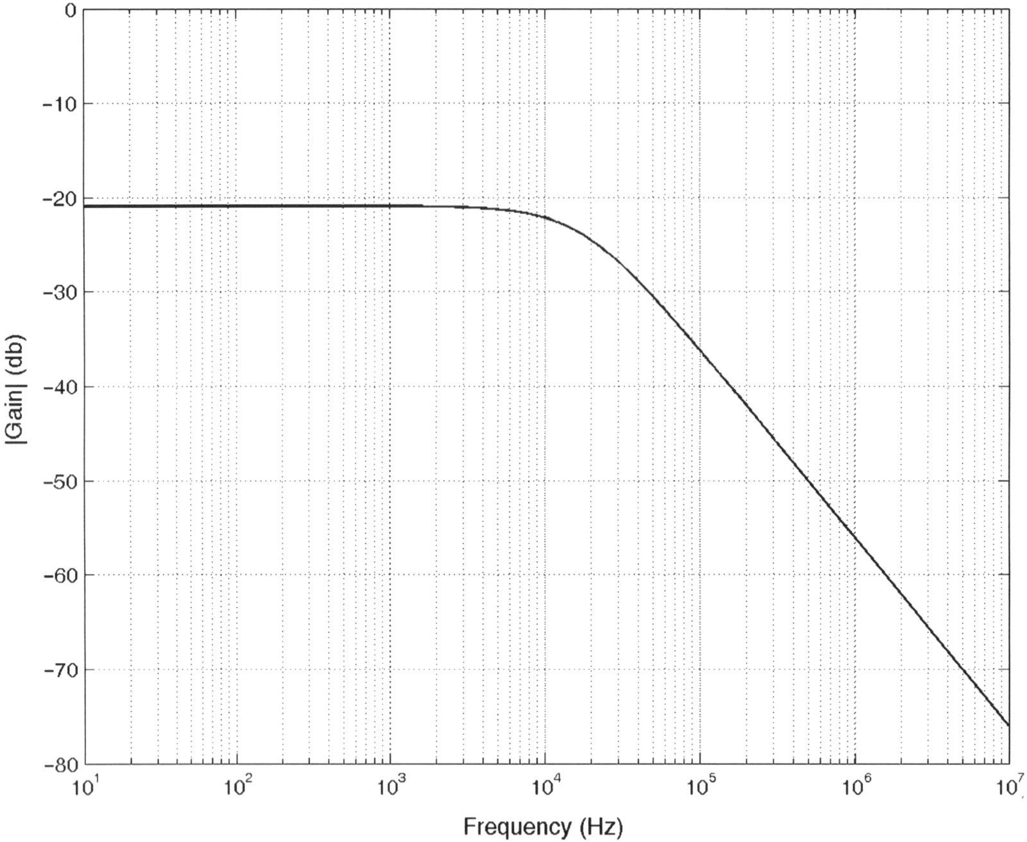

FIGURE 12.18 A) The magnitude frequency plot of the feedback gain, β, of the network shown in Figure 12.17. The equation for this curve (Equation 12.17) is found using standard phasor circuit analysis and plotted using Bode plot techniques.

This is just a low-pass filter with a cutoff frequency of 1.1×10^5 rad/sec or 17.5 kHz. Figure 12.18 shows the magnitude frequency plot of β plotted from Equation 12.17 using standard Bode plot techniques.

Figure 12.19 shows the $1/\beta$ curve, the inverse of the curve in Figure 12.18, plotted superimposed on the open-loop gain curve of the LF356. From Figure 12.19, the two curves intersect at around 85 kHz.

The phase shift contributed by the feedback network at 85 kHz can readily be determined from Equation 12.17. At 85 kHz, $\omega_1 = 2\pi\ (85 \times 10^3) = 5.34 \times 10^5$.

$$\beta = \frac{0.09}{1 + \frac{j\omega}{1.1 \times 10^5}} = \frac{0.09}{1 + \frac{j2\pi 85 \times 10^3}{1.1 \times 10^5}} = \frac{0.09}{4.96\angle 78} = 0.018\angle{-78}$$

Thus the feedback network contributes a 78-degree phase shift to the overall loop gain. While the phase shift of the op amp at 85 kHz is not known (detailed phase information is not often provided in op amp specifications), we can only be confident that the phase shift

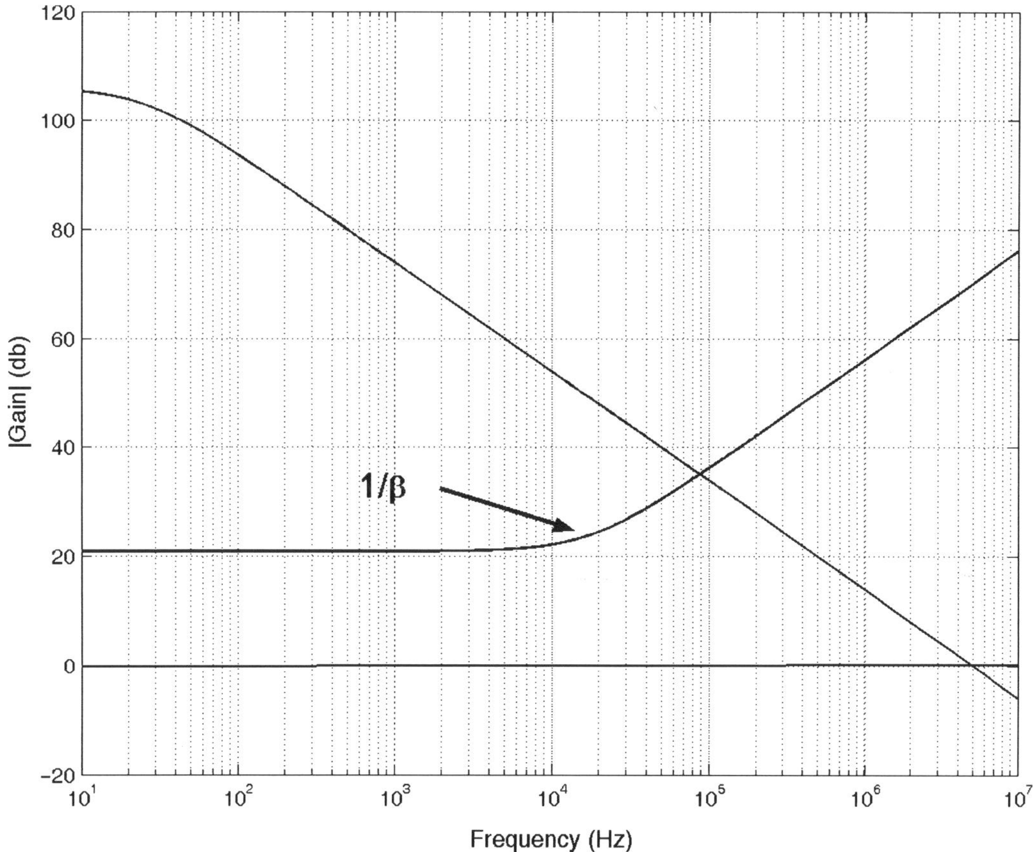

FIGURE 12.19 The inverse of the feedback gain given by Equation 12.17 (i.e., $1/\beta$) plotted superimposed against the A_V curve of the LF356 op amp.

is no more than 120 degrees. Adding the worst-case op amp phase shift to the feedback network phase shift (120 + 78) results in a total phase shift that is >180 degrees, so this circuit is likely to oscillate. A good rule of thumb is that the circuit will be unstable if the $1/\beta$ curve breaks upward before intersecting the A_V line of the op amp. The reverse is also true: The circuit will be stable if the $1/\beta$ line intersects the A_V line at a point where it is flat or going downward.

If the feedback network can make the circuit unstable, it stands to reason that it can also make the amplifier less unstable. This occurs when capacitance is added in parallel to the feedback resistor, the 100 kΩ resistor in Figure 12.17. In fact, the quick fix approach to oscillations in many op amp circuits is to add a capacitor to the feedback network in parallel with the feedback resistor. This usually works, although the capacitor might have to be fairly large. As shown in the section on filters (Section 12.7.5), adding feedback capacitance reduces bandwidth, the larger the capacitance the greater the reduction in bandwidth. Sometimes a reduction in bandwidth is desired to reduce noise, but often it is disadvantageous. The influence of feedback capacitance on bandwidth is explored in the section on filters, while its influence on stability is demonstrated in Problem 6.

Stability problems are all too frequent and when they occur, they can be difficult and frustrating to correct. Possible problems include the feedback network, excessive capacitive load (which modifies the op amp's phase shift characteristics), the use of an inappropriate op amp, and, most commonly, inadequate decoupling capacitors. The use of decoupling capacitors is discussed in Section 12.6.

12.5.2 Input Characteristics

The input characteristics of a real op amp can best be described as involved but not complicated. In addition to a large, but finite, input resistance, r_{in}, several voltage and current sources are found (see Figure 12.20). The values of these elements are given for the LF356 in parentheses. (The curious units for the voltage and current noise sources, $V/\sqrt{Hz}$ and $pA/\sqrt{Hz}$, are explained below.) These sources have very small values and can often be ignored, but again it is important for the bioengineer to understand the importance of these elements in order to make intelligent decisions.

12.5.2.1 Input Voltage Sources

The voltage source, v_o, is a constant voltage source termed the *input offset voltage*. It indicates that the op amp will have a nonzero output even if the input is zero; that is, the differential input $V_+ - V_- = 0$. The output voltage produced by this small voltage source depends on the gain of the circuit. To find the output voltage under zero input conditions (the output offset voltage) simply multiply the input offset voltage by the gain term $1/\beta$. This gain is also known as the *noise gain*, for reasons given later. Determination of the output offset voltage is demonstrated in the following example.

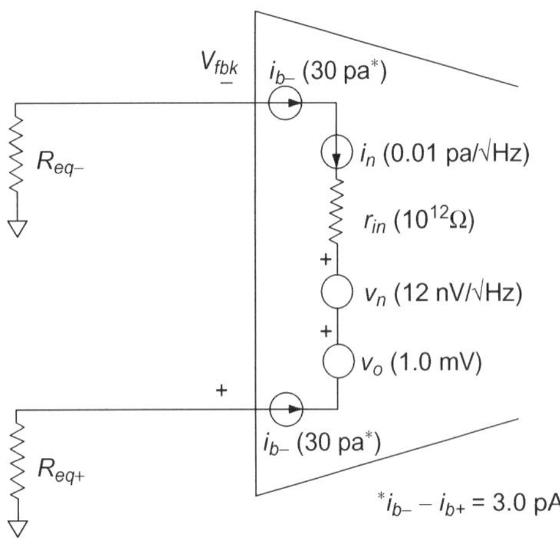

FIGURE 12.20 A schematic representation of the input elements of a practical op amp. The input can be thought of as containing several voltage and current sources as well as a large input impedance. Each of these sources is discussed.

EXAMPLE 12.6

Find the offset voltage at the output of the amplifier circuit shown in Figure 12.21. This is the same as asking for V_{out} when $V_{in} = 0$.

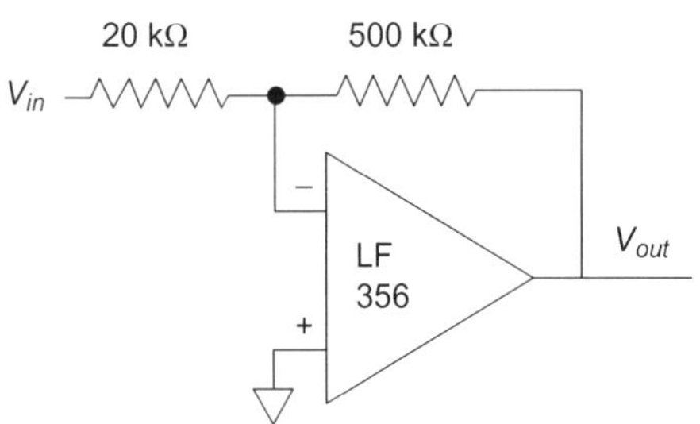

FIGURE 12.21 A typical inverting amplifier circuit. This amplifier will have a small output voltage even if $V_{in} = 0$. This offset voltage is determined in Example 12.6.

Solution: First find the noise gain, $1/\beta$, then multiply this by the input offset voltage, v_o (1.0 mV). The value of $1/\beta$ is:

$$\frac{1}{\beta} = \frac{R_f + R_1}{R_1} = \frac{500 + 20}{20} = 26$$

Given the typical value of v_o as 1.0 mV, V_{out} for zero input becomes:

$$V_{out} = v_o(26) = 26 \text{ mV}$$

The input offset voltage given in the LF356 specifications (see Appendix F) is a typical value, not necessarily the value of any individual operational amplifier chip. This means that the input offset voltage will be around the value shown, but for any given chip it could be larger or smaller. It could also be either positive or negative leading to a positive or negative output offset voltage. Sometimes a maximum or "worst case" value is also given.

The noise voltage source, v_n, specifies the noise normalized for bandwidth—more precisely, normalized to the square root of the bandwidth—which accounts for the strange units nV/$\sqrt{\text{Hz}}$. Like Johnson noise and shot noise described in Chapter 1, noise in an op amp is distributed over the entire bandwidth of the op amp. To determine the actual noise in an amplifier it is necessary to multiply v_n by the square root of the circuit bandwidth as determined using the methods described in Section 12.5.1.1. This value is multiplied by the noise gain (i.e., $1/\beta$) to find the noise at the output. We finally see why $1/\beta$ is also referred to as the noise gain, although it also applies to offset voltages. This procedure is demonstrated in Example 12.7.

EXAMPLE 12.7

Find the noise at the output of the amplifier used in Example 12.6 that is due only to the op amp's noise voltage. (The resistors in the feedback network also contribute Johnson noise as does the input current noise source, i_n. Calculation of total noise is covered in Example 12.8).

Solution: Find the noise gain, then determine the bandwidth from $1/\beta$ and the open-loop gain curve in Figure 12.13. Multiply the input noise voltage, v_n, by the square root of the bandwidth to find the noise at the input. Then multiply the result by the noise gain to find the value of noise at the output.

From Example 12.6, the noise gain is 26, or 28 dB. Referring to Figure 12.13, a $1/\beta$ line at 28 dB will intersect A_V at approximately 200 kHz. Hence, the bandwidth can be taken as 200 kHz and the input noise voltage becomes:

$$v_{n\ \mathrm{input}} = 12 \times 10^{-9}\sqrt{200 \times 10^3} = 5.4\ \mu\mathrm{V}$$

The noise at the output is this value times $1/\beta$:

$$V_{n\ \mathrm{output}} = v_{n\ \mathrm{input}}\left(\frac{1}{\beta}\right) = 5.4(26) = 140\ \mu\mathrm{V}$$

The value of input voltage noise used in Example 12.7 was again a typical value for frequencies above 100 Hz. Op amp noise generally increases at lower frequencies and many op amp specifications include this information, including those of the LF356. For the LF356, the input voltage noise increases from 12 $\mathrm{nV}/\sqrt{\mathrm{Hz}}$ at 200 Hz and up, to 60 $\mathrm{nV}/\sqrt{\mathrm{Hz}}$ at 10 Hz (see Appendix F). Presumably the voltage noise becomes even higher at lower frequencies, but values below 10 Hz are not given for this chip.

12.5.2.2 *Input Current Sources*

To evaluate the influence of input current sources, it is easiest to convert them to voltages by multiplying by the *equivalent resistance* at the terminals. Figure 12.20 notates the equivalent input resistances at the plus and minus terminals as R_{eq+} and R_{eq-}. These would have been determined by using the network reduction methods described in Chapter 11 as is demonstrated in a subsequent example. The two current sources at the inputs, i_{b+} and i_{b-}, are known as the *bias currents*. It may seem curious to show two current sources rather than one for both inputs, but this has to do with the fact that the bias currents at the two terminals, though not exactly equal, tend to be similar. Also as with the bias voltage, the bias currents could be in either direction, in or out of their respective terminals.

To determine how these bias currents contribute to the output, we convert them to voltages by multiplying by the appropriate R_{eq}, then multiplying that voltage by the noise gain. In most operational amplifiers, the two bias currents are approximately the same, so the influence on output offset voltage tends to cancel if the equivalent resistances at the two terminals are the same. Sometimes the op amp circuit designer tries to make the equivalent resistances at the two terminals the same just to achieve this cancellation. The amount that the two bias currents are different, the imbalance between the two currents, is called the *offset current* and is usually much less than the bias current. For example, in the LF356 typical bias currents are 30 pA while the offset current is only 3.0 pA, an order of magnitude less.

Figure 12.22 shows an inverting op amp circuit where a resistor has been added between the positive terminal and ground. The current flowing through this resistor is essentially zero (if you ignore the small bias currents) since the op amp's input impedance is quite large (recall rule 1 from Section 12.3). So there is negligible voltage drop across the resistor and the positive terminal is still at ground potential. This resistor performs no function in the circuit except to balance the voltage offset due to the bias currents. To achieve this balance, the resistor should be equal to the equivalent resistance at the op amp's negative terminal.

To determine the equivalent resistance at the negative terminal, we use the approaches described in Chapter 11 and make the usual assumption that the input to the op amp is from an ideal source (Figure 12.23). We also assume that the output of the op amp is essentially an ideal source. (Real op amp output characteristics are covered in Section 12.5.3, but the assumption is reasonable) The equivalent resistance hanging on the inverting terminal is the parallel combination of R_f and R_1 since both connected to ground as shown in Figure 12.23, right side:

$$R_{eq-} = \frac{R_f R_1}{R_f + R_1}; \quad \text{or more generally:} \quad Z_{eq-} = \frac{Z_f Z_1}{Z_f + Z_1} \tag{12.18}$$

So to balance the resistances (or impedances) at the input terminals, the resistance at the positive terminal should be set to the parallel combination of the two feedback resistors (Equation 12.18). Sometimes, as an approximation, the resistance at the positive terminal is set to equal the lower of the two feedback network resistors (usually R_1). Another strategy is to make this resistor variable, then adjust the resistance to cancel the output offset voltage from a given op amp. This has the advantage of removing any offset voltage

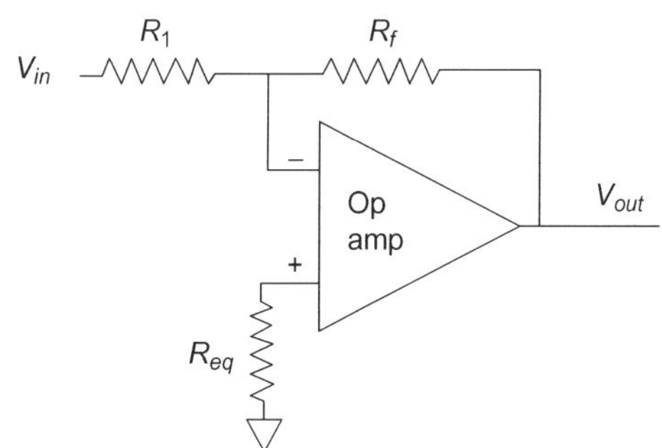

FIGURE 12.22 An inverting op amp with a resistor added to the noninverting terminal to balance the bias currents. R_{eq} should be set to equal the equivalent resistance on the inverting terminal. This circuit is also used in Example 12.8.

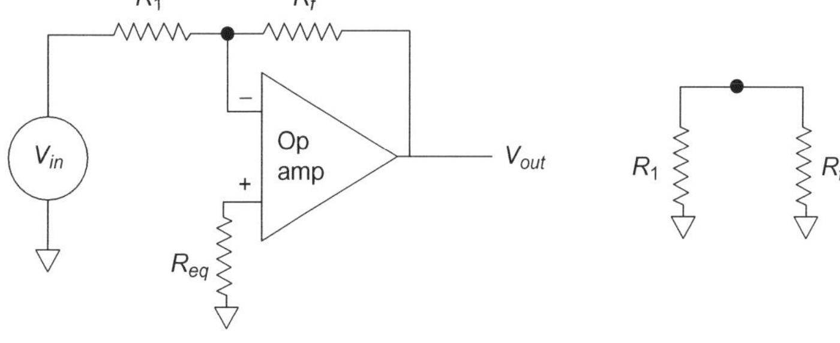

FIGURE 12.23 The left circuit is a typical inverting operational amplifier. If the op amp output, V_{out}, and the input, V_{in}, can be considered ideal sources, then they have effective impedances of 0 Ω. Since these two sources are connected to ground, the ends of the two resistors are connected to ground, one through the ideal input and the other through the op amp's ideal output.

due to the input offset voltage, v_o, as well. The primary downside to this approach is that the resistor must be carefully adjusted after the circuit is built. If the feedback circuit contains capacitors, then Z_{eq} will be an impedance equivalent to the parallel combination of the two feedback impedances.

The current noise source is treated the same way as the bias currents: it is multiplied by the two equivalent resistances to find the input current noise, then multiplied by the noise gain to find the output noise.

To find the total noise at the output, it is necessary to add in the voltage noise. Since the noise sources are independent, they add as the square root of the sum of the squares (Equation 1.7). In addition to the voltage and current noise of the op amp, the resistors will produce voltage noise as well. Repeating the equation for Johnson noise for a resistor from Chapter 1 (Equation 1.4):

$$V_J = \sqrt{4kT\,R\,BW} \text{ volts} \tag{12.19}$$

The three different noise sources associated with an operational amplifier circuit are all dependent on bandwidth. The easiest way to deal with these three different sources is to combine them in one equation that includes the bandwidth:

$$V_{n\ in} = [(v_n^2 + (i_n(R_{eq+} + R_{eq-}))^2 + 4kT(R_{eq+} + R_{eq-}))BW]^{\frac{1}{2}} \tag{12.20}$$

This equation gives the summed input noise. To find the output noise, multiply $V_{n\ in}$ by the noise gain.

$$V_{n\ out} = V_{n\ in}(Noise\ Gain) = V_{n\ in}\left(\frac{1}{\beta}\right) \tag{12.21}$$

Use of this approach to calculate the noise out of a typical op amp amplifier circuit is shown in Example 12.8.

EXAMPLE 12.8

Find the noise at the output of the operational amplifier circuit shown in Figure 12.22 where: $R_f = 500$ kΩ, $R_1 = 10$ kΩ, and $R_{eq} = 9.8$ kΩ.

Solution: First find the noise gain, $1/\beta$. From $1/\beta$ determine the bandwidth of the amplifier using the open-loop gain curves in Figure 12.13. Apply Equation 12.20 to find the total input voltage noise including resistor, current, and voltage noise. Then multiply this voltage by the noise gain to find output voltage noise (Equation 12.21).

The noise gain is:

$$\frac{1}{\beta} = \frac{R_f + R_1}{R_1} = \frac{500 + 10}{10} = 51$$

From the specifications of the LF356, $v_n = 12\ \text{nV}/\sqrt{\text{Hz}}$ and $i_n = 0.01\ \text{pA}/\sqrt{\text{Hz}}$. The equivalent resistance at the negative terminal is found from Equation 12.18 to be 9.8 kΩ (the parallel combination of 500 kΩ and 10 kΩ). For a $1/\beta$ of 51, the bandwidth is approximately 100 kHz (Figure 12.13). Using $T = 310°K$, $4kT$ is 1.7×10^{-20} J; and combining resistances Equation 12.20 becomes:

$$V_{n\ in} = [((12 \times 10^{-9})^2 + (0.01 \times 10^{-12}(19.6 \times 10^3))^2 + 1.7 \times 10^{-20}(19.6 \times 10^3))10^5]^{\frac{1}{2}}$$

$$V_{n\ in} = [(1.44 \times 10^{-16} + 3.8 \times 10^{-20} + 3.33 \times 10^{-16})10^5]^{\frac{1}{2}}$$

$$V_{n\ in} = [(4.77 \times 10^{-16})10^5]^{\frac{1}{2}} = 6.9\ \mu V$$

The noise at the output is found by multiplying by the noise gain:

$$V_{n\ out} = V_{n\ in}(Noise\ Gain) = 6.9 \times 10^{-6}(51) = 0.35\ mV$$

One advantage to including all the sources in a single equation is that the relative contributions of each source can be compared. After curents are converted to voltages, the current noise source is approximately four orders of magnitude less than the other two noise sources, so its contribution is negligible. The op amp's voltage noise does contribute to the overall noise, but most comes from the resistors. Finding an op amp with a lower noise voltage would imporve things, but of the 0.35-mV noise at the output, 0.29 mV is from the resistors. Of course, this is using the value of noise voltage for frequencies above 200 Hz. The noise voltage of the op amp at 10 Hz is 60 nV/$\sqrt{Hz}$, four times the value used in this example. If noise at the lower frequencies is a concern, another op amp should be considered. (For example, the OP27 op amp features a noise voltage of only 5.5 nV/$\sqrt{Hz}$ at 10 Hz.)

12.5.2.3 *Input Impedance*

While the input impedance of most op amps is quite large, the actual input impedance of the circuit depends on the configuration. The noninverting op amp has the highest input impedance, that of the op amp itself. In practice it may be difficult to attain the high impedance of many op amps due to leakage currents in the circuit board or wiring. Also the bias currents of an op amp will decrease its effective input impedance.

For an inverting amplifier, the input impedance is approximately equal to the input resistance, R_1 (see Figure 12.9). This is because the input resistor is connected to virtual ground in the inverting configuration. While this is much lower than the input impedance of the noninverting configuration, it is usually large enough for many applications. Where very high input impedance is required, the noninverting configuration should be used. If even higher input impedances are required, op amps with particularly high input impedances are available, but the limitations on impedance are usually set by other components of the circuit such as the lead-in wires and circuit boards.

12.5.3 Output Characteristics

Compared to the input characteristics, the output characteristics of an op amp are quite simple: a Thévenin source where the ideal source is $A_V\ (V_{in+} - V_{in-})$ and the resistance is r_{out} (Figure 12.24). For the LF356, A_V is given as a function of frequency in Figure 12.13 and r_{out} varies between 0.05 and 50 Ω depending on the frequency and closed-loop gain. The value of the output resistance is lowest at lower frequencies and when the closed-loop gain is 1 (i.e., $\beta = 1$).

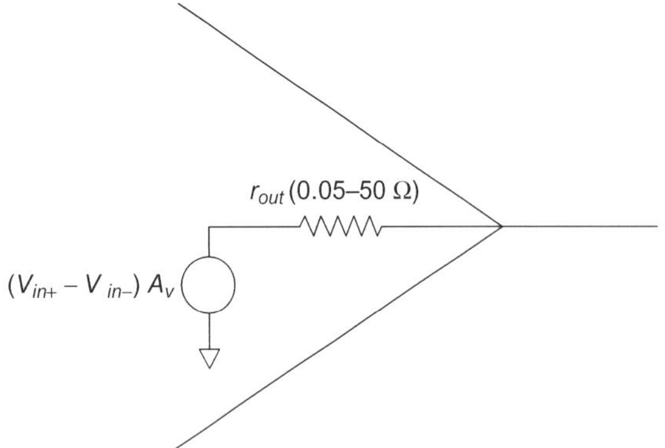

FIGURE 12.24 The output characteristics of an op amp are those of a Thévenin source including an ideal voltage source, $A_V\,(V_{in+} - V_{in-})$, and an output resistor, r_{out}.

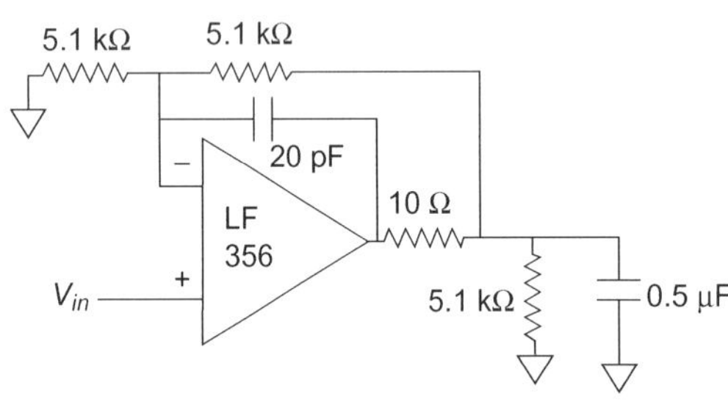

FIGURE 12.25 An op amp circuit that can be used to drive a large capacitive load. Note that this is a noninverting amplifier with a gain of 2. The main purpose of this circuit is to provide a drive to the capacitive load, not to amplify the signal.

In some circumstances, the output characteristics can become more complicated. Maximum voltage swing at the output must always be a few volts less than the voltage that powers the op amp (see next section), but the output signal range is also limited at higher frequencies. In addition, many op amps have stability problems when driving a capacitive load. Figure 12.25 shows a circuit taken from the spec sheet (Appendix F) that can be used to drive a large capacitive load. (A 0.5-µf capacitor is considered fairly large in electronic circuits.) One strategy illustrated in this figure is to place a resistor in parallel with the capacitor load. Another strategy that is shown is to place a small resistor at the output of the op amp before the feedback resistor. A final strategy is to add a small feedback capacitor which, as mentioned above, improves stability. These strategies are often implemented on an ad-hoc basis, but the design engineer should anticipate possible problems when capacitive (or inductive) loads are involved.

12.6 POWER SUPPLY

Op amps are active devices and require external power to operate. This external power is delivered as a constant voltage or voltages from a device known, logically, as a *power supply*. Power supplies are commercially available in a wide range of voltages and current capabilities. Many operational amplifiers are *bipolar*, that is, they handle both positive and negative voltages. (Unlike its use in psychology, in electronics the term bipolar has

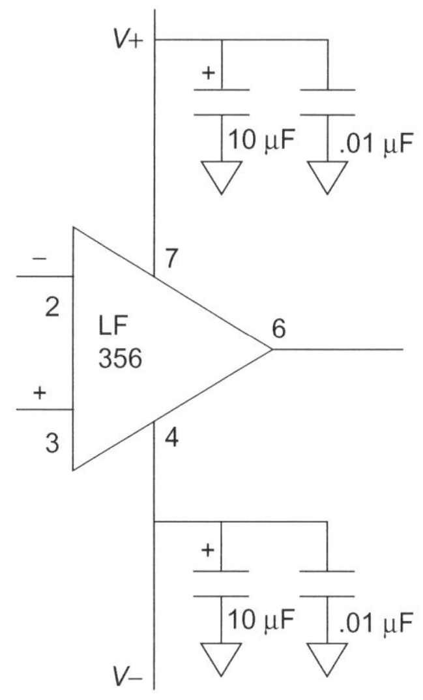

FIGURE 12.26 A LF356 op amp connected to positive and negative power supply lines. The number next to each line indicates the pin number of a common version of the chip (the 8-pin DIP package). The capacitors are used to remove noise or fluctuations on the power supply lines that may be produced by other components in the system. The capacitors are called *decoupling capacitors.*

nothing to do with stability.) Bipolar applications require both positive and negative power supply voltages, and values of ±12 volts or ±15 volts are common. The higher the power supply voltage, the larger the output voltages the op amp can produce, but all op amps have a maximum voltage limit. The maximum voltage for the LF356 is ±18 volts, but a special version, the LF356B, can handle ±22 volts. High-voltage op amps are available as are low-voltage op amps for battery use. The latter also feature lower current consumption. (The LF356 uses a nominal 5 to 10 mA and a few of them in a circuit will use up a 9-volt battery fairly quickly.)

The power supply connections are indicated on the op amp schematic by vertical lines coming from the side of the amplifier icon, as shown in Figure 12.26. Sometimes the actual chip pins are indicated on the schematic, as in this figure. Figure 12.26 also shows a curious collection of capacitors attached to the two power supply lines. Power supply lines often go to a number of different op amps or other analog circuitry. These common power supply lines make great pathways for spreading signal artifacts, noise, positive feedback signals, and other undesirable fluctuations. One op amp circuit might induce fluctuations on the power supply line(s), and these fluctuations then pass to all the other circuits. Practical op amps do have some immunity to power supply fluctuations, but this immunity falls off significantly with the frequency of the fluctuations. For example, the LF356 will attenuate power supply variations at 100 Hz by 90 dB (a factor of 31,623), but this attenuation falls to 10 dB (a factor of 3) at 1 MHz.

A capacitor placed right at the power supply pin will tend to smooth out voltage fluctuations and reduce artifacts induced by the power supply. Since such a capacitor tends to isolate or disconnect the op amp from power line noise it is called a *decoupling capacitor.* Figure 12.26 shows two capacitors on each supply line: a large 10 μf capacitor and a small 0.01 μf capacitor. Since the two capacitors are in parallel they are in theory equivalent to a single 10.01 μf capacitor. The small capacitor appears to be contributing very little. In fact, the small capacitor is there because large capacitors have poor high-frequency performance: They look more like inductors than capacitors at higher frequencies. The small capacitor serves to reduce high-frequency fluctuations while the large capacitor does the same at low frequencies. While a given op amp circuit may not need these decoupling capacitors, the 0.01 μF is routinely included by most design engineers on every power supply pin of every op amp. The larger capacitor is added if strong low-frequency signals are present in the network.

12.7 OP AMP CIRCUITS OR 101 THINGS TO DO WITH AN OP AMP

There are actually more than 101 different signal processing operations that can be performed by op amp circuits, but since this is an introductory course, only a handful are presented. For a look at the other 90 +, see *The Art of Electronics* by Horowitz and Hill (1989).

12.7.1 The Differential Amplifier

We have already shown how to construct inverting and noninverting amplifiers. Why not throw the two together to produce an amplifier that does both: a differential amplifier? As shown in Figure 12.27, a differential amplifier is a combination of inverting and noninverting amplifiers.

To derive the transfer function of the circuit in Figure 12.27, we once again employ the principle of superposition. Setting V_{in2} to zero effectively grounds the lower R_1 resistor, and the circuit becomes a standard inverting op amp with a resistance between the positive terminal and ground (Figure 12.28, left side). As stated previously, the only effect of this resistance is balanceing of the bias currents. For this partial circuit, the transfer function is:

$$V_{out} = -\frac{R_f}{R_1} V_{in1}$$

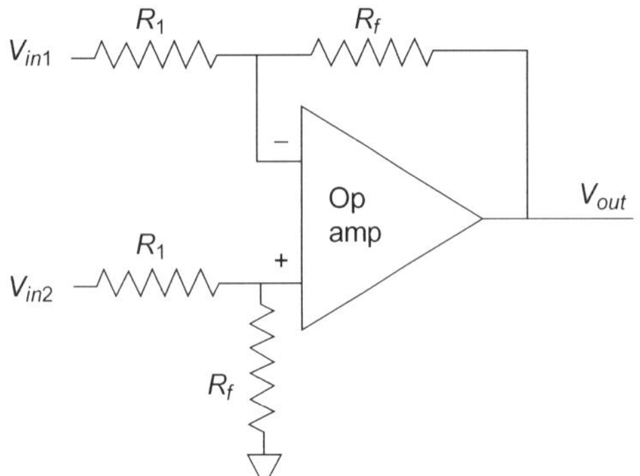

FIGURE 12.27 A differential amplifier circuit. This amplifier combines both inverting and noninverting amplifiers into a single circuit. As shown in the text, this circuit amplifies the difference between the two input voltages:

$$V_{out} = \frac{R_f}{R_{in}}(V_{in2} - V_{in1})$$

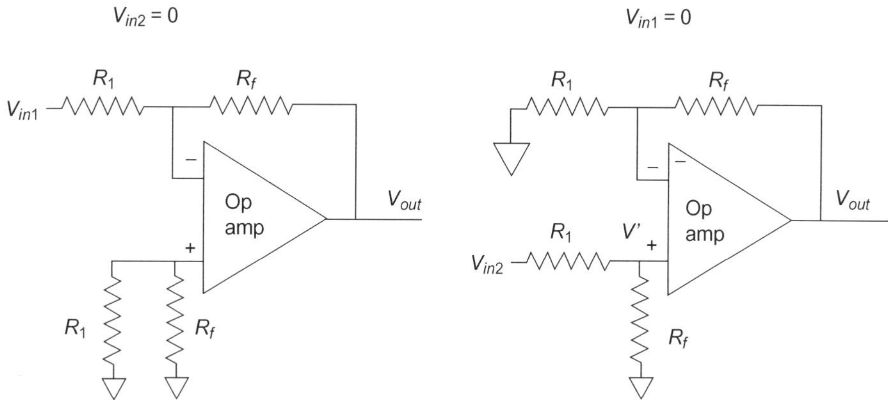

FIGURE 12.28 Superposition applied to the differential amplifier. Left circuit: The V_{in2} input is set to zero (i.e., grounded), leaving a standard inverting amplifier circuit. Right circuit: The V_{in1} input is grounded, leaving a noninverting operational amplifier with a voltage divider circuit on the input. Each one of these circuits will be solved separately and the overall transfer function determined by superposition.

Setting V_{in1} to zero grounds the upper R_1 resistor, and the circuit becomes a noninverting amplifier with a voltage divider on the input. With respect to the voltage V' (Figure 12.28, right side), the circuit is a standard noninverting op amp:

$$V_{out} = \frac{R_f + R_1}{R_1} V'$$

The relationship between V_{in2} and V' is given by the voltage divider equation:

$$V' = \frac{R_f}{R_f + R_1} V_{in2}$$

Substituting and solving for V_{in2}:

$$V_{out} = \frac{R_f + R_1}{R_1} V' = \frac{R_f + R_1}{R_1} \frac{R_f}{R_f + R_1} V_{in2} = \frac{R_f}{R_1} V_{in2}$$

By superposition, the two partial solutions can be combined to find the transfer function when both voltages are present.

$$V_{out} - \frac{R_f}{R_1} V_{in1} + \frac{R_f}{R_1} V_{in2} = \frac{R_f}{R_1}(V_{in2} - V_{in1}) \tag{12.22}$$

Thus, the circuit shown in Figure 12.27 amplifies the difference between the two input voltages.

12.7.2 The Adder

If the sum of two or more voltages is desired, the circuit shown in Figure 12.29 can be used.

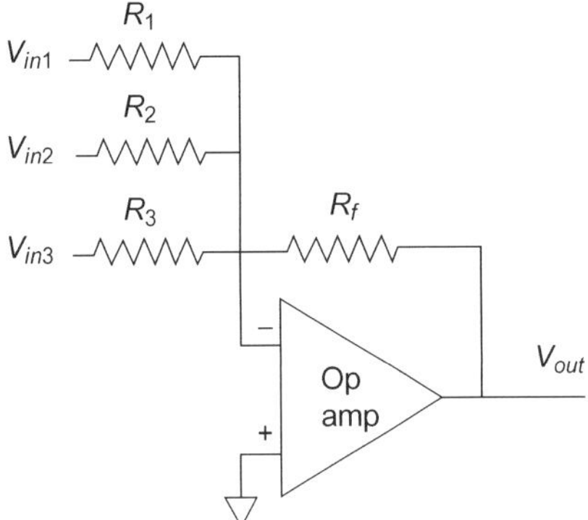

FIGURE 12.29 An op amp circuit that takes a weighted sum of three input voltages.

It is easy to show, using an extension of the approach used in Example 12.2, that the transfer function of this circuit is:

$$V_{out} = \left(\frac{R_f}{R_1}\right)V_{in1} + \left(\frac{R_f}{R_2}\right)V_{in2} + \left(\frac{R_f}{R_3}\right)V_{in3} \tag{12.23}$$

The derivation of this equation is requested in Problem 8 at the end of the chapter. This circuit can be extended to any number of inputs by adding more input resistors. If $R_1 = R_2 = R_3$, then the output is the straight sum of the three input signals amplified by R_f/R_1.

12.7.3 The Buffer Amplifier

At first glance the circuit in Figure 12.30 appears to be of little value. In this circuit all of the output is feedback to the inverting input terminal, so the feedback gain, β, equals 1. Since the gain of a noninverting amplifier is $1/\beta$, the gain of this amplifier is 1 and $V_{out} = V_{in}$. Although this amplifier does nothing to enhance the amplitude of the signal, it does a great deal when it comes to impedance. Specifically, the incoming signal sees a very large impedance, the input impedance of the operational amplifier ($>10^{12}\,\Omega$ for the LF356), while the output impedance is very low ($0.02\,\Omega$ at 10 kHz for the LF356), approaching that of an ideal source. This circuit can take a signal from a high-impedance Thévenin source and provide a low-impedance, nearly ideal, source that can be used to drive several other devices. While all noninverting op amp circuits have this impedance transformation function, the unity gain circuits are particularly effective and have the highest bandwidth, since $1/\beta = 1$. Since this circuit provides a buffer between the high-impedance source and the other devices, it is sometimes referred to as a *buffer amplifier*. The low-output impedance also reduces noise pick-up in the output wires, and this circuit can be invaluable whenever a signal is sent over long wires or just off the circuit board. Many design engineers routinely use a buffer amplifier as the output to any signal that will be sent any distance.

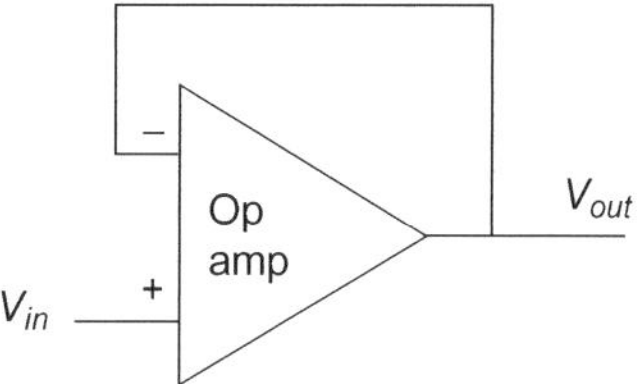

FIGURE 12.30 A *buffer amplifier* circuit. This amplifier provides no gain ($G = 1/\beta = 1$), but presents a very high impedance to the signal source (nearly an ideal load) and generates a low-impedance signal that approximates an ideal source.

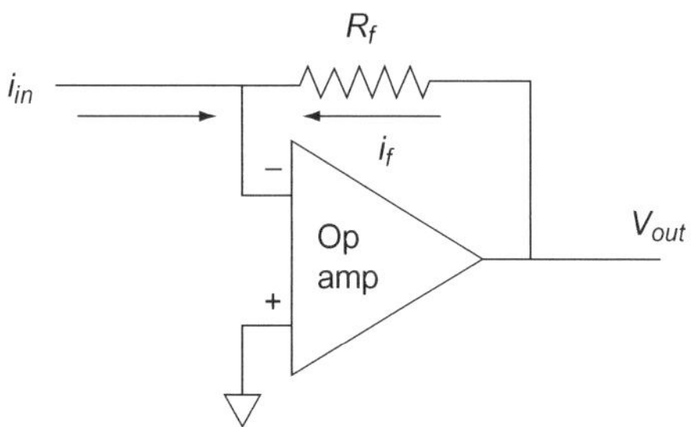

FIGURE 12.31 An op amp circuit used to convert a current to a voltage. Because of this current-to-voltage transformation, this circuit is also referred to as a *transconductance amplifier*.

12.7.4 The Transconductance Amplifier[1]

Figure 12.31 shows another simple circuit that looks like an inverting op amp circuit except that the input resistor is missing.

The input to this circuit is a current, not a voltage, and the circuit is used to convert this current into a voltage. Applying KCL to the negative input terminal, the transfer function for this circuit is easily determined.

$$i_n = i_f = \frac{V_{out}}{R_f};$$

Solving for V_{out}:

$$V_{out} = R_f\, i_n \tag{12.24}$$

Some transducers produce current as their output and this circuit can be used as the first stage to convert that current signal to a voltage. A common example in medical instruments is the photodetector transducer. These light-detecting transducers usually produce a current proportional to the light falling on them. These currents can be very small (in the nanoamps or picoamps) and R_f is chosen to be very large ($\approx$10–100 MΩ) so that a reasonable output voltage is produced.

Since the input to the op amp is current, noise depends only on current noise. This includes the noise current generated by both the op amp and the resistor. The net input current noise is then multiplied by R_f to find the output voltage noise. This approach is illustrated in the practical problem posed in Example 12.9.

[1]The tranconductance amplifier described here takes in a current and puts out a voltage. There are also tranconductance amplifiers that do the opposite: output a current proportional to an input voltage.

EXAMPLE 12.9

A transconductance amplifier shown in Figure 12.32 is used to convert the output of a photodetector into a voltage. The photodetector has a sensitivity of: $R = 0.01\ \mu A/\mu W$. The R stands for *responsivity*, which is the same as sensitivity. Sensitivity functions as the transfer function of a transducer, the output for a given input. The photodiode also has a noise current which is called *dark current*, of $i_d = 0.5\ pA/\sqrt{Hz}$. What is the minimum light flux, φ in μW, that can be detected with an SNR of 20 dB and a bandwidth of 1 kHz?

Solution: This problem requires a number of steps, but the heart of the problem is how much current noise is generated at the input of the op amp. Once this is determined, the minimum signal current can be determined as 20 dB, or a factor of 10, times this current noise. The responsivity defines the output for a given input ($\mathscr{R} = \varphi/i$), so, once the minimum signal current is found, the minimum light flux, φ_{min}, can be calculated as: $i_{min}/\mathscr{R}$. To find the noise current, use a version of Equation 12.20 for current rather than voltage:

$$i_{n\ Total} = \left[\left(i_d^2 + i_n^2 + \frac{4kT}{R_f}\right)BW\right]^{\frac{1}{2}} \tag{12.25}$$

Note that the value of current noise decreases for increased values of R_f, so a good design would use as large a value of R_f as possibly needed.

Solving into Equation 12.25 with the bandwidth of 1 kHz, a value of i_n from the LF356 spec sheet, the value of i_d from the photodetector, and a value of R_f of 10 MΩ:

$$i_{n\ Total} = \left[\left((0.05 \times 10^{-12})^2 + (0.01 \times 10^{-12})^2 + \frac{1.7 \times 10^{-20}}{10^7}\right)10^3\right]^{\frac{1}{2}}$$
$$= \left[(2.5 \times 10^{-27} + 1 \times 10^{-28} + 1.7 \times 10^{-27})10^3\right]^{\frac{1}{2}} = 2.07 \times 10^{-12}\ A = 2.07\ pA$$

Thus the minimum signal required for an SNR of 20 dB is $10 \times 2.07\ pA = 20.7\ pA$. From the sensitivity of the photodetector, the minimum light flux that can be detected is:

$$\phi_{min} = \frac{i_{min}}{\mathscr{R}} = \frac{20.7 \times 10^{-12}}{0.01} = 2.07 \times 10^{-9}\ W = 2.07 \times 10^{-3}\ \mu W$$

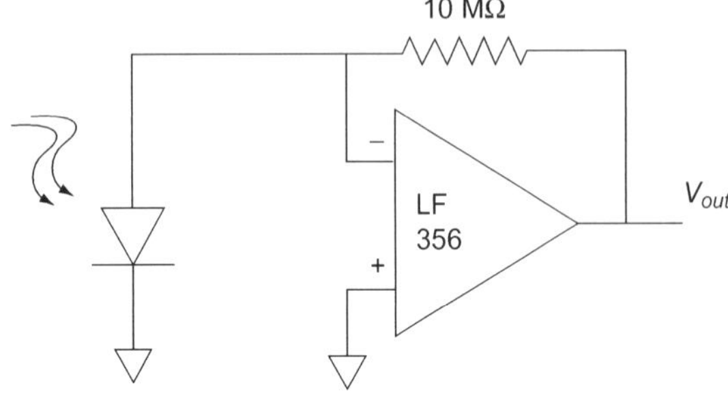

FIGURE 12.32 The output of a photodetector is fed to a transconductance amplifier. This circuit is used in Example 12.9 to determine the minimum light flux that can be detected with an SNR of 20 dB.

Note that since $\mathcal{R}$ is in μA/μW and these are scaled versions of A/W, it is not necessary to scale this number. To determine the output voltage produced by this minimum signal, or any other signal for that matter, simply multiply this input current by R_f:

$$V_{out} = i_{in} R_f = 2.07 \times 10^{-9}(10^7) = 20.7 \text{ mV}$$

This is a small signal, so it would be a good idea to increase the value of R_f. Since R_f is the second largest contributor of noise, increasing its value by a factor of 10 both increases the output signal by that amount and decreases the noise (since noise current is inversely related to the resistance). Improvement can be obtained by using an op amp with a lower current noise since, as seen in the calculations, this is the largest noise source.

12.7.5 Analog Filters

A simple single-pole low-pass filter can be constructed using an R-C circuit. An op amp can also be used to construct a low-pass filter with improved input and output characteristics and also provide increased signal amplitude. The easiest way to construct an op amp low-pass filter is to add a capacitor in parallel to the feedback resistor, as shown in Figure 12.33.

The equation for the transfer function of an inverting op amp with impedances in the feedback circuit is given in Equation 12.12 and repeated here:

$$\frac{V_{out}}{V_{in}} = -\frac{Z_f}{Z_1} \tag{12.26}$$

Applying Equation 12.26 to the circuit in Figure 12.33:

$$\frac{V_{out}}{V_{in}} = -\frac{Z_f}{Z_1} = -\frac{\frac{R_f \frac{1}{j\omega C_f}}{R_f + \frac{1}{j\omega C_f}}}{R_1} = \left(\frac{R_f}{R_1}\right)\frac{1}{(1 + j\omega R_f C_f)} \tag{12.27}$$

At frequencies well below the cutoff frequency, the second term goes to 1 and the gain is R_f/R_1. The second term is a low-pass filter with a cutoff frequency of $\omega = 1/(R_f C_f)$ rad/sec or $f = 1/(2\pi R_f C_f)$ Hz. Design of an active low-pass filter is found in Problem 12.

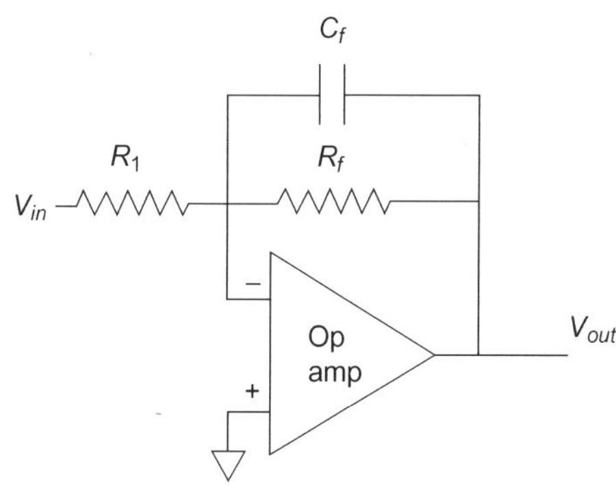

FIGURE 12.33 An inverting amplifier that also functions as a low-pass filter. As shown in the text, the low-frequency gain of this amplifier is R_f/R_1 and the cutoff frequency is $\omega = 1/(R_f C_f)$.

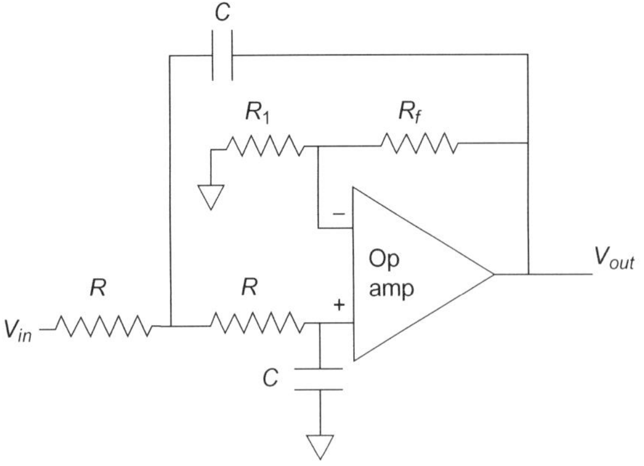

FIGURE 12.34 A 2-pole active low-pass filter constructed using a single op amp.

It is also possible to construct a second-order filter using a single op amp. A popular second-order op amp circuit is shown in Figure 12.34.

Derivation of the transfer function requires applying KCL to two nodes and is provided in Appendix A. The transfer function is:

$$\frac{V_{out}}{V_{in}} = \frac{\frac{G}{(RC)^2}}{s^2 + \frac{3-G}{RC}s + \frac{1}{(RC)^2}} \tag{12.28}$$

where:

$$G = \frac{R_f + R_1}{R_1}$$

where G is the gain of the noninverting amplifier and equals $1/\beta$. Equating coefficients of Equation 12.28 with the standard Laplace transfer function of a second-order system (Equation 5.45 and 6.31):

$$\omega_o = \frac{1}{RC}; \quad \delta = \frac{3-G}{2} \tag{12.29}$$

EXAMPLE 12.10

Design a second-order filter with a cutoff frequency of 5 kHz and a damping of 1.0.

Solution: Since there are more unknowns than equations, several component values may be set arbitrarily with the rest determined by Equation 12.29. To find R and C, pick a value for C that is easy to obtain, then calculate the value for R (capacitor values are more limited than resistor values). Assume $C = 0.001\ \mu f$, a common value. Then the value of R is:

$$\omega_o = 2\pi f = \frac{1}{RC} = 2\pi(5000) = 31,416 \text{ rad/sec}$$

$$R = \frac{1}{31,416(0.001 \times 10^{-6})} = 31.8 \text{ k}\Omega$$

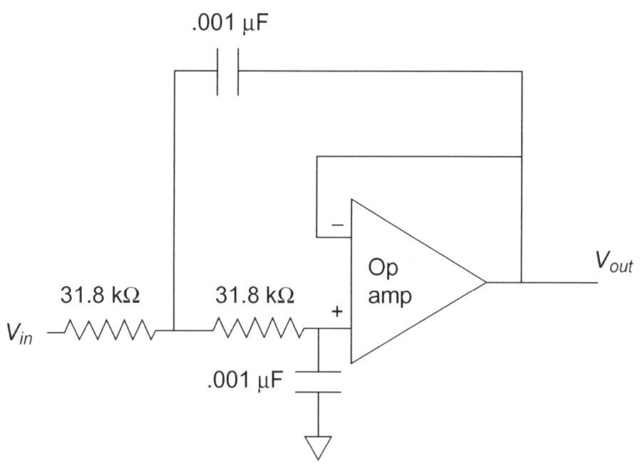

FIGURE 12.35 A 2-pole low-pass filter with a cutoff frequency of 5 kHz and a damping factor of 1.0.

The value for G, the gain of the noninverting amplifier, would be:

$$\delta = \frac{3 - G}{2} = 1.0; \quad G = 3 - 2\delta = 3 - 2 = 1$$

Hence for this particular damping, $G = 1/\beta = 1$ and $\beta = 1.0$. So all of the output is feedback to the noninverting input and a resistor divider network is not required. Other values of damping require the standard resistor divider network to achieve the desired gain. A second-order active filter having the desired cutoff frequency and damping is shown in Figure 12.35.

12.7.6 Instrumentation Amplifier

The differential amplifier shown in Figure 12.27 is useful in certain biomedical engineering applications, specifically to amplify signals from biotransducers that produce a differential output. Such transducers actually produce two voltages that move in opposite directions in response to their input. An example of such a transducer is the strain gage bridge shown in Figure 12.36.

Here the strain gages are arranged in such a way that when a force is applied to the gages, two of them (A-B and C-D) undergo tension while the other two (B-C and D-A)

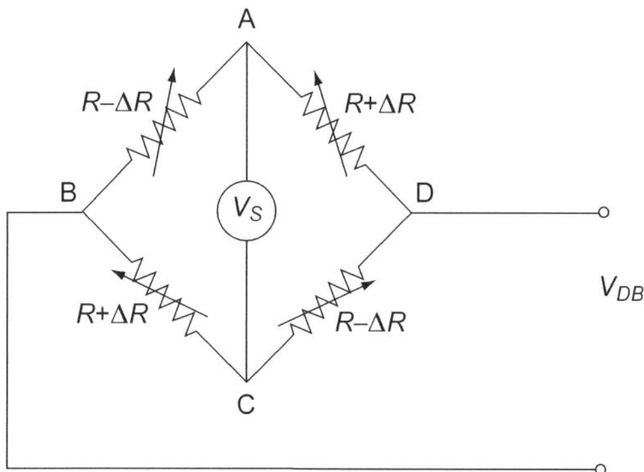

FIGURE 12.36 A bridge circuit that produces a differential output. The voltage at D moves in opposition to the voltage at B. The output voltage is best amplified by a differential amplifier.

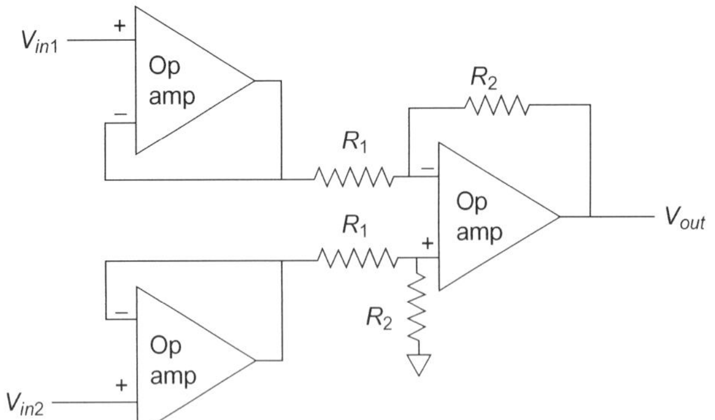

FIGURE 12.37 A deferential amplifier circuit with high input impedance. Resistor R_1 can be adjusted to balance the differential gain so that the two channels have equal but opposite gains. In the interest of symmetry, it is common to reverse the position of the positive and negative op amp inputs in the upper input op amp.

undergo compression. The two gages under tension decrease their resistance while the two under compression increase their resistance. The net effect is that the voltage at B increases while the voltage at D decreases an equal amount in response to the applied force. If the difference between these voltages is amplified using a differential amplifier such as the one shown in Figure 12.27, the output voltage will be the difference between the two voltages and reflect the force applied. If the force reverses, the output voltage will change sign.

One of the significant advantages of this differential operation is that much of the noise, particularly noise picked up by the wires leading to the differential amplifier, will be common to both of the inputs and will tend to cancel. To optimize this kind of noise cancellation, the gain of each of the two inputs must be exactly equal in magnitude (but opposite in sign, of course). Not only must the two inputs be balanced, but the input impedance should also be balanced and often it is desirable that the input impedance be quite high. An *instrumentation amplifier* is a differential amplifier circuit that meets these criteria: balanced gain along with balanced and high-input impedance. In addition, low noise is a common and desirable feature of instrumentation amplifiers.

A circuit that fulfills this role is shown in Figure 12.37. The output op amp performs the differential operation, and the two leading op amps configured as the unity gain buffer amplifier provide similar high-impedance inputs. If the requirements for balanced gain are high, one of the resistors is adjusted until the two channels have equal but opposite gains. It is common to adjust the lower R_1 resistor. Since the two input op amps provide no gain, the transfer function of this circuit is just the transfer function of the second stage, which is shown in Equation 12.22 to be:

$$V_{out} = \frac{R_1 + R_2}{R_1}(V_{in2} - V_{in1}) \tag{12.30}$$

There is one serious drawback to the circuit in Figure 12.37. To increase or decrease the gain it is necessary to change two resistors simultaneously: either both R_1s or both R_2s. Moreover, to maintain balance, they both have to be changed exactly the same amount. This can present practical difficulties. A differential amplifier circuit that requires only one resistor change for gain adjustment is shown in Figure 12.38. The derivation for the

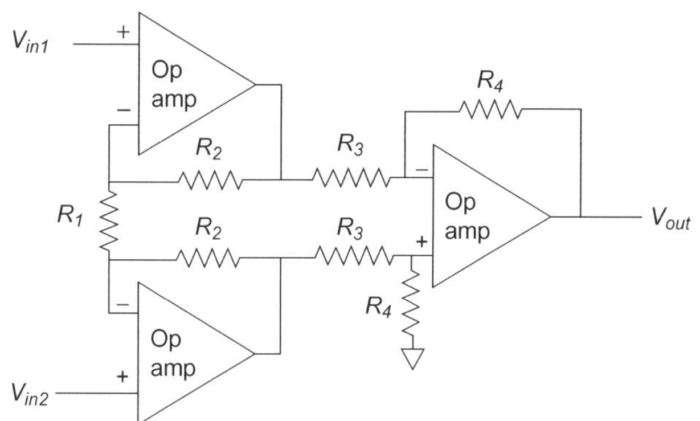

FIGURE 12.38 An instrumentation amplifier circuit. This circuit has all the advantages of the one in Figure 12.37 (i.e., balanced channel gains and high input impedance), but with the added advantage that the gain can be adjusted by modifying a single resistor, R_1.

input-output relationship of this circuit is more complicated than for the previous circuit, and is given in Appendix A:

$$V_{out} = \frac{R_4}{R_3}\left(\frac{R_1 + 2R_2}{R_1}\right)(V_{in2} - V_{in1}) \tag{12.31}$$

Since R_1 is a now a single resistor, the gain can be adjusted by modifying only this resistor. As this resistor is common to both channels, changing its value affects the gain of each channel equally and does not alter the balance between the gains of the two channels. It is possible to obtain integrated circuit instrumentation amplifiers that place all the components of Figure 12.38 on a single chip. Such packages generally have very good balance between the two channels, very high input impedance, and low noise. For example, an instrumentation amplifier made by Analog Devices, Inc., the ADC624, has an input impedance of $10^9\ \Omega$ and a noise voltage of 4.0 $nV/\sqrt{Hz}$ at 1.0 kHz. Such a chip also includes a collection of highly accurate internal resistors that can be used to set specific amplifier gains with no need of external components (just jumper wires between the appropriate pins).

The balance between the channels is measured in terms of V_{out} when the two inputs are at the same voltage. Voltage that is common (i.e., the same) to both input terminals is termed the *common mode voltage*. In theory, the output should be zero no matter what the input voltage is so long as it is the same at both inputs. However, any imbalance between the gains of the two channels will produce some output voltage, and this voltage will be proportional to the common mode voltage. Since the idea is to have the most cancellation and the smallest output voltage to a common mode signal, the common mode voltage is specified as inverse gain. This inverse gain is called the *common mode rejection ratio*, or *CMRR*, and is usually given in dB.

$$V_{out} = \frac{V_{CM}}{CMRR} \tag{12.32}$$

Hence the higher the CMRR, the smaller the output voltage due to common mode voltage and the better the noise cancellation. The ADC624 has a CMRR of 120 dB. This means that the common mode gain is -120 dB. For example, if 10 volts were applied to each of the input terminals (i.e., $V_{in1} = V_{in2} = 10$ v), V_{out} would be:

$$V_{out} = \frac{V_{CM}}{CMRR} = \frac{10}{120\ \text{dB}} = \frac{10}{10^{\frac{120}{20}}} = \frac{10}{10^6} = 10\ \mu\text{V}$$

While not zero, this value will be close to the noise level for most applications.

12.8 SUMMARY

There is more to analog electronics than just op amp circuits, but they do encompass the majority of analog applications. The ideal op amp is an extension of the concept of an ideal amplifier. An ideal amplifier has infinite input impedance, zero output impedance, and a fixed gain at all frequencies. An ideal op amp has infinite input impedance and zero output impedance, but has infinite gain. The actual gain of an op amp circuit is determined by the feedback network, which is generally constructed from passive devices. This provides great flexibility with a wide variety of design options and the inherent robustness and long-term stability of passive elements.

Real operational amplifiers come reasonably close to the idealization. They have very high input impedances and quite low output impedances. Deviations from the ideal fall into three categories: deviations in transfer characteristics, deviations in input characteristics, and deviations in output characteristics. The two most important transfer characteristics are bandwidth and stability where stability means the avoidance of oscillation. The bandwidth of an operational amplifier circuit can be determined by combining the frequency characteristics of the feedback network with the frequency characteristics of the op amp itself. The immunity of an op amp circuit from oscillation can also be estimated from the frequency characteristics of the particular op amp and those of the feedback network. Input errors include bias voltages and currents, and noise voltages and currents. The bias and noise currents are usually converted to voltages by multiplying them by the equivalent resistance at each of the input terminals. The effect of these input errors on the output can be determined by multiplying all the input voltage errors by the noise gain, $1/\beta$. Output deviations consist of small nonzero impedances and limitations on the voltage swing.

There is a wide variety of useful analog circuits based on the operational amplifier. These include both inverting and noninverting amplifiers, filters, buffers, adders, subtractors including differential amplifiers, transconductance amplifiers, and many more circuits not discussed here. The design and construction of real circuits that use op amps is fairly straightforward, although some care may be necessary to prevent noise and artifact from spreading through power supply lines. Decoupling capacitors, capacitors running from the power supply lines to ground, are placed at the op amp's power supply feed to reduce noise carried over the power lines.

PROBLEMS

1. Design a noninverting amplifier circuit with a gain of 500.
2. Design an inverting amplifier with a variable gain from 50 to 250.

3. What is the bandwidth of the noninverting amplifier below? If the same feedback network were used to design an inverting amplifier (by switching ground and V_{in}), what would be the bandwidth of this circuit?

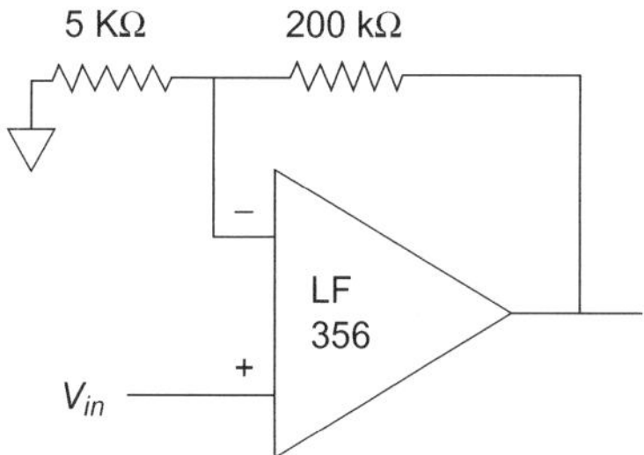

4. An amplifier has a gain bandwidth product (GBP) of 10 MHz. It is used in a noninverting amplifier where $\beta = 0.01$. What is the gain of the amplifier? What is the bandwidth?
5. An LF356 is used to implement the variable gain amplifier in Example 12.3. What are the bandwidths of this circuit at the two extremes of gain?
6. A 0.001 μf capacitor is added to the feedback circuit of the inverting op amp circuit shown below. You can assume that before the capacitor was added the phase shift due to the amplifier when $Av\beta = 1$ was 120 degrees. (The criterion for stability is that the phase shift induced by the op amp and the feedback network must be less than 180 degrees when $Av\beta = 1$, Equation 12.14.) After the capacitor is added, what is the phase shift of the op amp plus feedback network at the frequency where $Av\beta = 1$? Follow the example given in Section 12.5.1.2 on stability.

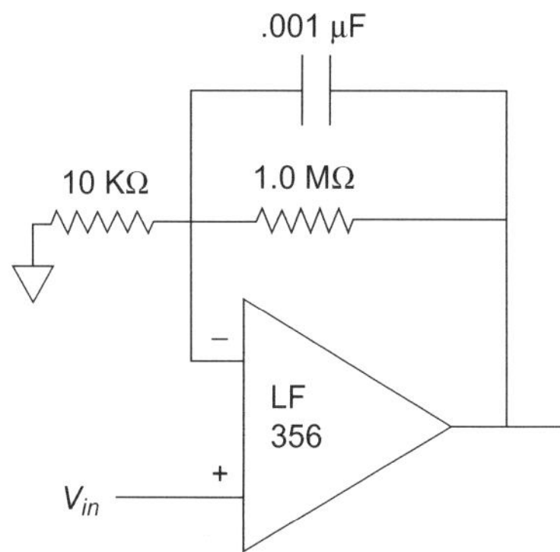

7. For the circuit below, what is the total offset voltage at the output? How much is this offset voltage increased if the 19-kΩ ground resistor on the positive terminal is replaced with a short circuit?

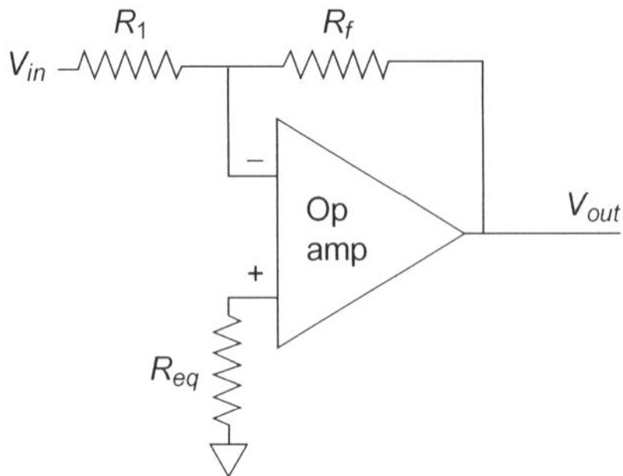

8. Derive the transfer function of the adder circuit below. Use KVL applied to node A. [Hint. Node A is virtual ground.]

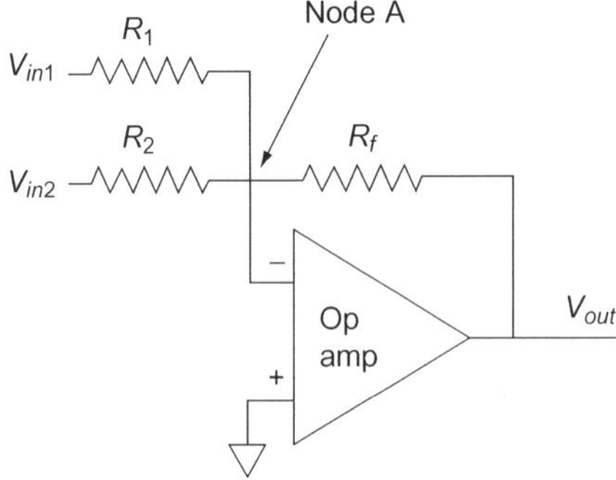

9. Find the total noise at the output of the circuit below. Identify the major source(s) of noise. What will the output noise be if the 50-kΩ ground resistor at the positive terminal is replaced with a short circuit? (Note that this resistor adds to both the Johnson noise from the resistors and to the voltage noise generated by the op amp's noise current.)

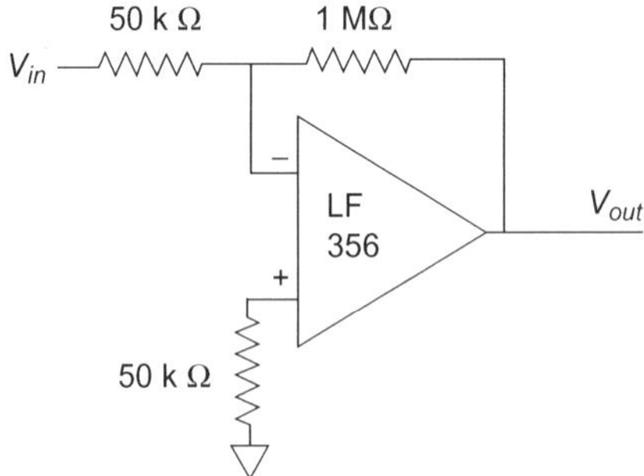

10. For the circuit in Problem 9, what is the minimum input signal that can be detected with a signal-to-noise ratio of 10 dB? What will be the voltage of such a signal at V_{out}?

11. An op amp has a noise current of 0.1 pA$\sqrt{}$Hz. This op amp is used as a transconductance amplifier (Figure 12.31). What should the minimum value of feedback resistance be so that the noise contribution from the resistor is less than the noise contribution from the op amp?
12. Design a 1-pole low-pass filter with a bandwidth of 1 kHz. Assume you have capacitor values of 0.001 μF, 0.01 μF, 0.05 μF, and 0.1 μF, and a wide range of resistors.
13. Design a 2-pole low-pass filter with a cutoff frequency of 500 Hz and a damping factor of 0.8. Assume the same component availability as in Problem 12.
14. Design a 2-pole high-pass filter with a cutoff frequency of 10 kHz and a damping factor of 0.707. (The circuit for a high-pass filter is the same as for a low-pass filter except that the capacitors and resistors are reversed, as shown in the figure below.)

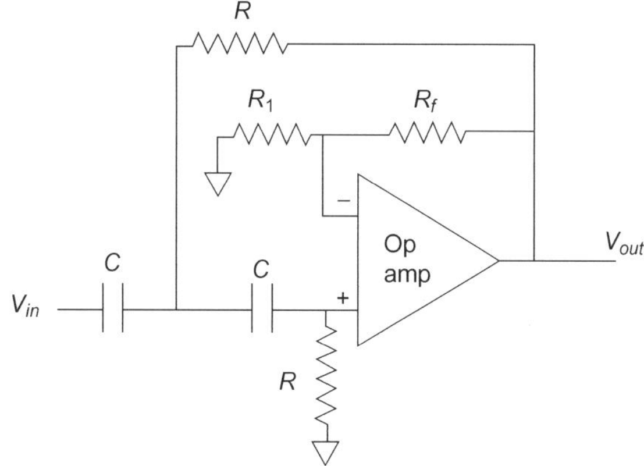

15. Design an instrumentation amplifier with a switchable gain of 10, 100, and 1000. [*Hint*: Switch the necessary resistors in or out of the circuit as needed.]

Appendix A
Derivations

A.1 DERIVATION OF EULER'S FORMULA

Assume a sinusoidal function:

$$x = \cos\theta + j\sin\theta \tag{A.1}$$

(where $j = \sqrt{-1}$ as usual)

Differentiating with respect to θ produces:

$$\frac{dx}{d\theta} = j(\cos\theta + j\sin\theta) = jx \tag{A.2}$$

Separating the variables gives:

$$\frac{dx}{x} = jd\theta \tag{A.3}$$

and integrating both sides gives:

$$\ln x = j\theta + K$$

where K is the constant of integration. To solve for this constant, note that in Equation A.1: $x = 1$ when $\theta = 0$. Applying this condition to Equation A.3:

$\ln 1 = 0 = 0 + K;\ K = 0;$

so Equation A.3 becomes: $\ln x = j\theta$
or

$$x = e^{j\theta} \tag{A.4}$$

but since x is defined in Equation A.1 as: $\cos\theta + j\sin\theta$

$$e^{j\theta} = \cos\theta + j\sin\theta \tag{A.5}$$

Alternatively,

$$e^{-j\theta} = \cos\theta - j\sin\theta \tag{A.6}$$

A.2 CONFIRMATION OF THE FOURIER SERIES

Fourier showed that a periodic function of period T can be represented by a series, possibly infinite, of sinusoids, or sine and cosines:

$$x(t) = \frac{a_0}{2} + \sum_{n=1}^{\infty}(a_n \cos n\omega_o t + b_n \sin n\omega_o t) \tag{A.7}$$

where $\omega_o = 2\pi/T$ and a_n and b_n are the Fourier coefficients.

To derive the Fourier coefficients, let us begin with the a_0 or DC term. Integrating both sides of Equation A.7 over a full period:

$$\int_0^T x(t)dt = \int_0^T \frac{a_0}{2}dt + \sum_{n=1}^{\infty}\int_0^T (a_n \cos n\omega_o t + b_n \sin n\omega_o t)dt \tag{A.8}$$

For $n = 0$, the second term on the right-hand side is zero since the sum begins at $n = 1$, and the equation becomes

$$\int_0^T x(t)dt = \int_0^T \frac{a_0}{2}dt; \quad \int_0^T x(t)dt = \frac{a_0 T}{2};$$
$$a_0 = \frac{2}{T}\int_0^T x(t)dt \tag{A.9}$$

To find the other coefficients, multiply both sides of Equation A.7 by $\cos(m\omega_o t)$, where m is an integer, and again integrate both sides.

$$\int_0^T x(t)\cos(m\omega_o t)dt = \int_0^T \frac{a_0}{2}\cos(m\omega_o t)dt + \sum_{n=1}^{\infty}\int_0^T (a_n\cos(m\omega_o t)\cos n\omega_o t + b_n\cos(m\omega_o t)\sin n\omega_o t)dt \tag{A.10}$$

Rearranging:

$$\int_0^T x(t)\cos(m\omega_o t)dt = \int_0^T \frac{a_0}{2}\cos(m\omega_o t)dt + \sum_{n=1}^{\infty} a_n \int_0^T \cos(m\omega_o t)\cos n\omega_o t dt$$
$$+ \sum_{n=1}^{\infty} b_n \int_0^T \cos(m\omega_o t)\sin n\omega_o t \tag{A.11}$$

Since m is an integer, the first term and third terms on the right-hand side integrate to zero for all m. The two terms integrate to zero for all m except $m = n$. At $m = n$, the second term becomes:

$$an\int_0^T \cos^2(n\omega_o t)dt = \frac{\pi}{\omega_o}a_n = \frac{T}{2}a_n; \tag{A.12}$$

so that:

$$\frac{T}{2}a_n = \int_0^T x(t)dt$$
$$an = \int_0^T \cos(n\omega_o t)dt; \quad m = 1, 2, 3, \ldots \tag{A.13}$$

The b_n coefficients are found in a similar fashion except Equation A.7 is multiplied by $\sin(m\omega_o t)$, then integrated. In this case, all but the third term integrate to zero and the third term in nonzero only for $m = n$.

$$bn\int_0^T \sin^2(n\omega_o t)dt = \frac{\pi}{\omega_o}b_n = \frac{T}{2}b_n;$$

so that:

$$\frac{T}{2}b_n = \int_0^T x(t)dt$$
$$b_n = \int_0^T \sin(n\omega_o t)dt; \qquad m = 1, 2, 3, \ldots$$

A.3 DERIVATION OF THE TRANSFER FUNCTION OF A SECOND-ORDER OP AMP FILTER

The op amp circuit for a second-order low-pass filter is shown in Figure A.1.

This derivation applies to the low-pass version shown below, but also applies to the high-pass version where the positions of R and C are reversed.

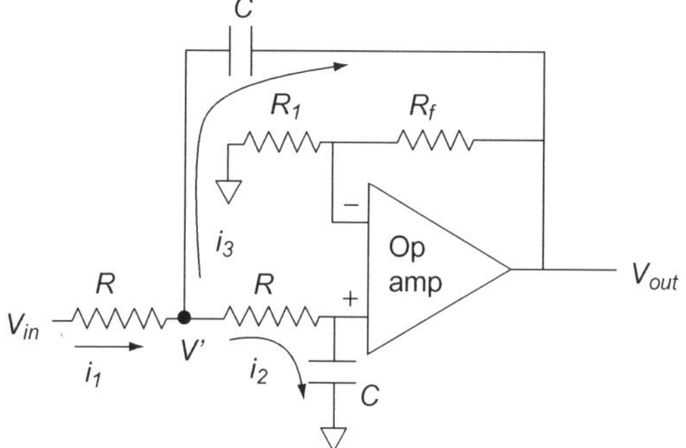

FIGURE A.1 A second-order op amp filter circuit.

Note that at node V' by KCL: $i_1 - i_2 - i_3 = 0$
This allows us to write a nodal equation around that node:

$$\frac{V_{in} - V'}{R} - \frac{V' - V_{out}}{\frac{1}{CS}} - i_3 = 0$$

where:

$$i_3 = \frac{V^+}{\frac{1}{Cs}} = V^+ Cs$$

Since the two terminals of an op amp must be at the same voltage, the voltage V^+ must be equal to V^-. Applying the voltage divider equation to the two feedback resistors, V^-, and hence V^+ can be found in terms of V_{out}:

$$V^+ = V^- = V_{out} G$$

where

$$G = \frac{R_1}{R_f + R_1}$$

So i_3 becomes: $V_{out}(Cs)/G$. Substituting i_3 into the nodal equation at V':

$$\frac{V_{in} - V'}{R} - (V_{out} - V')Cs - \frac{V_{out}(Cs)}{G} = 0$$

$$\frac{V_{in}}{R} - \frac{V'}{R} - V'Cs + V_{out}Cs - \frac{V_{out}Cs}{G} = 0$$

$$\frac{V_{in}}{R} - V'\left(\frac{1}{R} - V'Cs\right) + V_{out}Cs\left(1 - \frac{1}{G}\right) = 0$$

Note that V' can also be written in terms of just i_3:

$$V' = i_3\left(R + \frac{1}{CS}\right) = \frac{V_{out}}{G}\left(R + \frac{1}{CS}\right)$$

Substituting this for V' in the nodal equation:

$$\frac{V_{in}}{R} - \frac{V_{out}Cs}{G}\left(\frac{1}{R} + Cs\right)\left(R + \frac{1}{Cs}\right) + V_{out}Cs\left(1 - \frac{1}{G}\right) = 0$$

$$\frac{V_{in}}{R} = \frac{V_{out}Cs}{G}\left(3 + \frac{1}{RCs} + RCs - G\right)$$

Solving for V_{out}/V_{in}:

$$\frac{V_{out}}{V_{in}} = \frac{G}{RCs\left(1 + \frac{1}{RC} + RCs - g\right)} = \frac{G}{(RCs)^2 + (3 - G)RCs + 1}$$

$$\frac{V_{out}}{V_{in}} = \frac{\frac{G}{(RC)^2}}{S2 + \frac{3 - G}{RC}s + \frac{1}{(RC)^2}}$$

A.4 DERIVATION OF THE TRANSFER FUNCTION OF AN INSTRUMENTATION AMPLIFIER

The classic circuit for a three-op amp instrumentation amplifier is shown in Figure A-2.

To determine the transfer function, note that the voltage V_{in1} appears on both terminals of op amp 1 while V_{in2} appears on both terminals of op amp 2. The voltage out of op amp 1 will be equal to V_{in2} plus the voltage drop across the two resistors, R_2 and R_1:

$$V_{out1} = i_{12}(R_1 + R_2) + V_{in2};$$

but

$$i_{12} = \frac{V_{in1} - V_{in2}}{R_1}$$

$$V_{out1} = \frac{V_{in1} - V_{in2}}{R_1}(R_1 + R_2) + V_{in2}$$

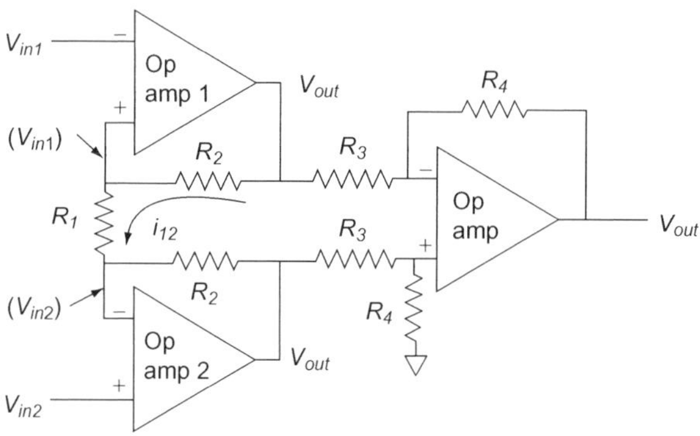

FIGURE A.2 Instrumentation amplifier circuit.

Applying the same logic to op amp 2, its output, V_{out2}, can be written as:

$$V_{out2} - V_{out1} = \frac{V_{in1} - V_{in2}}{R_1}(R_1 + R_2)2 - (V_{in1} - V_{in2})$$

$$= \left[\frac{2(R_1 + R_2)}{R_1} - 1\right](V_{in1} - V_{in2}) = \left[\frac{2R_1 + 2R_2 - R_1}{R_1}\right](V_{in1} - V_{in2})$$

$$V_{out2} - V_{out1} = \left(\frac{R_1 + 2R_2}{R_1}\right)(V_{in1} - V_{in2})$$

$$V_{out} = \left(\frac{R_4}{R_3}\right)(V_{out2} - V_{out1}) = \left(\frac{R_4}{R_3}\right)\left(\frac{R_1 + 2R_2}{R_1}\right)(V_{in1} - V_{in2})$$

The overall output, V_{out}, is equal to the difference of $V_{out2} - V_{out1}$ times the gain of the differential amplifier: R_4/R_3.

Appendix B
Laplace Transforms and Properties of the Fourier Transform

B.1 LAPLACE TRANSFORMS

Some useful Laplace transforms are given here. A more extensive list can be found in several tables available online.

1) 1	$\frac{1}{s}$
2) t^n	$\frac{n!}{s^{n+1}}$
3) $e^{-\alpha t}$	$\frac{1}{s+\alpha}$
4) $(1-e^{-\alpha t})$	$\frac{\alpha}{s(s+\alpha)}$
5) $\cos \beta t$	$\frac{s}{s^2+\beta^2}$
6) $\sin \beta t$	$\frac{\beta}{s^2+\beta^2}$
7) $\frac{1}{\beta^2}[1-\cos(\beta t)]$	$\frac{1}{s(s^2+\beta^2)}$
8) $t-\frac{1}{\beta}(1-e^{-\beta t})$	$\frac{\beta}{s^2(s+\beta)}$
9) $e^{-\alpha t}-e^{-\gamma t}$	$\frac{\gamma-\alpha}{(s+\alpha)(s+\gamma)}$
10) $t-\frac{1}{\alpha}(1-e^{-\alpha t})$	$\frac{\alpha}{s^2(s+\alpha)}$
11) $\left(\frac{b\beta-b\alpha+c}{2\beta}\right)e^{-(\alpha-\beta)t}+\left(\frac{b\beta+b\alpha-c}{2\beta}\right)e^{-(\alpha+\beta)t}$	$\frac{bs+c}{s^2+2\alpha s+\alpha^2-\beta^2}$
12) $e^{-\alpha t}t[b+(c-b\alpha)t]$	$\frac{bs+c}{(s+\alpha)^2}$
13) $^*e^{-\alpha t}\left(\frac{c-b\alpha}{\beta}\right)\sin \beta t+b\cos \beta t$	$\frac{bs+c}{s^2+2\alpha s+\alpha^2+\beta^2}$
14) $^*1-e^{-\alpha t}\left(\frac{\alpha-b}{\beta}\right)\sin \beta t+\cos \beta t$	$\frac{bs+\alpha^2+\beta^2}{s(s^2+2\alpha s+\alpha^2+\beta)}$
15) $^*\frac{\omega_n^2}{\sqrt{1-\delta^2}}\left[e^{-\delta\omega_n t}\sin\left(\omega_n\sqrt{1-\delta^2}\,t\right)\right]$	$\frac{\omega_n^2}{s^2+2\delta\omega_n+\omega_n^2}$

16) $*1 - \frac{e^{-\delta\omega_n t}}{\sqrt{1-\delta^2}} \sin\left(\omega_n \sqrt{(1-\delta^2)}t + \theta\right)$ $\qquad \frac{\omega_n^2}{s(s^2 + 2\delta\omega_n + \omega_n^2)}$

where

$$\theta = \tan^{-1}\left(\frac{\sqrt{1-\delta^2}}{-\delta}\right)$$

*Roots are complex.

B.2 PROPERTIES OF THE FOURIER TRANSFORM

The Fourier transform has a number of useful properties. A few of the properties discussed in the book are summarized here.

Linearity:

$$z(t) = ax(t) + by(t) \Rightarrow Z(\omega) = aX(\omega) + bY(\omega)$$

Differentiation:

$$\frac{dx(t)}{dt} \Rightarrow j\omega X(\omega)$$

Integration:

$$\int_{-\infty}^{t} x(\tau)d\tau \Rightarrow \frac{X(\omega)}{j\omega}$$

Time shift:

$$x(t-\tau) \Rightarrow X(\omega)e^{-j\omega\tau}$$

Time scaling:

$$x(at) \Rightarrow \frac{1}{a}X\left(\frac{\omega}{a}\right)$$

Convolution:

$$\int x(\tau)y(t-\tau)d\tau \Rightarrow X(\omega)Y(\omega)$$

Multiplication:

$$x(t)y(t) \Rightarrow \frac{1}{2\pi}\int X(\upsilon)Y(\omega-\upsilon)d\upsilon$$

where ω and υ are frequencies.

Appendix C
Trigonometric and Other Formulae

$\sin(-x) = -\sin\ x$

$\cos(-x) = \cos\ x$

$\tan\ x = \frac{\sin\ x}{\cos\ x}$

$\sin(\omega t) = \cos(\omega t - 90)$

$\cos(\omega t) = \sin(\omega t + 90)$

$\sin(x \pm y) = \sin\ x\ \cos\ x \pm \cos\ x\ \sin\ y$

$\cos(x \pm y) = \cos\ x\ \cos\ y \mp \sin\ x\ \sin\ y$

$\sin\ 2x = 2\ \sin\ x\ \cos\ x$

$\cos\ 2x = \cos^2 x - \sin^2 x = 2\cos^2 x - 1 = 1 - 2\sin^2 x$

$\sin^2 x = \frac{1 - \cos\ 2x}{2}$

$\cos^2 x = \frac{1 + \cos\ 2x}{2}$

$\sin^2 x + \cos^2 x = 1$

$\sin\ x \pm \sin\ y = 2\ \sin\frac{1}{2}(x \pm y)\cos\frac{1}{2}(x \mp y)$

$\cos\ x + \cos\ y = 2\ \sin\frac{1}{2}(x + y)\cos\frac{1}{2}(x - y)$

$\cos\ x - \cos\ y = -2\sin\frac{1}{2}(x + y)\cos\frac{1}{2}(x - y)$

$\sin\ x\ \cos\ y = \frac{1}{2}\left[\sin(x + y) + \sin(x - y)\right]$

$\cos\ x\ \cos\ y = \frac{1}{2}\left[\cos(x + y) + \cos(x - y)\right]$

$\sin\ x\ \sin\ y = \frac{1}{2}\left[\cos(x + y) - \cos(x - y)\right]$

$e^{\pm j\theta} = \cos\ \theta \pm j\ \sin\ \theta$

$\cos\ \theta = \frac{e^{j\theta} + e^{-j\theta}}{2}$

$\sin\ \theta = \frac{e^{j\theta} - e^{-j\theta}}{2j}$

Appendix D
Conversion Factors: Units

TABLE 1 Metric Conversions

1 cc (10^3 mm^3)	1×10^{-6} m^3
1 kg wt	9.80665 N
1 kg wt	9807×10^5 dynes
1 gm wt	980.665 dynes
1 kg-m	9.80665 joules
1 kg-m	8.6001 gm-cal
1 gm	0.001 kg
1 joule	1 watt-sec
1 joule	10^7 ergs
1 joule	2.778×10^{-7} kw hr
1 pascal	1 dyne/cm^2
1 rad	56.296 deg
1 rad/sec	57.296 deg/sec
1 N	10^5 dynes
1 dyne	0.0010197 gm wt
1 erg (1 dyne-cm)	1×10^{-7} joule
1 km	10^5 cm
1 m	10^{-10} Ang
1 m	10^{-12} microns
1 liter	1,000,027 (0.10) cm^3 (cc)
1 watt	1 joule/sec
1 watt	10^7 erg/sec
1 watt	10^7 dyne-cm/sec
1 amp	1 coul/sec
1 coul (amp/sec)	6.281×10^{18} electronic charges
1 coul	3×10^9 electrostatic units
1 ohm	1 volt/amp
1 volt	1 joule/coul
1 henry	1 volt-sec/amp
1 farad	1 coul/volt

TABLE 2 English-to-Metric Conversions

1 in (U.S.)	2.540 cm
1 ft	0.3048 m
1 yd	0.9144 m
1 fathom (6 ft)	1.829 m
1 mile	1.609 km
1 in^3	16.387 cm^3
1 in^2	6.4516 cm^2
1 pint (0.5 qt)	0.473 liter
1 gal (0.013368 ft^3)	3.7854 liter
1 gal	3785.4 cm^3
1 gal wt (8.337 lb)	3.785 kg
1 carat	6.2 gm
1 oz (troy)	31.134 gm
1 oz (avdps)	28.35 gm
1 lb (troy)	373.24 gm
1 lb (troy)	0.3782 kg
1 lb (avdps)	453.59 gm
1 lb (avdps)	0.45359 kg
1 lb wt	4.448×10^5 dynes
1 lb wt	4.448 N
1 BTU	1054.8 joules
1 horsepower	0.7457 kw
1 rpm	6 deg/sec
1 rpm	0.10472 rad/sec
Pressure Conversions	
1 cm Hg (0 °C)	1.333×10^4 dynes/cm^2
1 cm Hg (0 °C)	135.95 kg/m^2
1 cm Hg (4 °C)	980.368 dynes/cm^2
1 pascal	1 dyne/cm^2

TABLE 3 Metric-to-English Conversions

1 cm	0.3937 in
1 m	3.281 ft
1 km	0.6214 mile
1 km	3280.8 ft
1 km	1.0567×10^{-13} light year
1 gm	0.0322 oz (troy)
1 kg	2.2046 lbs (avdps)
1 deg/sec	0.1667 rpm
1 cm/sec	0.02237 mi/hr
1 cm/sec	3.728×10^{-4} mi/min

TABLE 4 Constants

Gravitational constant: g	32.174 ft/sec^2
Gravitational constant: g	980.665 cm/sec^2
Speed of light: vc	2.9986×10^{10} cm/sec
Dielectric constant (vacuum): ε_0	8.85×10^{-12} col^2/mm^2

TABLE 5 American Standard Wire Gage (AWG) or Brown and Sharpe (B&S) Gage

No.	Diameter (inches)
1	0.2893
2	0.2576
3	0.2294
4	0.2043
5	0.1819
6	0.1620
7	0.1443
8	0.1285
9	0.1144
10	0.1019
11	0.0907
12	0.0808
13	0.0720
14	0.0641
15	0.0571
16	0.0508
17	0.0453
18	0.0403
19	0.0359
20	0.0320
21	0.0285
22	0.0253
23	0.0226
24	0.0201
25	0.0179
26	0.0159
27	0.0142
28	0.0126
29	0.0113
30	0.100
31	0.00893
32	0.00795

TABLE 6 Scaling Prefixes

Scale	Prefix	Examples
10^{-15}	femto	femtoseconds
10^{-12}	pico	picoseconds, picoamps
10^{-9}	nano	nanoseconds, nanoamps, nanotechnology
10^{-6}	micro	microsecond, microvolts, microfarads
10^{-3}	milli	milliseconds, millivolts, milliamps, millimeters
10^{3}	kilo	kilohertz, kilohms, kilometers
10^{6}	mega	megahertz, megaohms, megabucks
10^{9}	giga	gigahertz, gigawatt
10^{12}	tera	terahertz

Appendix E
Complex Arithmetic

Complex numbers and complex variables consist of two numbers rolled into one. However, they are not completely independent as most operations affect both numbers in some manner. The two components of a complex number are the real part and the imaginary part, the latter so-called because it is multiplied by $\sqrt{-1}$. The imaginary part is denoted by the symbol i in mathematical circles, but the symbol j is used in engineering since i is reserved for current. A typical complex number would be written as: $a + jb$ where a is the real part and jb is the imaginary part.

Complex numbers are visualized as lying on a plane consisting of a real horizontal axis and an imaginary vertical axis, as shown in Figure E.1.

A complex number is one point on the real-imaginary plane and can be represented either in rectangular notation as a real and imaginary coordinate (Figure E.1, dashed lines), or in polar coordinates as a magnitude C and angle θ (Figure E.1, solid line). In this text the polar form is written using a shorthand notation: $C\angle\theta$.

To convert between the two representations, refer to the geometry of Figure E.1. To go from polar to rectangular, apply standard trigonometry:

$$a = C\cos\theta \quad b = C\sin\theta \tag{E.1}$$

To go in the reverse direction:

$$C^2 = a^2 + b^2$$

$$C = \sqrt{a^2 + b^2} \tag{E.2}$$

$$\tan\theta = \frac{b}{a}$$

$$\theta = \tan^{-1}\left(\frac{b}{a}\right) \tag{E.3}$$

These operations are very useful in complex arithmetic because addition and subtraction are done using the rectangular representation while multiplication and division are easier using the polar form. Care must be taken with regard to signs and the quadrant of the angle, θ.

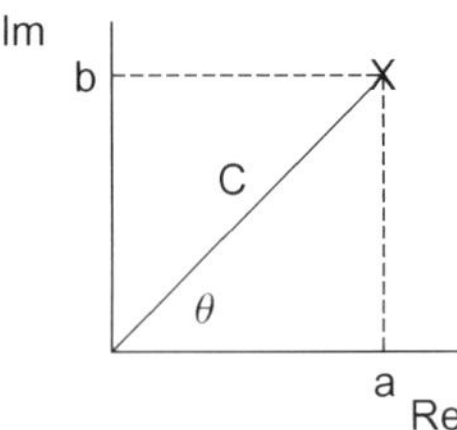

FIGURE E.1 A complex number can be visualized as one point on a plane.

E.1 ADDITION AND SUBTRACTION

To add two or more complex numbers, add real numbers to real, and imaginary numbers to imaginary:

$$(a + jb) + (c + jd) = (a + c) + j(b + d)$$

Subtraction follows the same strategy:

$$(a + jb) - (c + jd) = (a - c) + j(b - d)$$

If the numbers are in polar form (or mixed) covert them to rectangular form first.

EXAMPLE E.1

Add the following complex numbers:

$$8\angle -30 + 6\angle 60$$

Solution: Convert both numbers to rectangular form following the rules above. Note that the first term is in the fourth quadrant so it will have the general form: $a - jb$.

For the first term:

$a = 8\cos(-30)$; $b = 8\sin(-30)$;
$a = 6.9$; $b = -4$

For the second term:

$c = 6\cos(60) = 3$; $d = 6\sin(60) = 5.2$
Sum $= 6.9 - j4 + 3 + j5.2 = 9.9 + j1.2$

This could then be converted back to polar form if desired.

E.2 MULTIPLICATION AND DIVISION

These arithmetic operations are best done in polar form although they can be carried out in rectangular notation. For multiplication, multiply magnitudes and add angles:

$$C\angle\theta(D\angle\phi) = CD\angle(\theta + \phi)$$

For division, divide the magnitudes and subtract the angles:

$$\frac{C\angle\theta}{D\angle\phi} = \frac{C}{D}\angle(\theta - \phi)$$

EXAMPLE E.2

Perform the indicated multiplications or divisions:

A) $8\angle -30(6\angle 120)$;
B) $(10 - j6)(1 + j10)$;
C) $\dfrac{7\angle 305}{6\angle -80}$;
D) $\dfrac{6\angle 50}{8 - j6}$

Solution: For the numbers already in polar coordinates, simply follow the rules given above. Otherwise convert to polar form where necessary.

A) $8\angle -30(6\angle 120) = 48\angle 90$

B) $(10 - j6)(5 + j10) = 11.66\angle -31(11.2\angle + 63) = 130.5\angle 32$

C) $\dfrac{7\angle 305}{6\angle -80} = 1.166\angle 385 = 1.166\angle 25$

D) $\dfrac{6\angle 50}{8 - j6} = \dfrac{6\angle 50}{10\angle -36.9} = 0.6\angle 86.9$

More involved arithmetic operations can call for combinations of these conversions.

EXAMPLE E.3

Add the two complex fractions:

$$\frac{5 + j6}{3 - j7} + \frac{-8 + j6}{-3 - j8}$$

Solution: Convert all the rectangular representations to polar form, carry out the division, then convert back to rectangular form for the addition. When converting from rectangular to polar form, note that each number is in a different quadrant, so care must be taken with the angles.

Evaluate each fraction in turn:

$$\frac{5 + j6}{3 - j7} = \frac{7.8\angle 50}{7.6\angle -67} = 1.03\angle 117 = -0.47 + j0.92$$

$$\frac{-8 + j6}{-3 - j8} = \frac{10\angle 143}{8.5\angle -110} = 1.18\angle 253 = -0.34 - \text{j}1.13$$

So the sum becomes:

$$-0.47 + j0.92 - 0.34 - j1.13 = -0.81 - j0.21 = 0.83\angle -165$$

It is a good idea to visualize where each number falls in the real-imaginary plane, or at least what quadrant of the plane, to help keep angles and signs straight.

Multiplication or division by the number j has the effect of rotation of the complex point by ± 90 deg. This is apparent if the number j, which is in rectangular form, is converted to polar form: $j = 1\angle 90$. So multiplying or dividing by j adds or subtracts 90 degrees from a number:

$$j(C\angle\theta) = 1\angle 90(C\angle\theta) = C\angle(\theta + 90)$$

$$\frac{C\angle\theta}{j} = \frac{C\angle\theta}{1\angle 90} = C\angle(\theta - 90)$$

Appendix F
LF356 Specifications

Only those specifications useful for analyzing circuits and problems in Chapter 12 are given here. For a detailed set of specifications, see the various manufacturer spec. sheets. These specifications are for the LF356 except as noted, and typical values are given.

Description	**Specification**
Bias current	30 pA
Offset current	3.0 pA
Offset voltage	3.0 mV
Gain bandwidth product	5.0 MHz (LF356)
Gain bandwidth product	20 MHz (LF357; minimum $1/\beta = 5$)
Input impedance	$10^{12}\ \Omega$
Open loop gain (DC)	106 dB
CMRR	100 dB
Max. voltage	$\pm$18 volts (LF356); $\pm$22 volts (LF356B)
Noise current	0.01 pA/$\sqrt{\text{Hz}}$
Noise voltage (1000 Hz)	12 nV/$\sqrt{\text{Hz}}$
Noise voltage (100 Hz)	15 nV/$\sqrt{\text{Hz}}$
Noise voltage (10 Hz)	60 nV/$\sqrt{\text{Hz}}$

Appendix G
Determinants and Cramer's Rule

The solution of simultaneous equations can be greatly facilitated by matrix algebra. When the solutions must be done by hand, the use of determinants is helpful, at least when only two or three equations are involved. A determinant is a specific single value defined for a square array of numbers. Given a 2×2 array, the determinant would be found by the application of the so-called "diagonal rule" where the product of the main diagonal (solid arrow) is subtracted by the product of the off-diagonal (dotted arrow):

$$\text{det} = \begin{vmatrix} a_{11} & a_{12} \\ a_{21} & a_{22} \end{vmatrix}$$

This gives rise to the equation:

$$\text{det} = \begin{vmatrix} a_{11} & a_{12} \\ a_{21} & a_{22} \end{vmatrix} = a_{11}a_{22} - a_{12}a_{21} \qquad \text{(G.1)}$$

For a 3×3 array, the determinant is found by an extension of the diagonal rule. One way to visualize this extension is to repeat the first two columns at the right side of the array. Then the diagonals can be drawn directly:

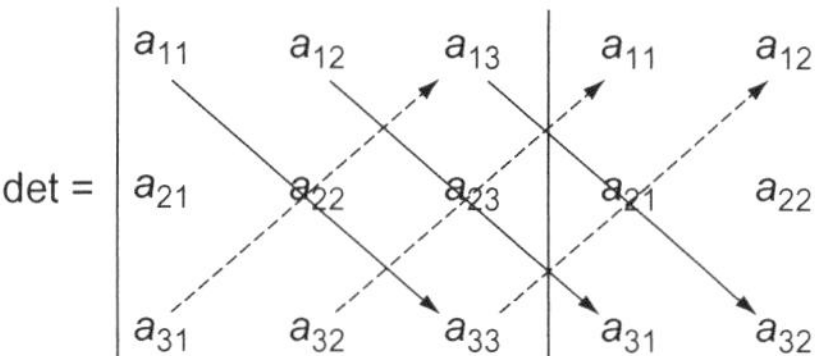

This procedure produces the equation:

$$\text{det} = \begin{vmatrix} a_{11} & a_{12} & a_{13} \\ a_{21} & a_{22} & a_{23} \\ a_{31} & a_{32} & a_{33} \end{vmatrix} = (a_{11}a_{22}a_{33} + a_{12}a_{23}a_{31} + a_{13}a_{32}a_{21}) - (a_{13}a_{22}a_{31} + a_{23}a_{32}a_{11} + a_{33}a_{21}a_{12}) \qquad \text{(G.2)}$$

It is a lot easier using MATLAB where the determinate is obtained by the command: `det(A)`, where `A` is the matrix.

Cramer's rule is used to solve simultaneous equations using determinants. The equations are first put in matrix format (show here using electrical variables):

$$\begin{vmatrix} v_1 \\ v_2 \end{vmatrix} = \begin{vmatrix} Z_{11} & Z_{12} \\ Z_{21} & Z_{22} \end{vmatrix} \begin{vmatrix} i_1 \\ i_2 \end{vmatrix} \tag{G.3}$$

The current i_1 is found using:

$$i_1 = \frac{\det Z_1}{\det Z} = \frac{\det\begin{vmatrix} v_1 & Z_{12} \\ v_2 & Z_{22} \end{vmatrix}}{\det\begin{vmatrix} Z_{11} & Z_{12} \\ Z_{21} & Z_{22} \end{vmatrix}} = \frac{v_1 Z_{22} - v_2 Z_{12}}{Z_{11}Z_{22} - Z_{21}Z_{12}} \tag{G.4}$$

And in a similar fashion, the current i_2 is found by:

$$i_2 = \frac{\det Z_1}{\det Z} = \frac{\det\begin{vmatrix} Z_{11} & v_1 \\ Z_{21} & v_2 \end{vmatrix}}{\det\begin{vmatrix} Z_{11} & Z_{12} \\ Z_{21} & Z_{22} \end{vmatrix}} = \frac{Z_{11} v_2 - Z_{21} v_1}{Z_{11}Z_{22} - Z_{21}Z_{12}} \tag{G.5}$$

Extending Cramer's rule to a 3×3 matrix equation

$$\begin{vmatrix} v_1 \\ v_2 \\ v_3 \end{vmatrix} = \begin{vmatrix} Z_{11} & Z_{12} & Z_{13} \\ Z_{21} & Z_{22} & Z_{23} \\ Z_{31} & Z_{32} & Z_{33} \end{vmatrix} \begin{vmatrix} i_1 \\ i_2 \\ i_3 \end{vmatrix} \tag{G.6}$$

the three currents are obtained as:

$$i_1 = \frac{\det\begin{vmatrix} v_1 & Z_{12} & Z_{13} \\ v_2 & Z_{22} & Z_{23} \\ v_3 & Z_{32} & Z_{33} \end{vmatrix}}{\det\begin{vmatrix} Z_{11} & Z_{12} & Z_{13} \\ Z_{21} & Z_{22} & Z_{23} \\ Z_{31} & Z_{32} & Z_{33} \end{vmatrix}} \quad i_2 = \frac{\det\begin{vmatrix} Z_{11} & v_1 & Z_{13} \\ Z_{21} & v_2 & Z_{23} \\ Z_{31} & v_3 & Z_{33} \end{vmatrix}}{\det\begin{vmatrix} Z_{11} & Z_{12} & Z_{13} \\ Z_{21} & Z_{22} & Z_{23} \\ Z_{31} & Z_{32} & Z_{33} \end{vmatrix}} \quad i_3 = \frac{\det\begin{vmatrix} Z_{11} & Z_{12} & v_1 \\ Z_{21} & Z_{22} & v_2 \\ Z_{31} & Z_{32} & v_3 \end{vmatrix}}{\det\begin{vmatrix} Z_{11} & Z_{12} & Z_{13} \\ Z_{21} & Z_{22} & Z_{23} \\ Z_{31} & Z_{32} & Z_{33} \end{vmatrix}} \tag{G.7}$$

where each determinant would be evaluated using Equation G.2 or using MATLAB.

Bibliography

Bruce, E.N., 2001. Biomedical Signal Processing and Signal Modeling. John Wiley and Sons, New York.

Davasahayam, S.R., 2000. Signals and Systems in Biomedical Engineering: Signal Processing and Physiological Systems Modeling. Kluwer Academic/Plenum Publishing, New York.

Horowitz, P., Hill, W., 1989. The Art of Electronics, second ed. Cambridge University Press, New York.

Johnson, D., 1998. Applied Multivariate Methods for Data Analysis. Brooks/Cole Publishing, Pacific Grove, CA.

Khoo, M.C.K., 2000. Physiological Control Systems: Analysis, Simulation, and Estimation. IEEE Press, Piscataway, NJ.

Northrop, R.B., 2003. Signals and Systems Analysis in Biomedical Engineering. CRC Press, Boca Raton, FL.

Northrop, R.B., 2004. Analysis and Application of Analog Electronic Circuits to Biomedical Instrumentation. CRC Press, Boca Raton, FL.

Rideout, V.C., 1991. Mathematical and Computer Modeling of Physiological Systems. Prentice Hall, Englewood Cliffs, NJ.

Semmlow, J.L., 2004. Biosignal and Biomedical Image Processing: MATLAB-Based Applications. Marcel Dekker, New York.

Smith, S., 1997. Digital Signal Processing. Calif. Tech. Pub., San Diego CA.

Index

N

O

P

Q

R